Bott, Hirzebruch, Singer, and Atiyah

Bott, Hirzebruch, Singer, and Atiyah

Frederick Hirzebruch

Michael Atiyah

At the banquet

At the banquet

The banquet hall

S.-T. Yau congratulating the honorees

Raoul Bott

Art by Milen Poenaru
Dedicated to the index theory founders

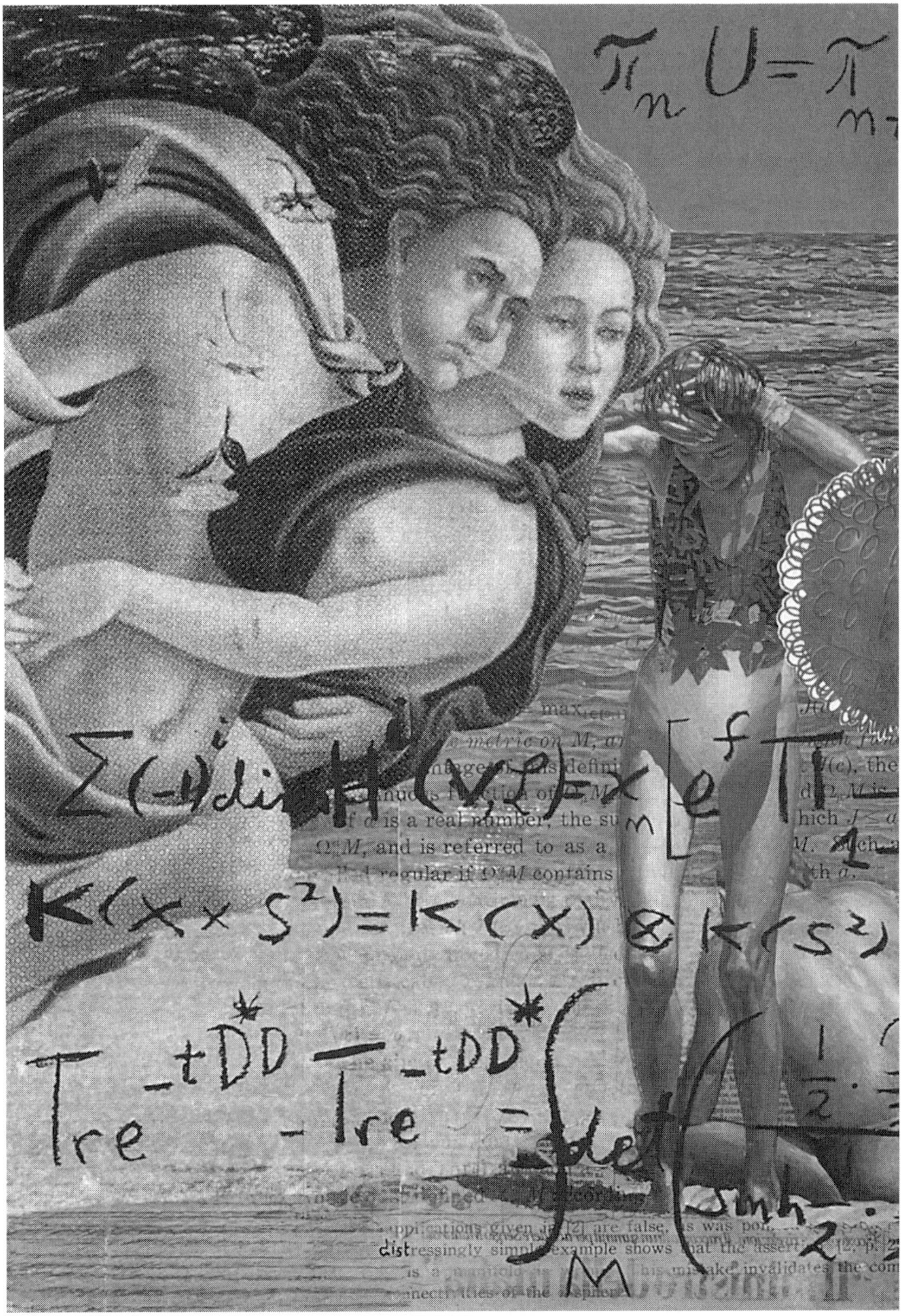

Art by Milen Poenaru
Dedicated to the index theory founders

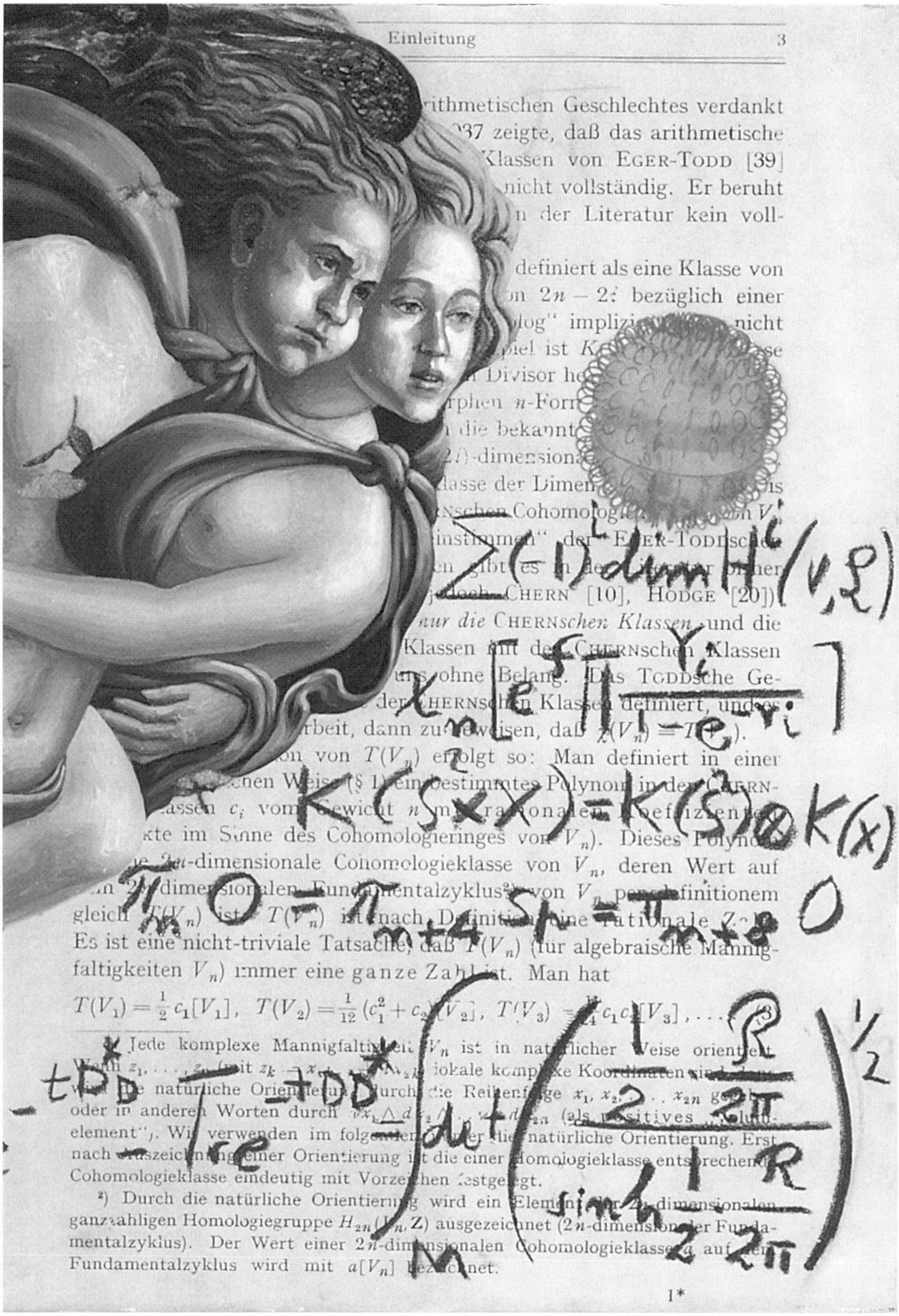

Art by Milen Poenaru
Dedicated to the index theory founders

Art by Milen Poenaru
Dedicated to the index theory founders

Preface

This volume arises out of the conference sponsored by the Journal of Differential Geometry and held at Harvard University to honor the four mathematicians who founded Index Theory. A large number of geometers gathered for this historic occasion which included numerous tributes and reminiscences which will be published in a separate volume. The four men who together created Index Theory: Michael Atiyah, Raoul Bott, Frederich Hirzebruch, and Isadore Singer, were sources of inspiration, mentors and teachers for the other speakers and participants at the conference. The larger than usual size of this volume derives directly from the tremendous respect and admiration felt for the honorees.

Along with this volume, we give our best wishes to each of the honorees that they might have many more years to continue their own research and to inspire and encourage future mathematicians.

Sincerely,

S.-T. Yau

SURVEYS IN DIFFERENTIAL GEOMETRY, 2000
Vol. VII ©2000, INTERNATIONAL PRESS
pp. 1-15

THE GEOMETRY OF CLASSICAL PARTICLES

MICHAEL ATIYAH

1. Introduction

In a recent paper [2] Berry and Robbins have described a classical approach to the spin-statistics theorem of quantum physics. In the course of their investigation they were led to a purely geometrical question in 3-dimensional Euclidean space. This paper grew out of an attempt to answer their question and to understand its significance.

The Berry-Robbins problem concerns two very well-known spaces:

(i) the configuration space $C_n(R^3)$, parametrizing n distinct ordered points in R^3;

(ii) the flag manifold $U(n)/T^n$, parametrizing 'flags' i.e., n ordered mutually orthogonal one-dimensional vector subspaces of C^n (here $U(n)$ is the unitary group, T^n the diagonal subgroup fixing a given flag).

The symmetric group Σ_n acts freely on both these spaces by permuting the points or the subspaces.

The problem is the following:

(1.1) Does there exist (for all n) a continuous map $f_n : C_n(R^3) \to U(n)/T^n$ which is compatible with the action of Σ_n?

In this paper I shall (in §6) give a positive answer to this problem using only elementary geometry. However, this solution has some unsatisfactory features and a much more elegant solution may exist. This

This research was supported in part by the Clay Mathematics Institute.

depends on a conjecture which is remarkably simple to state but appears to be difficult to settle. I shall explain this in §4 and I shall also describe a natural generalization to hyperbolic 3-space in §5.

There are clear indications that alternative approaches are possible which, though more complicated geometrically, have interesting physical interpretations. I hope to follow up on these in subsequent publications. There are also further generalizations in a number of directions which I hope to explore.

In §2 and §3 I make some preliminary comments on the topological aspects of the problem. These are helpful steps on the way to the solution.

2. Elementary comments

Before proceeding to study our problem seriously it is helpful to make a few preliminary remarks.

The case $n = 1$ is trivial, so the first interesting case arises from $n = 2$. By fixing the centre of mass we see that

$$C_2(R^3) = R^3 \times (R^3 - 0)$$

with Σ_2 being the antipodal map on the second factor. Since

$$U(2)/T^2 = P_1(C) = S^2$$

is the complex projective line, with Σ_2 as the antipodal map, it is clear how to define f_2. We simply use radial projection

$$(R^3 - 0) \to S^2.$$

This special case already shows two things. First it explains why the problem is a 3-dimensional one. We can define a configuration space $C_n(R^N)$ for all N, but only for $N = 3$ does the radial map

$$(R^N - 0) \to S^{N-1}$$

end up with S^2.

The second point, intimately related to the first, is that the complex numbers appear in our problem through the Riemann sphere "at ∞" in R^3. In fact the notable feature about the Berry-Robbins problems is that it relates **real** geometry in R^3 to linear independence of vectors over the **complex** numbers. Recalling that complex superposition is the hallmark of quantum theory we see here the germ of a link between classical geometry and quantum theory. This point (in a Minkowski framework, to which we shall return in a later paper) has been persistently emphasized by Roger Penrose.

Mathematically this link between R^3 and the Riemann sphere is best understood by regarding R^3 as the space of imaginary quaternions. This is systematically exploited in the theory of hyperkähler manifolds and in the related Penrose twistor theory.

Note that our map f_2 has the following obvious properties.

(1) f_2 is translation invariant.

(2) f_2 is scale invariant.

(3) f_2 is compatible with the action of $SO(3)$ on both sides.

In fact these properties are easily seen to characterize f_2 uniquely (modulo composition with a fixed element of $0(3)$).

Our task, for general n, is to search for maps f_n which are generalizations of f_2. It would in particular be reasonable to ask for analogues of properties (1), (2), (3), but for (3) we have to specify an action of $SO(3)$ on the flag manifold. Any representation of $SU(2)$ on C^n will induce a representation of $SO(3)$ on the complex projective space $P(C^n)$ and more generally on the flag manifold $U(n)/T^n$ (which can be viewed as a subspace of $P(C^n)^n$). We shall see later that the correct choice of representation of $SO(3)$ is (for each n) given by the (unique) irreducible representation of dimension n.

Clearly one would also like all the maps f_n, for different n, to be somehow related to each other. This can be made more precise by considering a "cluster decomposition" when the n points fall into two separate clusters of r and $n-r$ points far apart from each other. If we write $x = (x_1, ..., x_r)$ and $y = (y_1, ..., y_{n-r})$, we shall denote by $x * y$ the ordered n-tuple formed by putting x and y 'far apart'. We then expect an asymptotic formula.

$$f_n(x * y) \sim f_r(x) \times f_{n-r}(y) \tag{4}$$

where on the right we use the obvious product map

$$U(r)/T^r \times U(n-r)/T^{n-r} \to U(n)/T^n.$$

While properties (1)-(4) put constraints on f_n they will not (for $n \neq 2$) uniquely determine f_n. The Berry-Robbins problem is, in the general case, a problem in equivariant homotopy theory and only the homotopy class will be unique. However we may hope for some natural representative map which has special geometric or physical significance.

3. Homology calculations

A standard test in homotopy theory for a putative map is to check its possible effect on homology or cohomology. It is frequently possible to disprove the existence of a map by showing inconsistency in its effects on cohomology. For a continuous map

$$f : X \to Y$$

we get an induced map

$$f^* : H^*(Y) \to H^*(X)$$

on the integral cohomology rings. If a group Σ acts on both X, Y and is compatible with f, then f^* must also be compatible with the Σ action on $H^*(X)$, $H^*(Y)$. This gives strong restrictions.

As a preliminary test let us consider the (integral) cohomology rings of our two spaces

$$X = C_n(R^3) \qquad Y = U(n)/T^n.$$

These have well-known descriptions [1] [3]. Each is multiplicatively generated by its 2-dimensional elements, and

$$H^2(X) \text{ is generated by } x_{ij} (i \neq j) \text{ with } x_{ij} = -x_{ji}$$

$$H^2(Y) \text{ is generated by } y_i \text{ with } \Sigma y_i = 0.$$

The action of Σ_n in each case is by permutation of the indices.

Any map $f : X \to Y$ will induce on H^2 a map f^* with

$$f^*(y_i) = \sum_{j \neq i} a_{ij} x_{ij}.$$

Compatibility with Σ_n implies that

$$a_{ij} = \lambda \text{ for all } i, j$$

and some integer λ.

For $n = 2$, and $f = f_2$, we have $\lambda = 1$ and the asymptotic requirement (4) will then imply $\lambda = 1$ for all n. Note that, since $x_{ij} = -x_{ji}$, we do indeed have

$$\sum_i f^*(y_i) = \sum_i \sum_{j \neq i} x_{ij} = 0$$

as required by $\Sigma y_i = 0$.

One might expect some inconsistency to appear in the ring structure, but in fact f^* turns out to be fully compatible with the multiplication. This follows from the following known facts [3] [4] (for the analogous case of R^2):

(i) $H^*(Y)$ is the polynomial ring in $y_1, ..., y_n$ modulo the symmetric functions (of positive degree).

(ii) Both $H^*(Y)$ and $H^*(X)$, after tensoring with C, give the regular representation of Σ_n.

From (ii) it follows that the only invariant element in $H^*(X)$ is in $H^0(X)$ and so all the symmetric functions of the y_i (of positive degree) get mapped to zero by f^*. In view of (i) this shows that f^* does indeed extend from $H^2(Y)$ to the whole of $H^*(Y)$.

Despite (ii), which gives an abstract Σ_n-isomorphism between $H^*(X) \otimes C$ and $H^*(Y) \otimes C$, the map f cannot possibly give such an isomorphism for $n > 2$. In fact f^* is not even an isomorphism on H^2 because the ranks of $H^2(X)$ and $H^2(Y)$ are different (for $n > 2$). The way the pieces of the regular representation of Σ_n get their dimensions differs in the two cases as is clear from the explicit formulae for the Poincaré polynomials:

$$\begin{aligned} P(X) &= (1+t^2)(1+2t^2) \dots (1+(n-1)t^2) \\ P(Y) &= (1+t^2)(1+t^2+t^4) \dots (1+t^2+t^4+\dots+t^{2n-2}). \end{aligned}$$

In both cases, putting $t = 1$, gives $n!$, the order of Σ_n. Only for $n = 2$ do the two Poincaré polynomials coincide.

The conclusion of this little excursion into cohomology is that there appears to be no obvious obstruction to the existence of the required map f. On the other hand, algebraic topology alone has difficulty in giving a positive solution. With a bit more effort one can construct a map f having the desired cohomological properties but making it genuinely compatible with Σ_n (beyond its cohomological action) is far from easy. For this a direct geometrical construction must be sought. The cohomology calculations do provide a clue which will be followed up in the next section.

4. A candidate map

We shall now, by direct and elementary construction exhibit a map f_n which will be our first candidate for a solution. Its success depends however on the non-vanishing of a certain determinant and this, though highly probable, has not yet been established. First let us recall the polar decomposition

$$GL(n,C) = P \times U(n)$$

where P is the space of positive self-adjoint matrices.

If $g = pk$ is the decomposition of $g \in GL(n,C)$ then

$$\begin{aligned} gg^* &= pkk^*p^* \\ &= pp^* \text{ since } k \in U(n) \\ &= p^2 \text{ since } p = p^*. \end{aligned}$$

Hence p is the **positive square root** of the self-adjoint matrix gg^* and then $k = p^{-1}g$ gives the explicit retraction map

$$\phi : GL(n,C) \to U(n).$$

Note that this is compatible with the action of $U(n)$ on both left and right:

if $g = pk$ and $u \in U(n)$, then

$$\phi(gu) = \phi(pku) = ku = \phi(g)u$$

$$\text{and } \phi(ug) = \phi(upk) = \phi(upu^{-1}.uk) = uk = u\phi(g).$$

In particular, for any permutation σ,

$$\phi(g\sigma) = \phi(g)\sigma$$

so that ϕ is equivariant with respect to the action of the symmetric group Σ_n on the column vectors of the matrices. Moreover, factoring on the right by the action of the maximal torus T^n of $U(n)$, we get an induced Σ_n-equivariant retraction (still denoted by ϕ)

$$\phi : GL(n,C)/T^n \to U(n)/T^n.$$

What we shall try to do is to construct a Σ_n-equivariant map

$$F_n : C_n(R^3) \to GL(n,C)/T^n$$

and then follow this by the retraction ϕ, so that $f_n = \phi F_n$.

We now describe our construction. Given a configuration $x = (x_1, ..., x_n) \in C_n(R^3)$ we consider the points on the unit 2-sphere S^2 given by

$$t_{ij} = \frac{x_j - x_i}{|x_j - x_i|}.$$

In other words t_{ij} is the **direction** of the line $x_i x_j$ (equivalently $t_{ij} = f_2(x_i, x_j)$). Fixing i and taking all values $j \neq i$ we get $n - 1$ points on S^2 and hence a point

$$p_i \in S^{n-1}(S^2) = P_{n-1}(C)$$

in the symmetric product. If we identify S^2 with the Riemann sphere we can think of each t_{ij} as a complex number (or ∞), and then p_i is simply the polynomial whose roots are the t_{ij} for $j \neq i$. The coefficients of the polynomial are the homogeneous coordinates of $P_{n-1}(C)$. The polynomial p_i is only determined up to a non-zero scalar, but we can normalize so that $||p_i|| = 1$ in C^n and then p_i is only ambiguous up to a phase factor. The metric we use in C^n is the natural metric induced by a metric in C^2. This means it is invariant under $SU(2)$, the double cover of $SO(3)$. Since the representation is irreducible the metric is uniquely determined up to an overall scale. We now have n (normalized) polynomials, associated to $x \in C_n(R^3)$, namely

$$p_1, p_2, \ldots, p_n.$$

Let us assume that these are **linearly independent** (we shall discuss this question later). Then the matrix g, whose columns are $p_1, \ldots, p_n$, is well-defined in $GL(n, C)/T^n$, the factor T^n corresponding to the ambiguous phase factors. Clearly, from their construction, permuting the points $x_1, \ldots, x_n$ just leads to the corresponding permutation of $p_1, \ldots, p_n$. Hence

$$x \to (p_1, \ldots, p_n)$$

gives a Σ_n-equivariant map

$$F_n : C_n(R^3) \to GL(n, C)/T^n.$$

Following this by the retraction ϕ would then give us our map

$$f_n : C_n(R^3) \to U(n)/T^n.$$

This is compatible with Σ_n and also with the action of $SU(2)$ (or $SO(3)$). Note that $SU(2)$ acts on the left on $U(n)/T^n$.

Instead of working with matrices a more invariant way of thinking is to say that the p_i define a linear map $C^n \to V$, where V is the $(n-1)^{\text{th}}$ symmetric power of C^2. The symmetric group Σ_n acts by permuting the basis of C^n, while $SU(2)$ acts on V.

Clearly our map f_n would have the invariance properties (1)(2)(3) of §2.

So our problem is solved, provided we can establish the following.

(4.1) Conjecture. For any configuration of points in $C_n(R^3)$ the complex polynomials $p_1, ..., p_n$ are linearly independent.

We now turn to examine this point. At first glance one might expect the p_i would become dependent for certain degenerate configurations of points. In particular the worst case would seem to be when the points $x_1, ..., x_n$ are all **collinear**. Let us consider this case, and choose our complex coordinate t on S^2 so that the line of the x_i gives $t = 0$ and $t = \infty$. We can assume that $x_1, ..., x_n$ appear on the line in that order and let $x_1 x_n$ be the direction $t = \infty$. Then we see that our polynomials p_i are:

$$\begin{aligned} p_1 &= 1 \\ p_2 &= t \\ &\vdots \\ p_n &= t^{n-1} \end{aligned}$$

and these are clearly linearly independent. This is very encouraging!

Suppose next that our n points fall into 2 clusters x and y far apart, as discussed in §2. We again choose coordinates on S^2 so that $t = \infty$ is the direction connecting the clusters (from x to y). We then find that

$$\begin{aligned} p_i(x * y) &= p_i(x) && 1 \leq r \\ &= t^r p_{i-r}(y) && 1 > r. \end{aligned}$$

Since the $p_i(x)$ have degree $r - 1$, linear independence for $x * y$ would follow from linear independence for x and y separately. This also would establish property (4) of §2.

We have therefore reduced our problem to that of establishing the linear independence of the polynomials $p_i, ..., p_n$ for all configurations $x_i, ..., x_n$ in R^3. We have seen that linear independence holds in the collinear case (which includes the trivial case $n = 2$), so the first significant case is for $n = 3$. We shall now give a direct geometric proof for $n = 3$.

We can assume that the three points x_1, x_2, x_3 are not collinear (since this case is already covered) and so they lie in a definite plane. Starting with the triangle $1, 2, 3$, take 1 (i.e., x_1) as origin and draw the parallel to 23 through 1.

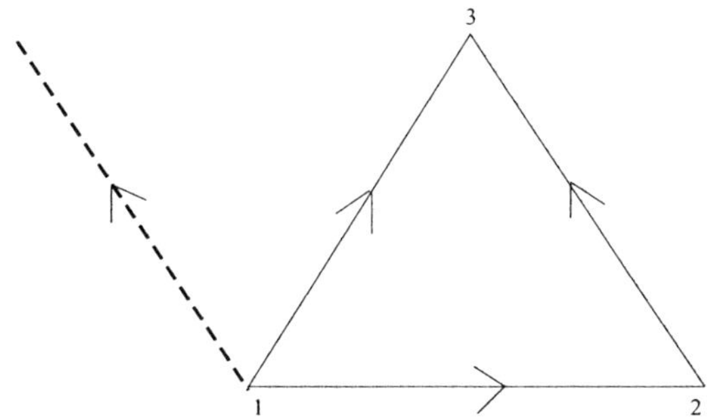

Taking intersections of the extended lines (both directions) with the unit circle we get the following picture, where t_{ij} is just the point denoted by ij on the circle:

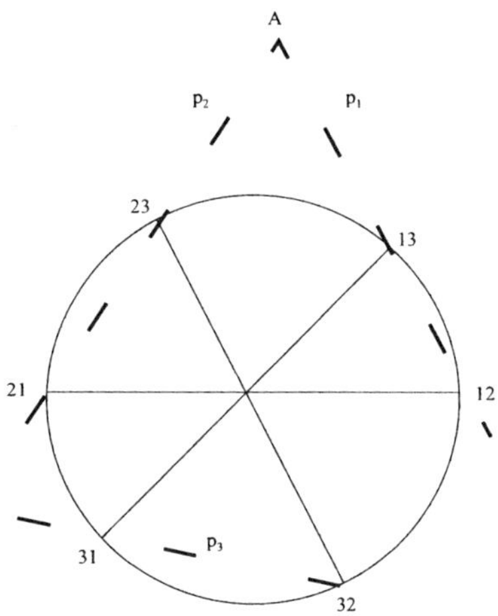

The essential point is the order in which the points i (which are all distinct) appear on the circle.

The three polynomials p_1, p_2, p_3 can be represented (linearly) by the three dotted lines in the picture. Linear independence means that the three lines should not be concurrent. But this is clear: p_1 and p_2 meet in a point A in the upper-half plane (above the diameter joining 1 and 2), and every line through A meets the circle either in no points or in a pair, one in each half-plane, while p_3 meets the circle in 2 points in the lower half-plane.

The case $n = 4$ is already much more complicated and I know of simple geometric proof. However computer calculations by Robbins appear to provide convincing evidence.

Needless to say, since $n = 4$ remains unproved, the same is true of the general case. It is possible that, with sufficient ingenuity, an elementary proof can be constructed for all n but despite my publicizing the problem on several occasions no solution has yet emerged.

Since a direct assault along these lines has reached an impasse we can try to think of ways round the problem. One well known mathematical procedure, when faced by an intractable problem, is to generalize it in the hope that new insight might follow. We shall follow this strategy in the next section and see that it does in fact lead to a solution of the Berry-Robbins problem. This solution, although explicit and elementary, does suffer from some aesthetic drawbacks and it should not be regarded as the end of the story, as I shall explain in subsequent papers.

5. The hyperbolic analogue

In this section I shall replace Euclidean space R^3 by hyperbolic space H^3. The curvature will play no role and it can be normalized to -1, though at a later stage we may wish to allow it to vary. We can therefore introduce the space $C_n(H^3)$ representing ordered configurations of n distinct points in H^3, and we can again ask for an analogue of the Berry-Robbins conjecture. **Is there a continuous map**

$$f : C_n(H^3) \to U(n)/T^n$$

compatible with the action of Σ_n? Since H^3 and R^3 are topologically equivalent this question, as it stands, is equivalent to the original conjecture concerning $C_n(R^3)$. The problem however becomes more natural and interesting if we ask for the existence of a continuous map

$$f : C_n(H^3) \to GL(n, C)/(C^*)^n \tag{5.1}$$

which is compatible with Σ_n **and with the action of $SL(2, C)$.** Here $SL(2, C)$ acts on H^3 as its group of isometries and on $GL(n, C)$ via the irreducible n-dimensional representation. Note that there is no invariant metric on C^n, so we use the full diagonal $(C^*)^n$, not just T^n.

The construction of §4, based on the polynomials, $p_1, ..., p_n$, can be repeated here in essentially the same way. Given two distinct points x_i, x_j of H^3 we define t_{ij} to be the "point at ∞" along the (oriented) geodesic $x_i x_j$. If we take the projective model of H^3, as the interior of the unit ball in R^3, the geodesics are just the usual straight lines, and t_{ij} is just the point where the (oriented) line $x_i x_j$ meets the unit sphere of R^3

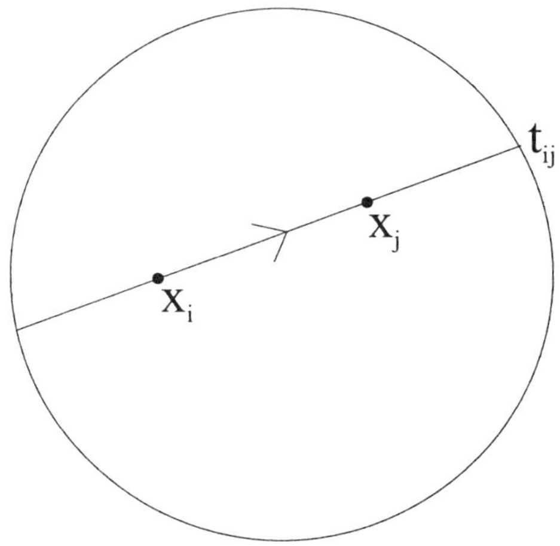

As before p_i is the polynomial with roots t_{ij} $(j \neq i)$, and is determined up to a scalar. Again we may

(5.2) Conjecture. For any configuration in $C_n(H^3)$ the complex polynomials $p_1, ..., p_n$ are linearly independent.

If this conjecture is true then the polynomials p_i define the required map (5.1).

Evidence[1] for the conjecture parallels that for the Euclidean case. The collinear case follows by the same argument as before, and the case $n = 3$ can be proved explicitly (see below). The cluster decomposition property also holds as before.

The fact that the p_i are constructed purely geometrically means, both for R^3 and for H^3, that the construction is compatible with the appropriate isometry group. In this respect the hyperbolic case is more interesting and more natural, because the whole group $SL(2, C)$ (modulo ± 1), acts effectively, whereas in the Euclidean case the translations act trivially and only the orthogonal group acts effectively.

As promised we now prove the conjecture for $n = 3$. The picture is very similar to the Euclidean case.

[1]**Added in proof:** Further computer calculations by P. Sutcliffe have now extended the evidence for the conjecture up to $n = 20$

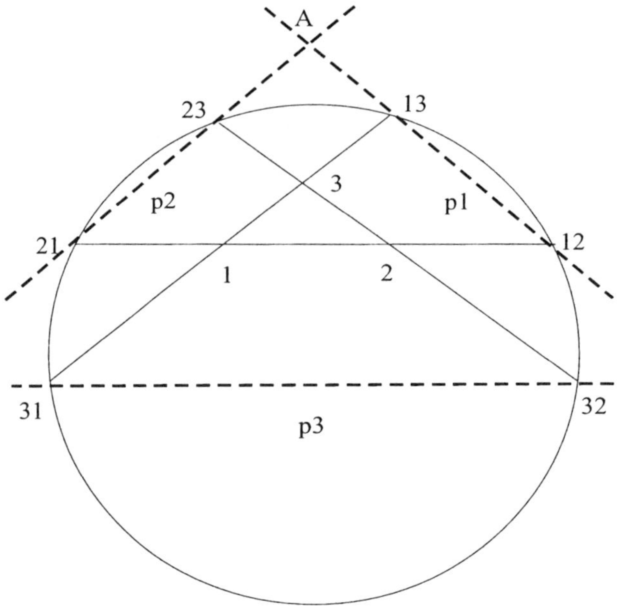

As before the lines p_1, p_2, p_3 are not concurrent for essentially the same reason as before. Lines through A which cut the circle do so at points on either side of the line joining 1 and 2, whereas p_3 cuts the circle at two points (31) and (32), both of which are below the line.

Any fixed configuration of R^3 lies inside some ball and, by rescaling, this can be taken as the unit ball, so that our configuration can now be thought of as in H^3. Varying the size of the ball corresponds to changing the curvature of H^3 and, if the size of the ball tends to ∞ the curvature tends to zero, so that we recover flat space. Moreover, taking a minimal ball, i.e., one which has a point of the configuration on its boundary corresponds to a configuration in H^3 with a point "at ∞". This case of the conjecture reduces easily to that for $(n-1)$ points, so this suggests a possible inductive proof of the conjecture for R^3 by using balls of different sizes, i.e., copies of H^3 with different curvatures. In fact, and perhaps surprisingly, such a naive procedure can actually be made to work and will be explained in detail in the next section. This will justify our belief in the advantage of generalizing difficult conjectures!

6. An explicit map

In this section we shall construct an explicit map

$$f_n : C_n(R^3) \to U(n)/T^n$$

compatible with the action of Σ_n and $SO(3)$.

We shall break the translation symmetry of the problem by fixing an origin. We shall also identify, by radial projection, all spheres with this origin as centre. Given a configuration $x_1, ..., x_n$ of R^3 we shall define polynomials, $p_1, ..., p_n$ with roots t_{ij} $(j \neq i)$ by a slight variant of the construction in §5. To define the roots of p_i we shall distinguish between the values of j:

(a) if $|x_j| \geq |x_i|$ we take $t_{ij} = x_j$(on the sphere of radius $|x_j|$).

(b) if $|x_j| \leq |x_i|$ we take t_{ij} to be the second intersection of the line $x_i x_j$ with the sphere of radius $|x_i|$.

Roughly speaking we treat points inside the ball of radius $|x_i|$ as in hyperbolic space while leaving alone the external points. Note that when $|x_i| = |x_j|$ (a) and (b) agree, so that t_{ij} and hence p_i is a continuous function of $(x_1, ..., x_n)$.

Although the points are treated differently in (a) and (b) our construction is still compatible with Σ_n. The action of Σ_n should be viewed as just altering the labels (suffixes) of the points, and the dichotomy leading to (a) or (b) does not depend on the labelling, but on the intrinsic geometry of the configuration.

The key claim is that **the polynomials** $p_1, ..., p_n$ **given by this new construction are linearly independent**. This is easily proved by induction on n. Given $(x_1, ..., x_n)$, choose an index j for which $|x_j|$ is maximal. For simplicity of notation we may take $j = n$. Now let $q_1, ..., q_{n-1}$ be the polynomials defined by the smaller configuration $(x_1, ..., x_{n-1})$. By the inductive hypothesis these are independent polynomials (of degree $n-2$). Now choose our complex parameter t on the 2-sphere so that x_n is $t = \infty$. Then $p_i = q_i$ for $i \leq n-1$, while p_n is a polynomial of genuine degree $n-1$ (i.e., none of its roots t_{nj} is ∞)

Thus adding p_n to the set $p_1, ..., p_{n-1}$ we still have linear independence, establishing the induction (which starts trivially with $n = 2$).

Just as in §4 we can normalize the p_i and then use the polar decomposition to end up with the required map

$$f_n : C_n(R^3) \to U(n)/T^n.$$

We have thus settled the original question posed by Berry and Robbins. As a bonus our map has $SO(3)$-invariance relative to our chosen origin (and also dilation-invariance). Unfortunately, and this is certainly a drawback, our map is definitely not translation invariant. We could make the origin depend on the configuration by choosing the "centre of mass", and this would restore translation invariance. However we would then lose the "cluster decomposition" property (4) of §2, which our construction with a fixed origin does satisfy.

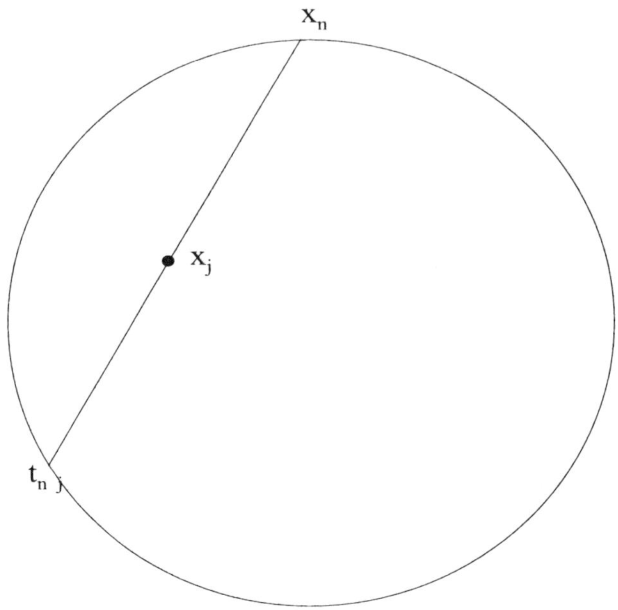

There is another variant of our construction which uses the upper half space model of H^3. We identify H^3 with $C \times R^+$, so that a point x is represented by a point (t, u) with $t \in C$ and $u > 0$. The geodesics are now circles orthogonal to $u = 0$. The dichotomy (a) versus (b) now depends on the u-component (with (a) corresponding to $u_j \leq u_i$).

The role of expanding concentric spheres exhausting R^3 is here played by parallel planes u = constant, and these are all identified with C through their t-component (we now take $R^3 = C \times R$).

This alternative construction is compatible with the subgroup of $SL(2, C)$ keeping a point $(t = \infty)$ fixed. This consists of transformations $t \to at + b$. It is also compatible with u-translations (the analogue of dilations).

All these constructions, though continuous, are not actually differentiable because of the sharp transition from (a) to (b). This can be overcome by a Σ_n-equivariant smoothing, but this is a little cumbersome.

To sum up, while the Berry-Robbins problem has been settled by constructing an explicit map there are some unsatisfactory features of this solution. One might hope for a more elegant geometric solution, for example by settling the conjecture of §4 (and its hyperbolic analogue in

§5). One might also ask for a solution which has some physical meaning. I hope to return to these questions in future publications.

References

[1] M. F. Atiyah & J. D. S. Jones, *Topological aspects of Yang-Mills theory*, Commun. Math. Phys. **61** (1978) 97–118.

[2] M. V. Berry & J. M. Robbins, *Indistinguishability for quantum particles: spin, statistics and the geometric phase*, Proc. R. Soc. London A **453** (1997) 1771–1790.

[3] A. Borel & F. Hirzebruch, *Characteristic classes and homogeneous spaces. I,* Amer. J. Math. **80** (1958) 458–538.

[4] G. I. Lehrer, *On the Poincáre series associated with Coxeter group actions on complements of hyperphones*, J. London Math. Soc. (2) (1987) 275–294.

DEPARTMENT OF MATHEMATICS & STATISTICS
UNIVERSITY OF EDINBURGH

SURVEYS IN DIFFERENTIAL GEOMETRY, 2000
Vol. VII ©2000, INTERNATIONAL PRESS
pp. 17-60

SINGULARITIES IN THE WORK OF FRIEDRICH HIRZEBRUCH

EGBERT BRIESKORN

My first love in mathematics was the theorem of Hirzebruch-Riemann-Roch. In my second year as an undergraduate in Munich I took a course on sheaf theory by Helmut Röhrl. Afterwards, Röhrl told us that if he went on for another year, he could tell us about the theorem of Riemann-Roch. I was deeply impressed, and when I asked Karl Stein where I should go to learn about this theorem, I was advised to go to Bonn.

The summer term 1959 was my first semester in Bonn. I enrolled for Hirzebruch's seminar on "Geometry and Topology" and for his course on "Algebraic Topology". I remember the first day of that course. Our teacher standing in front of the class was a very friendly, very young man, less formal than the German professors I had known so far. My first thought was, that it must be his assistant. However, it was Friedrich Hirzebruch himself. In my first letter to my mother from Bonn I wrote "Professor Hirzebruch ist mir sehr sympathisch. Er ist noch sehr jung". I became Hirzebruch's student, and since then my sympathy, my admiration and gratitude has grown continuously.

Hirzebruch had come to Bonn in 1956, after an offer of a chair in Göttingen had been withdrawn as the result of intervention by Carl Ludwig Siegel. Siegel had failed to recognize the significance of the new methods employed so successfully in Hirzebruch's Habilitationsschrift "Neue topologische Methoden in der algebraischen Geometrie", which appeared in 1956. This book, which culminates in the proof of the theorem of Riemann-Roch for complex projective algebraic manifolds, is dedicated to the teachers of Friedrich Hirzebruch, Heinrich Behnke and Heinz Hopf.

After the end of the war, Behnke had quickly restored his contacts with mathematicians in other countries, in particular with Henri Cartan, who together with Jean Pierre Serre applied the modern methods of sheaf theory introduced by Jean Leray in their investigation of Stein manifolds and of algebraic manifolds. Behnke's contacts had also made it possible for the young student Hirzebruch to visit Heinz Hopf in Zürich, who became his second teacher.

From 1952 to 1954 Friedrich Hirzebruch was at the Institute for Advanced Study in Princeton. This was certainly the most important period in his mathematical development, a period of learning, of intensive exchange and cooperation with Armand Borel, Kunihiko Kodaira and D. C. Spencer and, by letter, with René Thom and J. P. Serre. Many important results were obtained during this time, in particular the theorem of Riemann-Roch and large parts of the joint papers with A. Borel on characteristic classes and homogeneous spaces.

I believe that besides his great mathematical ability Friedrich Hirzebruch's personality, his friendly, open-minded, sincere character must have helped in establishing mathematical cooperation and in making friends in the mathematical world only a few years after the horrible crimes committed by Germans in the time of the Third Reich. As an example let me mention that Nicolaas Kuiper once told me that Friedrich Hirzebruch was the first German mathematician who he was able to speak to after the end of the German occupation of his country.

When Hirzebruch came to Bonn he began, of course, to build up a group of students. Both his lectures and his seminars played an important role in this. I have always admired his wonderful and unique style of lecturing. Every new idea appears at the same time spontaneously and naturally exactly at the right place, so much that one feels that one could almost have had the ideas oneself. I remember a talk by Hirzebruch in his seminar on the theorem of Riemann-Roch after which we almost had the impression that we could have discovered it ourselves. The clarity of these lectures becomes even more surprising when one looks at the notes made in preparation for them – just a few formulas scattered on one page or maybe nothing at all. Many lectures were prepared during the five-minute-walk from Hirzebruch's home to the mathematical institute.

For me, Hirzebruch's seminars were even more important than his lectures. One seminar was always called *"Seminar über Geometrie und Topologie"*. It dealt however with a wide variety of modern subjects. There we learnt about such notions as manifolds, fibre bundles, characteristic classes and theories such as homotopy theory, obstruction theory, Morse theory, the index theorem and much, much more. We learned at the same time modern conceptual forms of mathematical thought and the interplay between such general theories and the analysis of well-chosen interesting concrete problems and examples. It was exciting for us to have famous mathematicians like John Milnor in Bonn, lecturing to Hirzebruch's students on the latest theories. The most exciting week in the year was always the *Mathematische Arbeitstagung.*

The Arbeitstagung

The first Arbeitstagung took place in 1957. The participants were Michael Atiyah, Hans Grauert, Alexander Grothendieck, Friedrich Hirzebruch, Nicolaas Kuiper and Jacques Tits. In subsequent years, more names of first rank were added to the list of participants. Instead of trying to make a complete list, let me mention some of those who became particularly faithful friends of the Arbeitstagung. Raoul Bott, Michel Kervaire, John Milnor, Jean-Pierre Serre and René Thom were added to this list in 1958, Frank Adams, Armand Borel and Serge Lang in 1959. In the sixties, James Eells, Günter Harder, Wilfried Schmid and C.T.C. Wall were added to those who frequently contributed to the program of the Arbeitstagung. Of course, more names come to my mind: Palais, Quillen, Remmert, Smale, Van de Ven, Zagier ... Let me stop at this point. The program was decided on in a public program discussion chaired with subtle guidance by Friedrich Hirzebruch. The first lecture was usually given by Michael Atiyah, who contributed more to the Arbeitstagung than anybody else.

Altogether there were thirty meetings of the Arbeitstagung organized by Hirzebruch. The last one took place in 1991. There is now a second series, organized by G. Faltings, G. Harder, Y. Manin, and D. Zagier, but this is another story. Hirzebruch's Arbeitstagung was a unique phenomenon in the mathematics of the second half of the twentieth century. A large part of the history of mathematics of that period is reflected in the annals of the Arbeitstagung, and some of it was written during its meetings. For example, in his Arbeitstagung lecture given 16 July 1962 on "Harmonic Spinors and Elliptic Operators" Atiyah formulated the problem of expressing the index of elliptic operators in terms of topological invariants associated to their symbol and stated the fundamental conjecture for the Dirac operator *"that* $\mathrm{spin}(X,E) = \hat{A}(X,E)$, *where* $\hat{A}$ *is the so-called* $\hat{A}$*-genus (cf. Hirzebruch Ergebnisse book)."* He explained that this included as special cases the Hirzebruch index theorem and the theorem of Riemann-Roch for Kähler manifolds with zero first Chern class.

A few months later, in February 1963, Atiyah and Singer announced the general index formula for elliptic operators on closed manifolds and indicated the main steps of a proof in a note in the Bulletin of the American Mathematical Society. This first proof was modelled closely on Hirzebruch's proof of the Riemann-Roch theorem. K-theory, which gave the essential framework for the statement of the index theorem, had been introduced by Atiyah and Hirzebruch following Grothendieck's lead in their 1959 paper *Riemann-Roch theorems for differentiable manifolds.* In their paper *Vector bundles and homogeneous spaces* they had given the first systematic exposition of this new cohomology theory. The "central and deep point" of this new cohomology theory was the Bott isomorphism.

Bott's famous periodicity theorem $\pi_k(\mathbf{U}) \cong \pi_{k+2}(\mathbf{U})$ published October 1957 in the Proceedings of the National Academy of Sciences had been suggested to Bott by results of Borel and Hirzebruch published later in the paper *Homogeneous spaces and characteristic classes* and by computations of homotopy groups of Lie groups done by Toda. In the paper of Atiyah and Hirzebruch on Riemann-Roch for differentiable manifolds, Bott's theorem for the unitary group was reformulated as an isomorphism

$$K(X \times S^2) \cong K(X) \otimes K(S^2).$$

In this or similar forms, it was applied also in the subsequent paper of Atiyah and Hirzebruch on the Riemann-Roch for analytic embeddings and in the original proof of the Atiyah-Singer index theorem as well as in the later proof by embedding. Conversely, further generalization of the index theorem led Atiyah and Bott to a beautiful elementary proof of the periodicity theorem, which was presented by Hirzebruch during the Arbeitstagung 1963.

The fusion of analysis and topology in the development leading from the theorem of Riemann-Roch to the index theorem and the Lefschetz fixed point formula for elliptic differential operators was one of the most exciting achievements during the three decades of the Arbeitstagung organized by Friedrich Hirzebruch. It was characterized by a vivid interaction between a small group of leading mathematicians, and some part of that interaction happened during the Arbeitstagung in Bonn.

The work of Michael Atiyah presented in Bonn was not the only work discussed at these meetings that won a Fields Medal. Half of the medalists who won the award between 1950 and 1990 gave lectures at the meetings of the first series of the Arbeitstagung.

Of course, Hirzebruch's students tried to make themselves acquainted with the exciting new mathematics presented at the Arbeitstagung. Thus, in 1963, we had a seminar on the Atiyah-Singer index theorem, working hard on trying to understand the details of the proof. Two of us,

Karl Heinz Mayer and Klaus Jänich, wrote their PhD-thesis on related subjects. Mayer constructed certain elliptic differential operators and applied the index theorem in order to get an integrality theorem containing as special cases all the integrality theorems previously proved by Borel and Hirzebruch. The possibility of such a unified proof had been indicated by Atiyah in his talk in the Séminaire Bourbaki in May 1963.

Klaus Jänich constructed an isomorphism

$$[X, \mathcal{F}\,] \longrightarrow K(X)$$

where $[X, \mathcal{F}\,]$ is the ring of homotopy classes of maps of a compact space X into the space $\mathcal{F}$ of Fredholm operators of a separable Hilbert space. Jänich presented his result during the Mathematische Arbeitstagung 1964. In the proof he used a new theorem on which Nicolaas Kuiper had lectured during the same Arbeitstagung: "The unitary group of Hilbert space is k-connected". Klaus Jänich and Detlef Gromoll, who spoke on exotic spheres and metrics of positive curvature, were the first students of Hirzebruch to talk at the Arbeitstagung. The index map $[X, \mathcal{F}\,] \to K(X)$ was also constructed in a slightly more general form by Atiyah and was used in the definition of a map $K(S^2 \times X) \to K(X)$ leading to a new simple proof of Bott periodicity.

My own thesis written in 1962 dealt with subjects more in line with the previous work of my teacher. Its first part was a theorem on complex quadrics which was an analogue of a theorem on projective spaces proved by Hirzebruch and Kodaira in 1957. The proof was an application of the theorem of Riemann-Roch, and I had been given that problem because I was in love with this theorem. The second part of my thesis generalized work of one part of Hirzebruch's own thesis, in which he had investigated a particularly nice class of simply connected complex surfaces, namely $\mathbb{P}_1$-bundles over $\mathbb{P}_1$. My generalization dealt with $\mathbb{P}_n$-bundles over $\mathbb{P}_1$, which were also investigated from the new viewpoint of the deformation theory of Kodaira and Spencer.

The Thesis

The thesis of Friedrich Hirzebruch was written in 1950. In the year 2000 we celebrated the 50th anniversary of that event in Münster, the "Goldenes Doktorjubiläum", as it is called in Germany, and there Hirzebruch gave a talk on the other part of his thesis, which has been published under the title *Über vierdimensionale Riemannsche Flächen mehrdeutiger analytischer Funktionen von zwei komplexen Veränderlichen.* At this point returning to the beginning of Hirzebruch's work, I am finally approaching the subject given in the title: Singularities in the Work of Friedrich Hirzebruch.

The fame of great mathematicians is justly founded on their great achievements, the creation of new theories and the depth, originality and strength of their mind shown in formulating and solving problems of outstanding importance for the development of our science. In this way the achievements of Friedrich Hirzebruch have been described in the laudations given on the many occasions when he received awards of the highest rank. Instead of repeating such praise I shall try to understand some features in the work of my teacher by asking the questions: What were the objects that he liked? How did he look at them? What did he see?

These questions are not quite as harmless as they might appear, since any attempt to explain the meaning of the words "mathematical objects" must lead to deep philosophical problems. I remember discussions on such matters revealing the belief underlying a whole life devoted to mathematics.

Matthias Kreck has claimed that obviously manifolds are the central objects in Hirzebruch's work. Indeed, manifolds do occur in every work in the two volumes of his collected papers, and in one of these papers he himself writes

"Seit mehr als 30 Jahren beschäftige ich mich mit Mannigfaltigkeiten, besonders mit algebraischen Mannigfaltigkeiten."

But in the same place Hirzebruch mentions *"die Theorie der Singularitäten, die mich seit langem interessiert"*. This interest in singularities began with Hirzebruch's thesis. In the first volume of the collected papers the thesis is the only paper in which singularities play an essential role. However, for the second volume the situation is different; singularities appear in three out of four papers, and in some cases they even appear in the title. So singularities are obviously objects Hirzebruch is interested in. They were among the first objects which he studied, et l'on revient toujours à ses premiers amours.

I think that a case could also be made for yet another and more fundamental entity: number. Integrality problems, divisibility properties, the calculation of integral invariants, and relations between number theory and other fields such as topology, algebraic geometry and analysis on manifolds play a role in many ways in Hirzebruch's work. Finally, we must add to the list of things which Hirzebruch likes, things which are symmetric.

The most venerable symbols of symmetry are the platonic solids, and in particular the icosahedron. In Plato's Timaios the world was conceived as cosmos ordered and shaped by numbers and figures in the best possible way. Anything good had to be beautiful, and beauty was not possible without symmetry. The platonic solids were the elements of Plato's cosmology which has played a very important role in the evolution of European science. Today the icosahedron is to be seen at the entrance of the Max Planck Institute for Mathematics in Bonn. It was founded in 1980 with Friedrich Hirzebruch as its first director for the first fifteen years.

In Hirzebruch's papers singularities are mostly not studied as isolated objects for their own sake. Almost always they occur together with interesting manifolds, frequently in relation to certain symmetric configurations or group actions or in a number theoretic context. There is such a rich variety of beautiful constructions of modern context and relations to classical mathematics that it will be completely impossible to do justice to this work in a few pages. All I can do, is to present some of the themes. For the one which I know best, I shall also try to describe its evolution.

Let me begin with Hirzebruch's thesis. Adding our present knowledge about complex spaces we might summarize its contents as being a constructive resolution of the singularities of 2-dimensional complex spaces. However, at the time when this thesis was written, Heinrich Behnke, Karl Stein, and Henri Cartan had just begun to lay the foundations for the theory of complex spaces. In 1951, Behnke and Stein published a paper in the Mathematische Annalen entitled *"Modifikation komplexer Mannigfaltigkeiten und Riemannscher Gebiete"*, in which they introduced two new notions: the notion of complex space, defined by means of analytic coverings of domains in $\mathbb{C}^k$, and the notion of modification. Also in 1951, Cartan introduced his notion of complex space, modelled on normal analytic subsets of $\mathbb{C}^\ell$. In 1955, Serre allowed arbitrary analytic subsets, so that Cartan's spaces became what is now called normal complex spaces. The relation between the two notions of complex spaces was clarified by Hans Grauert and Reinhold Remmert. In their paper *"Komplexe Räume"*, published 1958 in Mathematische Annalen they proved that the notions of complex space in the sense of Behnke and Stein and in the sense of Cartan were coextensive.

Grauert and Remmert also clarified a question that Hirzebruch had to leave unanswered in his thesis. They proved that every k-dimensional normal complex space can be presented locally as an algebroid covering of a domain in $\mathbb{C}^k$. This means that locally it is the normalization of a Weierstraßcovering defined by an irreducible Weierstraßpolynomial in $\mathbb{C}\{z_1, \ldots, z_k\}[z_{k+1}]$. Hirzebruch's method of resolution uses such local presentations of 2-dimensional complex spaces as algebroid coverings of domains in $\mathbb{C}^2$. This is possible because normal singularities of 2-dimensional complex spaces are isolated so that in dimension two resolution is a local problem.

The discriminant of the Weierstraßpolynomial describing a 2-dimensional algebroid covering defines a curve in a domain of $\mathbb{C}^2$. The first step in Hirzebruch's resolution process consists in resolving the singularities of this curve by a sequence of σ-processes so that the total transform has only normal crossings. The notion of σ-process had been introduced by Hirzebruchs teacher Heinz Hopf in 1951 as a local process of modification of complex manifolds. It modifies a k-dimensional complex manifold X in a point p by replacing p by the $(k-1)$-dimensional complex projective space of tangent directions at p. Hopf knew that in algebraic geometry quadratic transformations were an old and successful method of modifying varieties. Zariski had introduced the notion of a quadratic transformation at a point p of a surface in 1939 in his paper *"The reduction of the singularities of an algebraic surface"*, and in 1943 he defined general monoidal transformations in his paper *"Foundations of a general theory of birational correspondences"*. In dimension two, the σ-process replaces a point by a Riemann sphere, and that is the reason for its name, σ being the first letter of the greek word $\sigma\varphi\alpha\tilde{\iota}\rho\alpha$.

Now let $\varphi : (X, x) \to (\mathbb{C}^2, 0)$ be an algebroid m-fold covering defined by an irreducible Weierstraßpolynomial of degree m in $\mathbb{C}\{z_1, z_2\}[z_3]$ such that the reduced discriminant curve has the equation $z_1 z_2 = 0$. The irreducibility implies that all points of $X' = X - \{x\}$ over $z_1 z_2 = 0$ are branch points of the same order $a-1$ over $z_1 = 0$ and of order $b-1$ over $z_2 = 0$, with integers $a, b > 1$. Define a ramified covering $\chi : \mathbb{C}^2 \to \mathbb{C}^2$ by $\chi(\zeta_1, \zeta_2) = (\zeta_1^a, \zeta_2^b)$. Let (Y, y) be the fibered product of $(\mathbb{C}^2, 0)$ and (X, x) with respect to χ and φ. The normalization of (Y, y) is a multigerm unbranched over $\mathbb{C}^2 - \{0\}$. So it consists of m germs isomorphic to $(\mathbb{C}^2, 0)$. Therefore we have a holomorphic lifting

$$\begin{array}{ccc} & & (X, x) \\ & \overset{\psi}{\nearrow} & \big\downarrow \varphi \\ (\mathbb{C}^2, 0) & \xrightarrow[\chi]{} & (\mathbb{C}^2, 0) \end{array}$$

such that ψ is an unramified covering over a suitable punctured neighbourhood $V_0 - \{x\}$ of x. If n is the degree of ψ, we have $m \cdot n = a \cdot b$. We want a precise description of ψ. Choose a polycylinder $W \subset \mathbb{C}^2$ sufficiently small so that $V = \varphi^{-1}(W) \subset V_0$. Let $U \subset \mathbb{C}^2$ be the polycylinder $\chi^{-1}(W)$. Let W', V', U' be the spaces obtained by removing the axes $z_1 z_2 = 0$ and their inverse images. These spaces have the homotopy type of $\mathbb{C}^* \times \mathbb{C}^*$. In particular, the fundamental groups are abelian. So we have a diagram of regular unramified coverings

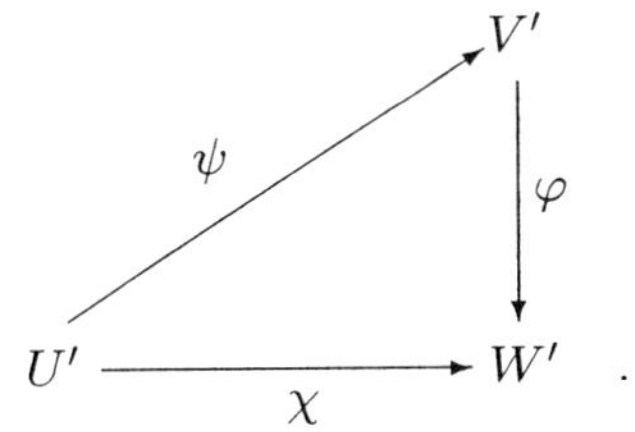

Let G and H be the groups of covering transformations of χ and ψ. The group $G \subset GL(2,\mathbb{C})$ is the group of diagonal matrices with diagonal entries the a-th and b-th roots of unity. The subgroup H of order n is also the group of covering tranformations of the regular unramified covering $\psi : U - \{0\} \to V - \{x\}$. Therefore it is a "small" subgroup of $GL(2,\mathbb{C})$ which means that its nontrivial elements have no eigenvalue 1. Therefore H must be one of the cyclic groups of order n

$$C_{n,q} = \left\langle \begin{pmatrix} e^{2\pi i q/n} & 0 \\ 0 & e^{2\pi i 1/n} \end{pmatrix} \right\rangle$$

where q is an integer $0 < q < n$ relatively prime to n. So we obtain the result that the germ (X, x) is isomorphic to the cyclic quotient singularity $(X_{n,q}, 0)$ where $X_{n,q} = \mathbb{C}^2/C_{n,q}$ may be described as the algebroid covering given by the Weierstraßpolynomial

$$z_3^n - z_1 z_2^{n-q}.$$

Hirzebruch borrows this result from an article by Heinrich W. E. Jung which appeared in 1908 in Crelles Journal. Of course, the modern terminology used above does not occur in Jung's paper. In particular the notion of the quotient of a complex space with respect to a properly discontinuous group of automorphisms was introduced not until 1953/54 when it appeared in the Séminaire Cartan.

In view of the result obtained above, all we have to do is to resolve the singularity of $X_{n,q}$. Hirzebruch constructs a resolution by means of an algorithm taken from the paper of Jung. I shall try to motivate this construction and present it so as to show its relation to the theory of toroidal embeddings developed by Kempf, Knudsen, Mumford and Saint-Donat in 1973. As a matter of fact, Mumford was partly motivated by later work of Hirzebruch on cusp singularities which may be seen as a natural continuation of his thesis.

Let T be the standard complex algebraic torus $\mathbb{C}^{*2} \subset \mathbb{C}^2$. The basic fact is that $X_{n,q}$ contains the algebraic torus $T_{n,q} = T/C_{n,q}$. We shall construct the resolution $\tilde{X}_{n,q} \to X_{n,q}$ by gluing several copies of $\mathbb{C}^2$ which map to $X_{n,q}$ so that T is mapped isomorphically onto $T_{n,q}$.

Let $\overline{X}_{n,q} \subset \mathbb{C}^3$ be the Weierstraßspace given by the equation $z_3^n - z_1 z_2^{n-q} = 0$. Let $X_{n,q} \to \overline{X}_{n,q}$ be the normalization map induced by the map $\mathbb{C}^2 \to \overline{X}_{n,q}$ given by $(z_1, z_2, z_3) = (t_1^n, t_2^n, t_1 t_2^{n-q})$. This maps $T_{n,q}$ isomorphically onto its image $\overline{T}_{n,q} \subset \overline{X}_{n,q}$. Therefore isomorphisms $T \to T_{n,q}$ can be given by

$$\begin{aligned} z_1 &= u^{\lambda} v^{\lambda'} \\ z_2 &= u^{\mu} v^{\mu'} \\ z_3 &= u^{\nu} v^{\nu'} \end{aligned}$$

where the exponents have to satisfy the conditions

$$|\mu\nu' - \mu'\nu| = 1$$

and λ, λ' are determined by the other exponents. Let N be the 2-dimensional lattice of algebraic homomorphisms $\mathbb{C}^* \to T$. The canonical homomorphisms $t \mapsto (t, 1)$ and $t \mapsto (1, t)$ form a canonical basis $(1, 0)$ and $(0, 1)$ of $N = \mathbb{Z}^2$. If we compose $T \to \overline{T}_{n,q}$ with the projection $\overline{T}_{n,q} \to T$ given by $(z_1, z_2, z_3) \to (z_2, z_3)$, we get an isomorphism $T \to T$, and $T \to T_{n,q}$ is determined by the induced map $N \to N$, i.e., by the images (μ, ν) and (μ', ν') of the canonical basis. These two vectors span a sector σ in $\mathbb{R}^3$, consisting of their linear combinations with nonnegative coefficients.

Consider the sector $S \subset \mathbb{R}^2$ spanned by $(0, 1)$ and $(n, n - q)$. If we identify N with $\mathrm{Hom}(\mathbb{C}^*, T_{n,q})$ via the projection $\overline{T}_{n,q} \to T$, the points in $S \cap N$ correspond to those algebraic isomorphisms $\mathbb{C}^* \to T_{n,q}$ which have a limit in $X_{n,q}$ for $t \to 0$. Therefore, an isomorphism $T \to T_{n,q}$ will extend to a holomorphic map $\mathbb{C}^2 \to X_{n,q}$ if and only if its sector σ is contained in S. If we try to construct the resolution $\tilde{X}_{n,q} \to X_{n,q}$ by gluing a finite number $s + 1$ of copies of $\mathbb{C}^2$ with maps $\mathbb{C}^2 \to X_{n,q}$, the condition that $\tilde{X}_{n,q} \to X_{n,q}$ has to be proper, means that the corresponding sectors $\sigma_0, \ldots, \sigma_s$ have to cover S. Such a covering is minimal if the sectors $\sigma_0, \ldots, \sigma_s$ form a subdivision of S, and if their number is minimal. As a matter of fact, there is a unique subdivision

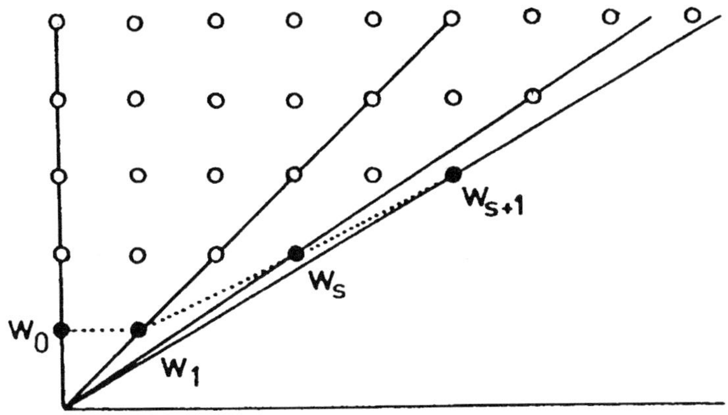

FIGURE 1

of S with that property. Let Σ be the convex hull of $S \cap N - \{0\}$, and let $w_0, w_1, \ldots, w_{s+1}$ be the points of N on $\partial\Sigma$ between $w_0 = (0,1)$ and $w_{s+1} = (n, n-q)$. Then S is subdivided by the rays $\mathbb{R}^+ \cdot w_k$ and the sectors between them. Figure 1 illustrates this for Jung's example $(n,q) = (5,2)$.

The vectors $w_k = (\mu_k, \nu_k)$ are computed recursively as follows:

$$\begin{aligned} w_{k+1} &= b_k w_k - w_{k-1} \quad , \quad k = 1, \ldots, s \\ w_0 &= (0,1) \\ w_1 &= (1,1). \end{aligned}$$

Here b_k are natural numbers larger than 1 computed from the continued fraction

$$\frac{n}{n-q} = b_1 - \cfrac{1}{b_2 - \cfrac{1}{\ddots - \cfrac{1}{b_s}}} \quad .$$

This is the Hirzebruch-Jung algorithm. $\tilde{X}_{n,q}$ is obtained by gluing $s+1$ copies of $\mathbb{C}^2$. The gluing transformation from the $(k-1)$-th copy to the k-th copy is

$$\begin{aligned} u_k &= u_{k-1}^{b_k} v_{k-1} \\ v_k &= u_{k-1}^{-1}. \end{aligned}$$

Gluing of these two copies gives the total space of the b_k-th tensor power of the Hopf line bundle over the Riemann sphere. Thus the inverse image

FIGURE 2

of the singular point of $X_{n,q}$ in $\tilde{X}_{n,q}$ is a chain of s nonsingular rational curves with selfintersection numbers $-b_k$, such that only subsequent curves intersect, and they intersect transversely.

Following Hopf's example, Hirzebruch describes the configuration of exceptional curves by a weighted dual graph

There are two interesting extreme cases: $q = 1$ and $q = n - 1$. For $q = 1$, the resolution graph is just one point with value n, and the exceptional curve in $X_{n,1}$ identifies with the zero section in the n-th power of the Hopf bundle. Compactifying that bundle by adding a point at infinity for each complex line of the bundle gives the Σ_n-surface treated in the other half of Hirzebruch's thesis.

The case $q = n - 1$ is characterized by the fact that in this case $C_{n,q}$ is a subgroup of $SL(2, \mathbb{C})$. It is also characterized by the fact that in this case all b_k are equal to two. This can be interpreted as follows. Up to a sign the intersection matrix of the exceptional curve in the resolution of $X_{n,n-1}$ is the Cartan matrix of the root system of type A_{n-1}, and the resolution graph is the Coxeter-Dynkin diagram of type A_{n-1}. This correspondence between $C_{n,n-1}$ and A_{n-1} is part of a perfect correspondence between conjugacy classes of subgroups $G \subset SL(2, \mathbb{C})$ and isomorphism classes of their quotient singularities $(\mathbb{C}^2/G, 0)$ on one hand and simple Lie algebras of type A_n, D_n, E_6, E_7, E_8 on the other hand. In their book "Compact Complex Surfaces" Barth, Peters and Van de Ven say the following about this:

The relation between simple singularities and simple Lie groups is one of the most beautiful discoveries in mathematics. It is impossible to attribute it to a single author.

Friedrich Hirzebruch is one of those who have a share in this discovery, and it is due to him that I too got involved in this on-going story of more than a century.

The importance of Hirzebruch's thesis from a historic point of view is perhaps not primarily to be seen in the fact that he proves the existence of a resolution of singularities of complex surfaces. As a matter of fact, Robert J. Walker had given the first rigorous proof for algebraic surfaces as early as 1935 in a paper in the Annals of Mathematics apparently not known to Hirzebruch at the time when he wrote his thesis. Walker had used essentially the same approach as Hirzebruch, quadratic transformations and Jung's algorithm. Hirzebruch's solution has the merit of clarity and simplicity made possible by his strictly local complex analytic approach as opposed to the projective algebraic methods of the previous proofs. However, the primary importance of this thesis probably is to be seen in the fact that it contained certain germs unfolded in future work of Hirzebruch and his students.

One of these germs is Hirzebruch's remark that the singular point $X_{n,q}$ has a neighbourhood in $X_{n,q}$ bounded by the lens space $L(n,q) = S^3/C_{n,q}$. Lens spaces, constructed by Poul Heegaard in 1898 and by Heinrich Tietze in 1908 were the first examples of closed orientable 3-manifolds not determined by their fundamental groups. In 1918 Alexander proved that $L(5,1)$ and $L(5,2)$ are not homeomorphic although both have the same fundamental group. In 1935 Kurt Reidemeister proved that $L(n,q)$ and $L(n',q')$ are homeomorphic if and only if $n = n'$ and $q' \equiv \pm q \mod n$ or $qq' \equiv \pm 1 \mod n$. On the other hand Hirzebruch proves that the singularities of $X_{n,q}$ and $X_{n',q'}$ are isomorphic if and only if $n = n'$ and $q = q'$ or $qq' \equiv 1 \mod n$.

The interest in the topology of singularities can be traced back to the last decade of the 19th and the first decade of the 20th century, when Poul Heegaard wanted to develop topological tools for the investigation of algebraic surfaces, and when Wilhelm Wirtinger adopted Felix Klein's geometric view of the theory of analytic functions and tried to understand the topology of the ramification of functions of two variables. The fascinating story how this led to the first result of modern knot theory, Tietze's proof that the trefoil knot is not trivial, is told in Moritz Epple's book "Die Entstehung der Knotentheorie". The story is too long to be told here. Let me just indicate in moderately unhistoric terms what Wirtinger did. He studied the algebraic function z of two variables x, y defined by the equation

$$z^3 + 3xz + 2y = 0.$$

In modern terms: the projection of the surface X with this equation to the (x,y)-plane is the semiuniversal unfolding of the 0-dimensional A_2-type singularity $z^3 = 0$. The discriminant curve $D \subset \mathbb{C}^2$ has the equation

$$x^3 + y^2 = 0.$$

The fundamental group of $\mathbb{C}^2 - D$ operates on the fibre over the base point by the monodromy representation. Wirtinger calculates $\pi_1(\mathbb{C}^2 - D)$

and finds a presentation with two generators and one relation $sts = tst$. In modern terms: π_1 is the braid group on 3 strings. The monodromy representation is the canonical homomorphism of this group to the symmetric group S_3.

The group S_3 is the Weyl group of A_2 operating on the plane $z_1 + z_2 + z_3 = 0$ by permutations. If we map this plane to the (x, y)-plane by means of the elementary symmetric functions σ_2, σ_3 and lift the covering of the (x, y)-plane by means of this base extension, we get $(z - z_1)(z - z_2)(z - z_3) = 0$. So we get a trivial covering over the complement of the discriminant $\Pi(z_i - z_j) = 0$. The fundamental group of that complement is the coloured braid group, i.e., the kernel of $B_3 \to S_3$. It is part of the beautiful relation between simple singularities and simple Lie algebras that all this generalizes to all types A_k, D_k, E_6, E_7, E_8.

In his computation of $\pi_1(\mathbb{C}^2 - D)$, Wirtinger used an idea of Heegaard. Heegaard reduced the complex geometry of an algebroid covering $(X, x) \to (\mathbb{C}^2, 0)$ with a singularity $(D, 0)$ of the discriminant to a situation of 3-dimensional topology. He considered a small 4-ball $B \subset \mathbb{C}^2$ centered at 0 with boundary $\partial B = S^3$, a 3-sphere. The intersection $L = D \cap S^3$ is a knot or link in S^3. In Wirtinger's example it is the trefoil knot. $D \cap B \subset B$ is homeomorphic to the cone over L. Therefore $B - D \cap B$ has the same fundamental group as the complement $S^3 - L$ of the link. Let $U \subset X$ be the inverse image of B and $\partial U = M$ the inverse image of S^3. Then M is a 3-manifold, which is a ramified covering of S^3 ramified over L. Moreover, M is a boundary of the neighbourhood U of x in X, and U is homeomorphic to the cone over M. This established a link between the geometry of singularities of complex surfaces and 3-dimensional topology which turned out to be very fruitful both for complex analytic geometry and topology.

The title of Hirzebruch's paper "Über vierdimensionale Riemannsche Flächen mehrdeutiger analytischer Funktionen von zwei komplexen Veränderlichen" and the 1928 paper of Wirtinger's student Brauner in the list of references are indications that Hirzebruch's thesis is to be seen in this context.

$\mathbf{E}_8$

The gist of the story that I want to tell now is expressed in the titles of a talk by Hirzebruch in the Séminaire Bourbaki, given November 1966, and of a paper by myself published in the same year in volume 2 of Inventiones Mathematicae. Hirzebruch's title was *"Singularities and Exotic Spheres"* and mine was *"Beispiele zur Differentialtopologie von*

Singularitäten".

The story begins in the year 1956 with John Milnor's sensational discovery that there are 7-dimensional differentiable manifolds which are homeomorphic but not diffeomorphic to the 7-dimensional sphere S^7. This discovery was certainly one of the most germinal achievements of mathematicians in the twentieth century. In less than seven years a theory describing these exotic differentiable structures was developed by an extraordinary meshing of the results of mathematicians working in diverse parts of topology.

Several fundamental tools and constructions had already been prepared during the two decades preceeding Milnor's discovery. Fibre bundles and the characteristic classes of Stiefel-Whitney, Chern and Pontryagin were at the disposition of differential topology. Cobordism and framed cobordism had been introduced by Thom and Pontryagin, and the signature theorem had been proved by Hirzebruch. In homotopy theory Freudenthal had proved the stability of $\pi_{d+n}(S^d)$ for $d > n+1$ in 1937, and the resulting stable groups Π_n had been proved to be finite for $n > 0$ and had been computed for low values by Serre and Toda. The stable homotopy groups of the classical groups had been computed by Bott. A link between these groups had been established by George Whitehead in 1942. Generalizing a construction by Heinz Hopf in his 1935 paper *"Über die Abbildung von Sphären auf Sphären niedrigerer Dimension"* Whitehead defined a homomorphism

$$J_n : \pi_n(\mathbf{SO}) \longrightarrow \Pi_n$$

from the stable homotopy groups of the orthogonal group to the stable n-stem of the homotopy groups of spheres. Work of Pontryagin culminating in his 1955 paper *"Smooth manifolds and their applications in homotopy theory"* identifies Π_n with the framed cobordism group of framed embedded n-manifolds. The canonical homomorphism from the framed cobordism group to Π_n is defined by means of the Thom-Pontryagin construction.

Some of the most outstanding results of the period following Milnor's discovery were the proofs of the Poincaré conjecture in dimensions greater than four by Stallings, Zeeman and Smale, the development of handlebody theory and the proof of the h-cobordism theorem by Smale and the determination of the image of the J-homomorphism by Frank Adams.

The h-cobordism theorem allows the identification of oriented diffeomorphism classes of topological n-spheres with h-cobordism classes for $n \geq 5$. Let Θ_n be the set of h-cobordism classes of closed oriented

C^∞-manifolds homotopy equivalent to S^n. This is a group with respect to the connected sum operation. Using Bott's calculation of $\pi_n(\mathbf{SO})$, Hirzebruch's signature theorem and the results of Adams, Kervaire and Milnor showed that homotopy spheres are stably parallelizable. Therefore, they can apply the Thom-Pontryagin construction in order to define a homomorphism

$$p : \Theta_n \longrightarrow \text{coker } J_n.$$

The kernel of p is the group bP_{n+1} of classes of oriented homotopy-spheres bounding parallelizable manifolds. The cokernel of p is trivial for $n \not\equiv 2 \mod 4$ and trivial or of order 2 if $n \equiv 2 \mod 4$. Kervaire and Milnor apply the technique of surgery developed by Milnor in order to determine the group bP_{n+1}. For n even bP_{n+1} is trivial. For n odd bP_{n+1} is finite cyclic. The order is 1 or 2 if $n = 4k + 1$. For $n = 4k - 1$, the order ist

$$\sigma_k/8 = 2^{2k-2}(2^{2k-1} - 1) \text{ numerator } (4B_k/k),$$

where B_k denotes the k-th Bernoulli number. Thus Θ_n, $n \neq 3$, is always a finite abelian group, and for n odd the calculation of its order is reduced to the calculation of the order of Π_n (up to a factor 2 if $n = 4k+1$). The first non-zero group Θ_n, $n \neq 3$, is Θ_7. In this case coker J_7 is trivial, and $\Theta_7 = bP_8$ is cyclic of order 28. The first nontrivial group bP_{n+1} with $n = 4k + 1$ is bP_{10}.

An isomorphism $bP_{4k} \to \mathbb{Z}/(\sigma_k/8\mathbb{Z})$ is obtained as follows. Let Σ be a homotopy sphere bounding a $(2k-1)$-connected parallelizable $4k$-manifold W with signature σ. The intersection form on $H_{2k}(W, \mathbb{Z})$ is symmetric, even and unimodular. Therefore, its signature σ is divisible by 8. The isomorphism maps the class of Σ in bP_{4k} to $\sigma/8 \mod \sigma_k/8$. In particular one obtains a generator for bP_{4k} if $\sigma = 8$. The minimal rank for an even unimodular quadratic form with signature 8 is 8, and up to isomorphism there is only one form with these properties, namely that of the root lattice of E_8.

In a mimeographed manuscript dated Princeton, January 23, 1959, Milnor constructed such a manifold W with this quadratic form. However, the choice of a basis of the lattice with which he begins his construction is not the simplest possible choice, since the graph describing the intersection matrix contains cycles. Up to isomorphism, there is only one choice where the graph is a tree and all intersection numbers are non-negative and 2 on the diagonal. The corresponding graph is the famous Coxeter-Dynkin diagram of E_8, shown in Figure 3.

Hirzebruch noticed the possibility of simplifying Milnor's construction and presented it in a colloquium lecture in Bonn in the winter

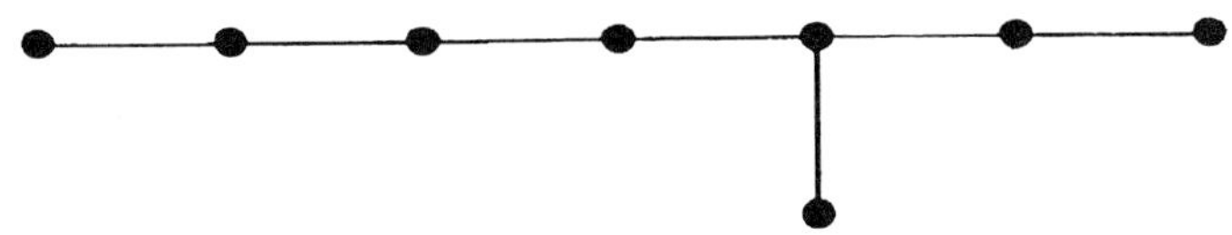

FIGURE 3

1960/61 and in a lecture in Vienna on November 18, 1960. The corresponding publication *"Zur Theorie der Mannigfaltigkeiten"* is Hirzebruch's shortest publication, just one page in the Internationale Mathematische Nachrichten. Milnor adopted Hirzebruch's "lovely" construction and used it in his essay *"Differential topology"* in the book *"Lectures on Modern Mathematics"* edited by Saaty.

Let me describe the construction in somewhat greater generality with a historical perspective. The first examples of exotic 7-spheres constructed by Milnor in 1956 were S^3-bundles over S^4. These may be viewed as boundaries of D^4-disk-bundles over S^4. Thus it may appear to be natural to consider more generally orientable D^m-bundles over S^m. In his paper *"Differentiable Structures on Spheres"* in the Annals of Mathematics 1959 Milnor took such bundles as basic building blocks for a certain construction of manifolds with boundary. For a suitable choice of the building blocks the boundary is an exotic sphere.

The construction is as follows. Take two D^m-bundles $p_i : W_i \to S^m$, $i = 1, 2$ with structure group $SO(m)$. Choose m-disks $U_i \subset S^m$ and trivializations $\varphi_i : D^m \times D^m \to p_i^{-1}(U)$. Let W be obtained from the disjoint union of W_1 and W_2 by identifying $\varphi_1(x, y)$ with $\varphi_2(y, x)$. The result of this is a bounded manifold with corners. Unbending of the corners finally gives a smooth compact orientable manifold W with boundary.

The boundary ∂W may be obtained by gluing two copies of $D^m \times S^{m-1}$ along their boundaries by means of a suitable diffeomorphism. For example, gluing $D^m \times S^{m-1}$ and $S^{m-1} \times D^m$ by means of the identity $S^{m-1} \times S^{m-1} \to S^{m-1} \times S^{m-1}$ gives S^{2m-1}. This was already observed by Hopf in 1935 in his paper mentioned above. The archetypical case of this construction is the Heegaard decomposition of S^3 into two solid tori. This is the case $m = 2$.

However, the construction is older yet. There are reasons to believe that the case $m = 1$ was known to Gauss. Some evidence for this is to be seen in the following two figures. The one on the left hand side

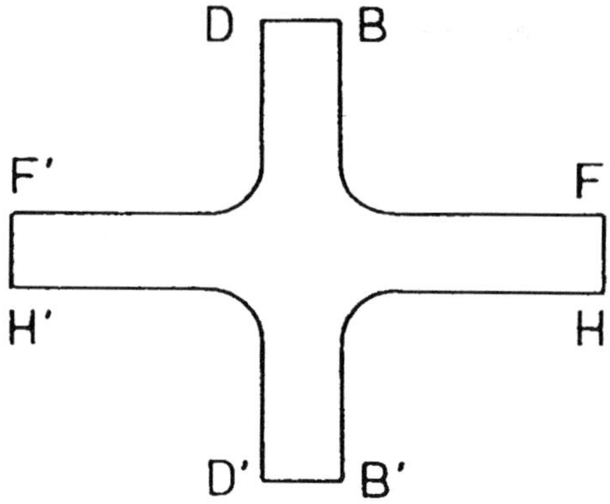

FIGURE 4

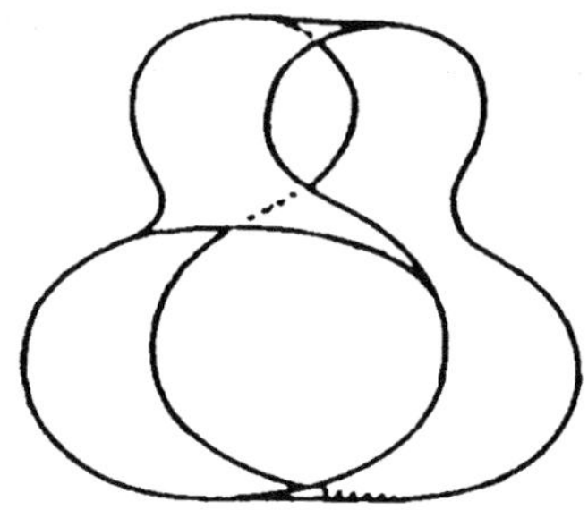

FIGURE 5

dates from the years 1858-60 and is to be found in the collected works of August Ferdinand Moebius, volume 2, page 541. The figure on the right hand side is to be found in the essay *"Der Census räumlicher Komplexe"* published in 1861 by Gauss' student Johann Benedikt Listing. So the simplest case of the construction introduced by Milnor in 1959 was at least one hundred years old and probably known to Gauss.

For a suitable choice of the disk bundles used in Milnor's construction the resulting manifold W will be parallelizable. In particular this will be so, if both copies are the unit-disc bundles in the tangent bundle of S^m. In 1960 Kervaire used the 10-dimensional manifold W obtained in this way for $m = 5$ in order to construct a manifold W_0 which does not admit any differentiable structure. W_0 is the union of W with the cone over ∂W. The boundary ∂W is homomorphic to S^9, but it follows from Kervaire's result that it is not diffeomorphic to it and it is thus the exotic Kervaire sphere generating the cyclic group of order two bP_{10}. An analogous construction can be done for all odd numbers m, and the resulting $(2m-1)$-dimensional Kervaire sphere is the nontrivial element of bP_{2m} whenever bP_{2m} is not trivial.

Milnor's construction can be generalized in various ways. In the first place, one may use more than two disc bundles in the construction, with identifications along disjoint copies of $D^m \times D^m$. The scheme for the construction may be given by a weighted graph. The vertices with weights specify the bundles, the edges are the prescription for the gluing. The graph has to be a tree if we want the resulting manifold M to be highly connected. A further generalization consists in admitting disk-bundles over more general bases such as for example Riemann surfaces of arbitrary genus instead of the Riemann sphere.

A construction of this kind was introduced in a germinal paper by David Mumford submitted to the Publications Mathématiques of the Institut des Hautes Etudes Scientifiques in May 1960 and published in 1961. The title of the paper was *"The topology of normal singularities of an algebraic surface and a criterion for simplicity"*. In this paper Mumford describes certain "good" neighbourhoods of normal singular points x of a complex algebraic surface X in two different ways leading to the same result. One way is to embed (X, x) in some affine space $\mathbb{C}^n$ and to intersect with a sufficiently small ball B^{2n} with center x. The resulting neighbourhood $V = X \cap B^{2n}$ has a boundary $K = \partial V$ which is a 3-dimensional closed orientable manifold. Later work of Hassler Whitney published in 1965 shows that this construction can be generalized to isolated singularities (X, x) of arbitrary dimensions, that the resulting neighbourhood boundary $K = \partial V$ is essentially uniquely determined by (X, x) and that V is homeomorphic to the cone over K.

The second description uses a good resolution $(Y, E) \to (X, x)$ of the 2-dimensional singularity. The exceptional curve E in the complex surface Y is a divisor with normal crossings. Its components E_i are Riemann surfaces intersecting transversely in at most one point. Mumford constructs a smooth boundary M of a tubular neighbourhood of E in Y from building blocks obtained from the normal S^1-bundles of the curves E_i by removing the inverse image of small disks around points where E_i intersects some E_j, $j \neq i$. These building blocks are "patched" by a *"standard plumbing fixture"*

$$\begin{aligned}\{(x,y,u,v)\ |(x^2+y^2) \leq\ 1/4, (u^2+v^2) \leq\ 1/4,\\ (x^2+y^2)^n(u^2+v^2)^m =\ \varepsilon < 1/8^{n+m}\}.\end{aligned}$$

The plumbing fixture is obviously homeomorphic to $S^1 \times S^1 \times [0, 1]$.

Mumford uses this description of the neighbourhood boundary M to derive a presentation of its fundamental group. He then proves the theorem that $\pi_1(M)$ is nontrivial if $x \in X$ is not a regular point. This implies that a normal complex surface which is a topological manifold must be nonsingular.

In the last paragraph of his paper Mumford studies an interesting example. He looks at surfaces in $\mathbb{C}^3$ defined by an equation

$$0 = x^p + y^q + z^r,$$

where p, q and r are pairwise relatively prime, $p < q < r$. Mumford does not resolve these singularities. Instead, he notices that the neighbourhood boundary K is an r-fold branched covering of S^3 branched over a torusknot of type (p, q). Mumford then refers to Herbert Seifert's paper *"Topologie dreidimensionaler gefaserter Räume"* in Acta mathematica 60, 1932, where it is proved that K is a homology 3-sphere. Among these homology spheres, there is only one with finite fundamental group, namely the one for $(p, q, r) = (2, 3, 5)$. Its fundamental group is the binary icosahedral group, and it is the spherical dodecahedral space, as proved by Seifert and Threlfall in part II of *"Topologische Untersuchung der Diskontinuitätsbereiche endlicher Bewegungsgruppen des dreidimensionalen sphärischen Raumes"*. Mumford studies the singular point x of the surface X defined by the equation

$$0 = x^2 + y^3 + z^5.$$

He proves that for a resolution $\pi : Y \to X$ we have $(R^1\pi\mathcal{O}_Y)_x = 0$. This is done without an explicit description of the exceptional divisor. In terms of Michael Artin's 1960 Harvard thesis this means that the singularity is rational. This, together with the fact that the neighbourhood boundary is a homology sphere, implies that the local ring $\mathcal{O}_{X,x} = \mathbb{C}\{x, y, z\}/(x^2 + y^3 + z^5)$ is a unique factorization domain. Actually it is the only nonregular two-dimensional analytic local ring with that property, as I was able to show some years later.

I have now described what was probably known to Hirzebruch when he found the beautiful construction of Milnor's exotic sphere generating bP_{4k}. The construction consists in gluing 8 copies of the tangent-disc-bundle of S^{2k} according to the E_8-scheme. The resulting parallelizable $4k$-manifold is $(2k-1)$-connected and has signature 8. Therefore, its boundary is the Milnor generator of bP_{4k}.

It would be interesting to know whether Hirzebruch's construction had its origin in the remarkable temporal coincidence of the constructions of Milnor and Mumford, one coming from differential topology and the other one from algebraic geometry. There is some evidence for such a fusion of ideas. In February 1963 Hirzebruch gave a talk in the Séminaire Bourbaki reporting on Mumford's paper with a final section "Further remarks", in which he mentions his E_8-construction of the Milnor sphere and points out that certain singularities given by equations have resolution graphs of type A_n, D_n, E_6, E_7, E_8. In this Bourbaki talk Hirzebruch adopts Mumford's term "plumbing" for the construction of manifolds by gluing disk bundles. However, the construction is presented in the way of Milnor, with bending of corners, instead of fitting in Mumford's "plumbing fixture".

In the one page paper *"Zur Theorie der Mannigfaltigkeiten"* Hirzebruch first refers to Milnor's mimeographed notes *"Differentiable manifolds which are homotopy spheres"*, but finishes with the sentences:

"Die Konstruktion wurde motiviert durch die Singularität der affinen algebraischen Fläche $z_1^2 + z_2^3 + z_3^5 = 0$ *in* $(0,0,0)$*. Löst man auf, dann wird der singuläre Punkt aufgeblasen in einen* E_8*-Baum von* 8 *nichtsingulären rationalen Kurven der Selbstschnittzahl* -2*."*

It is also interesting to take notice of Hirzebruch's commentary on this paper in his collected works. There he writes:

"In dem 2. Teil meiner Dissertation [...] *hatte ich zwar die Flächensingularitäten aufgelöst, aber leider, abgesehen von den Quotientensingularitäten* $A_{n,q}$ [...], *keine konkreten Beispiele behandelt. Um 1960 lernte ich die heute so berühmten "einfachen" Singularitäten kennen, deren Auflösungsbäume die aus der Theorie der Lieschen Gruppen bekannten Diagramme* $A_{n-1}, D_{n+2}, E_6, E_7, E_8$ *sind* $(n \geq 2;\ A_{n-1} = A_{n,n-1})$. *Ich benutzte die älteren Arbeiten von Patrick Du Val [...]. Später kamen dann sein Buch (Homographies, quaternions and rotations, Oxford University Press 1964) und eine interessante Korrespondenz mit Du Val hinzu, wodurch ich auch die Beziehungen zur Invariantentheorie nach F. Klein kennenlernte. Die Singularitäten wollte ich dann mittels "plumbing" in höheren Dimensionen "imitieren". So kam ich auf die* E_8*-Konstruktion der Milnorschen exotischen Sphäre."*

It is impossible to present in a few pages the historical development to which Hirzebruch alludes in these sentences. I shall restrict myself to a few comments on the names mentioned by Hirzebruch and to a narration of some part of the story in which Hirzebruch and I myself were involved.

Klein's invariant theory came into being in 1874. In his paper *"Über binäre Formen mit linearen Transformationen in sich selbst"* in Mathematische Annalen 9 we find among other things a relation between three invariants T, f and H of the binary icosahedral group acting on the ring of polynomials in two variables. In 1877 Klein writes this relation as

$$T^2 = 12f^5 - 12^4 H^3.$$

Essentially the same relation had been found a few years earlier by Hermann Amandus Schwarz. In his paper *"Über diejenigen Fälle, in welchen die Gaussische hypergeometrische Reihe eine algebraische Function ihres vierten Elementes darstellt"* published in Crelles Journal, vol. 75 (1872/73) Schwarz considers three polynomials $\varphi_{12}, \varphi_{20}$ and φ_{30} whose roots correspond to the vertices, the midpoints of the faces and the midpoints of the edges of an icosahedron inscribed in the Riemann sphere.

He obtains the identity

$$\varphi_{20}^3 - 4^3 \cdot 3^3 \varphi_{12}^5 = \varphi_{30}^2.$$

Today, we see this as the defining relation between three generators of the ring of invariants $\mathbb{C}[u,v]^G$ of the binary icosehedral groups G acting on $\mathbb{C}^2$, and we identify this ring with the ring of functions on the affine variety $\mathbb{C}^2/G$ imbedded in $\mathbb{C}^3$ and given by such an equation. However, it took a long time until it was possible to see things this way.

An important step relating singularities to root systems and groups generated by reflections was a paper by Patrick Du Val published in 1934 in the Proceedings of the Cambridge Philosophical Society. Du Val considered surface singularities which have a resolution such that all components of the exceptional curve are nonsingular with self-intersection number -2. He classified them by their resolution-graphs, which are, in modern terminology, the Coxeter-Dynkin diagrams of type A_n, D_n, E_6, E_7, E_8. He also describes these singularities as singularities of double coverings of the plane with a description of the singularity of the branch curve. This amounts to writing down equations of the form $z^2 = f(x,y)$. For E_8, or U_{10}^* in Du Val's notation $f(x,y) = y^3 - x^5$. Du Val notices the analogy between his classification and Coxeter's classification of finite groups generated by reflections obtained in the years 1931/34. He shows that the reflection groups, i.e., the Weylgroups of type A_n, D_n, E_6, E_7, E_8 can be used in a systematic discussion of exceptional curves of the first kind and of exceptional configurations of A-D-E-type on rational surfaces. Modern accounts of these matters were given by Manin in his book *"Cubic forms"* and by Demazure in his four talks on Del Pezzo surfaces in the *"Séminaire sur les Singularités des Surfaces"*, 1976/77, dedicated to P. Du Val.

Du Val's 1934 paper had established a link between singularities of type A-D-E and Weyl groups of type A-D-E. On the other hand, around 1960 Hirzebruch realized that these singularities have a relation to the finite subgroups G of $SU(2)$, since their neighbourhood boundaries have the same topological properties as the spherical space forms S^3/G. The exchange with Du Val finally clarified the situation. Du Val identified these singularities with the quotient singularities $\mathbb{C}^2/G$. Those of type A_n correspond to the cyclic groups, the ones of type D_n to the binary octahedral groups, and E_6, E_7, E_8 correspond to the binary tetrahedral, octahedral and icosahedral groups.

When Hirzebruch speaks of "simple" singularities, he is referring to a beautiful discovery of Vladimir Igorevich Arnold made in 1972. In a paper entitled *"Normal forms of functions near degenerate critical points,* *the Weyl groups A_k, D_k, E_k, and Lagrange singularities"* Arnold proved the following theorem:

Every 0-modal germ of an analytic function with an isolated singularity is stably equivalent to one of the germs of type A, D or E at zero; these germs are themselves 0-modal.

Two germs are stably equivalent, if they become equivalent when one adds a number of squares of new variables. Thus the germs equivalent to surface singularities of type A_2 or of type E_8 look like this:

$$z_1^3 + z_2^2 + z_3^2 + \cdots + z_n^2,$$
$$z_1^5 + z_2^3 + z_3^2 + \cdots + z_n^2.$$

These stabilized germs, which were characterized by Arnold by a property of their semiuniversal unfolding or deformation had already appeared in an entirely different context. In 1955 J. Herszberg had characterized them in his thesis by a property of their resolution: They are the only absolutely isolated double points on hypersurfaces. An isolated singular point is called absolutely isolated if it can be resolved by a sequence of monoidal transformations with 0-dimensional centre. For absolutely isolated double points of surfaces this theorem had already been obtained by D. Kirby and had been published in three parts in the Proceedings of the London Mathematical Society 1955-1957. The title was *"The Structure of an Isolated Multiple Point of a Surface"*. Herszberg and Kirby were aware of the earlier work of Du Val.

Later work of Hirzebruch, Milnor and myself was to show that these higher-dimensional singularities of type E_8 and A_2 have a very close relation to the Milnor and Kervaire spheres and to their plumbing construction, a relation going beyond the intentions of Hirzebruch when he wanted to "mimic" the 2-dimensional singularities by plumbing in higher dimensions. This development came as a surprise while I was struggling for the solution of another problem related to E_8.

It began when I asked Hirzebruch for a problem for my first postdoctoral work. This was at some time in 1963. Hirzebruch gave me a 7-page paper by Michael Atiyah published in 1958 in the Proceedings of the Royal Society. The title was *"On analytic surfaces with double points"*. Hirzebruch suggested that I might try to generalize this from ordinary double points, i.e., surface singularities of type A_1, to the other surface singularities of type A_n, D_n, E_6, E_7, E_8. At that time, there was correspondence between Hirzebruch and Du Val about these singularities, there were two Ph.D.-theses on plumbing written by two of Hirzebruch's students, Arlt and von Randow, and there were mimeographed notes of lectures by Hirzebruch at the University of California, Berkeley in 1962 entitled *"Differentiable manifolds and quadratic forms"*. In these notes the A-D-E singularities were treated as twofold algebroid coverings of the plane and resolved by Hirzebruch's method.

Atiyah's paper dealt with Kummer surfaces. The history of these surfaces is too long to be told here. I shall say only a few words about it. The first example of a Kummer surface appeared long before Kummer in the work of Fresnel between 1820 and 1830. He introduced a surface now called Fresnel surface describing the expansion of light in a crystal. Around 1860 this surface appeared in another context in the work of Kummer who investigated focal surfaces of algebraic ray systems. In 1865 Kummer proved that the focal surface of a ray system of order 2 in complex projective 3-space is a surface of degree 4 in $P_3(\mathbb{C})$ with 16 ordinary double points or a degeneration of such a surface. Conversely any surface of degree 4 in $P_3(\mathbb{C})$ with 16 ordinary double points is the focal surface of a ray system of order 2. Subsequent work of Weber, Borchardt, Rohn and Klein showed that these Kummer surfaces of degree 4 in $P_3(\mathbb{C})$ with 16 double points are exactly the surfaces $\mathrm{Jac}\,(C)/\{\pm 1\}$, where $\mathrm{Jac}\,(C)$ is the Jacobian of a Riemann surface of genus 2.

One of the results in Atiyah's 1958 paper is that the nonsingular complex surfaces obtained by the minimal resolution of the 16 double points of a Kummer surface are diffeomorphic to the nonsingular quartic surfaces in $P_3(\mathbb{C})$. This theorem was one of the starting points of the fabulous development of the theory of $K3$-surfaces which began at that time. André Weil refers to Atiyah's theorem in his final report on Contract No. AF 18(603)-57 and to his exchange with Atiyah. He says that he had observed independently that the minimal resolution of a surface in $P_3(\mathbb{C})$ with one ordinary double point and a nonsingular surface in $P_3(\mathbb{C})$ of the same degree are diffeomorphic. The reason for the name $K3$-surface introduced by Weil is given in his comment on that final report: "ainsi nommés en l'honneur de Kummer, Kähler, Kodaira et de la belle montagne $K2$ au Cachemire". Today, a $K3$-surface may be defined as a compact complex surface with trivial canonical bundle and first Betti number 0. The minimal resolutions of Kummer surfaces are very special $K3$-surfaces, which have remarkable symmetry properties, and which are used in the analysis of moduli problems of $K3$-surfaces. I refer to the chapter on $K3$-surfaces in the beautiful book *"Compact complex surfaces"* of Barth, Peters and Van de Ven.

Atiyah used the following basic facts. Consider the quadric cone V in $\mathbb{C}^4$ given by the equation

$$x_1x_2 - x_3x_4 = 0.$$

Let $Q \subset P_3(\mathbb{C})$ be the quadric of complex lines $\ell \subset V$. In the Grassmannian of complex planes in $\mathbb{C}^4$ there are two projective lines $P_\pm$ of planes $p \subset V$. Each line $\ell \in Q$ is contained in two uniquely determined planes $p_\pm(\ell) \in P_\pm$. This defines two projections $p_\pm : Q \to P_\pm$ and an isomorphism $Q \to P_+ \times P_-$.

Now define three modifications W_+, W_- and W of V as follows:

$$W_\pm = \{(x, p) \in X \times P_\pm \mid x \in p\},$$
$$W = \{(x, \ell) \in X \times Q \mid x \in \ell\}.$$

There is an obvious diagram of holomorphic maps

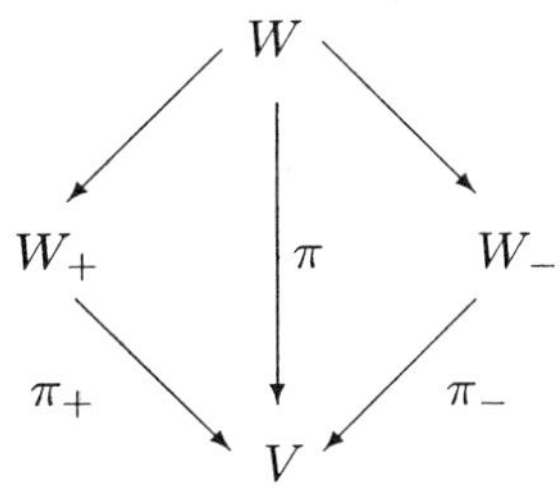

The modifications π, π_+, π_- are three different resolutions of the singularity $(V, 0)$. Whereas π replaces the singular point by the divisor Q in W, the resolutions π_+ and π_- are "small" resolutions. They replace the singular point only by the 2-codimensional curves $P_\pm$ in $W_\pm$. The resolution π is obtained by blowing up the maximal ideal of the local ring $\mathcal{O}_{V,0}$. This local ring is not a unique factorization domain. Its divisor class group is infinite cyclic. The divisors $p \in P_+$ represent one generator, those in P_- the other one. W_+ and W_- may be obtained by blowing up the nonprincipal ideals corresponding to these divisors, e.g. (x_1, x_3) and (x_1, x_4). The transition from W_+ to W_- is the simplest example of what people working on complex 3-manifolds nowadays call a flop.

Atiyah uses these modifications as follows. Let $f : (X, x) \to (S, s)$ be the germ of a map from a 3-dimensional complex manifold to a 1-dimensional complex manifold. Assume that the fibre has an ordinary double point of type A_1 at x. Let $\varphi : (T, t) \to (S, s)$ be a double covering by a smooth germ (T, t) ramified in t. Then the fibred product $(T \times_S X, t \times x)$ is isomorphic to the quadric cone $(V, 0)$. Choosing an isomorphism and choosing one of the two modifications W_+, W_-, we get a modification X' of $T \times_S X$. Choosing suitable representatives, we get a diagram of holomorphic maps of complex manifolds

$$\begin{array}{ccc} X' & \xrightarrow{\psi} & X \\ {\scriptstyle f'}\downarrow & & \downarrow{\scriptstyle f} \\ T & \xrightarrow[\varphi]{} & S \end{array}$$

with the following properties: (i) f' is a regular map, i.e., without singularities, (ii) φ is a ramified covering, (iii) ψ is proper and surjective, (iv) for each fibre X'_t of f' the map $X'_t \to X_{\varphi(t)}$ is a resolution of the singularities of the fibre $X_{\varphi(t)}$ of f. Let us call such a diagram a simultaneous resolution of the singularities of the fibres of f.

The construction of Atiyah indicated above gives the existence of simultaneous resolutions for maps $f : X \to S$ of 3-manifolds X to 1-manifolds S with only A_1-type singularities. His theorem on Kummer surfaces is an easy consequence of this.

I took it that my task was to generalize this to all surface singularities of type A_n, D_n, E_6, E_7, E_8. One difficulty in the beginning was that it was not quite clear what was meant by "the" A-D-E-singularities, since a priori the definition by the resolution graph was wider than the other definitions (i.e., by equations, or as quotient singularities or as absolutely isolated double points). Correspondence on this with Du Val and Kirby was not conclusive, but in 1964/65 the situation was clarified by means of Michael Artin's new work on rational singularities published in 1966 in the American Journal. The A-D-E-singularities were identified with Artin's rational double points and were determined up to isomorphism by the corresponding diagram.

When $f(x, y, z) = 0$ is the equation of such a singularity, the fibered product for a base extension by a covering of degree d will have the equation

$$f(x, y, z) - t^d = 0.$$

Thus, in the cases A_n, E_6 and E_8 this leads to equations of the form

$$x^a + y^b + z^c + t^d = 0.$$

I tried to find small modifications of these 3-dimensional singularities by mapping them to others such as the quadric cone and inducing the small modification from another one already constructed. For example mapping to the quadric cone V meant writing $F = f(x, y, z) - t^d$ in the form $\phi_1\phi_2 - \phi_3\phi_4$. With such methods and encouraged by my teacher I constructed in 1964 the simultaneous resolutions for A_n, D_n, E_6 and E_7. It also became clear that for maps $f : X \to S$ of 3-manifolds to 1-manifolds simultaneous resolutions could only exist for the A-D-E-singularities. So the only case in question was E_8.

My calculations in the other cases had shown that somehow the right number d for the base extension was the Coxeter number for the corresponding root system. Thus the equation to consider for E_8 was

$$x^2 + y^3 + z^5 + t^{30} = 0.$$

I was unable to treat this case with the methods which I had used for the other cases. During the Arbeitstagung in 1965 I talked about this with Heisuke Hironaka. He suggested that I should study the divisor class groups of the local rings of the 3-dimensional singularities which I wanted to modify by blowing up the ideals of nonprincipal divisors. In particular, I should study the cohomology of the neighbourhood boundaries of these singularities, since the divisor class group of the local ring of a 3-dimensional isolated Cohen Macaulay singularity injects into the second cohomology group of the neighbourhood boundary with integer coefficients.

Shortly afterwards I set sail for New England because following Hirzebruch's advice I had successfully applied for a C.L.E. Moore instructorship at MIT. This was a very good place for me, since Michael Artin was at MIT and David Mumford at Harvard. Michael was very friendly and always ready to help, and I learned a lot from the many discussions which we had. David was also very friendly and in a number of discussions gave me very valuable ideas. The two years in Boston and Cambridge are among the best in my mathematical life.

Shortly after my arrival in Boston I intended to calculate the divisor class group for

$$x^2 + y^3 + z^5 + t^{30} = 0.$$

However, Mumford suggested that I should first look at the simpler example

$$x^2 + y^3 + z^5 + t^2 = 0.$$

This is the 3-dimensional E_8-singularity. I decided to do first a much simpler case, namely, the 3-dimensional A_2-singularity

$$z_1^2 + z_2^2 + z_3^2 + z_4^3 = 0.$$

I discovered quickly that it was factorial because the second cohomology group of the neighbourhood boundary was zero. Then I did the E_8-case suggested by Mumford, which was much more tedious, since the resolution by a sequence of monoidal transforms with the singular points as centres leads to an exceptional divisor with a dozen components. Again I found that the second cohomology group was zero and the divisor class group trivial. I was not happy about this, since I wanted nontrivial divisor class groups. Anyhow, I had a closer look at the topology of the neighbourhood boundary of the 3-dimensional A_2-singularity, and in September 1965 I made the irritating discovery that this singularity was topologically trivial. Its neighbourhood boundary is homeomorphic to S^5. Thus, there was no analogue of Mumford's theorem for singularities of dimension higher than two. Of course, I told this immediately to Mumford, and I also wrote a letter to Hirzebruch. In that letter dated September 28, I speculated about the E_8-singularity and possible connections with Hirzebruch's E_8 plumbing contruction and exotic spheres.

At that time, Hirzebruch was at a conference in Rome. He gave a very nice talk entitled *"Über Singularitäten komplexer Flächen"*, where he explained the A-D-E surface singularities and many related subjects. Among other things, he reported on my work on simultaneous resolution, and on the recent discovery announced in my letter.

Meanwhile I continued my efforts to construct the missing simultaneous resolution for E_8. I tried to gather strength by looking at the beautiful crystal icosahedron on the mantelpiece of my apartment on Beacon-Hill, but I did not get anywhere, and I got more and more depressed. In December 1965 I wrote to my mother that I was abandoning E_8. I also wrote about an experience intensifying my melancholy mood, namely, meeting John Nash. I knew that he had done extraordinary things before he got ill. In 1965/66 he was in Brandais and back in MIT, and he was able to do mathematics. Sometimes late in the evening we met in the long high corridors of MIT and started to talk about mathematics. Nash was interested in the resolution of singularities of complex algebraic varieties. Some traces of our conversation may be seen in a draft of a paper entitled *"Arc structure of singularities"*, where absolutely isolated double points of dimensions 2 and 3 serve as examples illustrating his distinction between essential and inessential components of the exceptional set in a resolution. What made me sad to the extent of being terrified was the feeling that he had lost his strength. I felt that this once powerful mind had broken wings.

In February 1966 I gave a talk in Cornell on the topology of singularities showing my example

$$z_1^2 + \ldots + z_k^2 - z_0^3 = 0 \quad , \quad k > 1 \text{ odd},$$

for a singular normal complex space which is a topological manifold. I had been invited by my friend, Hirzebruch's student, Klaus Jänich. We had a good time together, thus feeling at home in a foreign country. I do not remember whether we talked about Jänich's work. If we did, we certainly didn't anticipate what happened next.

In a letter dated March 24, 1966 Hirzebruch told me that he had found out that there were very close connections between my work and that of Jänich, which were explained in a provisional manuscript of eight pages.

Jänich was studying actions of compact Lie groups G on connected differentiable manifolds X without boundary. Such a G-manifold X was called "special", if for each $x \in X$ the action of the isotropy group G_x on the normal space in x to the orbit Gx is the direct sum of a trivial and a transitive representation. "Transitive" means transitive on the set of rays. The orbit space X/G of a special G-manifold is canonically a differentiable manifold M with boundary. Let M^0 be the interior of M and A be the set of boundary components. The orbit structure of X associates to M^0 the conjugacy class of isotropy groups $H(x) = G_x$ of $x \in X$ over M_0 and to $\alpha \in A$ the conjugacy class of isotropy groups $U_\alpha(x) = G_x$ of points x over α. Jänich defined a notion of admissible fine orbit structures in terms of data $H \subset G$, $H_\alpha \subset G$, $\alpha \in A$. His main result was a classification of special G-manifolds X with quotient M in terms of these fine orbit structures. The result was published in Topology 5, 1966. Hirzebruch used the manuscript of this paper. At about the same time, there was independent closely related work of Wu-Chung Hsiang and Wu-Yi Hsiang announced in the Bulletin of A.M.S.

Hirzebruch applied Jänich's result to a very special class of examples. Motivated by my work he looked at the neighbourhood boundaries of absolutely isolated double points of type A_{d-1}. Thus, he considered the differentiable manifolds $W^{2n-1}(d)$ in $\mathbb{C}^{n+1}$ given by the equations

$$\begin{aligned} z_0^d + z_1^2 + \cdots + z_n^2 &= 0 \\ |z_0|^2 + |z_1|^2 + \cdots + |z_n|^2 &= 2. \end{aligned}$$

He noticed that there is an obvious operation of the orthogonal group $O(n)$ on $W^{2n-1}(d)$. The operation is obvious indeed, but only if you have the idea of looking for it. Hirzebruch proved that $W^{2n-1}(d)$ is a special $O(n)$-manifold with orbit space D^2, the 2-disk. The isotropy groups are conjugate to $O(n-1)$ for the special orbits, i.e., those with $|z_0| = 1$ and to $O(n-2)$ for the general orbits.

Hirzebruch applied Jänich's classification result to the special case of special $O(n)$-manifolds, $n \geq 2$, with this orbit structure $H = O(n-2)$, $U = O(n-1)$. According to Jänich, they are classified up to equivariant diffeomorphism by an integer $d \geq 0$. For $d = 0$ one has the diagonal action on $S^n \times S^{n-1}$. For $d > 0$ Hirzebruch proved that one gets exactly the $O(n)$-manifolds $W^{2n-1}(d)$, $d \geq 1$.

Certain $O(n)$-manifolds M_k^{2n-1} with orbit type $(O(n-2), O(n-1))$ had been studied by Bredon in a paper in Topology 3, 1965. Hirzebruch noticed that they are special, and thus he could identify M_k^{2n-1} with $W^{2n-1}(2k+1)$. Using Bredon's results, he could prove that $W^{2n-1}(d)$ is a homology sphere if and only if d is odd, and that for fixed n and different $d's$ one gets different knots.

But the most exciting result was derived from a result of Bredon on his M_1^9 derived from a result of Kosinski: M_1^9 is an exotic sphere. Hence Hirzebruch got

Theorem 3. *The manifold*

$$W^9(3) = \{(z_0, \ldots, z_5) \in \mathbb{C}^6 \mid z_0^3 + z_1^2 + \ldots + z_5^2 = 0 \ , \ \|z\| = 1\}$$

is an exotic 9-sphere.

$W^{2n-1}(d)$ is obviously embedded in S^{2n+1}. Kervaire had proved in a paper which appeared in the volume *"Differential and Combinatorial Topology"*:

A homotopy m-sphere can be imbedded in S^{m+2} if and only if it bounds a parallelizable manifold.

Thus it was clear that $W^9(3)$ is the 9-dimensional Kervaire sphere. But where was the highly connected parallelizable manifold obtained by plumbing two disc-bundles with boundary $W^9(3)$? And how about the absolutely isolated singularities of type E_8? The parallelizable manifolds were not expected to be found by resolution. Hirzebruch speculated on this question in a postscript referring to my own speculation on his E_8-construction. Hirzebruch expected to deal with the E_8-case by means of a certain generalization of Jänich's result to $O(n)$-manifolds with 3 types of orbits and to show in this way that the neighbourhood boundaries of the n-dimensional E_8-singularities were the Milnor generators of bP_{2n} for n even.

The joy about these wonderful results was so great that I was almost beside myself, and I wrote in a letter to Hirzebruch that I could not imagine a more beautiful interplay between teacher and students.

Two weeks later a new player appeared. On April 16, John Nash showed me a letter to him from John Milnor, dated Ann Arbor 4-13-66. As far as I know, this letter has not been published. I hope that it is not improper when I publish it here. It is also a sign of gratitude to John

Nash. The text is as follows.

Dear John,

I enjoyed talking to you last week. The Brieskorn example is fascinating. After staring at it a while I think I know which manifolds of this type are spheres, but the statement is complicated and the proof doesn't exist yet. Let $\Sigma(p_1, \dots, p_n)$ be the locus

$$z_1^{p_1} + \dots + z_n^{p_n} = 0 \ , \ |z_1|^2 + \dots + |z_n|^2 = 1$$

where $p_j \geq 2$. It is convenient to introduce the graph G which has one vertex for each p_i and one edge for each p_j, p_k which have $G.C.D.$ greater than one.

E.g. (4 , 6, 7, 15) $\longmapsto$ 

Assertion. $\Sigma(p_1, \dots, p_n)$ is a topological $(2n-3)$-sphere if and only if $n \neq 3$ and either

a) G has at least two isolated points

or b) G has one isolated point and one component consisting of an odd number of p_j's, any two of which have $G.C.D. = 2$.

For example $\Sigma(2,2,2,25)$ and $\Sigma(2,2,2,3,5)$ are topological spheres, but $\Sigma(4,6,7,15)$ or $\Sigma(2,2,2,2,3)$ is not.

In the case $2n-3 \equiv 1 \mod 8$ one can describe which are exotic spheres and which not; but I can't handle the other dimensions.

Are results of this type known to Brieskorn or Hirzebruch?

Note: The conjecture I mentioned about $(\text{disk}, \text{disk} \cap \Gamma) \cong (\text{slab}, \text{slab} \cap \Gamma)$ is true and not so difficult.

Regards

Jack

There was a little figure about 1 cm in diameter on the margin next to the last sentence. Figure 6 is a magnified facsimile.

Initially I must have overlooked this figure or failed to realize that it was a key for understanding Milnor's approach. For, when I sent a copy of the letter to Hirzebruch I wrote that I had no idea how he was going to prove his assertion.

But even without knowing Milnor's ideas I was able to prove his assertion by good luck in less than two weeks. On the shelf for the latest journals in the library of MIT I found an article by Frédéric Pham, submitted to the Bulletin de la Société Mathématique de France in March 1965. The title was *"Formules de Picard-Lefschetz généralisées et ramification des integrales"*. Pham, who at that time was working at the Service de Physique théorique in Saclay was motivated by problems which

FIGURE 6

at first sight seemed to be unrelated to what we were doing. The paper was a contribution to efforts of theoretical physicists aiming at a better understanding of the singularities and discontinuities of Feynman integrals by applying methods of algebraic topology developed for the topological analysis of algebraic manifolds. Thus these efforts had their mathematical roots in the two volume treatise *"Théorie des fonctions algébriques de deux variables indépendantes"* published by Picard-Simart in 1897 and 1906 and in Lefschetz' 1924 monograph *"L'analysis situs et la géométrie algébrique"*. Pham was also influenced by work of Leray and Thom. The first part of Pham's paper was a generalization of the classical Picard-Lefschetz formula. This formula describes the monodromy transformation on the homology of a general member of a pencil of algebraic varieties such that the singularities of the special members are at most ordinary double points. Pham generalized the Picard-Lefschetz formula exactly to the class of singularities considered in Milnor's letter to Nash.

Let $a = (a_0, \ldots, a_n)$ be a tuple of positive integers. Pham considers the pencil of affine hypersurfaces

$$\Xi^a(t) = \{(z_0, \ldots, z_n) \in \mathbb{C}^{n+1} \mid z_0^{a_0} + \ldots + z_n^{a_n} = t\}.$$

Let $\omega_k = e^{2\pi i/a_k}$ and C_{a_k} the cyclic group of unit roots generated by ω_k. Pham constructs a simplicial complex in $\Xi^a(1)$ which is a deformation retract of $\Xi^a(1)$. As an abstract complex it is the iterated join

$$C_{a_0} * \ldots * C_{a_n}.$$

$\Xi^a(0)$ is the only singular member of the pencil. Intersecting with the sphere $\|z\| = 1$ one gets Milnor's $\Sigma(a_0, \ldots, a_n)$. Removing the singular fibre on gets a fibre bundle $\mathbb{C}^{n+1} - \Xi^a(0) \to \mathbb{C} - \{0\}$. There is a geometric monodromy $(z_0, \ldots, z_n) \mapsto (\omega_0 z_0, \ldots, \omega_n z_n)$ acting in the obvious way on the complex $C_{a_0} * \ldots * C_{a_n}$. Using these facts Pham calculates the homology of $\Xi^a(1)$ and the monodromy transformation on the only nontrivial reduced homology group $H_n(\Xi^a(1))$ with integral coefficients. He also calculates the intersection form on the homology.

From these results I could painlessly deduce the homological part of Milnor's assertion. All I had to do was to calculate the characteristic polynomial of the monodromy

$$\Delta_a = \prod_{0<i_k<a_k} \left(x - \omega_0^{i_0} \ldots \omega_n^{i_n}\right)$$

and to show that $\Delta_a(1) = \pm 1$ is equivalent to Milnor's condition on the graph G_a associated to a. This was done on four pages by April 25 and sent to Milnor and Hirzebruch. Until the end of the month I had also proved on two more pages that $\pi_1(\Sigma(a_0, \ldots, a_n))$ is trivial for $n > 2$. Thus the proof of Milnor's assertion was complete and I went on to show that Pham's result also allowed to calculate the signature σ_a of $\Xi^a(1)$ for n even and thus to determine the differentiable structure when $\Sigma(a_0, \ldots, a_n)$ was an exotic sphere of dimension $4m - 1$. I did the case $\Sigma(2, 2, 2, 3, 5)$ explicitly and found that it was Milnor's generator of bP_8 constructed by Hirzebruch's E_8-plumbing construction.

Meanwhile by the end of April Milnor had completed a manuscript entitled *"On isolated singularities of hypersurfaces"*. It did not give a complete proof of the assertion about the class of singularities considered in his letter to Nash, but it contained foundational results on arbitrary isolated hypersurface singularities.

Consider holomorphic functions f defined in a neighbourhood U of 0 in $\mathbb{C}^{n+1}$ with an isolated singularity at 0 and $f(0) = 0$. Let $B_\varepsilon \subset U$ be the ball $\|z\| \leq \varepsilon$ and $S_\varepsilon = \partial B_\varepsilon$. For $\eta > 0$ let $D_\eta \subset \mathbb{C}$ be the disk $|t| < \eta$. Consider the "slab"

$$N_{\varepsilon,\eta} = B_\varepsilon \cap f^{-1}(D_\eta).$$

This is the slab shown by the figure in the letter to Nash. f defines a map

$$f : N_{\varepsilon,\eta} \longrightarrow D_\eta.$$

Let F_t denote the fibre of this map over $t \in D_\eta$. The fibre F_0 is singular for small ε. The intersection

$$K_\varepsilon = F_0 \cap S_\varepsilon$$

is a closed manifold and the diffeomorphism type of this neighbourhood boundary does not depend on ε. Fix ε and choose η sufficiently small. Then

$$f : N_{\varepsilon,\eta} - F_0 \longrightarrow D_\eta - \{0\}$$

is a differentiable locally trivial fibre bundle. The fibre F_t is parallelizable and $(n-1)$-connected with boundary diffeomorphic to K_ε. It has the homotopy type of a wedge of n-spheres. ∂F_t is $(n-2)$-connected. It is a homology sphere iff $\Delta(1) = \pm 1$ for the characteristic polynomial Δ of the fibre bundle. For n odd, the value $\Delta(-1) \mod 8$ determines the Arf invariant and hence the class of K in bP_{2n}, if K is a homotopy sphere. For homotopy spheres K with n even, the class of K in bP_{2n} is determined by the signature of F_t divided by 8.

There is a homeomorphism $N_{\varepsilon,\eta} \to B_\varepsilon$ keeping K pointwise fixed. It identifies $\partial N_{\varepsilon,\eta}$ with S_ε. The part of $\partial N_{\varepsilon,\eta}$ lying over ∂D_η is identified with the complement of a tubular neighbourhood of K in S_ε which therefore becomes a fibre bundle over a circle with typical fibre F_t. The fibration may be defined by $z \to f(z)/\|f(z)\|$. Thus $K \subset S_\varepsilon$ is a fibred knot. These fibrations are nowadays called *"Milnor fibrations"* and Milnor's results or analogues of them are fundamental for nearly all work on the topology of singularities.

Later on a local Picard-Lefschetz theory was developed which allows to represent certain bases of the homology of the Milnor fibre by vanishing cycles which are embedded n-spheres with tubular neighbourhoods isomorphic to their tangent disc bundles. In the case of the absolutely isolated double points of type A_k, D_k, E_6, E_7, E_8 a suitable choice of such a basis of vanishing cycles allows to identify the Milnor fibre directly with the corresponding parallelizable manifold constructed by plumbing.

Meanwhile Hirzebruch had been pursuing his idea of dealing with stabilized curve singularities

$$f(x,y) + z_1^2 + \ldots + z_n^2 = 0$$

via certain $O(n)$-manifolds with three types of orbits. Nearly simultaneously with Milnor's manuscript I got a manuscript from Hirzebruch dated May 1, 1966 with many beautiful results on these manifolds.

The general theory of such knot-G-manifolds was developed in the last paragraph of Jänich's article. Consider a closed connected $(2n+1)$-dimensional $O(n)$-manifold M with the following properties:

(i) The isotropy groups are conjugate to $O(n-2)$, $O(n-1)$ or $O(n)$.

(ii) The set F of fixed points is not empty and for any $x \in F$

the operation of $O(n)$ is the diagonal operation on $\mathbb{R}^{2n}$ plus a trivial operation on $\mathbb{R}$.

(iii) $M - F$ is a special $O(n)$-manifold.

These conditions imply that the orbit space $M/O(n)$ is a 4-dimensional manifold with boundary. M is called a knot $O(n)$-manifold, if there is a diffeomorphism of $M/O(n)$ with the 4-disk D^4 sending the image of F in the orbit space bijectively onto a knot $k \subset S^3 = \partial D^4$. Jänich proves that for each knot $k \subset S^3$ and each integer $n \geq 2$ there is a welldetermined $(2n+1)$-dimensional knot $O(n)$-manifold $\gamma_n(k)$. For $n = 1$ we define $\gamma_1(k)$ as the 2-fold covering of S^3 branched along k.

Hirzebruch announced the following result. $\gamma_n(k)$ is an $(n-1)$-connected manifold bounding a parallelizable manifold which can be constructed explicitly from a Seifert diagram of the knot. When $\gamma_n(k)$ is a homotopy sphere the invariants determining its class in bP_{2n+2} can be calculated from invariants of the knot.

Now let k be an algebraic knot, i.e., a knot associated to a plane curve singularity at the origin

$$k = \{(x,y) \in \mathbb{C}^2 \mid f(x,y) = 0\ ,\ |x|^2 + |y|^2 = 1\}.$$

Then $\gamma_n(k)$ is the neighbourhood boundary of the corresponding stabilized singularity and is imbedded in the standard sphere S^{2n+3} in $\mathbb{C}^{n+2}$:

$$\gamma_n(k) = \{(x,y,z_1,\ldots,z_n) \in S^{2n+3} \mid f(x,y) + z_1^2 + \ldots + z_n^2 = 0\}.$$

In particular for a torus knot $k = t(p,q)$ one obtains

$$\gamma_n(k) = \Sigma(p,q,2,\ldots,2).$$

Hirzebruch indicated a tentative way for calculating the signature and obtained an explicit formula for $t(p,q)$. In particular, he also concluded that for $\Sigma(3,5,2,\ldots,2)$ for n odd is the Milnor generator of bP_{2n+2}.

When Hirzebruch got my letter referring to Pham's paper, he saw very quickly how to calculate the signature of the evendimensional varieties of Pham by using Pham's description of the intersection form. He told me the result together with the proof in a letter dated May 9, 1966. The result is as follows. For $a = (a_0,\ldots,a_n)$, n even, the signature σ_a of $\Xi^a(1)$ is

$$\sigma_a = \sigma_a^+ - \sigma_a^-$$

where σ_a^+ and σ_a^- are the numbers of tuples $(j_0,\ldots,j_n)$ with $0 < j_k < a_k$ such that

$$\begin{aligned} 0 &< \Sigma j_k/a_k < 1 \mod 2 \quad \text{for} \quad \sigma_a^+, \\ -1 &< \Sigma j_k/a_k < 0 \mod 2 \quad \text{for} \quad \sigma_a^-. \end{aligned}$$

The proof given in my paper in Inventiones is Hirzebruch's proof.

Later on the formula for σ_a went through a remarkable metamorphosis. In March 1970 I got a letter from Don Zagier who had been a student of Atiyah and had attended lectures on singularities and exotic spheres which I gave in Oxford in 1969. Don Zagier had discovered a formula for σ_a resembling closely the form of the Atiyah-Bott fixed point theorems and Atiyah-Singer G-signature theorems. Here is the formula:

$$\sigma_a = \frac{(-1)^{n/2}}{N} \sum_{\substack{j=1 \\ j \text{ odd}}}^{2N} \cot\frac{\pi j}{2N} \cot\frac{\pi j}{2a_0} \cot\frac{\pi j}{2a_1} \ldots \cot\frac{\pi j}{2a_n}$$

where n is even and N is any common multiple of $a_0, \ldots, a_n$. I sent Zagier's letter to Hirzebruch, who found Zagier's result very interesting, since he had studied similar questions and had tried to get σ_a through the G-signature theorem. In March 1970 Hirzebruch had been taking part in the inauguration of the new Fine Hall in Princeton and had given a talk entitled *"The signature theorem: Reminiscences and recreations"*. The underlying theme was *"More and more number theory in topology"*. In that talk Hirzebruch dealt with Dedekind sums and reciprocity theorems and Markoff triples and tried to establish relations with the Atiyah-Bott-Singer index theorem and fixed point theorem. When he got Zagier's formula, he pointed out to Zagier that it could be deduced from a formula of Eisenstein, which Hirzebruch had found in Rademacher's lectures on Analytic Number Theory: Let $((x)) = x - [x] - \frac{1}{2}$ for $x \in \mathbb{R} - \mathbb{Z}$ and $((x)) = 0$ for $x \in \mathbb{Z}$. Then Eisenstein's formula expresses $((x))$ for rational $x = p/q$ with positive integers p, q by a trigonometric sum:

$$((p/q)) = \frac{i}{2q} \sum_{k=1}^{q-1} \cot\frac{\pi j}{q}\, e^{2\pi i k p/q}.$$

Zagier's formula can be deduced from this formula since σ_a can easily be expressed as a sum of values of $((\ ))$ for rational numbers.

Hirzebruch invited Zagier to discuss these matters with him in Bonn, and this was the beginning of a cooperation that led to Zagier's Lecture Notes *"Equivariant Pontrjagin Classes and Applications to Orbit Spaces"*, to the joint monograph *"The Atiyah-Singer theorem and elementary number theory"* and to joint papers on Hilbert modular surfaces. Zagier proves in his lecture notes a signature theorem expressing $\mathrm{Sign}(g, \Xi^a(1))$ as a trigonometric sum, where g is any element of the group $C_{a_0} \times \ldots \times C_{a_n}$ operating on Pham's $\Xi^a(1)$. For $g = 1$ this specializes to his formula for σ_a. Other specializations include a result of Hirzebruch and Jänich published in 1969 in their joint paper *"Involutions and*

Singularities". The description of exotic spheres bounding parallelizable manifolds as neighbourhood boundaries $\Sigma(a_0, \ldots, a_n)$ gives lots of exotic actions of finite groups on these manifolds. It is natural to try to distinguish them by invariants. One such invariant is the Browder-Livesay-invariant. Hirzebruch and Jänich identify it with an invariant introduced by Hirzebruch in his paper *"Involutions on manifolds"*. They calculate this invariant in certain cases for the involution T on $\Sigma(a_0, \ldots, a_n)$ given by $T(z) = -z$, where all a_j have the same parity. For even parity they have a general formula, and this formula turned out to be a special case of Zagier's theorem.

When the problem relating to the manifolds $\Sigma(a_0, \ldots, a_n)$ had been clarified by Hirzebruch, Milnor and myself, I returned to my old problem of constructing simultaneous resolutions for the rational double point E_8, the icosahedral singularity. Now I was finally able to solve it. I found that the number of solutions of the problem is

$$2^{14} \cdot 3^5 \cdot 5^2 \cdot 7.$$

This is the order of the Weyl group of type E_8. The divisor class group of the local ring of the singularity

$$x^2 + y^3 + z^5 + t^{30} = 0$$

has the structure of the lattice of weights of the root system E_8, and the different solutions correspond to the Weyl chambers. For each chamber one obtains a solution by blowing up any ideal class in the chamber. The construction of the solution used very classical algebraic geometry, an old paper of Max Noether on rational double planes from 1889 and properties of the exceptional curves on rational surfaces obtained by blowing up 8 points on a plane cubic. Some of these facts had been explained to Hirzebruch and to me by Du Val, and Hirzebruch had mentioned them in his talk in Rome.

In May 1969 Grothendieck read my papers on simultaneous resolution. He told me some interesting conjectures on related problems. Whereas I had been considering simultaneous resolutions of a very special kind of deformation of the A-D-E-singularities, he suggested to look at the semiuniversal deformation. He conjectured that this deformation was to be found in the adjoint quotient map of the simple Lie algebras of type A-D-E, and that a universal simultaneous resolution was to be obtained by means of a generalization of the Springer resolution of the nilpotent variety. I proved these conjectures with the help of Tits and explained it at the ICM in Nice 1970. Later developments were beautiful extensions to all simple Lie algebras by Slodowy and characterizations of the universal simultaneous resolution as universal deformation of the resolution of the rational double points by Michael Artin and Huikeshoven. Recently there has been very interesting work of Slodowy and Helmke on the relation between loop groups and elliptic singularities. But this is a different story.

Cusps

Let me return to the singularities in the work of Friedrich Hirzebruch. In the hierarchy of singularities as described by Arnold there is an interesting class of singularities lying between the simple singularities of A-D-E type and the simply elliptic singularities of type $\tilde{E}_6, \tilde{E}_7, \tilde{E}_8$. These are the singularities T_{pqr} with equation

$$x^p + y^q + z^r + xyz = 0,$$

where $1/p + 1/q + 1/r < 1$. These belong to a class of singularities which Hirzebruch discovered in 1970.

Hirzebruch has given four talks in the Séminaire Bourbaki. It is a remarkable fact that in three of them singularities played an important role. The first of these Bourbaki lectures was the report on the work of Mumford and the higher dimensional E_8-plumbing construction. The second was on singularities and exotic spheres. Finally the third lecture, delivered in June 1971, had the title: *"The Hilbert modular group, resolution of the singularities at the cusps and related problems"*.

This contribution of Hirzebruch is on one hand a direct continuation of work in his thesis and on the other hand has its origin in work of David Hilbert in 1893/94 and in the Habilitationsschrift of Hilbert's first student Otto Blumenthal. In his thesis Hirzebruch had considered surface singularities which are resolved by a chain of rational curves. Now the objects to be studied are surface singularities which are resolved by a cycle of rational curves. In the printed version of his thesis Hirzebruch had claimed without proof that there could be no cycles in the resolution graph of a surface singularity. He had soon noticed that this was wrong, and now singularities resolved by a cycle of rational curves became objects with which he occupied himself during a whole decade, from 1970 until 1980.

Let us consider a finite sequence of nonnegative integers $b_1, \ldots, b_q$. We want to construct a surface singularity with a cyclic resolution by nonsingular rational curves with self-intersection numbers $-b_k$. We want to do a construction in the style of Hirzebruch's thesis by using toroidal embeddings.

A cyclic configuration of exceptional curves with mutual intersection numbers 0 or 1 has at least 3 elements. So we assume $q \geq 3$. For $q > 3$ an exceptional curve of the first kind in such a configuration can be blown down, and the resulting configuration is still cyclic. So for $q > 3$ we can assume $b_k \geq 2$ for all k. Thus for $q \geq 3$ we admit all sequences such that $b_k \geq 2$ for all k, but not all b_k equal 2. For $q = 3$ we admit also sequences of the form $(a+3, 2, 1)$ with $a \geq 3$ and $(a_1+1, 1, a_2+1)$ with $a_1, a_2 \geq 2$ and $a_1 \geq 3$ or $a_2 \geq 3$. In these two cases blowing down of exceptional curves of the first kind leads to "reduced" sequences (a) and (a_1, a_2) of length 1 and 2 respectively. Hirzebruch gives two constructions for singularities with cyclic resolution. The first one is analogous to the construction in his thesis and uses the nonreduced sequence $b = (b_1, \ldots, b_q)$. The second one uses the reduced sequence, which I shall again denote by $(b_1, \ldots, b_q)$. I admit that this is an abuse of notation.

Here is the first construction. We define a doubly infinite sequence of integers b_k, $k \in \mathbb{Z}$ by $b_j \equiv b_k$ if $j \equiv k \mod q$. Now we proceed in strict analogy with the construction in Hirzebruch's thesis. We construct a 2-dimensional complex manifold Y_b by gluing an infinite number of copies of $\mathbb{C}^2$, one for each integer k. The transformation from the $(k-1)$-th copy to the k-th copy is

$$\begin{aligned} u_k &= u_{k-1}^{b_k} v_{k-1} \\ v_k &= u_{k-1}^{-1}. \end{aligned}$$

In Y_b we have an infinite chain of nonsingular rational curves with self-intersection numbers $-b_k$, and the complement of this system of curves is an algebraic torus $\mathbb{C}^* \times \mathbb{C}^*$. Because of the periodicity of the sequence b_k we have a transformation $T : Y_b \to Y_b$ identifying the k-th copy of $\mathbb{C}^2$ canonically with the $(k+q)$-th copy. T has a fixed point in $(1,1) \in \mathbb{C}^* \times \mathbb{C}^*$, but there is a T-invariant tubular neighbourhood Y_b^0 of the chain of exceptional curves on which T acts freely. $\tilde{X}_b = Y_b^0/\langle T \rangle$ is a complex manifold with a cyclic configuration of q nonsingular rational curves with self-intersection numbers $-b_1, \ldots, -b_q$. The conditions on these sequences imply that the intersection matrix is negative definite. So, according to Grauert, one can blow these curves down. Thus we get a normal complex space X_b with a singular point x, and we have constructed a singularity (X_b, x) with cyclic resolution $\tilde{X}_b \to X_b$.

The second construction is somehow analogous to the description of the $X_{n,q}$ in Hirzebruch's thesis as quotient singularities $\mathbb{C}^2/C_{n,q}$, where the group $C_{n,q}$ is constructed from the sequence of self-intersection numbers by means of a continued fraction. Now the singularity of X_b will be constructed as a partial compactification of a quotient $\mathbb{H}^2/G_b$, where $\mathbb{H}$ is the upper half plane and the group G_b acting on $\mathbb{H} \times \mathbb{H}$ is defined by means of infinite continued fractions.

We start with the doubly infinite sequence of integers $b_k \geq 2$ with reduced period q generated from the reduced sequence associated to b. (Recall that this is an abuse of notation.)

For any integer k we define a real number a_k by an infinite periodic continued fraction

$$a_k = b_k - \cfrac{1}{b_{k+1} - \cfrac{1}{b_{k+2} \ddots}} .$$

These numbers are totally positive algebraic numbers in the real quadratic field $K = \mathbb{Q}(a_1)$. In K we consider the lattice

$$M = \mathbb{Z}a_1 \oplus \mathbb{Z}.$$

In M we consider the sector of totally positive elements

$$M^+ = \{w \in M \mid w > 0, w' > 0\},$$

where w' is obtained from w by the nontrivial automorphism of K.

$$M^+ = \{y - xa_1 \mid (x, y) \in \mathbb{Z}^2, y - xa_1 > 0, y - xa_1' > 0\}.$$

The boundary of the convex hull of M^+ is an infinite polygon with vertices $w_k \in M^+$ which may be computed recursively by $w_0 = 1$ and $w_{k+1} = a_{k+1}^{-1} w_k$. Any pair (w_k, w_{k+1}) is a basis of the lattice M generating a sector contained in M^+, and the system of these sectors is a subdivision of M^+. The manifold constructed from these data by toroidal embeddings is exactly the manifold Y_b in the first construction, if the original sequence b was already reduced. The next figure illustrates the situation for the simplest reduced sequence $b_k = 3$ for all k and $q = 1$, where $a_k = (3 + \sqrt{5})/2$ for all k. This case corresponds to the singularity $T_{2,3,7}$. Figures of this kind occur already in Felix Klein's lectures *"Elementarmathematik vom höheren Standpunkte aus"*.

Now we define the group G_b. Let ε be the product $\varepsilon = a_1 \ldots a_q$. This is a totally positive unit in the ring of integers of the field K. Multiplication with ε is an automorphism of the abelian group M. Thus we may form the semidirect product

$$G_b = M \rtimes \langle \varepsilon \rangle$$

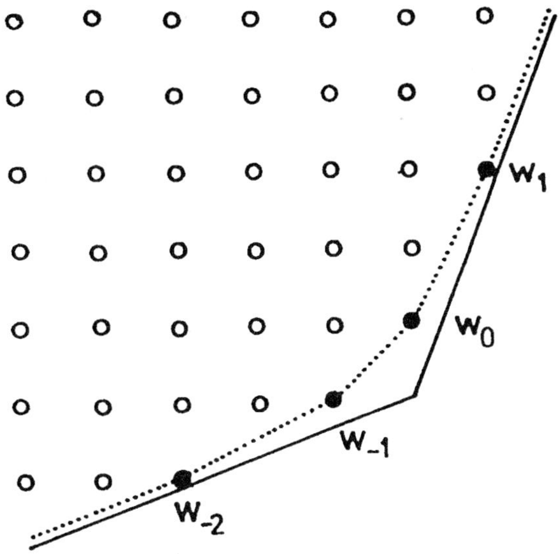

FIGURE 7

with the infinite cyclic group $\langle \varepsilon \rangle$ generated by ε. We identify G_b with a subgroup of $SL(2, K)$

$$G_b = \left\{ \begin{pmatrix} \varepsilon^n & \mu \\ 0 & 1 \end{pmatrix} \middle| \, n \in \mathbb{Z}, \mu \in M \right\}.$$

$SL(2, K)$ operates on $\mathbb{H} \times \mathbb{H}$ as follows:

$$\begin{pmatrix} a & b \\ c & d \end{pmatrix} (z_1, z_2) = \left(\frac{az_1 + b}{cz_1 + d}, \frac{a'\Pi z_2 + b'}{c'\coprod z_2 + d'} \right).$$

The action of G_b on $\mathbb{H} \times \mathbb{H}$ is properly discontinuous, so that the orbit space $\mathbb{H}^2/G_b$ is a 2-dimensional normal complex space. We define a partial compactification

$$\overline{\mathbb{H}^2/G_b} = \mathbb{H}^2/G_b \cup \{\infty\}.$$

A basis of neighbourhoods of ∞ is given by the sets $\operatorname{Im} z_1 \cdot \operatorname{Im} z_2 > d$, where d is any positive real number. A complex valued function on an open neighbourhood U of ∞ is holomorphic if it is continuous and holomorphic on $U - \{\infty\}$. With these definitions $\overline{\mathbb{H}^2/G_b}$ is a 2-dimensional normal complex space, and Hirzebruch proves

$$\overline{\mathbb{H}^2/G_b} \cong (X_b, x).$$

These singularities with cyclic resolution occur as cusp singularities of compactified orbit spaces $\overline{\mathbb{H}^2/G}$, where the groups G are certain discrete groups operating on $\mathbb{H}^2$ such as $SL(2, \underline{o})$, where $\underline{o}$ is the ring of integers in a real quadratic field K over $\mathbb{Q}$. By resolving all singularities of $\overline{\mathbb{H}^2/G}$, one gets the Hilbert modular surfaces. Hirzebruch has made a detailed investigation of such surfaces in a series of papers written between 1970 and 1980. Some of these papers were joint work with Don Zagier and with Van de Ven. As an example let me quote the main result of the joint paper with Van de Ven in Inventiones dedicated to Karl Stein on the occasion of his sixtieth birthday. Let $Y(p)$ be the Hilbert modular surface associated to $K = \mathbb{Q}(\sqrt{p})$, where p is a prime congruent 1 mod 4.

Theorem. *The surfaces* $Y(p)$ *are rational for* $p = 5, 13, 17$*; blown up elliptic* K_3*-surfaces for* $p = 29, 37, 41$*; honestly elliptic surfaces for* $p = 53, 61, 73$ *and surfaces of general type for* $p \geq 89$.

It is a pity that I am unable to render adequately the wealth of results in these papers on Hilbert modular surfaces. Let me mention at least one more beautiful result which I think is very typical of Hirzebruch's way of looking at mathematical objects. It is related to classical results of Clebsch and Klein. In 1873 Klein had proved that the famous diagonal surface of Clebsch, which is the surface in $P_4(\mathbb{C})$ with equations

$$\begin{aligned} x_0^3 + x_1^3 + x_2^3 + x_3^3 + x_4^3 &= 0 \\ x_0 + x_1 + x_2 + x_3 + x_4 &= 0, \end{aligned}$$

can be obtained from $P_2(\mathbb{C})$ by blowing up 6 points in $P_2(\mathbb{C})$ in a special position, namely the 6 points in $P_2(\mathbb{R}) = S^2/\{\pm 1\}$ corresponding to the 12 vertices of an icosahedron inscribed in S^2. Now Hirzebruch blows up 10 more points, namely those corresponding to the 20 vertices of the dual dodecahedron. The resulting surface Y can also be obtained from the Clebsch diagonal surface by blowing up 10 Eckhardt points, that is points, where 3 of the 27 lines on the surface meet. In a paper dedicated to P. S. Aleksandrov, this classical surface is identified with a Hilbert modular surface. Let $\underline{o} \subset \mathbb{Q}(\sqrt{5})$ be the ring of integers and $\Gamma \subset SL(2, \underline{o})$ the congruence subgroup mod 2. Hirzebruch proves: The icosahedral surface Y is the minimal resolution of $\overline{\mathbb{H}^2/\Gamma}$.

After his work on Hilbert modular surfaces Hirzebruch wrote a series of papers in which the problem of the existence of complex manifolds with invariants satisfying certain conditions was related to the problem of the existence of various types of geometric configurations and in particular to the problem of the existence of certain configurations of singularities

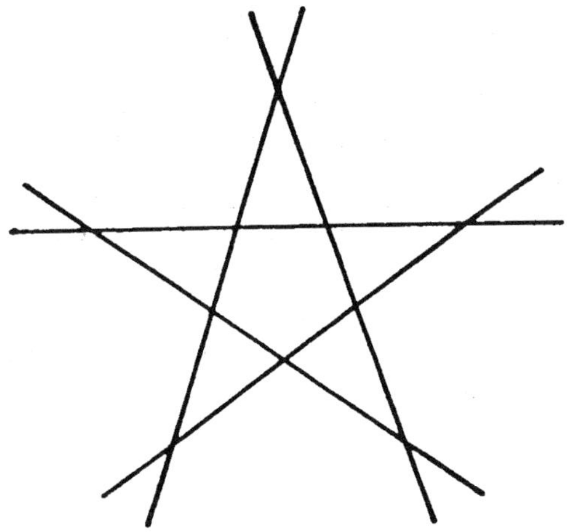

FIGURE 8

on certain algebraic manifolds. These papers contain such a wealth of beautiful geometry with relations to classical configurations of the 19-th century, but also to modern theoretical physics, that I am unable to produce an adequate summary. Instead, let me mention just one example which is taken from the last paper of the Collected Works published in 1987.

Consider hypersurfaces of degree d in complex projective n-space with singularities which are only ordinary double points of type A_1. Let $\mu_n(d)$ be the maximal number of double points that can occur on some hypersurface. For example $\mu_4(5) \leq 135$ by a theoretical estimate of Varchenko. It was not known whether this number is attained. In 1986 C. Schoen had constructed a quintic with 125 double points. In 1987 Hirzebruch constructed a quintic with one more double point. The construction is as follows. Consider a configuration of five lines in the real Euclidean plane forming a regular pentagram. Let $f(u, v)$ be a polynomial of degree 5 describing this configuration and invariant under its group of symmetries. f has 10 critical points of level 0 in the 10 intersection points of the five lines, 5 critical points of a certain level $a \neq 0$ in the triangles and one critical point of level b, $0 \neq b \neq a$, at the center of the pentagon. Now consider the quintic in $P_4(\mathbb{C})$ with the affine equation

$$f(u, v) - f(z, w) = 0.$$

Obviously this quintic has only ordinary double points and their number is

$$10 \cdot 10 + 5 \cdot 5 + 1 \cdot 1 = 126.$$

I have tried to show how singularities figure in the work of Friedrich Hirzebruch. I have also tried to show how much I owe to him. Many people, students and mathematicians from all parts of the world owe him thanks. He is always ready to listen, to give advice and to help. He has done an enormous amount of work organizing mathematical research and teaching and international cooperation. In the midst of all that he still has time and energy for wonderful mathematics When I asked him how he does it, he just said: I enjoy it.

MATHEMATISCHES INSTITUT DER UNIVERSITÄT BONN

SURVEYS IN DIFFERENTIAL GEOMETRY, 2000
Vol. VII ©2000, INTERNATIONAL PRESS
pp. 61-81

THE MODULI SPACE OF ABELIAN VARIETIES AND THE SINGULARITIES OF THE THETA DIVISOR

CIRO CILIBERTO & GERARD VAN DER GEER

Introduction

The object of study here is the singular locus of the theta divisor Θ of a principally polarized abelian variety (X, Θ). The special case when (X, Θ) is the Jacobian of a curve C shows that this is meaningful: singularities of Θ are closely related to the existence of special linear systems on the curve C and for curves of genus $g \geq 4$ the divisor Θ is always singular. But for the general principally polarized abelian variety the theta divisor Θ is smooth. In their pioneering work [1] Andreotti and Mayer introduced in the moduli space $\mathcal{A}_g$ of principally polarized abelian varieties the loci N_k of those principally polarized abelian varieties for which Θ has a k-dimensional singular locus:

$$N_k = \{[(X, \Theta)] \in \mathcal{A}_g : \dim \operatorname{Sing}(\Theta) \geq k\} \qquad k \geq 0.$$

(Warning: we shall use a slightly different definition of the N_k in this paper.) They were motivated by the Schottky problem of characterizing Jacobian varieties among all principally polarized abelian varieties. They showed that N_0 is a divisor and that the image of the moduli space of curves under the Torelli map is an irreducible component of N_{g-4} for $g \geq 4$. In a beautiful paper [18] Mumford calculated the cohomology class of the divisor N_0 and used it to show that the moduli space $\mathcal{A}_g$ is of general type for $g \geq 7$. Another reason for interest in the loci N_k might be the study of cycles on the moduli space $\mathcal{A}_g$. The theory of automorphic forms seems to suggest that there exist many algebraic cycles, but it seems very difficult to find them. The N_k yield many interesting examples of cycles. Unfortunately, for $k \geq 1$ our knowledge of the codimension of the N_k, let alone of the irreducible components of N_k, is very limited.

In this paper we give a new result on the codimension of the N_k and formulate a conjectural lower bound for the codimension of the N_k in $\mathcal{A}_g$.

The first author would like thank A. Verra for useful discussions and the second author would like to thank Roy Smith and Robert Varley for useful comments on a first draft of this paper.

It is a great pleasure for the authors to dedicate this paper to the illustrious four mathematicians Atiyah, Bott, Hirzebruch and Singer, who with their insight and personality changed mathematics in the second half of the 20th century in such a decisive way.

1. Review of known results

Let (X, Θ_X) be a principally polarized abelian variety of dimension g over $\mathbf{C}$. We shall assume that $g \geq 2$. Then $\Theta = \Theta_X$ is an ample effective divisor with $h^0(\Theta) = 1$ and Θ defines an isomorphism

$$\lambda : X \xrightarrow{\approx} \hat{X}, \qquad x \mapsto [\Theta - \Theta_x],$$

of X with its dual abelian variety $\hat{X}$; here Θ_x stands for the translate of Θ over x.

We denote by $\mathcal{A}_g$ the moduli space of principally polarized abelian varieties over $\mathbf{C}$. This is an orbifold of dimension $g(g+1)/2$.

The basic objects that we are interested in here are the loci

$$N_k := N_{g,k} = \overline{\{[(X, \Theta_X)] \in \mathcal{A}_g(\mathbf{C}) : \dim \mathrm{Sing}(\Theta_X) = k\}} \qquad (0 \leq k \leq g-2)$$

where the overline means Zariski closure. Note that this definition is slightly different from the definition of Andreotti-Mayer and the one used by Mumford. Andreotti and Mayer introduced these loci in 1967 in relation with the Schottky problem. To explain the relation, we consider the moduli space $\mathcal{M}_g$ of irreducible, smooth complete curves of genus g over $\mathbf{C}$. Then there is the Torelli map

$$t : \mathcal{M}_g(\mathbf{C}) \to \mathcal{A}_g(\mathbf{C})$$

given by

$$[C] \mapsto [(X = \mathrm{Jac}(C), \Theta \subset \mathrm{Pic}^{g-1}(C))],$$

where Θ is the divisor of effective line bundles of degree $g-1$ in $\mathrm{Pic}^{g-1}(C)$ and this defines a divisor up to translations, again denoted by Θ, on $\mathrm{Jac}(C)$. Riemann showed (see [1]) that for a Jacobian of genus $g \geq 4$ the singular locus $\mathrm{Sing}(\Theta)$ has dimension $g-4$ unless the curve C is hyperelliptic; in that case $\mathrm{Sing}(\Theta)$ has dimension $g-3$ and there is even an explicit description of $\mathrm{Sing}(\Theta)$:

$$\mathrm{Sing}(\Theta) = g_2^1 + W_{g-3}^0 \subset \mathrm{Pic}^{g-1}(C)$$

with W_{g-3}^0 the locus of effective divisor classes (line bundles) of degree $g-3$. We define

$$J_g := \overline{t(M_g)}, \quad \text{the Jacobian locus.}$$

It is an irreducible closed algebraic subset of $\mathcal{A}_g$ of dimension $3g-3$. We also need H_g, the hyperelliptic locus in $\mathcal{M}_g$, and we put

$$\mathcal{H}_g = \overline{t(H_g)}, \quad \text{the hyperelliptic locus in } \mathcal{A}_g.$$

It is irreducible and of dimension $2g-1$. Now Riemann's results on the dimension of $\mathrm{Sing}(\Theta)$ for Jacobians imply:

$$J_g \subseteq N_{g,g-4} \quad \text{and} \quad \mathcal{H}_g \subseteq N_{g,g-3}.$$

(1.1) Theorem. (Andreotti-Mayer). *i) The Jacobian locus J_g is an irreducible component of $N_{g,g-4}$. ii) The hyperelliptic locus $\mathcal{H}_g$ is an irreducible component of $N_{g,g-3}$.*

The fact that N_0 is a divisor was first noticed in an unpublished version of the Andreotti-Mayer paper. It was also proved by Beauville ([3]) for $g=4$ and Mumford observed that Beauville's proof works in the general case.

Mumford calculated in [18] the class of N_0 in the Chow group $CH^1(\tilde{\mathcal{A}}_g^{(1)})$ for the canonical partial compactification $\tilde{\mathcal{A}}_g^{(1)}$. If $\lambda = \lambda_1$ denotes the first Chern class of the determinant of the Hodge bundle (a line bundle the sections of which are modular forms) and δ is the class of the boundary then the result is this:

(1.2) Theorem. (Mumford). *The class of N_0 in $CH^1(\tilde{\mathcal{A}}_g^{(1)})$ is given by*

$$[N_0] = \left(\frac{(g+1)!}{2} + g!\right)\lambda - \frac{(g+1)!}{12}\,\delta.$$

A variation of the notion of Jacobian varieties is given by *Prym varieties*: take a double étale cover

$$\pi : \tilde{C} \to C$$

of smooth irreducible curves, where $\tilde{C}$ has genus $2g+1$ and C has genus $g+1$. The morphism π induces a norm map $\mathrm{Nm} = \pi_* : \mathrm{Pic}(\tilde{C}) \to \mathrm{Pic}(C)$. We now look at the restriction to the degree $2g$ part

$$\tilde{J} \cong \mathrm{Pic}^{2g}(\tilde{C}) \xrightarrow{\mathrm{Nm}} \mathrm{Pic}^{2g}(C) \cong J$$

and define the Prym variety P as the connected component of zero in the kernel of Nm:

$$P = P(\tilde{C}/C) = \ker \mathrm{Nm}(\tilde{J} \to J)^0.$$

As it turns out we can also write it as

$$P \cong \{L \in \mathrm{Pic}^{2g}(\tilde{C}) : \mathrm{Nm}(L) = K_C, h^0(L) \equiv 0 (\mathrm{mod}\, 2)\}.$$

and this is an abelian variety of dimension g. A principal polarization on P is provided by the divisor of effective line bundles

$$\Xi = \{L \in P : \mathrm{Nm}(L) = K_C, h^0(L) \equiv 0 (\mathrm{mod}\, 2), h^0(L) > 0\}.$$

Let $R\mathcal{M}_{g+1}$ be the moduli space of such double covers $\tilde{C} \to C$. It is an orbifold of dimension $3g$ with a natural map $R\mathcal{M}_{g+1} \to \mathcal{M}_{g+1}$ defined by forgetting the cover $\tilde{C}$ of C. The Torelli map has an analogue for this situation, the Prym-Torelli map:

$$p : R\mathcal{M}_{g+1} \longrightarrow \mathcal{A}_g \qquad (\tilde{C} \to C) \mapsto P(\tilde{C}/C).$$

(1.3) Theorem. (Friedman-Smith [15], Donagi [11], [12]). *The morphism p is dominant for $g \leq 5$; it is birational to its image for $g \geq 6$, but not injective.*

Note that the non-injectivity follows immediately from Mumford's description of the Prym varieties of hyperelliptic curves, see [17].

We define a locus in $\mathcal{A}_g$:

$$P_g := \overline{p(R\mathcal{M}_{g+1})}, \quad \text{the Prym locus in } \mathcal{A}_g$$

This new locus is of dimension $3g$ for $g \geq 5$ and it contains the Jacobian locus:

(1.4) Theorem. (Wirtinger, Beauville [3]). *The Prym locus contains the Jacobian locus:* $J_g \subseteq P_g$.

The classical result of Riemann on the singular locus of Θ for Jacobians has an analogue for Prym varieties. The singular points of Ξ are of two types. If $L \in P$ then we have $L \in \mathrm{Sing}(\Xi)$ if and only if

i) $h^0(L) \geq 4$, or

ii) L is of the form $\pi^*(E) + M$, where $M \geq 0$ and $h^0(E) \geq 2$.

The singularities of type i) are called *stable* and those of type ii) are called *exceptional.* Welters and Debarre proved that the singular locus of the divisor Ξ has dimension $\geq g-6$, see [25], [7]. It follows from their work and that of Debarre ([8]) that for a generic Prym variety every singular point of Ξ is stable and $\mathrm{Sing}(\Xi)$ is irreducible of dimension $g-6$ for $g \geq 7$, reduced of dimension 0 for $g=6$ and empty if $g \leq 5$. Mumford showed that if $\dim \mathrm{Sing}(\Xi) \geq g-4$ then $\mathrm{Sing}(\Theta)$ has an exceptional component and the curve C is either hyperelliptic, trigonal, bi-elliptic, a plane quintic or a genus 5 curve with an even theta characteristic. By work of Debarre we know that if C is not a 4-gonal curve then $\dim \mathrm{Sing}_{exc}(\Xi) \leq g-7$ for $g \geq 10$. He also gives a beautiful description of exceptional singular locus of 4-gonal curves.

(1.5) Theorem. (Debarre [8]). *The Prym locus P_g is an irreducible component of $N_{g,g-6}$ for $g \geq 7$.*

This shows once more that the components of $N_{g,k}$ give geometrically meaningful cycles on the moduli space.

What do we know about the structure of the loci $N_{g,k}$? Let us start with $k=0$.

(1.6) Theorem. (Debarre [9]). *The divisor $N_{g,0} = N_0$ consists of two irreducible components for $g \geq 4$:*

$$N_0 = \theta_{\mathrm{null}} + 2N_0'.$$

Some explanation is in order here. The generic point of the irreducible divisor θ_{null} corresponds to a polarized abelian variety (X, Θ) for which Θ has one singularity, a double point at a point of order 2 of X, while the generic point of the irreducible divisor N_0' corresponds to an abelian variety (X, Θ) where Θ has two singularities. Mumford has shown in [17] how the divisor N_0 can be defined scheme-theoretically so that it comes with multiplicities. The divisor θ_{null} is given as the zero divisor of the modular form given by the product of the $2^{g-1}(2^g+1)$ even thetanulls $\theta[{\epsilon_1 \atop \epsilon_2}](\tau, z)$.

(1.7) Example. If $g = 4$ the component N_0' is the Jacobian locus J_4 as Beauville showed. For $g = 5$ the component N_0' can be identified with the locus of intermediate Jacobians of double covers of $\mathbb{P}^3$ ramified along a quartic surface with 5 nodes, cf. [20], [9].

Mumford showed that for $k \geq 1$ none of the N_k have codimension 1:

$$\operatorname{codim}_{\mathcal{A}_g} N_k > 1 \quad \text{if} \quad k \geq 1.$$

At the other extreme we have $N_{g,g-2}$. We call a principally polarized abelian variety *decomposable* if it is a product of (positive-dimensional) principally polarized abelian varieties. The singular locus of Θ for a decomposable abelian variety has codimension 2. There are natural maps

$$\mathcal{A}_{i_1} \times \ldots \times \mathcal{A}_{i_r} \to \mathcal{A}_g, \qquad ([X_{i_1}], \ldots, [X_{i_r}]) \mapsto [X_{i_1} \times \ldots \times X_{i_r}].$$

We denote the image by $\mathcal{A}_{i_1,\ldots,i_r}$. Let

$$\Pi_g = \bigcup_{1 \leq i \leq g/2} \mathcal{A}_{i,g-i}$$

be the locus of decomposable abelian varieties in $\mathcal{A}_g$. This is a closed algebraic subset whose components have codimension $\geq g-1$ in $\mathcal{A}_g$.

(1.8) Proposition. *$N_{g,g-2} = \Pi_g$ with Π_g the locus of decomposable abelian varieties.*

This proposition is a corollary to a fundamental result by Kollár and a result of Ein-Lazarsfeld.

(1.9) Theorem. (Kollár [16]). *The pair (X, Θ) is log canonical.*

Kollár's result implies that

$$\Theta^{(r)} := \{x \in \Theta : \operatorname{mult}_x(\Theta) \geq r\}$$

has codimension $\geq r$ in X. Moreover, Ein and Lazarsfeld prove in [14]

$$\operatorname{codim}_X \Theta^{(r)} = r \iff X \text{ is decomposable as product of } r \text{ p.p.a.v.}$$

and this then implies the proposition, cf. also [22]. Ein and Lazarsfeld also proved that if Θ is irreducible then Θ is normal and the singularities are rational.

In the following table we collect what is known about components of the N_k. We do not give multiplicities.

Table for Low Genera

$g\backslash N_k$	$\mathcal{A}_g$	N_0	N_1	N_2	N_3	N_4	N_5
2	$A_2 = J_2$	Π_2	$\emptyset$	$\emptyset$	$\emptyset$	$\emptyset$	$\emptyset$
3	$A_3 = J_3$	$\mathcal{H}_3$	Π_3	$\emptyset$	$\emptyset$	$\emptyset$	$\emptyset$
4	$A_4 = P_4$	$\theta_{\text{null}} + J_4$	$\mathcal{H}_4$	Π_4	$\emptyset$	$\emptyset$	$\emptyset$
5	$A_5 = P_5$	$\theta_{\text{null}} + N_0'$	$J_5 + A + B + C$	$\mathcal{H}_5$	Π_5	$\emptyset$	$\emptyset$
6	$A_6 =?$	$\theta_{\text{null}} + N_0'$	?	$J_6+?$	$\mathcal{H}_6+?$	Π_6	$\emptyset$
7	$A_7 =?$	$\theta_{\text{null}} + N_0'$	$P_7+?$	?	$J_7+?$	$\mathcal{H}_7+?$	Π_7

It is known by work of Debarre that the three irreducible components of $N_{5,1}$ different from J_5 have dimensions 10, 9 and 9, cf. [5], [13]. Moreover, in [5] Debarre constructed components of $N_{g,g-4}$ (resp. $N_{g,g-6}$) (resp. $N_{g,g-8}$) of codimension $g'(g-g')$ for $2 \leq g' \leq g/2$ and $g \geq 5$ (resp. for $3 \leq g' \leq g/2$ and $g \geq 7$) (resp. for $4 \leq g' \leq g/2$ and $g \geq 9$) and these thus are part of the question marks in the table at positions $N_{5,1}$, $N_{6,2}$, $N_{7,3}$ and $N_{7,1}$.

2. Bounds and a conjecture on the codimension

As the review of the preceding section may show, very little is known about the components of the loci $N_{g,k}$. Apart from Mumford's estimate that $\operatorname{codim} N_k \geq 2$ for $k \geq 1$ we know almost nothing about the codimension of the $N_{g,k}$. Our new results give some lower bounds for the codimension. Debarre proved in an unpublished manuscript independently that $\operatorname{codim} N_{g,k} \geq k+1$ for $k \geq 1$.

(2.1) Theorem. *Let $g \geq 4$. Then for k with $1 \leq k \leq g-3$ we have* $\operatorname{codim} N_{g,k} \geq k+2$.

(2.2) Theorem. *Let $g \geq 5$. If k satisfies $g/3 < k \leq g-3$ then* $\operatorname{codim} N_{g,k} \geq k+3$.

The first theorem is sharp for $k = 1$ and $g = 4, 5$. However, we do not expect that this is an accurate description of reality and believe that Theorems (2.1) and (2.2) are never sharp for $k = 1$ and $g \geq 6$, or for $k \geq 2$. We conjecture the following much stronger bound.

(2.3) Conjecture. *If $1 \leq k \leq g-3$ and if M is an irreducible component of $N_{g,k}$ whose generic point corresponds to a simple abelian variety then* $\operatorname{codim} M \geq \binom{k+2}{2}$. *Moreover, equality holds if and only if $g = k+3$ (resp. $g = k+4$) and then $M = \mathcal{H}_g$ (resp. $M = J_g$).*

Note that by work of Beauville and Debarre ([3], [5]) the conjecture is true for $g = 4$ and $g = 5$.

We now describe some corollaries of this. Let $\pi : \mathcal{X}_g \to \mathcal{A}_g$ be the universal family of principally polarized abelian varieties. The reader should view this as a stack, or replace $\mathcal{A}_g$ by a fine moduli space, e.g. the moduli space of principally polarized abelian varieties with a level 3 structure. We can view $\mathcal{X}_g$ as the universal family of pairs (X, Θ). In it we can consider the algebraic subset S_g where the morphism $\pi|\Theta$ is not smooth. If we write $\mathcal{A}_g$ as the orbifold $\mathbf{H}_g/\mathrm{Sp}(2g, \mathbf{Z})$ with $\mathbf{H}_g$ the upper half plane and $\mathcal{X}_g$ as the orbifold

$$\mathbf{H}_g \times \mathbf{C}^g/\mathrm{Sp}(2g, \mathbf{Z}) \ltimes \mathbf{Z}^{2g}$$

then Θ is given in $\mathbf{H}_g \times \mathbf{C}^g$ by the vanishing of Riemann's theta function $\theta(\tau, z) = 0$, with

$$\theta(\tau, z) = \sum_{n \in \mathbf{Z}^g} e^{\pi i n^t \tau n + 2\pi i n^t z}$$

and S_g is defined in $\mathbf{H}_g \times \mathbf{C}^g$ by the $g+1$ equations

$$\theta = 0, \quad \frac{\partial \theta}{\partial z_i} = 0 \qquad i = 1, \ldots, g.$$

Therefore, S_g has codimension $\leq g + 1$. Theorem (2.1) implies that the codimension is equal to $g + 1$:

(2.4) Theorem. *Every irreducible component of* $S_g = \mathrm{Sing}(\Theta) \subset \mathcal{X}_g$ *has codimension* $g+1$ *in* $\mathcal{X}_g$, *hence* S_g *is locally a complete intersection.*

Proof. Take an irreducible component S of S_g and let N be its image under the natural map $\pi : \mathcal{X}_g \to \mathcal{A}_g$. We first assume that N is not contained in Π_g. Suppose now that N is contained in $N_{g,k}$ for some $k \geq 1$, and we may assume by (1.8) that $k \leq g - 3$. Then by Theorem (2.1) the codimension of N in $\mathcal{A}_g$ is at least $k + 2$. This implies that the codimension of S in $\mathcal{X}_g$ is at least $g + 2$, which is impossible. Hence generically, the fibres of $\pi_{|S} : S \to N$ are 0-dimensional and N must have codimension ≤ 1 in $\mathcal{A}_g$. So S maps dominantly to a component of $N_{g,0}$, a divisor and we get codim $\geq g + 1$. Finally, if N is contained in Π_g we observe that Π_g has codimension $g - 1$ and the fibres have dimension $g - 2$ leading also to codim $\geq g + 1$ and this concludes the proof.

This Corollary of Theorem (2.1) was obtained independently by Debarre in an unpublished note.

(2.5) Corollary. N_0 *is a divisor properly containing* $\cup_{k \geq 1} N_k$.

This raises the problem about the respective positions for higher N_k.

(2.6) Problem. Is it true that N_k properly contains $\cup_{i\geq k+1} N_i$?

For the generic point of a component N_0 we know the singularities of Θ. In general we know almost nothing about the nature of the singular locus $\mathrm{Sing}(\Theta)$ of a generic point of a component of $N_{g,k}$. For a discussion of the case N_1 we refer to Section 8.

3. Deformation theory and the heat equation

In this section we explain Welters' interpretation of the Heat Equation for the theta function, cf. [24]. The Heat Equation is one of the tools for obtaining our estimates on the codimension.

Let (X, Θ) be a principally polarized abelian variety of dimension g. We denote the invertible O_X-module $O_X(\Theta)$ associated to Θ by L. The space $\mathrm{Def}(X)$ of linear infinitesimal deformations of the algebraic variety X has a well-known cohomological interpretation:

$$\mathrm{Def}(X) \cong H^1(X, T_X),$$

where T_X denotes the tangent sheaf of X. The space of linear infinitesimal deformations of the pair (X, Θ) or equivalently of the pair (X, L), where we consider Θ or L up to translations on X, is given by

$$\mathrm{Def}(X, L) \cong H^1(X, \Sigma_L),$$

where Σ_L is the sheaf of germs of differential operators of order ≤ 1 on L (sums of functions and derivations). Given now a section $s \in \Gamma(L)$ we obtain a complex

$$0 \to \Sigma_L \xrightarrow{d^1 s} L \to 0, \quad \text{with } d^1 s : D \mapsto D(s)$$

on X given by associating to a differential operator D the section $D(s)$ of L. The cohomological interpretation of the space of linear infinitesimal deformations of the triple (X, L, s) is

$$\mathrm{Def}(X, L, s) \cong \mathbb{H}(d^1 s),$$

the hypercohomology of the complex. Explicitly, it can be given as follows: if $(X_\epsilon, L_\epsilon, s_\epsilon)$ is an infinitesimal deformation, then on a suitable open cover $U_j[\epsilon]$ of X_ϵ the section s_ϵ is given as $s_j + \epsilon\sigma_j$ with

$\sigma_j - \sigma_i = \eta_{ij}(s)$, where $\eta_{ij}(s)$ is a cocycle whose class in $H^1(X, \Sigma_L)$ determines the deformation (X_ϵ, L_ϵ). So we obtain a 1-cocycle $(\{\sigma_i\}, \{\eta_{ij}\}) \in C^0(U, L) \oplus C^1(U, \Sigma_L)$ of the total complex associated with

$$\begin{array}{ccccc} C^0(U,\Sigma_L) & \longrightarrow & C^1(U,\Sigma_L) & \longrightarrow & \dots \\ \big\downarrow{\scriptstyle d^1 s} & & \big\downarrow{\scriptstyle -d^1 s} & & \\ C^0(U,L) & \longrightarrow & C^1(U,L) & \longrightarrow & \dots \end{array}$$

and we thus have an element of $\mathbb{H}^1(d^1 s)$.

The central point is now the following:

(3.1) Claim. An element of $H^0(X, \mathrm{Sym}_2 T_X)$ determines canonically a linear infinitesimal deformation of (X, L) and (X, L, s).

This follows from the first connecting homomorphism of the exact sequence of hypercohomology of the short exact sequence of complexes

$$\begin{array}{ccccccccc} 0 & \longrightarrow & \Sigma_L & \longrightarrow & \Sigma_L^{(2)} & \longrightarrow & \mathrm{Sym}_2 T_X & \longrightarrow & 0 \\ & & \big\downarrow{\scriptstyle d^1 s} & & \big\downarrow{\scriptstyle d^2 s} & & \big\downarrow & & \\ 0 & \longrightarrow & L & \xrightarrow{\mathrm{id}} & L & \longrightarrow & 0 & \longrightarrow & 0, \end{array} \tag{1}$$

where $\Sigma_L^{(2)}$ stands for differential operators of order ≤ 2 on L and $\mathrm{Sym}_2 T_X$ is the *subspace* of elements fixed by the involution $(x_1, x_2) \mapsto (x_2, x_1)$ on $T_X \otimes T_X$. We thus have the connecting homomorphism of the upper exact sequence of (1)

$$b : H^0(X, \mathrm{Sym}_2 T_X) \longrightarrow H^1(X, \Sigma_L)$$

and the connecting homomorphism of the short exact sequence of complexes (1)

$$\beta : H^0(X, \mathrm{Sym}_2 T_X) \longrightarrow \mathbb{H}^1(d^1 s)$$

such that $b = f \cdot \beta$ with $f : \mathbb{H}^1(d^1 s) \longrightarrow H^1(X, \Sigma_L)$ the forgetful map, and we find the morphisms

$$H^0(X, \mathrm{Sym}_2 T_X) \longrightarrow \mathbb{H}^1(d^1 s) \longrightarrow H^1(X, \Sigma_L) \longrightarrow H^1(X, T_X). \tag{2}$$

But it is well-known that we can identify $H^0(X, \mathrm{Sym}_2 T_X)$ with $H^1(X, \Sigma_L)$: we have

$$H^1(X, T_X) \cong H^1(X, O_X) \otimes T_{X,0} \cong T_{\hat{X},0} \otimes T_{X,0}$$

and using the polarization $\lambda : X \xrightarrow{\sim} \hat{X}$ we see that the subspace corresponding to deformations preserving the polarizations is

$$\mathrm{Sym}_2 T_{X,0} \subset T_{X,0} \otimes T_{X,0} \cong T_{\hat{X},0} \otimes T_{X,0}.$$

The composition $H^0(X, \mathrm{Sym}_2 T_X) \to \mathbb{H}^1(d^1s) \to H^1(X, \Sigma_L)$ is therefore an isomorphism. The first spectral sequence for the hypercohomology gives us an exact sequence

$$H^0(\Sigma_L) \to H^0(L) \xrightarrow{\alpha} \mathbb{H}^1(d^1s) \to H^1(\Sigma_L) \to H^1(L),$$

where $\alpha(t) = (X = X_\epsilon, L = L_\epsilon, s + t\epsilon)$. So we get an exact sequence

$$0 \to H^0(L)/\mathbf{C} \cdot s \to \mathbb{H}^1(d^1s) \to H^1(\Sigma_L) \to 0$$

which shows that for principally polarized abelian varieties the forgetful map

$$f : \mathbb{H}^1(d^1s) \to H^1(X, \Sigma_L) \tag{3}$$

is also an isomorphism: *every deformation* (X_ϵ, L_ϵ) *of* (X, L) *canonically determines a deformation* s_ϵ *of* s. This is Welters' interpretation of the classical Heat Equations. If we represent X as a complex torus $X = \mathbf{C}^g/\Lambda$ with $\Lambda = \mathbf{Z}^g\tau + \mathbf{Z}^g$, $\tau \in \mathbf{H}_g$, $z \in \mathbf{C}^g$ and Θ as before by

$$\theta(\tau, z) = \sum_{n \in \mathbf{Z}^g} e^{\pi i n^t \tau n + 2\pi i n^t z}$$

then it satisfies the relation

$$2\pi i(1 + \delta_{ij}) \frac{\partial \theta}{\partial \tau_{ij}} = \frac{\partial^2 \theta}{\partial z_i \partial z_j}$$

where δ_{ij} denotes the Kronecker delta. These are the classical "Heat Equations" for Riemann's theta function.

4. Singularities of theta and quadrics

The tangent cone of a singular point x of Θ with multiplicity 2 defines after projectivization and translation to the origin a quadric Q_x in $\mathbf{P}^{g-1} = \mathbf{P}(T_{X,0})$. Another description is obtained as follows. The singular points of $\Theta \subset X$ are the points x where the map $d^1s : \Sigma_L \to L$ of Section 3 vanishes. Replace now in diagram (1) all sheaves by their fibres at x and denote the resulting maps by the suffix (x). Then $(d^1s)(x) = 0$ and diagram (1) implies then that at such points x the map $(d^2s)(x)$ factors through

$$(\mathrm{Sym}_2 T_X)_x \longrightarrow L_x. \tag{4}$$

This gives an element of $L_x \otimes \mathrm{Sym}_2(T_X)_x^\vee \cong H^0(\mathrm{Sym}^2(\Omega_X^1)) \otimes L_x$. We can view this as an equation q_x for the projectivized tangent cone Q_x of Θ at x (if the multiplicity of the point is 2; otherwise it is zero). Note that if $x \in \mathrm{Sing}(\Theta)$ then $\mathbb{H}^1((d^1s)(x))$ can be identified with L_x and we can identify (4) with

$$H^0(X, \mathrm{Sym}_2 T_X) \longrightarrow (\mathrm{Sym}_2 T_X)_x \longrightarrow \mathbb{H}^1((d^1s)(x)) = L_x.$$

The map $(d^2s)(x) : \mathrm{Sym}_2 T_X \to L_x$ sends an element w to 0 if and only if $q_x(w) = 0$, i.e., if and only if the quadric q_x and the dual quadric w are orthogonal.

Suppose that we have an element $w \in H^0(X, \mathrm{Sym}_2 T_X)$ determining by (3) an element of $\mathbb{H}^1(d^1s)$ with corresponding deformation $(X_\epsilon, L_\epsilon, s_\epsilon)$ of (X, L, s). This is given by a cocycle (σ_i, η_{ij}) representing an element of $\mathbb{H}^1(d^1s)$. With respect to a suitable covering $\{U_i\}$ of X_ϵ we can write the section s_ϵ as

$$s_i + \sigma_i\epsilon.$$

Identifying $\mathbb{H}^1((d^1s)(x))$ with L_x we see that the corresponding element of L_x is given by $\sigma_i(x)$. Suppose that $x \in X$ deforms to x_ϵ. The condition that $s_\epsilon(x_\epsilon) = 0$ can be translated as follows:

$$s_i(x) + (v_x s_i + \sigma_i(x))\epsilon = 0,$$

where v_x is the tangent vector to X at x corresponding to x_ϵ. Since $s_i(x) = 0$ and $v_x s_i = 0$ because x is a singular point, the condition is $\sigma_i(x) = 0$, i.e., $q_x(w) = 0$. We thus see:

(4.1) Lemma. *Let x be a quadratic singularity of Θ. The infinitesimal deformations of (X, Θ) which keep x on Θ are the deformations contained in $Q_x^\perp \subset \mathrm{Sym}_2(T_X)$. In particular, the deformations that keep x a singular point of Θ are contained in $Q_x^\perp$.*

Let R be an irreducible component of the locus $\mathrm{Sing}^{(2)}(\Theta)$ of quadratic singularities of Θ, where we are assuming that $\mathrm{Sing}^{(2)}(\Theta)$ is not empty. We now consider the map

$$\phi : R \to \mathbf{P}(\mathrm{Sym}_2(T_X)_x^\vee) = \mathbf{P}(T^\vee_{\mathcal{A}_g,[X]}) \qquad x \mapsto Q_x$$

given by associating to $x \in R$ the quadric $Q_x \subset \mathbf{P}^{g-1}$. We identified the space $\mathrm{Sym}_2(T_{X,0})$ with the tangent space $T_{\mathcal{A}_g,[X]}$ to the moduli space $\mathcal{A}_g$ at $[(X, \Theta)]$.

Since $\theta = 0$ and all derivatives $\partial_j\theta$ vanish on R the partial derivatives $\partial_i\partial_j\theta$ are sections of $O_R(\Theta)$:

(4.2) Proposition. *The map ϕ is given by sections of $O_R(\Theta)$.*

Another way to interpret this is using the exact sequence

$$0 \to \mathcal{T}_\Theta \longrightarrow \mathcal{T}_{X|\Theta} \longrightarrow \mathcal{N}_{\Theta,X} \longrightarrow \mathcal{T}^1_\Theta \to 0,$$

where $\mathcal{T}_\Theta \cong \mathcal{O}^g_\Theta$ is the tangent sheaf, $\mathcal{N}_{\Theta,X} \cong \mathcal{O}_\Theta(\Theta)$ is the normal sheaf, and $\mathcal{T}^1_\Theta \cong \mathcal{O}_{\mathrm{Sing}(\Theta)}(\Theta)$ is the first higher tangent sheaf of deformation theory and the middle arrow sends $\partial/\partial z_i$ to $\partial\theta/\partial z_i$. The induced Kodaira-Spencer map is

$$\delta : T_{\mathcal{A}_g,X} \longrightarrow H^0(\mathcal{T}^1_\Theta) = H^0(\mathcal{O}_{\mathrm{Sing}(\Theta)}(\Theta))$$

which maps $\partial/\partial\tau_{ij}$ to $\partial\theta/\partial\tau_{ij}$, cf. [23]. In this interpretation, for a singular point x the deformation $v \in T_{\mathcal{A}_g,X}$ keeps the point x on Θ if and only if $\delta(v)(x) = 0$. If we assume for simplicity that $\mathrm{Sing}(\Theta) = \mathrm{Sing}^{(2)}(\Theta)$ then the image of δ is a linear system on $\mathrm{Sing}(\Theta)$ and we thus find a map

$$\sigma : \mathrm{Sing}(\Theta) \longrightarrow \mathbb{P}(H^0(\mathcal{T}^1_\Theta)^\vee) \xrightarrow{\delta^\vee} \mathbb{P}(T^\vee_{\mathcal{A}_g,X}).$$

The Heat Equations tell us that this can be identified with the map ϕ that associates to $x \in \mathrm{Sing}(\Theta)$ the quadric defined by $\sum_{i,j}(\partial^2\theta/\partial z_i \partial z_j)\, x_i x_j$.

It might happen that all singularities of Θ are of higher order. In order to deal with this case we extend the approach to the partial derivatives of $s = \theta$ which are sections of L when restricted to singular points of Θ. We define

$$R^{(j)} := \{x \in X : m_x(s) \geq j\},$$

with m_x the multiplicity at x, the set of points of multiplicity $\geq j$ of Θ; so $R^{(0)} = X$, $R^{(1)} = \Theta$, etc. Suppose that s is a non-zero-section of L. Then any partial derivative $\eta = \partial_v\, s$ ($v \in \mathrm{Sym}^{(j)}(T_X)$) of weight j defines a section of $L|R^{(j)}$.

If η is a partial derivative of θ then it satisfies again the Heat Equation

$$2\pi i(1 + \delta_{ij})\frac{\partial\eta}{\partial\tau_{ij}} = \frac{\partial^2\eta}{\partial z_i \partial z_j}.$$

The algebraic interpretation is as follows. Given a partial derivative η of weight j we apply the formalism of Section 3 to η and find a map

$$\mathrm{Sym}_2 T_{X,0} \to H^0(R^{(j)}, \mathrm{Sym}_2 T_{R^{(j)}}) \to \mathbb{H}^1(d^1\eta) \to H^1(R^{(j)}, \Sigma_{R^{(j)}})$$

We claim that η satisfies the heat equation: any linear infinitesimal deformation of (X, L) determines canonically a deformation η_ϵ of η. This can be deduced in a way very similar to the earlier case by extending Welters' analysis.

5. The tangent space to N_k

Instead of looking at a component of N_0 (with its reduced structure) we may look at the space $\tilde{N}_0$ of triples defined by

$$S_g = \tilde{N}_0 = \{[(X, \Theta, x)] : x \in \mathrm{Sing}(\Theta)\} \subset \mathcal{X}_g,$$

where $\mathcal{X}_g$ is the universal abelian variety over $\mathcal{A}_g$. This has to be taken in the sense of stacks or one has to work with level structures. We have a natural map $\pi : \tilde{N}_0 \to N_0$. By Lemma (4.1) the image under $d\pi$ of the Zariski tangent space of $\tilde{N}_0$ at a point $[(X, x)]$ is contained in the space $q_x^{\perp} \subset \mathrm{Sym}_2(T_{X,0})$.

We shall call an abelian variety X simple if it does not contain abelian subvarieties $\neq X$ of positive dimension. The reason to consider simple abelian varieties is that we then can use the non-degeneracy of the *Gauss map*:

(5.1) Theorem. *If $Z \subset X$ is a positive-dimensional smooth subvariety of a simple abelian variety then the span of the tangent spaces to Z translated to the origin is not contained in a proper subspace of $T_{X,0}$.*

Suppose that for (X, Θ) we have $\mathrm{Sing}^{(2)}(\Theta) \neq \emptyset$ and that N_k is smooth at $[(X, \Theta)]$. The Zariski tangent space to N_k (with its reduced structure) at $[(X, \Theta)]$ is contained in the subspace of $\mathrm{Sym}_2(T_{X,0})$ orthogonal to the linear span in $\mathrm{Sym}^2(\Omega_X^1)$ of the quadrics q_x with $x \in R$ for some k-dimensional irreducible subvariety R of $\mathrm{Sing}(\Theta)$.

By sending $x \in R$ to the quadric Q_x we get a natural map

$$R \longrightarrow \mathbf{P}(N_{N_k}), \qquad x \mapsto Q_x$$

with $\mathbf{P}(N_{N_k})$ the projectivized conormal space to N_k. The image quadrics have rank $\leq g - k$ because of the following lemma.

(5.2) Lemma. *The Zariski tangent space to* $\mathrm{Sing}^{(2)}(\Theta)$ *at a point x equals* $\mathrm{Sing}(Q_x)$.

Proof. In local coordinates $z_1, \ldots, z_g$ a local equation of Θ at a point x is

$$f = q_x + \text{higher order terms}.$$

By putting $q_x = \sum a_{ij} z_i z_j$ we get for $v = (v_1, \ldots, v_g)$ that $f(z + v) = \sum_{i,j} a_{ij} v_j z_i + \ldots$ and we see $v \in \mathrm{Sing}(Q_x)$, i.e., $\sum_{i,j} a_{ij} v_j z_i = 0$, is equivalent to $f(x + v)$ having no linear term, i.e., $v \in T_{\mathrm{Sing}^{(2)}(\Theta), x}$.

(5.3) Proposition. *Let X be a simple principally polarized abelian variety and let S be an open part of a component of* $\mathrm{Sing}^{(2)}(\Theta)$ *where* *the rank of* Q_x *is constant, say* $g-d$. *Then the map* $S \to \mathrm{Gras}(d,g)$, $x \mapsto \mathrm{vertex}(Q_x)$ *has finite fibres.*

Proof. We first note by (5.2) that the tangent space at a point x to the reduced variety S_{red} is contained in the vertex of Q_x. If F denotes a fibre of the map $x \mapsto Q_x$ then the tangent spaces to F are contained in the subspace which is the vertex of the constant Q_x. The result then follows from (5.1).

(5.4) Proposition. *Let* X *be simple and let* x *be a quadratic singularity of* Θ *and a smooth point of* $\mathrm{Sing}^{(2)}(\Theta)$. *The general deformation* $w \in Q_x^{\perp}$ *preserves only finitely many singularities of* $\mathrm{Sing}^{(2)}(\Theta)$.

Proof. The deformations $w \in Q_x^{\perp}$ preserving $y \in \mathrm{Sing}(\Theta)$ are $Q_x^{\perp} \cap Q_y^{\perp}$. Hence all deformations preserving x preserve y if and only if $Q_x = Q_y$. But the map $x \mapsto \mathrm{vertex}(Q_x)$ has finite fibres. q.e.d.

(5.5) Example. Let C be a curve of genus g and $L \in \Theta \subset \mathrm{Jac}(C)$ a quadratic singularity of Θ. This means that the linear system $|L|$ defined by L is a g^1_{g-1}, i.e., has degree $g-1$ and projective dimension 1.

i) C is hyperelliptic. Then L is of the form $g^1_2 + D$, where D is a divisor of degree $g-3$ on C. The quadric Q_L is then the cone projecting the canonical image of C from the span of D. The image Σ of the map $\phi : \mathrm{Sing}(\Theta) \to \mathbb{P}^{\binom{g+1}{2}-1}$ can be identified with the quadratic Veronese $V : \mathbb{P}^{g-3} \longrightarrow \mathbb{P}^{\binom{g-1}{2}-1} \subset \mathbb{P}^{\binom{g+1}{2}-1}$. So the normal space to $N_{g,g-3}$ at $[\mathrm{Jac}(C), \Theta]$ is the subspace spanned by $\Sigma \cong \mathbb{P}^{\binom{g-1}{2}-1}$. Since $\mathrm{codim}_{\mathcal{A}_g} \mathcal{H}_g = \binom{g+1}{2} - \binom{g-1}{2}$ we see that $\mathcal{H}_g$ is a component of $N_{g,g-3}$.

ii) C is not hyperelliptic. By a theorem of M. Green the space spanned by the quadrics Q_x with $x \in \mathrm{Sing}^{(2)}(\Theta)$ is the space of quadrics containing the canonical curve. Since the space of quadratic differentials on C has dimension $3g-3$ it follows that the normal space to $N_{g,g-4}$ at $[\mathrm{Jac}(C), \Theta]$ has dimension $g(g+1)/2 - (3g-3)$. We thus see that J_g is a component of $N_{g,g-4}$.

6. A result on pencils of quadrics

One of the ingredients of our proofs is a classical result of Corrado Segre on pencils of quadrics. Judging from the reactions of experts this theorem seems to have been completely forgotten.

(6.1) Theorem. (C. Segre, 1883). *Let* L *be a linear pencil of singular quadrics of rank* $\leq n+1-r$ *in* $\mathbb{P}^n$ *with* $n \geq 2$ *whose generic member has rank* $n+1-r$ *(i.e., the vertex* $\cong \mathbb{P}^{r-1}$*). We assume that* *the vertex is not constant in this pencil. Then the Zariski closure of the generic vertex in this pencil*

$$V_L = \Big(\bigcup_{\mathrm{rk}(Q)=n+1-r} \mathrm{Vertex}(Q) \Big)^-$$

is a variety of dimension r and degree $m-r+1$ in a projective linear subspace $\mathbb{P}^m \subset \mathbb{P}^n$ with $m \leq (n+r-1)/2$ and $r \leq (n+1)/3$.

If L is a pencil of quadric cones whose generic member has rank n in $\mathbb{P}^n$ and such that the vertex does not stay fixed then the Zariski closure of the union of the vertices of the rank n quadrics is contained in the base locus of the family and it is a rational normal curve of degree m contained in a linear subspace $\mathbb{P}^m \subset \mathbb{P}^n$ with $m \leq n/2$.

For the proof we refer to Segre [19, p. 488-490]. It would be desirable to have an extension of this theorem to higher dimensional linear families of quadrics.

7. Sketch of the proof

We now sketch a proof of Theorem (2.1). Let M be an irreducible component of N_k. We choose a smooth point $\xi \in M$ corresponding to a pair (X, Θ). Note that we may assume that the abelian variety X is simple since the loci of non-simple abelian varieties have codimension $\geq g-1$ in $\mathcal{A}_g$ and $g-1 \geq k+2$ by our assumption on k. We choose a k-dimensional subvariety R of $\mathrm{Sing}(\Theta_X)$ which deforms. For simplicity we start with the case when the generic point x of R is a quadratic singularity of Θ.

The construction of the preceding section yields a rational map

$$\begin{array}{c} \phi : R \longrightarrow \mathbb{P}(N^\vee_{M/\mathcal{A}_g}), \quad x \mapsto Q_x \\ \| \\ \mathbb{P}^\nu \end{array}$$

Here $N^\vee_{M/\mathcal{A}_g}$ is the normal space of the component M in $\mathcal{A}_g$ and we are assuming that the codimension of M in $\mathcal{A}_g$ is $\nu+1$. We also have the *Gauss map*

$$\gamma : R_{\mathrm{smooth}} \longrightarrow \mathrm{Gras}(k, g), \quad x \mapsto \mathrm{vertex}(Q_x) = \mathbb{P}(T_{R,x}) \tag{5}$$

which associates to a smooth point of R its projectivized tangent space, or equivalently the vertex of the quadric Q_x. The non-degeneracy of the

Gauss map $\gamma : R \to \mathrm{Gras}(k, g)$ implies as in (5.3) that the fibres of ϕ must be zero-dimensional, and this gives immediately $\nu \geq k$.

To go further we assume that $\nu = k$. Again, using the Gauss map we see that ϕ maps dominantly to $\mathbb{P}^\nu$. We consider a pencil L of quadrics, i.e., a $\mathbb{P}^1 \subset \mathbb{P}^\nu$. By (5.2) these quadrics have rank $\leq g - k$. The theorem of Segre implies that the Gauss map γ restricted to $\phi^{-1}(L)$ is degenerate since the vertices lie in a linear subspace of $\mathbb{P}^{g-1}$, contradicting (5.1).

If the generic point of R has higher multiplicity, say r, then we apply the preceding to the partial derivatives $\partial_v \theta$ with $|v| = r - 2$ instead of to the section θ. These satisfy the Heat Equations and we proceed with these as with θ before. This completes the sketch of proof of Theorem (2.1).

For the proof of Theorem (2.2) we let M as before be a component of N_k, we pick a point $\xi \in M$ corresponding to (X, Θ) with X simple and let R be a k-dimensional subvariety of $\mathrm{Sing}(\Theta)$ which deforms. Assume then that $\mathrm{codim}(M) = k + 2$ in $\mathcal{A}_g$. We get again a rational map of R to a projective space

$$\phi : R \longrightarrow \mathbf{P}^{k+1}$$

whose image is a hypersurface Σ. We have to distinguish two cases:

i) Not every quadric corresponding to a point of $\mathbb{P}^{k+1}$ is singular.

ii) The general quadric corresponding to a point of $\mathbb{P}^{k+1}$ is singular, say of of rank $g - r$.

First we treat the case i). Consider the discriminant locus $\Delta \subset \mathbb{P}^{k+1}$. This is a hypersurface of degree g containing the image Σ of R with multiplicity at least k:

$$\Delta = k\Sigma + \Phi.$$

In order to be able to apply Segre's result we use the following well-known Lemma:

(7.1) Lemma. *A hypersurface of degree $\leq 2n - 3$ in $\mathbb{P}^n$ contains a line.*

So if the degree of the hypersurface Σ satisfies $\leq 2k - 1$ we have again a line and we can apply Segre's result. Note that since the degree of Δ equals g we have $\deg(\Sigma) \leq g/k$, so that $g/k \leq 2k - 1$ suffices and this follows from $g/3 < k$. This rules out case i) if $\dim(\mathrm{Sing}^{(2)}(\Theta)) \geq k$. If the generic singularity of Θ has higher order we apply the procedure to the higher derivatives as before.

To treat the remaining case ii), where all quadrics parametrized by $\mathbb{P}^{k+1}$ are singular note that $r \leq k$ because our quadrics generically have rank $g - k$ by (5.2) and this should be less than $g - r$, the generic rank of the whole family $\mathbb{P}^{k+1}$.

If $r = k$ then by Segre's result we get $k \leq g/3$, contrary to our assumption. If $r < k$ we shall use a refinement of Segre's theorem which says that the number of quadrics in a pencil of quadrics in $\mathbb{P}^n$ where the rank drops equals $n + r - 2m - 1$ in the notation of (6.1). Here one has to count a quadric with multiplicity d if the rank drops by d. In our case this yields

$$(k - r) \deg \Sigma \leq (g - 1) + r - 2m - 1 \leq g - 2 - r$$

since $m \geq r$. We get $\deg \Sigma \leq (g - 2 - r)/(k - r) \leq 2k - 1$ and this assures us that Σ contains a line. With this line we can apply Segre's theorem to get a contradiction to the non-degeneracy of the Gauss map. This concludes our sketch of proof.

A closer analysis shows that we can draw stronger conclusions from the proof. If $[X, \Theta]$ is a point of an irreducible component of N_1 such that X is simple and $\dim \mathrm{Sing}(\Theta) = 1$ then $[X, \Theta]$ admits a linear deformation in codimension 3 at most. If $\mathrm{codim}(N) = 3$ and $[X, \Theta] \in \mathrm{Sing}(N)$ then the singularities of Θ must "get worse."

This approach to getting estimates on the codimension of components of N_k is by no means exhausted. For example, if $\mathrm{codim}(N) = k + 3$ then the image of a component R of $\mathrm{Sing}(\Theta)$ under the Gauss map is a codimension 2 variety Σ in $\mathbb{P}^{k+2}$. We now can use the variety spanned by the secant lines instead of Σ and apply Segre's result to that. We hope to return to this point in the future (joint work with A. Verra)

8. An approach to the conjecture for N_1

We now restrict to the case of N_1. Then the codimension is at least 3 and this is sharp: the case when the codimension is 3 occurs.

(8.1) Example. Let $g = 4$. We consider the hyperelliptic locus $\mathcal{H}_4$. If $X = \mathrm{Jac}(C)$, the Jacobian of a hyperelliptic curve of genus 4, then the singular locus $\mathrm{Sing}(\Theta) = g_2^1 + W_1^0$ is a copy of C as explained above. The class of this curve in the cohomology is $\Theta^3/3!$. By associating to each point $x \in \mathrm{Sing}(\Theta)$ the vertex of the singular quadric $Q_x \subset \mathbb{P}^3$ we obtain the Gauss map $C \to \Gamma = \mathbb{P}^1$ and the image Γ is the rational normal curve of degree 3 in $\mathbb{P}^3$. The quadrics containing Γ form a net $\mathbb{P}^2$ of quadrics. In general for a net of quadrics the curve of vertices is a curve of degree 6 in $\mathbb{P}^3$ and the discriminant curve of singular quadrics in $\mathbb{P}^2$ is a curve Δ of degree 4. But in our case the map $\mathrm{Sing}(\Theta) \to \Gamma$ is of degree 2 to a rational curve and Δ is a conic with multiplicity 2.

(8.2) Example. Let $g = 5$. We consider the Jacobian locus J_5. If $X = \mathrm{Jac}(C)$, the Jacobian of a curve of genus 5, then the singular locus $\mathrm{Sing}(\Theta) = W_4^1$ is a smooth curve $\tilde{D}$ of genus 11 and class $\Theta^4/4!$ if C is not trigonal and with no semi-canonical pencils. The quotient of $\tilde{D}$ under the involution -1 is a curve D and $\tilde{D} \to D$ is a double unramified cover. The Gauss map $\tilde{D}$ is the Prym canonical map of $\tilde{D}$ to $\mathbb{P}^4$. The map $\phi : \tilde{D} \to \mathbb{P}^2$ is a map of degree 2 to a plane quintic Δ.

Our Conjecture says that a component of N_1 is of codimension 3 if and only if $g = 4$ (resp. $g = 5$) and the component in question is $\mathcal{H}_4$ (resp. J_5). A tentative approach to proving this might be the following.

Take an irreducible component N of N_1 and assume it has codimension 3 in $\mathcal{A}_g$. Let $[X, \Theta]$ be a general point of N and R a 1-dimensional component of $\mathrm{Sing}(\Theta)$.

Step i) Try to prove that the generic point of an irreducible component of R is a double points of Θ. If it is not, we should be able to prove that $\mathrm{codim}(N)$ is higher than 3.

Step ii) Assume that the general quadric Q_x of the span of Σ is smooth; otherwise use Segre's result. Let Δ be the discriminant locus of degree g in $\mathbf{P}^2$. The map $\phi : R \longrightarrow \Sigma \subset \Delta$ has degree ≥ 2 since it factors through -1. Prove that the degree is 2. This seems difficult. It would imply that $\Theta \cdot R \leq 2g$.

Step iii) We now assume that the class of R is a multiple $m\alpha$ of the minimal class $\alpha = \Theta^{g-1}/(g-1)! \in H^2(X, \mathbf{Z})$. If it is not, then $\mathrm{End}(X) \neq \mathbf{Z}$ and this implies that $\mathrm{codim}(N) \geq g - 1$. By the preceding step we now find $m \leq 2$.

Step iv) Apply now the Matsusaka-Ran criterion or a result of Welters. This implies that X is a Jacobian or a Prym variety. The cases $N \neq \mathcal{H}_4, J_5$ can then be ruled out by the results of Beauville.

References

[1] A. Andreotti & A. L. Mayer, *On period relations for abelian integrals on algebraic curves*, Ann. Sci. Norm. Pisa **3** (1967) 189–238.

[2] A. Beauville, *Variétés de Prym et jacobiennes intermédiaires*, Ann. Sci. École Norm. Sup. **10** (1977) 309–391.

[3] ———, *Prym varieties and the Schottky problem*, Invent. Math. **41** (1977) 149–196.

[4] ———, *Sous-variétés spéciales des variétés de Prym*, Compositio Math. **45** (1982) 357–383.

[5] O. Debarre, *Sur les variétés abéliennes dont le diviseur theta est singulier en codimension* 3, Duke Math. J. **57** (1988) 221–273.

[6] ———, *Sur les variétés de Prym des courbes tétragonales*, Ann. Sci. École Norm. Sup. **21** (1988) 545–559.

[7] ———, *Sur le problème de Torelli pour les variétés de Prym*, Amer. J. Math. **111** (1989) 111–134.

[8] ———, *Variétés de Prym et ensembles d'Andreotti et Mayer*, Duke Math. J. **60** (1990) 599–630.

[9] ———, *Le lieu des variétés abéliennes dont le diviseur thêta est singulier a deux composantes*, Ann. Sci. École Norm. Sup. **25** (1992) 687–707.

[10] ———, *Sur le théorème de Torelli pour les solides doubles quartiques*, Compositio Math. **73** (1990) 161-187.

[11] R. Donagi, *The tetragonal construction*, Bull. Amer. Math. Soc. **4** (1981) 181–185.

[12] ———, *The Schottky problem*, Theory of Moduli, Lecture Notes in Math., Springer, Berlin, Vol. 1337, 1988, 84–137.

[13] R. Donagi, *The fibers of the Prym map*, Curves, Jacobians, and abelian varieties (Amherst, MA, 1990), Contemp. Math., Amer. Math. Soc., Providence, RI, Vol. 136, 1992, 55–125.

[14] L. Ein & R. Lazarsfeld, *Singularities of the theta divisor and the birational geomtry of irregular varieties*, J. Amer. Math. Soc. **10** (1997) 243–258.

[15] R. Friedman & R. Smith, *The generic Torelli theorem for the Prym map*, Invent. Math. **67** (1982) 473–490.

[16] J. Kollár, *Shafarevich maps and automorphic forms*, Princeton Univ. Press, Princeton, 1995.

[17] D. Mumford, *Prym varieties. I*, Contributions to analysis (a collection of papers dedicated to Lipman Bers), Academic Press, New York, 1974, 325–350.

[18] ———, *On the Kodaira dimension of the Siegel modular variety*, Algebraic Geometry, Open Problems, Proc. Ravello, 1982, Lecture Notes in Math. (eds. C. Ciliberto, F. Ghione, F. Orecchia), Vol. 997, 348–376.

[19] C. Segre, *Ricerche sui fasci di coni quadrici in uno spazio lineare qualunque*, Atti della R. Accademia delle Scienze di Torino **XIX** (1883/4) 692–710.

[20] R. Smith & R. Varley, *Components of the locus of singular theta divisors of genus* 5, Algebraic geometry, Sitges (Barcelona), 1983, Lecture Notes in Math. Springer, Berlin, Vol. 1124, 1985, 338–416.

[21] ———, *Singularity theory applied to Θ-divisors,* Algebraic geometry (Chicago, IL, 1989), Lecture Notes in Math., Springer, Berlin, Vol. 1479, 1991, 238–257.

[22] ———, *Multiplicity g points on theta divisors*, Duke Math. J. **82** (1996) 319–326.

[23] ———, *Deformations of theta divisors and the rank 4 quadrics problem*, Comp. Math. **76** (1990) 367–398.

[24] G. E. Welters, *Polarized abelian varieties and the heat equation*, Compositio Math. **49** (1983) 173–194.

[25] ———, *A theorem of Gieseker-Petri type for Prym varieties*, Ann. Sci. École Norm. Sup. **18** (1985) 671–683.

UNIVERSITÀ DI ROMA II, ITALY
UNIVERSITEIT VAN AMSTERDAM, THE NETHERLANDS

Surveys in Differential Geometry, 2000
Vol. VII ©2000, International Press
pp. 83-106

HOLOMORPHIC SPHERES IN LOOP GROUPS AND BOTT PERIODICITY

RALPH L. COHEN, ERNESTO LUPERCIO & GRAEME B. SEGAL

Abstract

In this paper we study the topology of spaces of holomorphic maps from the Riemann sphere $\mathbb{P}^1$ to infinite dimensional Grassmanian manifolds and to loop groups. Included in this study is a complete identification of the homotopy types of $Hol_k(\mathbb{P}^1, BU(n))$ and of $Hol_k(\mathbb{P}^1, \Omega U)$, where the subscript k denotes the degree of the map. These spaces are shown to be homotopy equivalent to the kth Mitchell - Segal algebraic filtration of the loop group $\Omega U(n)$ [7], and to $BU(k)$, respectively.

Introduction

One of the most important theorems in Topology and Geometry is the "Bott Periodicity Theorem". In its most basic form it states that there is a natural homotopy equivalence,

$$\beta : \mathbb{Z} \times BU \xrightarrow{\simeq} \Omega U.$$

Here U is the infinite unitary group, $U = \varinjlim_n U(n)$, $BU = \varinjlim_n BU(n)$ is the limit of the classifying spaces, and $\Omega U = C^\infty(S^1, U)$ is the space of smooth, basepoint preserving loops. Here and throughout the rest of this paper all spaces will assumed to be equipped with a basepoint, and all maps and mapping spaces will be basepoint preserving.

If we input the fact that $U \simeq \Omega BU$, Bott periodicitiy states that there is a natural homotopy equivalence

$$\beta : \mathbb{Z} \times BU \xrightarrow{\simeq} \Omega^2 BU = C^\infty(S^2, BU).$$

First printed in Asian Journal of Mathematics, 1999. Used by permission.
The research of the first author was partially supported by a grant from the NSF, and the second author was partially support from CONACYT.

In a paper which first pointed to the deep relationship between the index theory of Fredholm operators and Algebraic Topology, Atiyah [1], defined a homotopy inverse to the Bott map β, which can be viewed as a map

$$\bar{\partial} : C^\infty(S^2, BU) \xrightarrow{\simeq} \mathbb{Z} \times BU.$$

This map was defined by studying the index of the family of operators obtained by coupling the $\bar{\partial}$ operator to a smooth map from S^2 to a Grassmannian.

Since the mapping spaces $C^\infty(S^2, BU)$ and ΩU both have path components naturally identified with the integers, we denote by $C^\infty(S^2, BU)_k$ the path components consisting of degree k - maps. Thus Bott periodicity, together with Atiyah's results says that for each integer k, there is a natural homotopy equivalence,

$$\bar{\partial} : C^\infty(S^2, BU)_k \xrightarrow{\simeq} BU\cdot$$

The goal of this paper is to prove a holomorphic version of this result. We first note however that the homotopy type of BU has many different models, several of which carry a holomorphic structure. For the purposes of this paper we think of BU as the colimit of the finite dimensional Grassmannians $Gr_m(\mathbb{C}^n)$ of m - dimensional subspaces of $\mathbb{C}^n$. We define $Hol_k(\mathbb{P}^1, BU)$ to be $\varinjlim_{m,n} Hol_k(\mathbb{P}^1, Gr_m(\mathbb{C}^n))$ topologized as a subspace of $C^\infty(S^2, BU) = \varinjlim_{m,n} C^\infty(S^2, Gr_m(\mathbb{C}^n))$, where these mapping spaces are given the compact open topology.

The following is the first result of this paper.

Theorem 1. *For each positive integer k, there is a natural homotopy equivalence*

$$Hol_k(\mathbb{P}^1, BU) \xrightarrow{\simeq} BU(k).$$

Remark. Observe that this theorem states implies the inclusion of holomorphic maps into all smooth maps (of degree k),

$$Hol_k(\mathbb{P}^1, BU) \hookrightarrow C^\infty(S^2, BU)_k$$

is homotopy equivalent to the inclusion of classifying spaces,

$$BU(k) \hookrightarrow BU.$$

In future work we will study consequences of this theorem to the understanding of holomorphic K - theory of a smooth, projective variety.

Our other main result in this paper has to do with an identification of the topology of subspaces $Hol_k(\mathbb{P}^1, BU(n))$ of $Hol_k(\mathbb{P}^1, BU)$. By Theorem 1, this latter space is homotopy equivalent to $BU(k)$, and so the inclusion $BU(n) \subset BU$ induces a map

$$j : Hol_k(\mathbb{P}^1, BU(n)) \to BU(k).$$

On the other hand, the inclusion of holomorphic maps into smooth maps, together with the identification of $\Omega^2 BU(n)$ with $\Omega U(n)$ induces a map

$$\iota : Hol_k(\mathbb{P}^1, BU(n)) \to \Omega_k U(n).$$

Furthermore, it is not difficult to see that the maps j and ι are compatible when composed to $BU \simeq \Omega U$.

Now S. Mitchell in [7] described an algebraic filtration of the loop group $\Omega SU(n)$ by compact, complex subvarieties:

$$F_{1,n} \subset F_{2,n} \subset \cdots \subset F_{k,n} \subset \cdots \subset F_{\infty,n} \simeq \Omega SU(n).$$

In this filtration, $F_{1,n} = \mathbb{CP}^{n-1}$ included in $\Omega U(n)$ via the usual complex J - map. $F_{k,n} \subset \Omega U(n)$ is the set of all k - fold products of elements of $F_{1,n} \subset \Omega U(n)$. See [11]. The subspaces $F_{k,n}$ have as their homology, precisely the intersection of $H_*(BU(k))$ and $H_*(\Omega SU(n))$, viewed as subgroups of $H_*(BU) = H_*(\Omega SU)$. In [8], Richter proved that this filtration stably splits, so that the loop group $\Omega SU(n)$ is stably homotopy equivalent to a wedge of the subquotients, $F_{k+1,n}/F_{k,n}$. The following theorem was conjectured by Mann and Milgram in [6] after an analysis of the holomorphic mapping spaces $Hol_k(\mathbb{P}^1, Gr_m(\mathbb{C}^n))$:

Theorem 2. *There is a natural homotopy equivalence*

$$Hol_k(\mathbb{P}^1, BU(n)) \simeq F_{k,n}.$$

These theorems are proven by using the identification of the loop group $\Omega U(n)$ with a certain moduli space of holomorphic bundles over $\mathbb{P}^1$, together with holomorphic trivializations on a disk. This identification was established in [9]. We then identify the homotopy type of $Hol_k(\mathbb{P}^1, BU(n))$ with a subspace of this moduli space consisting of bundles that are "negative" in the sense that they are (holomorphically) isomorphic to a direct sum of line bundles, each of which has nonpositive first Chern class. The topology of these moduli spaces are then studied in two ways: homologically, using calcluations of [6], and Morse theoretically, by analyzing the gradient flow of the Dirichlet energy functional on $\Omega U(n)$.

This paper is organized as follows. In Section 1 we review some results from the theory of loop groups. The main reference for this material is Pressley and Segal's book [9]. This theory will in particular allow us to define the terms and maps in the statements of the above theorems more carefully. In Section 2 we give a proof of Theorems 1 and 2, modulo a technical argument establishing that certain maps are (quasi)fibrations. This argument is carried out in Section 3.

The real Bott periodicity analogues of the above theorems (where the unitary groups are replaced by orthogonal and symplectic groups) were established in the Stanford University Ph.D thesis of the second author written under the direction of the first author. The authors are grateful to Paulo Lima-Filho, Steve Mitchell, Paul Norbury, and Giorgio Valli for helpful conversations regarding this work.

1. Loop groups

In this section we recall some of the basic constructions from the theory of loop groups as developed in [9]. We will use these constructions to define the holomorphic structures necessary to define the spaces and maps in the theorems described in the introduction. In what follows we will work with the Lie groups $U(n)$, but everything we use has obvious analogues for arbitrary compact semi-simple Lie groups. Again, we refer the reader to [9] for details.

As defined in the introduction, let LG denote the space of smooth maps from the circle S^1 to a Lie group G. The loop group $LGL(n, \mathbb{C})$ has the following important subgroups.

1. The group $L^+GL(n, \mathbb{C})$ of maps $\gamma : S^1 \to GL(n, \mathbb{C})$ that extend to holomorphic maps of the closed disk $D^2 \to GL(n, \mathbb{C})$.

2. The group $L_{pol}GL(n, \mathbb{C})$ of loops whose matrix entries are finite Laurent polynomials in z. That is, loops γ of the form

$$\gamma(z) = \sum_{k=-N}^{N} A_k z^k$$

for some N, where the A_k's are $n \times n$ matrices.

3. The based loop group $\Omega GL(n, \mathbb{C})$, and the corresponding subgroup of polynomial loops, $\Omega_{pol} GL(n, \mathbb{C})$.

In the theory of loop groups, there is an underlying Hilbert space H^n defined to be the space of square integrable functions from the circle to $\mathbb{C}^n$:

$$H^n = L^2(S^1, \mathbb{C}^n).$$

Notice that this Hilbert space has a natural polarization:

$$H^n \cong H^n_+ \oplus H^n_-$$

where H^n_+ consists of those functions whose negative Fourier coefficients are all zero; or equivalently those functions $f : S^1 \to \mathbb{C}^n$ that extend to a holomorphic map of the disk.

Observe that the loop group $LGL(n, \mathbb{C})$ has a natural representation on the Hilbert space $H^n = L^2(S^1, \mathbb{C}^n)$ given by matrix multiplication. In particular the Laurent polynomial ring $\mathbb{C}[z, z^{-1}]$ acts on H^n, via the action of the loops

$$z^k \to z^k \cdot I_{n \times n} \in LGL(n, \mathbb{C}).$$

Now recall the "restricted Grassmannian" $Gr(H^n)$ of H^n as defined in [9]. This is the space of all closed subspaces $W \subset H^n$ such that the orthogonal projections $pr_+ : W \to H^n_+$ and $pr_- : W \to H^n_-$ are Fredholm and Hilbert-Schmidt operators respectively.

Let $Gr_\infty(H^n) \subset Gr(H^n)$ be the dense submanifold consisting of elements $[W] \in Gr(H^n)$ such that the images of the projections $pr_+ : W \to H^n_+$ and $pr_- : W \to H^n_-$ consist of smooth functions. (See [9] §7.2.)

For ease of notation we denote $Gr_\infty(H^n)$ by Gr. Consider the following important submanifolds of Gr.

1. $Gr_0 = \{W \in Gr : \exists k \geq 0 \text{ such that } z^k H_+ \subset W \subset z^{-k} H^n_+\}$
2. $Gr^{(n)} = \{W \in Gr : zW \subset W\}$
3. $Gr^{(n)}_0 = Gr^{(n)} \cap Gr_0$.

The action of $LGL(n, \mathbb{C})$ on H^n induces an action on $Gr(H^n)$, and it is proved in [9] that the orbit of H^n_+ is precisely $Gr^{(n)}$. This is also its orbit under the subgroup $LU(n)$. The isotropy group of H^n_+ in $LGL(n, \mathbb{C})$

consists of those loops whose Fourier expansions contain only nonnegative powers of z. These loops are exactly the boundary values of holomorphic maps of the disk. In the above notation we called this subgroup $L^+GL(n, \mathbb{C})$. In fact there is a homeomorphism

$$LGL(n, \mathbb{C})/L^+GL(n, \mathbb{C}) \cong Gr^{(n)}.$$

Furthermore, if one restricts the action to $LU(n)$, one sees that the isotropy subgroup of H^n_+ is given by the subgroup $LU(n) \cap L^+GL(n, \mathbb{C})$. This is the subgroup of loops of $U(n)$ that are boundary values of holomorphic maps of the disk to $GL(n, \mathbb{C})$. A generalization of the maximum modulus principle, as proved in [9] shows that this subgroup consists only of the constant loops, $U(n) \subset LU(n)$. This, together with the analogous argument using $Gr_0^{(n)}$ and polynomial loops proves the following.

Theorem 1.1. *There are homeomorphisms*

$$\Omega U(n) \cong LU(n)/U(n) \cong LGL(n, \mathbb{C})/L^+GL(n, \mathbb{C}) \cong Gr^{(n)}$$

$$\Omega_{pol}U(n) \cong L_{pol}U(n)/U(n) \cong L_{pol}GL(n, \mathbb{C})/L^+_{pol}GL(n, \mathbb{C}) \cong Gr_0^{(n)}$$

given by $\gamma \to \gamma H^n_+$.

These homeomorphisms determine the complex structure on the infinite dimensional manifolds $\Omega U(n)$ and $\Omega_{pol}U(n)$ that are used in the statement of Theorem 1 in the introduction.

Given an element $W \in Gr^n$ define the **virtual dimension** of W, $vd(W)$, to be the Fredholm index of the projection of the projection map

$$pr_+ : W \to H^n_+.$$

So in particular we have

$$vd(W) = dim\left(W \cap H^n_-\right) - dim\left(W^\perp \cap H^n_+\right).$$

As we will see below, much of the analysis necessary to prove Theorem 1 comes from studying the situation when the projection operator $pr_+ : W \to H^n_+$ has trivial kernel, and hence $vd(W) = -dim(coker(pr_+))$.

Now notice that since every element $W \in Gr_0$ has the property that $z^k H^n_+ \subset W \subset z^{-k} H^n_+$ for some k, and hence by considering the projection $\bar{W} \subset z^{-k} H^n_+ / z^k H^n_+$, one has that Gr_0 is the union of finite dimensional Grassmannians,

$$Gr_0 = \bigcup_p Gr(z^{-p} H^n_+ / z^p H^n_+).$$

and is therefore topologically equivalent to $\mathbb{Z} \times BU$. The integer denotes the virtual dimension of W.

Via the identification of $Gr^{(n)}$ with the loop group, the virtual dimension determines the path component of the loop. In pariticular if $wdg(\gamma)$ denotes the winding number of the determinant of a loop in $U(n)$, then a straightforward exercise proves the following.

Proposition 1.1. *Let $\gamma \in \Omega U(n)$ correspond to $W \in Gr^{(n)}$. That is, $W = \gamma H_+^n$. Then the virtual dimension of W and the winding number of γ are related by*

$$v.d(W) = -wdg(\gamma).$$

Notice furthermore that if $\gamma \in \Omega_{pol} U(n)$ is a polynomial loop, then this composite $det \circ \gamma$ is also a polynomial loop in $S^1 \in \mathbb{C}$. The only such polynomial maps (i.e polynomials in one variable of constant unit length) are $z \to z^k$ for some $k \in \mathbb{Z}$. Therefore we may conclude that the polynomial loop group of $SU(n)$ consists precisely of those polynomial loops in $\Omega_{pol} U(n)$ with winding number zero. We again refer the reader to [9] for details.

Recall that there was another description of the loop group $\Omega U(n)$ in [9]that is given in terms of holomorphic bundles over the Riemann sphere $\mathbb{P}^1 = \mathbb{C} \cup \infty = D_0 \cup D_\infty$, where $D_0 = \{z : |z| \leq 1\}$ and $D_\infty = \{z : |z| \geq 1\}$.

Proposition 1.2. *Let $C_{k,n}$ denote the space $C_{k,n} = \{$isomorphism classes of pairs (E, ϕ), where $E \to \mathbb{P}^1$ is an n - dimensional bundle of Chern class $c_1(E) = -k$, and ϕ is a holomorphic framing of $E_{|D_\infty}\}$. Then there is a natural homeomorphism*

$$C_{k,n} \xrightarrow{\cong} \Omega_k U(n).$$

The homeomorphism in this proposition can be described as follows. Let $(E, \phi) \in C_{k,n}$. So $\phi : E|_{D_\infty} \to \mathbb{C}^n$ is a fixed trivialization. Let $\sigma : E|_{D_0} \to \mathbb{C}^n$ be any holomorphic trivialization of the restriction of E to the other disk, D_0. On the intersection $S^1 = D_0 \cap D_\infty$, the trivializations ϕ and σ differ by a loop $\gamma \in LGL(n, \mathbb{C})$. Of course the loop γ depends on the choice of trivialization σ, but if a different trivialization, say $\tilde{\sigma}$ were used to construct a different loop, say $\tilde{\gamma}$, then $\gamma \cdot (\tilde{\gamma})^{-1}$ would be a loop that extends to a holomorphic map $\sigma \cdot (\tilde{\sigma})^{-1} : D_0 \to GL(n, \mathbb{C})$. Thus $\gamma \cdot (\tilde{\gamma})^{-1} \in L^+GL(n, \mathbb{C})$. This procedure therefore gives a well defined map

$$C_{k,n} \to LGL(n, \mathbb{C})/L^+GL(n, \mathbb{C}) \cong \Omega U(n).$$

We refer the reader to [9] for details of the argument proving that this map is a homeomorphism.

We note that with respect to the diffeomorphisms

$$\mathcal{C}_{k,n} \cong \Omega_k U(n) \cong (Gr^{(n)})_k$$

the Chern class of the bundle in $\mathcal{C}_{k,n}$ corresponds to the virtual dimension of the subspace $W \in Gr^{(n)}$, which, as observed above, corresponds to the opposite of the winding number of the loop in $\Omega U(n)$.

These models of the loop groups will prove very important in our study of holomorphic mapping spaces. One aspect of these models that is quite useful is that they come equipped with filtrations by algebraic subvarieties. That is, there is a sequence of compact, complex subvarieties:

$$F_{1,n} \hookrightarrow \cdots F_{k,n} \hookrightarrow F_{k+1,n} \hookrightarrow \cdots F_{\infty,n} = \Omega_{pol} SU(n) \simeq \Omega SU(n).$$

The filtration $F_{k,n}$ can be defined as follows. As in [11], define the sub - semigroup $\tilde{\Omega}_n$ of the space of polynomial loops $\Omega_{pol} U(n)$ to consist of loops that only involve non-negative powers of z in their Fourier expansions. Furthermore, we write

$$\tilde{\Omega}_n = \coprod_{k \geq 0} \tilde{\Omega}_{k,n}$$

where $\tilde{\Omega}_{k,n}$ consists of loops of winding number k.

The space $F_{k,n}$ is homeomorphic to $\tilde{\Omega}_{k,n}$, but in order to see how these spaces give a filtration of $\Omega SU(n)$, we consider the loop

$$\lambda : S^1 \to U(n)$$

$$z \to \begin{pmatrix} z^{-1} & 0 & \cdots & 0 \\ 0 & 1 & 0 & \cdots \\ 0 & 0 & 1 & \cdots \\ & & \cdots & \\ & & & 1 \end{pmatrix}$$

Clearly λ is a polynomial loop of winding number -1. Therefore if we define

$$F_{k,n} = \lambda^k \tilde{\Omega}_{k,n}$$

then $F_{k,n}$ consists of polynomial loops of winding number zero. That is, $F_{k,n} \subset \Omega_{pol} SU(n)$. Moreover $F_{k,n} \subset F_{k+1,n}$ and

$$\Omega_{pol} SU(n) = \bigcup_{k \geq 1} F_{k,n}.$$

(See [7], [11].)

The homotopy theory of these filtrations has been studied in great detail by Mitchell [7], and by Richter in [8]. In particular one has the following properties (see [7])

Proposition 1.3. *If one takes the limit over the rank n, then there is a natural homotopy equivalence*

$$\lim_{n\to\infty} F_{k,n} \simeq BU(k).$$

Moreover the following diagram homotopy commutes:

$$\begin{array}{ccc} F_{k,n} & \xrightarrow{n\to\infty} & BU(k) \\ \cap\downarrow & & \downarrow\cap \\ \Omega U(n) & \underset{\hookrightarrow}{\longrightarrow} & \Omega U \simeq BU. \end{array}$$

Furthermore the homology of this filtration is given by

$$H_*(F_{k,n}) \cong H_*(\Omega U(n)) \cap H_*(BU(k)) \subset H_*(BU).$$

In [8] Richter also showed that this filtration stably splits. That is, there is a stable homotopy equivalence

$$\Omega SU(n) \simeq_s \bigvee_k F_{k,n}/F_{k-1,n}.$$

Now the spaces of loops $\tilde{\Omega}_{k,n}$ can also be interpreted in the above Grassmannian models in the following way. Since $\gamma \in \tilde{\Omega}_{k,n}$ has only positive terms in its Fourier expansion, the space $W = \gamma H^n_+$ is a subspace of H^n_+. Moreover since γ has winding number k, the subspace W has virtual dimension $-k$, (Proposition 1.2). This means that the dimension of H^n_+/W is equal to k. Thus we have the following.

Proposition 1.4. *Under the diffeomorphism described in Theorem 1.1, we have that*

$$\begin{aligned} \tilde{\Omega}_{k,n} &= \{W \in Gr_0^{(n)} : W \subset H^n_+ \text{ and } \dim H^n_+/W = k\} \\ &= \{W \in Gr_0^{(n)} : vd(W) = -k, \text{ and } ker(pr_+ : W \to H^n_+) = 0\}. \end{aligned}$$

It will be helpful to have a description of $\tilde{\Omega}_{k,n}$ (or equivalently $F_{k,n}$) in terms of holomorphic bundles as well. To do this we need to introduce the notion of a **negative** holomorphic bundle.

By a well known theorem of Grothendieck [4], every holomorphic bundle over $\mathbb{P}^1$ is isomorphic to a direct sum of holomorphic line bundles of the form $O(j)$, where $c_1(O(j)) = j \in \mathbb{Z}$. Furthermore this direct sum decomposition is unique (up to order). A holomorphic bundle $E \to \mathbb{P}^1$ is **negative** if

$$E \cong O(j_i) \oplus \cdots \oplus O(j_m)$$

with each $j_i \leq 0$. Notice that the dimension of E is m, and the first Chern class is given by

$$c_1(E) = \sum_{i=1}^{m} j_i.$$

It is a standard exercise that a holomorphic bundle E over $\mathbb{P}^1$ is negative if and only if it can be holomorphically embedded into a trivial bundle. Such bundles are obtained by pulling back the universal bundle over a Grassmannian via a holomorphic map. A **positive** holomorphic bundle over $\mathbb{P}^1$ has the analogous definition, and this property corresponds to a bundle being generated by its holomorphic sections.

Define the sub - moduli space

$$\mathcal{C}^-_{k,n} \subset \mathcal{C}_{k,n}$$

to consist of those $(E, \phi) \in \mathcal{C}_{k,n}$ such that E is negative.

Now an easy exercise verifies that with respect to the homeomorphism $\mathcal{C}_{k,n} \cong Gr^{(n)}$ described above, a holomorphic bundle E being negative corresponds to the projection operator $pr_+ : W \to H^n_+$ having zero kernel, and so $vd(W) = -dimcoker(pr_+)$. By Proposition 1.3 we therefore have the following.

Proposition 1.5. *With respect to the diffeomorphisms*

$$\mathcal{C}_{k,n} \cong \Omega_k U(n) \cong (Gr^{(n)})_k,$$

we have that

$$\begin{aligned}\tilde{\Omega}_{k,n} &\cong \mathcal{C}^-_{k,n} \cap \Omega_{pol} U(n)\\ &\cong \mathcal{C}^-_{k,n} \cap Gr^{(n)}_0.\end{aligned}$$

2. Proofs of Theorems 1 and 2

In this section we give proofs of Theorems 1 and 2, modulo a technical lemma whose proof we delay until the next section. We begin with Theorem 2, which asserts that $F_{k,n}$ is homotopy equivalent to the space $Hol_k(\mathbb{P}^1, BU(n))$.

In view of Proposition 1.5, it is sufficient to prove the following two theorems:

Theorem 2.1. *There is a natural homotopy equivalence*

$$Hol_k(\mathbb{P}^1, BU(n)) \simeq C^-_{k,n}.$$

Theorem 2.2. *The inclusion given by Proposition 1.5*

$$\tilde{\Omega}_{k,n} \subset C^-_{k,n}$$

is a homotopy equivalence.

We begin with a proof of Theorem 2.1.

Proof. We start by describing a model (up to homeomorphism) of $Hol_k(\mathbb{P}^1, BU(n))$.

Let V be an infinite dimensional complex vector space topologized as the union of its finite dimensional subspace. We take as our model for $BU(n)$ the Grassmannian $Gr_n(V)$ of n - dimensional subspaces of V. $Gr_n(V)$ is topologized as the limit

$$Gr_n(V) = \varinjlim_F Gr_n(F)$$

where the limit is taken over finite dimensional subspaces F of V.

Proposition 2.1. *Let $V_\infty \subset V$ be a fixed n - dimensional subspace. Define $\mathcal{M}_{k,n}$ to be the following moduli space:*

$\mathcal{M}_{k,n}$ ={*isomorphism classes of pairs (E, j), where $E \to \mathbb{P}^1$ is a negative n - dimensional holomorphic bundle of Chern class $c_1(E) = -k$, and $j : E \hookrightarrow \mathbb{P}^1 \times V$ is a holomorphic embedding of vector bundles taking the fiber at ∞, E_∞ to V_∞*}

Then there is a natural set bijection

$$h : Hol_k(\mathbb{P}^1, Gr_n(V)) \xrightarrow{\cong} \mathcal{M}_{k,n}.$$

Proof. Let $f \in Hol_k(\mathbb{P}^1, Gr_n(V))$. The universal bundle $\zeta \to Gr_n(V)$ is a holomorphic bundle embedded (holomorphically) in the trivial bundle $e : \zeta \hookrightarrow Gr_n(V) \times V$. We define

$$h(f) = (f^*(\zeta), g^*(e)) \in \mathcal{M}_{k,n}.$$

The fact that h is a bijection follows from the fact that a holomorphic bundle over $\mathbb{P}^1$ is (holomorphically) embeddable in a trivial bundle if and only if it is negative. q.e.d.

We give the set $\mathcal{M}_{k,n}$ the topology induced from $Hol_k(\mathbb{P}^1, Gr_n(V))$ via the bijection h. Theorem 2.1 will then be a consequence of the following.

Proposition 2.2. *The moduli spaces $\mathcal{C}^-_{k,n}$ and $\mathcal{M}_{k,n}$ are homotopy equivalent.*

Proof. We actually prove that the moduli spaces $\mathcal{M}_{k,n}$ and $\mathcal{C}^-_{k,n}$ are both homotopy equivalent to an intermediate space $\mathcal{X}_{k,n}$ defined by

$\mathcal{X}_{k,n} = \{$isomorphism classes of pairs (E, j, θ), where $(E, j) \in \mathcal{M}_{k,n}$, and $\theta : E_{|D_\infty} \to D_\infty \times V_\infty$ is a holomorphic trivialization$\}$.

Forgetting the trivialization θ defines a map

$$\pi_{k,n} : \mathcal{X}_{k,n} \to \mathcal{M}_{k,n}.$$

Similarly the projection map $(E, j, \theta) \to (E, \theta)$ defines a map

$$p_{k,n} : \mathcal{X}_{k,n} \to \mathcal{C}^-_{k,n}.$$

The following is rather technical, and so its proof will be delayed until the next section.

Lemma 2.3. *a. The map $\pi_{k,n} : \mathcal{X}_{k,n} \to \mathcal{M}_{k,n}$ is a locally trivial fibration.*

b. The map $p_{k,n} : \mathcal{X}_{k,n} \to \mathcal{C}^-_{k,n}$ is a quasifibration.

Assuming the validity of this lemma for now, we can easily complete the proof of Proposition 2.2. Notice that the fiber of the map $\pi_{k,n}$ is the space of (holomorphic) trivializations of $E_{|D_\infty}$, which is homeomorphic to $Hol(D^\infty, GL(n, \mathbb{C}))$, which is contractible. Similarly the fiber of the map $p_{k,n}$ is the space of holomorphic bundle embeddings $E \hookrightarrow \mathbb{P}^1 \times V$. Since every such embedding has image in a finite dimensional trivial subbundle, this space is given by the limit of the spaces of holomorphic embeddings of E into finite dimensional trivial subbundles of $\mathbb{P}^1 \times V$. Since V is infinite dimensional this limiting space is contractible. Since $\pi_{k,n}$ and $p_{k,n}$ both have contractible fibers, then by the lemma they are homotopy equivalences. This implies $\mathcal{C}_{k,n}^-$ and $\mathcal{M}_{k,n}$ are homotopy equivalent to each other.

This proves the proposition, and therefore completes the proof of Theorem 2.1, modulo Lemma 2.3. q.e.d.

We now proceed to prove Theorem 2.2. We will give two proofs of this theorem. The first is homological in nature and will rely on the calculations of the homology of the spaces $F_{k,n}$ by Mitchell [7] and of spaces of holomorphic maps from $\mathbb{P}^1$ to Grassmannians by Mann and Milgram [6]. The second proof will be Morse theoretic in nature, and will rely on the dynamics of a flow of a natural C^*- action on loop groups studied in [9]. This proof is more geometric in nature, and has the feature that it will give an alternative proof of Mann and Milgram's calculation of $H_*(Hol_k(\mathbb{P}^1, Gr_n(V)))$.

Homological Proof. Let $\beta : F_{k,n} \to Hol_k(\mathbb{P}^1, Gr_n(V))$ be the composition

$$\beta : F_{k,n} \cong \tilde{\Omega}_{k,n} \subset \mathcal{C}_{k,n}^- \simeq Hol_k(\mathbb{P}^1, Gr_n(V)).$$

Of course it suffices to prove that β induces an isomorphism in homology.

Notice from the construction of the map β that if one composes with the inclusion of holomorphic maps into smooth maps,

$$F_{k,n} \xrightarrow{\beta} Hol_k(\mathbb{P}^1, BU(n)) \hookrightarrow \Omega_k^2 BU(n) \simeq \Omega_k U(n) \simeq \Omega SU(n)$$

then this map is homotopic to the inclusion of the Mitchell filtration $F_{k,n} \subset \Omega SU(n)$.

Also, consider the composition of β with the map $Hol_k(\mathbb{P}^1, Gr_n(V)) \to BU(k)$ given by

$$\delta : Hol_k(\mathbb{P}^1, Gr_n(V)) \simeq \mathcal{C}_{k,n}^- \to Gr_k(H_+^n)$$

where the last map assigns to a loop $\gamma \in \Omega U(n)_k$ whose associated holomorphic bundle E_γ is negative (and therefore lies in $\mathcal{C}_{k,n}^-$) the cokernel of the projection operator pr_γ viewed as a subspace $coker(pr_\gamma) \subset H_+^n$. The composition

$$F_{k,n} \xrightarrow{\beta} Hol_k(\mathbb{P}^1, Gr_n(V)) \xrightarrow{\delta} Gr_k(H_+^n) \simeq BU(k)$$

is homotopic to the map described in Proposition 1.4. We therefore have the following homotopy commutative diagram:

$$\begin{array}{ccccc} F_{k,n} & \xrightarrow{\beta} & Hol_k(\mathbb{P}^1, Gr_n(V)) & \xrightarrow{\delta} & BU(k) \\ & & \downarrow & & \downarrow \cap \\ & & \Omega_k U(n) & \xrightarrow[\hookrightarrow]{} & \Omega U \simeq BU \end{array}$$

where the outer maps $\delta \circ \beta : F_{k,n} \to BU(k)$ and $F_{k,n} \to \Omega_k U(n)$ are the maps described in Proposition 1.3. Now by this proposition, the composite map $F_{k,n} \to BU$ maps injectively in homology, with image equal to the intersection $H_*(\Omega U(n)) \cap H_*(BU(k)) \subset H_*(BU)$. Thus by the commutativity of this diagram, the homomorphism

$$\beta_* : H_*(F_{k,n}) \to H_*(Hol_k(\mathbb{P}^1, Gr_n(V)))$$

is injective. To prove surjectivity, we use a result of Mann and Milgram [7] that says the inclusion of the holomorphic maps into all smooth maps

$$Hol_k(\mathbb{P}^1, Gr_n(\mathbb{C}^m)) \hookrightarrow \Omega^2_k Gr_n(\mathbb{C}^m)$$

induces an injection in homology for every Grassmannian $Gr_n(\mathbb{C}^m)$. This implies that the map

$$\begin{aligned} H_*(Hol_k(\mathbb{P}^1, Gr_n(V))) &= H_*(Hol_k(\mathbb{P}^1, BU(n))) \\ &\to H_*(\Omega^2 BU(n)) \\ &= H_*(\Omega U(n)) \end{aligned}$$

is injective. But since the inclusion

$$H_*(\Omega U(n)) \to H_*(\Omega U) \cong H_*(BU)$$

is injective, that means that the composition

$$H_*(Hol_k(\mathbb{P}^1, Gr_n(V))) \to H_*(\Omega U(n)) \to H_*(BU)$$

is injective. Also, by the commutativity of the diagram, we know that the image lies in the intersection

$$H_*(\Omega U(n)) \cap H_*(BU(k)) \subset H_*(BU).$$

But as remarked above this intersection is the image of (and isomorphic to) $H_*(F_{k,n})$. Hence both maps

$$H_*(Hol_k(\mathbb{P}^1, Gr_n(V))) \to H_*(\Omega U(n)) \cap H_*(BU(k))$$

and

$$H_*(F_{k,n}) \to H_*(\Omega U(n)) \cap H_*(BU(k))$$

are isomorphsms. Thus by the commutativity of the above diagram

$$\beta_* : H_*(F_{k,n}) \to H_*(Hol_k(\mathbb{P}^1, Gr_n(V)))$$

is an isomorphism. Now since both these spaces are known to be simply connected ([7], [6]), this implies β is a homotopy equivalence. q.e.d.

Morse theoretic proof. The second proof that the inclusion $\tilde{\Omega}_{k,n} \hookrightarrow \mathcal{C}^-_{k,n}$ is a homotopy equivalence is by studying the dynamics of a $\mathbb{C}^*$ - action on $\Omega U(n)$ described in [9].

To be more specific, consider the natural circle action on $H^n = L^2(S^1, \mathbb{C}^n)$. This action preserves the polarization and therefore induces an action on the Grassmannian $Gr(H^n)$. It was shown in [9] that this action extends to a smooth action of the units in $\mathbb{C}$ lying in the unit disk

$$\mathbb{C}^*_{\leq 1} \times Gr^{(n)} \to Gr^{(n)}$$

and to all of $\mathbb{C}^*$ on the polynomial Grassmannian,

$$\mathbb{C}^* \times Gr_0^{(n)} \to Gr_0^{(n)}.$$

The induced flow given by the action of the reals was shown to be the gradient flow of the energy function on $\Omega U(n)$:

$$E : \Omega U(n) \to \mathbb{R}$$
$$\gamma \to \frac{1}{4\pi} \int_0^{2\pi} |\gamma(t)^{-1}\gamma'(t)|^2 dt.$$

The critical points of this action are homomorphisms $\lambda : S^1 \to U(n)$, and hence the critical levels are indexed by the conjugacy class of the homomorphism; namely a partition $\bar{a} = (a_1, \cdots, a_n)$. We write $\lambda_{\bar{a}}$ to denote the critical level given by conjugates of the homomorphism

$$z \to (z^{a_1}, \cdots, z^{a_n}) \in U(1) \times \cdots \times U(1) \subset U(n).$$

The resulting Morse decomposition of $\Omega U(n) = LGL(n, \mathbb{C})/L^+$ was studied in great detail in [9]. In particular it was shown that the stable and unstable manifolds of the critical levels $\lambda_{\bar{a}}$ are given by the orbits under the (left) action of L^- and L^+ respectively. (Here, as above, L^+ denotes those loops that are boundary values of holomorphic maps of the disk D_0, and L^- denotes those loops that are boundary values of holomorphic maps of the disk around infinity, D_∞.)

Recall that with respect to a generic metric, the unstable and stable manifolds of a Morse function intersect each other transversally. In this case the critical points have a partial ordering, where one says that $a \geq b$ if there is a flow line emanating from a and converging to b. The resulting ordering of critical points of this flow on $\Omega U(n)$ was described in [9] in the following way.

Let $S \subset \mathbb{Z}$ be a set of integers whose symmetric difference with the nonnegative integers, $\mathbb{Z}^+$ is finite. That is,

$$S \Delta \mathbb{Z}^+ = \{S - \mathbb{Z}^+\} \cup \{\mathbb{Z}^+ - S\}$$

is a finite set. The *virtual cardinal* of such a set S, $v.c(S)$ is defined to be

$$v.c(S) = \#\{S - \mathbb{Z}^+\} - \#\{\mathbb{Z}^+ - S\}.$$

A set S with virtual cardinal $v.c(S) = k$ will be of **type** n, if for every $s \in S$,

$$s + n \subset S.$$

It was shown in [9] that the critical manifolds of the above flow when restricted to the component $\Omega_k U(n)$, are indexed by sets S of type n having virtual cardinal $= -k$. The relationship with the above indexing of critical points via conjugacy classes of homomorphisms $\lambda : S^1 \to U(n)$, of winding number k, and thus partitions $\bar{a} = (a_1, \cdots, a_n)$, with $\sum_{i=1}^n a_i = k$, is given as follows.

Given such a partition $\bar{a} = (a_1, \cdots, a_n)$, then let $b_i = na_i + i$, and let $S_{\bar{a}}$ be the type n set genererated by $\{b_1, \cdots, b_n\}$. That is,

$$S_{\bar{a}} = \bigcup_{q \in \mathbb{Z}^+} \{b_1 + qn, b_2 + qn, \cdots, b_n + qn\}.$$

One reason that this indexing of the critical levels is useful, is that, as was observed in [9], every set S with virtual cardinal $v.c(S) = m$, can be written as an ordered sequence

$$S = \{s_{-m}, s_{-m+1}, \cdots, s_q, \cdots\}$$

with $s_{-m} < s_{-m+1} < \cdots, s_q, \cdots$, and $s_m = m$ for all m sufficiently large. When written this way, the sets of virtual cardinal k have a natural partial ordering given by

$$S \geq S' \quad \text{if} \quad s_j \leq s'_j \quad \text{for all } j.$$

The following was proved in [9].

Lemma 2.4. *The above partial ordering on the type n sets S of virtual cardinal $-k$ corresponds to the Morse - Smale partial ordering of the critical levels of the energy functional resticted to the component $\Omega_k U(n)$.*

As we did with holomorphic bundles, we refer to a type n set S of virtual cardinal $-k$ as *negative*, if $S \subset \mathbb{Z}^+$. This terminology might be somewhat confusing , but we use it because in the definition $v.c(S) = \#\{S - \mathbb{Z}^+\} - \#\{\mathbb{Z}^+ - S\}$, a *negative* set S is one for which $\#\{S - \mathbb{Z}^+\} = 0$ and hence $v.c(S) = -\#\{\mathbb{Z}^+ - S\}$.

For a given k and n, define the type n set of virtual cardinal $-k$, S_k, to be

$$S_k = \{k, k+1, k+2, \cdots\}.$$

Write the integer k in the form $k = mn + i$, where $0 \leq i \leq n-1$. Then it is seen from the definition that the partition that S_k corresponds to is $(m, m, \cdots, m, m+1, \cdots, m+1)$, where there are $n-i$ copies of m and i copies of $m+1$ in this partition of k.

S_k therefore corresponds to the homomorphism

$$\begin{aligned} \lambda_k : S^1 &\to U(1) \times \cdots \times U(1) \subset U(n) \\ z &\to (z^m, \cdots, z^m, z^{m+1}, \cdots, z^{m+1}) \end{aligned}$$

or, equivalently to the holomorphic bundle given by the direct sum

$$\lambda_i = \oplus_{n-i} O(-m) \quad \oplus \quad \oplus_i O(-(m+1)).$$

The following is an exercise with the above definitions.

Lemma 2.5 *The sets S_k are minimal in the sense that if S is a type n set of virtual cardinal $-k$ with*

$$S_k \geq S,$$

then $S = S_k$. Furthermore, a type n set S of virtual cardinal $-k$ satisfies

$$S \geq S_k$$

if and only if S is negative.

This leads quite quickly to the following result.

Proposition 2.6. *The moduli space $C^-_{k,n}$, when viewed as a subspace of $\Omega U(n)_k$, is equal to the union of the stable manifolds of the critical levels indexed by type n sets S of virtual cardinal $-k$, satisfying $S \geq S_k$. That is, $C^-_{k,n}$ is the closure of the stable manifold of the critical level consisting of homomorphisms conjugate to λ_k, as defined above.*

Proof. As described above, the stable manifold of the critical level of the space of homomorphisms conjugate to $\lambda_{\bar{a}}$ is given by the orbit under the left action of $L^-(GL(n, \mathbb{C}))$. Now the Bruhat factorization (see [9]) of a loop says that every loop $\gamma \in LGL(n, \mathbb{C})$ has a decomposition as

$$\gamma = \gamma_- \cdot \lambda \cdot \gamma_+$$

where $\gamma_- \in L^-$ and $\gamma_+ \in L^+$, and $\lambda : S^1 \to U(n)$ is a homomorphism. This decomposition is unique up to the conjugacy class of λ. The corresponding holomorphic bundle E_γ is isomorphic to E_λ which is a direct sum of line bundles, and hence gives the Grothendieck decomposition. This then says that under the left action of L^-, the Grothendieck type of the corresponding holomorphic bundle is preserved. Thus the stable manifold of a critical set $\lambda_{\bar{a}}$ is exactly the space of loops γ whose corresponding holomorphic bundles E_γ are isomorphic to the direct sum of line bundles given by the partition $\bar{a}$. The proposition then follows from Lemma 2.4. q.e.d.

The action of $L^+GL(n, \mathbb{C})$ on the polynomial loop group $\Omega_{pol}U(n)_k$ was studied in detail in [9] as well. As mentioned above, the orbits of this action give the unstable manifolds of the flow of energy functional, when restricted to the polynomial loop group. The following result can be viewed as the dual of Proposition 2.6.

Proposition 2.7. *The space $\tilde{\Omega}_{k,n}$, when viewed as a subspace of $\Omega_{pol}U(n)_k$, is equal to the union of the unstable manifolds of the critical levels indexed by type n sets S of virtual cardinal $-k$, satisfying $S \geq S_k$.*

Theorem 2.2 now follows from Propositions 2.5 and 2.6 and standard Morse theory arguments, using the fact that the energy functional on $\Omega U(n)$ satisfies the Palais - Smale condition. q.e.d.

Now that we have Theorem 2, we have identified the homotopy type of the holomorphic mapping space, $Hol_k(\mathbb{P}^1, BU(n))$ in terms of the Mitchell filtration, $F_{k,n}$. By taking the limit over the rank n and using Proposition 1.4, we get a homotopy equivalence

$$\phi_k : Hol_k(\mathbb{P}^1, BU) \xrightarrow{\simeq} BU(k).$$

This is the assertion of Theorem 1. q.e.d.

3. Proof of Lemma 2.3

In the last section we completed the proofs of Theorems 1 and 2 modulo a technical result (Lemma 2.3) establishing that certain maps are (quasi)fibrations. We will prove this lemma in this section.

Proof. (Lemma 2.3 part (a)). Recall that the space $\mathcal{X}_{k,n}$ is defined by $\mathcal{X}_{k,n} = \{$isomorphism classes of triples (E, j, θ), where $(E, j) \in \mathcal{M}_{k,n}$, and $\theta : E_{|D_\infty} \to D_\infty \times V_\infty$ is a holomorphic trivialization$\}$, where V is an infinite dimensional complex vector space (topologized as the limit of its finite dimensional subspaces), and $V_\infty \subset V$ is a fixed n - dimensional subspace. Also recall that $\mathcal{M}_{k,n}$ is naturally homeomorphic to the holomorphic mapping space $Hol_k(\mathbb{P}^1, Gr_n(V))$ (propositon 2.1). It is therefore clear that we can identify the elements of $\mathcal{X}_{k,n}$ as a set with the family of commutative diagrams,

$$\mathcal{X}_{k,n} = \left\{ \begin{array}{ccc} D_\infty & \xrightarrow{\theta} & Fr_n(V) \\ \cap\downarrow & & \downarrow \\ \mathbb{CP}^1 & \xrightarrow{f} & Gr_n(V) \end{array} \quad : \theta, f \text{ are holomorphic} \right\}$$

where $Fr_n(V)$ is the usual $GL(n, \mathbb{C})$ principal frame bundle over the Grassmannian $Gr_n(V)$. This in turn establishes the topology on $\mathcal{X}_{k,n}$, where the above family of commutative diagrams is topologized as a subspace of the obvious product of mapping spaces (given the compact open topology).

Observe that $\mathcal{M}_{k,n}$ is the limit of the connected complex manifolds

$$\mathcal{M}^m_{k,n} = Hol_k(\mathbb{CP}^1, Gr_n(\mathbb{C}^{n+m})).$$

We now define $\mathcal{X}^m_{k,n}$ in the same way as $\mathcal{X}_{k,n}$, except the vector space V is replaced by $\mathbb{C}^{n+m}$. Namely,

$$\mathcal{X}^m_{k,n} = \left\{ \begin{array}{ccc} D_\infty & \xrightarrow{\theta} & Fr_n(\mathbb{C}^{n+m}) \\ \cap\downarrow & & \downarrow \\ \mathbb{P}^1 & \xrightarrow{f} & Gr_n(\mathbb{C}^{n+m}) \end{array} \quad : \theta, f \text{ are holomorphic} \right\}.$$

We now prove that

$$\pi_m : \mathcal{X}^m_{k,n} \to \mathcal{M}^m_{k,n}$$

is a locally trivial fibration. This is enough to conclude that π is a locally trivial fibration.

To do this, let $f : \mathbb{P}^1 \to Gr_n(\mathbb{C}^{n+m})$ lie in $\mathcal{M}^m_{k,n}$, and let U be an open neighborhood of f. Recall that $\mathcal{M}^m_{k,n}$ is a complex manifold of dimension $k(n+m)$, so we can take U to be a holomorphic disk of that dimension. Consider the adjoint of the inclusion map

$$\iota : U \times \mathbb{P}^1 \to Gr_n(\mathbb{C}^{n+m}).$$

This is a holomorphic map. When restricted to the holomorphic disk $U \times D_\infty$, $\iota^*(E_n)$ is holomorphically trivial. Here $E_n \to Gr_n(\mathbb{C}^{n+m})$ is the universal bundle. Let

$$\Psi : U \times D_\infty \times \mathbb{C}^n \xrightarrow{\cong} \iota^*(E_n)_{|U \times D_\infty}$$

be such a trivialization. For each $(x, y) \in U \times D_\infty$,

$$\Psi(x, y) : \mathbb{C}^n \cong \iota(x, y) \subset \mathbb{C}^{n+m}$$

is a linear embedding. This gives a lift

$$\begin{array}{ccc} U \times D_\infty & \xrightarrow{\Psi} & Fr_n(\mathbb{C}^{n+m}) \\ {\scriptstyle =}\downarrow & & \downarrow \\ U \times D_\infty & \xrightarrow[\iota]{} & Gr_n(\mathbb{C}^{n+m}). \end{array}$$

By definition, this defines a section $\Psi : U \to \mathcal{X}^m_{k,n}$. To get an induced trivialization, let $\gamma : D_\infty \to GL(n, \mathbb{C})$ be any holomorphic map, and define

$$\Psi_\gamma : U \times D_\infty \to Fr_n(\mathbb{C}^{n+m})$$

by

$$\Psi_\gamma(x, y) = \Psi(x, y) \circ \gamma(y) : \mathbb{C}^n \xrightarrow{\gamma(y)} \mathbb{C}^n \xrightarrow{\Psi(x,y)} \iota(x, y) \subset \mathbb{C}^{n+m}.$$

This then defines

$$\bar{\Psi} : U \times D_\infty \times Hol(D_\infty, GL(n, \mathbb{C})) \to Fr_n(\mathbb{C}^{n+m})$$

or, equivalently,

$$\bar{\Psi} : U \times Hol(D_\infty, GL(n, \mathbb{C})) \to \pi_m^{-1}(U)$$

given by $\bar{\Psi}(x,\gamma) =$

$$\begin{array}{ccc} D_\infty & \xrightarrow{\bar{\Psi}(x,-,\gamma)} & Fr_n(\mathbb{C}^{n+m}) \\ \cap\downarrow & & \downarrow \\ \mathbb{P}^1 & \xrightarrow[\iota(x,-)]{} & Gr_n(\mathbb{C}^{n+m}). \end{array}$$

This gives our local trivialization of π_m and therefore completes the proof of part (a) of Lemma 2.3. q.e.d.

Proof. (Lemma 2.3 part (b)). The first step in proving that $p_{k,n}$ (which we abbreviate simply as p) is a quasifibration is to prove that it is continuous. To do this we keep the notation as above.

Recall that the moduli space $C^-_{k,n}$ is a subspace of

$$\Omega U(n) \cong LGL(n,\mathbb{C})/L^+GL(n,\mathbb{C}).$$

Now given a point $(f,\theta) \in \mathcal{X}_{k,n}$ then there exists a neighborhood $\tilde{\mathcal{U}}$ around it in $\mathcal{X}_{k,n}$ so that $p|_{\tilde{\mathcal{U}}}$ can be factored as

$$\tilde{\mathcal{U}} \xrightarrow{\Theta} LGL(n,\mathbb{C}) \to LGL(n,\mathbb{C})/L^+GL(n,\mathbb{C})$$

and hence we only need to prove that the map Θ is continuous.

We choose $\tilde{\mathcal{U}} = \pi^{-1}(\mathcal{U})$ as above, then with respect to the above local trivialization, Θ has the formula

$$\Theta : \mathcal{U} \times Hol(D_\infty, GL(n,\mathbb{C})) \to LGL(n,\mathbb{C})$$

$$\Theta(f(z),\tau(z)) = \tau(z)\cdot P(z).$$

This is just the attaching map induced by the canonical trivialization given by the local section ψ. Clearly Θ is continuous.

Finally, we need to show that p is a quasifibration. To prove this we will use the following proposition about quasifibrations taken directly from [12], which in turn is a consequence of Theorem 2.15 in the classical paper by A. Dold and R. Thom [13].

Proposition 3.1. *Let B be a space filtered by subspaces*

$$B_1 \hookrightarrow B_2 \hookrightarrow \cdots \hookrightarrow B_n \hookrightarrow \cdots \hookrightarrow B$$

with $B = \bigcup_n B_n$ having the topology of the union. Suppose

$$p : E \longrightarrow B$$

is a surjective map, and that there are trivializations on the strata, $E_{n+1} - E_n \cong (B_{n+1} - B_n) \times F$*, where* $E_n = E_{|B_n}$*. Suppose also that every* B_n *is a strong deformation retract of a neighborhood* $V_n \subset B_{n+1}$*, with retraction* $r_n : V_n \to B_n$*. Suppose furthermore that the retractions are covered by maps* $\tilde{r}_n$*,*

$$\begin{array}{ccc} p^{-}(V_n) & \xrightarrow{\tilde{r}_n} & E_n \\ {\scriptstyle p}\downarrow & & \downarrow{\scriptstyle p} \\ V_n & \xrightarrow[r_n]{} & B_n \end{array}$$

that induce homotopy equivalences on the fibers. Then p *is a quasifibration.*

Now in order to apply this proposition in our situation we consider the Birkhoff and Bruhat factorization theorems proved in chapter 8 of [9].

For a given $\gamma \in \mathcal{C}^{-}_{k,n} \subset \Omega U(n)$, write $\gamma = (E, \theta)$ as above. Then the proofs of these theorems give a canonical isomorphism with its Grothendieck decomposition,

$$E \cong \mathcal{O}(k_1) \oplus \cdots \oplus \mathcal{O}(k_n).$$

Moreover this isomorphism clearly extends over the neighborhood U_γ of γ in $\mathcal{C}^{-}_{k,n}$ in which the Grothendieck type of the holomorphic bundle is constant. Hence when restricted to this neighborhood, the fiber of $p_{|U_\gamma} : \mathcal{X}_{k,n_{|U_\gamma}} \to U_\gamma$ is canonically identified with the space of linear holomorphic embeddings of $\mathcal{O}(k_1) \oplus \cdots \oplus \mathcal{O}(k_n)$ into the infinite dimensional trivial bundle $\mathbb{P}^1 \times V$, $Emb(\mathcal{O}(k_1) \oplus \cdots \oplus \mathcal{O}(k_n), \mathbb{P}^1 \times V)$.

Now in Section 8.4 of [9] a "Bruhat" stratification of $\Omega U(n)$ is studied, which as observed earlier coincides with the Morse stratification of the energy functional

$$E : \Omega U(n) \to \mathbb{R}$$

described in Section 2 above. As seen there (Proposition 2.8), this stratification restricts to a Morse stratification of $\mathcal{C}^{-}_{k,n}$, where the strata consist of those loops $\gamma = (E, \theta)$, where E has fixed Grothendieck type. Thus

$$p : \mathcal{X}_{k,n} \longrightarrow \mathcal{C}^{-}_{k,n}$$

is surjective, and has canonical local trivializations on the strata. Again, the fiber of p on the stratum corresponding to loops $\gamma = (E, \theta)$ with $E \cong \mathcal{O}(k_1) \oplus \cdots \oplus \mathcal{O}(k_n)$ is given by the space of linear embeddings

$Emb(\mathcal{O}(k_1) \oplus \cdots \oplus \mathcal{O}(k_n), \mathbb{P}^1 \times V)$. This space is nonempty because the Grothendieck type is negative, and it is contractible because V is infinite dimensional.

As studied in Section 2 above, these strata are partially ordered in a way corresponding to the Morse - Smale partial ordering of the critical levels of the energy functional restricted to $\mathcal{C}^-_{k,n}$ (Lemma 2.6). This then defines a partially ordered filtration of $\mathcal{C}^-_{k,n}$, indexed by the Grothendieck types, where p restricted to the strata has canonical trivialization. Moreover standard Morse theory tells us that the inclusion of one filtration another has a neighborhood deformation retract, where the retractions are given by following flow lines of the energy functional. These retractions are then clearly covered by maps of the restrictions of $\mathcal{X}_{k,n}$ to these spaces. These are again defined by following flow lines. The fact that these maps induce homotopy equivalences on fibers follows from the fact that all the fibers (i.e the embedding spaces described above) are contractible. Hence by Proposition 3.1 the map $p : \mathcal{X}_{k,n} \to \mathcal{C}^-_{k,n}$ is a quasifibration, as claimed. This completes the proof of Lemma 2.3. q.e.d.

References

[1] M. F. Atiyah, *Bott Periodicity and the index of elliptic operators*, Quart. J. Math. **19** (1968) 113–140.

[2] ———, *Instantons in two and four dimensions*, Comm. Math. Phys. **93** (1984) 437–451.

[3] S. Donaldson, *Instantons and geometric invariant theory*, Comm. Math. Phys. **93** (1984) 456–460.

[4] A. Grothendieck, *Sur la classification des fibres holomorphes sur la sphere de Riemann*, Amer. J. Math. **79** (1957) 121–138.

[5] F. Kirwan, *Geometric invariant theory and the Atiyah Jones conjecture*, Proc. S. Lie Mem. Conf., Scand. Univ. Press, 1994, 161–188.

[6] B. Mann & R. J. Milgram, *Some spaces of holomorphic maps to complex Grassmann manifolds*, J. Differential Geom. **33** (1991) 301–324.

[7] S. Mitchell, *The filtration of the loops on $SU(n)$ by Schubert varieties*, Math Z. **193** (1986) 347–362.

[8] W. Richter, *Splitting of loop groups*, To appear.

[9] A. Pressley & G. Segal, *Loop Groups*, Oxford Math. Monographs, Clarendon Press, 1986.

[10] M. Sanders, *Classifying spaces and Dirac operators coupled to instantons*, Trans. of Amer. Math. Soc. 347 (1995) 4037–4072.

[11] G. B. Segal, *Loop groups and harmonic maps*, in London Math Soc Lect. Notes **139** (1989) 153–164.

[12] M. Aguilar & C. Prieto, *Quasifibrations and Bott periodicity*, To appear.

[13] A. Dold & R. Thom, *Quasifaserungen und unendliche symmetrische produkte*, Ann. of Math. **67** (1958) 239–281.

Stanford University, California
University of Michigan, Ann Arbor
Cambridge University, Cambridge, England

SURVEYS IN DIFFERENTIAL GEOMETRY, 2000
Vol. VII ©2000, INTERNATIONAL PRESS
pp. 107-127

MOMENT MAPS AND DIFFEOMORPHISMS

S. K. DONALDSON

Atiyah and Bott pointed out, in [1], that the curvature of a connection on a bundle over a surface can be viewed as the "momentum" corresponding to the action of the gauge group. This observation, together with various extensions, has stimulated a great deal of work and provides a conceptual framework to understand many phenomena in Yang-Mills theory. Our purpose in this paper is to explore some similar ideas in the framework of diffeomorphism groups. We begin by identifying a moment map in a rather general setting, and then see how the ideas work in some more specific situations. We hope to show that the moment map point of view is useful, both in understanding certain established results and also in suggesting new problems in geometry and analysis. While these analytical questions are the main motivation for the work, we will concentrate here on the formal aspects and will not make any serious inroads on the analysis.

1. Identification of moment maps

1.1. Volume-preserving diffeomorphisms. Suppose a Lie group G acts on a symplectic manifold $(\mathcal{M}, \Omega)$, preserving the symplectic form. A moment, or momentum, map for the action is a map $\mu : \mathcal{M} \to \mathrm{Lie}(G)^*$ with the following property. For each element ξ in the Lie algebra $\mathrm{Lie}(G)$ the function $\langle \mu, \xi \rangle$ on $\mathcal{M}$ has derivative

$$d\langle \mu, \xi \rangle = i_{X(\xi)}(\Omega),$$

where $X(\xi)$ is the vector field on $\mathcal{M}$ defined by the infinitesimal action of ξ. This is to say that $\langle \mu, \xi \rangle$ is a Hamiltonian function for the 1-parameter subgroup generated by ξ.

First published in The Asian Journal of Mathematics, 1999. Used by permission.
The research is supported in part by NSF grant DMS 98-03192.

Now consider the following set-up. Let S be a compact k-manifold with a fixed volume form $\sigma \in \Omega^k(S)$. Let (M, ω) be a symplectic manifold and let $\mathcal{M}$ be the infinite-dimensional space of smooth maps from S to M, in some fixed homotopy class. This may be considered as an infinite-dimensional manifold in the usual way: the tangent space to $\mathcal{M}$ at a map $f : S \to M$ is the space of sections of the bundle $f^*(TM)$ over S. (We will ignore, in this paper, any foundational questions about infinite dimensional manifolds.) Now $\mathcal{M}$ carries a natural symplectic form Ω: for sections v, w of $f^*(TM)$ we define

$$\Omega(v, w) = \int_S \omega(v, w)\sigma.$$

The Lie group $\mathcal{G}$ of volume-preserving diffeomorphisms of S acts on $\mathcal{M}$ by composition on the right, preserving Ω, so we may seek a moment map for this action. Let us suppose first that $f^*([\omega])$ is zero in the de Rham cohomology $H^2(S)$ and that $H^1(S) = 0$. Then for each $f \in \mathcal{M}$ we can make the following construction. We can, by hypothesis, choose a 1-form $a \in \Omega^1(S)$ such that $f^*(\omega) = da$. Then for any vector field ξ on S we can define a pairing

$$\langle a, \xi \rangle = \int_S a(\xi)\sigma.$$

Suppose that ξ is in the Lie algebra of $\mathcal{G}$: that is,

$$L_\xi(\sigma) = d(i_\xi(\sigma)) = 0,$$

(here L_ξ denotes the Lie derivative). Then the pairing $\langle a, \xi \rangle$ is independent of the choice of a. For, since $H^1(S) = 0$, any other choice a' differs by the derivative of a function, $a' - a = dg$ say, and

$$\int_S (dg)(\xi)\sigma = \int_S dg \wedge (i_\xi(\sigma) = \int_S d(gi_\xi(\sigma)) - gd(i_\xi(\sigma)) = 0.$$

So in sum we have a well-defined linear map $\xi \mapsto \langle a, \xi \rangle$ on the Lie algebra of $\mathcal{G}$ which we denote by $\mu(f)$. Another way of expressing this is to say that the Lie algebra of $\mathcal{G}$ is identified with ker $d : \Omega^{k-1}(S) \to \Omega^k(S)$ and we have a dual pairing of this space and coker $d : \Omega^0(S) \to \Omega^1(S)$ which is isomorphic to the space ker $d : \Omega^2(S) \to \Omega^3(S)$ in which $f^*(\omega)$ lives.

Proposition 1. *If $H^1(S) = 0$ and $[\omega] = 0$ in $H^2(S)$ then the above construction gives a map $\mu : \mathcal{M} \to Lie(\mathcal{G})^*$ which is a moment map for the action of $\mathcal{G}$ on $(\mathcal{M}, \Omega)$.*

To verify this, consider a map $f : S \to M$, a section v of $f^*(TM)$ over S and a volume-preserving vector field ξ on S. Then, with μ defined as above, $\langle \mu, \xi \rangle$ is a function on $\mathcal{M}$ and $d\langle \mu, \xi \rangle$ is a 1-form on $\mathcal{M}$, which we can evaluate on the tangent vector $v \in T\mathcal{M}_f$. This is the derivative $\frac{d}{dt}\mu(f_t)(\xi)$, where f_t is a 1-parameter family of maps from S to M with $f_0 = f$ and t-derivative v (all time- derivatives being evaluated at $t = 0$). Now

$$\frac{d}{dt} f_t^*(\omega) = d(i_V(\omega)),$$

so $\frac{d}{dt}a_t = i_V(\omega)$ and

$$\frac{d}{dt}\mu(f_t)(\xi) = \int_S i_v(\omega)(f_*(\xi)\sigma = \int_S \omega(v, f_*(\xi)) = \Omega(v, X(\xi)),$$

and this is precisely the identity required for a moment map.

We may vary the topological hypotheses in force above somewhat. First, if $f^*([\omega]) \neq 0$ in $H^2(S)$ we may fix a reference form ν in the de Rham cohomology class and choose a so that $f^*(\omega) - \nu = da$. The calculation goes through as before and we do get a moment map, but not an *equivariant* moment map, with respect to the co-adjoint action. If $H^1(S) \neq 0$ we may proceed in two ways. On the one hand there is a Calabi homomorphism C from the group of volume-preserving diffeomorphisms to the torus $H^{n-1}(S; \mathbf{R})/H^{n-1}(S; \mathbf{Z})$. The kernel of C is a Lie group $\mathcal{G}_0$ whose Lie algebra can be identified with the exact $(n-1)$-forms, and our construction yields a moment map for the action of this group. On the other hand, suppose that the class $[\omega/2\pi]$ is an integral class in $H^2(M)$ so there is a unitary line bundle L over M having a connection with curvature $-i\omega$. Suppose also that $f^*(L)$ is trivial as a complex line bundle over S. Let $\tilde{\mathcal{M}}$ be the covering space of $\mathcal{M}$ consisting of pairs (f, τ) where τ is a homotopy class of trivialisations of $f^*(L)$, a Galois covering with group $H^1(S; \mathbf{Z})$. Then for each point of $\tilde{\mathcal{M}}$ we can choose a trivialisation of $f^*(L)$ within the given homotopy class. The connection form in this trivialisation yields a 1-form a with $da = f^*(\omega)$ and our construction gives a moment map for the natural action of $\mathcal{G}$ on $\tilde{\mathcal{M}}$.

1.2. The symplectic case. Now consider the case when S is also a symplectic manifold of dimension $k = 2l$, with symplectic form ρ. This gives rise to a volume form $\sigma = \rho^l/l!$ and so fits into the framework above, but we can write the relevant moment map in a different way. The group of interest is now the group $\mathcal{G}^{\mathrm{Sp}}$ of symplectomorphisms of S.

For simplicity let us again assume that $H^1(S) = 0$, so all symplectic vector fields on S are Hamiltonian and the Lie algebra of the symplectomorphism group can be identified with the functions on S modulo the constants, or equally with the functions of integral zero. This carries an invariant L^2 inner product which embeds the Lie algebra naturally in its dual. Now let F be a function on S and ξ_F be the Hamiltonian vector field which it generates. Following the notation of the previous section, if $f^*(\omega) = da$ the pairing $\langle a, \xi_F \rangle$ can be written

$$\begin{aligned}\langle a, \xi_F \rangle &= 1/l! \int_S a(\xi_F)\rho^l \\ &= 1/l! \int_S a \wedge dF \wedge \rho^{l-1}, \\ &= 1/l! \int_S F da \wedge \rho^{l-1} \\ &= 1/l! \int_S F f^*(\omega) \wedge \rho^{l-1} \\ &= \int_S F H_f \sigma = \langle F, H_f \rangle,\end{aligned}$$

where H_f is the function on S defined by

$$H_f \sigma = f^*(\omega) \wedge \rho^{l-1}. \tag{2}$$

This means that the moment map for the action of $\mathcal{G}^{\mathrm{Sp}}$ is given simply by $\mu(f) = H_f$. Notice that H_f is defined locally, without any topological assumptions. In the case when $H^1(S) \neq 0$ the relevant group consists of the *exact* symplectomorphisms: the kernel of the restriction of the Calabi map. To sum up we have

Proposition 3. *The map $f \mapsto H_f$ is an equivariant moment map for the action of the group of exact symplectomorphisms of S on $\mathcal{M}$.*

2. Brief recap of moment map geometry

We will quickly recall some standard constructions, as background for the rest of the paper. For details we refer to [7], [8]. In this exposition we have in mind primarily the case of finite-dimensional manifolds, although of course all the formal aspects will go over immediately to infinite-dimensional situations, of the kind we are considering in the body of this paper.

2.1. If μ is an equivariant moment map for the action of G on $(\mathcal{M}, \Omega)$ one may form the *symplectic quotient*

$$\mathcal{M}//G = \mu^{-1}(0)/G.$$

This is a manifold if G acts freely on $\mathcal{M}$ and it inherits an induced symplectic form. Suppose now that $\mathcal{M}$ is a Kahler manifold, and Ω is the Kahler form. Suppose that the action of G extends to an action of the complexified group G^c. Then, at least on an open set of "stable points", one has an identification

$$\mathcal{M}//G = \mathcal{M}/G^c.$$

This says that on an open set of stable points, each G^c-orbit meets the zero-set $\mu^{-1}(0)$ in a unique G-orbit. Using this identification, one sees that the symplectic quotient has a natural Kahler structure.

2.2. The relation between the symplectic and complex quotients, and the role of stability, is clarified by a flow that one can define on $\mathcal{M}$ in this Kahler situation, provided that the Lie algebra of G has an invariant inner product. This means that the moment map can be regarded as a map into the Lie algebra, rather than its dual. This flow is defined by the equation

$$\frac{dx}{dt} = IX(\mu(x)), \tag{4}$$

for $x(t) \in \mathcal{M}$. Here I denotes the usual action of complex multiplication on tangent vectors in $\mathcal{M}$. This is the gradient flow of the function $\|\mu\|^2$ on $\mathcal{M}$, one has:

$$\frac{d}{dt}\|\mu(x)\|^2 = -\|X(\mu(x))\|^2.$$

This gradient flow clearly preserves the G^c-orbits in $\mathcal{M}$. The stable points, which one expects will form a dense open set, are those which flow down to the minimum of the function: the zero set of μ. More generally one can study the stratification of $\mathcal{M}$ defined by infimum of $\|\mu\|^2$ on the G^c orbits.

2.3. Now consider the case when $\mathcal{M}$ is a *hyperkahler* manifold: so we have three complex structures I, J, K satisfying the algebraic relations of the quaternions and corresponding Kahler forms $\Omega_1, \Omega_2, \Omega_3$ defining the same Riemannian metric on $\mathcal{M}$. Suppose that the action of G preserves all this structure and we have equivariant moment maps μ_1, μ_2, μ_3 for the three symplectic forms. We can put these together into a single map

$$\underline{\mu} : \mathcal{M} \to \mathrm{Lie}(G)^* \otimes \mathbf{R}^3.$$

The *hyperkahler quotient* [7] is $\underline{\mu}^{-1}(0)/G$ and this inherits a hyperkahler structure. The final fact we wish to record is less well-known: it is essentially implicit in the work of Taubes [11], who studied the case of Yang-Mills theory over $\mathbf{R}^4$ where the hyperkahler quotient of the space of all connections is the instanton moduli space. In general, suppose that the Lie algebra of G has an invariant inner product and consider the G-invariant function $E = \|\underline{\mu}\|^2$ on $\mathcal{M}$. The gradient flow of this function is given by

$$\frac{dx}{dt} = IX(\mu_1(x)) + JX(\mu_2(x)) + KX(\mu_3(x)).$$

The minimum of E is obviously given by the zero-set of $\underline{\mu}$, and we want to focus now on the other critical points. We associate to the problem an index d equal to the dimension of the hyperkahler quotient, if this is nonempty. Thus, in finite dimensions:

$$d = \dim(\mathcal{M}) - 4 \dim G.$$

In suitable infinite-dimensional problems one interprets this as minus the Fredholm index of a linear operator L. Given any point $x \in \mathcal{M}$ the infinitesimal action is a linear map $\xi \mapsto X_x(\xi)$ from Lie(G) to $T\mathcal{M}_x$. The operator $L = L_x : \mathbf{H} \otimes \mathrm{Lie}G \to T\mathcal{M}_x$ is formed from this using the action of the quaternions:

$$L(\xi_0 + i\xi_1 + j\xi_2 + k\xi_3) = X_x(\xi_0) + IX_x(\xi_1) + JX_x(\xi_2) + KX_x(\xi_3).$$

If $\mathcal{M}$ is connected the index of L_x will not depend upon x. If x lies in the zero set of $\underline{\mu}$ the tangent space of the hyperkahler quotient at x can be identified (modulo a suitable implicit function theorem) with the kernel of the adjoint of L_x.

Proposition 6. *If $d \geq 0$ then there are no strictly stable critical points of E on $\mathcal{M}$ outside the minimum set $\underline{\mu}^{-1}(0)$.*

We outline a proof of this proposition. The equation defining a critical point x of E is

$$L_x(\underline{\mu}(x)) = 0.$$

So if $\mu(x) \neq 0$, L_x has a non-trivial kernel. The hypothesis $d \geq 0$ then imples that the adjoint operator L^* has a non-trivial kernel. The operator is $\mathbf{H}$-linear, so the kernel is a quaternionic vector space. Let v be an element of this kernel and H be the Hessian of E at x, a quadratic form on $T\mathcal{M}_x$. A calculation shows that

$$H(v) + H(Iv) + H(Jv) + H(Kv) = 0.$$

So H cannot be positive definite, as asserted.

One can hope, at least in particular cases, to strengthen this statement to show that H has a non-trivial negative subspace and further to get a lower bound on the dimension, as in the work of Taubes.

3. Moduli spaces

3.1. Special Lagrangian submanifolds. Suppose M is a complex n-manifold with a non-vanishing holomorphic n- form $\theta \in \Omega^{n,0}(M)$. Suppose that in addition M has a Kahler metric ω. Then a *special Lagrangian submanifold* $P \subset M$ is, by definition, a submanifold of (real) dimension n such that

1. the restriction of θ to P is a real n-form:
2. the restriction of ω to P is zero, i.e. P is a Lagrangian submanifold in the ordinary sense of symplectic geometry.

These submanifolds were introduced by Harvey and Lawson [5] and have been studied intensively, following the work of Strominger, Yau and Zaslow [10], in the context of mirror symmetry [4], [6]. They may be fitted into our general picture as follows. Consider a fixed compact n-manifold S with a volume form σ, and suppose first that $H^1(S) = 0$. Then the group of volume-preserving diffeomorphisms of S acts on the space $\mathcal{M}$ of maps from S to M in a given homotopy class and we have identified a moment map for the action above. A map $f : S \to M$ is a zero of the moment map precisely when $f^*(\omega) = 0$. Now the complex structure on M means that we can regard $\mathcal{M}$ as an infinite dimensional complex manifold, via the complex structure on the bundles $f^*(TM)$, moreover Ω becomes a Kahler form on $\mathcal{M}$. Consider now the subset $\mathcal{N} \subset \mathcal{M}$ consisting of maps $f : S \to M$ with $f^*(\theta) = \sigma$. Note that such maps are necessarily immersions. This subset $\mathcal{N}$ is clearly preserved by the action of $\mathcal{G}$ on $\mathcal{M}$. Moreover $\mathcal{N}$ is a complex submanifold of $\mathcal{M}$. For, by definition, it is the zero-set of the map $f \mapsto f^*(\theta) - \sigma$, which we can regard as mapping to the vector space of complex n-forms on S of integral 0. The derivative of this map, at a point $f_0 \in \mathcal{N}$, is the linear map D,

$$D(v) = L_v(\theta) = d(i_v(\theta)),$$

for $v \in T\mathcal{M}_{f_0} = \Gamma(f^*(TM))$. Here we have used the fact that $d\theta = \overline{\partial}\theta = 0$. Now D maps *onto* the forms of integral zero, so $\mathcal{N}$ is a submanifold, and moreover D is complex linear, since θ has type $(n, 0)$, so $\mathcal{N}$ is a complex submanifold. The zeros of the moment map for the action of $\mathcal{G}$ on $\mathcal{N}$ are the "parametrised" immersed special Lagrangian submanifolds of the given topological type, and the moduli space of special Lagrangian submanifolds appears in this framework as the symplectic quotient $\mathcal{N}//\mathcal{G}$.

As it stands the discussion above misses the cases of most interest, because if $H^1(S) = 0$ the special Lagrangian submanifolds are isolated, and the moduli space is just a discrete set. We can extend the set-up in the manner of 1.1 to allow non-trivial H^1 using the kernel $\mathcal{G}_0$ of the Calabi map. The symplectic quotient $\mathcal{N}//\mathcal{G}_0$ is a torus bundle over the moduli space, V say, of special Lagrangian submanifolds, with fibre $T = H_1(S; \mathbf{R})/H_1(S; \mathbf{Z})$. The general moment map theory yields a Kahler metric on $\mathcal{N}//\mathcal{G}_0$. Unfortunately this is not quite the same as the space considered in the Mirror symmetry literature, which is a bundle over V with fibre the dual torus $T^* = H^1(S; \mathbf{R})/H^1(S; \mathbf{Z})$.(However Hitchin has shown how to modify the construction to fit in with the literature on the geometry of this latter space). In the special case when $n = 2$, which we will discuss further below, the two spaces are the same since T is then isomorphic to T^*.

3.2. The symplectic case. There is a parallel discussion in the case when M is a *complex* symplectic manifold, so there is a holomorphic symplectic form $\Theta \in H^{2,0}(M)$. We call an (immersed) submanifold $P \subset M$ (with $\dim_{\mathbf{R}} P = \dim_{\mathbf{C}} M = n$) an*LS-submanifold* if it is *Lagrangian* with respect to the (real) symplectic form $\mathrm{Re}(\Theta)$ and *symplectic* with respect to the (real) symplectic form $\mathrm{Im}(\Theta)$. (Of course we can always replace Θ by $i\Theta$, reversing the roles of the real and imaginary parts.) We digress to point out the following example:

Example 7. Let Z be any complex manifold with $H^1(Z) = 0$ and let M be the total space of the cotangent bundle T^*Z, with the canonical complex structure and holomorphic symplectic form Θ. We consider those LS submanifolds $P \subset T^*Z$ which are graphs of sections of the fibration $T^*Z \to Z$. The real part of Θ is the canonical 2-form on T^*Z, regarded as the real cotangent bundle, so the first condition says that P is the graph of an exact 1-form $d\phi$, where ϕ is a real-valued function on Z. A litle thought shows that second condition asserts that $i\overline{\partial}\partial\phi > 0$ on Z, i.e., ϕ is a Kahler potential. Thus the LS-graphs in T^*Z can be identified with the Kahler potentials modulo constants. There is a global variant of this which applies to any complex manifold Z with a holomorphic line bundle $L \to Z$. Let $p : J_1(L) \to Z$ be the bundle of 1-jets of sections of L and let $U \subset J_1(L)$ be the subset of jets of non-vanishing sections. Let M_L be the fibrewise quotient of U by the natural action of $\mathbf{C}^*$. Locally in Z we can trivialise L which identifies M_L with the cotangent bundle, and two different trivialisations induce the same 2-form on M_L, so M_L has a canonical complex symplectic structure. Then the LS-sections of M_L can be identified with the Kahler metrics on M in the cohomology class $c_1(L)$.

Returning to the main theme, fix a real symplectic manifold (S,ρ) and consider the set $\mathcal{N}$ of maps $f : S \to M$ with $f^*(\Theta) = i\rho$. Then, just as in the previous case, $\mathcal{N}$ is an infinite-dimensional complex manifold and the group $\mathcal{G}_0^{\rm Sp}$ of exact symplectomorphisms of S acts on $\mathcal{N}$. Now suppose that M has in addition a Kahler form ω. So we have three different symplectic forms on M: $\omega_1 = \mathrm{Re}(\Theta), \omega_2 = \mathrm{Im}(\Theta), \omega_3 = \omega$. Then $\mathcal{N}$ gets an induced Kahler structure and the Kahler quotient $\mathcal{N}//\mathcal{G}_0^{\rm Sp}$ is a torus bundle over the moduli space of submanifolds $P \subset M$ which satisfy the three conditions

$$(7) \qquad \omega_1|_P = 0 \ , \ \omega_2|_P \text{ is nondegenerate}, \quad \omega_3|_P \wedge \omega_2^{n-1} = c\omega_2^n.$$

Here c is a constant determined by the homotopy class of the map, which we are allowed to include since the constants act trivially as Hamiltonians. The general theory tells us that this torus bundle over the moduli space inherits a natural Kahler structure.

3.3. The hyperkahler case. In the case when M has real dimension 4 the two discussions co-incide. It is natural also then to suppose that M is a hyperkahler 4-manifold, with three complex structures I, J, K giving an action of the quaternions, and three Kahler forms $\omega_1, \omega_2, \omega_3$, with a symmetry under the group $SO(3)$. From either point of view the objects we are studying are, after a suitable rotation of the complex structures, complex curves in M. We obtain then

Proposition 8. *Suppose M is a hyperkahler 4-manifold and let α be a homology class in $H_2(M;\mathbf{Z})$. Suppose that $\omega_2(\alpha) = \omega_3(\alpha) = 0$ and $\omega_1(\alpha) > 0$. There is a hyperkahler metric on the moduli space of pairs (C,λ) where C is a smooth I-holomorphic curve in the homology class α and λ is a holomorphic line bundle of degree 0 over C.*

This structure arises because the moduli space can be regarded as the hyperkahler quotient of $\mathcal{M}$ by the group $\mathcal{G}_0^{\rm Sp}$.

4. Minimising the norm of the moment map

Suppose again that S is a symplectic manifold and M is Kahler. Then we are in the familar formal picture with a mapping space $\mathcal{M}$

which is Kahler and a symmetry group $\mathcal{G}^{\mathrm{Sp}}$ whose Lie algebra admits an invariant inner product—the L^2 norm. So we may ask how the circle of ideas sketched in II.2 works in this context. On the one hand we may try to identify "orbits" of the complexification and search for zeros of the moment map in these orbits. On the other hand we can look at the gradient flow (4) of the norm of the moment map. It is not clear whether this programme is sensible in general: for example the gradient flow equation is not usually parabolic and one cannot be sure if solutions exist even for a short time with smooth initial data. In this section we will examine a number of cases when we do arrive at apparently sensible differential geometric problems.

4.1. Diffeomorphisms of surfaces. Suppose here that M is a compact Riemann surface with a fixed metric ω, and that S is diffeomorphic to M. There is no loss in supposing that the total areas of M, S are equal. Restrict attention to the open set in $\mathcal{M}$ of oriented diffeomorphisms $f : S \to M$. Any such f defines an area form $(f^*)^{-1}(\rho)$ on the Riemann surface M which we can write as $J_f \omega$, where J_f is a positive function on S. By definition, J_f determines f up to the action of $\mathcal{G}^{\mathrm{Sp}}$. The gradient flow equation is

$$\frac{d}{dt} f_t = I f_*(\xi_{H(f_t)}), \tag{9}$$

where $H(f)$ is the moment map—the function $f^*(\omega)/\rho$ on S —and ξ_H is the Hamiltonian vector field of H. Thus $H(f)$ is the composite of f with the real-valued function J_f^{-1} on S. The evolution equation can be written as an evolution equation for J_f. For the image of the vector field ξ_H under f_* is $I\mathrm{grad}(J_f^{-1})$ so

$$\frac{d}{dt} J_{f_t} = \mathrm{div}(I f_*(\xi_{H(f_t)})) = \mathrm{divgrad}(J_f^{-1}),$$

and $J(t) = J_{f_t}$ satisfies the equation

$$\frac{d}{dt} J = \Delta(J^{-1}), \tag{10}$$

where Δ is the ordinary Laplace operator on M. Conversely, given a positive solution $J(t)$ of this equation (10), with $J(0) = J_{f_0}$, we can define a time-dependent vector field on M,

$$X_t = \mathrm{grad}(J(t)),$$

and let $\psi_t : M \to M$ be the family of diffeomorphisms obtained from the integral curves of X_t. Then the composites $f_t = \psi_t \circ f_0$ satisfy (9). Now equation (10) is parabolic and the maximimum principle implies that $\max(J)$ is decreasing and $\min(J)$ is increasing. It is a straightforward exercise to show that, with any initial data, solutions exist for all time and converge to constant functions. But maps with J_f constant are just the area-preserving maps from S to M which are precisely the zeros of the moment map (since the constants act trivially)—just what we would expect in the general picture.

4.2. The reverse porous-medium equation. Here we ask what happens if we vary the set-up above to allow maps which are not diffeomorphisms. Consider the case when $M = \mathbf{C}$, the complex plane, with its standard symplectic structure, and let S be a compact surface with area form ρ. A map $f : S \to \mathbf{C}$ can be written as $f_1 + i f_2$ for real valued functions f_1, f_2 on S. We can also think of these maps as elements of the complexification of the Lie algebra of functions on S under Poisson bracket, and the equation (9) becomes

$$\begin{aligned} \frac{d}{dt} f_1 &= \{\{f_1, f_2\}, f_2\} \\ \frac{d}{dt} f_2 &= \{\{f_2, f_1\}, f_1\} \end{aligned} \tag{11}$$

(Note that we may study the corresponding ODE for pairs of elements of any Lie algebra.) The zeros of the moment map are pairs (f_1, f_2) with $\{f_1, f_2\} = 0$, and these are just the maps which have a 1-dimensional image in $\mathbf{C}$. Suppose that S is the double of a manifold with boundary S_+ and so has a fixed involution $\sigma : S \to S$. Assume this is compatible with the symplectic form, so $\sigma^*(\rho) = -\rho$. Restrict attention to the set $\mathcal{U}$ of maps $f : S \to \mathbf{C}$ with $f \circ \sigma = f$ and such that f maps the interior of S_+ diffeomorphically to its image in $\mathbf{C}$, a domain $\Omega_f \subset \mathbf{C}$. We can define a function J_f on Ω_f in the same fashion as in (4.1), using the restriction of f to Σ_+. The equation (11) corresponds to the same equation (10) in the interior of Ω_f, but this must be supplemented by boundary conditions. Since the domain Ω_f depends upon f we encounter a "free boundary problem". One can show that the appropriate equation is just (10) where J^{-1} is viewed as a distribution on $\mathbf{C}$—extended by zero outside Ω_f. Now the function J_f is typically unbounded: for a generic map f of the kind we are considering $J_f = O(d^{-1/2})$ where d is the distance to the boundary of Ω_f. If we write $U = J_f^{-2}$ then, generically at least, U is smooth up to the boundary and vanishes transversally there. Our equation can now be written in terms of U, and becomes

$$\frac{dU}{dt} = -U \, \Delta U - \frac{1}{2} |\nabla U|^2, \tag{12}$$

where we have in mind that U_t is a positive function on a domain Ω_t, vanishing on the boundary, and the evolution of Ω_t is determined by saying that the boundary moves inwards with normal velocity $|\nabla U|$. This can be compared with the *porous medium equation*, much studied in applied mathematics, which is

$$\frac{dU}{dt} = -U\Delta U + \frac{1}{2}|\nabla U|^2.$$

In this latter case the domain expands; the boundary moving *outwards* with normal velocity $|\nabla U|$.

Now, modulo questions of smoothness on the boundary, if we are given a solution of (12) we can recover a solution of (11) by integrating a time-dependent vector field just as before. The zeros of the moment map do not lie inside the open set $\mathcal{U}$ we are considering at present but clearly there are sequences f_i of maps in $\mathcal{U}$ which converge to zeros of the moment map. In terms of the functions J_f this corresponds to $J_{f_i} \to \Gamma$ as measures on $\mathbf{C}$, where Γ is a measure supported on a 1-dimensional set $|\Gamma| \subset \mathbf{C}$. So we are lead to propose the following

Problem 13. For which initial data U_0 is there a solution of the the free-boundary problem (12) defined for $t \in [0, \infty)$, such that $U_t^{-1/2}$ converges to a measure supported on a 1-dimensional subset of $\mathbf{C}$, as $t \to \infty$?

If such a solution does exist we interpret it as an integral curve of the gradient equation (4), converging to a zero of the moment map.

For some initial data this question does have an affirmative answer. Let S be the standard 2-sphere in $\mathbf{R}^3$ with the induced area form, and let x, y, z be the standard Euclidean co-ordinates, so

$$\{x, y\} = z \quad \text{etc.},$$

generating a copy of the Lie algebra $\mathfrak{so}(3) = \mathfrak{su}(2)$ inside $C^\infty(S)$, and we can look for solution of (11) inside $\mathfrak{so}(3)$. The simplest solution is to take:

$$f_1 = (2t)^{-1/2}x, \quad f_2 = (2t)^{-1/2}y,$$

which yields a solution of (12) with U_t supported on a disc of radius $(2t)^{-1/2}$, and given in this disc by:

$$U(w, t) = (2t)^{-1}((4t)^{-1} - |w|^2).$$

In this case the map f_t converges to zero, and $U^{-1/2}$ converges to $4\pi\delta_0$. More generally we have solutions:

$$f_1 = (\tanh(t/2) + \coth(t/2))\, x \ , \ f_2 = (\tanh(t/2) - \coth(t/2))y$$

which yields functions U_t supported on ellipses and with $U_t^{-1/2}$ converging to a measure supported on the segment $[-2,2] \subset \mathbf{C}$. One can also show, by reducing the equation to a linear, parabolic, equation, that (13) has an affirmative answer for any circularly symmetric initial data.

However we cannot expect to find these solutions for all initial data. To see this let G be the usual Green's operator on $\mathbf{C} = \mathbf{R}^2$

$$G(\phi)(w) = (2\pi)^{-1} \int_{\mathbf{R}^2} \phi(w') \log|w - w'| dw'.$$

Then the equation (10) implies that

$$\frac{d}{dt} G(J_t) = J^{-1}; \tag{14}$$

in particular $G(J_t)$ is *non-decreasing* everywhere and is *constant* outside the support of J_0. Suppose (13) has an affirmative answer for initial data U_0, with support Ω_0. The restriction of $G(U_0^{-1/2})$ is a harmonic function h_0 on $\mathbf{C} \setminus \Omega_0$ and this *extends* to a harmonic function h_∞—the restriction of $G(\Gamma)$— on $\mathbf{C} \setminus |\Gamma|$. Moreover $h_\infty \geq G(U_0^{-1/2})$ on its domain $\mathbf{C} \setminus |\Gamma|$. One can construct examples of initial data where the function h_0 cannot be extended in this way, and other examples when the harmonic extension is not bounded below by $G(U_0^{-1/2})$, so for these examples the question has a negative answer. A possibility is that these constraints on $G(U_0^{-1/2})$ are both necessary and sufficient conditions for a positive answer to (13).

4.3. Kahler metrics. Now suppose that M is a compact Kahler manifold of complex dimension n and that the symplectic manifold S is diffeomorphic to M. We restrict attention to the open set of diffeomorphisms $f : S \to M$, so the case $n = 1$ was the topic of (4.1). As in that case, we can rewrite the gradient flow equation as an equation for the 1-parameter family of symplectic forms $\chi_t = (f_t^*)^{-1}(\rho)$ on M. The equation is

$$\frac{d}{dt}\chi_t = L_{I\,\xi_t}\chi_t,$$

where ξ_t is the Hamiltonian vector field of the function H_t defined by

$$H_t\, \chi^n = \omega \wedge \chi^{n-1},$$

with respect to the symplectic form χ_t. The case we wish to discuss now is when the initial data χ_0 is a positive form of type $(1,1)$–that is, a second Kahler structure on the complex manifold M. In this case the quantity $L_{I\,\xi_0}\chi_0$ can be written as $i\bar{\partial}\partial H_0$ so the t-derivative of χ_t is again of type $(1,1)$. That is, the set of Kahler forms is preserved by the gradient flow. The gradient equation can now be written as an equation for a Kahler potential,

$$\chi_t = \chi_0 + i\bar{\partial}\partial\phi_t,$$

and becomes

$$\frac{d\phi_t}{dt} = H_t = \frac{\omega \wedge (\chi_0 + i\bar{\partial}\partial\phi_t)^{n-1}}{(\chi_0 + i\bar{\partial}\partial\phi_t)^n}, \tag{15}$$

which is parabolic. We arrive here at the point of view explained in [3]. We interpret the set $\mathcal{H}_0$ of Kahler forms cohomologous to χ_0 as the images of the fixed form ρ under the maps in a single complexified orbit in $\mathcal{M}$. So the problem of finding a zero of the moment map in a complexified orbit leads us to the following

Problem 16. Let (M, ω) be a Kahler manifold and let $[\chi_0]$ be another Kahler metric on M. Can one find a Kahler metric χ in the cohomology class $[\chi_0]$ such that

$$\chi^{n-1} \wedge \omega = c\chi^n, \tag{17}$$

where c is a constant[1]?

The equation (17) is a nonlinear elliptic PDE of Monge-Ampere type for the Kahler potential of χ. If the cohomology class of χ_0 is the same as $[\omega]$ there is a trivial solution, $\chi = \omega$, and for nearby cohomology classes one can use a the implicit function theorem to find a solution. Moreover an argument using the maximum principle shows that if a solution exists it is unique. However, the problem does not always have a solution. To see this, notice first that if a solution does exist then the constant c is determined topologically by the cohomology classes of the forms. Now consider a solution of (17) at a fixed point in M. We can choose an basis for the tangent space at this point which is orthonormal with respect to ω and in which χ is diagonal, with diagonal entries λ_i. The condition is:

$$\sum \lambda_i^{-1} = nc,$$

so $nc\lambda_i \geq 1$ for each i. This means that $nc\chi - \omega$ is a *positive* $(1,1)$ form on M. Conversely if $[nc\chi - \omega]$ is not a Kahler class then no solution can

[1]Further discussion of this equation can be found in [12]

exist. For example if M is a complex surface, $n = 2$, we may suppose that $[\omega]^2 = 1$ and write

$$[\chi] = s[\omega] + t\eta,$$

where $\eta \in H^2(M)$ with $\eta^2 = -1, \eta.[\omega] = 0$. Then the necessary condition is that:

$$2c[\chi] - [\omega] = (s^2 - t^2)^{-1}(\ (s^2 + t^2)[\omega] + 2st\ \eta\ ) > 0.$$

Now let R be the supremum of the set of parameters r such that $[\omega] + r\eta$ is a Kahler class, and suppose that $R < 1$. Then taking $\chi = [\omega] + t\eta$ where t is slightly less than than R, we find that $2c[\chi] - [\omega]$ is not a Kahler class. Thus we conclude that *if (16) has a solution for all Kahler classes on a complex surface M then the Kahler cone of M is a component of the entire positive cone for the intersection form* i.e. the Kahler cone consists of the (1,1) classes χ with $\chi^2 > 0, \chi.\omega > 0$. By the Nakai criterion this is true precisely when M does not contain any curves of negative self-intersection. Of course the obvious conjecture is that this necessary condition, $[nc\chi] - [\omega] > 0$, is also sufficient for the existence of a solution to (16). One would expect that in cases when this condition is violated the solution to the parabolic equation (15) will blow up over some curves of negative self-intersection.

4.4. Surfaces in 4-manifolds. Here we will discuss cases when the maps involved are immersions, rather than diffeomorphisms. We suppose that M is a complex symplectic manifold of complex dimension $2m$, with a holomorphic symplectic form Θ, and consider the space $\mathcal{N}$ of maps f from our symplectic $2m$ manifold S to M with $f^*(\Theta) = \rho$. We consider the moment map gradient equation

$$\frac{df}{dt} = If_*(\xi_f) \tag{18}$$

where ξ_f is the Hamiltonian vector field of the function $\omega \wedge \rho^{m-1}/\rho^m$ on S, with respect to the fixed symplectic form ρ. This evolution preserves $\mathcal{N}$, by the discussion of (3.2). We may regard the equation as an evolution equation for a 1-parameter family of (immersed) LS submanifolds $P_t \subset M$: instantaneously the normal velocity of P_t is the normal component of IX where X is the vector field on P_t obtained as the Hamiltonian vector field on the symplectic manifold $(P_t, \Theta|_{P_t})$. Thought of in this way, the equation (18) becomes parabolic, so short-time solutions exist and one might hope to prove that solutions exist for all time and converge to zeros of the moment map: i.e. to submanifolds satisfying (7).

Now let us suppose that M has real dimension 4 and is hyperkahler, as in (3.3). Then the potential limits of the flow (18) on the LS surfaces in M are complex curves for an appropriate complex structure. Let us consider, more generally, the hyperkahler picture. We return to the full space $\mathcal{M}$ of maps from S to M on which we have three moment maps μ_1, μ_2, μ_3 for the action of $\mathcal{G}^{\mathrm{Sp}}$. In line with the general theory, we consider the functional

$$E(f) = \|\mu_1\|^2 + \|\mu_2\|^2 + \|\mu_3\|^2,$$

on the space $\mathcal{M}$. The gradient flow of this functional is

$$\frac{df}{dt} = If_*(\xi_1) + Jf_*(\xi_2) + Kf_*(\xi_3). \tag{19}$$

Here the moment maps are functions on S:

$$H_i = f^*(\omega_i)/\rho,$$

and ξ_i is the Hamiltonian vector field of H_i on S. So, when restricted to the set of maps $\mathcal{N}$, this hyperkahler flow coincides with the flow (18) we considered before since H_2, H_3 are constants on S in this case, and ξ_2, ξ_3 vanish. To clarify the geometrical meaning of (19), recall that the Grassmannian of oriented 2-planes in a tangent space TM_p can be identified with a product of 2-spheres:

$$Gr_2 = S(\Lambda^2_+) \times S(\Lambda^2_-),$$

and the three Kahler forms $\omega_1, \omega_2, \omega_3$ provide a standard orthnormal basis for Λ^2_+. So if p lies in the image of a map f and if we write $\lambda^2 = H_1^2(p) + H_2^2(p) + H_3^2(p)$ then H_i/λ are the three standard co-ordinates specifying the $S(\Lambda^2_+)$-component $Tf(S)^+_p$ of the tangent space $Tf(S)_p \in Gr_2$. The Riemannian area form dA induced on S from the metric on M is simply $dA = \lambda\rho$. So the total area of the image surface is

$$\mathrm{Area}(f(S)) = \int_S \lambda\rho,$$

while the hyperkahler energy is

$$E(f) = \int_S \lambda^2\rho.$$

(More precisely, the energy functional as we have defined it is given by the sum of L^2 norms of the functions $H_i - \hat{H}_i$, where $\hat{H}_i$ are their average values, determined by the homotopy class of the map. This is because the constant Hamiltonians act trivially. However the two functionals differ by a constant.) Comparing the two integrals we see that

$$\text{Area}(f(S)) \leq \sqrt{I}\sqrt{E}(f), \tag{20}$$

where I is the integral of ρ, a constant in the problem. Equality holds in (20) if and only if λ is a constant. The relation between area and hyperkahler energy is thus much the same as the familar relation between length and energy of paths in a Riemannian manifold. In particular we have an immediate corollary

Proposition 21. *An immersion $f : S \to M$ is a critical point of the energy functional E if and only if its image is a minimal surface in the ordinary sense and the Riemannian area dA is a constant multiple of ρ.*

In view of this, it is not surprising that the flow (19) is related to the mean-curvature flow studied in Riemannian geometry. To see this, consider an immersion $f : S \to M$ and the resulting "Gauss map" $\gamma_f : S \to S^2$, given by $\gamma_f(x) = Tf(S)^+_{f(x)}$. Suppose, without loss of generality, that at a given point $x \in S$, $\gamma(x) = \omega_1$, so there is a preferred complex structure I on $TM_{f(x)}$, which has the property that the tangent space $Tf(S_x)$ is a complex subspace. So we have an induced complex structure on TS_x. The derivative of the Gauss map γ_f at x is a $\mathbf{R}$-linear map $D : TS_x \to TS^2_{\omega_1} = \mathbf{R}\omega_2 + \mathbf{R}\omega_3$. We can decompose D into a sum of a complex linear and complex anti-linear part $D = D' + D''$, using the standard complex structure on S^2 and the complex structure induced by I on TS_x. Some calculation shows that mean curvature h of the image surface $f(S)$ at $f(x)$ can be identified with D' when we use the natural identification of the normal bundle of $f(S)$ at $f(x)$

$$\nu_{f(S)} \cong TS^* \otimes_{\mathbf{C}} TS^2_{\omega_1}. \tag{22}$$

Using this point of view one sees that the deformation vector field $I\xi_1 + J\xi_2 + K\xi_3$ can be decomposed into components tangential and normal to the surface $f(S)$ where the normal component is λh and the tangential component is the gradient in the ordinary Riemannian sense of the function λ on $f(S)$. So we may think of our flow as generating a 1-parameter family of pairs (P_t, λ_t), where P_t is an immersed surface and λ_t is a positive function on P_t: instantaneously P_t evolves by the mean curvature vector, scaled by λ, while λ evolves by a variant of the equation studied in (4.1) above.

We may apply our general discussion from (2.3) of the stability of non-minimal critical points of the norm of the hyperkahler moment map in this case. A critical point is a parametrised minimal surface, and it is easy to see from the relation between energy and area that this index is the same as the usual index in minimal surface theory. The relevant linear operator is a map:

$$D : C^{\infty}(S)_0 \otimes \mathbf{R}^4 \to \Gamma(f^*TM), \tag{23}$$

where C_0^{∞} denotes the functions of integral zero. This is simply

$$D(h_0, h_1, h_2, h_3) = f_*(\xi_0) + If_*(\xi_1) + Jf_*(\xi_2) + Kf_*(\xi_3), \tag{24}$$

where ξ_i is the Hamiltonian vector field of the function h_i on the symplectic surface (S, ρ). This is an elliptic operator, in fact the symbol is the same as that of the Cauchy-Riemann operator on Σ with values in a vector bundle

$$V = \underline{\mathbf{C}} \oplus T\Sigma^* \otimes_{\mathbf{C}} \nu_f^*,$$

where ν_f is the normal bundle of the immersion, regarded as a complex line bundle. The index of this Cauchy-Riemann operator in the ordinary sense, but taking *real* dimensions, is

$$2(c_1(V) + 2(1-g)) = -2e,$$

where e is the Euler number of the normal bundle. The index of D is thus -2e-4, where we substract 4 for the constants C^{∞}/C_0^{∞}. Hence the index d in our problem is $d = (2e+4)$. We obtain, from the general hyperkahler theory (Proposition 6), a result which is related to a theorem of Micallef and Wolfson [9]

Proposition 26. *Any compact, immersed, minimal surface in a hyperkahler 4-manifold with normal Euler number $e \geq -2$, and which is not a complex curve for some complex structure on M, is not strictly stable.*

Finally, we point out that although we have derived the energy functional and evolution equation from the hyperkahler point of view the formulae above show that they can be defined for maps from a surface into any oriented Riemannian 4-manifold. This is rather similar to the case of Yang-Mills theory on a 4-manifold: when the manifold is hyperkahler the Yang-Mills functional can be viewed as the norm of the hyperkahler moment map, but the functional makes sense for general manifolds.

4.5. Symplectic forms on 4-manifolds. We will now discuss a hyperkahler version of the case studied in (4.3), so we let M be a (compact) hyperkahler 4-manifold and suppose that S is a symplectic 4-manifold diffeomorphic to M. For simplicity we suppose that ρ is cohomologous to $f^*(\omega_1)$. We consider the set $\mathcal{M}$ of diffeomorphisms from S to M. This is hyperkahler, with a hyperkahler moment map for the action of the symplectomorphism group $\mathcal{G}^{\mathrm{Sp}}$. On $\mathcal{M}$ we have an energy functional E in the scheme of (2.3). As before this can be expressed in terms of the induced symplectic form $\chi = (f^*)^{-1}(\rho)$ on M. If we write $\chi.\omega_i = g_i$, using the standard inner product given by the fixed metric on M, then the energy is given by:

$$E(\chi) = \int_M \frac{g_1^2 + g_2^2 + g_3^2}{|\chi \wedge \chi|} dV, \tag{27}$$

where dV is the standard Riemannian volume form on M. That is

$$E(\chi) = \int_M \frac{|\chi_+|^2}{|\chi \wedge \chi|} dV. \tag{28}$$

Here, as usual, χ_+ is the self-dual part of χ with respect to the fixed metric on M. The absolute minimum of E is attained when $\chi = \omega$, and we know that any higher critical points cannot be strictly stable, since one readily sees that the relevant index d is 0. The hyperkahler gradient equation goes over to a certain parabolic evolution equation on the space of symplectic forms on M. This suggests the possibility of applications to symplectic topology. Let $\mathcal{S}$ be the space of symplectic forms on M, in the cohomology class $[\omega_1]$. *A priori* this could be disconncted, i.e. there could be different deformation classes of symplectic structure. If one could build up a calculus of variations for the energy functional E on $\mathcal{S}$ one would hope to show that each connected component contains a local minimum of E. But we know that the only strictly stable critical point is the standard structure ω_1, suggesting that in fact $\mathcal{S}$ should be connected.

Of course there are a great many ways in which this programme could fail: one would certainly need to consider critical points "at infinity" in $\mathcal{S}$. Let us just observe that the energy functional does give some control of the symplectic form. If we write

$$|\chi_+| = |\chi \wedge \chi|^{1/2} \frac{|\chi_+|}{|\chi \wedge \chi|^{1/2}},$$

we obtain

$$\begin{aligned}\int_M |\chi_+| dV &\leq \left(\int_M |\chi \wedge \chi| dV. \int_M \frac{|\chi_+|^2}{|\chi \wedge \chi|} dV \right)^{1/2} \\ &= \left(\int_M \chi \wedge \chi \,.\, E(\chi) \right)^{1/2} \\ &= ([\omega_1]^2 \, E(\chi)^{1/2}.\end{aligned}$$

On the other hand $|\chi_-| < |\chi_+|$ pointwise, since $\chi \wedge \chi > 0$. So we deduce that the L^1 norm of χ is bounded by a fixed multiple of $E(\chi)^{1/2}$. Thus any minimising sequence for E has a subsequence which converges weakly, to some closed current on M.

Notice that again the final expression (28) for the functional E does not involve the hyperkahler structure explicitly, so one can try to extend at least some of the ideas to more general 4-manifolds.

References

[1] M. F. Atiyah & R. Bott, *The Yang-Mills equations over Riemann surfaces*, Philosophical Trans. Roy. Soc. London, Ser. A **308** (1982) 523–615.

[2] S. K. Donaldson , *Remarks on gauge theory, complex geometry and 4-manifold topology*, The Field's Medal Volume, World Scientific, 1998, 384–403.

[3] ———, *Symmetric spaces, Kahler geometry and Hamiltonian dynamics*, Proc. Northern California Seminar on Symplectic Geometry, (Weinstein and Eliashberg, eds.), To appear.

[4] M. Gross, *Special Lagrangian fibrations. II: Geometry*, Preprint.

[5] R. Harvey & H. B. Lawson, *Calibrated geometries*, Acta Math. **148** (1982) 47–157.

[6] N. J. Hitchin, *The moduli space of special Lagrangian submanifolds*, Preprint (dg-ga/9711002).

[7] N. J. Hitchin, A. Karlhede, U. Lindstrom & M. Rocek, *Hyperkahler metrics and supersymmetry*, Commun. Math. Phys. **108** (1987) 535–89.

[8] F. C. Kirwan, *Cohomology of quotients in symplectic and algebraic geometry,* 1984, Princeton U.P.

[9] M. J. Micallef & J. Wolfson , *The second variation of area of minimal surfaces in 4-manifolds*, Math. Ann. **295** (1993) 245–67.

[10] A. Strominger, S. T. Yau & F. Zaslow , *Mirror symmetry is T-duality*, Nucl. Phys. **B 479** (1996) 243–59.

[11] C. H. Taubes , *Stability in Yang-Mills theories*, Commun. Math. Phys. **91** (1983) 235–263.

[12] X. Chen, *On the lower bound of the Mabuchi energy and its appliacation*, Int. Math. Res. Notices **12** (2000) 607-623.

STANFORD UNIVERSITY

SURVEYS IN DIFFERENTIAL GEOMETRY, 2000
Vol. VII ©2000, INTERNATIONAL PRESS
pp. 129-194

DIRAC CHARGE QUANTIZATION AND GENERALIZED DIFFERENTIAL COHOMOLOGY

DANIEL S. FREED

To the Gang Of Four[1]

The *classical* Maxwell equations, which describe electricity and magnetism in four-dimensional spacetime, may be generalized in many directions. For example, nonabelian generalizations play an important role in both geometry and physics. The equations also admit abelian generalizations in which differential forms of degree greater than two come into play. Such forms enter into high dimensional supergravity theories, so also into string theory and M-theory. There are analogs of electric and magnetic currents for these higher degree forms. In the classical theory these are also differential forms, and their de Rham cohomology classes in real cohomology (with support conditions) are the corresponding electric and magnetic charges.

In *quantum* theories *Dirac charge quantization* asserts that these charges are constrained to lie in a lattice in real cohomology. In many examples this lattice is the suitably normalized image of integer cohomology, but recently it was discovered that Ramond-Ramond charges[2] in Type II superstring theory lie in the suitably normalized image of complex K-theory instead. (See [39] and the references contained therein.) Furthermore, physical arguments suggest that there is a refined Ramond-Ramond charge in K-theory whose image in real cohomology is the cohomology class of the Ramond-Ramond current. Inspired by this example,

---

[1]The author is supported by NSF grant DMS-0072675.

[2]These are often called "D-brane charges," but that is a misnomer. After all, in ordinary electromagnetism the notion of charge is attached to the abelian gauge field, not to the point particles, monopoles, etc. which are charged with respect to it. Similarly, Ramond-Ramond fields have associated charges. D-branes are Ramond-Ramond charged, just as point particles are electrically charged.

we argue in §2 that the group of charges associated to *any* abelian gauge field is a generalized cohomology group. The rationale is that the group of charges attached to a manifold X should depend locally on X, and generalized cohomology groups are more or less characterized as being topological invariants which satisfy locality (in the form of the Mayer-Vietoris property). The choice of generalized cohomology theory and its embedding into real cohomology affects both the lattice of charges measured by the gauge field and also the possible torsion charges. Both integral cohomology and K-theory (in many of its variations) occur in examples; I do not know an argument to rule out more exotic cohomology theories. Particular physical properties—decay processes, anomalies, etc.—are used to determine which generalized cohomology theory applies to a particular gauge field. We do not review such arguments in this paper.

Our concern instead is a more formal question: How do we implement generalized Dirac charge quantization in a functional integral formulation of the quantum theory? The quantization of charge means that the currents have the local degrees of freedom of a differential form yet carry a global characteristic class in a generalized cohomology theory. Furthermore, the fact that currents and gauge fields couple means that gauge fields are the same species of geometric object. We answer this query using *generalized differential cohomology theories*. The marriage of integral cohomology and differential forms, which we term ordinary differential cohomology, appears in the mathematics literature in two guises: as Cheeger-Simons differential characters [13] and as smooth Deligne cohomology [16]. For field theory we must go beyond differential cohomology groups and use cochains and cocycles. Again this is due to locality—gauge fields have automorphisms (gauge transformations) and we cannot cut and paste equivalence classes. For more subtle reasons electric and magnetic currents must also be refined to cocycles. Many aspects of a cocycle theory for ordinary differential cohomology are developed in [28], and for generalized cohomology theories it is an ongoing project of the author, M. Hopkins, and I. M. Singer. That theory is the mathematical foundation for the discussion in this paper; we give a provisional summary in §1. The application to abelian gauge theory is one motivation for the development of generalized differential cohomology theory, and indeed the presentation here will help shape the theory. There are other mathematical motivations for generalized differential cohomology as well.

The heart of the paper is §2, where we write the action for an abelian gauge field in the language of generalized differential cohomology. Both electric and magnetic currents are cocycles for a differential cohomology class. The gauge field is a cochain which trivializes the magnetic current; this is a geometric version of the Maxwell equation $dF = j_B$. The electric current appears in the action; in classical electromagnetism the other Maxwell equation is the Euler-Lagrange equation. That term in the action is anomalous if there is both electric and magnetic current, and the anomaly has a natural expression (2.30) in the language of differential cohomology. It is a bilinear form in the electric and magnetic currents j_E and j_B. The various ingredients which enter the discussion are collected in Summary 2.32. We also describe how twistings of differential cohomology enter; they are closely related to orientation issues.

Our illustrations in §2 are mostly for 0-form and 1-form gauge fields. In §3 we turn to theories of more current interest, where higher degree gauge fields occur. After a brief comment on Chern-Simons theory, we focus on superstring theories in 10 dimensions. There is a new theoretical ingredient: *self-dual* gauge fields. In Definition 3.11 we specify the additional data we need to define a self-duality constraint. The main ingredient is a quadratic form, whose use in defining the partition function of a self-dual field was elucidated in [38]. Here we also observe that the same quadratic form is used to divide the usual electric coupling term by two; see (3.26) for the action of a self-dual gauge field. Therefore, the quadratic form enters into the formula for the anomaly as well. Note that for self-dual fields the electric and magnetic currents are (essentially) equal. In this paper we do not explain how the data which define the self-duality constraint are used in the quantum theory. These ideas were applied to D-branes in Type II superstring theory in [21], where the focus is anomaly cancellation. We review that argument briefly here. For the theory with nonzero Neveu-Schwarz B-field twistings of generalized cohomology play a crucial role. Indeed, the Ramond-Ramond fields are cochains in B-twisted differential K-theory. In this language a certain restriction on D-branes (equation (3.34)) appears naturally. We also explain a puzzle [4] about the formula for Ramond-Ramond charge with nonzero B-field.

At the end of §3 we treat the Green-Schwarz anomaly cancellation in the low energy limit of Type I superstring theory, including global anomalies. (As far as we know these global anomalies have not previously been discussed.) Since the charges in Type I have been shown to live in KO-theory, the 2-form gauge field is naturally interpreted in differential KO-theory. As with Type II the formulation is self-dual. But here there are background electric and magnetic currents which are present even in the absence of D-branes. Their presence is most naturally explained in our framework by the observation that the KO quadratic form which defines the self-duality constraint is not symmetric about the origin. Rather, the center is a differential KO class which determines the background charges. The theory of this center is discussed in Appendix B, written jointly with M. Hopkins. The gauge field in Type I is a trivialization of the background magnetic current, and this leads to a constraint (3.46) in KO-theory which generalizes the usual cohomological constraint.[3] For spacetimes of the form Minkowski spacetime cross a compact r-dimensional manifold the KO constraint is no new information if $r \leq 7$. The computational aspects of the anomaly cancellation in our treatment are not different than the original, though novel computations are required to relate our self-dual KO-formulation with the standard formulation in terms of a 2-form field. We also verify the local and global anomaly cancellation for D1- and D5-branes. The case of Type I theories "without vector structure" also fits naturally into our approach—it involves a twisted version of KO—but we do not develop the underlying mathematical ideas. These global anomaly cancellations are further evidence that KO-theory is the correct generalized cohomology theory for the gauge field in Type I.

The Atiyah-Singer index theorem, in a geometric form, computes the pfaffian of a Dirac operator as an integral in differential KO-, KSp-, or K-theory, depending on the dimension. In quantum field theories it appears as the anomaly of the fermionic functional integral. Sometimes these fermion anomalies cancel among each other. In the Green-Schwarz mechanism these fermion anomalies cancel against a boson anomaly: the anomaly in the electric coupling of an abelian gauge field in the presence of nonzero magnetic current. The gauge field is quantized by some flavor of K-theory, and so the anomaly in the electric coupling is also an integral in a version of differential K-theory. This idea was first presented in [21]. It indicates that gauge fields involved in this type of anomaly cancellation will always be quantized by some variation of K-theory.

Each factor in an exponentiated (effective) action is a section of a complex line bundle with metric and connection. That geometric line bundle is called the "anomaly," and to say the anomaly cancels between two factors is to say that the tensor product of the corresponding geometric line bundles is isomorphic to the trivial bundle. To define the product of those factors as a function—so to define the partition function—one needs a choice of isomorphism. In this paper we do not address the

[3]Let E be the rank 32 real bundle over spacetime X^{10}. The usual constraint asserts that both X and E are spin, and that $\lambda(E) = \lambda(X)$, where $2\lambda = p_1$.

construction of such isomorphisms. It is undoubtedly true that the geometric form of the index theorem gives a canonical isomorphism between the pfaffian line bundle of a family of Dirac operators and an integral in some differential K-theory. The definition of the partition function in cases where the Green-Schwarz mechanism operates depends on this.[4]

There are two appendices. The first is a heuristic discussion of Wick rotation. We include it since some elementary points, especially in the context of self-dual gauge fields, cause confusion. As mentioned above, the second (with M. Hopkins) contains mathematical arguments needed for the anomaly cancellation in Type I.

It is a pleasure to dedicate this paper to Michael Atiyah, Raoul Bott, Fritz Hirzebruch, and Is Singer. I hope they enjoy seeing the full-blown K-theory form of the index theorem for families of Dirac operators appear in physics. Discussions with many mathematicians and physicists over a long period of time contributed to the presentation here. I particularly thank Jacques Distler and Willy Fischler for clarifying many aspects and Mike Hopkins for his collaboration on a variety of topological issues in the main text and in Appendix B.

1. Generalized differential cohomology

In differential geometry we encounter the *real* cohomology of a manifold via representative closed differential forms. In this section we describe differential geometric objects which represent *integral* generalized cohomology classes. For example, a principal circle bundle with connection is a differential geometric representative of a degree two integral cohomology class. A detailed development of the ideas outlined here is the subject of ongoing work with M. Hopkins and I. M. Singer. The treatment here is only a sketch, offered as background for the discussion of abelian gauge fields in §2.

Let Γ be a multiplicative generalized cohomology theory.[5] We give examples shortly, but in brief Γ obeys the axioms of ordinary cohomology H except that the ring[6] $\Gamma^\bullet(pt)$ may differ from $H^\bullet(pt) \cong \mathbb{Z}$. We

---

[4]I thank Ed Witten for emphasizing this point.

[5]For simplicity of exposition we assume that the cohomology theory Γ is multiplicative—all of our examples are—but much of what we say does not require this hypothesis.

[6]Throughout, $A^\bullet$ denotes a $\mathbb{Z}$-graded abelian group $A^\bullet = \oplus_{q\in\mathbb{Z}} A^q$. Often it has a graded ring structure as well. If $A^\bullet, B^\bullet$ are graded groups, then $A^\bullet \otimes B^\bullet$ is double graded. We denote the associated simply graded group as $(A \otimes B)^\bullet$.

introduce the notation

$$\pi_{-n}\Gamma = \Gamma^0(S^{-n}) = \Gamma^n(pt), \qquad n \in \mathbb{Z},$$

for the cohomology of a point. (Another typical notation for this graded ring is '$\Gamma^\bullet$'.) The most important property of a generalized cohomology theory is the Mayer-Vietoris exact sequence, which we view as asserting the locality of the assignment $X \mapsto \Gamma^\bullet(X)$, where X ranges over a suitable category of finite dimensional spaces. Now after tensoring with the reals, Γ is isomorphic to ordinary real cohomology. More precisely, there is for each X a *canonical* map

$$\begin{array}{ccc} \Gamma^\bullet(X) & \longrightarrow & \big(H(X;\mathbb{R}) \otimes \Gamma(pt)\big)^\bullet \\ \lambda & \longmapsto & \lambda_{\mathbb{R}} \end{array} \tag{1.1}$$

It is natural to introduce the notation

$$\pi_{-n}\Gamma_{\mathbb{R}} = \Gamma^n_{\mathbb{R}}(pt) = \Gamma^n(pt) \otimes \mathbb{R}.$$

Then the codomain of (1.1) is the (hyper)cohomology of X with coefficients in the graded ring $\pi_{-\bullet}\Gamma_{\mathbb{R}}$. The image of (1.1) is a full lattice $\overline{\Gamma}^\bullet(X) \subset H\big(X;\pi\Gamma_{\mathbb{R}}\big)^\bullet$; the kernel is the torsion subgroup of $\Gamma^\bullet(X)$.

Example 1.2 (integral cohomology). There are many cochain models for integral cohomology: singular, Čech, Alexander-Spanier, etc. Such cochains have integral coefficients, and on the cochain level the map (1.1) is the standard inclusion $\mathbb{Z} \hookrightarrow \mathbb{R}$. A class in the image of (1.1) is represented by a closed differential form ω on X such that $\int_Z \omega$ is an integer for all cycles Z in X.

Example 1.3 (K-theory). Historically, this is the first example of a generalized cohomology theory [8], [3]. For X compact we can represent an element of $K^0(X)$ by a $\mathbb{Z}/2\mathbb{Z}$-graded vector bundle $E = E^0 \oplus E^1$, thought of as the formal difference $E^0 - E^1$. The cohomology ring of a point is $\pi_{-\bullet}K \cong \mathbb{Z}[[u,u^{-1}]]$, where $\deg u = 2$. The element $u^{-1} \in K^{-2}(pt) \cong K^0(S^2)$ is called the *Bott element*; multiplication by u^{-1} is the Bott periodicity map. The element u^{-1} is represented by the hyperplane (Hopf) complex line bundle over $\mathbb{CP}^1 \cong S^2$. The map (1.1) is the Chern character

$$\mathrm{ch}\ K^\bullet(X) \longrightarrow H\big(X;\mathbb{R}[[u,u^{-1}]]\big)^\bullet.$$

Example 1.4 (KO- and KSp-theory). These are the variations of K-theory for real and quaternionic bundles, respectively. Whereas

$KO^\bullet(X)$ is a ring—the tensor product of real bundles is real—$KSp^\bullet(X)$ is not. In fact, $KSp^\bullet(X)$ is a module over $KO^\bullet(X)$ and there is also a tensor product $KSp^\bullet(X) \otimes KSp^\bullet(X) \to KO^\bullet(X)$. So it is natural to consider the $(\mathbb{Z}/2\mathbb{Z} \times \mathbb{Z})$-graded theory $KOSp^\bullet = KO^\bullet \times KSp^\bullet$, which does have a multiplicative structure. Notice that the ring $\pi_{-\bullet}KOSp$ has torsion in this case. Over the reals there is an isomorphism $\pi_{-\bullet}KOSp_{\mathbb{R}} \cong \mathbb{R}[[u^2, u^{-2}]]$. The element $u^{-2} \in KSp^{-4}(pt) \cong KSp^0(S^4)$ is represented by the hyperplane (Hopf) quaternionic line bundle over $\mathbb{HP}^1 \cong S^4$. Odd powers of u^{-2} are quaternionic; even powers are real. Note also that twice a quaternionic bundle (e.g. $2u^{-2}$) is real.

Differential Γ*-theory*, which we denote $\check{\Gamma}$, combines Γ with closed differential forms Ω_{cl}. It is defined on the category of smooth manifolds. Loosely speaking, it is the pullback in the diagram

$$\begin{array}{ccc} \check{\Gamma}^\bullet(\,\cdot\,) & \longrightarrow & \Omega_{\text{cl}}(\,\cdot\,;\pi\Gamma_{\mathbb{R}})^\bullet \\ \big\downarrow & & \big\downarrow \\ \Gamma^\bullet(\,\cdot\,) & \longrightarrow & H(\,\cdot\,;\pi\Gamma_{\mathbb{R}})^\bullet \end{array} \tag{1.5}$$

The northeast corner is the set of closed differential forms with coefficients in $\pi_{-\bullet}\Gamma_{\mathbb{R}}$. As a first approximation to $\check{\Gamma}$, for a manifold X define the group $A^\bullet_\Gamma(X)$ by the pullback diagram

$$\begin{array}{ccc} A^\bullet_\Gamma(X) & \longrightarrow & \Omega_{\text{cl}}(X;\pi\Gamma_{\mathbb{R}})^\bullet \\ \big\downarrow & & \big\downarrow \\ \Gamma^\bullet(X) & \longrightarrow & H(X;\pi\Gamma_{\mathbb{R}})^\bullet \end{array}$$

In other words, for each $q \in \mathbb{Z}$

$$A^q_\Gamma(X) = \Big\{ (\lambda,\omega) \in \Gamma^q(X) \times \Omega_{\text{cl}}(X;\pi\Gamma_{\mathbb{R}})^q : \lambda_{\mathbb{R}} = [\omega]_{\text{dR}} \Big\}. \tag{1.6}$$

Here $[\omega]_{\text{dR}}$ is the de Rham cohomology class of the form ω. But $\check{\Gamma}$—the pullback in (1.5)—is a pullback *as a cohomology theory*.[7] So a class

[7] There is a subtlety which we avoid in the main text. Namely, the cohomology theory whose q^{th} cohomology is Ω^q_{cl} depends on q. Precisely, on a manifold X we use the cochain complex

$$\Omega(X;\pi\Gamma_{\mathbb{R}})^q \xrightarrow{d} \Omega(X;\pi\Gamma_{\mathbb{R}})^{q+1} \xrightarrow{d} \Omega(X;\pi\Gamma_{\mathbb{R}})^{q+2} \to \cdots$$

to define the theory in the northeast corner of (1.5). So for each $q \in \mathbb{Z}$ we have a pullback diagram (1.5). This leads to a *bi*graded cohomology theory in the northwest corner: the pth cohomology in the qth theory is denoted $\check{\Gamma}(q)^p$. The groups we call $\check{\Gamma}^q$ are the diagonal groups $\check{\Gamma}(q)^q$ in the bigraded theory.

in $\check{\Gamma}^q(X)$ is a pair (λ, ω) with $\lambda_{\mathbb{R}} = [\omega]_{\mathrm{dR}}$ as in (1.6), together with an "isomorphism" of $\lambda_{\mathbb{R}}$ and $[\omega]_{\mathrm{dR}}$ in $H\big(X; \pi\Gamma_{\mathbb{R}}\big)^q$. If we understand cohomology classes on X to be homotopy classes of maps from X into some universal space B, then an "isomorphism" is an explicit choice of homotopy (up to homotopies of the homotopy). Even when $\lambda = \omega = 0$ there may be nontrivial isomorphisms, and this is the sense in which $\check{\Gamma}$ carries topological information beyond Γ. Equivalence classes of nontrivial isomorphisms appear as the kernel torus in the exact sequence

$$0 \longrightarrow \frac{H\big(X; \pi\Gamma_{\mathbb{R}}\big)^{q-1}}{\overline{\Gamma}^{q-1}(X)} \longrightarrow \check{\Gamma}^q(X) \xrightarrow{\;c\;} A^q_{\Gamma}(X) \longrightarrow 0. \tag{1.7}$$

In some situations the kernel torus sometimes captures topological information not detected by the topological group $\Gamma^q(X)$. If $\check{\lambda} \in \check{\Gamma}^q(X)$ with $c(\check{\lambda}) = (\lambda, \omega)$ it is natural to call λ the *characteristic class* of $\check{\lambda}$ and ω the *curvature* of $\check{\lambda}$. We can rewrite this exact sequence as

$$0 \longrightarrow \frac{\Omega(X; \pi\Gamma_{\mathbb{R}})^{q-1}}{\Omega_{\mathrm{cl}}(X; \pi\Gamma_{\mathbb{R}})^{q-1}_{\Gamma}} \longrightarrow \check{\Gamma}^q(X) \longrightarrow \Gamma^q(X) \longrightarrow 0, \tag{1.8}$$

where $\Omega_{\mathrm{cl}}(X; \pi\Gamma_{\mathbb{R}})^{\bullet}_{\Gamma}$ is the set of closed differential forms whose cohomology class lies in $\overline{\Gamma}^{\bullet}(X)$. The second map is the characteristic class. The curvature of a differential cohomology class defined by a global $(q-1)$-form B is the exact q-form dB. A third way to present (1.7) and (1.8) is the exact sequence

$$0 \longrightarrow \Gamma^{q-1}(X; \mathbb{R}/\mathbb{Z}) \longrightarrow \check{\Gamma}^q(X) \longrightarrow \Omega_{\mathrm{cl}}(X; \pi\Gamma_{\mathbb{R}})^q_{\Gamma} \longrightarrow 0. \tag{1.9}$$

The second map is the curvature. The kernel is the set of "flat" differential cohomology classes, an abelian group whose identity component is the kernel torus in (1.7) and whose group of components is the torsion subgroup of $\Gamma^q(X)$.

As with topological cohomology theories there are many possible ways to represent classes in differential cohomology theories. In computations we are free to use whichever model is most convenient. We use the usual notations $\check{C}^{\bullet}_{\Gamma}(X), \check{Z}^{\bullet}_{\Gamma}(X), \check{B}^{\bullet}_{\Gamma}(X)$ for cochains, cocycles, and coboundaries in a given model for $\check{\Gamma}$. (This is schematic, as models do not necessarily involve cochain complexes.) In any model we construct a category[8] whose set of equivalence classes is $\check{\Gamma}^{\bullet}(X)$. The homotopy

[8] We work in the bigraded theory. Then the objects in the category form a set, the set of cocycles $\check{Z}_{\Gamma}(q)^q(X)$. If $\check{a}', \check{a} \in \check{Z}_{\Gamma}(q)^q(X)$, then a morphism $\check{b}\; \check{a}' \to \check{a}$ is a cochain $\check{b} \in \check{C}_{\Gamma}(q)^{q-1}(X)$ such that $\check{a} = \check{a}' + d\check{b}$ in $\check{Z}_{\Gamma}(q)^q(X)$, but we take such

theory neatly encodes the categorical (and multi-categorical) structure in cochain complexes, or better in spaces of maps. We need the notion of a "trivialization" of a cocycle $\check{a} \in \check{Z}^q_\Gamma(X)$. For our purposes[9] we take it to be a cochain $\check{b} \in \check{C}^{q-1}_\Gamma(X)$ such that $d\check{b} = \check{a}$. The meaning of 'd' in this equation depends on the model. Associated to $\check{b}$ is a differential form $\eta \in \Omega^{q-1}(X; \pi\Gamma_{\mathbb{R}})$—the *covariant derivative* of $\check{b}$—such that $d\eta = \omega$, where $\omega \in \Omega^q_{\mathrm{cl}}(X; \pi\Gamma_{\mathbb{R}})$ is the curvature of $\check{a}$.

We next give some explicit models for $\Gamma = H$ (integral cohomology) and $\Gamma = K$ (complex K-theory). The differential cohomology groups $\check{H}^\bullet(X)$ are also known as the groups of *Cheeger-Simons differential characters* [13] or as the *smooth Deligne cohomology* groups [16].

Example 1.10 (differential cohomology [28]). We represent an element of $\check{H}^q(X)$ by a triple

$$(c, h, \omega) \in C^q(X; \mathbb{Z}) \times C^{q-1}(X; \mathbb{R}) \times \Omega^q(X)$$

of differentiable singular cochains [36, §5.31] and differential forms which satisfy

$$\begin{aligned} \delta c &= 0, \\ d\omega &= 0, \\ \delta h &= \omega - c_{\mathbb{R}}. \end{aligned}$$

In the last equation we view the differential form ω as a singular cochain by integration over (smooth) chains. This last equation very directly expresses the pullback diagram (1.5); h is the isomorphism of the images of c, ω in a set of cochains representing $H^\bullet(X; \mathbb{R})$. We also have maps of such triples:

$$(s, t)\ (c, h, \omega) \longrightarrow (c', h', \omega'),$$

cochains up to equivalence. An equivalence $\check{c}\,\check{b}' \to \check{b}$ is a cochain $\check{c} \in \check{C}_\Gamma(q)^{q-2}(X)$ with $\check{b} = \check{b}' + d\check{c}$ in $\check{Z}_\Gamma(q)^{q-1}(X)$. The group of automorphisms of any cocycle is $\check{\Gamma}(q)^{q-1}(X) \cong \Gamma^{q-1}(X) \otimes \mathbb{R}/\mathbb{Z}$. The construction of a category from a cochain complex is standard. It may be continued to construct higher categories as well.

[9]There are different notions of trivialization, and they appear naturally in the bigraded theory. The most useful notion makes precise the one mentioned in the text: A trivialization of a cocycle $\check{a} \in \check{Z}_\Gamma(q)^q(X)$ is a cochain $\check{b} \in \check{C}_\Gamma(q-1)^{q-1}(X)$ such that $d\check{b} = \check{a}$ in $\check{C}_\Gamma(q-1)^q(X)$. A map $\check{c}\,\check{b}' \to \check{b}$ of trivializations is then a cochain $\check{c} \in \check{C}_\Gamma(q-1)^{q-2}(X)$ with $\check{b} = \check{b}' + d\check{c}$ in $\check{C}_\Gamma(q-1)^{q-1}(X)$. One can go on to discuss equivalence classes of such maps and make a category of trivializations of $\check{a}$, analogous to the discussion in the previous footnote.

where $s \in C^{q-1}(X;\mathbb{Z})$, $t \in C^{q-2}(X;\mathbb{R})$, and

$$\begin{aligned} c' &= c + \delta s, \\ \omega' &= \omega, \\ h' &= h - s_{\mathbb{R}} - \delta t. \end{aligned}$$

In the category representing $\check{H}^q(X)$ we equate maps (s,t) and $(s+\delta e, t-e_{\mathbb{R}} - \delta f)$ for $e \in C^{q-2}(X;\mathbb{Z})$, $f \in C^{q-3}(X;\mathbb{R})$.

This may be more neatly formulated as a cochain complex whose q^{th} cohomology is $\check{H}^q(X)$.

Example 1.11 (differential cohomology). Whereas the last model was based on singular cochains, this model is based on Čech theory. Fix $q \geq 0$. Let $\{U_i\}_{i\in I}$ be an open cover of X with ordered index set I, and set

$$\check{C}^{r,s}(X) = \begin{cases} 0, & s < -1 \text{ or } s > q-1; \\ \prod\limits_{i_0<\cdots<i_p} C^0(U_{i_0} \cap \cdots \cap U_{i_r} \to \mathbb{Z}), & s = -1; \\ \prod\limits_{i_0<\cdots<i_p} \Omega^s(U_{i_0} \cap \cdots \cap U_{i_r}), & 0 \leq s \leq q-1. \end{cases}$$

Then $\check{C}^{\bullet,\bullet}$ is a double complex with the Čech differential δ of degree (1,0) and a differential $\check{d}$ of degree (0,1) defined by

$$\check{d} = \begin{cases} \text{inclusion} & \text{on } \check{C}^{r,-1}; \\ d & \text{on } \check{C}^{r,s},\ 0 \leq s \leq q-1. \end{cases}$$

The degree q cohomology of the total complex is $\check{H}^q(X)$. This is the model which is described, for example, in [23, §6].

Example 1.12 (differential K-theory). Here we only discuss models for $\check{K}^0$. Our first model requires fixing an infinite dimensional manifold B whose homotopy type is the classifying space $\mathbb{Z} \times BU$ of K^0. There are many possibilities, for example the space of Fredholm operators on a separable complex Hilbert space. For complex K-theory we have $\pi_{-\bullet}K_{\mathbb{R}} = K^{\bullet}_{\mathbb{R}}(pt) \cong \mathbb{R}[[u,u^{-1}]]$. Fix a closed differential form $\omega_B \in \Omega_{\text{cl}}(B;\pi K_{\mathbb{R}})^0$ which represents the Chern character of the universal K-theory class on B. Then

$$\omega_B = \omega_B^0 + \omega_B^1 u^{-1} + \omega_B^2 u^{-2} + \dots,$$

where $\omega_B^i \in \Omega^{2i}(B)$ represents the ith universal Chern character class. A representative of an element in $\check{K}^0(X)$ is a triple

$$(f,\eta,\omega) \in \text{Map}(X,B) \times \Omega(X;\pi K_{\mathbb{R}})^{-1} \times \Omega(X;\pi K_{\mathbb{R}})^0$$

with

$$\begin{aligned} d\omega &= 0 \\ d\eta &= \omega - f^*\omega_B. \end{aligned}$$

Let $\pi\ [0,1] \times X \to X$ be projection. Then a map of triples is

$$(F,\sigma)\ (f,\eta,\omega) \longrightarrow (f',\eta',\omega'),$$

where $F\ [0,1] \times X \to B$ and $\sigma \in \Omega(X;\pi K_{\mathbb{R}})^{-2}$ satisfy

$$\begin{aligned} F_0 &= f \\ F_1 &= f' \\ \omega' &= \omega \\ \eta' &= \eta + \pi_* F^*\omega_B + d\sigma. \end{aligned}$$

There is an equivalence relation on maps (F,σ): the maps (F_0,σ_0) and (F_1,σ_1) are equivalent if there exists a homotopy $\tilde{F}\ [0,1] \times [0,1] \times X \to B$ from F_0 to F_1 and a form $\phi \in \Omega(X;\pi K_{\mathbb{R}})^{-3}$ such that $\sigma_1 = \sigma_0 + \Pi_* \tilde{F}^*\omega_B + d\phi$, where $\Pi\ [0,1] \times [0,1] \times X \to X$ is projection.

Example 1.13 (differential K-theory). Next we give a more geometric picture of elements in $\check{K}^\bullet(X)$, though we do not give a complete "cochain model" which computes differential K-theory. In other words, we do not specify maps between representatives.

Simply stated: A vector bundle $E \to X$ with connection ∇ represents an element of $\check{K}^0(X)$. Certainly (E,∇) determines a pair $(\lambda,\omega) \in A^0_K(X)$. Namely, $\lambda \in K^0(X)$ is the equivalence class of E and $\omega = \mathrm{ch}(\nabla)$ is the Chern-Weil representative of the Chern character using the connection ∇. To make contact with our previous model, assume $\omega_B = \mathrm{ch}(\nabla_B)$ is the Chern character form of a universal vector bundle with connection (E_B,∇_B) on B. Choose a classifying map $\tilde{f}\ E \to E_B$ and let $f\,X \to B$ be the induced map. Then $\tilde{f}^*\nabla_B$ is connection on E, and there is a secondary (Chern-Simons) form $\eta = \eta(\tilde{f}^*\nabla_B, \nabla)$ with $d\eta = f^*\omega_B - \omega$. This gives a triple (f,ω,η) as in our previous model.

We can also use Quillen's superconnections [33] to represent elements of $\check{K}^0(X)$. A superconnection on $E = E^0 \oplus E^1$ has a 0-form piece which is a pair of maps $E^0 \rightleftarrows E^1$. We can allow E^0, E^1 to be infinite dimensional if we restrict these maps to be Fredholm. Such infinite dimensional superconnections play a prominent role in Bismut's treatment of index theory [6]. The expression of that work and of other geometric developments in index theory in terms of differential K-theory is part of ongoing research. We learned recently that some versions of differential K-theory and close relatives, together with applications to index theory, appear in the literature. See [30] and the references therein.

There are other, more geometric, models for differential cohomology in low degree. First, we have

$$\begin{aligned}\check{H}^0(X) &\cong H^0(X;\mathbb{Z}),\\ \check{H}^1(X) &\cong \mathrm{Map}(X,\mathbb{R}/\mathbb{Z}).\end{aligned}$$

A circle bundle with connection represents an element of $\check{H}^2(X)$. There are various concrete models for elements of $\check{H}^3(X)$, often called "circle gerbes with connection". See [10], [27], [11], [24] for example.

Multiplication and pushforward on $\check{\Gamma}$ are induced from the corresponding operations on Γ and Ω_{cl}. Explicit formulas for these operations depend on the particular cochain model. Multiplication in $\check{\Gamma}$ combines multiplication in Γ and Ω_{cl}. Thus the map c in (1.7) is a ring homomorphism. In particular, if $\check{\lambda},\check{\lambda}'$ are differential cohomology classes with curvatures ω,ω', and we have a (locally defined) form α with $d\alpha=\omega$, then the curvature of the product $\check{\lambda}\cdot\check{\lambda}'$ is (locally) the differential of $\alpha\wedge\omega$. Pushforward is defined for suitably oriented maps. In this paper we encounter fiber bundles $X\to T$ with compact fibers and inclusions $i\, W\hookrightarrow X$ of submanifolds. For a fiber bundle $X\to T$ we need at least to orient the tangent bundle along the fibers in topological Γ-cohomology. For ordinary cohomology this suffices. For the various forms of K-theory we also need a Riemannian structure on the family, i.e., a Riemannian metric on the relative tangent bundle $T(X/T)$ and a distribution of horizontal planes on X. This data determines a Levi-Civita connection on $T(X/T)$. (See [20, §1].) We call $X\to T$ a "Riemannian fiber bundle." Pushforward is integration along the fibers

$$\int_{X/T}\check{\Gamma}^{\bullet}(X)\longrightarrow\check{\Gamma}^{\bullet-n}(T), \tag{1.14}$$

where the relative dimension is n. This map refines to a map on cochain representatives, and suitable versions of Stokes' theorem hold for this extension. For example, if the fibers are closed and a cochain $\check{b}$ is a trivialization of a cocycle $\check{a}$, then $\int_{X/T}\check{b}$ is a trivialization of $\int_{X/T}\check{a}$. For an inclusion $i\, W\hookrightarrow X$ we must orient the normal bundle to W in X in Γ-cohomology and also choose a smooth closed differential form Poincaré dual to W. Then pushforward

$$i_*\,\check{\Gamma}^{\bullet}(W)\longrightarrow\check{\Gamma}^{\bullet+r}(X)$$

is defined, where r is the codimension of W in X. Curvature does not commute with pushforward. For example, if $X \to T$ is a $\check{\Gamma}$-oriented fiber bundle, then there is a closed differential form $\hat{A}_\Gamma(X/T)$ on X so that if $\check{\lambda} \in \check{\Gamma}^\bullet(X)$ has curvature ω, then the curvature of $\int_{X/T} \check{\lambda}$ is

$$\int_{X/T} \hat{A}_\Gamma(X/T) \wedge \omega. \tag{1.15}$$

For integral cohomology this form is the constant $\hat{A}_H(X/T) \equiv 1$; for K-theory it is $\hat{A}(X/T) \wedge e^{\eta/2}$, where $\hat{A}$ is the usual $\hat{A}$-genus of the curvature and $-2\pi i\eta$ the curvature of a spin^c connection on X. For KO- and KSp-theory it is[10] $\hat{A}(X/T)$.

Generalized cohomology theories admit twistings, and so too do generalized differential cohomology theories. For example, if $F \to X$ is a flat real vector bundle, then $H^\bullet(X; F)$ is a twisted version of real cohomology. It may be computed by an extension of the de Rham complex to F-valued forms. Quite generally, for any cohomology theory E (which could be a topological theory $E = \Gamma$ or a differential theory $E = \check{\Gamma}$) a real vector bundle $V \to X$ determines a one-dimensional twisting $\zeta(V) = \zeta_E(V)$ of $E^\bullet(X)$. We denote the $\zeta(V)$-twisted E-cohomology as $E^{\bullet+\zeta(V)}(X)$. Then there is a Thom homomorphism

$$E^{\bullet+\zeta(V)}(X) \longrightarrow E^{\bullet+r}_{\mathrm{cv}}(V), \tag{1.16}$$

where $\operatorname{rank} V = r$. In the codomain we use cohomology with compact vertical support. In topological theories (1.16) is an isomorphism, but in differential theories it only has a left inverse. For a manifold X we use the notation $\zeta(X) = \zeta(TX)$ for the twisting derived from its tangent bundle. An *E-orientation* of V is a trivialization of the twisting $\zeta(V)$, which then induces an isomorphism $E^{\bullet+\zeta(V)}(X) \cong E^\bullet(X)$. For example, in ordinary

[10] Write the curvature of an element $\check{\lambda} \in (\check{KO})^0(X)$ as $\omega = \omega_0 + \omega_4 u^{-2} + \omega_8 u^{-4} + \cdots$, where ω_i are the Chern character forms of the complexification $\check{\lambda}_{\mathbb{C}}$. If, for example, $\dim X/T = 4$, then since $\hat{A}(X/T) = 1 - p_1(X/T)/24 + \cdots$ we have

$$\int_{X/T} \hat{A}(X/T) \wedge \omega = \int_{X/T} u^{-2}(\omega_4 - \omega_0\, p_1(X/T)/24) + \cdots,$$

and the curvature of $\int_{X/T} \check{\lambda}$ is

$$(2u^{-2}) \int_{X/T} \frac{1}{2}(\omega_4 - \omega_0\, p_1(X/T)/24) + \cdots.$$

Since $2u^{-2}$ is the generator of $KO^{-4}(pt) \cong \mathbb{Z}$, the coefficient (with the factor 1/2) computes the KO index.

cohomology $\zeta(V) = w_1(V) \in H^1(X;\mathbb{Z}/2\mathbb{Z})$ is the first Stiefel-Whitney class, the characteristic class of the real line bundle $\operatorname{Det} V$. In topological K-theory $\zeta(V) = \bigl(w_1(V), W_3(V)\bigr) \in H^1(X;\mathbb{Z}/2\mathbb{Z}) \times H^3(X;Z)$, so a K-theory orientation of V is an orientation in the usual sense (trivialization of $w_1(V)$) together with a spin^c structure (trivialization of $W_3(V)$). In *differential* K-theory the twisting class is

$$\check{\zeta}(V) = \bigl(w_1(V), \check{w}_2(V)\bigr) \in H^1(X;\mathbb{Z}/2\mathbb{Z}) \times \check{H}^3(X), \tag{1.17}$$

where we use the map $H^2(X;\mathbb{Z}/2\mathbb{Z}) \to H^2(X;\mathbb{R})/H^2(X;\mathbb{Z}) \to \check{H}^3(X)$ (cf. (1.7)) to regard the second Stiefel-Whitney class as a differential cohomology class ("flat gerbe") of order two. A twisting in a generalized differential cohomology theory induces a twisting of differential forms. If $\check{\zeta}(V)$ is the twisting of a real vector bundle V, the induced twisting on forms is by the real line bundle $\operatorname{Det} V \to X$. When V is the tangent bundle to X, then $\operatorname{Det} V$ is the orientation bundle. A twisted n-form is simply a density. (See [18, §2.2] for more about twisted forms and densities.)

Pushforward is defined using the Thom isomorphism, so without choice of topological orientation makes sense in twisted cohomology. For example, for suitable[11] fiber bundles $X \to T$ we have

$$\int_{X/T} \check{\Gamma}^{\bullet-\check{\zeta}(X/T)}(X) \longrightarrow \check{\Gamma}^{\bullet-n}(T), \tag{1.18}$$

where $\check{\zeta}(X/T) = \check{\zeta}\bigl(T(X/T)\bigr)$ is the twisting class of the tangent bundle along the fibers.

2. Gauge fields and quantization

In a classical nonrelativistic formulation Maxwell's equations concern a time-varying electric field $E \in \Omega^1(\mathbb{R}^3)$, a time varying magnetic field $B \in \Omega^2(\mathbb{R}^3)$, a time-varying electric current $J_E \in \Omega^2(\mathbb{R}^3)$, and a time varying electric charge density[12] $\rho_E \in \Omega^3(\mathbb{R}^3)$. The relativistic invariance is manifest if we work instead on Minkowski spacetime $M^4 = \mathbb{R}_t \times \mathbb{R}^3$, where t is the time coordinate and the speed of light

[11] For the various forms of K-theory we need a Riemannian fiber bundle; for ordinary cohomology (1.18) is true for any fiber bundle.

[12] We write ρ_E, J_E as forms, rather than densities, using the canonical orientation of $\mathbb{R}^3$. We discuss the role of orientation at the end of this section. Using the standard metric and volume form on $\mathbb{R}^3$ as well, we can write E, B, J_E as vector fields and ρ_E as a function.

has been set to unity. Introduce

$$\begin{aligned} F &:= B - dt \wedge E && \in \Omega^2(M^4), \\ j_E &:= \rho_E - dt \wedge J_E && \in \Omega^3(M^4). \end{aligned}$$

Then Maxwell's equations assert

$$\begin{aligned} dF &= 0, \\ d * F &= j_E. \end{aligned} \tag{2.1}$$

With an eye towards generalizations we introduce a magnetic current $j_B \in \Omega^3(M^4)$ and allow dF to be nonzero:

$$\begin{aligned} dF &= j_B, \\ d * F &= j_E. \end{aligned} \tag{2.2}$$

This version of Maxwell's equations is our starting point.

The form F is called the *field strength*, j_E the *electric current*, and j_B the *magnetic current*. We assume that on any spacelike slice j_E, j_B have compact support (or more generally satisfy some integrability condition). The integral of j_E (j_B) over a spacelike slice $N \cong \mathbb{R}^3$ is the total *electric* (*magnetic*) *charge*. Maxwell's equations (2.2) imply that the currents j_E, j_B are closed, and as a consequence the total charge is constant in time. Letting j denote either of the closed forms j_E or j_B we have a de Rham cohomology interpretation of the charges $\overline{Q}_E$ and $\overline{Q}_B$:

$$\overline{Q} = \left[j\big|_N\right]_{dR} \in H^3_c(N;\mathbb{R}). \tag{2.3}$$

The subscript 'c' indicates that the cohomology is taken with compact support. Notice that the field strength F need not have compact support, so equation (2.2) does not imply that the charge vanishes. However, it does imply

$$\overline{Q} \in \ker\big(H^3_c(N;\mathbb{R}) \longrightarrow H^3(N;\mathbb{R})\big). \tag{2.4}$$

When $N \cong \mathbb{R}^3$ this refinement is vacuous, but for more general manifolds N (of higher dimension, for example) it may be nontrivial. For example, if N is compact (2.4) implies that $\overline{Q} = 0$. For a similar discussion of charge, see [32, §2].

So far we have presented the equations of *classical* electromagnetism. A new feature enters in the *quantum* theory: charge is quantized. *Dirac charge quantization* asserts that in appropriate units the total charge is an integer. Equivalently, the cohomology class $\overline{Q}$ in (2.3), (2.4) lies in the image of the map $H^3_c(N;\mathbb{Z}) \to H^3_c(N;\mathbb{R})$. This is the correct quantization condition for Maxwell theory. For general abelian gauge fields integral cohomology may be replaced by a generalized cohomology theory, as we now explain.

We work in arbitrary dimensions and allow space N to be any oriented Riemannian manifold of dimension $n-1$. Let F be an (abelian) field strength and j_E, j_B currents. These are real differential forms on $\mathbb{R}\times N$. We allow F to have arbitrary degree; it may or may not be homogeneous. (Shortly we will consider F as a form with coefficients, as in §1.) The degrees of the currents are then determined from (2.2). The currents are assumed to have compact support on spacelike slices, but we do not make the support condition explicit in the notation. Quantization of the charge associated to F means that the integral of a current j over a closed cycle in N is not an arbitrary real number, but rather takes discrete values. Therefore, the charge $\overline{Q} = [j]$ is restricted to lie in a lattice $\overline{\Gamma}^\bullet(N) \subset H^\bullet(N;\mathbb{R})$. It is natural from a mathematical point of view—there are physical arguments which motivate this—to postulate an abelian group $\Gamma^\bullet(N)$ and a map $\Gamma^\bullet(N) \to H^\bullet(N;\mathbb{R})$ with image $\overline{\Gamma}^\bullet(N)$ such that the charges $\overline{Q} \in \overline{\Gamma}^\bullet(N)$ are refined to charges $Q \in \Gamma^\bullet(N)$. Furthermore, the locality of quantum field theory implies that the group of possible charges $\Gamma^\bullet(N)$ should depend locally on N. As stated at the beginning of §1, locality is a characteristic feature of generalized cohomology theories whose expression is the Mayer-Vietoris exact sequence. We are led, then, to postulate that the group of charges $\Gamma^\bullet(N)$ assigned to a space N is a generalized cohomology group. We will not discuss the physical motivation behind the choice of Γ and the choice of map to real cohomology. There are detailed discussions of particular cases in the physics literature (most recently concerning Ramond-Ramond fields in Type II superstring theory and its close cousins). Notice that different choices of Γ and the map to real cohomology lead to different lattices $\overline{\Gamma}^\bullet(N)$, so to different quantization conditions on charges measured around cycles. Also, different choices of Γ lead to different torsion phenomena for charges.

Now we Wick rotate to Euclidean field theory and formulate the theory on oriented[13] Riemannian manifolds X of dimension n. (Appendix A reviews Wick rotation in general terms, so provides the setting for our discussion here.) We will not specify explicit support conditions on currents, though the reader should keep in mind the compact spatial

[13]At the end of this section we relax the orientation assumption.

support condition above on manifolds of the form $X = \mathbb{R} \times N$. Correlation functions in the Euclidean theory are defined (formally) by a functional integral over Euclidean fields using a Euclidean action. Our task is to describe the Euclidean fields and Euclidean action precisely. We implement our conclusion in the previous paragraph in the Euclidean setting by choosing: (i) a generalized cohomology theory Γ, and (ii) a map

$$\Gamma^\bullet(X) \to H(X; \pi\Gamma_{\mathbb{R}})^\bullet \tag{2.5}$$

to real cohomology. This is precisely the data we need to define differential Γ-theory $\check{\Gamma}$. Now given a generalized cohomology theory Γ there is a canonical map (1.1), and any other map (2.5) is obtained by multiplying by an invertible element in $H(X; \pi\Gamma_{\mathbb{R}})^0$. In gauge theory there is an invertible closed differential form

$$\omega_X \in \Omega_{\mathrm{cl}}(X; \pi\Gamma_{\mathbb{R}})^0 \tag{2.6}$$

which represents this class; it depends locally on X in a suitable sense. Then as we will see shortly, it is then natural to lift the currents j to differential cohomology classes $\check{j} \in \check{\Gamma}^\bullet(X)$ whose image in $A^\bullet_\Gamma(X)$ is[14]

$$c(\check{j}) = (Q, \frac{j}{\omega_X}). \tag{2.7}$$

(Recall the definition of $A^\bullet_\Gamma$ from (1.6).) Note that $\overline{Q}$ is $[\omega_X]_{dR}$ times the image of Q under (2.5). A particularly good choice for the normalizing factor (2.6) is

$$\omega_X = 2\pi\sqrt{\hat{A}_\Gamma(X)}, \tag{2.8}$$

where $\hat{A}_\Gamma(X)$ is defined in (1.15). (The 2π is convention; the $\sqrt{\hat{A}_\Gamma(X)}$ is to make bilinear pairings in Γ compatible with integration of curvatures.) To make sense of the first Maxwell equation $dF = j_B$ we must refine the field strength F to differential Γ-theory as well. We will see that in fact we must lift F and j to *cocycles* representing generalized differential cohomology classes. (If $j_B \neq 0$ then F is lifted to a cochain rather than a cocycle.) In this setting the differential forms F and j have coefficients in $\pi_{-\bullet}\Gamma_{\mathbb{R}}$. We proceed with the construction of the Euclidean theory after describing two motivating examples.

[14]We continue to use "charges" $Q, \overline{Q}$ in the Euclidean setting. The physical interpretation of these quantities as charges is in the Hamiltonian situation $X = \mathbb{R} \times N$ after rotating back to real time.

Example 2.9 (typical p-form gauge field). Suppose F is a homogeneous form of degree $p+1$, so that $\deg j_B = p+2$ and $\deg j_E = n-p$. Typically the quantization law asserts that charges lie in integral cohomology $\Gamma = H$ with $\omega_X = 2\pi$. In other words, the pair (Q_E, Q_B) lives in $H^{p+2}(X;\mathbb{Z}) \oplus H^{n-p}(X;\mathbb{Z})$.

Example 2.10 (Ramond-Ramond fields).These occur in the low energy field theory description of the Type II superstring. Here X^{10} is a spin Riemannian manifold. If the B-field vanishes, then the Ramond-Ramond charges naturally live in K-theory. (For a physical discussion of the choice of K-theory, see [39] and the references therein.) The charge is a homogeneous class in $K^\bullet(X)$, and by Bott periodicity only the parity of the degree matters. The parity is odd in Type IIA and even in Type IIB. For definiteness we suppose the charge lives in $K^1(X)$ in Type IIA and $K^0(X)$ in Type IIB. (For Type IIB—and probably for any theory—the charge is in $\epsilon^{-1}(0)$ for $\epsilon\ K^0(X) \to H^0(X)$ the augmentation.) Thus the Ramond-Ramond field strengths and currents are refined to differential K-theory classes of degree 0 and -1, respectively. In this case (2.8) is $\omega_X = 2\pi\sqrt{\hat{A}(X)}$. Also, there is a self-duality condition which enters in the construction of the functional integral. We discuss self-dual fields in §3.

The B-field, which heretofore was assumed zero, is locally a 2-form. Its field strength, usually denoted H, is a closed 3-form on X which obeys integrality constraints corresponding to integral cohomology. In other words, we postulate a class $\zeta \in H^3(X)$ with $\zeta_{\mathbb{R}} = [H]_{dR}$, and suppose that globally the B-field is a cocycle $\check{\zeta}$ for a class in the differential cohomology group $\check{H}^3(X)$. Then the Ramond-Ramond charges live in twisted K-theory $K^{\bullet+\zeta}(X)$, and currents are lifted to twisted differential K-theory $\check{K}^{\bullet+\check{\zeta}}(X)$.

There is a lagrangian formulation of the classical Maxwell equations (2.1) (with no magnetic current) in the classical Lorentzian theory. The field variable is a gauge field. The first Maxwell equation is the Bianchi identity and holds off-shell. The second Maxwell equation is the variational equation of a classical action for the gauge field. As stated above, our task is to incorporate Dirac charge quantization into the Wick rotated Euclidean theory. (We summarize our answer in Summary 2.32.) As we have seen charge quantization means choosing a generalized cohomology theory Γ and an embedding (2.5), refined to a differential form (2.6). We begin with the case where the currents j_E

and j_B both vanish. Fix a degree[15] $\mathbf{d}$. Then we: (i) refine the characteristic class $[F/\omega_X]_{dR}$ of the normalized field strength $F \in \Omega_{\text{cl}}(X; \pi\Gamma_{\mathbb{R}})^{\mathbf{d}}$ to a class $\lambda \in \Gamma^{\mathbf{d}}(X)$, and (ii) refine the field strength itself to a cohomology class $\check{F} \in \check{\Gamma}^{\mathbf{d}}(X)$ such that

$$c(\check{F}) = \Big(\lambda, \frac{F}{\omega_X}\Big).$$

Here we use a multi-normalization $\omega_X = \big((\omega_X)_1, \dots, (\omega_X)_k\big)$ corresponding to the components of $F = (F_1, \dots, F_k)$. The differential cohomology group $\check{\Gamma}^{\mathbf{d}}(X)$ is the space of abelian gauge fields (or gauge potentials) *up to gauge transformations*. It is the space over which one integrates in the Euclidean functional integral. Cocycles in $\check{Z}^{\mathbf{d}}_{\Gamma}(X)$ representing a class $\check{F} \in \check{\Gamma}^{\mathbf{d}}(X)$ are particular *gauge fields*. If $\check{A} \in \check{Z}^{\mathbf{d}}_{\Gamma}(X)$ is such a cocycle, we denote its cohomology class in $\check{\Gamma}^{\mathbf{d}}(X)$ as $\check{F}_A$. Cocycles are the proper variables for field theory—they are local. We are led, then, to a Euclidean theory in which the space of fields is the category $\check{Z}^{\mathbf{d}}_{\Gamma}(X)$ of cocycles (of particular degrees). The Wick rotated version of the classical Lorentzian action makes sense for our refined fields: If $\check{A} \in \check{Z}^{\mathbf{d}}_{\Gamma}(X)$ is a field with curvature F_A/ω_X, then the Euclidean action is

$$S(\check{A}) = \frac{1}{2e^2} \int_X F_A \wedge *F_A. \tag{2.11}$$

Here $e = (e_1, \dots, e_k)$ is a set of coupling constants, and the notation implies a sum over components:

$$S(\check{A}) = \sum_{i=1}^{k} \frac{1}{2e_i^2} \int_X (F_A)_i \wedge *(F_A)_i.$$

Since the curvature depends only on the cohomology class $\check{F}_A$ of $\check{A}$ in $\check{\Gamma}^{\mathbf{d}}(X)$, the action is gauge-invariant. Of course, in writing (2.11) we are implicitly assuming either that X is compact or some support condition on the fields.

Example 2.12 (1-form gauge field). Let X^n be an oriented n-manifold and suppose $F \in \Omega^2_{\text{cl}}(X)$ obeys quantization using integral cohomology (with $\omega_X = 2\pi$). As a model for $\check{Z}^2_H(X)$ we take the category of connections on principal circle bundles over X. A gauge field $\check{A}$

[15]We use a multi-index notation

$$\mathbf{d} = (d_1, \dots, d_k).$$

Then if $A^\bullet$ is a graded group, an element $a \in A^{\mathbf{d}}$ is a k-tuple $(a_1, \dots, a_k)$ with $a_i \in A^{d_i}$. Arithmetic is done componentwise. For example, $\mathbf{d} + 1 = (d_1 + 1, \dots, d_k + 1)$.

is such a connection and $\check{F}_A$ its equivalence class under isomorphisms of circle bundles with connection. The field strength F_A is $\sqrt{-1}$ times the curvature of $\check{A}$; the characteristic class is the first Chern class. If the characteristic class vanishes, then the gauge field may be represented by a global 1-form, uniquely up to differentials of circle-valued functions (see (1.8)).

Example 2.13 (periodic scalar field). Again X^n is an oriented n-manifold. Suppose $F \in \Omega^1_{\mathrm{cl}}(X)$, with quantization specified by integer cohomology. Now $\check{Z}^1_H(X)$ may simply be taken to be the *set* $\mathrm{Map}(X, \mathbb{R}/\mathbb{Z})$ of periodic real scalar fields on X. Taking $\omega_X = 2\pi$ a gauge field is a map $\phi X \to \mathbb{R}/2\pi\mathbb{Z}$ and its field strength is $d\phi$. Then (2.11) is the usual action $\frac{1}{2e^2}\int_X d\phi \wedge *d\phi$. It is convenient in this case to write formulas in terms of the exponentiated circle-valued scalar field $e^{i\phi}\, X \to \mathbb{T}$.

Next, we allow the currents to be nonzero. First, consider $j_E \neq 0$. In the classical Lorentzian theory there is an additional term in the action whose variation gives the right-hand side of the second Maxwell equation[16] in (2.1). To write its Wick rotation in our framework we need to postulate maps[17]

$$\begin{aligned} \Gamma^1 &\longrightarrow H^1, \\ \Gamma^2 &\longrightarrow H^2, \\ \pi_{-\bullet}\Gamma_{\mathbb{R}} &\longrightarrow \pi_{-\bullet}H_{\mathbb{R}} \cong \mathbb{R}. \end{aligned} \tag{2.14}$$

Using the pullback square (1.5) there are induced maps

$$\begin{aligned} \check{\Gamma}^1 &\longrightarrow \check{H}^1 \\ \check{\Gamma}^2 &\longrightarrow \check{H}^2. \end{aligned} \tag{2.15}$$

We must also assume that X is $\check{\Gamma}$-oriented so that integration over X in $\check{\Gamma}$-theory is defined. Recall that the closed differential form j_E is refined to a differential cohomology class $\check{j}_E \in \check{\Gamma}^{n-\mathbf{d}+1}(X)$. (See (2.7).) The additional term in the Euclidean action is the purely imaginary expression

$$2\pi i \int_X \check{j}_E \cdot \check{F}_A. \tag{2.16}$$

[16]The coupling constant is $e^2 = 2\pi$ in (2.1).

[17]For complex K-theory there are natural "determinant" maps $K^1 \to H^1$ and $K^2 \to H^2$. The map $\pi_{-\bullet}K_{\mathbb{R}} \cong \mathbb{R}[[u, u^{-1}]] \to \pi_{-\bullet}H_{\mathbb{R}} \cong \mathbb{R}$ sends u to zero. We will also make use of the natural "pfaffian" map $KSp^2 \to H^2$. (See [20, §3]; note $KSp^2 \cong KO^{-2}$ by Bott periodicity.) The map in (2.14) in degree 1 is used in (2.17); that in degree 2 is used later in (2.30). For general Γ maps to low degree cohomology with coefficients exist canonically, using the Postnikov tower, and with some choice one produces (2.14).

The product takes place in $\check{\Gamma}^{\bullet}(X)$, and the degrees are such that the integrand is an element of[18] $\check{\Gamma}^{n+1}(X)$. Hence the integral lands in $\check{\Gamma}^{1}(pt)$, and using (2.15) we map to $\check{H}^1(pt) \cong \mathbb{R}/\mathbb{Z}$. Therefore, the exponentiated action

$$e^{-S}(\check{A}) = \exp\Big(-\frac{1}{2e^2}\int_X F_A \wedge *F_A\Big)\,\exp\Big(-2\pi i \int_X \check{\jmath}_E \cdot \check{F}_A\Big) \tag{2.17}$$

is a well-defined complex number.

Several comments are in order. First, it is more illuminating to work with a family of gauge fields $\check{A}$ parametrized by a manifold T. Then the exponentiated action is a function $e^{-S}\, T \to \mathbb{C}$ and we can use Stokes' theorem to differentiate the second term of the action. Also, the fact that (2.17) depends on $\check{A}$ only through $F_A, \check{F}_A$ means that the exponentiated action e^{-S} is gauge-invariant. Finally, for gauge fields quantized by integer cohomology, as in Example 2.12 and Example 2.13, the electric coupling term (2.16) is usually written as "$\frac{i}{2\pi}\int_X j_E \wedge A$". Indeed, (2.16) reduces to this if $F_A = dA$ for some form A; otherwise this is only valid locally. The correct global expression is (2.16).

Example 2.18 (1-form gauge field). Continuing Example 2.12, a typical electric current j_E is induced from point charges. For manifolds of the form $\mathbb{R} \times N$ the point charges are described by a finite set P of points in N with integers attached. If the particles are static, then $\mathbb{R} \times P$ is the set of their "worldlines"; if they move the worldline is the graph of a function $\mathbb{R} \to N$. More generally, let $W \subset X$ be a 1-dimensional oriented submanifold and

$$q_E\, W \longrightarrow \mathbb{Z} \tag{2.19}$$

a (locally constant) function.[19] Let $i\, W \hookrightarrow X$ denote the inclusion. Then the electric charge is the pushforward of q_E in cohomology:

$$Q_E = i_* q_E \in H^{n-1}(X;\mathbb{Z}). \tag{2.20}$$

We can regard q_E as a class in $\check{H}^0(X)$ (or better as a cocycle in $\check{Z}^0_H(W)$). Then the refined electric current is the pushforward of q_E in differential cohomology:

$$\check{\jmath}_E = i_* q_E \in \check{H}^{n-1}(X). \tag{2.21}$$

[18]If $\mathbf{d} = (d_1, \dots, d_k)$ is a multi-degree, then (2.16) is the sum of k terms, one for each component of $\check{\jmath}_E$ and $\check{F}_A$. For the signs to work out properly, we should regard $\check{\jmath}_E$ as a form twisted by the orientation bundle, so of degree $-(d-1)$. (A density—twisted n-form—has degree 0.) This is formalized in (2.33) below.

[19]For manifolds of the form $\mathbb{R} \times N$ we require $W \cap (\{\tau\} \times N)$ to be compact for all $\tau \in \mathbb{R}$. For general X we do not specify support conditions, and in any case omit them from the notation as usual.

Recall from §1 that this depends on choosing a smooth closed differential form Poincaré dual to W. Without making any choices one could define a distributional electric current—a current in the sense of de Rham—supported on W. But we prefer to remain in the smooth category.[20] The second factor in the exponentiated action (2.17) may be rewritten as

$$\exp\Big(-2\pi i \int_W \check{F}_A\Big)^{q_E}. \tag{2.22}$$

This is the product over components of W of the q_E^{th} power of the holonomy of the connection A.

Example 2.23 (periodic scalar field). There is an analogous story for the periodic scalar $e^{i\phi}\ X \to \mathbb{T}$, continuing Example 2.13. In this case W is 0-dimensional—a "(-1)-brane"—and (2.19)–(2.21) hold with $n-1$ replaced by n. Expression (2.22) becomes a product over the points of W:

$$\prod_{w\in W} e^{-iq_E(w)\phi(w)} \tag{2.24}$$

This factor is often viewed as a local operator inserted into the functional integral, but in our context it is the electric coupling in the exponentiated action.

Next, we consider nonzero magnetic current $j_B \neq 0$. As before, we suppose j_B is refined to a differential cohomology class $\check{j}_B \in \check{\Gamma}^{\mathbf{d}+1}(X)$, and assume that we have a fixed cocycle representative, also denoted $\check{j}_B$. Recall the first Maxwell equation in (2.2). If the magnetic current is nonzero, then the field strength is not closed and it makes no sense to refine it to a differential cohomology class. Rather, we postulate that the gauge field $\check{A}$ is a trivialization of the refined magnetic current $\check{j}_B$, in the sense explained before Example 1.10. This means that $\check{A} \in \check{C}^{\mathbf{d}}_{\Gamma}(X)$ with $d\check{A} = \check{j}_B$. The equivalence class of $\check{A}$ (under the equivalence of trivializations of $\check{j}_B$ discussed in a footnote preceding Example 1.10) is denoted $\check{F}_A$ as before. The covariant derivative F_A/ω_X of $\check{A}$ is a global differential form, and $dF_A = j_B$. Notice that the existence of a global trivialization of $\check{j}_B$ implies that the magnetic charge Q_B vanishes in the global cohomology group (no support condition), which is consistent with (2.4). Also, if $\check{j}_B = 0$ we recover the previous definitions of $\check{A}, \check{F}_A, F_A$.

Example2.25 (periodic scalar field). We continue Example 2.23. In this case $\deg j_B = 2$ and in our model for $\check{H}^2$ the refined magnetic

[20]Smoothness plays a greater role when we come to magnetic currents. For example, it was a key idea in [22]. The point is to avoid illegal products of distributions.

current $\check{\jmath}_B \in \check{H}^2(X)$ is a principal circle bundle with connection on X whose curvature is $\sqrt{-1}$ times j_B. The exponentiated gauge field $e^{i\phi}$ is now a global section of this bundle, and therefore the bundle is topologically trivial. The field strength $F_A = d\phi$ is the "covariant derivative" of this section, that is, the pullback to X of the connection form on the total space of the circle bundle. Notice that in case $\check{\jmath}_B = 0$—the cocycle $\check{\jmath}_B$ is the trivial circle bundle with product connection—we recover our previous description of the gauge field as a map $X \to \mathbb{T}$.

Specialize now to $n = \dim X = 2$. Then we can consider a magnetically charged (-1)-brane. Thus suppose $i\, W \hookrightarrow X$ is the inclusion of an oriented 0-manifold and

$$q_B\, W \longrightarrow \mathbb{Z}, \tag{2.26}$$

which we regard as a class in $\check{H}^0(W)$. Then we set

$$\check{\jmath}_B = i_* q_B \in \check{H}^2(X). \tag{2.27}$$

Recall that the pushforward depends on a choice of Poincaré dual form, which in this case is a closed bump 2-form localized near points $w \in W$ and whose integral near w is $q_B(w)$. A cocycle representative of the refined magnetic current is a circle bundle with connection whose curvature is this Poincaré dual form. This construction should be compared to the construction in complex geometry of a holomorphic line bundle from a divisor.

Example 2.28 (1-form gauge field). In this case $\check{\jmath}_B \in \check{Z}^3_H(X)$ is intuitively a "gerbe with connection" and $\check{A}$ is a trivialization—a "translated version" of a circle bundle with connection. The theory outlined in §1 is a natural home for these notions and their generalizations to higher degrees.

If the electric current j_E vanishes, then the action (2.11) is well-defined and gauge-invariant. In case both j_E and j_B are nonzero we must re-examine the second factor in the exponentiated action (2.17), whose form does not change, but whose geometric nature does. As before it is gauge-invariant, since it only depends on $\check{A}$ through $\check{F}_A$. In a family of gauge fields parametrized by T, we compute the action as a function on T. Suppose now that the refined electric current $\check{\jmath}_E$ has been lifted to a cocycle. Since $\check{F}_A$ is a trivialization of $\check{\jmath}_B$, by Stokes' theorem (see the text following (1.14))

$$\begin{aligned}&\exp\bigl(-2\pi i\int_{X/T}\check{\jmath}_E\cdot\check{F}_A\bigr)\\ (2.29)\qquad &\qquad\text{is a trivialization of}\quad \exp\bigl(-2\pi i\int_{X/T}\check{\jmath}_E\cdot\check{\jmath}_B\bigr).\end{aligned}$$

The degrees work out so that

$$\exp\bigl(-2\pi i\int_{X/T}\check{\jmath}_E\cdot\check{\jmath}_B\bigr) \tag{2.30}$$

lives in[21] $\check{Z}^2_H(T)$, i.e., is a circle bundle with connection over T. Equivalently, the first expression in (2.29) is a section of a hermitian line bundle with connection, as then is the entire exponentiated action (2.17). Actions which are sections of hermitian line bundles with connection are potentially anomalous; the anomaly is the obstruction to trivializing the line bundle with connection. Here (2.30) is the formula for the anomaly, where we interpret the integral as a differential cohomology class in $\check{H}^2_H(T)$.

Example 2.31 (periodic scalar field). We continue Example 2.25 in any dimension n. Recall that $i\ W \hookrightarrow X$ is the inclusion of a 0-manifold—which we assume to be compact, i.e., a finite set of points—and $q_E\ W \to \mathbb{Z}$ encodes the electric charges of the points of W. The refined magnetic current $\check{\jmath}_B = i_*q_B$ is a circle bundle with connection over X and the gauge field $e^{i\phi}$ is a section of $\check{\jmath}_B$. Let $L \to X$ denote the associated hermitian line bundle with connection. The electric coupling in the exponentiated action is (2.24), and the anomaly formula reduces to the obvious assertion that it is an element of the hermitian line

$$\bigotimes_{w\in W}(L_w)^{\otimes(-q_E(w))}.$$

As we vary over a family T of connections (and also embeddings $W \hookrightarrow X$) these lines assemble into a smooth hermitian line bundle with connection over T.

The anomaly (2.30) is nonzero in this example since W is both electrically and magnetically charged. The example generalizes to higher dimensional submanifolds which are both electrically and magnetically

[21] Note we use the degree 2 map in (2.14) to define (2.30) as an element in $\check{Z}^2_H(T)$, though we do not make it explicit in the notation.

charged. In these higher dimensional cases the Euler class $\chi_{\check{\Gamma}}(\nu)$ in $\check{\Gamma}$-theory of the normal bundle ν to W in X enters.

For convenience we collate the various parts of the discussion.

Summary 2.32. The data needed to define an abelian gauge field is:

(i) a generalized cohomology theory Γ;

(ii) maps (2.14) to ordinary cohomology;

(iii) a multidegree $\mathbf{d} = (d_1, \dots, d_k)$;

(iv) normalizing differential forms $\omega_X = \big((\omega_X)_1, \dots, (\omega_X)_k\big)$ which depend functorially and locally on X; and

(v) coupling constants $e = (e_1, \dots, e_k)$.

Then the gauge field, magnetic current, and electric current live in:

$$\begin{aligned} \check{A} &\in \check{C}_{\Gamma}^{\mathbf{d}}(X), \\ \check{\jmath}_B &\in \check{Z}_{\Gamma}^{\mathbf{d}+1}(X), \\ \check{\jmath}_E &\in \check{Z}_{\Gamma}^{n-\mathbf{d}+1}(X). \end{aligned}$$

The gauge field $\check{A}$ is a nonflat trivialization of the magnetic current $\check{\jmath}_B$. The exponentiated action is (2.17), and the electric coupling has an anomaly given by (2.30).

Finally, we relax the orientability assumption on X. For this we use the discussion of twistings and orientation at the end of §1. Thus let X be a Riemannian manifold which is not oriented and possibly not orientable. The only change[22] from Summary 2.32 is that the refined electric current $\check{\jmath}_E$ lives in $-\check{\zeta}(X)$-twisted $\check{\Gamma}$-theory:

$$\check{\jmath}_E \in \check{Z}_{\Gamma}^{n-\mathbf{d}+1-\check{\zeta}(X)}(X). \tag{2.33}$$

The twisting refers to differential $\check{\Gamma}$-theory. Then the integral in (2.16) is well-defined (cf. (1.18)). The refined magnetic current $\check{\jmath}_B$ still lives in the untwisted theory. Suppose these currents are induced from submanifolds $i\, W \hookrightarrow X$ and cocycles $\check{q}_E, \check{q}_B$ on W, as in (2.19) and (2.27), but now we

[22] There is another possible scenario in which the gauge field is twisted, hence the magnetic current is twisted, and the electric current is untwisted. This occurs in M-theory, for example.

allow twisted cocycles. In other words, postulate twistings $\check{\tau}_E, \check{\tau}_B$ on W such that

$$\begin{aligned} \check{q}_E &\in \check{Z}_\Gamma^{\bullet+\check{\tau}_E}(W), \\ \check{q}_B &\in \check{Z}_\Gamma^{\bullet+\check{\tau}_B}(W). \end{aligned}$$

Then if $\check{j}_E = i_*\check{q}_E$ and $\check{j}_B = i_*\check{q}_B$ the twistings must satisfy

$$\begin{aligned} \check{\tau}_E &= \check{\zeta}(\nu) - i^*\check{\zeta}(X), \\ \check{\tau}_B &= \check{\zeta}(\nu), \end{aligned} \tag{2.34}$$

where $\nu \to W$ is the normal bundle to W in X.

In some theories the gauge fields and currents live in twisted versions of differential cohomology. Precisely, we have for some twisting $\check{\zeta}$:

$$\begin{aligned} \check{A} &\in \check{C}_\Gamma^{\mathbf{d}+\check{\zeta}}(X), \\ \check{j}_B &\in \check{Z}_\Gamma^{\mathbf{d}+1+\check{\zeta}}(X), \\ \check{j}_E &\in \check{Z}_\Gamma^{n-\mathbf{d}+1-\check{\zeta}-\check{\zeta}(X)}(X). \end{aligned} \tag{2.35}$$

Note the sign change in the twisting for the electric current. This makes the electric coupling (2.16) well-defined. Equations (2.34) are now

$$\begin{aligned} \check{\tau}_E &= \check{\zeta}(\nu) + i^*\check{\zeta} - i^*\check{\zeta}(X), \\ \check{\tau}_B &= \check{\zeta}(\nu) + i^*\check{\zeta}. \end{aligned} \tag{2.36}$$

3. Applications

Chern-Simons class

Recall that the origin of differential cohomology lies in the work of Cheeger-Simons [13]. Their primary motivation is the application to secondary characteristic classes. We focus on 4-dimensional characteristic classes.

Fix a compact Lie group G, and suppose BG is a *smooth* classifying space. The odd real cohomology of BG vanishes, so from (1.7) we conclude

$$\begin{aligned} \check{H}^4(BG) &\cong A^4_H(BG) \\ &= \{(\lambda, \omega) \in H^4(BG) \times \Omega^4_{\mathrm{cl}}(BG) : \lambda_{\mathbb{R}} = [\omega]_{dR}\}. \end{aligned} \tag{3.1}$$

Fix a connection A_{univ} on the universal bundle $EG \to BG$ and suppose $\lambda \in H^4(BG)$ is a characteristic class. Let $\omega_{\text{univ}} \in \Omega^4_{\text{cl}}(BG)$ be its Chern-Weil representative. Then $(\lambda, \omega_{\text{univ}}) \in A^4_H(BG)$, and by (3.1) this data determines a universal *Chern-Simons class* in $\check{H}^4(BG)$. Fix a cocycle representative $\check{\alpha}_{\text{univ}} \in \check{Z}^4_H(BG)$.

Now suppose $P \to X$ is a principal G-bundle over a smooth manifold M. Let A be a connection on P. A *classifying map* for A is a G-equivariant map $f\ P \to EG$ such that $f^*A_{\text{univ}} = A$. It is well-known that classifying maps exist. Let $\bar{f}\ M \to BG$ be the map induced from a classifying map f. Then

$$\check{\alpha}(A) = \bar{f}^*\alpha_{\text{univ}} \in \check{Z}^4_H(M) \tag{3.2}$$

is the Chern-Simons cocycle of A. Note that the curvature of $\check{\alpha}(A)$ is the Chern-Weil 4-form $\omega(A)$ of A. As stated here $\check{\alpha}(A)$ depends on the classifying map f. Any two classifying maps are homotopic through classifying maps, so the cohomology class of $\check{\alpha}(A)$ in $\check{H}^4(M)$ is well-defined. In fact, there is a more refined context in which we can work so that $\check{\alpha}(A)$ is canonically defined as a cocycle.

In 3-dimensional Chern-Simons theory [19] one considers a family of connections A on a compact oriented manifold X parametrized by T, where $n = \dim X \leq 3$. Then the associated classical Chern-Simons invariant is

$$\int_{(X\times T)/X} \check{\alpha}(A) \in \check{C}^{4-n}_H(T).$$

If X is closed the result is a cocycle, so represents a differential cohomology class. For example, if $\dim X = 3$ this cocycle is a map $T \to \mathbb{R}/\mathbb{Z}$, the usual classical Chern-Simons action. For $\dim X < 3$ we obtain other geometric invariants.

If a field theory (in arbitrary dimensions) contains a nonabelian gauge field A, we can couple a "2-form field" B to it in a nontrivial way using the Chern-Simons cocycle (3.2). Let the underlying manifold be X. Then we interpret the field B globally as an element $B \in \check{C}^3_H(X)$ which trivializes $\check{\alpha}(A)$, as explained before Example 1.10. Schematically, we write

$$d\check{B} = \check{\alpha}(A). \tag{3.3}$$

There are theories of A, B alone which use this coupling [5], and it appears in more complicated theories as well, as we explain next.

Type I B-field: differential cohomology

The coupling (3.3) between a nonabelian gauge field A and an abelian 2-form field B occurs in type I supergravity in 10 dimensions [12]. The B-field occurs in *pure* supergravity, where it may be interpreted as a cocycle in $\check{Z}^3_H(X)$. But in classical Type I supergravity coupled to super Yang-Mills there is also a nonabelian gauge field A. Suppose $\check{\alpha}(A) \in \check{Z}^4_H(X)$ is its Chern-Simons cocycle, as in (3.2). Then the B-field is a cochain $\check{B} \in \check{C}^3_H(X)$ such that $\check{B}$ trivializes $\check{\alpha}(A)$:

$$d\check{B} = \check{\alpha}(A). \tag{3.4}$$

The *Green-Schwarz anomaly cancellation mechanism* in the low energy description of Type I superstring theory [25] is a modification to the global geometric nature of B and an additional term in the action. Here we interpret it in differential cohomology, along the lines of the standard story. Namely, (3.4) is replaced by

$$d\check{B} = \check{\alpha}(A) - \check{\alpha}(g), \tag{3.5}$$

where $\check{\alpha}(g)$ is the Chern-Simons cocycle of the Levi-Civita connection. The additional term in the action has the form

$$2\pi i \int_X \check{\gamma}(g,A) \cdot \check{B}, \tag{3.6}$$

where the curvature of the differential cocycle $\check{\gamma}(g,A) \in \check{Z}^8_H(X)$ is an 8-form $P_8(g,A)$ which occurs in the anomaly computation from the fermionic functional integrals. (The ability to write the fermion anomaly in this form restricts the gauge algebra to a few possibilities. The precise formula for $P_8(g,A)$ may be found in (3.38).) The exponential of (3.6) is a section of a hermitian line bundle whose curvature is

$$2\pi i \int_X P_8(g,A) \wedge \big[\omega(A) - \omega(g)\big].$$

(Recall that ω denotes the Chern-Weil 4-form.) This cancels the curvature from the fermion Pfaffian line bundles. Note that the existence of the trivialization $\check{B}$ in (3.5) implies the topological constraint

$$\lambda(A) = \lambda(g), \tag{3.7}$$

where λ is the *integral* characteristic class used to define the Chern-Simons cocycle. Equation (3.7) plays an important role in heterotic string theory, for example in the cancellation of worldsheet anomalies. It also appears in the Type I superstring.

In the scenario presented here the local anomaly (curvature) cancels, but there remains a global anomaly. Later we revise this discussion for the Type I superstring (and one of the heterotic strings). We replace integral cohomology by KO-theory; then the global anomaly cancels as well. Equation (3.7) is refined to equation (3.46) in KO-theory.

Self-dual gauge fields[23]

Often these are termed *chiral gauge fields* or *gauge fields with self-dual field strength*; for simplicity we call them *self-dual gauge fields*. In *Lorentzian* field theory on $\mathbb{R} \times N$ the self-duality condition makes sense *classically*, and it states $F_A = *F_A$. This is not an equation of motion from an action principle, but rather is an auxiliary condition imposed by hand. The set of self-dual solutions (up to equivalence) to the classical equations of motion is a symplectic submanifold of the set of all solutions, and so gives a well-defined classical system. Note that electric and magnetic currents and charges are equal for a self-dual gauge field: $j_E = j_B$. In this section we outline the additional data needed to define self-dual gauge fields—including charge quantization—in Euclidean quantum field theory. We assume throughout that the Riemannian manifold X is compact; if not, one should add convergence conditions on the integrals over X.

There are three main examples we have in mind.

Example 3.8 (doubling). Here $n = \dim X$ is arbitrary. Let $\check{A} \in \check{Z}^{\mathbf{d}}_{\Gamma}(X)$ be any gauge field, and consider now $\tilde{\mathbf{d}} = (\mathbf{d}, n - \mathbf{d})$ and $\check{\mathcal{A}} = (\check{A}, \check{A}') \in \check{Z}^{\tilde{\mathbf{d}}}_{\Gamma}(X) = \check{Z}^{\mathbf{d}}_{\Gamma}(X) \times \check{Z}^{n-\mathbf{d}}_{\Gamma}(X)$. The self-duality condition asserts that $\check{A}'$ is the electromagnetic dual of $\check{A}$, and it allows us to recover the theory of $\check{A}$ from the theory of the pair $\check{\mathcal{A}} = (\check{A}, \check{A}')$.

Example 3.9 (integer cohomology). Here $n = \dim X = 4\ell + 2$ for some integer ℓ. Then on a middle-dimensional gauge field $\check{A} \in \check{Z}^{2\ell+1}_{H}(X)$ we can impose the self-duality constraint. Returning momentarily to the Lorentzian framework on Minkowski spacetime M^n, free theories

[23] We take this opportunity to point out a conceptual mistake in [21]. It occurs in the paragraph following equation (6), and also in the footnote which follows. In fact, there is no change in the quantization law of the gauge field, but rather it is the quadratic form introduced in (3.13) below which is needed to make sense of the electric coupling. See Example 3.27 for an analogous case. Also, there is no constraint on the Ramond-Ramond gauge field as proposed before equation (15) in [21]; the factor of 1/2 is implemented by the quadratic form (3.13).

correspond to representations of the Poincaré group. In this case we obtain an irreducible massless representation induced from the action of the little group $\mathrm{Spin}_{n-2} = \mathrm{Spin}_{4\ell}$ on self-dual 2ℓ-forms.

Example 3.10 (K-theory). We continue Example 2.10. Here $n = \dim X = 10$ and X is spin. Then for vanishing B-field the self-duality is imposed on $\check{A} \in \check{Z}^d_K(X)$. This occurs in the low energy description of the Type II superstring—$\check{A}$ is the *Ramond-Ramond gauge field.*

We list the data and constraints necessary to define a self-dual gauge field.

Definition 3.11. Fix a dimension n, a cohomology theory Γ, a multi-degree $\mathbf{d} = (d_1, \dots, d_k)$, a multi-coupling constant $e^2 = (e_1^2, \dots, e_k^2)$, and maps (2.14) from Γ to integer cohomology H in low degrees. A *self-duality constraint* on the corresponding gauge field $\check{A}$ is the additional data:

(i) an automorphism $\theta\, \check{\Gamma}^\bullet \to \check{\Gamma}^{n+2-\bullet}$ of a product of k copies of generalized differential cohomology so that for any $\check{\Gamma}$-oriented fiber bundle $X \to T$ with closed fibers of dimension n, the bilinear form

$$\begin{aligned} B_{X/T}\, \check{Z}^{\mathbf{d}+1}_\Gamma(X) \times \check{Z}^{\mathbf{d}+1}_\Gamma(X) &\longrightarrow \check{Z}^2_\Gamma(T) \\ \check{a}_1 \times \check{a}_2 &\longmapsto \int_{X/T} \theta(\check{a}_1) \cdot \check{a}_2 \end{aligned} \tag{3.12}$$

is symmetric;

(ii) for each fiber bundle $X \to T$ as above a quadratic map

$$q_{X/T}\, \check{Z}^{\mathbf{d}+1}_\Gamma(X) \longrightarrow \check{Z}^2_\Gamma(T) \tag{3.13}$$

which refines the bilinear map (3.12) in the sense that there is a natural isomorphism

$$\begin{aligned} q_{X/T}(\check{a}_1 + \check{a}_2) - q_{X/T}(\check{a}_1) - q_{X/T}(\check{a}_2) + q_{X/T}(0) \\ \cong B_{X/T}(\check{a}_1, \check{a}_2), \qquad \check{a}_1, \check{a}_2 \in \check{Z}^{\mathbf{d}+1}_\Gamma(X). \end{aligned} \tag{3.14}$$

If $\check{b} \in \check{C}^{\mathbf{d}}_\Gamma(X)$ is a trivialization of $\check{a} \in \check{Z}^{\mathbf{d}+1}_\Gamma(X)$, then there is a canonically induced trivialization $q_{X/T}(\check{b})$ of $q_{X/T}(\check{a})$.

In addition, we take the specific normalizing form $\omega_X = \big((\omega_X)_1, \dots, (\omega_X)_k\big)$ (see (2.6)) defined by

$$(\omega_X)_i = \sqrt{\pi e_i^2 \hat{A}_\Gamma(X)}, \tag{3.15}$$

where $\hat{A}_\Gamma(X)$ is the form in (1.15). Finally, the electric and magnetic currents are constrained to satisfy

$$\check{j}_E = \theta(\check{j}_B). \tag{3.16}$$

The definition requires several comments. The quadratic form q is modeled after [28], who treat Example 3.9 in great detail. They specify a more precise set of axioms for q which we do not state explicitly here but implicitly require.[24] One statement we will use is: The quadratic form (3.13) is a map of categories. (Recall the discussion before Example 1.10.) The quadratic form q is determined by an analogous quadratic refinement in Γ-cohomology, so is really a topological choice. Typically, q has no constant term—$q_{X/T}(0) = 0$—and often q has no linear term—$q_{X/T}(-\check{a}) = q_{X/T}(\check{a})$. However, we will encounter one example (involving KO-theory) where the symmetry is of the form $q_{X/T}(\check{\lambda} - \check{a}) = q_{X/T}(\check{a})$ for some class $\check{\lambda}$. If the gauge field and magnetic current live in $\check{\zeta}$-twisted differential cohomology, in which case the electric current lives in $(-\check{\zeta})$-twisted differential cohomology (see (2.35)), then $\theta \check{\Gamma}^\bullet_{\check{\zeta}} \to \check{\Gamma}^{n+2-\bullet}_{-\check{\zeta}}$ and the domains of $B_{X/T}$ and $q_{X/T}$ are suitably twisted. In some cases (e.g. Example 3.22 below) the codomain of $q_{X/T}$ does not involve $\check{\Gamma}$ but rather a different differential theory which maps into $\check{\Gamma}$. In fact, we really only use the quadratic form obtained by composing $q_{X/T}$ with the map $\check{Z}^2_\Gamma(T) \to \check{Z}^2_H(T)$ obtained from (2.14). The constraint (3.16) on the currents means that there is only one current and one charge in the theory. This is the basic meaning of self-duality. Finally, we note that the quadratic refinement q of B is used twice in the self-dual theory: it determines the partition function (see [38]) and is used in the electric coupling term in the action (see (3.25) below).

Example 3.17 (doubling). The map θ is

$$\theta(\check{a}, \check{a}') = ((-1)^{(\mathbf{d}+1)(n-\mathbf{d}+1)}\check{a}', \check{a}), \qquad \check{a}, \check{a}' \in \check{\Gamma}^\bullet.$$

Let $\tilde{a}_1 = (\check{a}_1, \check{a}_1')$ and $\tilde{a}_2 = (\check{a}_2, \check{a}_2')$ be elements of $\check{\Gamma}^{\mathbf{d}+1} \times \check{\Gamma}^{n-\mathbf{d}+1}$. Then the bilinear form

$$\tilde{a}_1, \tilde{a}_2 \longmapsto \theta(\tilde{a}_1) \cdot \tilde{a}_2 = \check{a}_1\check{a}_2' + \check{a}_2\check{a}_1' \tag{3.18}$$

[24] They extend to maps $q_{X/T}\ \check{Z}^{\mathbf{d}+1}_\Gamma(Y) \to \check{Z}^i_\Gamma(T)$ over fiber bundles $Y \to T$ with relative dimension $n+2-i$ for $i = 0, 1, 2$. The additional axioms involve functoriality under base change and composition of fiber bundles.

has a canonical quadratic refinement

$$\tilde{a} = (\check{a}, \check{a}') \longmapsto \check{a} \cdot \check{a}'. \tag{3.19}$$

The required $q_{X/T}$ is obtained from this by integration.

Example 3.20 (integer cohomology). Set $n = 4\ell + 2$ and $d = 2\ell + 1$. The map θ is the identity. Hopkins and Singer [28], following Browder [9], explain that on the category of compact oriented manifolds with "Wu structure" there is a functorial quadratic refinement defined. In the familiar case $n = 2$ of a self-dual scalar field, a Wu structure is simply a spin structure. The normalization (3.15) corresponds to the "free fermion radius". Namely, with $\omega_X = 2\pi$ so that the gauge field ϕ has periodicity 2π, the kinetic action is

$$S_{\text{kin}}(\phi) = \frac{1}{8\pi} \int_X d\phi \wedge *d\phi. \tag{3.21}$$

Example 3.22 (K-theory). Recall $n = 10$ and X is spin. Then

$$\begin{aligned} \theta\, \check{K}^q(X) &\longrightarrow \check{K}^{12-q}(X) \\ \check{a} &\longmapsto u^{6-q}\bar{\check{a}} \end{aligned} \tag{3.23}$$

where $u \in K^2(pt)$ is the inverse Bott element. If $\check{a} \in \check{K}^0(X)$ is represented by a complex vector bundle $E \to X$ with connection, then $\bar{\check{a}}$ is represented by the complex conjugate vector bundle $\overline{E} \to X$ with the conjugate connection. The quadratic refinement is defined by Witten in[25] [40], and it uses the fact that we integrate over Riemannian *spin* fiber bundles $X \to T$. Namely, for $\check{a} \in \check{Z}^{\bullet}_K(X)$ the element $\theta(\check{a}) \cdot \check{a}$ has a canonical lift to $\check{Z}^{12}_{KSp}(X)$, and topologically $q_{X/T}(a)$ is the pushforward of this lift in KSp -theory.

In the quantum theory the Euclidean partition function and correlation functions of self-dual fields are not defined by the usual functional integral, but rather a special procedure is needed which accounts for the self-duality constraints. The reader should keep in mind that the interpretation of this action is not that of the usual functional integral.

We work in the situation of Definition 3.11. Let $\check{\jmath} \in \check{Z}^{\mathbf{d}+1}_{\Gamma}(X)$ and set $\check{\jmath}_B = \check{\jmath}$, $\check{\jmath}_E = \theta(\check{\jmath})$. The gauge field $\check{A} \in \check{C}^{\mathbf{d}}_{\Gamma}(X)$ is a trivialization of the refined magnetic current $\check{\jmath}$. The kinetic term in the action is unchanged

[25] Witten actually defines a related quadratic form, but the basic idea is the same.

from (2.11), but we do use the normalization (3.15). For a gauge field quantized by integer cohomology with $\omega_X = 2\pi$ we find

$$S_{\text{kin}}(A) = \frac{1}{8\pi} \int_X F_A \wedge *F_A,$$

generalizing (3.21). In the presence of the self-duality constraint the electric coupling term is half[26] the usual term (2.16). Recall first the anomaly (2.30); in the self-dual case it is the exponential of $2\pi i$ times

$$\text{“} -\frac{1}{2} \int_{X/T} \theta(\check{\jmath}) \cdot \check{\jmath} = -\frac{1}{2}\, B_{X/T}(\check{\jmath}, \check{\jmath})\text{”}$$

where we use the degree 2 map in (2.14) to land in $\check{H}^2(T)$. Taking half the integral is precisely what the quadratic form $q_{X/T}$ does, so the self-dual anomaly is

$$\exp\Big(-2\pi i \; q_{X/T}(\check{\jmath})\Big). \tag{3.24}$$

The electric coupling term is formally the exponential of $2\pi i$ times

$$\text{“} -\frac{1}{2} \int_{X/T} \theta(\check{\jmath}) \cdot \check{F}_A = -\frac{1}{2}\, B_{X/T}(\check{\jmath}, \check{F}_A),\text{”}$$

which is

$$\exp\Big(-2\pi i \; q_{X/T}(\check{F}_A)\Big). \tag{3.25}$$

This is a trivialization of the circle bundle (3.24). The entire exponentiated action is

$$e^{-S}(\check{A}) = \exp\big(-\frac{1}{8\pi} \int_X F_A \wedge *F_A\big) \exp\big(-2\pi i \; q_{X/T}(\check{F}_A)\big). \tag{3.26}$$

Example 3.27 (doubling). It is instructive to work out the quadratic map (3.13) in the case of a pair of $\mathbb{T}$-valued scalar fields, a special case of Example 3.8 and Example 3.17. Thus $\Gamma = H$ is integer cohomology and the degree is $\mathbf{d} = (1, 1)$. Let the dimension be $n = 2$. Assume $X \to T$ has fibers which are closed oriented surfaces. We use formulas (3.18) and (3.19) for the bilinear form and its quadratic refinement. Consider first $\check{\jmath} = 0$ so that the gauge field $\check{A} = (e^{i\phi}, e^{i\phi'})$ is a pair of maps $X \to \mathbb{T}$. Then the anomaly (3.24) vanishes so the electric coupling (3.25) is an

[26] Some justification for the factor 1/2 comes by considering the classical Lorentzian field theory; see [21, (6)].

ordinary function $T \to \mathbb{T}$. The square of this function is identically 1 as it is the exponential of $2\pi i$ times $-B_{X/T}(\check{\jmath}, \check{F}_A)$. Thus (3.25) is a locally constant function with value ± 1. In fact, it is the exponential of $2\pi i$ times

$$-\frac{1}{2}\int_{X/T}\frac{d\phi}{2\pi}\wedge\frac{d\phi'}{2\pi}. \tag{3.28}$$

Note that $\frac{d\phi}{2\pi}\wedge\frac{d\phi'}{2\pi}$ represents an integral cohomology class on X, the characteristic class of the product $e^{i\phi}\cdot e^{i\phi'}$ in $\check{H}^2(X)$.

Now consider $\check{\jmath} = (0, \check{k}')$ where $\check{k}' \in \check{H}^2(X)$ is represented by a circle bundle $P' \to X$ with connection. Then the gauge field is $\check{A} = (e^{i\phi}, f')$, where $e^{i\phi}\, X \to \mathbb{T}$ and f' is a section of P'. The exponential of $2\pi i$ times $B_{X/T}(\check{\jmath}, \check{F}_A) = \int_{X/T} e^{i\phi}\cdot \check{k}'$ is again an ordinary function $T \to \mathbb{T}$, the anomaly $\exp\bigl(-2\pi i\; q_{X/T}(\check{\jmath})\bigr)$ vanishes, and $\exp\bigl(-2\pi i\; q_{X/T}(\check{F}_A)\bigr)$ is an ordinary function—a square root of $\exp\bigl(-2\pi i\int_{X/T} e^{i\phi}\cdot\check{k}'\bigr)$. Let $\theta \in \Omega^1(X)$ be the covariant derivative of f'; then $d\theta$ is the curvature of P'. A computation in differential cohomology yields

$$\exp\bigl(-2\pi i\; q_{X/T}(\check{F}_A)\bigr) = \exp\bigl(-\frac{2\pi i}{2}\int_{X/T}\frac{d\phi}{2\pi}\wedge\frac{\theta}{2\pi}\bigr), \tag{3.29}$$

a generalization of (3.28). As a special singular case let P' be flat outside a "divisor" of degree 0 on each component of X. For example, let P' have curvature $\delta_p - \delta_q$ where p and q are sections of $X \to T$ whose images lie in the same component and δ_p and δ_q are distributional 2-forms supported on these images. Then (compare (2.24))

$$\exp\bigl(-2\pi i\; B_{X/T}(\check{F}_A, \check{\jmath})\bigr) = \frac{e^{i\phi(q)}}{e^{i\phi(p)}}.$$

(3.29) is a square root of this function, but as f' is discontinuous at p, q it is not easy to describe geometrically.

Type II Ramond-Ramond fields ($B = 0$)

As described in Example 2.10, in the low energy description of the Type II superstring X is a spin 10-manifold. If the B-field vanishes, then the Ramond-Ramond gauge field $\check{A}$ lives in $\check{C}^{\bullet}_K(X)$ with $\bullet = 0$ for Type IIA and $\bullet = -1$ for Type IIB. Furthermore, the gauge field is self-dual and the extra data (see Definition 3.11) needed to describe it is specified in Example 3.22. Our goal here is to make everything a bit more explicit for the current $\check{\jmath}$ induced from a submanifold $i\, W \hookrightarrow X$, the worldvolume of a *D-brane*.

Suppose W has codimension r in X; r is odd in Type IIA and even in Type IIB. We assume given $\check{q} \in \check{Z}^0_{K,\check{\tau}}(W)$, analogous to (2.19) and (2.26), but where $\check{\tau}$ is a twisting of differential K-theory. (Twistings are discussed at the end of §1 and in this context at the end of §2.) Define the magnetic current as

$$\check{\jmath} = u^{-[\frac{p}{2}]}\, i_* \check{q}. \tag{3.30}$$

Concretely (see Example 1.13) $\check{q}$ is usually described as a complex vector bundle with connection $Q \to W$, the "Chan-Paton vector bundle". For the pushforward to be well-defined we have equation (2.34) for the twistings. Recall from (1.17) that the twisting class in $\check{K}$-theory of a real vector bundle V is the pair $\big(w_1(V), \check{w}_2(V)\big)$. Now since X is assumed spin, its twisting class vanishes. Hence (2.34) asserts that $\check{q}$ is a twisted cocycle on W, the twisting being $\big(w_1(\nu), \check{w}_2(\nu)\big)$, where $\nu \to W$ is the normal bundle to W in X. Usually one assumes W is oriented, in which case the twisting is by $\check{w}_2(\nu)$. For example, a (locally) rank one element $\check{q}$ is not represented by a complex line bundle over W with connection, but rather by a spin^c connection on ν. This was derived from perturbative string theory in [23].

Next we work out the electric coupling term (3.25) and the associated anomaly. First, the electric current is

$$\theta(\check{\jmath}) = u^{e-[\frac{p}{2}]}\, i_* \bar{\check{q}},$$

where $e = 5$ in Type IIA and $e = 6$ in Type IIB. The electric coupling term is a section of a hermitian line bundle with connection over a parameter space T; the line bundle represents the anomaly. We compute the Chern class of this line bundle from (3.24) as

$$-q_{X/T}(\check{\jmath}) = -\int_{X/T} u^{6-p}\, i_* \bar{\check{q}} \cdot i_* \check{q},$$

where the integral is in KSp . We can write this as an integral over W/T

$$\begin{aligned} -q_{X/T}(\check{\jmath}) &= -\int_{W/T} u^{6-p}\, \bar{\check{q}} \cdot i^* i_* \check{q} \\ &= -\int_{W/T} u^{6-p}\, \bar{\check{q}} \cdot \check{q} \cdot \mathrm{Euler}_{\check{K}}(\nu), \end{aligned}$$

but the integrals can no longer be interpreted in KSp . The last section in [21] explains how to interpret this computation depending on the dimension of W, and also gives a formula for the $\check{K}$-theory Euler class. (The subtle point is while for r odd the Euler class vanishes in topological K-theory, it is an element of order two in differential $\check{K}$-theory.) As explained there, this anomaly cancels the anomaly from the fermions on W. The electric coupling (3.25) may be written formally as

$$(3.31) \qquad 2\pi i\, q_{X/T}(\check{F}_A) \text{ “}=\text{” } \frac{2\pi i}{2}\int_{X/T} i_*\bar{\check{q}}\cdot\check{F}_A = \frac{2\pi i}{2}\int_{W/T}\bar{\check{q}}\cdot i^*\check{F}_A.$$

The integrals are in $\check{K}$-theory and the factor of $1/2$ is because of the quadratic form. The electric coupling appears in this form in [21, (15)]. To convert to a formula with differential forms, we assume that $\check{q}$ is defined by a complex vector bundle $Q \to W$ with connection. Suppose also that W is spin^c and the curvature of the spin^c connection is $-2\pi i\eta \in \Omega^2(W)$. Finally, suppose the Ramond-Ramond field is determined by a differential form $A/\big(2\pi\sqrt{\hat{A}(X/T)}\big)$ with $dA = F_A$, at least over W (see (1.8)). Then (3.31) reduces to

(3.32)

$$2\pi i\, q_{X/T}(\check{F}_A) = \pi i\int_{W/T}\hat{A}(W/T)\wedge e^{\eta/2}\wedge \mathrm{ch}(\overline{Q})\wedge i^*\left(\frac{A}{2\pi\sqrt{\hat{A}(X/T)}}\right).$$

The formula appears in roughly this form in the physics literature (e.g. [31], [14]). Notice that we ignore the magnetic current in writing this expression. We have already given a precise definition of the electric coupling; (3.32) is included to make contact with the literature.

Type II Ramond-Ramond fields ($B \neq 0$)

The Type II B-field is a cocycle $\check{B} \in \check{Z}^3_H(X)$. In other words, it is a “usual” 2-form gauge field quantized by integer cohomology. As explained at the end of §1 it determines a twisted version $\check{K}^{\bullet+\check{B}}(X)$ of differential K-theory, and the Ramond-Ramond fields are cochains in this twisted theory. The Ramond-Ramond charges take values in the twisted K-group $K^{\bullet+\zeta}(X)$, where $\zeta \in H^3(X)$ is the characteristic class of the B-field $\check{B}$.

The previous discussion may be reconsidered with this twist. The automorphism θ has the same formula (3.23) as before, but it reverses the twisting:

$$\theta\ \check{K}^{q+\check{B}}(X) \longrightarrow \check{K}^{12-q-\check{B}}(X).$$

By (2.36) the twisting $\check{\tau}$ of the Chan-Paton bundle $\check{q} \in \check{Z}^{0+\check{\tau}}_K(W)$ satisfies

$$(3.33) \qquad \check{\tau} = \check{w}_2(\nu) + i^*\check{B}.$$

This equation was deduced from perturbative open string theory in [23]; it is one of many pieces of evidence that Ramond-Ramond charge lives in K-theory. Equation (3.33) is a nontrivial constraint on D-branes. Traditionally one thinks of the Chan-Paton vector bundle on a "single" D-brane as having rank one. The concept of rank does not make sense in every twisted K-theory, and the only reasonable interpretation is that a rank one element is a cochain in $\check{C}^2_H(W)$ which trivializes $\check{\tau} \in \check{Z}^3_H(W)$. Such trivializations exist if and only if the topological class of the twisting vanishes:

$$W_3(\nu) + i^*\lambda(\check{B}) = 0. \tag{3.34}$$

This constraint on a single D-brane was derived from different points of view in [41], [39], and [23]. If the class $W_3(\nu) + i^*\lambda(\check{B})$ is torsion of order N, then again it makes sense to talk about twisted K-theory elements of finite rank, but the rank is constrained to be a multiple of N. Elements of virtual rank zero exist for any twisting; one might instead formally consider them to have infinite rank [37].

Explicit formulas for twisted $\check{K}$-theory are difficult to write in general, but can be written if we assume $\check{B}$ to be defined by a global real 2-form $B \in \Omega^2(X)$ (cf. (1.8)). Note that the characteristic class ζ of this $\check{B}$ vanishes and the curvature is dB. The form B induces a map

$$\psi\ \check{Z}^2_H(X) \longrightarrow \check{C}^2_H(X) \tag{3.35}$$

whose image consists of trivializations of $\check{B}$. In the model of Example 1.10 the triple $(0, B, dB)$ represents $\check{B}$ and the map (3.35) is

$$\psi\ (c, h, \omega) \longmapsto (c, h, \omega + B).$$

In our current notation, if $\check{A} \in \check{Z}^2_H(X)$ has field strength F_A, then the field strength (covariant derivative) of $\psi(A)$ is $F_A + B$. Note that trivializations of $\check{B}$ are "rank one" cocycles for $\check{K}^{\bullet+\check{B}}(X)$.

We can apply these remarks to construct $\check{q} \in \check{Z}_K^{0+i^*(\check{B})}(W)$ in case W is spin. Thus we suppose $\check{A} \in \check{Z}^2_H(W)$ is an ordinary 1-form gauge field on W and set $\check{q} = \psi(\check{A})$. In explicit formulas like (3.32) the field strength F_A is replaced by $F_A + B$, for example in the factor $\mathrm{ch}(\overline{Q})$. Also, these remarks explain a puzzle [4] about Ramond-Ramond charge which was resolved in [35], [1]. Namely, since the characteristic class of $\check{B}$ vanishes the Ramond-Ramond charges take values in ordinary K-theory: $\check{B}$-twisted topological K-theory is not twisted. The remarks above tell that the $\check{B}$-twisted K-theory class of $\check{\jmath} = u^{-[\frac{p}{2}]}\, i_*\check{q} \in \check{Z}_K^{\bullet+\check{B}}(X)$

(see (3.30)) is the ordinary K-theory class of $i_*\check{A}$. So in explicit formulas for the Ramond-Ramond charge—as in the papers just cited—one finds the field strength F_A, not $F_A + B$.

Type I B-field: differential KO-theory

We have already discussed the B-field in Type I superstring theory and the Green-Schwarz local anomaly cancellation from the point of view of integral cohomology. But according to [39] the charges in Type I superstring theory lie in KO-theory. Hence we expect the B-field to be interpreted in differential KO-theory. In fact, this 2-form field is related to the Ramond-Ramond field in Type IIB, and this also leads us to expect that the corresponding charge is quantized in terms of some form of K-theory. Since the Ramond-Ramond fields are self-dual, we expect that in the differential KO formulation the Type I B-field is also self-dual. Finally, the Atiyah-Singer index theorem computes the fermion anomaly as an integral in differential KO-theory. For this anomaly to cancel against local and global anomalies involving bosonic gauge fields, we expect the gauge fields to be cochains in differential KO-theory. We develop these ideas in this section. Proofs of some mathematical assertions made in this discussion are deferred to Appendix B, written jointly with M. Hopkins.

Let $X \to T$ be a Riemannian spin fiber bundle with fibers closed 10-manifolds. Recall this means that there is a Riemannian metric on the relative tangent bundle $T(X/T)$ and a distribution of horizontal planes on X, as well as a spin structure on $T(X/T)$. In Type I superstring theory there is a real rank 32 vector bundle $E \to X$ with connection A. The fermion anomaly has three contributions: a chiral spinor field with values in $\bigwedge^2 E$, the adjoint bundle to E; a chiral Rarita-Schwinger field, which is a spinor field coupled to $T(X/T) - 1$; and a chiral spinor field of the opposite chirality. (The trivial bundle 1 is subtracted from the relative tangent bundle to obtain the pure spin-3/2 field.) The fermion anomaly is a complex line bundle $\mathcal{L} \to T$ with connection, and it is computed by a geometric form of the index theorem [7]. We express[27] the answer in differential KO-theory:

$$\mathcal{L} = \operatorname{pfaff} \int_{X/T} \bigwedge^2 \check{E} + \check{T}(X/T) - 2. \tag{3.36}$$

Here $\check{E} \in \check{Z}^0_{KO}(X)$ is the cocycle corresponding to the real vector bundle

[27] As mentioned at the end of Example 1.13, the rigorous derivation of formulas like (3.36) is part of an ongoing project with M. Hopkins and I. Singer.

with connection E; similarly, $\check{T}(X/T) \in \check{Z}^0_{KO}(X)$; the integral is a map $\int_{X/T} \check{Z}^0_{KO}(X) \to \check{Z}^{-10}_{KO}(T)$; and pfaff $\check{Z}^{-10}_{KO}(T) \to \check{Z}^2_{KO}(T)$ is the pfaffian line bundle. The standard formula in the physics literature ([26, §13.5], for example) is for the curvature of $\mathcal{L}$, which we write as

$$\operatorname{curv} \mathcal{L} = 2\pi i \int_{X/T} \frac{1}{2} P_8(g, A) \wedge \left[p_1(g) - \mathrm{ch}_2(A)\right], \tag{3.37}$$

where

$$P_8(g, A) = -\mathrm{ch}_4(A) + \frac{1}{48} p_1(g)\,\mathrm{ch}_2(A) - \frac{1}{64} p_1(g)^2 + \frac{1}{48} p_2(g). \tag{3.38}$$

The integrand in (3.37) is a rational combination of Chern-Weil differential forms for the Pontrjagin classes of $T(X/T)$ and the Chern character classes of the complexification of E. The extra factor of $1/2$ is due to the fact that $\mathcal{L}$ is the *pfaffian* line bundle, a square root of the determinant line bundle. The factorization of the integrand in (3.37) is a crucial ingredient in the standard story. Usually $P_8(g, A)$ is expressed in terms of characteristic forms of $\bigwedge^2 E$ rather than E; in that case there is a term $\left[\mathrm{ch}_2(\bigwedge^2 A)\right]^2$. The fact that (3.38) is affine linear in A is important to our argument. Set

$$\check{\mathcal{T}} = \check{T}(X/T) + 22.$$

When $\check{E} = \check{\mathcal{T}}$ the curvature of $\mathcal{L}$ vanishes, as the first factor in the integrand does. We claim, and provide a proof in Proposition B.1, that $\mathcal{L}$ itself is trivial for $\check{E} = \check{\mathcal{T}}$, and so in general we can rewrite the formula (3.36) for the fermion anomaly $\mathcal{L}$ as

$$\mathcal{L} \cong \operatorname{pfaff} \int_{X/T} \textstyle\bigwedge^2 \check{E} - \bigwedge^2 \check{\mathcal{T}}. \tag{3.39}$$

We turn now to the B-field, a local 2-form field whose global description we now make precise. The charges associated to this gauge field lie in[28] $KO^0(X)$, so the gauge field $\check{B}$ is at first glance a cocycle for $(KO\ )^{-1}(X)$. In fact, there is a background magnetic current, which we have already seen in a different scenario in (3.5). We now give the self-duality data of Definition 3.11. The cohomology theory underlying this example is the $(\mathbb{Z}/2\mathbb{Z}\times\mathbb{Z})$-graded theory $KOSp$ which was described in Example 1.4. The automorphism θ in the degree we need is

$$\begin{aligned} \theta\ (KO\ )^0(X) &\longrightarrow (KSp\ )^{12}(X) \\ \check{a} &\longmapsto u^6 \check{a}. \end{aligned}$$

[28] As in Type IIB, charges lie in the *augmentation ideal* $\epsilon^{-1}(0)$, where $\epsilon\ KO^0(X) \to H^0(X)$ is the augmentation.

(Recall that u^6 is quaternionic.) For the quadratic refinement (3.13) of the bilinear form (3.12) we set

$$\begin{aligned} q_{X/T}\ \check{Z}^0_{KO}(X) &\longrightarrow \check{Z}^2_{KSp}(T) \\ \check{a} &\longmapsto \int_{X/T} u^6\, \lambda^2(\check{a}), \end{aligned} \tag{3.40}$$

where $\lambda^2\, \check{Z}^0_{KO}(X) \to \check{Z}^0_{KO}(X)$ is the second exterior power operation and the integral is a map $\int_{X/T} \check{Z}^{12}_{KSp}(X) \to \check{Z}^2_{KSp}(T)$. If $\check{a}$ is the KO -cocycle of a real vector bundle with connection E, then $\lambda^2(\check{a})$ is the KO -cocycle of $\bigwedge^2 E$ with the induced connection. The normalizing form is

$$\omega_X = 2\pi\sqrt{\hat{A}(X)},$$

and the coupling constant is $e^2 = 4\pi$.

The quadratic form $q_{X/T}$ satisfies $q_{X/T}(0) = 0$, but it is not symmetric about the origin. In Appendix B we define a $\mu(X) \in KO^0(X)$ canonically associated to spin 10-manifolds X. The class $2\mu(X) \in KO^0(X)$ is a KO-theoretic analog of the Wu class in cohomology. If X has a Riemannian structure then there is a canonical lift of $2\mu(X)$ to differential KO theory (and a canonical cocycle representative), but there may be many lifts of $\mu(X)$ to a differential KO class $\check{\mu}(X)$. (An analogy: Square roots of the canonical bundle of a Riemann surface exist, but none is canonically picked out.) We term a choice of $\check{\mu}(X)$ a *$\check{\mu}$-structure*. Furthermore, we only consider families $X \to T$ in which an appropriate class $\check{\mu}(X/T)$ is defined. These ideas are developed in Appendix B, where the following facts are part definition, part proposition:

- The quadratic form pfaff $q_{X/T}$ has a symmetry:

  $$\operatorname{pfaff} q_{X/T}(2\check{\mu} - \check{a}) \cong \operatorname{pfaff} q_{X/T}(\check{a}), \qquad \check{a} \in \check{Z}^0_{KO}(X). \tag{3.41}$$

  In other words, $\check{\mu}$ is a center for pfaff $q_{X/T}$.

- Set

  $$\check{\epsilon} = \check{\mu} - \check{T}.$$

  Then $\check{\epsilon}$ restricts to zero on the 7-skeleton of X. For example, the Chern character of $\check{\epsilon}$ contains only forms of degree ≥ 8. There is an explicit formula for $\check{\epsilon}$ after inverting 2:

  $$\check{\epsilon} = \frac{1}{64}\big(7\lambda^2(\check{T}) - 5\,\mathrm{Sym}^2(\check{T}) + 4\check{T} - 80\big) + \cdots, \tag{3.42}$$

  where $\check{T} = \check{T}(X/T)$ is the relative tangent bundle.

- We have

$$\text{pfaff}\, q_{X/T}(\check{\mu}) \cong \text{pfaff}\, q_{X/T}(\check{\mathcal{T}}) \cong \text{pfaff} \int_{X/T} \bigwedge{}^2 \check{\mathcal{T}}. \tag{3.43}$$

From the formula (3.42) for $\check{\mu}$ we compute the Chern character

$$\text{ch}(\check{\mu}) = 32 + p_1(g)\, u^{-2} + \left[\frac{1}{192} p_1(g)^2 + \frac{1}{48} p_2(g)\right] u^{-4} + \cdots.$$

Also, from (3.14) and (3.41) we deduce

$$\text{pfaff}\, 2q_{X/T}(\check{a}) \cong \text{pfaff}\, B_{X/T}(\check{a}, \check{a} - 2\check{\mu}), \qquad \check{a} \in \check{Z}^0_{KO}(X). \tag{3.44}$$

If $X^{10} \cong \mathbb{R}^3 \times Y^7$ then according to the second point above there is a canonical $\check{\mu}$-structure $\check{\mu}(X) = \check{\mathcal{T}}(X)$. Furthermore, in this case the characteristic class $[\check{\mathcal{T}}] \in KO^0(X)$ is determined by: $\text{rank}(\check{\mathcal{T}}) = 32$; $w_1(\check{\mathcal{T}}) = w_2(\check{\mathcal{T}}) = 0$; and $\lambda(\check{\mathcal{T}}) \in H^4(X)$, where λ is the canonical $\frac{1}{2}p_1$ for spin bundles.

Our postulate for the 2-form field in Type I on a Riemannian spin 10-manifold X is:

$$\check{B} \text{ is a nonflat isomorphism } \check{\mu}(X) \longrightarrow \check{E}. \tag{3.45}$$

The notion of nonflat trivialization was explained in the paragraph following (1.9); it is the same notion which underlies equation (3.5). A nonflat isomorphism is similar. One version is this: $\check{\mu}(X)$ and $\check{E}$ are elements of a category and $\check{B}$ is a morphism between $\check{\mu}(X)$ and $\check{E} - H$ for some global differential form H (see (1.8)). Another version uses the bigraded theory referred to in the footnotes of §1. The form H has components in degrees 3 and 7 on a 10-manifold. Assertion (3.45) means that there is a background magnetic current equal to $\check{E} - \check{\mu}(X)$. A necessary and sufficient condition for $\check{B}$ to exist is that

$$[E] = [\mu(X)] \qquad \text{in } KO^0(X). \tag{3.46}$$

From the remark in the previous paragraph, we see that (3.46) is equivalent to the standard condition (3.7) if $X = \mathbb{R}^3 \times Y^7$; in many cases it is a stronger condition. There is also a background electric current, manifested through an electric coupling term of the form (3.25). We write it for a family $X \to T$ of Riemannian spin 10-manifolds for which $\check{\mu} = \check{\mu}(X/T)$ exists. Recall that in (3.25) the map $(KSp\ )^2 \to \check{H}^2$ is omitted from the notation and the factor $2\pi i$ is part of the identification of $\check{H}^2$

with connections on circle bundles. Now since $\check{B}$ is a nonflat isomorphism $\check{\mu} \to \check{E}$, by applying the quadratic form, which is a map of categories, we obtain a nonflat isomorphism $q_{X/T}(\check{B})\, q_{X/T}(\check{\mu}) \to q_{X/T}(\check{E})$. Thus the background electric coupling term—the *inverse*[29] pfaffian applied to $q_{X/T}(\check{B})$—is a nonflat trivialization of

$$\text{pfaff } -\{q_{X/T}(\check{E}) - q_{X/T}(\check{\mu})\}. \tag{3.47}$$

In other words, the anomaly in the background electric coupling term is (3.47). Using (3.40), (3.43), and Bott periodicity in KO-theory we see that (3.47) precisely cancels the anomaly (3.39) from the fermions.

There is also a version of Type I theories in which E is a projective bundle with a nontrivial cocycle $w \in H^2(X;\mathbb{Z}/2\mathbb{Z})$.[30] This means that $\check{E}$ lies in a twisted version of KO , namely $\check{E} \in \check{Z}_{KO}^{0+w}(X)$. Note that w is a torsion version of the 2-form field (called 'B') in Type II. It seems that arguments parallel to those in Appendix B yield a "twisted" KO class $\mu_w(X) \in KO^{0+w}(X)$ associated to a spin 10-manifold, and so a parallel discussion with $\check{\mu}$ replaced by $\check{\mu}_w$, but we do not discuss such twisted classes in this paper.

Our motivation for (3.45) is not simply the anomaly cancellation. After all, because of (3.43) we could substitute $\check{\mathcal{T}}$ for $\check{\mu}(X)$ in (3.45) and still cancel the anomaly. An additional motivation for the choice of $\check{\mu}(X)$ is that it is the choice which renders the magnetic current equal to the electric current, which we require for a self-dual field. Another motivation is the presumed twisted analog of $\check{\mu}$ just described; there is no such twisted analog of $\check{\mathcal{T}}$, for example.

To make contact with the usual presentation of the *local* anomaly cancellation, we now relate the electric coupling to the standard formula for the Green-Schwarz term. We write $\check{B}$ as a differential form $(B_2u^{-2} + B_6u^{-4})/(2\pi\sqrt{\hat{A}(X/T)})$ relative to a fixed trivialization of the background magnetic current $\check{E} - \check{\mu}$. The differential of the covariant derivative $(H_3u^{-2} + H_7u^{-4})/(2\pi\sqrt{\hat{A}(X/T)})$ is the Chern character of

[29] due to the minus sign in (3.25).

[30] It is usually asserted that the gauge group of Type I is $\mathrm{Spin}_{32}/(\mathbb{Z}/2\mathbb{Z})$; the cocycle w is the obstruction to lifting the associated $\mathrm{Spin}_{32}/(\mathbb{Z}/2\mathbb{Z} \times \mathbb{Z}/2\mathbb{Z})$ bundle to an SO_{32} bundle.

the background magnetic current, and so

$$
\begin{aligned}
d\left(\frac{H_3u^{-2}+H_7u^{-4}}{2\pi}\right) &= \sqrt{\hat{A}(X/T)}\ \mathrm{ch}(\check{E}-\check{\mu}) \\
&= \Big[\mathrm{ch}_2(A)-p_1(g)\Big]u^{-2} \\
&\quad + \Big[\mathrm{ch}_4(A)-\frac{1}{48}p_1(g)ch_2(A) \\
&\qquad +\frac{1}{64}p_1(g)^2-\frac{1}{48}p_2(g)\Big]\,u^{-4}.
\end{aligned}
\tag{3.48}
$$

The fixed global trivialization of the background magnetic current gives a fixed trivialization of the anomaly (3.47), relative to which the electric coupling term may be written as an integral of differential forms. Set $\check{a}=\check{E}-\check{\mu}$. Then

$$
\begin{aligned}
q_{X/T}(\check{E})-q_{X/T}(\check{\mu}) &= q_{X/T}(\check{\mu}+\check{a})-q_{X/T}(\check{\mu}) \\
&= q_{X/T}(\check{a})+B_{X/T}(\check{\mu},\check{a}),
\end{aligned}
$$

so combining with (3.44) we find

$$
\mathrm{pfaff}\ 2\big[q_{X/T}(\check{E})-q_{X/T}(\check{\mu})\big]=\mathrm{pfaff}\ B_{X/T}(\check{a},\check{a}). \tag{3.49}
$$

Thus the electric coupling in the unexponentiated action is

$$
\begin{aligned}
&(2\pi i)\,\frac{1}{2}u^6\int_{X/T}\left[\hat{A}(X/T)\wedge\mathrm{ch}(\check{a})\wedge\left(\frac{B_2u^{-2}+B_6u^{-4}}{2\pi\sqrt{\hat{A}(X/T)}}\right)\right]_{(10)} \\
&=(2\pi i)\,\frac{1}{2}u^6\int_{X/T}\left[\sqrt{\hat{A}(X/T)}\wedge\big[\mathrm{ch}(\check{E}-\check{\mu})\big]\wedge\left(\frac{B_2u^{-2}+B_6u^{-4}}{2\pi}\right)\right]_{(10)}.
\end{aligned}
\tag{3.50}
$$

The factor of 1/2 in (3.50) is due to the factor of 2 in (3.49); the u^6 is in (3.40). We expand $\sqrt{\hat{A}(X/T)}\wedge\mathrm{ch}(\check{E}-\check{\mu})$ using (3.48).

Now (B_2,B_6) is a self-dual pair of gauge fields (with no Dirac quantization condition). The self-dual Euclidean action has the form

$$
\frac{1}{2}\ \mathrm{kinetic}(B_2)+\frac{1}{2}\ \mathrm{kinetic}(B_6)+\frac{2\pi i}{2}\int\frac{j_E}{2\pi}\wedge\frac{B_2}{2\pi}+\frac{2\pi i}{2}\int\frac{j_B}{2\pi}\wedge\frac{B_6}{2\pi}
$$

and the magnetic currents are determined by

$$
\begin{aligned}
dH_3 &= j_B, \\
dH_7 &= j_E.
\end{aligned}
$$

The "$\frac{1}{2}$ kinetic" terms are one-half the value for non-self-dual gauge fields. If we eliminate B_6 and write the same system in terms of a single gauge field $\tilde{B}_2$, then the corresponding Euclidean action is

$$\text{kinetic}(\tilde{B}_2) + 2\pi i \int \frac{j_E}{2\pi} \wedge \frac{\tilde{B}_2}{2\pi} \tag{3.51}$$

with magnetic current determined by

$$d\tilde{H}_3 = j_B. \tag{3.52}$$

From (3.48) and (3.50) we read off

$$\begin{aligned} j_B &= 2\pi\Big[\operatorname{ch}_2(A) - p_1(g)\Big], \\ j_E &= 2\pi\Big[\operatorname{ch}_4(A) - \frac{1}{48}p_1(g)\operatorname{ch}_2(A) + \frac{1}{64}p_1(g)^2 - \frac{1}{48}p_2(g)\Big], \end{aligned}$$

Thus (3.52) yields

$$d\left(\frac{\tilde{H}_3}{2\pi}\right) = \operatorname{ch}_2(A) - p_1(g),$$

in agreement with (3.5). The electric coupling in (3.51) is the Green-Schwarz term

$$-2\pi i \int_{X/T} P_8(g, A) \wedge \frac{\tilde{B}_2}{2\pi},$$

where $P_8(g, A)$ is given in (3.38). Therefore, our electric coupling term is indeed a refinement of the standard Green-Schwarz term.

Finally, we consider the anomaly for D1- and D5-branes in Type I; the discussion is similar to the treatment [21] of D-brane anomalies in Type II. A D1-brane is a compact spin submanifold $i\, W^2 \hookrightarrow X^{10}$. It is endowed with a real vector bundle Q with connection, which as usual we write as a cocycle $\check{q} \in \check{Z}^0_{KO}(W)$. The corresponding contribution to the magnetic current is

$$\check{\jmath} = u^{-4}\, i_* \check{q}.$$

As usual magnetic current modifies the geometric nature of the gauge field, so when added to the background magnetic current the appropriate modification of (3.45) is:

$$\check{B} \text{ is a nonflat isomorphism } \check{\mu} \longrightarrow \check{E} + \check{\jmath}.$$

The electric coupling $\operatorname{pfaff}\big(-q_{X/T}(\check{B})\big)$ is now the inverse pfaffian of a nonflat isomorphism

$$q_{X/T}(\check{\mu}) \;\longrightarrow\; q_{X/T}(\check{E} + \check{\jmath}) \;\cong\; q_{X/T}(\check{E}) + q_{X/T}(\check{\jmath}) + B_{X/T}(\check{E}, \check{\jmath}).$$

Recall that $q_{X/T}$ is defined in (3.40) and $B_{X/T}(\breve{a}, \breve{a}') = \int_{X/T} u^6 \breve{a} \cdot \breve{a}'$ for $\breve{a}, \breve{a}' \in \check{Z}^0_{KO}(X)$. Thus the new contribution to the anomaly (beyond (3.47)) is

$$\text{pfaff } -\{q_{X/T}(\breve{\jmath}) + B_{X/T}(\check{E}, \breve{\jmath})\}. \tag{3.53}$$

Next, we rewrite the expression in braces as an integral over W. The second term is immediate using the push-pull formula:

$$B_{X/T}(\check{E}, \breve{\jmath}) = u^2 \int_W i^* E \cdot \check{q}. \tag{3.54}$$

For the first term we claim, and prove in Proposition B.40, that

$$q_{X/T}(\breve{\jmath}) = u^2 \int_W \Delta^+(\nu) \cdot \lambda^2(\check{q}) \;-\; \Delta^-(\nu) \cdot \mathrm{Sym}^2(\check{q}), \tag{3.55}$$

where $\nu \to W$ is the normal bundle and $\Delta^\pm$ are the half-spin bundles.

The low energy theory on the D1-brane W has fermions[31] whose anomaly exactly cancels (3.53). Namely, the D1-D9 strings give massless positive chirality spinor fields with coefficients in the bundle $\mathrm{Hom}(Q, i^* E)$, and the D1-D1 strings give positive chirality spinor fields with coefficients in $\Delta^+(\nu) \otimes \bigwedge^2(Q)$ as well as negative chirality spinor fields with coefficients in $\Delta^-(\nu) \otimes \mathrm{Sym}^2(Q)$. As before, a geometric form of the Atiyah-Singer index theorem computes the anomaly to be the pfaffian of (3.54) times the pfaffian of (3.55), and this cancels the anomaly (3.53) from the electric coupling.

The story for the D5-brane is parallel, except that Q is quaternionic.

Appendix: Wick rotation

We include these well-known remarks since some of these issues caused the author confusion from which we hope to spare others. The exposition has no pretense to rigor.

On Minkowski spacetime there are both classical and quantum versions of field theory. A classical theory consists of a symplectic manifold of fields (often there are classical field equations which define it) and a symplectic action of the Poincaré group. A quantum theory consists of a Hilbert space and operators (in particular for the Lie algebra of the Poincaré group), and from this data one defines correlation functions of operators. Such classical and quantum theories exist as well on

[31]I warmly thank Jacques Distler for computing the low energy theory on Type I D-branes.

Lorentzian spacetimes of the form $\mathbb{R}_t \times N$, where (N, g_N) is Riemannian and $\mathbb{R}_t$ represents time; the Lorentz metric is $dt^2 - g_N$. *Wick rotation* occurs in the quantum theory as follows. Correlation functions depend on the positions $(t,n) \in \mathbb{R}_t \times N$ of local operators. Assuming[32] they extend to holomorphic functions of t, one restricts to purely imaginary values of $t = \sqrt{-1}\,\tau$ to obtain correlation functions on $\mathbb{R}_\tau \times N$, which is a Riemannian manifold with metric $d\tau^2 + g_N$. In good cases one can functorially define correlation functions on all Riemannian manifolds X (of fixed dimension) not necessarily of the form $\mathbb{R}_\tau \times N$. This, then, is the substance of Euclidean field theory: quantum correlation functions. There is not a Hilbert space interpretation, nor does one try to make physical sense of classical field theory.

In the Euclidean context one often introduces "classical fields" and an action functional and then writes correlation functions as a functional integral over fields. We review that process briefly, but issue the warning that despite the language of classical fields one is *not* doing classical field theory. In particular, the classical field equations derived mathematically from the Euclidean action have no physical meaning in Euclidean field theory.[33]

The Euclidean functional integral on $\mathbb{R}_\tau \times N$ is derived formally from a corresponding (formal) functional integral in the Lorentzian theory on the spacetime $\mathbb{R}_t \times N$ in case the quantum theory on $\mathbb{R}_t \times N$ is obtained by quantizing a classical theory which has a lagrangian description. The derivation of the Lorentzian functional integral is given in standard texts.[34] The main result has the schematic form

$$\langle \mathcal{O} \rangle = \iint D\phi\, D\psi\; e^{iS(\phi,\psi)}\, \mathcal{O}(\phi,\psi), \tag{A.1}$$

[32]In axiomatic formulations of quantum field theory on Minkowski spacetime (see [29, §2], for example) this is a consequence of more basic axioms.

[33]For example, the Euler-Lagrange equation derived from the action (2.17) is

$$d * F_A = -\sqrt{-1}\, C_X\, j_E,$$

for some real form C_X. As both F_A and j_E are real (see below), this equation clearly has no solutions with nonzero j_E. Of course, variational equations do have a distinguished place in Riemannian geometry. We also remark that Euclidean functional integrals have other applications in physics. But here our concern is strictly Euclidean quantum field theory.

[34]One subtlety, which already occurs in quantum mechanics ($N = pt$), is that the action should be at most quadratic in the first time derivative of the fields (see [34, p.163], for example).

where

$$S(\phi, \psi) = \int_{\mathbb{R}_t \times N} L(\phi, \psi). \tag{A.2}$$

In these formulas ϕ stands for a collection of Bose fields and ψ for a collection of Fermi fields. The fields are defined on $\mathbb{R}_t \times N$, and in (A.1) the integral is over fields with finite action. It is an important principle of unitary quantum field theory that the fields and action are *real*. (Complex or quaternionic notation may be used, but the point remains.) In the quantum theory this leads to the fact that operators corresponding to real observables are symmetric. The integrand which defines the action S is the lagrangian density L. The symbol $\mathcal{O}$ in (A.1) denotes a functional (or finite product of functionals) of the fields. The left-hand side $\langle \mathcal{O} \rangle$ is the quantum correlation function. Now a typical operator

$$\mathcal{O}_{(t,n)}\phi = \phi(t,n) \tag{A.3}$$

evaluates the field at a point. It is the t-dependence of the left-hand side of (A.1) which one (formally) analytically continues to complex values of t. We now discuss the corresponding Wick rotation of the right-hand side. For this one simultaneously rotates both the finite dimensional integral in (A.2) and the functional integral in (A.1). Bose fields and Fermi fields in (A.1) are treated differently. We emphasize that our presentation is formal and algebraic. In any case the Lorentzian functional integral (A.1) is usually oscillatory and badly behaved, and one should view this heuristic argument as motivation for the definition of correlation functions by Euclidean functional integrals.[35]

First, we discuss Wick rotation in the space of fields. Partially complexify the fields to spaces of complex-valued functions which depend holomorphically on a complex variable t. Real fields on $\mathbb{R}_t \times N$ satisfy the reality condition

$$\phi(\bar{t}, n) = \overline{\phi(t,n)}. \tag{A.4}$$

Operators such as (A.3) extend to complex operators defined on complexified fields; they also depend holomorphically on t. The restriction of complexified fields to purely imaginary values of $t = \sqrt{-1}\,\tau$, i.e., to $\mathbb{R}_\tau \times N$, we term *Euclidean fields*. Here is where the treatment of Bose and Fermi fields differs. There is no reality constraint imposed on Euclidean Fermi fields ψ_E: the fermionic functional integral is algebraic,

[35]As Coleman [15, p.148] says, the Lorentzian functional integral is "ill-defined, even by our sloppy standards."

and extending the coefficients from $\mathbb{R}$ to $\mathbb{C}$ does not affect the answer. For example, the pfaffian of a real operator equals the pfaffian of its complexification. For Euclidean Bose fields ϕ_E, on the other hand, (A.4) is "rotated" to the reality constraint

$$\phi_E(-\bar{t}, n) = \overline{\phi_E(t, n)}. \tag{A.5}$$

In particular, Euclidean Bose fields are real-valued on $\mathbb{R}_\tau \times N$. This rotation in the complexified function space is the change of integration domain from the Lorentzian functional integral to the Euclidean functional integral.

In the finite dimensional integral (A.2) rotate the domain of integration from $\mathbb{R}_t \times N$ to $\mathbb{R}_\tau \times N$. The Euclidean lagrangian density L_E on $\mathbb{R}_\tau \times N$ is defined by analytic continuation from L as a function of Euclidean fields:

$$L_E(\phi, \psi) = \frac{1}{\sqrt{-1}} L(\phi, \psi).$$

The reality constraint (A.5) is not used in defining L_E; one essentially substitutes $t = \sqrt{-1}\,\tau$ into L and divides by $\sqrt{-1}$. This has the usual effect of changing the sign of potential energy terms, etc.[36] Notice that imposing (A.5) does *not* render L_E real in general. The integral of L_E over $\mathbb{R}_\tau \times N$ is the Euclidean action S_E. By a similar change of variables, operators $\mathcal{O}$ are re-expressed as Euclidean operators $\mathcal{O}_E$.

With this understood the Euclidean correlation functions are

$$\langle \mathcal{O}_E \rangle_E = \iint D\phi_E \, D\psi_E \; e^{-S_E(\phi_E, \psi_E)} \, \mathcal{O}_E(\phi_E, \psi_E).$$

The functional integral is over Euclidean fields of finite Euclidean action.

Usually the Euclidean theory may be formulated on Riemannian manifolds X not necessarily of the form $\mathbb{R}_\tau \times N$. The reality condition on the classical fields and action in the Lorentzian case is replaced by (i) the reality condition on Euclidean Bose fields, and (ii) the condition that the Euclidean action undergo complex conjugation when the orientation is reversed.[37] The corresponding reality condition on the quantum Euclidean correlation functions is called *reflection positivity*.

[36] See [18, §7] for typical examples. However, be warned that there is a crucial notational mistake in the second paragraph: the analytic continuation of a real scalar field on Minkowski spacetime is not real when restricted to Euclidean space. This confusion—which extends to more complicated fields—was one of our motivations to include this appendix.

[37] Orientation reversal should be construed locally so that the condition makes sense on unorientable manifolds.

(See [17, p.690] for a formal derivation of reflection positivity from the reality condition (ii) on the Euclidean action.)

Appendix: KO and anomalies in type I
Daniel S. Freed & Michael J. Hopkins

In this appendix we provide proofs of several assertions needed in the anomaly cancellation arguments for Type I given in §3. We begin in Proposition B.1 with a special computation of a pfaffian line bundle, stated before (3.39). Then we turn to the theory of the quadratic form (B.31) in differential KO-theory which is relevant to the self-dual field in Type I. It is not symmetric about the origin; rather, the symmetry involves a (differential) KO-theoretic analog of the Wu class. We restrict to situations where this center exists. In Proposition B.40 we carry out a computation needed for D-brane anomaly cancellation in Type I. At the end of the appendix we prove that the Adams operation ψ_2 deloops once, a fact used earlier in this appendix. We do not resolve all questions about the quadratic form—for example, existence and uniqueness questions concerning the center—so view this account as provisional.

Proposition B.1. *Let $X \to T$ be a Riemannian spin fiber bundle with fibers smooth closed spin 10-manifolds. Let $\mathcal{L}$ be the pfaffian line bundle of the family of Dirac operators on the fibers of X coupled to*

$$\bigwedge\nolimits^2 (T(X/T) + 22) + T(X/T) - 2. \qquad (B.2)$$

Then $\mathcal{L}$, together with its natural metric and connection, is trivial.

The natural connection is defined, and the curvature and holonomy computed, in [7]. We do not claim here to give a *canonical* trivialization, so as explained at the end of the introduction while this proposition can be used to prove the cancellation of anomalies, it is not strong enough to construct correlation functions.

Proof. The proof is similar to the proof of [23, Theorem 4.7].

The curvature of $\mathcal{L}$ was computed in (3.37), and was seen to vanish. To compute the holonomy we pull back over a loop in T, so consider a family $X \to S^1$. Let the circle have its bounding spin structure; then X is a closed spin 11-manifold. The holonomy of $\mathcal{L}$ around S^1 is[38] $\exp(-2\pi i \xi_X/2)$, where $\xi_X = \frac{1}{2}(\eta_X + \dim\ker D_X)$ is the Atiyah-Patodi-Singer invariant for the Dirac operator coupled to (B.2). Now

[38]The extra factor of 2 is due to the pfaffian (as opposed to a determinant); the absence of an adiabatic limit is due to the vanishing of the curvature.

any closed spin 11-manifold is the boundary $X = \partial M$ of a compact spin 12-manifold, and we take M to have a Riemannian metric which is a product near the boundary. We compute $\xi_X/2 \pmod 1$ via the Atiyah-Patodi-Singer theorem, but for that we need to extend (B.2) to M, which is not necessarily fibered over S^1. A short computation shows that

$$\mathcal{E} = \bigwedge^2(TM + 20) + TM - 4$$

restricts on ∂M to (B.2). Then

$$\xi_X/2 \equiv \int_M \left[\frac{1}{2}\hat{A}(M)\,\mathrm{ch}(\mathcal{E})\right]_{(12)} \pmod 1,$$

and a straightforward computation shows that the integrand vanishes. Hence the holonomy of $\mathcal{L}$ is trivial. q.e.d.

As a preliminary to the rest of this appendix we recall some facts about KO-theory. Let X be any manifold. First, there is a canonical filtration, the *Atiyah-Hirzebruch filtration*. A class $a \in KO(X)$ has filtration q if q is the largest integer such that the pullback of a via any smooth map $\Sigma^k \to X$ of a k-dimensional manifold Σ into X vanishes for all $k < q$. The product of classes of filtration q and filtration q' has filtration $\geq q + q'$. Second, suppose $\nu \to X$ is a real spin bundle of even rank $2r$. The *K-theory Thom class U* is an element in $K^{2r}_{\mathrm{cv}}(\nu)$, where 'cv' denotes 'compact vertical support.' Let $i\,X \hookrightarrow \nu$ be the zero section. By the splitting principle we formally write

$$(B.3) \qquad \nu \otimes \mathbb{C} = \bigoplus_{i=1}^{r} (\ell_i + \ell_i^{-1}).$$

Then

$$(B.4) \qquad i^*U = u^r\,\left[\Delta^+(\nu) - \Delta^-(\nu)\right] = u^r \prod_{i=1}^{r} (\ell_i^{1/2} - \ell_i^{-1/2}),$$

where $\Delta^{\pm}$ are the half-spin representations. If $r \equiv 0 \pmod 4$, then U is real; if $r \equiv 2 \pmod 4$, then r is quaternionic. When $r \equiv 0 \pmod 4$ there is a KO-theory Thom class (also called 'U') whose complexification is the K-theory Thom class. Next, we state the form of Poincaré duality in KO-theory which we need. Let X be a spin manifold of dimension n. Then

$$(B.5) \qquad \mathrm{Hom}\big(KO^q_c(X;\mathbb{R}/\mathbb{Z}),\mathbb{R}/\mathbb{Z}\big) \cong KO^{n+4-q}(X),$$

where 'c' denotes 'compact support.' The analogous statement in ordinary cohomology does not have the shift by 4. Ordinarily, a generalized cohomology theory does not contain such a duality statement. Finally, we make use of the Adams operation ψ_2. It is a natural *ring endomorphism* of $KO^0(X)$ for any manifold X, and is related to the exterior square λ^2 (which is not a ring homomorphism) by

$$2\lambda^2(a) = a^2 - \psi_2(a). \tag{B.6}$$

The operation ψ_2 is defined on line bundles ℓ by the formula $\psi_2(\ell) = \ell^2$; it extends to arbitrary elements of KO using the splitting principle and the fact that ψ_2 is a ring homomorphism. The same definitions for ψ_2 work on complex K-theory, and then ψ_2 extends to K^{-q} for $q \geq 0$. Its action on the Bott element u^{-1} is[39]

$$\psi_2(u^{-1}) = 2u^{-1}. \tag{B.7}$$

If we invert 2, then ψ_2 extends to K^q for all q. Furthermore, after inverting 2 there is an inverse operation $\psi_{1/2}$. If ℓ is a line bundle, and $x = 1 - \ell$, then

$$\psi_{1/2}(\ell) = (1-x)^{1/2} = 1 - \frac{1}{2}x - \frac{1}{8}x^2 + \dots.$$

Note that x^q has filtration $\geq q$, so on a finite dimensional manifold the infinite series terminates. On an infinite dimensional manifold we must work in a certain completion $\hat{K}$ of K-theory. We also need a single delooping of ψ_2.

Proposition B.8. *There exists an operation $\psi_2 KO^1(X) \to KO^1(X)$ which is compatible with the standard ψ_2 under suspension, i.e., the diagram*

$$\begin{array}{ccc} KO^0(X) & \xrightarrow{\psi_2} & KO^0(X) \\ \downarrow & & \downarrow \\ KO^1(\Sigma X) & \xrightarrow{\psi_2} & KO^1(\Sigma X) \end{array} \tag{B.9}$$

commutes. Furthermore, ψ_2 extends to an operation on the differential group $(KO\ )^1(X)$ and so restricts to an operation on $KO^0(X;\mathbb{R}/\mathbb{Z})$.

[39]Since $u^{-1} = H - 1$ for H the hyperplane bundle on $S^2 \cong \mathbb{CP}^1$, we compute $\psi_2(u^{-1}) = H^2 - 1 = 2(H-1)$ in the reduced K-theory of S^2, which is isomorphic to $K^{-2}(pt)$.

The proof, which involves homotopy-theoretic techniques, is deferred to the end of the appendix. More elementary is the extension of the operations ψ_2 and λ^2 to the differential group $(\widehat{KO})^0$. For that we use the fact that the topological operations are defined on the level of cochains, not just cohomology, and there are compatible operations on differential forms. Then such operations on $\widehat{KO}$ are defined from the basic pullback square (1.5).

Concerning the Atiyah-Hirzebruch filtration, we have the following easy statement.

Lemma B.10. *Suppose $F \to X$ is a real vector bundle with $w_1(F) = w_2(F) = 0$. Then the class of $(F - \operatorname{rank} F)$ in $KO(X)$ has filtration ≥ 4.*

Proof. The classifying map $X \to \mathbb{Z} \times BO$ of $(F - \operatorname{rank} F)$ lifts to $B\,\mathrm{Spin}$, and the 3-skeleton of $B\,\mathrm{Spin}$ is trivial. q.e.d.

We introduce a characteristic class $\hat{\rho}(F) \in \widehat{KO}[\frac{1}{2}](X)$ associated to a real vector bundle $F \to X$. Let $i\,X \to F$ be the zero section. Suppose first that F is spin of even rank $2r$. Let ℓ_i, $i = 1, \dots, r$ as in (B.3) and set

$$(B.11) \qquad \hat{\rho}(F) = i^*\left(\frac{U}{\psi_{1/2}(U)}\right) = \prod_{i=1}^{r} \frac{\ell_i^{1/4} + \ell_i^{-1/4}}{2},$$

where we use (B.4) and (B.7). Here $\psi_{1/2}(U) \in \widehat{KO}[\frac{1}{2}]^{2r}_{\mathrm{cv}}(F)$ and under the Thom isomorphism it corresponds to a class $i^*\big(\psi_{1/2}(U) \,/\, U\big) \in \widehat{KO}[\frac{1}{2}]^0(X)$ which has the form $1 + z$ for z of filtration ≥ 1. The characteristic class $\hat{\rho}(F)$ is its inverse. From the last expression in (B.11) we see that $\hat{\rho}(F)$ is defined for any real vector bundle F. To compute a formula for $\hat{\rho}(F)$, write $\ell_i = 1 - x_i$ and $\ell_i^{-1} = 1 - y_i$. Expand (B.11) using the binomial theorem, take the log, and write the result in terms of $s_i = x_i + y_i = x_i y_i$ using the Newton polynomials for $x_i^n + y_i^n$:

$$\log \hat{\rho}(F) = \sum_{i=1}^{r}\left(-\frac{1}{32}s_i - \frac{3}{1024}s_i^2 - \frac{5}{12288}s_i^3 + \cdots\right).$$

Note $\sum s_i$ is the reduced bundle $2r - F$, so has filtration ≥ 1. We then express power sums in s_i in terms of the elementary symmetric polynomials $p_1, p_2, \cdots$ in s_i and exponentiate:

$$(B.12) \qquad \begin{aligned} \hat{\rho}(F) = 1 &- \left(\frac{1}{32}p_1\right) + \left(\frac{3}{512}p_2 - \frac{5}{2048}p_1^2\right) \\ &+ \left(-\frac{5}{4096}p_3 + \frac{17}{16384}p_1 p_2 - \frac{21}{65536}p_1^3\right) + \cdots . \end{aligned}$$

Finally, we compute

$$\begin{aligned} p_1 &= 2r - F \\ p_2 &= (2r^2 - 3r) - (2r-2)F + \lambda^2 F \\ p_3 &= \left(\frac{4r^3 - 18r^2 + 20r}{3}\right) - (2r^2 + 7r - 5)F \\ &\quad + 2(r-2)\lambda^2 F - \lambda^3 F, \end{aligned}$$

so find

$$\begin{aligned} \hat{\rho}(F) = 1 &+ \frac{1}{2^5}[F - 2r] \\ &+ \frac{1}{2^{11}}\Big[7\lambda^2(F) - 5\operatorname{Sym}^2(F) + (24 - 4r)F + (4r^2 - 36r)\Big] \\ &+ \frac{1}{2^{16}}\Bigg[33\lambda^3(F) - 26R(F) + 21\operatorname{Sym}^3(F) \\ &\qquad - (14r - 184)\lambda^2 F + (10r - 136)\operatorname{Sym}^2(F) \\ &\qquad + (4r^2 + 1036r - 400)F \\ &\qquad - \left(\frac{8r^3 - 216r^2 + 1600r}{3}\right)\Bigg] + \cdots . \end{aligned} \tag{B.13}$$

Here $R(F)$ is the associated bundle to F which satisfies

$$F^{\otimes 3} \cong \operatorname{Sym}^3 F \oplus 2R(F) \oplus \lambda^3 F, \qquad F \otimes \lambda^2 F \cong \lambda^3 F \oplus R(F).$$

Note from Lemma B.10 and (B.12) that if $w_1(F) = w_2(F) = 0$, then the second term in (B.13) has filtration ≥ 4, the third term has filtration ≥ 8, etc.

For a finite dimensional manifold Y define $\hat{\rho}(Y) = \hat{\rho}(TY) \in KO(Y)[\frac{1}{2}]$.

Proposition B.14. *Let Y^{8n+4} be a spin manifold. Then $2^{4n+2}\hat{\rho}(Y)$ is the image in $KO[\frac{1}{2}](Y)$ of a canonical class $\lambda_n(Y) \in KO(Y)$.*

The class $\lambda_n(Y)$ is a KO-theoretic analog of a Wu class. It is defined for any spin manifold Y^d of dimension $\leq 8n+4$; apply Proposition B.14 to $Y^d \times \mathbb{R}^{8n+4-d}$.

Proof. By Poincaré duality (B.5) the functional

$$\begin{aligned} KO^0_c(Y; \mathbb{R}/\mathbb{Z}) &\longrightarrow KO^{-(8n+4)}(pt; \mathbb{R}/\mathbb{Z}) \cong \mathbb{R}/\mathbb{Z} \\ a &\longmapsto \int_Y \psi_2(a) \end{aligned} \tag{B.15}$$

is represented by a class $u^{4n+4}\lambda_n(Y) \in KO^{8n+8}(Y)$:

$$(B.16) \qquad \int_Y \psi_2(a) = \int_Y \lambda_n(Y) \cdot a, \qquad a \in KO^0_c(Y;\mathbb{R}/\mathbb{Z}).$$

Note that the existence of the functional (B.15) relies on Proposition B.8. Now integration in KO-theory is defined using an embedding $i\,Y \hookrightarrow \mathbb{R}^N$ with $N = (8n+4)+8k$ for some k; then

$$i_*\, KO^0(Y;\mathbb{R}/\mathbb{Z}) \longrightarrow KO^{8k}_c(\mathbb{R}^N;\mathbb{R}/\mathbb{Z}) \cong \mathbb{R}/\mathbb{Z}$$

is the integral. Let U be the KO Thom class of the normal bundle $\nu \to Y$ of Y in $\mathbb{R}^N$. Then we compute

$$\begin{aligned} i_*\psi_2(a) &= \psi_2(a)\cdot U \\ &= \psi_2\big(a\cdot\psi_{1/2}(U)\big) \\ &= 2^{-4k}a\cdot\psi_{1/2}(U) \qquad\qquad (B.17) \\ &= 2^{-4k}\left(a\cdot\frac{\psi_{1/2}(U)}{U}\right)\cdot U. \end{aligned}$$

In the first equation we pull $\psi_2(a)$ back to ν and extend $\psi_2(a)\,U$ to $\mathbb{R}^N$ using the fact that U has compact vertical support. In the second equation we regard U in KO with 2 inverted. In the third equation we use the fact that $KO^{8k}_c(\mathbb{R}^N)$ is generated by u^{-4k}. Thus from (B.7) we know that ψ_2 acts on $KO^{8k}_c(\mathbb{R}^N)$ and $KO^{8k}_c(\mathbb{R}^N;\mathbb{R})$ as multiplication by 2^{-4k}; it now follows from (B.9) that ψ_2 also acts on $KO^{8k}_c(\mathbb{R}^N;\mathbb{R}/\mathbb{Z})$ as multiplication by 2^{-4k}. Let V be the KO Thom class of $TY \to Y$. Since $TY \oplus \nu$ is trivial of rank N, we deduce

$$\frac{\psi_{1/2}(U)}{U}\,\frac{\psi_{1/2}(V)}{V} = \frac{\psi_{1/2}(u^N)}{u^N} = 2^{N/2}.$$

This is an equation in $KO(Y)$; the factors on the left-hand side are implicitly restricted to Y. Substituting into (B.17) we find

$$(B.18) \qquad i_*\psi_2(a) = a\cdot\left(2^{4n+2}\,\frac{V}{\psi_{1/2}(V)}\right)\cdot U = i_*\big(a\cdot 2^{4n+2}\hat{\rho}(V)\big).$$

Thus

$$\int_Y \psi_2(a) = \int_Y 2^{4n+2}\hat{\rho}(Y)\cdot a, \qquad a \in KO^0_c(Y;\mathbb{R}/\mathbb{Z}).$$

Comparing with (B.16), and using the Poincaré duality isomorphism (B.5), we deduce that the image of $\lambda_n(Y)$ in $KO[\frac{1}{2}](Y)$ is $2^{4n+2}\hat{\rho}(Y)$, as desired. q.e.d.

On a spin manifold of dimension $\leq 8n+3$, the class λ_n is canonically divisible by 2. (Compare with a similar assertion about Wu classes in [28].)

Proposition B.19. *Let X^{8n+3} be a spin manifold. Then there is a canonically associated class $\mu_n(X) \in KO(X)$ with $2\mu_n(X) = \lambda_n(X)$.*

The proposition applies to manifolds of dimension $< 8n+3$ by taking the product with a vector space as before.

Proof. The operation λ^2 loops to an operator $\Omega\lambda^2$ on $KO^{-1}(X)$. It is linear since products of suspended classes vanish. Similarly, there is a linear operator $\Omega\lambda^2$ on $KO^{-1}(X;\mathbb{R}/\mathbb{Z})$ compatible with $\Omega\lambda^2$ on $KO^{-1}(X;\mathbb{R})$ and λ^2 on $KO^0(X)$ in the long exact sequence. From (B.6) we have

$$2\,\Omega\lambda^2 = -\psi_2. \tag{B.20}$$

Now Poincaré duality (B.5) implies that the linear functional

$$\begin{aligned} KO_c^{-1}(X;\mathbb{R}/\mathbb{Z}) &\longrightarrow KO^{-(8n+4)}(pt;\mathbb{R}/\mathbb{Z}) \cong \mathbb{R}/\mathbb{Z} \\ a &\longmapsto \int_X \Omega\lambda^2(a) \end{aligned}$$

is represented by a class $-u^{4n+4}\mu_n(X) \in KO^{8n+8}(X)$:

$$\int_X \Omega\lambda^2(a) = \int_X -\mu_n(X)\cdot a, \qquad a \in KO_c^{-1}(X;\mathbb{R}/\mathbb{Z}). \tag{B.21}$$

From (B.20) and (B.16) we have

$$\int_X 2\,\Omega\lambda^2(a) = \int_X -\psi_2(a) = \int_X -\lambda_n(X)\cdot a. \tag{B.22}$$

Comparing (B.22) and (B.21) we deduce $2\mu_n(X) = \lambda_n(X)$. q.e.d.

Turning to *differential* KO-theory we have the following.

Proposition B.23. *Let Y^{8n+4} be a Riemannian spin manifold. Then there is a canonical lift $\check{\lambda}_n(Y) \in (KO\ )^0(Y)$ of $\lambda_n(Y)$ such that*

$$\int_Y \psi_2(\check{a}) = \int_Y \check{\lambda}_n(Y)\cdot\check{a} \quad \in \mathbb{R}/\mathbb{Z}$$

for all $\check{a} \in (KO\ )^1_c(Y)$.

The proof is parallel to the proof of Proposition B.14. It relies on Poincaré duality for $KO\ $, which on an n-dimensional Riemannian spin manifold Y states that there is an "almost perfect" pairing

$$\begin{aligned} (KO\ )_c^{q+1}(Y) \quad &\otimes (KO\ )^{n+4-q}(Y) &&\longrightarrow \mathbb{R}/\mathbb{Z} \\ \breve{a} \qquad &\otimes \qquad \breve{b} &&\longmapsto \int_Y \breve{a}\cdot\breve{b}. \end{aligned}$$

Note that the integral lands in $(KO\ )^5(pt) \cong KO^4(pt;\mathbb{R}/\mathbb{Z}) \cong \mathbb{R}/\mathbb{Z}$. This duality combines the topological duality (B.5) with a duality on differential forms—see (1.8) and (B.18). The "almost perfect" refers to the fact that the dual of a differential form is a de Rham current. Thus the application of $KO\ $-theory Poincaré duality to the functional $\breve{a} \mapsto \int_Y \psi_2(\breve{a})$, $\breve{a} \in (KO\ )_c^1(Y)$ only gives a distributional class $\check{\lambda}_n(Y)$. Then the computation of its image in $\big(KO[\frac12]\big)(Y)$, parallel to the computation in the proof of Proposition B.14, shows that its curvature is in fact smooth.

There is no *canonical* lift of $\mu_n(X)$ to a differential class on a Riemannian spin $(8n+3)$-manifold, though lifts do exist. We define suitable lifts below.

Specialize to $n = 1$, so to the classes $\lambda(Y) = \lambda_1(Y) \in KO^0(Y)$ and $\check{\lambda}(Y) = \check{\lambda}_1(Y) \in (KO\ )^0(Y)$ canonically associated to a Riemannian spin 12-manifold Y. We also have the topological class $\mu(X) = \mu_1(X) \in KO^0(X)$ canonically associated to a spin 11- or 10-manifold X. Note by (B.13) that

$$\lambda(Y) = 2(TY + 22) + (\text{classes of filtration} \geq 8) \qquad \in KO[\tfrac12]^0(Y);$$

there are similar equations for $\check{\lambda}(Y)$ and $\mu(X)$ after inverting 2.

Lemma B.24. *Let Y be a Riemannian spin 12-manifold and X a Riemannian spin 11- or 10-manifold. Then* (without inverting 2)

$$\lambda(Y) = 2(TY + 22) + \epsilon_1 \tag{B.25}$$

$$\check{\lambda}(Y) = 2(\check{T}Y + 22) + \check{\epsilon}_1 \tag{B.26}$$

$$\mu(X) = (TX + 22) + \epsilon_2 \tag{B.27}$$

where $\epsilon_1, \check{\epsilon}_1, \epsilon_2$ have filtration ≥ 8.

Here $\check{T}Y \in (KO\ )^0(Y)$ is the class of the tangent bundle of Y with its Levi-Civita connection. Also, we induce a filtration on $KO\ (X)$ from the Atiyah-Hirzebruch filtration on $KO(X)$ via the characteristic class $KO\ (X) \to KO(X)$.

Proof. The first assertion (B.25) is equivalent to

$$f(Y,a) := \int_Y \psi_2(a) - 2(TY+22)a = 0, \qquad a \in KO^0(Y;\mathbb{R}/\mathbb{Z}), \quad \text{filtration}\, a \geq 8. \tag{B.28}$$

(Note that an element of $KO^0(Y;\mathbb{R}/\mathbb{Z})$ of filtration ≥ 5 has filtration ≥ 8.) There is a similar rewriting of the other two assertions. As a first step we argue that it suffices to assume that Y is compact. Namely, using a proper Morse function we can find a compact manifold with boundary Y' contained in Y such that the support of a lies in the interior of Y'; then replace Y by the double of Y'. This does not change the value of (B.28). Then our proof of (B.28) relies on the fact that $f(Y,a)$ depends only on the bordism class of (Y,a): If $Y^{12} = \partial Z^{13}$ for a compact spin manifold Z, and a extends to a class on Z, then (B.28) vanishes. Since $f(Y,0) = 0$ it suffices to consider the bordism class of $(Y,a)-(Y,0)$. The spectrum which classifies such reduced pairs is $M\,\mathrm{Spin}\wedge BO\langle 8\rangle$; an element of $\pi_n(M\,\mathrm{Spin}\wedge BO\langle 8\rangle)$ represents a spin n-manifold which bounds together with a class in KO^0 of filtration ≥ 8. One computes

$$\begin{aligned} \pi_{10}(M\,\mathrm{Spin}\wedge BO\langle 8\rangle) &\cong \mathbb{Z}/2\mathbb{Z}, \\ \pi_{11}(M\,\mathrm{Spin}\wedge BO\langle 8\rangle) &= 0, \\ \pi_{12}(M\,\mathrm{Spin}\wedge BO\langle 8\rangle) &\cong \mathbb{Z}\times\mathbb{Z}, \\ \pi_{13}(M\,\mathrm{Spin}\wedge BO\langle 8\rangle) &= 0. \end{aligned} \tag{B.29}$$

From these facts one deduces that the reduced bordism group of pairs (Y,a) with $a \in KO^0(Y;\mathbb{R}/\mathbb{Z})\langle 8\rangle$ is isomorphic to $\mathbb{R}/\mathbb{Z}\times\mathbb{R}/\mathbb{Z}$. In particular, it is arbitrarily divisible. Since $f(Y,32b) = 0$ for all b (see (B.13)), and any $(Y,a)-(Y,0)$ is bordant to $(Y,32b)-(Y,0)$ for some b by the divisibility, we obtain the desired result $f(Y,a) = 0$.

The proof of (B.27) is similar; the relevant bordism group of reduced pairs $(X,a)-(X,0)$, $a \in KO^{-1}(X;\mathbb{R}/\mathbb{Z})\langle 8\rangle$ is again isomorphic to $\mathbb{R}/\mathbb{Z}\times \mathbb{R}/\mathbb{Z}$.

For (B.26) we must show

$$g(Y,\check{a}) := \int_Y \psi_2(\check{a}) - 2(\check{T}Y+11)\cdot\check{a}$$

vanishes for all $\check{a} \in (KO\ )^1(Y)$ of filtration ≥ 8. From the exact sequence (1.7), the topological result (B.25), and the fact that differential forms are divisible we conclude that $g(Y,\check{a})$ depends only on the characteristic class $[a] \in KO^1(Y)$ of $\check{a}$. But the assertion about π_{11} in (B.29) implies that $(Y,[a])-(Y,0)$ vanishes in the appropriate reduced bordism group, whence g vanishes.

We remark that Lemma B.24 follows formally from the stronger Lemma B.36 below using more bordism theory. q.e.d.

The following definition is analogous to the definition of a square root of the canonical bundle of a Riemann surface.

Definition B.30. Let X be a Riemannian spin 11- or 10-manifold. Then a *$\check{\mu}$-structure* on X is a class $\check{\mu}(X) \in (KO\ )^0(X)$ such that

(i) $2\check{\mu}(X) = \check{\lambda}(X)$;

(ii) The cohomology class of $\check{\mu}(X)$ is $\mu(X)$;

(iii) $\check{\mu}(X)$ differs from $\check{T}X + 22$ by an element of filtration ≥ 8.

The preceding shows that $\check{\mu}$-structures exist; differences of $\check{\mu}$-structures are certain points of order 2 on the torus $KO^{-1}(X;\mathbb{R})/KO^{-1}(X)$.

Let $X \to T$ be a Riemannian spin fiber bundle with fibers closed manifolds of dimension 10. Recall from (3.40) the quadratic form

$$\begin{aligned} q = q_{X/T}\ \check{Z}^0_{KO}(X) &\longrightarrow \check{Z}^2_{KSp}(T) \\ \check{a} &\longmapsto \int_{X/T} u^6\, \lambda^2(\check{a}) \end{aligned} \tag{B.31}$$

which refines the bilinear form

$$\begin{aligned} B = B_{X/T}\ \check{Z}^0_{KO}(X)\ \times \check{Z}^0_{KO}(X) &\longrightarrow \check{Z}^2_{KSp}(T) \\ \check{a}\ \times\ \check{a}' &\longmapsto \int_{X/T} u^6\, \check{a}\cdot\check{a}'. \end{aligned}$$

Note that $q(0) = 0$. The quadratic form q does not necessarily have a symmetry; we restrict to fiber bundles for which a symmetry exists for the pfaffian.

Definition B.32. A *$\check{\lambda}$-structure* on a Riemannian spin fiber bundle $X \to T$ of closed 10-manifolds is a cocycle $\check{\lambda} = \check{\lambda}(X/T) \in \check{Z}^0_{KO}(X)$ and isomorphisms

$$\operatorname{pfaff} q(\check{a}) \cong \operatorname{pfaff} q(\check{\lambda} - \check{a}) \tag{B.33}$$

natural in $\check{a} \in \check{Z}^0_{KO}(X)$.

An easy computation shows that (B.33) is equivalent to natural isomorphisms

$$(B.34)\qquad \operatorname{pfaff}\int_{X/T}\psi_2(\check{a})\cong\operatorname{pfaff}\int_{X/T}\check{\lambda}\cdot\check{a}$$

together with an isomorphism pfaff $q(\check{\lambda})\cong 0$. (Note (B.34) implies pfaff $2q(\check{\lambda})\cong 0$.) Also, Proposition B.23 implies that the equivalence class of the restriction of $\check{\lambda}$ to the fiber is canonically determined. A computation parallel to that in the proof of Proposition B.14, now for fiber bundles and in differential KO, computes the image of $\check{\lambda}$ in $\left(KO[\frac{1}{2}]\right)^0(X)$ as

$$(B.35)\qquad \check{\lambda}=2\check{\mathcal{T}}+\frac{1}{32}\left(7\lambda^2(\check{T})-5\operatorname{Sym}^2(\check{T})+4\check{T}-80\right)+\cdots,$$

where $\check{T}=\check{T}(X/T)$ and $\check{\mathcal{T}}=\check{T}+22$. More precisely, analogous to Lemma B.24 we have the following.

Lemma B.36. *$\check{\lambda}\cong 2\check{\mathcal{T}}$ modulo cocycles of filtration ≥ 8.*

Proof. As in the proof of Lemma B.24 we must show

$$\operatorname{pfaff}\int_{X/T}\psi_2(\check{a})\cong\operatorname{pfaff}\int_{X/T}2(\check{\mathcal{T}}+22)\cdot\check{a}$$

for all $\check{a}$ of filtration ≥ 8. It follows from (B.29) that the universal family of spin 10-manifolds (up to bordism) together with a class in KO of filtration ≥ 8 is simply-connected. Thus to prove (B.39)—an isomorphism of circle bundles with connection—it suffices to prove that some powers are isomorphic over the universal parameter space, since there are no flat circle bundles there (cf. (1.7) for $\check{H}^2$). But this follows from (B.35). q.e.d.

As for a single manifold, there is no canonical division of $\check{\lambda}$ by 2, so no canonical center for q. We restrict to fiber bundles for which a center exists.

Definition B.37. A *$\check{\mu}$-structure* on a Riemannian spin fiber bundle $X\to T$ with $\check{\lambda}$-structure is a cocycle $\check{\mu}=\check{\mu}(X/T)\in\check{Z}^0_{KO}(X)$ and an isomorphism $2\check{\mu}\cong\check{\lambda}$ such that $\check{\mu}=\check{\mathcal{T}}$ modulo terms of filtration ≥ 8.

These are the fiber bundles used in §3. We leave to the future an investigation of existence and uniqueness questions for $\check{\lambda}$- and $\check{\mu}$-structures.

Next, we prove a fact used in (3.39).

Proposition B.38. *Let $X \to T$ be a fiber bundle with a $\check{\mu}$-structure, as in Definition B.37. Then*

$$\operatorname{pfaff} q(\check{\mu}) = \operatorname{pfaff} q(\check{\mathcal{T}}) = \operatorname{pfaff} \int_{X/T} \bigwedge\nolimits^2 \check{\mathcal{T}}.$$

Proof. Set $\check{\epsilon} = \check{\mu} - \check{\mathcal{T}}$. Then $\check{\epsilon}$ has filtration ≥ 8, whence $\operatorname{pfaff} B(\check{\epsilon}, \check{\epsilon}) = 0$. Thus

$$q(\check{\mathcal{T}}) = q(\check{\mu} - \check{\epsilon}) = q(\check{\mu}) - q(\check{\epsilon}) + B(\check{\epsilon}, \check{\epsilon} - \check{\mu}),$$

so it suffices to prove

$$(B.39) \qquad \operatorname{pfaff} q(\check{\epsilon}) \cong \operatorname{pfaff} B(\check{\epsilon}, -\check{\mu}).$$

In fact, (B.39) holds for *any* class $\check{\epsilon}$ of filtration ≥ 8. As in the proof of Lemma B.36, we must only prove some power of (B.39). Now

$$\operatorname{pfaff} B(\check{\epsilon}, -2\check{\mu}) \cong \operatorname{pfaff} B(\check{\epsilon}, \check{\epsilon} - 2\check{\mu}) \cong \operatorname{pfaff} B(\check{\epsilon}, \check{\epsilon} - \check{\lambda}),$$

and from (B.33) or (B.34) it follows that this is isomorphic to $\operatorname{pfaff} 2q(\check{\epsilon})$, which gives the square of (B.39). q.e.d.

We now prove (3.55), which we restate as follows.

Proposition B.40. *Let $X \to T$ be a fiber bundle of 10-manifolds with Riemannian, spin, and $\check{\mu}$-structures, and $W \to T$ a fiber bundle of 2-dimensional spin submanifolds. Denote the inclusion map as $i : W \hookrightarrow X$. Then for $\check{q} \in \check{Z}^0_{KO}(W)$,*

$$(B.41) \qquad q(u^{-4}\, i_* \check{q}) = u^2 \int_W \Delta^+(\nu) \cdot \lambda^2(\check{q}) \;-\; \Delta^-(\nu) \cdot \operatorname{Sym}^2(\check{q}),$$

where $\nu \to W$ is the normal bundle and $\Delta^{\pm}$ are the half-spin bundles.

Proof. Quite generally, for any manifold W let $\pi\, \nu \to W$ be a rank 8 real spin bundle[40] and $Q \to W$ a real vector bundle of rank r. Denote the zero section of ν as $iW \hookrightarrow \nu$. We first compute the element $x \in KO^0(W)$ defined by

$$(B.42) \qquad x := u^4\, \pi_* \lambda^2(u^{-4} i_* Q).$$

We claim that

$$(B.43) \qquad x = \Delta^+(\nu) \cdot \lambda^2(Q) - \Delta^-(\nu) \cdot \operatorname{Sym}^2(Q).$$

[40]The computation holds for any even rank over the complexes. For rank 8 the half-spin bundles associated to ν are real; for rank 4 they are quaternionic.

Let $U \in KO^8_{\text{cv}}(\nu)$ be the Thom class. Then (B.42) implies

$$U \cdot \pi^* x = u^4 \, \lambda^2\big(U \cdot \pi^*(u^{-4}Q)\big).$$

Apply i^* to conclude

$$(B.44) \qquad i^*U \cdot x = u^4 \, \lambda^2(i^*U \cdot u^{-4}Q).$$

This equation, and its solution (B.43), may be viewed as equations in the representation ring $RSpin_8 \times RSO_r$; the corresponding relations in $KO^0(W)$ then follow by passing to the principal bundles underlying ν, Q and the vector bundles associated to the representations. Note (B.44) is an equation of real representations, but we prove it by working in the complex representation ring. To compute the right-hand side of (B.44) we use the Adams operation ψ_2 in the representation ring. Use the splitting principle—i.e., restrict to the maximal torus of $Spin_8$—to write $\nu \otimes \mathbb{C} = \bigoplus_{i=1}^{4} (\ell_i \oplus \ell_i^{-1})$. From (B.4) we compute (the factors of u cancel)

$$\begin{aligned} \psi_2(i^*U \cdot u^{-4}Q) &= \prod_{i=1}^{4} (\ell_i - \ell_i^{-1}) \cdot \psi_2(Q) \\ &= \prod_{i=1}^{4} (\ell_i^{1/2} - \ell_i^{-1/2}) \cdot (\ell_i^{1/2} + \ell_i^{-1/2}) \cdot \psi_2(Q) \\ &= \frac{i^*U}{u^4} \cdot \big[\Delta^+(\nu) + \Delta^-(\nu)\big] \cdot \psi_2(Q). \end{aligned}$$

Hence from (B.6)

$$(B.45) \qquad \begin{aligned} 2u^4 \, \lambda^2(i^*U \cdot u^{-4}Q) &= i^*U \cdot \Big\{\big[\Delta^+(\nu) - \Delta^-(\nu)\big] \cdot Q^2 \\ &\qquad - \big[\Delta^+(\nu) + \Delta^-(\nu)\big] \cdot \psi_2(Q)\Big\} \\ &= 2\, i^*U \cdot \big\{\Delta^+(\nu) \cdot \lambda^2(Q) - \Delta^-(\nu) \cdot \mathrm{Sym}^2(Q)\big\}, \end{aligned}$$

where we use $Q^2 = \lambda^2(Q) + \mathrm{Sym}^2(Q)$. We deduce the desired result (B.43) from (B.44) and (B.45) using the fact that the ring $RSpin_8 \times RSO_r$ has no zero divisors. Finally, this universal relation in the representation ring applies to bundles with connection, so to differential KO-theory, whence (B.41) holds. q.e.d.

Finally, we provide the proof of the delooping of ψ_2.

Proof of Proposition B.8. The construction is based on Atiyah's construction of Adams operations [2]. Start with $x \in KO^0(X)$ and square it, remembering the $\mathbb{Z}/2$-action, to get $P(x) \in KO^0_{\mathbb{Z}/2}(X)$. Since the group $\mathbb{Z}/2$ is not acting on X, there is an isomorphism

$$KO^0_{\mathbb{Z}/2}(X) \approx RO(\mathbb{Z}/2) \otimes KO^0(X),$$

where

$$RO(\mathbb{Z}/2) = \mathbb{Z}[t]/(t^2 - 1)$$

is the real representation ring of $\mathbb{Z}/2$. The Adams operation is the image of $P(x)$ in

$$\mathbb{Z} \underset{RO(\mathbb{Z}/2)}{\to} \otimes KO^0_{\mathbb{Z}/2}(X) = KO^0(X),$$

where the ring homomorphism $RO(\mathbb{Z}/2) \cong \mathbb{Z}[t]/(t^2 - 1) \to \mathbb{Z}$ sends t to -1.

This whole discussion would make sense for X a *spectrum*, provided we had an equivariant map $X \to X \wedge X$ to play the role of the diagonal. We'll define the operation ψ_2 on $KO^1(X)$ by defining it on $KO^0(S^{-1} \wedge X)$. So start with $x \in KO^0(S^{-1} \wedge X)$, form the equivariant external square, and restrict along the diagonal $X \to X \wedge X$ to define

$$P(x) \in KO^0_{\mathbb{Z}/2}(S^{-1} \wedge S^{-1} \wedge X).$$

For $a, b \geq 0$, let S^{a+bt} be the 1-point compactification of the representation of $\mathbb{Z}/2$ on $\mathbb{R}^a_{(+1)} \times \mathbb{R}^b_{(-1)}$. (The subscript indicates the eigenvalue of action of the non-trivial element of $\mathbb{Z}/2$.) By forcing the exponents to add under smash product, we define equivariant spectra S^{a+bt} for all $a, b \in \mathbb{Z}$. The shearing isomorphism implies the sphere $S^{-1} \wedge S^{-1}$ with the flip action is isomorphic to S^{-1-t}, so we can regard

$$P(x) \in KO^0_{\mathbb{Z}/2}(S^{-t} \wedge S^{-1} \wedge X).$$

We'll produce below, for any spectrum Y with trivial $\mathbb{Z}/2$ action, a canonical isomorphism

$$KO^0_{\mathbb{Z}/2}(S^{-t} \wedge Y) \approx KO^0(Y). \tag{$B.46$}$$

In particular, this gives an isomorphism

$$KO^0_{\mathbb{Z}/2}(S^{-1} \wedge S^{-1} \wedge X) \approx KO^0(S^{-1} \wedge X). \tag{$B.47$}$$

We then define

$$\psi_2(x) \in KO^0(S^{-1} \wedge X)$$

to be the image of $P(x)$ under (B.47).

To construct (B.46) consider the cofibration

$$S^0 \to S^t \to S^t \wedge \mathbb{Z}/2_+$$

in which the first map is map of suspension spectra gotten by suspending the inclusion of the fixed points. Smash this with S^{-t} to get

$$S^{-t} \to S^0 \to S^0 \wedge \mathbb{Z}/2_+.$$

Passing to equivariant KO-groups leads to a sequence

$$0 \to \mathbb{Z} \xrightarrow{1+t} RO(\mathbb{Z}/2) \to KO^0_{\mathbb{Z}/2}(S^{-t}) \to 0,$$

from which it follows that

$$\begin{aligned} KO^0_{\mathbb{Z}/2}(S^{-t}) &\cong RO(\mathbb{Z}/2)/(1+t) \cong \mathbb{Z} \\ KO^1_{\mathbb{Z}/2}(S^{-t}) &= 0. \end{aligned}$$

Smashing this sequence with Y then leads to a short exact sequence

$$0 \to KO^0(Y) \xrightarrow{1+t} RO(\mathbb{Z}/2) \otimes KO^0(Y) \to KO^0_{\mathbb{Z}/2}(S^{-t} \wedge Y) \to 0,$$

which gives the desired result (B.46).

It is useful to note that the map $S^{-t} \to S^0$ is also the one derived from the diagonal map $S^1 \to S^1 \wedge S^1$ in

$$(B.48) \qquad \left(S^{-1} \wedge_{\text{flip}} S^{-1}\right) \wedge S^1 \longrightarrow \left(S^{-1} \wedge S^1\right) \wedge_{\text{flip}} \left(S^{-1} \wedge S^1\right) = S^0$$

with the $\mathbb{Z}/2$ action as indicated.

Now suppose that $X = S^1 \wedge Y$. We need to show that the diagram

$$\begin{array}{ccc} KO^0(Y) & \xrightarrow{\psi_2} & KO^0(Y) \\ \downarrow & & \downarrow \\ KO^0\left(S^{-1} \wedge \left(S^1 \wedge Y\right)\right) & \xrightarrow[\psi_2]{} & KO^0\left(S^{-1} \wedge \left(S^1 \wedge Y\right)\right) \end{array}$$

commutes. The main thing to check is that the map

$$\left(S^{-1} \wedge_{\text{flip}} S^{-1}\right) \wedge S^1 \longrightarrow \left(S^{-1} \wedge S^1\right) \wedge_{\text{flip}} \left(S^{-1} \wedge S^1\right) = S^0,$$

derived from the diagonal map of S^1, leads to a factorization

$$\begin{array}{ccc} RO(\mathbb{Z}/2) & \longrightarrow & KO^0_{\mathbb{Z}/2}(S^{-1} \wedge S^{-1} \wedge S^1) \\ {\scriptstyle t \mapsto -1}\downarrow & & \downarrow{\scriptstyle \approx} \\ \mathbb{Z} & \longleftarrow & KO^0(S^{-1} \wedge S^1) \end{array}$$

in which the isomorphism labeled "$\approx$" is the one of (B.47) with $X = S^1$. But this follows immediately from the previous discussion, especially (B.48). q.e.d.

References

[1] A. Alekseev, A. Mironov & A. Morozov, *On B-independence of RR charges*, hep-th/0005244.

[2] M. F. Atiyah, *Power operations in K-theory*, Quart. J. Math. **17** (1966) 165–193.

[3] M. F. Atiyah & F. Hirzebruch, *Vector bundles and homogeneous spaces*, Proc. Symp. Pure Math. **3** (1961) 7–38.

[4] C. Bachas, M. Douglas & C. Schweigert, *Flux stabilization of D-branes*, JHEP **0005:048** (2000), hep-th/0003037.

[5] C. Bizdadea, L. Saliu & S. O. Saliu, *On Chapline-Manton couplings: a cohomological approach*, Phys. Scripta **61** (2000) 307–310, hep-th/0008022.

[6] J. M. Bismut, *The Atiyah-Singer index theorem for families of Dirac operators: two heat equation proofs*, Invent. Math. **83** (1986) 91–151.

[7] J. M. Bismut & D. S. Freed, *The analysis of elliptic families. I: Metrics and connections on determinant bundles*, Commun. Math. Phys. **106** (1986) 159–176; *The analysis of elliptic families. II: Dirac operators, eta invariants, and the holonomy theorem of Witten*, Commun. Math. Phys. **107** (1986) 103–163.

[8] A. Borel & J.-P. Serre, *Le théorème de Riemann-Roch*, Bull. Soc. Math. France **86** (1958) 97–136.

[9] W. Browder, *The Kervaire invariant of framed manifolds and its genearlization*, Ann. of Math. **90** (1969) 157–186.

[10] J.-L. Brylinski, *Loop spaces, characteristic classes and geometric quantization*, Birkhäuser, Boston, 1993.

[11] A. L. Carey, M. K. Murray, B. L. Wang, *Higher bundle gerbes and cohomology classes in gauge theories*, J. Geom. Phys. **21** (1997) 183–197, hep-th/9511169.

[12] G. F. Chapline, & N. S. Manton, *Unification of Yang-Mills theory and supergravity in ten dimensions*, Phys. Lett. B **120** (1983) 105–109.

[13] J. Cheeger & J. Simons, *Differential characters and geometric invariants*, Geom. Topology (College Park, Md., 1983/84), Lecture Notes in Math. Vol. 1167, Springer, Berlin, 1985, 50–80.

[14] Y.-K. E. Cheung & Z. Yin, *Anomalies, branes, and currents*, Nucl. Phys. **B517** (1998) 185-196, hep-th/9803931.

[15] S. Coleman, *Aspects of symmetry*, Cambridge Univ. Press, Cambridge, 1985.

[16] P. Deligne, *Théorie de Hodge. II*, Inst. Hautes Études Sci. Publ. Math. **40** (1971) 5–57.

[17] Quantum Fields and Strings: A Course for Mathematicians, (eds. P. Deligne, P. Etingof, D. S. Freed, L. C. Jeffrey, D. Kazhdan, J. W. Morgan, D. R. Morrison, E. Witten), Amer. Math. Soc., 1999, Providence, RI, Vol. 1,.

[18] P. Deligne & D. S. Freed, *Classical field theory*, Quantum Fields and Strings: A Course for Mathematicians, (eds. P. Deligne, P. Etingof, D. S. Freed, L. C. Jeffrey, D. Kazhdan, J. W. Morgan, D. R. Morrison, E. Witten), Amer. Math. Soc., 1999, Providence, RI, Vol. 1, 137–225.

[19] D. S. Freed, *Classical Chern-Simons theory*, Adv. Math. **113** (1995) 237–303, `hep-th/9206021`.

[20] ———, *On determinant line bundles*, Mathematical Aspects of String Theory, (ed. S. T. Yau) World Scientific Publishing, 1987.

[21] D. S. Freed & M. J. Hopkins, *On Ramond-Ramond fields and K-theory*, J. High Energy Phys. (2000) Paper 44, `hep-th/0002027`.

[22] D. S. Freed, J. A. Harvey, R. Minasian & G. Moore, *Gravitational anomaly cancellation for M-theory fivebranes*, Adv. Theor. Math. Phys. **2** (1998) 601–618, `hep-th/9803205`.

[23] D. S. Freed, E. Witten, *Anomalies in string theory with D-branes*, Asian J. Math. **3** (1999) 819–851; `hep-th/9907189`.

[24] P. Gajer, *Geometry of Deligne cohomology*, Invent. Math. **127** (1997) 155–207.

[25] M. B. Green, J. H. Schwarz, *Anomaly cancellations in supersymmetric $D = 10$ gauge theory and superstring theory*, Phys. Lett. B **149** (1984) 117–122.

[26] M. B. Green, J. H. Schwarz, E. Witten, *Superstring theory*, Vol. 2, Cambridge Univ. Press, Cambridge, 1987.

[27] N. J. Hitchin, *Lectures on special lagrangian submanifolds*, `math.DG/9907034`.

[28] M. J. Hopkins & I. M. Singer, *Quadratic functions in geometry, topology, and M-theory*, to appear.

[29] D. Kazhdan, *Introduction to QFT*, notes by Roman Bezrukavnikov, Quantum Fields and Strings: A Course for Mathematicians, (eds. P. Deligne, P. Etingof, D. S. Freed, L. C. Jeffrey, D. Kazhdan, J. W. Morgan, D. R. Morrison, E. Witten), Amer. Math. Soc., 1999, Providence, RI, Vol. 1, 377–418.

[30] J. Lott, *$\mathbb{R}/\mathbb{Z}$ index theory*, Comm. Anal. Geom. **2** (1994) 279–311.

[31] R. Minasian & G. Moore, *K-theory and Ramond-Ramond charge*, J. High Energy Phys. (1998) No. 11, Paper 2, 7, `hep-th/9710230`.

[32] G. Moore & E. Witten, *Self-duality, Ramond-Ramond fields, and K-theory*, J. High Energy Phys. **0005** (2000) 032, `hep-th/9912279`.

[33] D. Quillen, *Superconnections and the Chern character*, Topology **24** (1985) 89–95.

[34] L. H. Ryder, *Quantum field theory*, Cambridge Univ. Press, Cambridge, 1985.

[35] W. Taylor, *D2-branes in B fields*, JHEP **0007** (2000) 039, `hep-th/0004141`.

[36] F. W. Warner, *Foundations of differentiable manifolds and Lie groups*, Springer, New York, 1983.

[37] E. Witten, *Overview of K-theory applied to strings*, `hep-th/0007175`.

[38] ———, *Five-brane effective action in M-theory*, J. Geom. Phys. **22** (1997) 103–133, `hep-th/9610234`.

[39] ———, *D-branes and K-theory*, JHEP **812:019** (1998) `hep-th/9810188`.

[40] ———, *Duality realtions among topological effects in string theory*, JHEP **0005** (2000) 031, `hep-th/9912086`.

[41] ———, *Baryons and branes in anti de Sitter space*, JHEP **9807** (1998) 006, `hep-th/9805112`.

DEPARTMENT OF MATHEMATICS, UNIVERSITY OF TEXAS AT AUSTIN

SURVEYS IN DIFFERENTIAL GEOMETRY, 2000
Vol. VII ©2000, INTERNATIONAL PRESS
pp. 195-219

THE HOLOMORPHIC KERNEL OF THE RANKIN–SELBERG CONVOLUTION

DORIAN GOLDFELD & SHOUWU ZHANG

1. Introduction

Fix positive integers k, ℓ, N, D. Let $S_k(\Gamma_0(N))$ denote the $\mathbb{C}$–vector space of holomorphic cusp forms of weight k for the congruence subgroup

$$\Gamma_0(N) = \left\{ \begin{pmatrix} a & b \\ c & d \end{pmatrix} \in SL(2,\mathbb{Z}) \;\middle|\; c \equiv 0 \pmod{N} \right\}.$$

For a Dirichlet character ϵ of $(\mathbb{Z}/D\mathbb{Z})^\times$, let $M_\ell(\Gamma_0(D), \epsilon)$ denote the $\mathbb{C}$–vector space of holomorphic modular forms of weight ℓ with character ϵ for the congruence group $\Gamma_0(D)$

Let $f \in S_k(\Gamma_0(N))$ and $g \in M_\ell(\Gamma_0(D), \epsilon)$ have Fourier expansions of the form

$$f(z) = \sum_{n=1}^{\infty} a(n)\, n^{\frac{k-1}{2}} e^{2\pi i n z}, \qquad g(z) = b(0) + \sum_{n=1}^{\infty} b(n)\, n^{\frac{\ell-1}{2}} e^{2\pi i n z}.$$

Rankin and Selberg [4], [6] proved that the convolution L–function (in the case of equal weights $k = \ell$)

$$L(s, f \otimes g) = \sum_{n=1}^{\infty} \frac{a(n)b(n)}{n^s}$$

converges absolutely for complex s with $Re(s) > 1$, has a meromorphic continuation in s with at most a simple pole at $s = 1$, and satisfies a functional equation $s \to 1 - s$. This result was later generalized [L] to

First printed in Asian Journal of Mathematics, 1999. Used by permission.

more general situations, and in particular, to arbitrary pairs of weights k, ℓ.

The proof of the meromorphic continuation and the functional equation of$L(s, f \otimes g)$ was obtained by expressing $L(s, f \otimes g)$ as an inner product of $f \cdot g$ with a nonholomorphic Eisenstein series. We shall give a new proof of this result which does not use Eisenstein series at all, but instead expresses the Rankin–Selberg convolution L–function as an inner product of f with a holomorphic kernel function which depends on g and s. The main result of the paper is the Fourier expansion of the kernel function (when D is squarefree) which is given in Theorem 6.5. In the case where ϵ is a quadratic Dirichlet character $\pmod{D}$, a simpler and more explicit version of this result is given in Theorem 9.1. The functional equation of the kernel is stated and proved in various important cases in sections §10, §11.

In the special case that g is a theta function attached to the imaginary quadratic extension $\mathbb{Q}(\sqrt{-D})$, the value of the holomorphic kernel function (or its derivative) at $s = \frac{1}{2}$ coincides with the kernel function computed by Gross and Zagier [1] in their celebrated formula relating the derivative of an L–function of an elliptic curve with the height of a certain Heegner point. Thus, our method simultaneously gives a new simplified proof of the L–value computation in the Gross–Zagier formula together with a new proof of the meromorphic continuation and functional equation of the Rankin–Selberg convolution. The original method of Gross–Zagier used non–holomorphic Eisenstein series defined on a smaller group. The kernel was then obtained by a trace map and a holomorphic projection. In our method, all calculations are done directly on $S_k(\Gamma_0(N))$ and it is not necessary to go outside the holomorphic space with different level and then project back in later.

2. Poincaré series

For $\gamma = \begin{pmatrix} a & b \\ c & d \end{pmatrix} \in SL(2, \mathbb{Z})$ let

$$j(\gamma, z) = cz + d$$

denote the one–cocycle which satisfies $j(\gamma \cdot \gamma', z) = j(\gamma, \gamma' z) \cdot j(\gamma', z)$ for all matrices $\gamma, \gamma' \in SL(2, \mathbb{Z})$. Fix positive integers m, k. For $Re(s) > 1 - \frac{k}{2}$, the series

$$P_m(z, s) = m^{\frac{k-1}{2}} \sum_{\gamma \in \Gamma_\infty \backslash \Gamma_0(N)} e^{2\pi i m \gamma z} \, j(\gamma, z)^{-k} \, (\mathrm{Im} \gamma z)^s .$$

converges absolutely and uniformly to an automorphic form of weight k on $\Gamma_0(N)$. This series was first introduced by Selberg [7] and shown to have a meromorphic continuation to the entire complex s–plane. We define the holomorphic Poincaré series

$$P_m(z) = \lim_{s\to 0} P_m(z,s)$$

by analytic continuation.

Fourier Expansion. The Fourier coefficients $p_m(n)$ of

$$P_m(z) = \sum_{n=1}^{\infty} p_m(n)\, n^{\frac{k-1}{2}} e^{2\pi i n z}$$

are given by the formula (see [5]),

$$p_m(n) = \delta_{m,n} + 2\pi i^k \sum_{\substack{c=1 \\ c\equiv 0 \pmod N}}^{\infty} \frac{S(m,n;c)}{c} J_{k-1}\left(\frac{4\pi\sqrt{mn}}{c}\right), \tag{2.1}$$

where $\delta_{m,n}$ (Kronecker's delta function) is 1 if $m = n$ and zero otherwise; $S(m,n;c)$ is the Kloosterman sum

$$S(m,n;c) = \sum_{ad\equiv 1 \pmod c} e^{\frac{2\pi i}{c}(ma+nd)}, \tag{2.2}$$

and

$$J_{k-1}(y) = \frac{1}{2\pi i} \int_{\epsilon-\frac{k-1}{2}-i\infty}^{\epsilon-\frac{k-1}{2}+i\infty} \frac{\Gamma\left(\frac{k-1}{2}+w\right)}{\Gamma\left(\frac{k+1}{2}-w\right)} \left(\frac{y}{2}\right)^{-2w} dw \tag{2.3}$$

is the Bessel function.

Petersson Formula. By unfolding the integral, one can show that for any cusp form

$$h(z) = \sum_{m=1}^{\infty} c(m) m^{\frac{k-1}{2}} e^{2\pi i m z} \in S_k(\Gamma_0(N)),$$

we have

$$c(m) = \frac{(4\pi)^{k-1}}{(k-2)!} < h, P_m > . \tag{2.4}$$

3. Outline of the method

Let $(\cdot,\cdot)$ denote the Weil–Petersson inner product on $S_k(\Gamma_0(N))$. Now, fix $s \in \mathbb{C}$ and

$$g = b(0) + \sum_{n=1}^{\infty} b(n)\, n^{\frac{\ell-1}{2}} e^{2\pi i n z} \quad \in \quad M_\ell(\Gamma_0(D), \epsilon).$$

Consider the linear map

$$f \longrightarrow L(s, f \otimes g) \ = < \Phi_{s,g}, f > \tag{3.1}$$

for a unique holomorphic Riesz kernel (cusp form) $\Phi_{s,g} \in S_k(\Gamma_0(N))$ with Fourier expansion

$$\Phi_{s,g}(z) = \sum_{n=1}^{\infty} \phi_{s,g}(n) n^{\frac{k-1}{2}} e^{2\pi i n z}. \tag{3.2}$$

We now use the properties of the Poincaré series (Fourier expansion and Petersson formula) to obtain a formula for the Fourier coefficients of the Riesz kernel $\Phi_{s,g}$ given in formula (3.2).

Set $h = \Phi_{s,g}$ in formula (2.4). It immediately follows from (3.1) that

$$\phi_{s,g}(m) = \frac{(4\pi)^{k-1}}{(k-2)!} L(s, P_m \otimes g). \tag{3.3}$$

By the Fourier expansion (2.1) for the Poincaré series, we have for complex s with $Re(s) > 1 + \frac{k-1}{2}$ that

$$L(s, P_m \otimes g) = b(m) m^{-s} + 2\pi i^k\, T_m(s) \tag{3.4}$$

where

$$T_m(s) = \sum_{\substack{c=1 \\ N|c}}^{\infty} \sum_{n=1}^{\infty} \frac{S(m,n:c)}{c} \frac{b(n)}{n^s}$$

$$\cdot \frac{1}{2\pi i} \int_{\epsilon - \frac{k-1}{2} - i\infty}^{\epsilon - \frac{k-1}{2} + i\infty} \frac{\Gamma(\frac{k-1}{2} + w)}{\Gamma(\frac{k+1}{2} - w)} \left(\frac{2\pi\sqrt{mn}}{c} \right)^{-2w} dw$$

$$= \sum_{\substack{c=1 \\ N|c}}^{\infty} \sum_{r \in (\mathbb{Z}/c\mathbb{Z})^\times} e^{\frac{2\pi i m r}{c}}$$

$$\cdot \frac{1}{2\pi i} \int_{\epsilon - \frac{k-1}{2} - i\infty}^{\epsilon - \frac{k-1}{2} + i\infty} \frac{\Gamma(\frac{k-1}{2} + w)}{\Gamma(\frac{k+1}{2} - w)} \frac{(2\pi\sqrt{m})^{-2w}}{c^{1-2w}} L_g\left(s + w, \frac{\bar{r}}{c}\right) dw, \tag{3.5}$$

and

$$L_g\left(s, \frac{r}{c}\right) = \sum_{n=1}^{\infty} b(n) e^{\frac{2\pi i n r}{c}} n^{-s}$$

with $r\bar{r} \equiv 1 \pmod{c}$.

In the remainder of this section we briefly illustrate our method in the special case $N = D = 1$, $k = \ell$, and g is a cusp form. Complete details for the more general case are given in §4 through §10.

Our assumptions imply that in this case $L_g(s, \frac{a}{c})$ has holomorphic continuation to all $s \in \mathbb{C}$ and satisfies the functional equation (see Proposition 4.2)

$$L_g(s, \frac{a}{c}) = i^{-k} \left(\frac{c}{2\pi}\right)^{1-2s} \frac{\Gamma\left(\frac{k+1}{2} - s\right)}{\Gamma\left(\frac{k-1}{2} + s\right)} L_g\left(1 - s, -\frac{\bar{a}}{c}\right)$$

where $\bar{a}$ is the inverse of $a \pmod{c}$. If we apply this functional equation to the formula for $T_m(s)$, given in (3.5), we obtain

$$T_m(s) = \frac{(2\pi)^{2s-1}}{i^k} \sum_{n=1}^{\infty} \frac{b(n)}{n^{1-s}} S(s, m-n)\, I_s\left(\frac{m}{n}\right)$$

where

$$S(s, B) = \sum_{c=1}^{\infty} \frac{1}{c^{2s}} \sum_{r \in (\mathbb{Z}/c\mathbb{Z})^\times} \exp\left(\frac{2\pi i B}{c} r\right),$$

is the classical Ramanujan sum, and

$$I_s(y) = \frac{1}{2\pi i} \int_{\epsilon - \frac{k-1}{2} - i\infty}^{\epsilon - \frac{k-1}{2} + i\infty} \frac{\Gamma\left(\frac{k-1}{2} + w\right) \Gamma\left(\frac{k+1}{2} - s - w\right)}{\Gamma\left(\frac{k+1}{2} - w\right) \Gamma\left(\frac{k-1}{2} + s + w\right)} y^{-w}\, dw$$

is a hypergeometric function.

Formula for $S(s, B)$. The formula

$$S(s, B) = \frac{1}{\zeta(2s)} \sum_{d \mid B} d^{1-2s}$$

was first given by Ramanujan [3]. When $B = 0$,

$$S(s, 0) = \frac{\zeta(2s-1)}{\zeta(2s)} = \pi^{2s-3/2} \frac{\Gamma(1-s)}{\Gamma\left(s - \frac{1}{2}\right)} \frac{\zeta(2(1-s))}{\zeta(2s)}.$$

Formula for $I_s(x)$. We will show in Proposition 8.3 that

$$I_s(x) = \begin{cases} \frac{\Gamma(k-s)}{\Gamma(k)\Gamma(s)} x^{\frac{k-1}{2}} (1-x)^{s-1} F\left(1-s, s, k; \frac{x}{x-1}\right), & \text{if } 0 < x < 1 \\ \frac{\Gamma(k-s)}{\Gamma(s)} \frac{2^{2s-2}\Gamma\left(s-\frac{1}{2}\right)}{\sqrt{\pi}\Gamma(k+s-1)}, & \text{if } x = 1 \\ \frac{\Gamma(k-s)}{\Gamma(k)\Gamma(s)} x^{\frac{-k+1}{2}} (x-1)^{s-1} F\left(1-s, s, k; \frac{1}{1-x}\right), & \text{if } x > 1, \end{cases}$$

where $F(\alpha, \beta, \gamma; z)$ denotes the Gauss hypergeometric function defined for $|z| < 1$ by the absolutely convergent series

$$F(\alpha, \beta, \gamma; z) = 1 + \frac{\alpha \cdot \beta}{\gamma \cdot 1} z + \frac{\alpha(\alpha+1)\beta(\beta+1)}{\gamma(\gamma+1) \cdot 1 \cdot 2} z^2 + \cdots$$

and for all values of z by analytic continuation.

Combining these formulas we obtain:

Proposition 3.6. *Define $\gamma(s) = \frac{\Gamma(s)\Gamma(k)}{\Gamma(k-s)} \frac{\zeta(2s)}{(2\pi)^{2s}}$. Then we have*

$$\gamma(s) L(s, P_m \otimes g) = \sum_{n=1}^{\infty} b(n)\, \gamma(m, n; s),$$

where

$$\gamma(m, n; s) = \begin{cases} \left(\frac{n}{m}\right)^{\frac{k-1}{2}} \sigma^*_{1-2s}(m-n)\, F\left(1-s, s, k; \frac{n}{n-m}\right) & \text{if } n < m, \\ \gamma(s) m^{-s} + \gamma(1-s) m^{s-1} & \text{if } n = m, \\ \left(\frac{m}{n}\right)^{\frac{k-1}{2}} \sigma^*_{1-2s}(n-m)\, F\left(1-s, s, k; \frac{n}{m-n}\right) & \text{if } n > m, \end{cases}$$

and $\sigma^*_\nu(n) = n^{-\frac{\nu+1}{2}} \sum_{d|n} d^\nu$ for positive integers n and complex ν.

Note that Proposition (3.6) (for the group $\Gamma = \Gamma_0(1)$) is also easily obtained by the standard Rankin–Selberg method. By unfolding the Poincaré series P_m instead of the Eisenstein series $E(z,s)$ we obtain

$$\begin{aligned} &\int_{\Gamma\backslash\mathfrak{h}} y^k \overline{P_m(z)} g(z)\, E(z,s)\, \frac{dxdy}{y^2} \\ &\quad = m^{\frac{k-1}{2}} \int_{\Gamma\backslash\mathfrak{h}} y^k\, \overline{e^{2\pi i m z}}\, g(z)\, E(z,s)\, \frac{dxdy}{y^2} \\ &\quad = \sum_{n=1}^{\infty} (mn)^{\frac{k-1}{2}} b(n) \int_0^\infty y^{k-2}\, e^{-2\pi(m+n)y} e_{m-n}(s,y) dy, \end{aligned}$$

where $e_r(s,y)$ denotes the coefficient of $e^{2\pi irx}$ in the Fourier expansion of $E(z,s)$. The formula for these Fourier coefficients is well–known: $e_r(s,y)$ is the product of $\sigma^*_{1-2s}(|r|)$ and a simple analytic function of $|r|y$ for $r \neq 0$, and a linear combination of $\zeta(2s)y^s$ and $\zeta(2s-1)y^{1-s}$ for $r=0$. Substituting this into the above unfolding identity immediately gives (3.6).

Remarks. The expression for $\gamma(s)L(s, P_m \otimes g)$ (on the right hand side in Proposition (3.6)) is absolutely convergent for all s and each term is invariant under $s \to 1-s$ except the first two, which are interchanged, so one immediately deduces the meromorphic continuation and functional equation. It follows that $L(s, P_m \otimes g)$ is holomorphic everywhere except for a simple pole at $s=1$ with residue proportional to $b(m)$. The classical results of Rankin [R] and Selberg [S1] are immediately recovered.

The classical Rankin–Selberg proof is simpler than our new method if f and g are the same level and if g is a cusp form. Otherwise, unfolding P_m will force one to take the trace of gE first, and it will be necessary to truncate $\mathrm{Tr}(gE)$ in order to make the integral convergent. This is more complicated than our new method given here and is very close to the original Gross–Zagier method. Our method was discovered by trying to simplify the proof of the Gross–Zagier formula. In that case g is a theta function (not a cusp form) of different level than f and our method avoids taking the trace and doing a holomorphic projection.

The formula (3.6) may yield new applications. For example, the rapid convergence of this formula, and the fact that that it is true also for s outside the region of convergence of the original Dirichlet series $L(s, f \otimes g)$, might make it suitable for certain theoretical or computational applications. Also, the fact that $F(1-s, s, k; x)$ becomes a polynomial for integral values of s might be useful for obtaining new results, or new proofs of known results, about special values of $L(s, f \otimes g)$ at such arguments. It would also be of interest to see if our new method can be used to obtain higher convolutions of Rankin–Selberg type.

4. Functional equation for $L_g(s, \frac{r}{c})$

In this section we derive the functional equation for

$$L_g\left(s, \frac{r}{c}\right) = \sum_{n=1}^{\infty} b(n) e^{\frac{2\pi i n r}{c}} n^{-s}$$

where $g \in M_\ell(\Gamma_0(D), \epsilon)$. Here, we assume that ℓ is a fixed positive integer and that ϵ is a Dirichlet character of $(\mathbb{Z}/D\mathbb{Z})^\times$.

Let $\gamma = \begin{pmatrix} a & b \\ c & d \end{pmatrix}$ be a matrix with real entries and positive determinant. Given, $F(z)$ a holomorphic function on the upper half plane, define

$$F\big|_\gamma(z) = (ad - bc)^{\frac{\ell}{2}}(cz + d)^{-\ell} F(z)$$

which satisfies $F\big|_\gamma\big|_{\gamma'} = F\big|_{\gamma\gamma'}$.

Assume now that D is square free. Let $\epsilon = \prod_{p|D} \epsilon_p$ be the decomposition of ϵ. Set

$$\delta = \frac{D}{(c, D)}, \qquad \delta' = (c, D).$$

Since $(\delta, \delta') = 1$ it follows that there exist $x, y \in \mathbb{Z}$ such that $x\delta - y\delta' = 1$. Define a matrix w_δ by the formula

$$w_\delta = \begin{pmatrix} x & y \\ \delta' & \delta \end{pmatrix} \begin{pmatrix} \delta & 0 \\ 0 & 1 \end{pmatrix}.$$

Then w_δ normalizes the subgroup $\Gamma_0(D)$.

Define

$$g^\delta(z) = g\big|_{w_\delta}(z).$$

Then g^δ belongs to $M_\ell(\Gamma_0(D), \epsilon^\delta)$ where

$$\epsilon^\delta = \epsilon_\delta^{-1} \cdot \epsilon_{\delta'} = \prod_{p|\delta} \epsilon_p^{-1} \cdot \prod_{p|\delta'} \epsilon_p. \tag{4.1}$$

Proposition 4.2. *The function $L_g(s, \frac{a}{c})$ has a meromorphic continuation to the entire complex s–plane with simple poles at $s = \frac{\ell+1}{2}, \frac{1-\ell}{2}$ (with residue $-b(0)$ at $s = \frac{1-\ell}{2}$) and satisfies the functional equation*

$$L_g(s, \frac{a}{c}) = \epsilon\left(\frac{a}{c}\right)\left(\frac{\delta c^2}{4\pi^2}\right)^{\frac{1}{2}-s} \frac{\Gamma\left(\frac{\ell+1}{2} - s\right)}{\Gamma\left(\frac{\ell-1}{2} + s\right)} L_{g^\delta}(1 - s, -\frac{\bar{a}\bar{\delta}}{c})$$

where

$$\epsilon\left(\frac{a}{c}\right) = i^{-\ell}\epsilon_\delta\left(\frac{c}{\delta'}\right)\epsilon_{\delta'}^{-1}(a\delta)$$

and $\bar{a}$ is the inverse of a (mod c).

Proof. Since $(c, \delta) = 1$, there exists a matrix $\gamma = \begin{pmatrix} a & b \\ c & d \end{pmatrix} \in \mathrm{SL}(\mathbb{Z})$ with $\delta | d$. Write

$$\gamma' = \gamma \begin{pmatrix} x & y \\ \delta' & \delta \end{pmatrix}^{-1} = \begin{pmatrix} a\delta - b\delta' & -ay + bx \\ c\delta - d\delta' & -cy + dx \end{pmatrix}.$$

Then $\gamma' \in \Gamma_0(D)$ since $\delta|d$, $\delta'|c$, and $\delta\delta' = D$. Since

$$\gamma = \gamma' w_\delta \begin{pmatrix} \delta^{-1} & 0 \\ 0 & 1 \end{pmatrix},$$

we obtain

$$g\big|_\gamma(z) = \epsilon(-cy+dx)\delta^{-\frac{\ell}{2}} g^\delta\left(\frac{z}{\delta}\right).$$

Here

$$\epsilon(-cy+dx) = \epsilon_\delta(-cy)\epsilon_{\delta'}(dx) = \epsilon_\delta\left(\frac{c}{\delta'}\right)\epsilon_{\delta'}^{-1}(a\delta).$$

Write

$$\frac{az+b}{cz+d} = \frac{a}{c} - \frac{1}{c(cz+d)}$$

and make the substitution $z \to -\frac{1}{c^2 z} - \frac{d}{c}$. We have

$$g\left(\frac{a}{c}+z\right)(cz)^\ell = \epsilon_\delta\left(\frac{c}{\delta'}\right)\epsilon_{\delta'}^{-\ell}(a\delta)\delta^{-\frac{\ell}{2}} g^\delta\left(-\frac{1}{\delta c^2 z} + \frac{a'}{c}\right)$$

where $a' = -d/\delta$.

Let $L_g^*(s, \frac{a}{c})$ denote the Mellin transformation

$$L_g^*(s, \frac{a}{c}) = \int_0^\infty \left[g\left(\frac{a}{c}+iy\right) - b(0)\right] y^{\frac{\ell-1}{2}+s}\,\frac{dy}{y}$$

where $b(0) = 0$ if g is a cusp form. Then we have

$$L_g^*(s, \frac{a}{c}) = \frac{\Gamma(\frac{\ell-1}{2}+s)}{(2\pi)^{\frac{\ell-1}{2}+s}}\, L_g(s, \frac{a}{c}).$$

Now

$$L_g^*\left(s, \frac{a}{c}\right) = \left(\int_0^{\frac{1}{c\sqrt{\delta}}} + \int_{\frac{1}{c\sqrt{\delta}}}^\infty\right)\left[g\left(\frac{a}{c}+iy\right) - b(0)\right] y^{\frac{\ell-1}{2}+s}\,\frac{dy}{y}.$$

On the other hand, from the functional equation of g and g^δ, we have

$$\begin{aligned}
&\int_0^{\frac{1}{c\sqrt{\delta}}} \left[g\left(\frac{a}{c}+iy\right) - b(0)\right] y^{\frac{\ell-1}{2}+s}\,\frac{dy}{y} \\
&= A\int_0^{\frac{1}{c\sqrt{\delta}}} \left[g^\delta\left(-\frac{1}{\delta c^2 iy} + \frac{a'}{c}\right) - b(0)\right](ciy)^{-\ell} y^{\frac{\ell-1}{2}+s}\,\frac{dy}{y} \\
&\quad + Ab(0)(ci)^{-\ell}(c\sqrt{\delta})^{-s+\frac{\ell+1}{2}} \cdot \frac{1}{s-\frac{\ell+1}{2}} \\
&\quad - b(0)(c\sqrt{\delta})^{-s-\frac{\ell-1}{2}} \cdot \frac{1}{s+\frac{\ell-1}{2}}
\end{aligned}$$

where

$$A = \epsilon_\delta\left(\frac{c}{\delta'}\right)\epsilon_{\delta'}^{-1}(a\delta)\delta^{-\frac{\ell}{2}}.$$

If we make the substitution $y \to \frac{1}{c^2\delta y}$, we obtain

$$\begin{aligned}
&\int_0^{\frac{1}{c\sqrt{\delta}}}\left[g\left(\frac{a}{c}+iy\right)-b(0)\right]y^{\frac{\ell-1}{2}+s}\,\frac{dy}{y}\\
&\quad = A\int_{\frac{1}{c\sqrt{\delta}}}^{\infty}\left[g^\delta\left(\frac{a'}{c}+iy\right)-b(0)\right](ci)^{-\ell}(c^2\delta y)^{\frac{\ell+1}{2}-s}\,\frac{dy}{y}\\
&\quad + \;Ab(0)(ci)^{-\ell}(c\sqrt{\delta})^{-s+\frac{\ell+1}{2}}\cdot\frac{1}{s-\frac{\ell+1}{2}}\\
&\quad - \;b(0)(c\sqrt{\delta})^{-s-\frac{\ell-1}{2}}\cdot\frac{1}{s+\frac{\ell-1}{2}}.
\end{aligned}$$

$$\epsilon_\delta\left(\frac{c}{\delta'}\right)\epsilon_{\delta'}^{-1}(a\delta)\delta^{-\frac{\ell}{2}}\int_0^\infty g^\delta\left(-\frac{1}{\delta c^2 iy}+\frac{a'}{c}\right)(ciy)^{-\ell}y^{\frac{\ell-1}{2}+s}\,\frac{dy}{y}.$$

The functional equation

$$L_g^*\left(s,\frac{a}{c}\right) = i^{-\ell}\epsilon_\delta\left(\frac{c}{\delta'}\right)\epsilon_{\delta'}^{-1}(a\delta)(\delta c^2)^{1/2-s}L_{g^\delta}^*\left(1-s,\frac{a'}{c}\right)$$

and Proposition (4.2) immediately follow. q.e.d.

5. Generalized Ramanujan sums

In this section we fix a decomposition $D = \delta\cdot\delta'$ of the square–free integer D. For any integer A we decompose

(5.1) $$A = A_1A_2$$

so that A_1 is positive with prime factors dividing δ' and A_2 is prime to δ'.

Definition 5.2. We define

$$G(\delta) = \prod_{p\mid\delta'}\epsilon_p^{-1}\left(\frac{\delta'}{p}\right)\sum_{r\in(\mathbb{Z}/p\mathbb{Z})^\times}\epsilon_p^{-1}(r)\,e^{\frac{2\pi ir}{p}}.$$

Let c be a positive integer and $B\in\mathbb{Z}$. The sum

$$\sum_{r\in(\mathbb{Z}/c\mathbb{Z})^\times}\epsilon_{\delta'}(r)e^{\frac{2\pi iBr}{c}}$$

is a generalized Ramanujan sum. We evaluate it in the next lemma using the notation $e^x = \exp(x)$.

Lemma 5.3. *Let c, B positive integers with $c > 0$, $(c, D) = \delta'$, and $B \neq 0$. Set $c = c_1c_2$, $B = B_1B_2$ as in (5.1). Then the sum*

$$\sum_{r\in(\mathbb{Z}/c\mathbb{Z})^\times} \epsilon_{\delta'}(r)\exp\left(\frac{2\pi iB}{c}r\right)$$

is equal to (with $G(\delta)$ given in Definition 5.2)

$$G(\delta)B_1\epsilon_{\delta'}\left(\frac{B_2}{c_2}\right)\sum_{d|(c_2,B_2)}\mu\left(\frac{c_2}{d}\right)d$$

if $c_1 = B_1\delta'$; otherwise it is zero.

Proof. Let $c = \prod_{i=1}^s p_i^{n_i}$ be the prime decomposition of c with $n_i > 0$. Then every $r \in (\mathbb{Z}/c\mathbb{Z})^\times$ can be uniquely written as $\sum_i r_i(c/p_i^{n_i})$ with $r_i \in (\mathbb{Z}/p_i^{n_i}\mathbb{Z})^\times$. Since

$$\epsilon_{\delta'}\left(\sum r_ic/p_i^{n_i}\right) = \prod_{p_i|\delta'}\epsilon_{p_i}((c/p_i^{n_i})r_i),$$

one has

$$\sum_{r\in(\mathbb{Z}/c\mathbb{Z})^\times}\epsilon_{\delta'}(r)^{-1}\exp\left(\frac{2\pi iB}{c}r\right)$$
$$=\prod_{p_i|\delta'}\sum_{r\in(\mathbb{Z}/p_i^{n_i}\mathbb{Z})^\times}\epsilon_{p_i}^{-1}((c/p_i^{n_i})r)\exp\left(\frac{2\pi iB}{p_i^{n_i}}r\right)\cdot$$
$$\cdot\prod_{p_i\nmid\delta'}\sum_{r\in(\mathbb{Z}/p_i^{n_i}\mathbb{Z})^\times}\exp\left(\frac{2\pi iB}{p_i^{n_i}}r\right).$$

Let us evaluate the two products separately. If $p_i|\delta'$, then every element in $\mathbb{Z}/p_i^{n_i}\mathbb{Z}$ can be uniquely written as $r + tp_i$ with $r \in (\mathbb{Z}/p_i\mathbb{Z})^\times$ and $t \in \mathbb{Z}/p_i^{n_i-1}\mathbb{Z}$. It follows that

$$\sum_{r\in(\mathbb{Z}/p_i^{n_i}\mathbb{Z})^\times}\epsilon_{p_i}^{-1}((c/p_i^{n_i})r)\exp\left(\frac{2\pi iB}{p_i^{n_i}}r\right)$$
$$=\sum_{r\in(\mathbb{Z}/p_iF)^\times}\epsilon_{p_i}^{-1}((c/p_i^{n_i})r)\exp\left(\frac{2\pi iB}{p_i^{n_i}}r\right)\sum_{t\in\mathbb{Z}/p_i^{n_i-1}\mathbb{Z}}\exp\left(\frac{2\pi iB}{p_i^{n_i-1}}t\right).$$

If $\mathrm{ord}_{p_i}(B) < n_i - 1$, the last sum is zero; otherwise it is

$$p_i^{n_i-1} \sum_{r \in (\mathbb{Z}/p_i\mathbb{Z})^\times} \epsilon_{p_i}^{-1}((c/p_i^{n_i})r) \exp\left(\frac{2\pi i B}{p_i^{n_i}} r\right).$$

Again this sum is 0 if $\mathrm{ord}_{p_i}(B) \geq n_i$. Otherwise, replace r by $r(B/p_i^{n_i-1})^{-1} \pmod{p_i}$ to obtain

$$p_i^{n_i-1} \epsilon_{p_i}(Bp_i/c) \sum_{r \in (\mathbb{Z}/p_i\mathbb{Z})^\times} \epsilon_{p_i}^{-1}(r) \exp\left(\frac{2\pi i}{p_i} r\right).$$

It follows that

$$\prod_{p_i | \delta'} \sum_{r \in (\mathbb{Z}/p_i^{n_i}\mathbb{Z})^\times} \epsilon_{p_i}^{-1}((c/p_i^{n_i})r) \exp\left(\frac{2\pi i B}{p_i^{n_i}} r\right)$$

is nonzero only if $B_1\delta' = c_1$; in this case, it is equal to

$$B_1 \epsilon_{\delta'}(B_2/c_2) G(\delta).$$

Now, we assume that $p_i \not| \delta'$, then

$$\begin{aligned}
\sum_{r \in (\mathbb{Z}/p_i^{n_i}\mathbb{Z})^\times} \exp\left(\frac{2\pi i B}{p_i^{n_i}} r\right) &= \sum_{r \in \mathbb{Z}/p_i^{n_i}\mathbb{Z}} \exp\left(\frac{2\pi i B}{p_i^{n_i}} r\right) \\
&\quad - \sum_{r \in \mathbb{Z}/p_i^{n_i-1}\mathbb{Z}} \exp\left(\frac{2\pi i B}{p_i^{n_i}} rp\right) \\
&= \sum_{d | (B, p_i^{n_i})} \mu\left(\frac{p_i^{n_i}}{d}\right) d.
\end{aligned}$$

It follows that

$$\prod_{p_i | \delta'} \sum_{r \in (\mathbb{Z}/p_i^{n_i}\mathbb{Z})^\times} \exp\left(\frac{2\pi i B}{p_i^{n_i}} r\right) = \sum_{d | (c_2, B)} \mu\left(\frac{c}{d}\right) d.$$

q.e.d.

This completes the proof of Lemma 5.3.

6. The holomorphic kernel $\Phi_{s,g}$

We recall formulas (3.4), (3.5) which we now relabel as (6.1), 6.2).

$$L(s, P_m \otimes g) = b(m)m^{-s} + 2\pi i^k T_m(s), \tag{6.1}$$

$$\begin{aligned}
T_m(s) &= \sum_{\substack{c=1 \\ N|c}}^{\infty} \sum_{n=1}^{\infty} \frac{S(m,n:c)}{c} \frac{b(n)}{n^s} \\
&\quad \cdot \frac{1}{2\pi i} \int_{\epsilon-\frac{k-1}{2}-i\infty}^{\epsilon-\frac{k-1}{2}+i\infty} \frac{\Gamma(\frac{k-1}{2}+w)}{\Gamma(\frac{k+1}{2}-w)} \left(\frac{2\pi\sqrt{mn}}{c}\right)^{-2w} dw \\
&= \sum_{\substack{c=1 \\ N|c}}^{\infty} \sum_{r\in(\mathbb{Z}/c\mathbb{Z})^\times} e^{\frac{2\pi i m r}{c}} \\
&\quad \cdot \frac{1}{2\pi i} \int_{\epsilon-\frac{k-1}{2}-i\infty}^{\epsilon-\frac{k-1}{2}+i\infty} \frac{\Gamma(\frac{k-1}{2}+w)}{\Gamma(\frac{k+1}{2}-w)} \frac{(2\pi\sqrt{m})^{-2w}}{c^{1-2w}} L_g\left(s+w, \frac{\bar{r}}{c}\right) dw,
\end{aligned} \tag{6.2}$$

and

$$L_g\left(s, \frac{r}{c}\right) = \sum_{n=1}^{\infty} b(n) e^{\frac{2\pi i n r}{c}} n^{-s}$$

with $r\bar{r} \equiv 1 \pmod c$. Since $L_g\left(s, \frac{r}{c}\right)$ is holomorphic in s, formula (6.2) holds for all s.

In (6.2) we will apply the functional equation given in Proposition 4.2. The Mellin–Barnes integral (for $x > 0, s \in \mathbb{C}, Re(s) > 1$)

$$I_s(x) = \frac{1}{2\pi i} \int_{\epsilon-\frac{k-1}{2}-i\infty}^{\epsilon-\frac{k-1}{2}+i\infty} \frac{\Gamma\left(\frac{k-1}{2}+w\right)\Gamma\left(\frac{\ell+1}{2}-s-w\right)}{\Gamma\left(\frac{k+1}{2}-w\right)\Gamma\left(\frac{\ell-1}{2}+s+w\right)} x^{-w}\, dw \tag{6.3}$$

naturally appears. This integral is evaluated in Proposition 8.3. Further, the Kloosterman sums then turn into generalized Ramanujan sums (here $B \in \mathbb{Z}, s \in \mathbb{C}$ with $Re(s) > 1$)

$$S^\delta(s,B) = \sum_{\substack{c=1 \\ N|c (c,D)=\delta'}}^{\infty} \frac{\epsilon_\delta(c/\delta')}{c^{2s}} \sum_{r\in(\mathbb{Z}/c\mathbb{Z})^\times} \epsilon_{\delta'}^{-1}(r) \exp\left(\frac{2\pi i B}{c} r\right). \tag{6.4}$$

These sums are evaluated by Lemma 5.3 in Proposition 7.1. This is the key idea for obtaining the final formula for the holomorphic kernel as given in Theorem 6.5 which is the main Theorem of this paper.

Theorem 6.5. *Fix positive integers k, ℓ, N, D and $g(z) = b(0) + \sum_{n=1}^{\infty} b(n)\, n^{\frac{\ell-1}{2}} e^{2\pi i n z}$ in $M_\ell(\Gamma_0(D), \epsilon)$. Assume that D is square free, ϵ a Dirichlet character (mod D), and $s \in \mathbb{C}$. Then we have:*

(a) *The kernel function $\Phi_{s,g}(z)$ defined in (3.1), (3.2) has the Fourier expansion*

$$\Phi_{s,g}(z) = \frac{(4\pi)^{k-1}}{(k-2)!} \sum_{m=1}^{\infty} L(s, P_m \otimes g) e^{2\pi i m z}.$$

(b) *The function $L(s, P_m \otimes g)$ is given by*

$$L(s, P_m \otimes g) = b(m) m^{-s} + 2\pi i^k \sum_{\delta | D} T_m^\delta(s)$$

with

$$T_m^\delta(s) = i^{-\ell} \left(\frac{\delta}{4\pi^2}\right)^{\frac{1}{2}-s} \epsilon_{\frac{D}{\delta}}^{-2}(\delta) \sum_{n=1}^{\infty} \frac{b^\delta(n)}{n^{1-s}}\, S^\delta(s, m\delta - n)\, I_s\left(\frac{\delta m}{n}\right),$$

where $b^\delta(n)$ are fourier coefficients of g^δ defined in §3, $I_s(x)$ is the Mellin–Barnes integral (6.3), and $S^\delta(s, B)$ is the generalized Ramanujan sum 6.4.

Proof. It follows from the functional equation given in Proposition 4.2 that

$$\begin{aligned} &\frac{(2\pi\sqrt{m})^{-2w}}{c^{1-2w}} L_g\left(s + w, \frac{\bar{r}}{c}\right) \\ &\quad = \epsilon\left(\frac{a}{c}\right) \frac{(\delta m)^{-w}}{c^{2s}} \left(\frac{\delta}{4\pi^2}\right)^{\frac{1}{2}-s} \frac{L\left(\frac{\ell+1}{2} - s - w\right)}{L\left(\frac{\ell-1}{2} + s + w\right)} \\ &\qquad \cdot L_{g^\delta}\left(1 - s - w, -\frac{\bar{a}\bar{\delta}}{c}\right). \end{aligned}$$

If we use this identity in equation (6.2) and recall that

$$g^\delta(z) = \sum_{n \geq 0} b^\delta(n) n^{\frac{\ell-1}{2}} e^{2\pi i n z},$$

and

$$L_{g^\delta}\left(s, \frac{r}{c}\right) = \sum_{n=1}^{\infty} b^\delta(n) e^{\frac{2\pi i n r}{c}} n^{-s},$$

it follows that

$$T_m(s) = \sum_{\delta | D} T_m^\delta(s) \tag{6.6}$$

where

$$T_m^\delta(s) = i^{-\ell} \left(\frac{\delta}{4\pi^2}\right)^{\frac{1}{2}-s} \epsilon_{\delta'}^{-2}(\delta) \sum_{n=1}^{\infty} \frac{b^\delta(n)}{n^{1-s}}\, S^\delta(s, m\delta - n)\, I_s\left(\frac{\delta m}{n}\right). \tag{6.7}$$

q.e.d.

7. Evaluation of $S^\delta(s, B)$

As before, we work with a fixed decomposition $D = \delta \cdot \delta'$ of the square–free integer D and ϵ is a Dirichlet character (mod D). Recall the definition of ϵ^δ given in (4.1):

$$\epsilon^\delta = \epsilon_\delta^{-1} \cdot \epsilon_{\delta'} = \prod_{p|\delta} \epsilon_p^{-1} \cdot \prod_{p|\delta'} \epsilon_p.$$

For any given number e prime to D, let $L_e^\delta(s)$ denote the Dirichlet L-function

$$L_e^\delta(s) = \sum_{(n,De)=1} \frac{\epsilon^\delta(\frac{1}{n})}{n^s}.$$

When $e = 1$ we simply denote it by $L^\delta(s)$.

Proposition 7.1. *Let B be an integer with decomposition $B = B_1 B_2$ as in (5.1). Let $N = N_1 N_2$ as in (5.1). Define*

$$S_e^\delta(s, B) = \begin{cases} \frac{G(\delta)\epsilon_\delta(B_1)\epsilon_{\delta'}(B_2)}{(\delta')^{2s} B_1^{2s-1}} \sum_{\substack{d | \frac{B_2}{e} \\ (d,D)=1}} d^{1-2s} \epsilon^\delta\left(\frac{1}{d}\right). & B \neq 0,\ N | B_1 \delta' \\ L^\delta(2s-1) & B = 0,\ \delta = D,\ (N, D) = 1 \\ 0 & \text{otherwise} \end{cases}$$

Then

$$S^\delta(s, B) = \frac{\epsilon^\delta(1/N_2) N_2^{1-2s}}{L_{N_2}^\delta(2s)} \sum_{e | (B_2, N_2)} \mu\left(\frac{N_2}{e}\right) \frac{e}{N_2} S_e^\delta(s, B).$$

Proof. Assume that $B \neq 0$ first. By Lemma 5.3, if $S^\delta(s,B) \neq 0$ then there is a positive integer c such that $N|c$, $(c,D) = \delta'$, and $c_1 = B_1\delta'$. This implies that $N_1|B_1\delta'$. Assuming this, Lemma 5.3 then gives

$$S^\delta(s,B) = \sum_{\substack{N_2|c_2\\(c_2,D)=1}} \frac{\epsilon_\delta(B_1c_2)}{(B_1\delta'c_2)^{2s}} G(\delta)B_1\epsilon_{\delta'}(B_2/c_2) \sum_{d|(c_2,B_2)} \mu(c_2/d)d.$$

Interchanging the summation, we obtain

$$S^\delta(s,B) = \frac{G(\delta)\epsilon_\delta(B_1)\epsilon_{\delta'}(B_2)}{(B_1\delta')^{2s}} \sum_{d|B_2} d \sum_{\substack{N_2|c_2\\d|c_2\\(c_2,D)=1}} \frac{\epsilon^\delta(1/c_2)\mu(c_2/d)}{c_2^{2s}}$$

If $S^\delta(s,B) \neq 0$ then $(N_2,D) = 1$. Assume this and let $e|N_2$ be a factor such that

$$\left(\frac{c_2}{d}, N_2\right) = \frac{N_2}{e}.$$

Substituting c_2 by $d\frac{N_2}{e}\ell$, then $(e,\ell) = 1$, and we obtain

$$S^\delta(s,B) = \frac{G(\delta)\epsilon_\delta(B_1\delta')\epsilon_{\delta'}(B_2)}{(\delta')^{2s}B_1^{2s-1}} \sum_{\substack{d|B_2\\(d,D)=1}} \sum_{\substack{e|N_2\\e|d}} d\frac{\epsilon^\delta(1/(dN_2e^{-1}))\mu(N_2/e)}{(dN_2/e)^{2s}}.$$

$$\cdot \sum_{\substack{(\ell,D)=1\\(\ell,e)=1\\(\ell,N_2/e)=1}} \frac{\epsilon^\delta(1/\ell)\mu(\ell)}{\ell^{2s}}.$$

Interchange the sums over e and d and replace d by $d \cdot e$. The Proposition follows in the case $B \neq 0$. The case $B = 0$ can be treated similarly.

q.e.d.

8. Evaluation of I_s

Let

$$F(\alpha,\beta,\gamma;z) = 1 + \frac{\alpha\cdot\beta}{\gamma\cdot 1}z + \frac{\alpha(\alpha+1)(\beta(\beta+1)}{\gamma(\gamma+1)\cdot 1\cdot 2}z^2 + \frac{\alpha(\alpha+1)(\alpha+2)\beta(\beta+1)(\beta+2)}{\gamma(\gamma+1)(\gamma+2)\cdot 1\cdot 2\cdot 3}z^3 + \cdots$$

denote the hypergeometric function. It is well known that the hypergeometric function F satisfies the following identities:

$$F(\alpha,\beta,\gamma;z) = F(\beta,\alpha,\gamma;z) \tag{8.1}$$

$$F(\alpha,\beta,\gamma;z) = (1-z)^{-\alpha}F\left(\alpha,\gamma-\beta,\gamma;\frac{z}{z-1}\right). \tag{8.2}$$

We use these identities to prove the following:

Proposition 8.3. *Assume that* $Re(s) < \frac{k+\ell}{2}$. *Then* $I_s(x)$ *is given by the following formulae:*

$$x^{\frac{k-1}{2}}\frac{\Gamma\left(\frac{\ell+k}{2}-s\right)}{\Gamma(k)\Gamma\left(\frac{\ell-k}{2}+s\right)}(1-x)^{\frac{\ell-k}{2}-1+s}F\left(\frac{k-\ell}{2}+s,\frac{k-\ell}{2}+1-s,k;\frac{x}{x-1}\right),$$

if $0 < x < 1$;

$$\frac{\Gamma\left(\frac{\ell+k}{2}-s\right)\Gamma(2s-1)}{\Gamma\left(\frac{k-\ell}{2}+s\right)\Gamma\left(\frac{\ell-k}{2}+s\right)\Gamma\left(\frac{\ell+k}{2}+s-1\right)},$$

if $x = 1$;

$$\frac{\Gamma\left(\frac{\ell+k}{2}-s\right)}{\Gamma(\ell)\Gamma\left(\frac{k-\ell}{2}+s\right)}x^{\frac{-k+1}{2}}(x-1)^{\frac{k-\ell}{2}+s-1}F\left(\frac{\ell-k}{2}+1-s,\frac{\ell-k}{2}+s,k;\frac{1}{1-x}\right),$$

if $x > 1$.

Proof. Recall formula (6.3)

$$I_s(x) = \frac{1}{2\pi i}\int_{\epsilon-\frac{k-1}{2}-i\infty}^{\epsilon-\frac{k-1}{2}+i\infty}\frac{\Gamma\left(\frac{k-1}{2}+w\right)\Gamma\left(\frac{\ell+1}{2}-s-w\right)}{\Gamma\left(\frac{k+1}{2}-w\right)\Gamma\left(\frac{\ell-1}{2}+s+w\right)}x^{-w}\,dw.$$

For $0 < x < 1$, we compute the integral by shifting the line of integration to the left. The integrand has poles at $w = -\frac{k-1}{2}-n$ with $n = 0, 1, 2, \ldots$. Consequently

$$I_s(x) = \sum_{n=1}^{\infty}\frac{(-1)^n}{n!}\frac{\Gamma\left(\frac{\ell+k}{2}-s+n\right)}{\Gamma(k+n)\Gamma\left(\frac{\ell-k}{2}+s-n\right)}x^{\frac{k-1}{2}+n}.$$

Lemma 8.4. *For* $x > 0$

$$\sum_{n=0}^{\infty}\frac{(-1)^n}{n!}\frac{\Gamma(a+n)}{\Gamma(b+n)\Gamma(c-n)}x^n = \frac{\Gamma(a)}{\Gamma(b)\Gamma(c)}F(a,1-c,b;x).$$

Proof. By the properties $x\Gamma(x) = \Gamma(x+1)$, $\Gamma(1) = 1$ of the Gamma function, and the definition of the hypergeometric function F, we have

$$\begin{aligned}\sum_{n=0}^{\infty} \frac{(-1)^n}{n!} & \frac{\Gamma(a+n)}{\Gamma(b+n)\Gamma(c-n)} x^n \\ &= \frac{\Gamma(a)}{\Gamma(b)\Gamma(c)}\left(1 - \frac{a\cdot(c-1)}{1!\, b}x + \frac{a(a+1)(c-1)(c-2)}{2!\, b(b+1)}x^2 - \cdots\right) \\ &= \frac{\Gamma(a)}{\Gamma(b)\Gamma(c)} F(a, 1-c, b; x),\end{aligned}$$

which concludes the proof of Lemma 8.4. q.e.d.

It follows from Lemma 8.4 that

$$I_s(x) = x^{\frac{k-1}{2}} \frac{\Gamma\left(\frac{\ell+k}{2} - s\right)}{\Gamma(k)\Gamma\left(\frac{\ell-k}{2} + s\right)} F\left(\frac{k+\ell}{2} - s, \frac{k-\ell}{2} + 1 - s, k; x\right).$$

We apply to this the first transformation (8.1) and then the functional equation (8.2) with $x \to x/(x-1)$. The first formula in Proposition 8.3 immediately follows.

For $x > 1$, we must shift the line of integration to the right. The integrand has poles at $w = \frac{\ell+1}{2} - s + n$ with $n = 0, 1, 2, \ldots$. We have

$$I_s(x) = \sum_{n \geq 0} \frac{(-1)^n}{n!} \frac{\Gamma\left(\frac{k+\ell}{2} - s + n\right)}{\Gamma\left(\frac{k-\ell}{2} + s - n\right)\Gamma(\ell+n)} x^{-\frac{\ell+1}{2}+s-n}.$$

Applying Lemma 8.4, we have

$$I_s(x) = x^{-\frac{\ell+1}{2}+s} \frac{\Gamma\left(\frac{k+\ell}{2} - s\right)}{\Gamma\left(\frac{k-\ell}{2} + s\right)\Gamma(\ell)} F\left(\frac{k+\ell}{2} - s, \frac{\ell-k}{2} + 1 - s, \ell; x^{-1}\right).$$

Again, the transformation (8.1) and the functional equation (8.2) $\frac{1}{x} \to \frac{1}{x}/(\frac{1}{x} - 1) = 1/(x-1)$ gives the formula in Proposition 8.3 in the case $x > 1$.

In the remaining case when $x = 1$, we require the following lemma.

Lemma 8.5.

$$F(a, b, c; 1) = \frac{\Gamma(c)\Gamma(c-a-b)}{\Gamma(c-a)\Gamma(c-b)}.$$

Proof. Using the identity

$$\frac{\Gamma(x)\Gamma(y)}{\Gamma(x+y)} = \int_0^1 t^{x-1}(1-t)^{y-1}dt$$

and the Tayler expansion of $(1-tz)^a$ at $z=0$, we obtain:

$$F(a,b,c;z) = \frac{\Gamma(c)}{\Gamma(b)\Gamma(c-b)} \int_0^1 (1-tz)^{-a}t^{b-1}(1-t)^{c-b-1}dt.$$

This gives the formula in lemma 8.5 after setting $z=1$. q.e.d.

The formula for $I_s(1)$ in Proposition 8.3 follows by applying lemma 8.5 to the case of the first formula for $I_s(x)$ when $0 < x < 1$. This completes the proof of Proposition 8.3. q.e.d.

Proposition 8.6. *Define*

$$\tilde{I}_s(x) = \frac{\Gamma\left(\frac{k-\ell}{2}+s\right)}{\Gamma\left(\frac{k+\ell}{2}-s\right)} I_s(x).$$

Then for $x \neq 1$ we have the functional equation

$$\frac{\tilde{I}_{1-s}(x)}{|x-1|^{1-s}} = sgn(x-1)^{k-\ell}\frac{\tilde{I}_s(x)}{|x-1|^s}.$$

Proof. Assume first that $0 < x < 1$. It follows from Proposition 8.3 that

$$\frac{\tilde{I}_{1-s}(x)}{|x-1|^{1-s}}$$
$$= \frac{\Gamma\left(\frac{k-\ell}{2}-s\right)}{\Gamma(k)\Gamma\left(\frac{\ell-k}{2}+s\right)}(1-x)^{\frac{\ell-k}{2}-1} F\Big(\frac{k-\ell}{2}+s, \frac{k-\ell}{2}+1-s, k; \frac{x}{x-1}\Big).$$

By property (8.1), the hypergeometric function F above is invariant under the transformation $s \to 1-s$. Further, since $k \equiv \ell \pmod 2$, we may set $a = \frac{k-\ell}{2} \in \frac{1}{2}\mathbb{Z}$. Then we must have

$$\frac{\Gamma(a+s)}{\Gamma(-a+s)} = (-1)^{2a}\frac{\Gamma(a+1-s)}{\Gamma(-a+1-s)},$$

since

$$\begin{aligned}\Gamma(a+s)\Gamma(1-a-s) &= \frac{\pi}{\sin(\pi(a+s))}\\ &= (-1)^{2a}\frac{\pi}{\sin(\pi(-a+s))}\\ &= (-1)^{2a}\Gamma(-a+s)\Gamma(1-(-a+s)).\end{aligned}$$

The functional equation immediately follows. In the case $x > 1$, the proof is even easier since the gamma factors cancel out. q.e.d.

9. The holomorphic kernel $\widetilde{\Phi}_{s,g}$ for real characters

In general, $\Phi_{s,g}$, does not have a simple functional equation. However, in the case ϵ is a real quadratic character, then we can replace $\Phi_{s,g}$ by a new function $\widetilde{\Phi}_{s,g}$ which has simpler Fourier coefficients.

Proposition 9.1. *Fix positive integers k, ℓ, N, D and $g(z) = b(0) + \sum_{n=1}^{\infty} b(n)\, n^{\frac{\ell-1}{2}} e^{2\pi i n z}$ in $M_\ell(\Gamma_0(D), \epsilon)$ with ϵ a real quadratic Dirichlet character* (mod D). *For $s \in \mathbb{C}$ define*

$$\widetilde{\Phi}_{s,g}(z) = \sum \tilde{\phi}_{s,g}(m) m^{\frac{k-1}{2}} e^{2\pi i m z}$$

with

$$\begin{aligned}\tilde{\phi}_{s,g}(m) = & \frac{b(m)}{m^s} \sum_{e|N_2} \mu\left(\frac{N_2}{e}\right) \frac{e}{N_2} \cdot \frac{\Gamma\left(\frac{k-\ell}{2}+s\right) L_e(2s)}{\Gamma\left(\frac{k+\ell}{2}-s\right)(2\pi)^{2s}\epsilon(e) e^{1-2s}} \\ & + \; 2\pi i^k T_{m,N_2}(s)\end{aligned}$$

where

$$T_{m,N_2}(s) = \frac{\Gamma\left(\frac{k-\ell}{2}+s\right)}{\Gamma\left(\frac{k+\ell}{2}-s\right)} \sum_{\delta|D} \frac{\delta^{\frac{1}{2}-s}}{2\pi i^\ell} \sum_{n=1}^{\infty} \frac{b^\delta(n)}{n^{1-s}}\, S_e^\delta\left(s, m\delta - n\right)\, I_s\left(\frac{\delta m}{n}\right),$$

$S_{N_2}^\delta(s, m\delta - n)$ is given in Proposition 6.5, and $I_s\left(\frac{\delta m}{n}\right)$ is given in (6.3).

Then $\widetilde{\Phi}_{s,g}(z)$ is a cusp form of weight k for $\Gamma_0(N)$. Further, for any newform f of weight k for $\Gamma_0(N)$, we have

$$L(s, f \otimes g) = \frac{(4\pi)^{k-1}}{(k-2)!} \frac{\Gamma\left(\frac{k+\ell}{2}-s\right)(2\pi)^{2s}\epsilon(N_2) N_2^{1-2s}}{\Gamma\left(\frac{k-\ell}{2}+s\right) L_{N_2}(2s)} \cdot < \widetilde{\Phi}_{s,g}, f > .$$

Proof. Since ϵ is real we have $\epsilon^2 = 1$ and $\epsilon^\delta = \epsilon$. By Proposition 6.5, $S_e^\delta(s, B) \neq 0$ only if $N | B\delta'$ and $(N, \delta) = 1$. In the decomposition $N = N_1 N_2$ (as in (5.1)) we may, therefore, assume that N_2 is maximal and prime to D.

For any factor e of N_2 define

$$T_{m,e} = \sum_{\delta|D} T_{m,e}^\delta$$

with $T^{\delta}_{m,e}$ given by the formula

$$T^{\delta}_{m,e}(s) = \frac{\delta^{\frac{1}{2}-s}}{2\pi i^{\ell}} \sum_{n=1}^{\infty} \frac{b^{\delta}(n)}{n^{1-s}} \, S^{\delta}_{e}(s, m\delta - n) \, \tilde{I}_s\left(\frac{\delta m}{n}\right)$$

where $\tilde{I}_s$ is defined by

$$\tilde{I}_s(x) = \frac{\Gamma\left(\frac{k-\ell}{2}+s\right)}{\Gamma\left(\frac{k+\ell}{2}-s\right)} I_s(x).$$

Define $\tilde{\phi}_{s,g}(m)$ by the formula

$$\tilde{\phi}_{s,g}(m) = b(m)m^{-s}B_s(N_2) + 2\pi i^k T_{m,N_2}(s)$$

where

$$B_s(N_2) = \sum_{e|N_2} \mu\left(\frac{N_2}{e}\right) \frac{e}{N_2} A_s(e)^{-1}$$

and

$$A_s(e) = \frac{\Gamma\left(\frac{l+k}{2}-s\right)(2\pi)^{2s}}{\Gamma\left(\frac{k-l}{2}+s\right)} \frac{\epsilon(e)e^{1-2s}}{L_e(2s)}.$$

By definition, $T_{m,e}(s)$ depends only on $N_1 e$ and m. It follows from (6.6), (6.7), and Proposition 6.5 that

$$T_m(s) = A_s(N_2) \sum_{e|N_2} \mu\left(\frac{N_2}{e}\right) \frac{e}{N_2} T_{m,e}(s). \tag{9.2}$$

It further follows from (6.1) and (9.2) that

$$\tilde{\phi}_{s,g}(m) = \sum_{e|N_2} \mu\left(\frac{N_2}{e}\right) \frac{e}{N_2} A_s(e) L(s, P_{m,e} \otimes g)$$

where $P_{m,e}$ denotes the mth Poincaré series for $\Gamma_0(N_1 e)$. q.e.d.

10. The functional equation of $\widetilde{\Phi}_{s,g}$ when $D = 1$

Theorem 10.1. *Fix positive integers k, ℓ, N with $k \equiv \ell \pmod 2$. Fix a modular form $g \in M_\ell(\Gamma_0(1))$. Define $\widetilde{\Phi}_{s,g}$ as in Theorem 9.1 with* *the choice $D = 1$. Then $\widetilde{\Phi}_{s,g}$ has the Fourier coefficients*

$$\begin{aligned}\tilde{\phi}_{s,g}(m) =& i^{k-\ell}N^{1-s}\frac{\Gamma(1-s)\Gamma(s)}{(2\pi)^{2s}\Gamma\left(\frac{\ell-k}{2}+1-s\right)\Gamma\left(\frac{\ell+k}{2}-s\right)}\zeta(2s)\\ &+ i^{k-\ell}N^{1-s}\frac{\Gamma(1-s)\Gamma(s)}{(2\pi)^{2-2s}\Gamma\left(\frac{\ell+k}{2}+s\right)\Gamma\left(\frac{\ell+k}{2}-1+s\right)}\zeta(2-2s)\\ &+ i^{k-\ell}N^{1-s}\sum_{\substack{n\geq 1, n\neq mD\\ n\equiv m \pmod N}} b(n)\\ &\cdot \sum_{\left|\frac{m-n}{N}\right| = d_1\cdot d_2} \frac{1}{d_1^s d_2^{1-s}}\left|\frac{m}{n}-1\right|^{1-s}\tilde{I}_s\left(\frac{m}{n}\right)\end{aligned}$$

and we have the functional equation

$$\widetilde{\Phi}_{1-s,g} = N^{1-2s}\widetilde{\Phi}_{s,g}.$$

Proof. The formula follows from Proposition 9.1 by taking $D = 1$ and $N_2 = 1$. The functional equation follows from the functional equation of $\zeta(s)$ and Proposition 8.6. q.e.d.

11. The functional equation of $\widetilde{\Phi}_{s,g}$ when g is a theta function

We now assume that N is prime to D and $\epsilon(-1) = -1$. Extend ϵ to a character on $\mathbb{A}^\times/\mathbb{Q}^\times$. Assume that

$$g(z) = b(0) + \sum_{n=1}^{\infty} b(n)\, n^{\frac{\ell-1}{2}} e^{2\pi i n z}$$

transforms like a theta function attached to an imaginary quadratic field $\mathbb{Q}(\sqrt{-D})$: In this case the Fourier coefficients $b(n)$ (with $n > 0$) satisfy the following properties.

$$\text{For any } \delta | D,\ \ b(\delta) = \pm 1 \tag{11.1}$$

$$b(n) \neq 0 \text{ only if } \epsilon(n) = 1. \tag{11.2}$$

$$\text{For } \delta | d,\ \ b(n\delta) = b(n)\, b(\delta). \tag{11.3}$$

(11.4) For $\delta|D$, let $\kappa(\delta) = \sqrt{\epsilon_\delta(-1)}$. Then $b^\delta(n) = \kappa(\delta)^{-1} b(n\delta)\epsilon_\delta(n)$.

Theorem 11.5. *The function $\widetilde{\Phi}_{s,g}$ defined in §9 has Fourier coefficients given by*

$$\begin{aligned}\tilde{\phi}_{s,g}(m) =& i^{k-\ell} L(2s,\epsilon)\frac{D^s N^{2s-1}}{(2\pi)^s}\frac{\Gamma\left(\frac{1}{2}-s\right)\Gamma\left(\frac{1}{2}+s\right)}{\left(\frac{\ell-k}{2}+1-s\right)\Gamma\left(\frac{\ell+k}{2}-s\right)}\frac{b^D(m)}{m^s}\\ &+ i^{k-\ell} L(2-2s,\epsilon)\frac{D^{2-3s}}{(2\pi)^{2-2s}}\frac{\Gamma\left(s-\frac{1}{2}\right)\Gamma\left(\frac{3}{2}-s\right)}{\Gamma\left(\frac{\ell-k}{2}+s\right)\Gamma\left(\frac{\ell+k}{2}-1+s\right)}\frac{b^D(m)}{m^{1-s}}\\ &+ i^{k-\ell} D^{1/2-s} b(D)\kappa(D)^{-1}\\ &\cdot \sum_{\substack{n\geq 1, n\neq mD\\ mD\equiv n \pmod N}} \frac{b(n)}{n^{1-s}} J_s\left(\frac{mD-n}{N}\right)\tilde{I}_s\left(\frac{Dm}{n}\right)\end{aligned}$$

with

$$J_s(t) = \epsilon^*(mD-Nt)\sum_{\substack{d|t\\(d,D)=1}}\epsilon(d)d^{1-2s}\cdot\sum_{\delta'|(D,t)}\epsilon_{\delta'}(Nt(Nt-mD))t_{\delta'}^{1-2s}$$

where $\epsilon^* = \epsilon\cdot\epsilon_D^{-1}$ and $t_{\delta'}$ denotes the maximal positive divisor of t whose prime factors are those of δ'.

Further, we have the functional equation,

$$\widetilde{\Phi}_{1-s,g} = \epsilon(-N)(DN^{-1})^{2s-1}\widetilde{\Phi}_{s,g}.$$

Proof. By definition, we have

$$\tilde{\phi}_{s,g}(m) = B_s(N)\frac{b(m)}{m^s} + 2\pi i^k T_{m,N}(s)$$

with

$$T_{m,N}(s) = \sum_{\delta|D}\frac{\delta^{1/2-s}}{2\pi i^\ell}\sum_{n=1}^{\infty}\frac{b^\delta(n)}{n^{1-s}}S_N^\delta(m\delta-n)\tilde{I}_s\left(\frac{\delta m}{n}\right).$$

One precise computation of $B_s(N)$ will give the first term in the formula of Theorem 11.5. Replacing n by n/δ', interchanging the sums, and using the formula for b^δ (equation (11.4)), we have

$$T_{m,N}(s) = \frac{D^{1/2-s}b(D)}{2\pi i^\ell\kappa(D)}\sum_{\substack{n\geq 1\\ mD\equiv n \pmod N}}\frac{b(n)}{n^{1-s}}J_s^*(n)\tilde{I}_s\left(\frac{Dm}{n}\right)$$

where

$$J_s^*(n) = \sum_{\delta'|(D,n)} \sqrt{\delta'}\kappa(\delta')\epsilon_\delta(n/\delta')S_N^\delta\left(\frac{mD-n}{\delta'}\right).$$

If $n = mD$, then

$$J_s^*(mD) = \epsilon(mD)L(2s-1,\epsilon).$$

Applying the functional equation of $L(s,\epsilon)$, the term $n = mD$ in the last formula of $T_{m,N}(s)$ will give the second term in the formula of the Theorem. Now we assume that $n \neq mD$. Then

$$J_s^*(n) = \sum_{\delta'|(D,n)} \sqrt{\delta'}\kappa(\delta')\epsilon_\delta(n/\delta')S_N^\delta\left(\frac{mD-n}{\delta'}\right).$$

Notice that S_N^δ is nonzero only if $mD \equiv n \pmod N$. Write $t = \frac{mD-n}{N}$, then

$$S_N^\delta(Nt/\delta') = \frac{G(\delta)\epsilon_{\delta'}(Nt/\delta')}{(\delta')^{2s}(t_{\delta'}/\delta')^{2s-1}} \sum_{\substack{d|t \\ (d,D)=1}} \epsilon(d)d^{1-2s}.$$

Since $G(\delta) = \kappa(\delta')\sqrt{\delta'}$, and

$$\epsilon_{\delta'}(-1)\epsilon_\delta(n/\delta')\epsilon_{\delta'}(Nt/\delta') = \epsilon^*(n)\epsilon_{\delta'}(-nNt),$$

we have $J_s^*(n) = J_s(t)$ as in the last term in the formula of Theorem 11.5.

We now obtain the functional equation for

$$J_s(t) = \epsilon^*(mD-Nt) \sum_{\substack{d|t \\ (d,D)=1}} \epsilon(d)d^{1-2s} \cdot \sum_{\delta'|(D,t)} \epsilon_{\delta'}(Nt(Nt-mD))t_{\delta'}^{1-2s}.$$

Replace d by $|t|/(t_D d)$ in the first sum, and replace δ' by $(D,t)/\delta'$ in the second sum to obtain

$$J_{1-s}(t) = t^{2s-1}\epsilon_D(|t|/t_D)\epsilon_{(D,n)}(-Ntn)J_s(t).$$

Notice that $\epsilon_D(|t|/t_D) = \mathrm{sgn}(t)\epsilon_D(t)$, and for any p not dividing (t,D),

$$\epsilon_p(-Ntn) = 1$$

because $-Ntn \equiv (Nt)^2 \pmod p$. We, therefore, obtain that

$$\epsilon(|t|/t_D)\epsilon_{(D,t)}(-Ntn) = \mathrm{sgn}(t)\epsilon_D(t)\epsilon_D(-Ntn) = \mathrm{sgn}(t)\epsilon(-N)\epsilon(n).$$

It follows that $J_s(t)$ satisfies the functional equation

$$J_{1-s}(t) = t^{2s-1}\mathrm{sgn}(t)\epsilon(-N)\epsilon(n)J_s(t).$$

Combining this with the functional equation for $\tilde{I}_s$ in Proposition 8.6, we obtain the functional equation for $\tilde{\Phi}_{s,g}$. q.e.d.

Remark. In the case that g is a theta series attached to an imaginary quadratic field, and m is prime to N, Gross and Zagier [1] have computed the value (when $\epsilon(N) = 1$) and the derivative (when $\epsilon(N) = -1$) of $\tilde{\phi}_{s,g}(m)$. It is not difficult to see that our results coincide with those of Gross–Zagier in this case. Our results go beyond [1] in that we give the whole kernel (not only the special value or derivatives) in terms of divisor functions and hypergeometric functions.

References

[1] B. Gross & D. B. Zagier, *Heegner points and derivatives of L-series*, Invent. Math. **84** (1986) 225–320.

[2] W. C. W. Li, *L-series of Rankin type and their functional equations,* Math. Ann. **244** (1979), no. 2, 135–166.

[3] S. Ramanujan, *On certain trigonometrical sums and their applications in the theory of numbers*, Trans. Cambridge Philos. Soc. XXII, No. 13 (1918) 259–276.

[4] R. Rankin, *Contributions to the theory of Ramanujan's function $\tau(n)$ and similar arithmetic functions. I and II,* Proc. Cambridge Phil. Soc. **35** (1939) 351–356; 357–372.

[5] P. Sarnak, *Some applications of modular forms*, Cambridge Tracts in Math., Cambridge Univ. Press, Cambridge, Vol. 99, 1990, 22–25.

[6] A. Selberg, *Bemerkungen über eine Dirichletsche reihe, die mit der theorie der modulformer nahe verbunden ist,* Arch. Math. Naturvid. **43** (1940) 47–50.

[7] ———, *On the estimation of Fourier coefficients of modular forms,* Proc. Sympos. Pure Math. (Cal Tech, Pasadena, Cal. 1963), Amer. Math. Soc., Providence, RI, Vol VIII, 1–15.

DEPARTMENT OF MATHEMATICS, COLUMBIA UNIVERSITY

Surveys in Differential Geometry, 2000
Vol. VII ©2000, International Press
pp. 221-257

EQUIVARIANT DE RHAM THEORY AND GRAPHS

V. GUILLEMIN & C. ZARA

Abstract

Goresky, Kottwitz and MacPherson have recently shown that the computation of the equivariant cohomology ring of a G-manifold can be reduced to a computation in graph theory. This opens up the possibility that many of the fundamental theorems in equivariant de Rham theory may, on closer inspection, turn out simply to be theorems about graphs. In this paper we show that for some familiar theorems, this is indeed the case.

1. Introduction

This article will consist of two essentially disjoint parts. Part 1 is an exposition of (mostly) well-known results about G-manifolds. In Section 1.1–1.3 we review the definition of the equivariant de Rham cohomology ring of a G-manifold and recall the statements of the two fundamental "localization theorems" in equivariant de Rham theory: the Atiyah-Bott-Berline-Vergne theorem and the Jeffrey-Kirwan theorem. In Section 1.4 we discuss the "Smith" problem for G-manifolds (which is concerned with the question: Given a G-manifold with isolated fixed points, what kinds of representations can occur as isotropy representations at the fixed points?) Then in Sections 1.5–1.7 we report on some very exciting recent results of Goresky-Kottwitz-MacPherson which have to do with the tie-in between "equivariant de Rham theory" and "graphs" alluded to in our title. These results show that for a large class of G-manifolds, M, with M^G finite, the equivariant cohomology ring of M is isomorphic to the equivariant cohomology ring of a pair (Γ, α), where Γ

First published in the Asian Journal of Mathematics, Volume 3, Number 1 (March) 1999. Used by permission.
The first author was supported in part by NSF grant DMS 890771.

is the intersection graph of a necklace of embedded S^2's, each of which is equipped with a circle action (i.e., an axis of symmetry), and α is an "axial" function which describes the directions in which the axes of these S^2's are tilted. Finally, in Section 1.8 we discuss a Morse theoretic recipe for computing the Betti numbers of M in terms of the pair (Γ, α).

The second part of this article is concerned with the combinatorial invariants of a pair (Γ, α), Γ being *any* finite simple d-valent graph and α an abstract analogue of the axial function alluded to above. In particular, for such a pair we will prove combinatorial versions of the theorems described in Sections 1.2-1.3 and 1.8. These combinatorial "localization" theorems help to shed some light on the role of the localization theorems in Smith theory : From the localization theorems one can generate a lot of complicated identities among the weights of the isotropy representations. However, the question of whether one can extract from these identities any new information about the isotropy representations themselves has been an open question for a long time. Our graph theoretical results seem to indicate that one can't.

This article is the first of a series of two articles on graphs and equivariant cohomology. In the second article in this series we will discuss K-theoretical analogues of the results above and give a purely combinatorial proof of the so-called "quantization commutes with reduction" conjecture.

1.1 Equivariant de Rham theory

Let G be an n-dimensional Lie group which is compact, connected and abelian, i.e., an n-dimensional torus. Let $\mathfrak{g}$ be its Lie algebra and $\mathfrak{g}^*$ the vector space dual of $\mathfrak{g}$. We will fix a basis $\xi^1, \ldots, \xi^n$ of $\mathfrak{g}$ and let $x_1, \ldots, x_n$ be the dual basis. Using this basis, the symmetric algebra $\mathbb{S}(\mathfrak{g}^*)$ can be identified with the polynomial ring $\mathbb{C}[x_1, \ldots, x_n]$.

Let M be a $2d$-dimensional manifold and τ an action of G on M. From τ one gets an infinitesimal action, $\partial\tau$, of $\mathfrak{g}$ on M which associates to every element ξ of $\mathfrak{g}$ a vector field ξ_M. Let $\Omega(M)$ be the usual complex of de Rham forms on M and $\Omega(M)^G$ the subcomplex of G-invariant de Rham forms. One defines the *equivariant* de Rham complex of M to be the tensor product

$$\Omega_G(M) = \Omega(M)^G \otimes \mathbb{S}(\mathfrak{g}^*) \tag{1.1}$$

with the coboundary operator

$$d_G(\omega \otimes f) = d\omega \otimes f + \sum \iota(\xi^i_M)\omega \otimes x_i f\,. \tag{1.2}$$

The *equivariant cohomology ring* of M, $H_G(M)$, is the cohomology ring of this complex. A few properties of this ring which we will need below are:

1. $H_G(M)$ is an $\mathbb{S}(\mathfrak{g}^*)$-module. (This follows from the fact that $\Omega_G(M)$ is an $\mathbb{S}(\mathfrak{g}^*)$ module by (1.1) and d_G is an $\mathbb{S}(\mathfrak{g}^*)$ module morphism by (1.2).)

2. $H_G(pt) = \mathbb{S}(\mathfrak{g}^*)$.

3. Suppose M is compact and oriented. Then there is an integration operation

(1.3) $$\int : \Omega_G(M) \longrightarrow \mathbb{S}(\mathfrak{g}^*)$$

defined by
$$\int (\omega \otimes f) = f \int \omega .$$

It is easily checked that $\int d_G = 0$ and hence that this integration operation induces an integration operation on cohomology

(1.4) $$\int : H_G(M) \longrightarrow \mathbb{S}(\mathfrak{g}^*) .$$

4. One can write d_G as a sum $d_1 + d_2$, d_1 and d_2 being the first and second terms on the right hand side of (1.2). Thus $\Omega_G(M)$ is a *bi-complex* and the additive structure of $H_G(M)$ can be computed by the spectral sequence of this bi-complex. The E_1 term in this spectral sequence is the d_1-cohomology of $\Omega_G(M)$, namely

(1.5) $$H(M) \otimes \mathbb{S}(\mathfrak{g}^*) .$$

One says that M is *equivariantly formal* if the spectral sequence is trivial, i.e., if, as vector spaces,

(1.6) $$H_G(M) = H(M) \otimes \mathbb{S}(\mathfrak{g}^*) .$$

One can show, by the way, that if (1.6) holds as an identity of vector spaces, it also holds as an identity of $\mathbb{S}(\mathfrak{g}^*)$-modules. However, (1.6) does not, in general, tell one very much about the *ring* structure of $H_G(M)$ (about which we will have more to say in §1.6).

The property of being equivariantly formal is a bit technical; however, there are a number of interesting assumptions on M which will imply this property. (See [8].) Of these assumptions, the one that will be of most interest to us is the following:

Theorem (Kirwan). *If M is a symplectic manifold and the action τ is Hamiltonian, then M is equivariantly formal.*

1.2 The Atiyah-Bott-Berline-Vergne localization theorem

Let M be compact and oriented and, also, to simplify the statement of the localization theorem, let M^G be finite. For $p \in M^G$ one has an isotropy representation τ_p of G on T_p and we will denote the weights of this representation by $\alpha_{i,p}, i = 1, \ldots, d$. Since τ_p is a *real* representation, these weights are, strictly speaking, only defined up to sign; however, since M is oriented, the product

$$\alpha_{1,p} \cdots \alpha_{d,p}$$

is well-defined as an element of $\mathbb{S}^d(\mathfrak{g}^*)$. Let

$$j_p : pt \longrightarrow M$$

be the mapping "pt" onto p and note that if c is in $H_G(M)$, $j_p^* c$ is in $H_G(pt)$ and thus in $\mathbb{S}(\mathfrak{g}^*)$. The localization theorem asserts that, for every equivariant cohomology class $c \in H_G(M)$,

$$\int c = \sum_{p \in M^G} \frac{j_p^* c}{\prod \alpha_{i,p}}. \tag{1.7}$$

There are many deep and beautiful applications of (1.7) but the focus of our interest in this article is that (1.7) implies a lot of complicated identities among the weights $\alpha_{i,p}$. For instance, for $c = 1$, it implies

$$\sum \left(\prod \alpha_{i,p} \right)^{-1} = 0 \,. \tag{1.8}$$

What are these identities? In particular, are there simpler identities of which they are formal consequences ? We will show in P art 2 of this article that one can shed some light on these questions by looking at a graph-theoretical analogue of (1.7).

1.3 The Jeffrey-Kirwan theorem

Another interesting source for identities of type (1.7) is the Jeffrey-Kirwan theorem ([11]): Let K be a one-dimensional connected closed subgroup of G with Lie algebra $\mathfrak{k}$ and let $\xi \in \mathfrak{k}$ be a basis vector of the group lattice of K. Suppose M possesses a G-invariant symplectic form ω and that the action of K on M is Hamiltonian, i.e.

$$\iota(\xi_M)\omega = -df, \tag{1.9}$$

f being a G-invariant function. In addition, suppose that

$$M^K = M^G$$

and hence that the critical points of f coincide with the fixed points of G. Let a be a regular value of f and let $M_a = f^{-1}(a)$. By the remark above, M_a contains no K-fixed points, so the action of K on M_a is locally free and the quotient space

$$M_a/K =: M_{red}$$

is an orbifold. Moreover, from the action of G on M_a one gets an inherited action of the quotient group

$$G/K =: G_1$$

on M_{red}. Let j be the inclusion of M_a into M and π the projection of M_a onto M_{red}. By the Marsden-Weinstein theorem there exists a symplectic form ω_{red} on M_{red} satisfying $\pi^*\omega_{red} = j^*\omega$. In particular, M_{red} is *oriented.* If $\mathfrak{g}_1$ is the Lie algebra of G_1, its vector space dual, $\mathfrak{g}_1^*$, can be identified with the annihilator $\mathfrak{g}_\xi^*$ of ξ in $\mathfrak{g}$; so there is an integration operator

$$\int : H_{G_1}(M_{red}) \longrightarrow \mathbb{S}(\mathfrak{g}_\xi^*). \tag{1.10}$$

Also, since the action of K on M_a is locally free, the map π induces an isomorphism

$$\pi^* : H_{G_1}(M_{red}) \longrightarrow H_G(M_a)\ ,$$

so one gets a map

$$(\pi^*)^{-1} j^* =: \mathcal{K}$$

of $H_G(M)$ into $H_{G_1}(M_{red})$. The Jeffrey-Kirwan theorem asserts that for every equivariant cohomology class $c \in H_G(M)$,

$$\int \mathcal{K}(c) = \mathrm{Res}_\xi \left(\sum_{f(p)>a} \frac{j_p^* c}{\prod \alpha_{j,p}} \right), \quad p \in M^G, \tag{1.11}$$

Res_ξ being the residue of the rational function in brackets with respect to the "ξ-coordinate" on $\mathfrak{g}^*$, the other coordinates being held fixed. (This residue can be defined intrinsically to be an element of $\mathbb{S}(\mathfrak{g}_\xi^*)$. See §2.6)

1.4 The Smith problem

The Smith conjecture asserts that if M^G consists of two points p and q, the isotropy representation of G at p is isomorphic (as a representation over $\mathbb{R}$) to the isotropy representation of G at q. The first complete proof of this theorem (for G an arbitrary compact Lie group) is due to Atiyah, Bott and Milnor (See [3], Theorem 3.83). If the cardinality of M^G is greater than two, the question of how the isotropy representations of G at distinct fixed points are related to each other is still open and is known as the "Smith problem". In this section we will describe a few of the more obvious relations:

Relations of type J. Suppose that M admits a G-invariant almost-complex structure. Then for every $p \in M^G$, the isotropy representations of G on T_p is a complex representation, so the weights of this representation,

$$\alpha_{i,p}, \quad i = 1, \dots, d \tag{1.12}$$

are unambiguously defined (not just defined up to sign).

For every closed subgroup H of G let $\mathfrak{h}$ be the Lie algebra of H and

$$\rho_H : \mathfrak{g}^* \longrightarrow \mathfrak{h}^*$$

the transpose of the inclusion map $\mathfrak{h} \longrightarrow \mathfrak{g}$. Let X be a connected component of M^H and p and q elements of X^G. We claim that the weights (1.12) can be ordered so that

$$\rho_H \alpha_{i,p} = \rho_H \alpha_{i,q} \,. \tag{1.13}$$

Proof. Let x be an arbitrary point of X and consider the isotropy representation of H on the normal space to X at x. This representation is a complex representation, so the weights of this representation are unambiguously defined and can't vary as x varies in X. Thus, in particular, they have to be the same at p and at q, implying (1.13). q.e.d.

Relations of type ω. Assuming that M admits a G-invariant almost-complex structure J is equivalent to assuming that M admits a G-invariant "almost-symplectic" structure, i.e. a two-form ω which is everywhere of maximal rank. Suppose that ω is actually a symplectic form and the action τ is *Hamiltonian*, or, in other words, that there exists a moment map

$$\Phi : M \longrightarrow \mathfrak{g}^* .$$

From the convexity theorem ([1], [9]) one gets

Theorem. *Let Δ be the set of regular values of Φ in $\Phi(M)$. Then Δ is a disjoint union of open convex polytopes. Moreover, the vertices of these polytopes are the images of the fixed points $p \in M^G$ and the edges going out of these vertices are pointing in the directions of the vectors $\alpha_{i,p}$.*

The one-skeleton, Γ, of this configuration is called the *moment graph* of M. (See [10].) It exhibits a lot of relations among the $\alpha_{i,p}$'s which are probably not much simpler than the relations (1.7) but have the virtue of being of a more geometric character.

1.5 The Goresky-Kottwitz-MacPherson graph

We will assume from now on that M admits a G-invariant almost-complex structure. Thus, for very $p \in M^G$, the weights $\alpha_{i,p} \in \mathfrak{g}^*$ are unambiguously defined. In addition we will assume: if $i \neq j$, then $\alpha_{i,p}$ and $\alpha_{j,p}$ are linearly independent. This "GKM hypothesis" has the following consequence: Let $\mathfrak{h} = \mathfrak{h}_i$ be the annihilator of $\alpha_{i,p}$ in $\mathfrak{g}$ and let H be the $(n-1)$-dimensional subtorus of G with Lie algebra $\mathfrak{h}$. Clearly, $p \in M^G \subseteq M^H$.

Proposition. *Let X be the connected component of M^H containing p. Then*

$$X \cong \mathbb{C}P^1 \cong S^2$$

and the action of G on X is the standard action of the circle S^1 on S^2 by "rotation about the z-axis". In particular X contains just two G-fixed points (one of which is p).

Proof. The tangent space to X at p is the 2-dimensional subspace of T_pM on which G acts with weight $\alpha_{i,p}$, so X itself is 2-dimensional. Since X is compact and the action of G/H is non-trivial, X is diffeomorphic to S^2, and this action is the standard action of S^1 on S^2 by the Korn-Lichtenstein theorem. q.e.d.

Let q be the other fixed point of G in X. Let $\alpha_{p,e} := \alpha_{i,p}$ be the weight of the isotropy representation of G on T_pX and let $\alpha_{q,e} := \alpha_{j,q}$ be the weight of the isotropy representation of G on T_qX. From the fact that the action of G on X is diffeomorphic to the standard action of S^1 on S^2 it follows that

$$\alpha_{p,e} = -\alpha_{q,e} . \tag{1.14}$$

For each of the weights $\alpha_{i,p}$ one gets an embedded $\mathbb{C}P^1$ of the type above, and we can represent these $\mathbb{C}P^1$'s graphically by d lines issuing from p. Each of these lines joins p to another fixed point, q, and the $\mathbb{C}P^1$'s associated with the weights $\alpha_{j,q}$ can also be represented graphically by d lines issuing from q. One of these will be the line from p to q, but the remaining $d-1$ lines will join q to other fixed points. By repeating this construction over and over until one runs out of fixed points, one obtains a finite d-valent graph, Γ, the vertices of which are the fixed points of G and the edges of which correspond to embedded $\mathbb{C}P^1$'s, each of these $\mathbb{C}P^1$'s being a connected component of the fixed point set of an $(n-1)$-dimensional subtorus of G. We will call Γ the *Goresky-Kottwitz-MacPherson (GKM) graph of M.*

Example. Suppose M is a Hamiltonian G-manifold whose moment map Φ maps M^G injectively into $\mathfrak{g}^*$. Then Φ embeds the GKM graph into $\mathfrak{g}^*$, and its image is the *moment graph.*

1.6 The Goresky-Kottwitz-MacPherson theorem

The graph Γ is equipped with an additional piece of structure. Namely, let I_Γ be the incidence relation of this graph: the set of all pairs (p, e), p being a vertex and e an edge containing p. Then one has a map

$$\alpha : I_\Gamma \longrightarrow \mathfrak{g}^* , \tag{1.15}$$

sending (p, e) to the weight $\alpha_{p,e}$. We will call this the *axial function* of Γ. It has the following properties (the first two of which we have already commented on):

1. If e is an edge and p and q are the vertices joined by e, then

$$\alpha_{p,e} = -\alpha_{q,e}. \tag{1.16}$$

2. If p is a vertex and $e_1, \ldots, e_d$ are the edges containing p, the vectors

$$\alpha_{p,e_i}, \quad i = 1, \ldots, d \tag{1.17}$$

are pair-wise linearly independent.

3. Let e be an edge and p and q the vertices joined by e. Let

$$\mathfrak{g}_e = \{\xi \in \mathfrak{g}, \alpha_{p,e}(\xi) = 0\}$$

and let

$$\rho_e : \mathfrak{g}^* \longrightarrow \mathfrak{g}_e^*$$

be the transpose of the inclusion map $\mathfrak{g}_e \longrightarrow \mathfrak{g}$. Let $e_i, i = 1, \ldots, d$ and $e'_i, i = 1, \ldots, d$ be the edges containing p and q respectively, with $e_d = e'_d = e$. Then the e_i's can be ordered so that

$$\rho_e \alpha_{p,e_i} = \rho_e \alpha_{q,e'_i} \,. \tag{1.18}$$

Proof. (1.18) is just a special case of (1.13), H being the subtorus of G with Lie algebra $\mathfrak{g}_e$. q.e.d.

From the data (Γ, α) one can construct a graded ring

$$H(\Gamma, \alpha) = \bigoplus H^{2k}(\Gamma, \alpha) \tag{1.19}$$

as follows. For each edge e, the map $\rho_e : \mathfrak{g}^* \longrightarrow \mathfrak{g}_e^*$ extends to a ring morphism

$$\rho_e : \mathbb{S}(\mathfrak{g}^*) \longrightarrow \mathbb{S}(\mathfrak{g}_e^*) \,. \tag{1.20}$$

Let V_Γ be the set of vertices of Γ and let $H^{2k}(\Gamma, \alpha)$ be the set of all maps

$$f : V_\Gamma \longrightarrow \mathbb{S}^k(\mathfrak{g}^*)$$

satisfying the compatibility condition:

$$\rho_e f(p) = \rho_e f(q) \tag{1.21}$$

for all vertices p and q and edges e joining p to q. Then $H(\Gamma, \alpha)$ can be given a ring structure by pointwise multiplication

$$(f_1 f_2)(p) = f_1(p) f_2(p) \,.$$

(Notice that if f_1 and f_2 satisfy (1.21) so does $f_1 f_2$ since ρ_e is a ring morphism.) In addition, $H(\Gamma, \alpha)$ contains $\mathbb{S}(\mathfrak{g}^*)$ as a subring: the ring of constant maps of V_Γ into $\mathbb{S}(\mathfrak{g}^*)$. In particular, $H(\Gamma, \alpha)$ is a module over $\mathbb{S}(\mathfrak{g}^*)$.

Theorem([8]). *If M is equivariantly formal, then $H_G(M)$ is isomorphic, as a graded ring, to $H(\Gamma, \alpha)$.*

We will sketch a proof of this at the end of §1.8.

1.7 Holonomy

Let $\mathbb{P}$ be the complex projective line, let z be the standard coordinate function on $\mathbb{C} = \mathbb{P}_0 = \mathbb{P} - \{\infty\}$ and $w = z^{-1}$ the corresponding coordinate function on $\mathbb{P}_\infty = \mathbb{P} - \{0\}$. The multiplicative group, $\mathbb{C}^* = \mathbb{C} - \{0\}$, acts on $\mathbb{C}$ by homotheties

$$(a, z) \longrightarrow az$$

and this extends to a holomorphic action, ρ, of $\mathbb{C}^*$ on $\mathbb{P}$. Let E be a holomorphic, rank r vector bundle over $\mathbb{P}$ and suppose ρ lifts to an action ρ of $\mathbb{C}^*$ on E by vector bundle automorphisms. Let E_0 and E_∞ be fibers of E over 0 and ∞. The isotropy representations of $\mathbb{C}^*$ at 0 and ∞ decompose these spaces into invariant one-dimensional subspaces

$$E_0 = V_1 \oplus \cdots \oplus V_r \tag{1.22}$$

and

$$E_\infty = V_1' \oplus \cdots \oplus V_r' \,. \tag{1.23}$$

Let $m_1, \ldots, m_r$ be the weights of the representations of $\mathbb{C}^*$ on $V_1, \ldots, V_r$. We will assume that these weights are all distinct and that $m_1 < m_2 < \cdots < m_r$. Similarly we will assume that the weights m_i' of the representations of $\mathbb{C}^*$ on $V_1', \ldots, V_r'$ are all distinct (but we won't require that $m_1' < \cdots < m_r'$). By a theorem of Birkhoff-Grothendieck ([12]) there is an equivariant decomposition of E into line-bundles

$$E = \mathbb{L}_1 \oplus \cdots \oplus \mathbb{L}_r \tag{1.24}$$

such that the fiber of $\mathbb{L}_i$ over 0 is V_i. Moreover this decomposition is *unique up to isomorphism.* (To see this, let

$$E = \mathbb{L}_1' \oplus \cdots \oplus \mathbb{L}_r'$$

be another decomposition with these properties. Over $\mathbb{P}_0$, one can find a trivializing section s_i of $\mathbb{L}_i$ which transforms under $\mathbb{C}^*$ according to the law

$$\rho_a^* s_i = a^{m_i} s_i \tag{1.25}$$

and a trivializing section s_i' of $\mathbb{L}_i'$ with the same transformation property. Moreover, one can assume that $s_i(0) = s_i'(0)$. We claim that $s_1 \equiv s_1'$ (and hence $\mathbb{L}_1 = \mathbb{L}_1'$). To see this we note that because of the transformation law (1.25),

$$s_1' = s_1 + \sum_{i>1} c_i z^{m_1 - m_i} s_i$$

on $\mathbb{P}_0$. However, since s_1' is holomorphic near 0 and $m_1 - m_i$ is strictly negative, the constants c_i are all zero.

Applying this argument to the quotient bundle $E/\mathbb{L}_1$ one concludes by induction that $\mathbb{L}_i \cong \mathbb{L}_i'$ for all i. Q.E.D.)

In particular, coming back to the isotropy decompositions (1.22) and (1.23), one has a canonical map of the r-element set

$$\{V_1, \dots, V_r\} \tag{1.26}$$

onto the r-element set

$$\{V_1', \dots, V_r'\} \tag{1.27}$$

which maps $V_i = (\mathbb{L}_i)_0$ onto $(\mathbb{L}_i)_\infty$; and we can relabel the V_i''s so that the element V_i of the set (1.26) corresponds under this map to the element V_i' of the set (1.27). We will call the map the *holonomy* map.

Let us apply these remarks to one of the $\mathbb{C}P^1$'s in §1.5; i.e., let H be a subtorus of G of codimension one and $X = \mathbb{C}P^1$ a 2-dimensional connected component of M^H. Let NX be the normal bundle of X and let $\beta_1, \dots, \beta_k$ be the weights of the isotropy representation of H on a typical fiber of X. The fiber of NX over $x \in X$ splits into a direct sum of vector subspaces

$$(E_1)_x \oplus \cdots \oplus (E_k)_x \tag{1.28}$$

such that the weight of the representation of H on $(E_i)_x$ is β_i. Since the β_i's can occur with multiplicities, the dimension of $(E_i)_x$ doesn't have to be one; however, the assignment $x \longrightarrow (E_i)_x$ defines a complex vector bundle E_i over X. Moreover, this bundle can be given a holomorphic structure, and the action of $S^1 = G/H$ on E_i can be extended to a holomorphic action of $\mathbb{C}^*$. In particular one can define for E_i a *holonomy* map of the type we described at the beginning of this section.

Now let $X^G = \{p, q\}$, and let $\alpha_{i,p}$ and $\alpha_{i,q}$ be the weights of the isotropy representations of G on T_pM and T_qM. By assumption these weights occur with multiplicity one, so one gets decompositions of T_pM and T_qM into direct sums of one-dimensional weight spaces

$$T_p^{\alpha_{1,p}} \oplus \cdots \oplus T_p^{\alpha_{d,p}}$$

and

$$T_q^{\alpha_{1,q}} \oplus \cdots \oplus T_q^{\alpha_{d,q}} .$$

Let ρ be the projection of $\mathfrak{g}^*$ onto $\mathfrak{h}^*$. By (1.18) we can reorder the $\alpha_{i,q}$'s so that $\rho(\alpha_{i,p}) = \beta_i = \rho(\alpha_{i,q})$; so, if the β_i's are all distinct, one has a *canonical* map of the set

$$\{\alpha_{1,p}, \dots, \alpha_{d,p}\} \tag{1.29}$$

onto the set

$$\{\alpha_{1,q}, \dots, \alpha_{d,q}\} \tag{1.30}$$

mapping $\alpha_{i,p}$ onto $\alpha_{i,q}$. However, this canonical map can even be defined when the β_i's are *not* distinct (that is, when the vector bundles E_i are *not* of rank one) by using the holonomy structure on these bundles. In other words there exists a canonical holonomy mapping from the set (1.29) to the set (1.30) which, when the β_i's are distinct, is defined trivially by the recipe

$$\alpha_{i,p} \longrightarrow \rho(\alpha_{i,p}) = \beta_i = \rho(\alpha_{i,q}) \longrightarrow \alpha_{i,q},$$

but, when the β_i's are not distinct, involves some topological properties of the bundles E_i. We will denote this map by $\theta_{p,e}$, where e is the edge corresponding to X.

One can give a slightly more "graphical" description of this map. As in §1.6, let Γ be the GKM graph, let V_Γ be its vertices and let I_Γ be the incidence relation. The projection $I_\Gamma \longrightarrow V_\Gamma$ can be regarded as a fiber bundle over V_Γ, the fiber, E_p, over $p \in V_\Gamma$ being the set of all pairs (p, e) in I_Γ (i.e., the set of all oriented edges of Γ pointing out from p). In this "fiber bundle" picture, $\theta_{p,e}$ is just a map

$$\theta_{p,e} : E_p \longrightarrow E_q \tag{1.31}$$

with the properties

$$\theta_{p,e}(p, e) = (q, e) \tag{1.32}$$

and

$$\theta_{q,e} = \theta_{p,e}^{-1}, \tag{1.33}$$

e being the edge joining p to q. A family of maps, θ, with the properties (1.32)–(1.33) is called a *connection* on Γ (cf. [6]). In terms of this connection, one can reformulate (1.18) more precisely:

Theorem 1.1. *Let e be an edge of Γ joining the vertex p to the vertex q. Then for every $(p, e_i) \in E_p$*

$$\rho_e(\alpha_{p,e_i}) = \rho_e(\alpha_{q,e_i'}), \tag{1.34}$$

where $(q, e_i') = \theta_{p,e}(p, e_i)$.

1.8 Betti numbers

The theorem of Goresky-Kottwitz-MacPherson described in § 1.6 implies that the odd Betti numbers of M are zero. The even Betti numbers can be computed as follows: As in §1.3 let K be a one-dimensional closed connected subgroup of G with $M^G = M^K$ and let ξ be a basis vector of $\mathfrak{k}$. For every $p \in V_\Gamma$ let σ_p be the number of edges e with $\alpha_{p,e}(\xi) < 0$.

Theorem. *The $2k$-th Betti number, β_k, of M is equal to the number of points $p \in V_\Gamma$ with $\sigma_p = k$.*

If M possesses a G-invariant symplectic form having the properties described in §1.3, this theorem can be proved by Morse theory : Let f be the function defined by (1.9). The critical points of this function coincide with the fixed points of G and it is not difficult to show that the index of the Hessian of f at $p \in M^G$ is just $2\sigma_p$.

We will now sketch a proof of the Goresky-Kottwitz-MacPherson theorem. It is clear that

$$H_G(M^G) = H^0(M^G) \otimes \mathbb{S}(\mathfrak{g}^*) = \mathrm{Maps}(V_\Gamma, \mathbb{S}(\mathfrak{g}^*));$$

so if $i : M^G \longrightarrow M$ is the inclusion, there is an induced map

$$i^* : H_G(M) \longrightarrow \mathrm{Maps}(V_\Gamma, \mathbb{S}(\mathfrak{g}^*)).$$

It is easy to see that the image of i^* is contained in $H(\Gamma, \alpha)$. Moreover, by a classical theorem of Borel, i^* is an isomorphism modulo torsion; in particular, the kernel of i^* is a torsion submodule of $H_G(M)$. However, by (1.6), $H_G(M)$ is a *free* $\mathbb{S}(\mathfrak{g}^*)$-module; so $i^* : H_G^{2k}(M) \longrightarrow H^{2k}(\Gamma, \alpha)$ is injective and

$$\dim\, H_G^{2k}(M) = \sum \beta_r \,\dim\, \mathbb{S}^{k-r}(\mathfrak{g}^*).$$

Therefore, in order to prove that

$$\dim\, H^{2k}(\Gamma, \alpha) = \sum \beta_r \,\dim\, \mathbb{S}^{k-r}(\mathfrak{g}^*), \tag{1.35}$$

it suffices to show that the dimension of $H^{2k}(\Gamma, \alpha)$ is less than or equal to the right hand side of (1.35), and we will prove this (by a very simple algebraic argument) in Section 2.9.

2.1 Abstract one-skeletons

In Part 2 of this article, $\mathfrak{g}$ will simply be a vector space over $\mathbb{R}$ of dimension n and $\mathfrak{g}^*$ its vector space dual. (In particular $\mathfrak{g}$ will *not* necessarily be the Lie algebra of a group G.) Let Γ be a finite simple d-valent graph and let I_Γ be its incidence relation, which can be identified with E_Γ, the set of oriented edges of Γ: $(p,e) \in I_\Gamma$ corresponds to the *oriented* edge e, oriented such that $p = i(e)$, the initial vertex of e. For $e \in E_\Gamma$, let $\bar{e}$ be the same edge with the reversed orientation; for a vertex $p \in V_\Gamma$, let E_p be the set of oriented edges issuing from p.

Definition. An *axial function* on Γ is a map $\alpha : E_\Gamma \longrightarrow \mathfrak{g}^*$ having the properties

1. If $e \in E_\Gamma$, then

(2.1) $$\alpha_{\bar{e}} = -\alpha_e .$$

2. If p is a vertex and $E_p = \{e_1, \dots, e_d\}$, then the vectors

(2.2) $$\alpha_{e_i}, \quad i = 1, \dots, d$$

are pair-wise linearly independent.

3. Let e be an edge and p and q the vertices joined by e. Let

$$\mathfrak{g}_e = \{\xi \in \mathfrak{g}, \alpha_e(\xi) = 0\}$$

and let

$$\rho_e : \mathfrak{g}^* \longrightarrow \mathfrak{g}_e^*$$

be the transpose of the inclusion map $\mathfrak{g}_e \longrightarrow \mathfrak{g}$. Let $e_i, i = 1, \dots, d$ and $e'_i, i = 1, \dots, d$ be the edges issuing from p and q respectively, with $e_d = e$ and $e'_d = \bar{e}$. Then the e_i's can be ordered so that

(2.3) $$\rho_e \alpha_{e_i} = \rho_e \alpha_{e'_i} .$$

(See (1.16)-(1.18)).

Definition. A *connection*, θ, on Γ is a collection of maps

(2.4) $$\theta_e : E_p \longrightarrow E_q ,$$

indexed by $e = (p,q) \in E_\Gamma$, with the properties

(2.5) $$\theta_e(e) = \bar{e}$$

and

(2.6) $$\theta_{\bar{e}} = \theta_e^{-1} .$$

(See (1.32)–(1.33)). An axial function α and a connection θ are *compatible* if for every oriented edge $e = (p, q) \in E_\Gamma$ and every $e' \in E_p$,

$$\rho_e(\alpha_{e'}) = \rho_e(\alpha_{\theta_e(e')}). \tag{2.7}$$

(See (1.34)).

For instance, suppose that for every vertex p and every edge $e \in E_p$, no two of the vectors (2.3) are equal. Then there is a *unique* connection on Γ which is compatible with α.

Definition. An *abstract one-skeleton* is a pair (Γ, α) consisting of a finite simple d-valent graph Γ and an axial function $\alpha : E_\Gamma \longrightarrow \mathfrak{g}^*$.

We will list below a few examples of abstract one-skeletons and describe some of their functorial properties:

Example 1 (*The complete one-skeleton on N vertices*). The vertices of the graph are the elements of the N-element set $\{1, \ldots, N\}$ and each pair of elements, (i, j), $i \neq j$, is joined by an edge. Thus the set of oriented edges is just the set

$$\{(i, j); 1 \leq i, j \leq N,\ i \neq j\}.$$

A connection θ can be defined as follows: For every oriented edge (i, j) of Γ, let $\theta_{i,j}$ be the map of the set of edges of Γ issuing from i onto the set of edges of Γ issuing from j which sends (i, j) onto (j, i) and, for $k \neq i, j$, sends (i, k) onto (j, k).

Let α_i, $i = 1, \ldots, N$ be non-zero elements of $\mathfrak{g}^*$. Then the function $(i, j) \longrightarrow \alpha_i - \alpha_j$ satisfies (2.1) and (2.3) and hence it is an axial function iff, for every i, the $N - 1$ vectors $\alpha_i - \alpha_j$, $i \neq j$, are pairwise linearly independent. Conversely, we claim that *every* axial function compatible with θ is of this form. (Proof : Let $(i, j) \longrightarrow \alpha_{i,j}$ be an axial function. Then, if i, j and k are distinct,

$$\alpha_{i,k} = \alpha_{j,k} + c_{i,j}\alpha_{i,j} \qquad c_{i,j} = c_{j,i}.$$

Hence

$$\alpha_{k,i} = \alpha_{j,i} + c_{k,j}\alpha_{k,j} = \alpha_{k,j} + c_{i,j}\alpha_{j,i}$$

so $(1 - c_{i,j})\alpha_{j,i} = (1 - c_{k,j})\alpha_{k,j}$ and $c_{i,j} = c_{k,j} = 1$. Now let $\alpha_1 = 0$ and let $\alpha_i = \alpha_{i,1}$ for $i > 1$.)

Example 2 (*Sub-one-skeletons.*) Let Γ_1 be an r-valent sub-graph of Γ and let i be the embedding of V_{Γ_1} into V_Γ. From i one gets an embedding $i : E_{\Gamma_1} \longrightarrow E_\Gamma$ and one can pull-back the axial function α to E_{Γ_1}. In general $i^*\alpha := \alpha_1$ won't be an axial function; however if it is, we will say that Γ_1 is *compatible* with α and call (Γ_1, α_1) a *sub-one-skeleton* of (Γ, α).

Example 3 (*Sub-one-skeletons which are "totally geodesic" with respect to a connection*). Let θ be a connection on Γ and let Γ' be an r-valent subgraph of Γ. For every vertex p of Γ let E_p and E'_p be the oriented edges of Γ and of Γ' issuing from p. We will say that Γ' is totally geodesic with respect to θ if, for every oriented edge $e = (p, q)$ of Γ', the restriction of the holonomy map $\theta_e : E_p \longrightarrow E_q$ to E'_p maps E'_p onto E'_q.

If this happens, this restriction defines an induced connection, θ', on Γ'. Moreover, if $\alpha : E_\Gamma \longrightarrow \mathfrak{g}^*$ is an axial function which is compatible with θ, the restriction of α to $E_{\Gamma'}$ is an axial function and is compatible with θ'.

Example 4 (*The totally geodesic sub-one-skeletons of* (Γ_N, α)). Let Γ_N be the complete graph on N vertices and, for every subset S of $\{1, \dots, N\}$, let Γ_S be the graph whose vertices are the elements of S and whose oriented edges are the pairs (s_1, s_2), $(s_i \in S)$, $s_1 \neq s_2$. It is obvious that Γ_S is a totally geodesic with respect to the connection we defined in Example 1; and, in fact, every totally geodesic sub-one-skeleton of Γ_N is a Γ_S for some subset S. (To see this, let Γ' be a connected totally geodesic subgraph and let (p_1, p_2) be an oriented edge of Γ'. If q is a vertex of Γ' distinct from p_1 and p_2 and (p_1, q) is an oriented edge of Γ', then (p_2, q) has to be an oriented edge of Γ', so it follows from the connectivity of Γ' that, for every pair of vertices p and q of Γ', (p, q) is an oriented edge of Γ'.)

Example 5 (*The one-skeleton* $(\Gamma_\mathfrak{h}, \alpha)$). Let $\mathfrak{h}$ be a vector subspace of $\mathfrak{g}$ and let $\rho_\mathfrak{h} : \mathfrak{g}^* \longrightarrow \mathfrak{h}^*$ be the transpose of the inclusion map $\mathfrak{h} \longrightarrow \mathfrak{g}$. Let $\Gamma_\mathfrak{h}$ be the subgraph of Γ whose edges are the edges e of Γ for which

$$\rho_\mathfrak{h}\alpha_e = -\rho_\mathfrak{h}\alpha_{\bar{e}} = 0. \tag{2.8}$$

Each connected component of this graph is k-valent for some k and is a sub-one-skeleton in the sense of item 2. Moreover, if p and q are joined by an edge $e \in E_\Gamma$ and are in the same connected component of $\Gamma_\mathfrak{h}$, and if e_i and e'_i, $i = 1, ..., d$, are the edges of Γ issuing from p and q, then one can order the e_i's so that

$$\rho_\mathfrak{h}\alpha_{e_i} = \rho_\mathfrak{h}\alpha_{e'_i} \tag{2.9}$$

(compare with (1.13)).

Example 6 (*Product one-skeletons*). Let Γ_1 be a graph of valence d_1 and Γ_2 a graph of valence d_2. The vertices of the *product* graph $\Gamma_1 \times \Gamma_2$ are the pairs (p, q), $p \in V_{\Gamma_1}$ and $q \in V_{\Gamma_2}$; two vertices (p, q) and (p', q') are joined by an edge if either $p = p'$ and q and q' are joined by an edge in Γ_2 or $q = q'$ and p and p' are joined by an edge in Γ_1. Thus this product graph is a $d_1 + d_2$-valent graph and its set of oriented edges is the disjoint union

$$E_{\Gamma_1 \times \Gamma_2} = E_{\Gamma_1} \amalg E_{\Gamma_2} \tag{2.10}$$

of the set of oriented edges of Γ_1 and the set of oriented edges of Γ_2, and for a vertex (p, q) of $\Gamma_1 \times \Gamma_2$,

$$E_{p,q} \simeq E_p \amalg E_q \ .$$

If

$$\alpha_i : E_{\Gamma_i} \longrightarrow \mathfrak{g}^*, \qquad i = 1, 2$$

is an axial function on Γ_i, one defines the product axial function α on $\Gamma_1 \times \Gamma_2$ to be the function which is equal to α_1 on the first summand of (2.10) and equal to α_2 on the second summand. Then α satisfies (2.1) - (2.3) and it is called *the product axial function.* If in addition, θ_1 and θ_2 are connections on Γ_1 and Γ_2, then one can define a product connection θ on $\Gamma_1 \times \Gamma_2$ by letting

$$\theta_{(p,q),(p',q)} = \theta_{1;p,p'} \amalg (\mathrm{Id})_{E_q}$$

and

$$\theta_{(p,q),(p,q')} = (\mathrm{Id})_{E_p} \amalg \theta_{2;q,q'} \ .$$

If α_i and θ_i are compatible, for $i = 1$ and 2, then the product axial function which we defined above is compatible with θ.

Example 7 (*Blowing-up*). This operation can be defined for any sub-one-skeleton; however, for simplicity, we will only consider here the special case of a point. Let Γ be a finite simple d-valent graph and let p_0 be an arbitrary vertex of Γ. Let e_i, $1 = 1, ..., d$, be the edges of Γ issuing from p_0 and let q_i be the vertex joined by e_i to p_0. From this data one can construct a new graph, $\Gamma^{\#}$, as follows. Replace the vertex p_0 by d new vertices, p_i, $i = 1, ..., d$, (which one should think of as being the "baricenters" of the edges e_i) and to each of these new vertices adjoin d

edges; one edge going from p_i to q_i (which one should think of as being a replacement for the old edge e_i) and one edge going from p_i to each of the p_j's, $j \neq i$. Let

$$\beta : V_{\Gamma^\#} \longrightarrow V_\Gamma \tag{2.11}$$

be the map which sends $\{p_1, ..., p_d\}$ to p_0 and is the identity on the complement of $\{p_1, ..., p_d\}$. We will call this map the *blowing-down* map. The set $\{p_1, ..., p_d\}$, which is the pre-image of p_0 with respect to β, is the set of vertices of a sub-graph of $\Gamma^\#$, Γ_0 (the complete graph on d vertices), which we will call *the singular locus* of β.

Now let $\alpha : E_\Gamma \longrightarrow \mathfrak{g}^*$ be an axial function and let α_i, $i = 1, ..., d$, be the values of α on the edges e_i issuing from p_0. Let us assume that for each i, the $d-1$ vectors $\alpha_j - \alpha_i$, $j \neq i$, are pairwise linearly independent. We can then define an axial function, $\alpha^\#$, on $\Gamma^\#$, as follows:

1. On the oriented edges e of Γ, not containing p_0, $\alpha^\#(e) = \alpha(e)$.

2. On the oriented edges $e = (p_i, q_i)$, $\alpha^\#(e) = \alpha_i$.

3. On the oriented edges $e = (p_i, p_j)$, $\alpha^\#(e) = \alpha_j - \alpha_i$.

This defines $\alpha^\#$ on *all* edges of $\Gamma^\#$ and it is easy to check that $\alpha^\#$ satisfies the axioms (2.1) - (2.3).

Finally, if θ is a connection on Γ, then there is a unique connection $\theta^\#$ on $\Gamma^\#$ with the following properties.

1. The restriction of $\theta^\#$ to Γ_0 is the connection described in Example 1.

2. $\theta^\#_{(p_i,p_j)}$ maps the oriented edge (p_i, q_i) to the oriented edge (p_j, q_j).

3. $\theta^\#_{(p_i,q_i)}$ maps the oriented edge (p_i, p_j) to the oriented edge $\theta_{(p_0,q_i)}(e_j)$.

4. If q is not equal to one of the p_i's and q and q_i are joined by an edge, then
$$\theta^\#_{(q_i,q)}(q_i, p_i) = \theta_{(q_i,q)}(q_i, p)\,.$$
On the other edges of $\Gamma^\#$ issuing from q_i, $\theta^\#_{(q_i,q)} = \theta_{(q_i,q)}$.

5. If q and q' are not equal to one of the q_i's or one of the p_i's, then $\theta^\#_{(q,q')} = \theta_{(q,q')}$.

Example 8 (*The case d=dim* $\mathfrak{g}^*$ *=2*). Let Γ be a finite connected 2-valent graph with N vertices, $\mathfrak{g}^*$ a 2-dimensional vector space and α an axial function. Let $p_1, ..., p_N, p_{N+1} = p_1$ be an enumeration of the vertices of Γ and let α_i be the value of α on the oriented edge (p_i, p_{i+1}). Then (1.18) is equivalent to

$$\alpha_{i+1} \wedge \alpha_i = \alpha_i \wedge \alpha_{i-1} \tag{2.12}$$

for all i. (For example for $N = 4k$, let $\{\alpha_1, \alpha_2\}$ be a basis of $\mathfrak{g}^*$. Then a solution of (2.12) is obtained by letting $\alpha_1 = -\alpha_3 = \alpha_5 = ...$ and $\alpha_2 = -\alpha_4 = \alpha_6 = ...$).

2.2 Orientations

Let (Γ, α) be an abstract one-skeleton and let

$$\mathcal{P} = \{\xi \in \mathfrak{g}, \alpha_e(\xi) \neq 0 \text{ for all } e \in E_\Gamma\}.$$

Then for every $\xi \in \mathcal{P}$, the axial function α defines an *orientation* of Γ; in other words, for each edge e, it fixes a new ordering of the vertices of e. Namely if e joins p to q, one can order p and q so that

$$p \prec q \Leftrightarrow \alpha_e(\xi) > 0. \tag{2.13}$$

It is clear that this orientation doesn't depend on ξ but only on the connected component of $\mathcal{P}$ in which ξ is contained. On the other hand it is clear that different components will give rise to different orientations (for instance, replacing ξ by $-\xi$ reverses all the orientations). We will say that Γ satisfies the *no-cycle condition* if, for at least one of these orientations, Γ has no cycles.

Definition. Given $\xi \in \mathcal{P}$, a function $f : V_\Gamma \longrightarrow \mathbb{R}$ is *positively oriented with respect to* ξ if, for every pair of vertices p and q and edge e joining p to q, the ratio of $f(q) - f(p)$ to $\alpha_e(\xi)$ is positive.

If f is positively oriented with respect to ξ, then the orientation of Γ associated with ξ can't have closed cycles since f has to be strictly increasing along any oriented path. We will prove that the converse is true:

Theorem. *If the orientation of* Γ *associated with* ξ *has no cycles, then there exists a function* $f : V_\Gamma \longrightarrow \mathbb{R}$ *which is positively oriented with respect to* ξ.

Proof. Given $p \in V_\Gamma$, consider the *longest* oriented path with initial point p, i.e., the longest sequence

$$p = p_1, p_2, \dots, \quad p_i \in V_\Gamma \tag{2.14}$$

with the property that p_i and p_{i+1} are the vertices of a common edge and, relative to the orientation on this edge, $p_i \prec p_{i+1}$. If Γ has no cycles this longest path has to be of finite length, i.e., has to terminate at some point p_N. Now set $f(p) = -N$. It is easy to check that this function is positively oriented with respect to ξ. q.e.d.

Remarks.

1. The vertices, p, where $f(p) = 0$ have the property that all edges containing p are pointing "into" p, i.e. p is a "maximum" of the oriented graph Γ. In particular, if $f(p) = -N$, this is true of the vertex p_N in the sequence (2.14); so the argument above shows that every vertex can be joined by an ascending path to a maximal vertex.

2. One can perturb f so that it remains positively oriented with respect to ξ and, in addition, takes on distinct values at distinct vertices. Namely, suppose that

$$f^{-1}(k) = \{p_1, \dots, p_r\}.$$

Redefine f on the set $\{p_1, \dots, p_r\}$ by setting f equal to $k + \epsilon_i$ on p_i where $\epsilon_i \neq \epsilon_j$ for $i \neq j$ and the ϵ_i's are small. This redefined function is still positively oriented with respect to ξ but now takes distinct values at $p_1, \dots, p_r$.

2.3 The cohomology ring of (Γ, α)

We define the *cohomology ring* of (Γ, α) to be the ring $H(\Gamma, \alpha)$ which we defined in §1.6. As (Γ, α) is no longer the GKM data associated with a G-manifold, it is, perhaps, a misnomer to refer to this ring as a "cohomology ring"; however, there are other reasons for using this terminology. For instance, if Γ is the one-skeleton of a simplicial polytope, $H(\Gamma, \alpha)$ is just the Stanley-Reissner cohomology ring of the dual polytope. (We are indebted to Mark Goresky for this observation.) We will describe below a few properties of this ring.

1. As we pointed out in §1.6, $H(\Gamma, \alpha)$ contains $\mathbb{S}(\mathfrak{g}^*)$ as a subring.

2. Chern classes: For each $p \in V_\Gamma$, let $E_p = \{e_1, \dots, e_d\}$ and let $c_k(p)$ be the k-th elementary symmetric function in the monomials $\alpha_{e_1}, \dots, \alpha_{e_d}$. The function $p \longrightarrow c_k(p)$ defines an element c_k of $H^{2k}(\Gamma, \alpha)$, which can be thought of as the k-th Chern class of the "tangent bundle" of Γ.

3. Symplectic structures: An element of $H^2(\Gamma, \alpha)$ is just a map $c : V_\Gamma \longrightarrow \mathfrak{g}^*$ satisfying

$$c(q) - c(p) = \lambda_e \alpha_e \tag{2.15}$$

for every pair of vertices p, q and edge e joining p to q. We will call c *symplectic* if, for every edge e, λ_e is positive. The existence of a symplectic structure implies that for every $\xi \in \mathcal{P}$, the orientation of Γ associated with ξ has the no-cycle property. (Proof: It follows from (2.15) that the ξ-component of c is an $\mathbb{R}$-valued function on V_Γ which is positively oriented with respect to ξ.)

4. Thom classes: Fix a vertex p and let $E_p = \{e_1, \dots, e_d\}$. Let $\tau : V_\Gamma \longrightarrow \mathbb{S}^d(\mathfrak{g}^*)$ be the map which is zero at $q \neq p$ and at p is equal to the product $\alpha_{e_1} \dots \alpha_{e_d}$. Then τ is in $H^{2d}(\Gamma, \alpha)$.

5. Sub-one-skeletons: Let (Γ_1, α_1) be a sub-one-skeleton of (Γ, α). Then the inclusion map $i : V_{\Gamma_1} \longrightarrow V_\Gamma$ induces a map

$$i^* : H(\Gamma, \alpha) \longrightarrow H(\Gamma_1, \alpha_1), \quad \alpha_1 = i^* \alpha\,.$$

6. Gysin maps: Suppose that (Γ_1, α) is a sub-one-skeleton of (Γ, α). The *Thom class* of Γ_1 is the map $\tau : V_\Gamma \longrightarrow \mathbb{S}^{d-r}(\mathfrak{g}^*)$ which is zero on the vertices of Γ which are not vertices of Γ_1 and on vertices p of Γ_1, is equal to

$$\tau(p) = \alpha_{e_1} \dots \alpha_{e_s}\ ,$$

where $s = d - r$ and $e_1, \dots, e_s$ are the edges of Γ starting at p which don't belong to Γ_1. From τ one gets a Gysin map

$$H^{2k}(\Gamma_1, \alpha_1) \longrightarrow H^{2(k+s)}(\Gamma, \alpha)\ ,$$

sending f to τf. (Since τ is supported on V_{Γ_1}, this map is well-defined.)

7. The cohomology of blow-ups: Let (Γ, α) be a d-valent abstract one-skeleton. Let p_0 be a vertex of Γ and let $(\Gamma^\#, \alpha^\#)$ be the blow-up of (Γ, α) at p_0 (See §2.1). From the blowing down map (2.11) one gets a pull-back map on cohomology

$$\beta^* : H(\Gamma, \alpha) \longrightarrow H(\Gamma^{\#}, \alpha^{\#}) \ ,$$

which embeds $H(\Gamma, \alpha)$ as a subring of $H(\Gamma^{\#}, \alpha^{\#})$. Moreover the singular locus, Γ_0, of β, is a sub-one-skeleton of $\Gamma^{\#}$ (in the sense of Example 2 in Section 2.1) and its Thom class,

$$\tau \in H^2(\Gamma^{\#}, \alpha^{\#})$$

generates $H(\Gamma^{\#}, \alpha^{\#})$ over the sub-ring $H(\Gamma, \alpha)$, subject to the relation

$$\tau^d - c_1 \tau^{d-1} + c_2 \tau^{d-2} - \ldots \pm c_d,$$

the c_i's being the Chern classes of Γ for $i < d$ and c_d being the Thom class of p_0.

2.4 The Atiyah-Bott-Berline-Vergne localization theorem

We have just discussed functoriality for sub-one-skeletons of (Γ, α). What about quotient objects? To take the most extreme case, let "*pt*" be the trivial zero-valent graph consisting of one vertex, *pt*, and no edges, and let

$$\pi : V_\Gamma \longrightarrow pt$$

be the constant map. We have already seen (see §2.3 item 2.3) that there is a functorial map

$$\pi^* : H^{2k}(pt) \longrightarrow H^{2k}(\Gamma, \alpha) \, .$$

However, does there exist a Gysin map

$$\pi_* : H^{2k}(\Gamma, \alpha) \longrightarrow H^{2(k-d)}(pt) \, ?$$

Such a map, if it existed, would have to have the following property. Let p be a vertex of Γ and let $j_p : pt \longrightarrow V_\Gamma$ be the map $pt \longrightarrow p$. Then, by functoriality, π_* would have to satisfy

$$\pi_*(j_p)_* = \text{ identity}$$

and, by items 2.3 and 2.3 of §2.3, π_* would have to have the form

$$\pi_* f = \sum f(p) (\prod_{e \in E_p} \alpha_e)^{-1}. \tag{2.16}$$

However, it is by no means obvious that this map is well defined, i.e. that the right hand side of (2.16) is in $\mathbb{S}(\mathfrak{g}^*)$. We will prove that it is :

Theorem 2.2. *π_* maps $H^{2k}(\Gamma, \alpha)$ into $\mathbb{S}^{k-d}(\mathfrak{g}^*)$.*

Proof. Let $f \in H^{2k}(\Gamma, \alpha)$; then $\pi_* f$ can be written as

$$\pi_* f = \frac{g}{\prod_{j=1}^{N} \alpha_j}, \tag{2.17}$$

where $g \in \mathbb{S}^{k-d+N}(\mathfrak{g}^*)$ and $\alpha_1, \cdots, \alpha_N$ are pair-wise linearly independent. We will show that α_1 divides g.

The vertices of Γ can be divided into two categories:

1. The first subset, V_1, contains the vertices $p \in V_\Gamma$ for which none of the α_e's, $e \in E_p$, is a multiple of α_1;
2. The second subset, V_2, contains the vertices $p \in V_\Gamma$ for which there exists an edge $e \in E_p$ such that α_e is a multiple of α_1. (Notice that (2.2) implies that there will be exactly one such edge.)

The part of (2.16) corresponding to vertices in the first category will then be of the form

$$\sum_{p \in V_1} \frac{f(p)}{\prod \alpha_e} = \frac{g_1}{\prod_{j=2}^{N} \alpha_j}, \tag{2.18}$$

with $g_1 \in \mathbb{S}(\mathfrak{g}^*)$.

If $p \in V_2$ then there exists an edge $e \in E_p$ such that $\alpha_e = \lambda \alpha_1$, with $\lambda \in \mathbb{R} - \{0\}$; let q be the terminal vertex of e. Since $\alpha_{\bar{e}} = -\alpha_e$, it results that $q \in V_2$ as well and thus the vertices in V_2 can be paired as above.

Let $e_i, i = 1, \ldots, d$ and $e'_i, i = 1, \ldots, d$ be the edges issuing from p and q respectively, with $e_d = e$ and $e'_d = \bar{e}$. Then the e_i's can be ordered (cf. 2.3) so that

$$\alpha_{e_i} \equiv \alpha_{e'_i} \pmod{\alpha_1}. \tag{2.19}$$

Also (1.21) implies that

$$f(q) \equiv f(p) \pmod{\alpha_1}. \tag{2.20}$$

The part of (2.16) corresponding to p and q is given by

$$\begin{aligned} &\frac{f(p)}{\prod_{j=1}^{d} \alpha_{e_j}} + \frac{f(q)}{\prod_{j=1}^{d} \alpha_{e'_j}} \\ &= \frac{f(p)\alpha_{e'_1} .. \alpha_{e'_{d-1}} - f(q)\alpha_{e_1} .. \alpha_{e_{d-1}}}{\lambda \alpha_1 \alpha_{e_1} ... \alpha_{e'_{d-1}}}. \end{aligned} \tag{2.21}$$

But multiplying together the congruences (2.19) and (2.20) we obtain that α_1 divides the numerator of (2.21) so that

$$\frac{f(p)}{\prod_{j=1}^{d}\alpha_{e_j}}+\frac{f(q)}{\prod_{j=1}^{d}\alpha_{e'_j}}=\frac{g_{p,q}}{\prod_{j=2}^{N}\alpha_j}, \tag{2.22}$$

with $g_{p,q}\in\mathbb{S}(\mathfrak{g}^*)$. Therefore

$$\sum_{p\in V_2}\frac{f(p)}{\prod\alpha_e}=\frac{g_2}{\prod_{j=2}^{N}\alpha_j}, \tag{2.23}$$

with $g_2\in\mathbb{S}(\mathfrak{g}^*)$. Adding (2.18) and (2.23) we get that

$$\frac{g}{\prod_{j=1}^{N}\alpha_j}=\frac{g_1+g_2}{\prod_{j=2}^{N}\alpha_j},$$

with $g_1+g_2\in\mathbb{S}(\mathfrak{g}^*)$, i.e. that α_1 divides g. The same argument can be used to show that each α_j divides g and therefore $\pi_*f\in\mathbb{S}^{k-d}(\mathfrak{g}^*)$, as desired. q.e.d.

2.5 The Kirwan map

From now on we will assume that (Γ,α) satisfies the no-cycle condition. Let ξ be an element of $\mathcal{P}$ which gives an orientation of Γ without cycles and let $\phi: V_\Gamma\longrightarrow\mathbb{R}$ be positively oriented with respect to ξ; without loss of generality we can assume that ϕ is injective. For $c\in\mathbb{R}-\phi(V_\Gamma)$, we define the *c-cross section*, Γ_c, of Γ to be the set of edges $e=(p,q)\in E_\Gamma$ with the property that $\phi(p)<c<\phi(q)$. Let $\mathfrak{g}_\xi^*$ be the annihilator of ξ in $\mathfrak{g}^*$ and let $M^{2k}(\Gamma_c)$ be the set of all maps

$$f:\Gamma_c\longrightarrow\mathbb{S}^k(\mathfrak{g}_\xi^*).$$

The sum

$$M(\Gamma_c)=\bigoplus M^{2k}(\Gamma_c)$$

is a graded ring under point-wise multiplication and we will define a morphism of graded rings

$$\mathcal{K}_c: H(\Gamma,\alpha)\longrightarrow M(\Gamma_c)$$

as follows: Let $e=(p,q)\in\Gamma_c$. The projection $\mathfrak{g}^*\longrightarrow\mathfrak{g}_e^*$ maps $\mathfrak{g}_\xi^*$ bijectively onto $\mathfrak{g}_e^*$, since $\alpha_e(\xi)\neq 0$, so one has a composite map

$$\mathfrak{g}^*\longrightarrow\mathfrak{g}_e^*\leftrightarrow\mathfrak{g}_\xi^*$$

and hence an induced morphism of graded rings

$$\mathbb{S}(\mathfrak{g}^*) \longrightarrow \mathbb{S}(\mathfrak{g}_e^*) \leftrightarrow \mathbb{S}(\mathfrak{g}_\xi^*) .$$

If f is in $H(\Gamma, \alpha)$, then the images of f_p and f_q in $\mathbb{S}(\mathfrak{g}_e^*)$ are the same by (1.21) and hence so are their images in $\mathbb{S}(\mathfrak{g}_\xi^*)$. We define $\mathcal{K}_c(f)(e)$ to be this common image and call the map $\mathcal{K}_c$ *the Kirwan map*.

Next we will define a morphism, γ_c, of $\mathbb{S}(\mathfrak{g}_\xi^*)$-modules, mapping $M(\Gamma_c)$ into the quotient field of $\mathbb{S}(\mathfrak{g}_\xi^*)$. To define γ_c, let $e = (p, q) \in \Gamma_c$. Let

$$e_i \,, \quad i = 1, \ldots, d-1$$

be the other edges of Γ (other than e) issuing from p and let

$$e_i' \,, \quad i = 1, \ldots, d-1$$

be the other edges of Γ intersecting at q. By the compatibility axiom we can assume that α_{e_i} and $\alpha_{e_i'}$ have the same image in $\mathfrak{g}_e^*$ and, hence, under the identification $\mathfrak{g}_e^* \leftrightarrow \mathfrak{g}_\xi^*$, have the same image in $\mathfrak{g}_\xi^*$. This implies that

$$\alpha_{e_i} - m_{i,e}\alpha_e = \alpha_{e_i'} - m_{i,e}'\alpha_{\bar{e}} =: \alpha_{i,e}^{\#} \,,$$

where

$$m_{i,e} = \frac{\alpha_{e_i}(\xi)}{\alpha_e(\xi)} \qquad \text{and} \qquad m_{i,e}' = \frac{\alpha_{e_i'}(\xi)}{\alpha_{\bar{e}}(\xi)}.$$

Let

$$m_e = \alpha_e(\xi) = -\alpha_{\bar{e}}(\xi)$$

and note that, since $\phi(p) < \phi(q)$, $m_e = |m_e| > 0$.

We now define, for $f \in M^{2k}(\Gamma_c)$,

$$\gamma_c f = \sum_{e \in \Gamma_c} \frac{1}{m_e} \frac{f(e)}{\prod_i \alpha_{i,e}^{\#}} \tag{2.24}$$

and we define $\mathfrak{p}_c$ to be the composition

$$\mathfrak{p}_c = \gamma_c \mathcal{K}_c \,. \tag{2.25}$$

Theorem 2.3. *$\mathfrak{p}_c$ is a map of $H^{2k}(\Gamma, \alpha)$ into $\mathbb{S}^{k-d+1}(\mathfrak{g}_\xi^*)$.*

We will prove this by obtaining an explicit "residue formula" for $\mathfrak{p}_c(f)$, $f \in H^{2k}(\Gamma, \alpha)$. This residue formula can be viewed as a kind of combinatorial version of the Jeffrey-Kirwan theorem described in §1.3. It is also closely related to the localization theorem proved by Jaap Kalkman and the first author in [7], and the residue results which we will describe in the next section are mostly taken from [7].

2.6 Residues

Let $\alpha_1, \dots \alpha_d$ be elements of $\mathfrak{g}^*$ and ξ be an element of $\mathfrak{g}$ with the property that $\alpha_i(\xi) \neq 0$ for all i. Given $f \in \mathbb{S}^k(\mathfrak{g}^*)$ we define an element

$$\operatorname{Res}_\xi \frac{f}{\prod \alpha_i} \tag{2.26}$$

of $\mathbb{S}^{k-d+1}(\mathfrak{g}^*_\xi)$ as follows: Choose a basis $x, y_1, .., y_{n-1}$ of $\mathfrak{g}^*$ such that $y_1, .., y_{n-1}$ is a basis of $\mathfrak{g}^*_\xi$ and $x(\xi) = 1$. Let

$$\alpha_i = m_i x - \sum_{j=1}^{n-1} a_{ij} y_j \tag{2.27}$$

and let

$$f(x) = \sum_{r=0}^{k} x^r f_r(y)$$

and

$$\beta_i = m_i^{-1} \sum a_{ij}\, y_j ; \tag{2.28}$$

then

$$\frac{f}{\alpha_1 \dots \alpha_d} = \left(\prod m_i\right)^{-1} x^{-d} \sum x^r f_r \left(1 - \frac{\beta_1}{x}\right)^{-1} \cdots \left(1 - \frac{\beta_d}{x}\right)^{-1} .$$

Now replace $(1 - \frac{\beta_i}{x})^{-1}$ by the power series

$$\sum_{k=0}^{\infty} x^{-k} \beta_i^k$$

and define (2.26) to be the coefficient of x^{-1} in the product on the right. It is easy to see that this definition doesn't depend on the choice of $x, y_1, \dots, y_n$. If the α_i's are pairwise linearly independent (i.e., if, for $i \neq j$, α_i and α_j are not multiples of each other) there is a relatively simple formula for (2.26).

Lemma 1. *Let A be a graded commutative algebra over $\mathbb{C}$ and let $f = f(x)$ be a polynomial in x with coefficients in A. Then for indeterminants $z_1, \dots, z_d$*

$$Res_x \frac{f(x)}{(x - z_1) \cdots (x - z_d)} = \sum_{i=1}^{d} \frac{f(z_i)}{\prod_{j \neq i} (z_i - z_j)} . \tag{2.29}$$

Proof. The conclusion follows immediately from the decomposition in simple fractions

$$\frac{f(x)}{(x-z_1)\cdots(x-z_d)} = F(x) + \sum_{i=1}^{d} \frac{f(z_i)}{\prod_{j\neq i}(z_i - z_j)} \frac{1}{x - z_i}$$

where $F(x)$ is a polynomial term in x. q.e.d.

Let

$$h = \frac{f}{\prod_{j=1}^{d}(x - z_j)} \quad \text{and} \quad h_j = \frac{f(z_j)}{\prod_{r\neq j}(z_j - z_r)}, \quad \text{for all } j = 1,..,d.$$

Lemma 2. *$h \in A[x]$ if and only if $Res_x(x^k h) = 0$, for all $k \geq 0$.*

Proof. From (2.29) we get that

$$\mathrm{Res}_x(x^k h) = \sum_{j=1}^{d} (z_j)^k h_j.$$

Then the fact that $\mathrm{Res}_x(x^k h) = 0$ for all $k = 1, ..., d$ can be written as

$$\begin{pmatrix} z_1^1 & \cdots & z_j^1 & \cdots & z_d^1 \\ \vdots & & \vdots & & \vdots \\ z_1^k & \cdots & z_j^k & \cdots & z_d^k \\ \vdots & & \vdots & & \vdots \\ z_1^d & \cdots & z_j^d & \cdots & z_d^d \end{pmatrix} \begin{pmatrix} h_1 \\ \vdots \\ h_j \\ \vdots \\ h_d \end{pmatrix} = 0.$$

Since the corresponding Vandermonde determinant is non-zero we deduce that $h_1 = \cdots = h_d = 0$, i.e. $f(z_j) = 0$, for all $j = 1, ..., d$, from which we obtain that $h \in A[x]$; the other implication is clear. q.e.d.

We will now apply Lemma 1 to the evaluation of (2.26). Let

$$m_i = \alpha_i(\xi)$$

and for $i \neq j$ let

$$\alpha_{j,i}^{\#} = \alpha_j - m_{j,i}\alpha_i$$

with

$$m_{j,i} = \frac{\alpha_j(\xi)}{m_i}.$$

Note that $\alpha^{\#}_{j,i} \in \mathfrak{g}^*_\xi$, since $\alpha^{\#}_{j,i}(\xi) = \alpha_j(\xi) - \alpha_j(\xi) = 0$. Let

$$\mathfrak{g}^*_i = \mathfrak{g}^* / \{c\alpha_i \, ; \, c \in \mathbb{R}\} \, .$$

The projection map $\mathfrak{g}^* \longrightarrow \mathfrak{g}^*_i$ is bijective on $\mathfrak{g}^*_\xi$, so one gets a composite map

$$\mathfrak{g}^* \longrightarrow \mathfrak{g}^*_i \xrightarrow{\simeq} \mathfrak{g}^*_\xi$$

as in §2.5 and hence a ring morphism

$$\mathcal{K}_i \, : \, \mathbb{S}(\mathfrak{g}^*) \longrightarrow \mathbb{S}(\mathfrak{g}^*_\xi) \, .$$

Theorem 2.4. *For* $f \in \mathbb{S}^k(\mathfrak{g}^*)$

$$Res_\xi \frac{f}{\alpha_1 \dots \alpha_d} = \sum_i \frac{1}{m_i} \frac{\mathcal{K}_i f}{\prod_{j \neq i} \alpha^{\#}_{j,i}} \, . \tag{2.30}$$

Proof. With the notations (2.27)-(2.28)

$$\frac{f}{\prod \alpha_i} = (\prod m_k)^{-1} \frac{f(x,y)}{\prod (x - \beta_k(y))} \, .$$

Thus by Lemma 1

$$\begin{aligned} \mathrm{Res}_\xi \frac{f}{\prod \alpha_i} &= (\prod m_k)^{-1} \sum_i \frac{f(\beta_i, y)}{\prod_{k \neq i} (\beta_i - \beta_k)} \\ &= \sum_i \frac{1}{m_i} \frac{f(\beta_i, y)}{\prod_{k \neq i} m_k (\beta_i - \beta_k)} . \end{aligned}$$

But $m_k(\beta_i - \beta_k) = \alpha^{\#}_{k,i}$ and $\mathcal{K}_i$ maps x to β_i and y_k to itself, so $\mathcal{K}_i f(x,y) = f(\beta_i, y)$. Thus the sum on the right is identical with the right hand side of (2.29). q.e.d.

2.7 The Jeffrey-Kirwan theorem

We will prove Theorem 2.3 by deducing it from the following result:

Theorem 2.5. *For* $f \in H^{2k}(\Gamma, \alpha)$

$$\mathfrak{p}_c(f) = \sum_{\phi(p) < c} Res_\xi \frac{f_p}{\prod \alpha_e} \, . \tag{2.31}$$

In particular, $\mathfrak{p}_c(f)$ *is in* $\mathbb{S}^{k-d+1}(\mathfrak{g}^*_\xi)$.

Proof. Choose values $c = c_0 > c_1 > c_2 > ... > c_N$ in $\mathbb{R} - \phi(V)$ so that $\phi(V) \subset (c_N, \infty)$ and for all $r \geq 0$ there is exactly one vertex, p_r, with $\phi(p_r) \in (c_{r+1}, c_r)$. Inspection of (2.25) shows that

$$\mathfrak{p}_{c_r}(f) - \mathfrak{p}_{c_{r+1}}(f) = \sum \frac{1}{m_i} \frac{\mathcal{K}_i f}{\prod_{j \neq i} \alpha^{\#}_{j,i}}, \tag{2.32}$$

where e_i, $i = 1, \ldots, d$, are the edges of Γ starting at p_r and $\alpha_i = \alpha_{e_i}$. On the other hand, by Theorem 2.4, the right hand side of (2.32) is just

$$\mathrm{Res}_\xi \frac{f(p_r)}{\prod \alpha_e}.$$

The conclusion follows since $\mathfrak{p}_{c_N}(f) = 0$. q.e.d.

Corollary 2.1. *If π_* is the map given by (2.16) then, for $f \in H^{2k}(\Gamma, \alpha)$,*

$$Res_\xi(\pi_* f) = 0. \tag{2.33}$$

We conclude this section by observing that combining Corollary 2.1 and Lemma 2 we obtain a new proof of Theorem 2.2 for graphs that satisfy the no-cycle condition:

Let $f \in H^{2k}(\Gamma, \alpha)$. Then, as in (2.17),

$$\pi_* f = \frac{g}{\prod_{j=1}^N \alpha_j},$$

where $g \in \mathbb{S}^{k-d+N}(\mathfrak{g}^*)$ and $\alpha_1, \ldots, \alpha_N$ are pair-wise linearly independent.

Let ξ generate an orientation of Γ with no cycles and choose $\theta \in \mathfrak{g}^*$ such that $\theta(\xi) = 1$ and θ is not equal to any of $\alpha_1, \cdots, \alpha_N$. Then $\theta^r f \in H^{2(k+r)}(\Gamma, \alpha)$ and

$$\pi_*(\theta^r f) = \frac{\theta^r g}{\prod_{i=1}^N \alpha_i}.$$

But (2.33) implies that $\mathrm{Res}_\xi(\pi_*(\theta^r f)) = 0$ for all $r \geq 0$ and it follows now from Lemma 2 that $\pi_* f \in \mathbb{S}^{k-d}(\mathfrak{g}^*)$.

2.8 The Betti numbers of abstract one-skeletons

For $\xi \in \mathcal{P}$ and $p \in V_\Gamma$, let $\sigma_p = \sigma_p(\xi)$ be the number of edges $e \in E_p$ for which $\alpha_e(\xi) < 0$. Let β_k be the number of vertices p with $\sigma_p = k$. Since σ_p depends on ξ, it is surprising to find that these "Betti numbers" don't.

Theorem 2.6. *The Betti numbers, β_k's, are combinatorial invariants of (Γ, α), i.e. they don't depend on ξ.*

Proof. Let $\mathcal{P}_i$, $i = 1, ..., N$, be the connected components of $\mathcal{P}$ and consider an $(n-1)$-dimensional wall separating two adjacent $\mathcal{P}_i$'s. This wall is defined by an equation of the form

$$\alpha_e(\xi) = 0 \tag{2.34}$$

for some $e = (p, q) \in E_\Gamma$. Lets compute the changes in σ_p and σ_q as ξ passes through this wall: Let e_i, $i = 1, ..., d$ be the edges starting at p and e_i', $i = 1, ..., d$ be the edges starting at q, with $e_d = e$ and $e_d' = \bar{e}$. By (2.3) we can order the e_i's so that, for $i \leq d-1$,

$$\alpha_{e_i} = \alpha_{e_i'} + c_i \alpha_e.$$

From (2.2) follows that for every $i = 1, ..., d-1$,

$$\dim(\ker \alpha_e \cap \ker \alpha_{e_i}) = n - 2.$$

Therefore there exists ξ_0 such that $\alpha_e(\xi_0) = 0$ but $\alpha_{e_i}(\xi_0) \neq 0 \neq \alpha_{e_i'}(\xi_0)$, for all $i = 1, ..., d-1$.

Then there exists a neighborhood U of ξ_0 in $\mathfrak{g}$ such that for all $i = 1, ..., d-1$ and $\xi \in U$, $\alpha_{e_i}(\xi)$ and $\alpha_{e_i'}(\xi)$ have the same sign and this common sign doesn't depend on $\xi \in U$. Such a neighborhood will intersect both regions created by the wall (2.34). Now suppose that $\xi \in U$ and that r of the numbers $\alpha_{e_i}(\xi)$, $i = 1, ..., d-1$, are negative. Since $\alpha_e(\xi) = -\alpha_{\bar{e}}(\xi)$, it follows that for $\alpha_e(\xi)$ positive

$$\sigma_p = r \quad \text{and} \quad \sigma_q = r + 1$$

and for $\alpha_e(\xi)$ negative

$$\sigma_p = r + 1 \quad \text{and} \quad \sigma_q = r.$$

In either case, as ξ passes through the wall (2.34), the Betti numbers don't change. q.e.d.

(For this simple and beautiful proof of the well-definedness of the Betti numbers we are indebted to Ethan Bolker.)

2.9 Betti numbers and cohomology

Simple examples show that the formula (1.35) is not true for an arbitrary abstract one-skeleton (Γ, α). However, we will prove that if (Γ, α) has the no-cycle property for some $\xi \in \mathcal{P}$ then the equality (1.35) can be replaced by the inequality

$$\dim H^{2k}(\Gamma,\alpha) \leq \sum \beta_r \dim \mathbb{S}^{k-r}(\mathfrak{g}^*). \tag{2.35}$$

In addition we will show that, for k large,

$$\dim H^{2k}(\Gamma,\alpha) = \sum \beta_r \dim \mathbb{S}^{k-r}(\mathfrak{g}^*) + O(k^{n-3}). \tag{2.36}$$

(Note that since

$$\begin{aligned}\dim \mathbb{S}^k(\mathfrak{g}^*) &= \binom{k+n-1}{n-1} \\ &= \frac{1}{(n-1)!}\left(k^{n-1} + \binom{n}{2} k^{n-2} + O(k^{n-3})\right),\end{aligned}$$

the first term on the right hand side is strictly greater than the error term.) In particular, if $n = 2$, the formula (2.36) asserts that

$$\dim H^{2k}(\Gamma,\alpha) = \sum \beta_r \dim \mathbb{S}^{k-r}(\mathfrak{g}^*) \tag{2.37}$$

for all k greater than some fixed k_0.

Proof. Let $\alpha_i \in \mathfrak{g}^*$, $i = 1, ..., N$, be a pairwise linearly independent set of vectors with the property that every one of the vectors α_e, $e \in E_\Gamma$, is a multiple of a vector in this set. Let I be the graded ideal in $\mathbb{S}(\mathfrak{g}^*)$ generated by the monomials

$$g_i = \alpha_1 \cdots \widehat{\alpha_i} \cdots \alpha_N.$$

Lemma 3. *The algebraic dimension of the quotient ring* $\mathbb{S}(\mathfrak{g}^*)/I$ *is* $n-2$.

Proof. This follows trivially from the fact that the algebraic variety defined by I is the union of the sets $\alpha_i = \alpha_j = 0$, $i \neq j$. q.e.d.

As a corollary of this lemma we get the bound

$$\dim \mathbb{S}^k(\mathfrak{g}^*)/I^k = O(k^{n-3}). \tag{2.38}$$

Now let $\phi : V_\Gamma \longrightarrow \mathbb{R}$ be a function which is positively oriented with respect to ξ and let $H_c(\Gamma,\alpha)$ be the subring of $H(\Gamma,\alpha)$ consisting of all maps $f : V_\Gamma \longrightarrow \mathbb{S}(\mathfrak{g}^*)$ with support on the set $\phi \geq c$. Let $p \in V_\Gamma$ with

$\phi(p) = c$ and suppose that there are no points $q \in V_\Gamma$ with $\phi(q)$ on the interval (c, c'). Let $\sigma_p = r$; we will prove the Morse inequality

$$\dim H_c^{2k}(\Gamma, \alpha) - \dim H_{c'}^{2k}(\Gamma, \alpha) \leq \dim \mathbb{S}^{k-r}(\mathfrak{g}^*) \tag{2.39}$$

and an inequality in the opposite direction:

$$\dim H_c^{2k}(\Gamma, \alpha) - \dim H_{c'}^{2k}(\Gamma, \alpha) \geq \dim I^{k-r}. \tag{2.40}$$

To prove (2.39) let e_i, $i = 1, ..., d$, be the edges of Γ issuing from p and let $\alpha_i = \alpha_{e_i}$. We will order the α_i's so that $\alpha_i(\xi) < 0$ for $1 \leq i \leq r$ and $\alpha_i(\xi) > 0$ for $r + 1 \leq i \leq d$. If $f \in H_c^{2k}(\Gamma, \alpha)$, then $f(p)$ must be a multiple of $\alpha_1, ..., \alpha_r$, so the image of the restriction map

$$H_c^{2k}(\Gamma, \alpha) \longrightarrow \mathbb{S}^k(\mathfrak{g}^*), \quad f \longrightarrow f(p), \tag{2.41}$$

is contained in $\mathbb{S}^{k-r}(\mathfrak{g}^*)\alpha_1 \cdots \alpha_r$. Since the kernel of this map is $H_{c'}^{2k}(\Gamma, \alpha)$, this proves (2.39). We will prove the inequality (2.40) by showing that if $h \in I^{k-r}$, then $h\alpha_1 \cdots \alpha_r$ is in the image of (2.41). Indeed, if $h \in I^{k-r}$, then $h\alpha_1 \cdots \alpha_r$ can be written as a sum

$$\sum_{i=r+1}^{N} h_i \alpha_1 \cdots \widehat{\alpha_i} \cdots \alpha_N.$$

Let p_j be the vertex joined to p by e_j for $j = r + 1, \cdots, d$, and, for fixed $j_0 \in \{r + 1, ..., d\}$, define $f : V_\Gamma \longrightarrow \mathbb{S}^k(\mathfrak{g}^*)$ to be the map which takes the value $h\alpha_1 \cdots \alpha_r$ at p, the value $h_{j_0}\alpha_1 \cdots \widehat{\alpha_{j_0}} \cdots \alpha_N$ at p_{j_0} and zero elsewhere. It is easily checked that $f \in H_c^{2k}(\Gamma, \alpha)$ and $f(p) = h\alpha_1 \cdots \alpha_r$. This proves (2.40).

Next let c and c' be any pair of real numbers with $c < c'$. From (2.39)-(2.40) one gets, by a simple induction, the Morse inequalities

$$\dim H_c^{2k}(\Gamma, \alpha) - \dim H_{c'}^{2k}(\Gamma, \alpha) \leq \sum \beta_r(c, c') \dim \mathbb{S}^{k-r}(\mathfrak{g}^*)$$

and

$$\dim H_c^{2k}(\Gamma, \alpha) - \dim H_{c'}^{2k}(\Gamma, \alpha) \geq \sum \beta_r(c, c') \dim I^{k-r},$$

where $\beta_r(c, c')$ is the number of vertices $p \in V_\Gamma$ with $c \leq \phi(p) < c'$ and $\sigma_p = r$. In particular, for $c' >> 0$ and $c << 0$, one gets the inequalities (2.35) and (2.36) from these estimates and from (2.38). q.e.d.

2.10 The role of the zeroth Betti number

An example of an abstract one-skeleton that fails to satisfy (1.35) is the $d = n = 2$ example described at the end of §2.1. The graph in this example is a connected graph; so its topological zeroth Betti number, defined as dim $H^0(\Gamma, \alpha)$, is 1. However, its graph theoretical zeroth Betti number, β_0, is N. A simple computation shows that, for this example, the identity (1.35) holds for all $k > 0$. But for $k = 0$ the left hand side of (1.35) is 1 (since the graph is connected) whereas the right hand side, β_0, is N. From this example one can generate examples of abstract one-skeletons (Γ, α) for which the estimate (2.36) is "best possible" by taking Cartesian products of this graph with graphs which do satisfy (1.35).

However, by making some additional assumptions on the pair (Γ, α) one can considerably improve (2.36). The assumptions we will make are of two kinds:

1. To avoid the problem posed by the example we have just described, we will assume that the graph theoretical zeroth Betti numbers of certain connected subgraphs of Γ are equal to 1.

2. For every $p \in V_\Gamma$, we will make certain "general position" hypothesis about the vectors α_e, $e \in E_p$. To formulate these hypothesis we introduce the following refinement of the notion of "pairwise linearly independent":

Definition. A collection of vectors $\alpha_i \in \mathfrak{g}^*$, $i = 1, ..., N$, is l-*independent* if, for every sequence $1 \leq i_1 < i_2 < ... < i_l \leq N$, the vectors $\alpha_{i_1}, ..., \alpha_{i_l}$ are linearly independent.

Now let (Γ, α) be an abstract one-skeleton which satisfies the no-cycle condition for some $\xi \in \mathcal{P}$. The main result of this section is the following sharpening of (2.36):

Theorem 2.7. *Suppose the following hypotheses hold:*

1. *For every subspace $\mathfrak{h}$ of $\mathfrak{g}$ of codimension strictly less than l, the zeroth Betti numbers of the connected components of $\Gamma_{\mathfrak{h}}$ are equal to 1.*

2. *For every vertex p of Γ, the vectors α_e, $e \in E_p$, are l-independent.*

Then:

$$\text{(2.42)} \qquad dim\ H^{2k}(\Gamma, \alpha) = \sum \beta_r \ dim\ \mathbb{S}^{k-r}(\mathfrak{g}^*) + O(k^{n-1-l}).$$

Remark. For $l = 2$ the above conditions are always satisfied; (2.36) is the particular case of (2.42) corresponding to $l = 2$.

For $l = n$ this theorem says that the left hand side of (2.42) is equal to the first term on the right for k greater than some fixed k_0. This result can be slightly improved.

Theorem 2.8. *If the hypotheses of theorem 2.7 hold for $l = n$, then*

$$dim\ H^{2k}(\Gamma, \alpha) = \sum \beta_r\ dim\ \mathbb{S}^{k-r}(\mathfrak{g}^*) \tag{2.43}$$

for $k > d - n$.

We will prove these two results by refining the Morse inequalities (2.40). For this we will need the following generalization of Lemma 3 of §2.9:

Lemma 4. *Let $\gamma_1, ..., \gamma_N$ be a collection of vectors in $\mathfrak{g}^*$ which are l-independent and let I_l be the ideal in $\mathbb{S}(\mathfrak{g}^*)$ generated by the monomials*

$$\frac{\gamma_1 \cdots \gamma_N}{\gamma_{i_1} \cdots \gamma_{i_{l-1}}}, \qquad 1 \leq i_1 < \cdots i_{l-1} \leq N. \tag{2.44}$$

Then the algebraic dimension of $\mathbb{S}(\mathfrak{g}^*)/I_l$ is $n - l$. Moreover, if $N \geq n$ and $n = l$ then $\mathbb{S}^m(\mathfrak{g}^*) = I_l^m$ for $m > N - n$.

(For the proof of this lemma see the appendix at the end of this section.)

Proof of theorem 2.7. Let $\phi : V_\Gamma \longrightarrow \mathbb{R}$ be a strictly monotone function which is positively oriented with respect to ξ. Let p be an arbitrary vertex of Γ, let $c = \phi(p)$ and assume that there are no points $q \in V_\Gamma$ with $c < \phi(q) < c'$. Let $e_1, ..., e_d$ be the edges of Γ starting at p and let $\alpha_i = \alpha_{e_i}$. We can order these vectors so that $\alpha_i(\xi) < 0$ for $1 \leq i \leq r$ and $\alpha_i(\xi) > 0$ for $r + 1 \leq i \leq d$, where $r = \sigma_p$. Let $N = d - r$.

Suppose $l \leq N$. We will prove that

$$\dim\ H_c^{2k}(\Gamma, \alpha) - \ \dim\ H_{c'}^{2k}(\Gamma, \alpha) \geq\ \dim\ I_l^{k-r}\ , \tag{2.45}$$

where I_l is the ideal of $\mathbb{S}(\mathfrak{g}^*)$ constructed as in Lemma 4 using the $N = d - r$ vectors $\alpha_{r+1}, ..., \alpha_d$ (which are l-independent, by hypothesis 2), i.e. I_l is the ideal in $\mathbb{S}(\mathfrak{g}^*)$ generated by the monomials

$$\frac{\alpha_{r+1} \cdots \alpha_d}{\alpha_{i_1} \cdots \alpha_{i_{l-1}}}, \quad r + 1 \leq i_1 < ... < i_{l-1} \leq N.$$

To show (2.45), consider (as in §2.9) the restriction map

$$H_c^{2k}(\Gamma,\alpha) \longrightarrow \mathbb{S}^{k-r}(\mathfrak{g}^*)\alpha_1\cdots\alpha_r \tag{2.46}$$

given by $f \longrightarrow f(p)$. The kernel of this map is $H_{c'}^{2k}(\Gamma,\alpha)$; so, to prove (2.45) it suffices to show that the image of this map contains $I_l^{k-r}\alpha_1\cdots\alpha_r$. Consider the set of vectors $\alpha_{i_1},\ldots,\alpha_{i_{l-1}}$, $r+1 \leq i_1 < \ldots < i_{l-1} \leq d$. These vectors are linearly independent; so their annihilator, $\mathfrak{h}$, is of codimension $l-1$ in $\mathfrak{g}$. Let Γ_1 be the connected component of $\Gamma_{\mathfrak{h}}$ containing p. Since the numbers $\alpha_{i_1}(\xi),\ldots,\alpha_{i_{l-1}}(\xi)$ are greater than zero, p is a local minimum point for the restriction of ϕ to $\Gamma_{\mathfrak{h}}$; so, by hypothesis 1, p is also a global minimum. Therefore the vertices of Γ_1 are contained in the set $\phi \geq c$, hence the Thom class τ_1 of Γ_1 is supported on this set, i.e. is an element of $H_c(\Gamma,\alpha)$. However, at p, τ_1 is equal to

$$\left(\frac{\alpha_{r+1}\cdots\alpha_d}{\alpha_{i_1}\cdots\alpha_{i_{l-1}}}\right)\alpha_1\cdots\alpha_r.$$

This argument shows that, for all generators, f, of I_l, $f\alpha_1\cdots\alpha_r$ is in the image of the restriction map (2.46). Hence the image of this restriction map contains $I_l^{k-r}\alpha_1\cdots\alpha_r$, which proves (2.45).

Now suppose that $l > N$. In this case we can simply take $\mathfrak{h}$ to be the annihilator of $\alpha_{r+1},\ldots,\alpha_d$, and, by the same argument as above, conclude that the image of the restriction map is equal to $\mathbb{S}^{k-r}(\mathfrak{g}^*)\alpha_1\cdots\alpha_r$. Hence for $l > N$,

$$\dim H_c^{2k}(\Gamma,\alpha) - \dim H_{c'}^{2k}(\Gamma,\alpha) = \dim \mathbb{S}^{k-r}(\mathfrak{g}^*). \tag{2.47}$$

The proof of the estimate (2.42) via (2.45), (2.47) and Lemma 4 is the same as the proof of the estimate (2.36) via (2.40) and Lemma 3. We will omit details. q.e.d.

Proof of Theorem 2.8. Let $l = n$. If $N < n$ the equality (2.47) holds for all k (as we have just seen). If $N \geq n$ then, by Lemma 4 and by (2.45), the equality (2.47) holds if $k - r > N - n$, i.e. if $k > d - n$. Thus for $l = n$, (2.42) can be sharpened to (2.43). q.e.d.

Appendix: The proof of Lemma 4

We will prove by induction that the algebraic variety defined by I_l is the union of the sets

$$\gamma_{i_1} = \cdots = \gamma_{i_l} = 0, \quad 1 \leq i_1 < \cdots < i_l \leq N.$$

Let x be a point on this variety. Since $I_{l-1} \subset I_l$, the variety defined by I_{l-1} contains the variety defined by I_l; so it follows by induction that there exists a sequence $1 \leq i_1 < \cdots < i_{l-1} \leq N$ with $\gamma_{i_1} = \cdots = \gamma_{i_{l-1}} = 0$ at x. However, since I_l contains the quotient of $\gamma_1 \cdots \gamma_N$ by $\gamma_{i_1} \cdots \gamma_{i_{l-1}}$, there exists some $j \neq i_1, ..., i_{l-1}$ such that $\gamma_j = 0$ at x. This proves the first assertion of lemma 4. Now let $l = n$. We will prove that

$$\mathbb{S}^m(\mathfrak{g}^*) = I_n^m$$

for $m > N - n$ by a double induction on n and N.

The equality above is true if $N = n$ or if $n = 1$, as can be easily checked. Consider now a pair (N, n) with $N > n$. We now assume that the assertion is true for $(N-1, n-1)$ and for $(N-1, n)$.

Let $\mathfrak{h}$ be the annihilator of γ_N in $\mathfrak{g}$. The restriction map

$$\mathbb{S}(\mathfrak{g}^*) \longrightarrow \mathbb{S}(\mathfrak{h}^*)$$

maps $\gamma_1, ..., \gamma_{N-1}$ onto vectors $\beta_1, ..., \beta_{N-1}$, which are $(n-1)$-independent in $\mathfrak{h}^*$; so, by induction, every element of $\mathbb{S}^m(\mathfrak{h}^*)$, $m > N - n$, is in the ideal generated by the monomials

$$\frac{\beta_1 \cdots \beta_{N-1}}{\beta_{i_1} \cdots \beta_{i_{n-2}}}, \quad 1 \leq i_1 < ... < i_{n-2} \leq N - 1.$$

Since the kernel of the map $\mathbb{S}(\mathfrak{g}^*) \longrightarrow \mathbb{S}(\mathfrak{h}^*)$ is the ideal generated by γ_N, it follows that for $m > N - n$, every element of $\mathbb{S}^m(\mathfrak{g}^*)$ can be written as a linear combination of

$$\frac{\gamma_1 \cdots \gamma_{N-1}\gamma_N}{\gamma_{i_1} \cdots \gamma_{i_{n-2}}\gamma_N}, \quad 1 \leq i_1 < ... < i_{n-2} \leq N - 1,$$

with polynomial coefficients, plus a term of the form $f\gamma_N$, $f \in \mathbb{S}^{m-1}(\mathfrak{g}^*)$. By induction the theorem is true in dimension n for the vectors $\gamma_1, ..., \gamma_{N-1}$. Then f is in the ideal generated by the monomials

$$\frac{\gamma_1 \cdots \gamma_{N-1}}{\gamma_{i_1} \cdots \gamma_{i_{n-1}}}, \quad 1 \leq i_1 < ... < i_{n-1} \leq N - 1$$

and hence $f\gamma_N$ is in the ideal generated by

$$\frac{\gamma_1 \cdots \gamma_N}{\gamma_{i_1} \cdots \gamma_{i_{n-1}}}, \quad 1 \leq i_1 < ... < i_{n-1} \leq N - 1.$$

References

[1] M. F. Atiyah, *Convexity and commuting Hamiltonians*, Bull. London Math. Soc., **14** (1982) 1–15.

[2] M. F. Atiyah & R. Bott, *The moment map and equivariant cohomology*, Topology **23** (1984) 1–28.

[3] ———, *A Lefschetz fixed point formula for elliptic complexes. I*, Ann. of Math. **86** (1967) 374–407.

[4] N. Berline & M. Vergne, *Classes caractéristiques équivariantes*, C.R. Acad. Sci., Paris, **295** (1982) 539–541.

[5] M. Brion, *Equivariant cohomology and equivariant intersection theory*, Representation Theory and Algebraic Geometry, Kluwer Acad. Publ., Dordrecht, 1998, 1–37.

[6] F. R. K. Chung & S. Sternberg, *Laplacian and vibrational spectra for homogeneous graphs*, J. Graph Theory **16** (1992) 605–627.

[7] V. Guillemin & J. Kalkman, *The Jeffrey-Kirwan localization theorem and residue operations in equivariant cohomology*, J. Reine Angew. Math. **470** (1996) 123–140.

[8] M. Goresky, R. Kottwitz & R. MacPherson, *Equivariant cohomology, Koszul duality and the localization theorem*, Invent. Math. **131** (1998) 25–83.

[9] V. Guillemin & S. Sternberg, *Convexity properties of the moment mapping*, Invent. Math. 67 (1982) 491–513.

[10] V. Guillemin, *Deformations of a Hamiltonian action of a compact Lie group*, in Integrable systems (Luminy 1991), Progr. Math. Birkhäuser, Boston, Vol. 115 1993, 227–233.

[11] L. Jeffrey & F. Kirwan, *Localization for non-abelian group actions*, Topology **34** (1995) 291–327.

[12] A. A. Klyachko, *Equivariant bundles on toric varieties*, Math USSR Izvestiya, **35:2** (1990) 337–375.

Department of Mathematics, MIT
Department of Mathematics, MIT

SURVEYS IN DIFFERENTIAL GEOMETRY, 2000
Vol. VII ©2000, INTERNATIONAL PRESS
pp. 259-311

MORSE THEORY AND STOKES' THEOREM

F. REESE HARVEY & H. BLAINE LAWSON, JR.

Abstract

We present a new, intrinsic approach to Morse Theory which has interesting applications in geometry. We show that a Morse function f on a manifold determines a submanifold T of the product $X \times X$, and that (in the sense that Stokes theorem is valid) T has boundary consisting of the diagonal $\Delta \subset X \times X$ and a sum

$$P = \sum_{p \in Cr(f)} U_p \times S_p$$

where S_p and U_p are the stable and unstable manifolds at the critical point p. In the language of currents,

(Stokes Theorem) $$\partial T = \Delta - P.$$

This current (or kernel) equation on $X \times X$ is equivalent to an operator equation

(Chain Homotopy) $$d \circ \mathbf{T} + \mathbf{T} \circ d = \mathbf{I} - \mathbf{P},$$

where $\mathbf{P}$ is a chain map onto the finite complex of currents $\mathcal{S}_f$ spanned by (integration over) the stable manifolds of f. The operator $\mathbf{P}$ can be defined on an exterior form α by

$$\mathbf{P}(\alpha) = \lim_{t \to \infty} \varphi_t^* \alpha$$

First printed in Asian Journal of Mathematics, 1999. Used by permission.
The research of both authors was partially supported by the NSF.

where φ_t is a gradient flow for f. The de Rham differential in the complex $\mathcal{S}_f$ is easily computed in terms of the flow lines. The chain homotopy equation also holds on certain integral chain complexes. Poincaré duality over $\mathbb{Z}$ follows from time-reversal in our operator equations. The method has many generalizations and applications. Residue theorems are established for functions with critical manifolds of higher dimension. The methods apply immediately to equivariant cohomology. Cup product formulas and a Lefschetz-type theorem are proved for the Thom-Smale Complex. Other applications include a new proof of the Carrell-Lieberman Theorem and a proof of a local version of the MacPherson Formula for characteristic classes and bundle maps.

Introduction

We present in this paper a new approach to Morse Theory which is stronger than the classical theory and has some interesting applications. It leads to formulas relating characteristic forms and singularities, and it unifies a body of results on holomorphic actions. It applies directly to the equivariant case. It also has the virtue of fitting neatly into the modern theory of invariants arising from topological quantum field theory.

This work resulted from addressing the following.

Question. Consider a flow $\varphi_t : X \to X$ generated by a smooth vector field on a compact manifold X. Under what cicumstances does the limit

$$\alpha_\infty \equiv \lim_{t\to\infty} \varphi_t^* \alpha$$

exist for a given smooth differential form α on X?

We do not demand that α_∞ be smooth. Even so, one would expect the answer to be "rarely, if ever". However, we shall prove that for generic gradient dynamical systems, this limit does exist and has a beautiful, simple structure.

In fact we shall show that setting

$$\mathbf{P}(\alpha) \stackrel{\text{def}}{=} \lim_{t\to\infty} \varphi^* \alpha \tag{0.1}$$

defines a continuous operator of degree 0

$$\mathbf{P} : \mathcal{E}^*(X) \longrightarrow \mathcal{D}'^*(X)$$

from smooth forms to generalized forms, i.e., currents. Furthermore, we shall show that this operator is chain homotopic to the inclusion $\mathbf{I} : \mathcal{E}^*(X) \hookrightarrow \mathcal{D}'^*(X)$. That is, using the flow we shall construct a continuous operator

$$\mathbf{T} : \mathcal{E}^*(X) \longrightarrow \mathcal{D}'^*(X)$$

of degree -1 such that

$$d \circ \mathbf{T} + \mathbf{T} \circ d \;=\; \mathbf{I} - \mathbf{P}. \tag{0.2}$$

By de Rham [11], $\mathbf{I}$ induces an isomorphism in cohomology. Hence so does $\mathbf{P}$.

The existence of $\mathbf{P}$ and $\mathbf{T}$ satisfying (0.1) and (0.2) is established for any flow of finite volume. This concept, which is central to our paper, is introduced in §2. A flow φ_t is said to have **finite volume** if the graph of the relation $x \overset{o}{\prec} y$, defined by the forward motion of the flow, has finite $(n+1)$-dimensional volume in $X \times X$, where $n = \dim(X)$. Any flow whose space-time graph

$$\mathcal{T}_\varphi \;\equiv\; \{(t, \varphi_{\frac{1}{t}}(x), x) \;:\; 0 < t \le 1 \;\text{ and }\; x \in X\} \;\subset\; \mathbf{R} \times X \times X$$

is of finite volume has this property.

Consider now a Morse function $f : X \to \mathbf{R}$ with (finite) critical set $Cr(f)$. Suppose there is a riemannian metric on X for which the gradient flow φ_t of f has the following properties:

1. φ_t is of finite volume.

2. The stable and unstable manifolds, S_p, U_p for $p \in Cr(f)$, are of finite volume in X.

3. $p \prec q \;\Rightarrow\; \lambda_p < \lambda_q$ for all p, $q \in Cr(f)$, where λ_p denotes the index of p and where $\prec$ is the closure of the relation $\overset{o}{\prec}$.

Note that $p \prec q$ means there is a piecewise flow line connecting p in forward time to q.

We shall prove in §14 that such metrics always exist. They can be taken to be *canonically flat* in some neighborhood of $Cr(f)$ (see §2) thereby making the gradient flow particularly tractible. In fact properties (1)–(3) are shown to be generic in the space of such metrics.

Under these assumptions the operator $\mathbf{P}$ is shown to have the following simple form:

$$\mathbf{P}(\alpha) \;=\; \sum_{p \in Cr(f)} r_p(\alpha)[S_p] \tag{0.3}$$

for all $\alpha \in \mathcal{E}^*(X)$, where

$$r_p(\alpha) = \begin{cases} \displaystyle\int_{U_p} \alpha & \text{if } \deg \alpha = \lambda_p \\ 0 & \text{otherwise} \end{cases}$$

and where $[S_p]$ denotes the current defined by integration over S_p. Note that the image of $\mathbf{P}$ is the finite dimensional vector subspace

$$\mathcal{S}_f = \text{span}_{\mathbf{R}}\{[S_p]\}_{p\in Cr(f)}.$$

From (0.2) we see that $\mathcal{S}_f$ is d-invariant, i.e., that $(\mathcal{S}_f, d)$ is a complex, and we show that the inclusion $\mathcal{S}_f \subset \mathcal{D}'^*(X)$ induces an isomorphism

$$H^*(\mathcal{S}_f) \cong H^*_{\text{de Rham}}(X)$$

This immediately yields the classical strong Morse inequalities.

The exterior derivative restricted to $\mathcal{S}_f$ has the form

$$d[S_p] = \sum_{q\in Cr(f)} n_{pq}[S_q].$$

By a result of Federer the constants n_{pq} are integers, and so

$$\mathcal{S}_f^{\mathbb{Z}} = \text{span}_{\mathbb{Z}}\{[S_p]\}_{p\in Cr(f)}$$

is a finite-rank subcomplex of the integral currents $\mathcal{I}_*(X)$ on X. Using (0.2) we show that the inclusion $\mathcal{S}_f^{\mathbb{Z}} \subset \mathcal{I}_*(X)$ induces an isomorphism

$$H^*\left(\mathcal{S}_f^{\mathbb{Z}}\right) \cong H^*(X;\mathbb{Z}).$$

In the special case of Morse-Smale flows the integers n_{pq} can be computed by counting flow lines from p to q. This follows from Stokes' Theorem (cf. [25] and §4).

Poincaré duality (over $\mathbb{Z}$) is now directly deduced from time-reversal in the flow (§5).

The analogous relative theorems for a Morse exhaustion function on a non-compact manifold, including Alexander-Lefschetz duality, are proved in §7.

Our method of proof involves converting the operator equation (0.2) to a kernel equation

$$(0.4) \qquad \partial T = [\Delta] - P$$

on $X \times X$, where Δ denotes the diagonal. There is a general correspondence between operators $\mathbf{K} : \mathcal{E}^*(X) \to \mathcal{D}'^*(X)$ of degree ℓ and currents K of dimension $n - \ell$ on $X \times X$ ([23], See Appendix A.) Under this transformation: $\mathbf{I}$ corresponds to $[\Delta]$, the pull-back of forms by φ_t corresponds to the graph of φ_t, and the chain homotopy $d \circ \mathbf{K} + \mathbf{K} \circ d$

corresponds to the current boundary ∂K. Thus equation (0.4) carries back directly to equation (0.2).

The current T in (0.4) is simply defined by integration over the graph of the relation $\overset{o}{\prec}$. If the space-time graph $\mathcal{T}_\varphi$ has finite volume, then $T = \mathrm{pr}_* \mathcal{T}_\varphi$ where $\mathrm{pr} : \mathbf{R} \times X \times X \to X \times X$ is the projection. Our finite-volume assumption on T implies that T is a rectifiable current and therefore that its current boundary is flat in the sense of Federer [13]. Applying the Federer Support Lemma, we conclude that

$$(0.5) \qquad P = \sum_{p \in Cr(f)} [U_p] \times [S_p].$$

Thus, T provides a homology between the diagonal and the sum of Künneth currents in $X \times X$ given by products of the unstable and stable manifolds of the flow. In other words, our Morse function gives a canonical chain approximation to the diagonal *together with an explicit homology.* This "transgression current" $T = T_f$ plays a role in defining more subtle invariants of manifolds and knots.

The entire procedure outlined above can be applied to functions with non-degenerate critical manifolds, i.e., functions of Bott-type. Suppose f is such a function and F_j, $j = 1, ..., \nu$ are the connected components of the critical set. In this case the kernel P in (0.4) is written as a sum of fibre-products

$$(0.6) \qquad P = \sum_{j=1}^{\nu} [U_j \times_{F_j} S_j].$$

of the stable and unstable manifolds of the flow. Here the subcomplex *image*($\mathbf{P}$) is not finite-dimensional. However, for each smooth form α we have

$$\mathbf{P}(\alpha) = \sum_{j=1}^{\nu} \mathrm{Res}_j(\alpha)[S_j]$$

where $\mathrm{Res}_j(\alpha)$ is a smooth, integrable *residue form* on the manifold S_j computed directly in terms of α.

This approach works nicely in the case of holomorphic $\mathbb{C}^*$-actions with fixed-points on Kähler manifolds. Here there is an underlying function of Bott-type. One finds a complex analogue of the current T and replaces equation (0.4) with a $\partial\bar{\partial}$-equation. General results of Sommese imply that T and all of the stable and unstable manifolds of the flow are subvarieties of finite-volume. One retrieves, in particular, classical results of Bialynicki-Birula [5] and of Carrell-Lieberman-Sommese [8],

[9]. The approach also fits directly into MacPherson's Grassmann graph construction and Gillet-Soulé's construction of transgression classes appearing in the refined Riemann-Roch Theorem [16].

The arguments apply literally without change to the case of equivariant cohomology. It yields rapid calculations in certain cases and has been used by J. Latschev to derive a spectral sequence associated to functions in the equivariant case.

The method can be applied to derive an equation of forms and currents which relates the singularities of a smooth bundle map $A : E \to F$ to characteristic forms of E and F. For *normal* bundle maps, i.e., those which are transversal to the universal singularity sets in $\mathrm{Hom}(E, F)$, this yields local proof of a classical topological formula of MacPherson [27],[28]. However, the method also yields formulas in many other cases, such as generic direct sum and tensor product mappings [19], [20]. This work, which is discussed briefly in §9, began in [17] and led to the Morse Theory presented here.

Much has been written about assigning topological invariants of manifolds and knots to "Feynman graphs". Primary invariants of this type, such as those discussed in [3] and [4], can be constructed using operators $\mathbf{P}$. The invariants of Kontsevich and Vasseliev (cf. [26], [7]) involve the currents $\mathbf{T}$.

It should be remarked that while the Morse Theory due to Ed Witten [36] involves the de Rham complex, it is distinctly different from the approach presented here. Witten considers the conjugates d_t of exterior differentiation d by the function e^{-tf} for $t \geq 0$ and examines the asymptotics of the associated Hodge laplacians. Elliptic operators do not enter the story in our approach. Moreover, a crucial simplifying component of our approach (namely, the calculus of [23] for the operator d) is not available for other operators such as the Laplacian or d_t. It would be interesting to find a more direct connection between the two theories.

The idea of using the stable manifolds of a generic gradient flow to give a cell structure to a manifold goes back to R. Thom [35] (See also [32] and [31].) F. Laudenbach was the first to consider the stable manifolds as de Rham currents [25]. He studied Morse-Smale flows and computed the boundary operator in the Thom-Smale complex by using Stokes' Theorem as we do here. The concept of finite volume flows and its use in Morse Theory were introduced in [21]. This article contains an expanded and more leisurely account of the basic ideas presented there. Furthermore, several additional results are included here.

The authors would like to thank Janko Latschev for many useful comments during the preparation of this work.

1. Flows of finite volume

Let X be a compact smooth manifold of dimension n, and let $\varphi_t : X \to X$ be the flow generated by a smooth vector field V on X. Consider the operator $\mathbf{P}_t : \mathcal{E}^*(X) \to \mathcal{E}^*(X)$ on the space of smooth differential forms which is given by pull-back

$$\mathbf{P}_t(\alpha) \equiv \varphi_t^*(\alpha).$$

We will exhibit a chain homotopy operator $\mathbf{T}_t : \mathcal{E}^k(X) \to \mathcal{E}^{k-1}(X)$ satisfying

$$\textbf{(A)} \qquad d \circ \mathbf{T}_t + \mathbf{T}_t \circ d = \mathbf{I} - \mathbf{P}_t,$$

and show that under a "finite volume" condition it is possible to take the limit as $t \to \infty$. We thereby obtain operators

$$\textbf{(B)} \qquad \mathbf{T} = \lim_{t\to\infty} \mathbf{T}_t \quad \text{and} \quad \mathbf{P} = \lim_{t\to\infty} \mathbf{P}_t$$

satisfying:

$$\textbf{(C)} \qquad d \circ \mathbf{T} + \mathbf{T} \circ d = \mathbf{I} - \mathbf{P}.$$

Now operator equations such as (**A**) and (**C**) are difficult to solve directly, but as we shall see in §3, they can be converted into current equations on the product manifold $X \times X$ which are much more tractable. For example, equation **(A)** becomes the current equation

$$\text{(A)} \qquad \partial T_t = [\Delta] - P_t \qquad \text{on } X \times X$$

where $\Delta \subset X \times X$ is the diagonal and

$$P_t = [\text{graph}(\varphi_t)] = \{(\varphi_t(x), x) : x \in X\}$$

is the (reverse) graph of the diffeomorphism φ_t. The remaining conditions are that

$$\text{(B)} \qquad T = \lim_{t\to\infty} T_t \quad \text{and} \quad P \equiv \lim_{t\to\infty} P_t$$

exist and satisfy

$$\text{(C)} \qquad \partial T = [\Delta] - P \qquad \text{on } X \times X.$$

Here T_t, T, and P are currents on $X \times X$ yet to be determined. Details of the above correspondence are discussed in §3. The remainder of this section is devoted to solving the second set of equations (A), (B), (C).

Equation (A) is particularly easy to solve. Consider the family of compact manifolds with boundary:

$$\mathcal{T}_t \equiv \{(s, \varphi_s(x), x) : 0 \leq s \leq t \text{ and } x \in X\} \tag{1.1}$$

contained in $\mathbf{R}\times X\times X$. Obviously, $\mathcal{T}_t$ has the two boundary components $\{0\} \times \Delta$ and $\{t\} \times P_t$. Assume X is oriented (this condition will be dropped later) and orient $\mathcal{T}_t$ so that

$$\partial \mathcal{T}_t = \{0\} \times \Delta - \{t\} \times P_t. \tag{1.2}$$

Let $\mathrm{pr} : \mathbf{R} \times X \times X \to X \times X$ denote projection and set

$$T_t \equiv (\mathrm{pr})_*(\mathcal{T}_t). \tag{1.3}$$

Since ∂ commutes with $(\mathrm{pr})_*$, the push-forward of (1.2) by pr_* gives equation (A).

The current $\mathcal{T}_t$ can be equivalently defined by

$$T_t = \Phi_*([0,t] \times X) \tag{1.3$'$}$$

where $\Phi : \mathbf{R} \times X \to X \times X$ is the smooth mapping given by $\Phi(s,x) = (\varphi_s(x), x)$. This mapping is an immersion exactly on the subset $\mathbf{R}\times(X - Z(V))$ where $Z(V) = \{x \in X : V(x) = 0\}$. Thus if we fix a riemannian metric g on X, then $\Phi^*(g\times g)$ is a symmetric positive semi-definite tensor whose associated volume is > 0 exactly on the subset $\mathbf{R} \times (X - Z(V))$.

This brings us to one of the central concepts of the paper.

Definition 1.1. A flow φ_t on X is called a **finite volume flow** if $\mathbf{R}^+ \times (X - Z(V))$ has finite volume with respect to the metric induced by the immersion Φ. (This concept is independent of the choice of riemannian metric on X.)

Theorem 1.2. *Let φ_t be a finite volume flow on a compact manifold X. Then both the limits*

$$P \equiv \lim_{t\to\infty} [\mathrm{graph}\, \varphi_t] \quad \textit{and} \quad T \equiv \lim_{t\to\infty} T_t \tag{B}$$

exist as currents, and by taking the boundary of T we obtain the equation of currents

$$\partial T = [\Delta] - P \quad \textit{on } X \times X. \tag{C}$$

relating P to the diagonal Δ in $X \times X$.

Proof. Since φ_t is a finite-volume flow, the current $T \equiv \Phi_*((0,\infty) \times X)$ is the limit in the mass norm of the currents $T_t = \Phi_*((0,t) \times X)$ as $t \to \infty$. The continuity of the boundary operator and equation (A) imply the existence of $\lim_{t\to\infty} P_t$ and also establish equation (C). q.e.d.

Remark 1.3. Since $(\varphi_t(x), x) = (y, \varphi_{-t}(y))$ if $y = \varphi_t(x)$, it follows that

$$T^* = \Phi_*((-\infty, 0) \times X)$$

is also a well-defined current for a finite-volume flow. It corresponds to the push-forward of T under the flip $(y,x) \mapsto (x,y)$ on $X \times X$.

Remark 1.4. The immersion $\Phi : \mathbf{R} \times (X - Z(V)) \to X \times X$ is an embedding outside the subset $\mathbf{R} \times Per(V)$ where

$$Per(V) = \{x \in X \ : \ \varphi_t(x) = x \ \text{ for some } t > 0\}$$

are the non-trivial periodic points of the flow. Thus, if $Per(V)$ has measure zero, then T_t is given by integration over the embedded finite-volume submanifold $\Phi(R_t)$, where $R_t = (0,t) \times (X - Z(V) \cup Per(V))$. If furthermore the flow has finite volume, then T is given by integration over the embedded, finite-volume submanifold $\Phi(R_\infty)$.

There is evidence that any flow with periodic points cannot have finite volume. Now a gradient flow never has periodic points, and many such flows are of finite volume (§14.). However, finite-volume flows are more general than gradient flows. For a first example, note that any flow with fixed points on S^1 has finite volume.

Remark 1.5. If we define a relation on $X \times X$ by setting $x \overset{o}{\prec} y$ if $y = \varphi_t(x)$ for some $0 \le t < \infty$, then T is just the (reversed) graph of this relation. This relation is always transitive and reflexive, and it is antisymmetric if and only if φ_t has no periodic orbits (i.e., $\overset{o}{\prec}$ is a partial ordering precisely when φ_t has no periodic orbits).

Remark 1.6. A standard method for showing that a given flow is finite volume can be outlined as follows. Pick a coordinate change $t \mapsto \rho$ which sends $+\infty$ to 0 and $[t_0, \infty]$ to $[0, \rho_0]$. Then show that

$$\begin{aligned} \mathcal{T} \ \equiv \ &\{(\rho, \varphi_{t(\rho)}(x), x) \ : \ 0 < \rho < \rho_0\} \\ &\qquad \text{has finite volume in } \ \mathbf{R} \times X \times X. \end{aligned} \tag{1.4}$$

Pushing forward to $X \times X$ then yields the current T with finite mass.

Perhaps the most natural such coordinate change is $r = 1/t$. Another natural choice (if the flow is considered multiplicatively) is $s = e^{-t}$. Of course finite volume in the r coordinate insures finite volume in the s coordinate since $r \mapsto s = e^{-1/r}$ is a C^∞-map.

Many interesting flows can be seen to be finite volume as follows.

Proposition 1.7. *If X is analytic and $\mathcal{T} \subset \mathbf{R} \times X \times X$ is contained in a real analytic subvariety of dimension $n+1$, then φ_t is a finite volume flow.*

Proof. The manifold points of a real analytic subvariety have (locally) finite volume. q.e.d.

A flow need not be a gradient flow to be a finite volume flow.

Example 1.8. (The standard degenerate flow on S^n) Consider the translational flow $\varphi_t(y) = y + tu$ on $\mathbf{R}^n$ where $u \in \mathbf{R}^n$ is a unit vector. We can identify $\mathbf{R}^n$ with $S^n - \{\infty\}$ so that φ_t extends to S^n as a finite volume flow. To do this choose coordinates $x = y/|y|^2$ on $\mathbf{R}^n \cong S^n - \{0\}$. Then

$$\varphi_t(x) = \frac{|x|^2}{|x + t|x|^2u|^2}(x + t|x|^2u) \tag{1.5}$$

(The vector field $V = \dot{\varphi}_t\big|_{t=0}$ is given by $V(y) = u$ on $\mathbf{R}^n = S^n - \{\infty\}$, and by $V(x) = |x|^2u - 2\langle x, u\rangle x$ on $\mathbf{R}^n = S^n - \{0\}$.)

The flow φ_t is finite volume flow on S^n. To see this let $r = 1/t$ and note that

$$\mathcal{T} = \{(r, z, x) \ : \ z = \varphi_{1/r}(x),\, 0 < r < \infty \ \text{ and } \ x \in \mathbf{R}^n = S^n - \{0\}\}$$

is defined by

$$z|rx + |x|^2u|^2 = |x|^2r(rx + |x|^2u)$$

so that Proposition 1.7 is applicable. Note that ∞ ($x = 0$) is the only zero of V. Although V is not a gradient vector field, it is the limit of gradient vector fields.

Our next problem is to explicitly compute the current P and its associated operator $\mathbf{P}$ under additional assumptions on the flow. We shall show that when V is a "good" gradient vector field for a Morse function, the operator $\mathbf{P}$ is projection onto the finite complex of currents spanned by the stable manifolds of the flow. Furthermore, for any given Morse function the "good" gradients are generic.

2. Morse-Stokes flows

Let $f \in C^\infty(X)$ be a Morse function on a compact n-manifold X, and let $\mathrm{Cr}(f)$ denote the (finite) set of critical points of f. Recall that f is a **Morse function** if its Hessian at each critical point is non-degenerate. The standard Morse Lemma asserts that in a neighborhood of each $p \in \mathrm{Cr}(f)$ of index λ, there exist **canonical local coordinates** $(u_1, ..., u_\lambda, v_1, ..., v_{n-\lambda})$ for $|u| < r$, $|v| < r$ with $(u(p), v(p)) = (0, 0)$ such that

$$f(u, v) = f(p) - |u|^2 + |v|^2. \tag{2.1}$$

Choose a Riemannian metric on X and let φ_t denote the flow associated to ∇f. Suppose that this metric has the form $|du|^2 + |dv|^2$ in some canonical coordinate system (u, v) about each $p \in \mathrm{Cr}(f)$. Then in these coordinates the gradient flow is given by

$$\varphi_t(u, v) = (e^{-t}u, e^t v)$$

Metrics with this property will be called **canonically flat near** $\mathrm{Cr}(f)$ and the flow will be called **f-tame**.

Now to each $p \in \mathrm{Cr}(f)$ are associated the **stable** and **unstable manifolds** of the flow, defined respectively by

(2.2)
$$S_p = \{x \in X : \lim_{t\to\infty} \varphi_t(x) = p\} \quad \text{and} \quad U_p = \{x \in X : \lim_{t\to-\infty} \varphi_t(x) = p\}.$$

For coordinates (u, v) at p, chosen as above, we consider the disks

$$S_p(\epsilon) = \{(u, 0) : |u| < \epsilon\} \qquad \text{and} \qquad U_p(\epsilon) = \{(0, v) : |v| < \epsilon\}$$

and observe that

$$S_p = \bigcup_{-\infty<t<0} \varphi_t\left(S_p(\epsilon)\right) \qquad \text{and} \qquad U_p = \bigcup_{0<t<+\infty} \varphi_t\left(U_p(\epsilon)\right). \tag{2.3}$$

Hence, S_p and U_p are submanifolds (but not closed subsets) of X with

$$\mathrm{dim} S_p = \lambda_p \qquad \text{and} \qquad \mathrm{dim} U_p = n - \lambda_p \tag{2.4}$$

where λ_p is the index of the critical point p. For each p we choose an orientation on U_p. This gives an orientation on the normal bundle of S_p via the splitting $T_pX = T_pU_p \oplus T_pS_p$. We thereby obtain a kernel $[U_p] \times [S_p]$ on $X \times X$ (See A.9.).

The flow φ_t induces a partial ordering on X by setting $x \prec y$ if there is a continuous path consisting of a finite number of forward-time orbits, which begins with x and ends with y. We shall see in the proof of Lemma 2.4 that this is the closure of the partial ordering of Remark 1.5.

Definition 2.1. The gradient flow of a Morse function f on a riemannian manifold X is called **Morse-Stokes** if it is f-tame and:

(i) It is a finite-volume flow.

(ii) Each of the stable and unstable manifolds S_p and U_p for $p \in \mathrm{Cr}(f)$ has finite volume.

(iii) If $p \prec q$ and $p \neq q$ then $\lambda_p < \lambda_q$, for all p, $q \in \mathrm{Cr}(f)$.

Remark 2.2. In Section 14 we shall prove that if the gradient flow of f is Morse-Smale, then it is Morse-Stokes. Furthermore, for any Morse function f on a compact manifold X there exist Riemannnian metrics on X for which the gradient flow of f is Morse-Stokes. In fact these metrics are dense in the set of all metrics which are canonically flat near $Cr(f)$.

Theorem 2.3. *Let $f \in C^\infty(X)$ be a Morse function on a compact riemannian manifold whose gradient flow is Morse-Stokes. Then there is an equation of integral currents*

$$\partial T \ = \ [\Delta] - P \tag{2.5}$$

on $X \times X$, where T is an embedded submanifold of finite volume, $\Delta \subset X \times X$ is the diagonal, and where

$$P \ = \ \sum_{p \in Cr(f)} [U_p] \times [S_p] \tag{2.6}$$

Proof. For each critical point $p \in \mathrm{Cr}(f)$ we define

$$\widetilde{U}_p \stackrel{\mathrm{def}}{=} \bigcup_{p \prec q} U_q = \{x \in X : p \prec x\}.$$

Lemma 2.4. *Let $f \in C^\infty(X)$ be any Morse function whose gradient flow is of finite volume, and let* spt $P \subset X \times X$ *denote the support of the current P defined in Theorem 1.2. Then*

$$\mathrm{spt}\, P \subset \bigcup_{p \in \mathrm{Cr}(f)} \widetilde{U}_p \times S_p.$$

Proof. Since $P = \lim_{t\to\infty} P_t$ and $P_t = \{(\varphi_t(x), x) : x \in X\}$, it is clear that $(y,x) \in \operatorname{spt} P$ only if there exist sequences $x_i \to x$ in X and $s_i \to \infty$ in $\mathbb{R}$ such that $y_i \equiv \varphi_{s_i}(x_i) \to y$. Let $L(x_i, y_i)$ denote the oriented flow line from x_i to y_i. Since the lengths of these lines are bounded, an elementary compactness argument shows that a subsequence converges to a piecewise flow line $L(x,y)$ from x to y. By the continuity of the boundary operator on currents, $\partial L(x,y) = [y] - [x]$. Finally, since $s_i \to \infty$ there must be at least one critical point on $L(x,y)$, and we define $p = \lim_{s\to\infty} \varphi_s(x)$. q.e.d.

Consider now the subset

$$\Sigma \equiv \bigcup_{\substack{p\prec q \\ p\neq q}} U_q \times S_p \subset X \times X.$$

and set $\Sigma' = \Sigma \cup \{(p,p) : p \in Cr(f)\}$. Note that from (2.4) and Axiom (iv) in Definition 2.1 that

(2.7) Σ' is a disjoint union of submanifolds of dimension $\leq n-1$.

Lemma 2.5. *Let $f \in C^\infty(X)$ be a Morse function and φ_t an f-tame gradient flow. Consider the embedded submanifold*

$$T = \{(y,x) : x \notin Cr(f), \text{ and } y = \varphi_t(x) \text{ for some } 0 < t < \infty\}.$$

of $X \times X - \Sigma'$. Then the closure $\overline{T}$ of T is a proper C^∞-submanfold with boundary

$$\partial\overline{T} = \Delta - \sum_{p\in \mathrm{Cr}(f)} U_p \times S_p$$

in $X \times X - \Sigma'$

Proof. We first show that it will suffice to prove the assertion in a neighborhood of $(p,p) \in X \times X$ for $p \in Cr(f)$. Consider $(\bar{y},\bar{x}) \in \overline{T} - T$. If $(\bar{y},\bar{x}) \notin \Sigma'$, then the proof of Lemma 2.4 shows that either $(\bar{y},\bar{x}) \in \Delta$ or $(\bar{y},\bar{x}) \in U_p \times S_p$ for some $p \in Cr(f)$. Near points $(\bar{x},\bar{x}) \in \Delta$, $\bar{x} \notin Cr(f)$, one easily checks that $\overline{T}$ is a submanifold with boundary Δ. If $(\bar{y},\bar{x}) \in U_p \times S_p$, then for sufficiently large $s > 0$, the diffeomorphism $\psi_s(y,x) \equiv (\varphi_{-s}(y), \varphi_s(x))$ will map $(\bar{y},\bar{x})$ into any given neighborhood of (p,p). Note that ψ_s leaves the subset $U_p \times S_p$ invariant, and that ψ_s^{-1}

maps T into T. Hence, if $\partial\overline{T} = \Delta - U_p \times S_p$, in a neighborhood of (p,p), then $\partial\overline{T} = -U_p \times S_p$ near $(\bar{y}, \bar{x})$.

Now in a neighborhood $\mathcal{O}$ of (p,p) we may choose coordinates as in (2.1) so that T consists of points $(y,x) = (\bar{u}, \bar{v}, u, v)$ with $\bar{u} = e^{-t}u$ and $\bar{v} = e^t v$ for some $0 < t < \infty$. Consequently, in $\mathcal{O}$ the set $\overline{T}$ is given by the equations

$$\bar{u} = su \quad \text{and} \quad v = s\bar{v} \qquad \text{for some } \ 0 \le s \le 1.$$

This obviously defines a submanifold in $\mathcal{O} - \{(p,p)\}$ with boundary consisting of Δ and the set $\{\bar{u} = 0,\ v = 0\} \cong (U_p \times S_p) \cap \mathcal{O}$. q.e.d.

Lemma 2.5 has the following immediate consequence

$$\text{spt}\left\{P - \sum_{p \in \text{Cr}(f)} U_p \times S_p\right\} \subset \Sigma' \tag{2.8}$$

We now apply the following important result of Federer.

Proposition 2.6. ([13, 4.1.20]) *Let S be a flat current of dimension k in $\mathbb{R}^n$. If the Hausdorff k-measure of the support of S is zero, then $S = 0$.*

Recall that if S is a current defined by integration over an oriented submanifold of locally finite volume, then both S and dS are flat. (In fact for such currents having the further property that dS is supported in a submanifold N with $\dim(N) < \dim(dS)$, Proposition 2.6 has an elementary proof [13, 4.1.15].) Combining Proposition 2.6 with (2.7) and (2.8) proves (2.6) and completes the proof of Theorem 2.3. q.e.d.

Related to Proposition 2.6 is the following result of Federer ([13, 4.1.15], [13, 4.2.16(2)]).

Proposition 2.7. *Let S be a flat current of dimension k with $dS = 0$ defined in a convex open subset U of $\mathbb{R}^n$. If $\text{spt}\,(S) \subset U \cap \mathbb{R}^k$, then $S = c[U \cap \mathbb{R}^k]$ for some $c \in \mathbb{R}$. Moreover, if $S = dT$ where T is locally rectifiable, then c is an integer.*

Propositions 2.6 and 2.7 are philosophically central to our paper.

Remark 2.8. Using the flow given in Example 1.8 and the methods above, one constructs an $(n+1)$-current T on $S^n \times S^n$ with the property that

$$\partial T \ = \ S^n \times \{*\} + \{*\} \times S^n.$$

This is a singular analogue of the form used by Bott and Taubes to study knot invariants [7].

3. From current equations to operator equations

We now explain how to pass from the current equations (A), (B), (C) to the operator equations **(A)**, **(B)**, **(C)** discussed in §1. This kernel calculus was introduced in [23]. There is a brief appendix on currents with definitions and notation at the end of the paper. The discussion here includes the non-orientable case.

Let X and Y be compact manifolds, and let π_Y and π_X denote projection of $Y \times X$ onto Y and X respectively. Then each partially twisted current (or **kernel**) $K \in \mathcal{D}'^*(Y \times \widetilde{X})$, determines an operator $\mathbf{K} : \mathcal{E}^*(Y) \to \mathcal{D}'^*(X)$ by the formula

$$\mathbf{K}(\alpha) = (\pi_X)_*(K \wedge \pi_Y^* \alpha). \tag{3.1}$$

The formula (3.1) can be rewritten as

$$\mathbf{K}(\alpha)(\beta) = K(\pi_Y^* \alpha \wedge \pi_X^* \beta)$$

where $\beta \in \widetilde{\mathcal{E}}^*(X)$ is a twisted form on X. This definition is motivated by the following example.

Example 3.1. Suppose $\varphi : X \to Y$ is a smooth map, and let $\mathbf{P}_\varphi(\alpha) \equiv \varphi^*(\alpha)$ be the pull-back operator on on differential forms. Now a differential form $\varphi^* \alpha$ defines a current by setting

$$(\varphi^* \alpha)(\beta) \equiv \int_X (\varphi^* \alpha) \wedge \beta \tag{3.2}$$

for all twisted forms $\beta \in \widetilde{\mathcal{E}}^*(X)$. Consider the **graph** of φ given by $\operatorname{graph} \varphi = \{(\varphi(x), x) : x \in X\} \subset Y \times X$. Integration over the graph of φ determines a kernel or (partially twisted) current on $Y \times X$, by this same formula (3.2). Namely,

$$[\operatorname{graph} \varphi](\pi_Y^* \alpha \wedge \pi_X^* \beta) \equiv \int_X \varphi^* \alpha \wedge \beta. \tag{3.3}$$

The left hand side equals

$$([\operatorname{graph} \varphi] \wedge \pi_Y^* \alpha)(\pi_X^* \beta) = \big((\pi_X)_*([\operatorname{graph} \varphi] \wedge \pi_Y^* \alpha)\big)(\beta).$$

Therefore,

$$\mathbf{P}_\varphi(\alpha) = (\pi_X)_*([\operatorname{graph} \varphi] \wedge \pi_Y^* \alpha). \tag{3.4}$$

To complete the transfer of the operator equations (**A**) and (**C**) to current equations on $X \times X$, we need the following result.

Lemma 3.2. *Suppose the operator* $\mathbf{K} : \mathcal{E}^*(Y) \to \mathcal{D}'^*(X)$ *has kernel* $K \in \mathcal{D}'^*(Y \times \widetilde{X})$. *Suppose that* $\mathbf{K}$ *lowers degree by one, or equivalently, that* $deg(K) = dimY - 1$. *Then*

$$\text{(3.5)} \qquad \textit{The operator } d \circ \mathbf{K} + \mathbf{K} \circ d \ \textit{ has kernel } \partial K.$$

Proof. The boundary operator ∂ is the dual of exterior differentiation. That is ∂K is defined by $(\partial K)(\pi_Y^*\alpha \wedge \pi_X^*\beta) = K(d(\pi_Y^*\alpha \wedge \pi_X^*\beta))$. Also, by definition, $(\mathbf{K}(d\alpha), \beta) \equiv K(\pi_Y^*(d\alpha) \wedge \pi_X^*\beta)$, and

$$(d(\mathbf{K}(\alpha)), \beta) \equiv (-1)^{\deg \alpha}\mathbf{K}(\alpha)(d\beta) = (-1)^{\deg \alpha} K(\pi_Y^*(\alpha) \wedge \pi_X^* d\beta),$$

since $\mathbf{K}(\alpha)$ has degree equal to $\deg \alpha - 1$. q.e.d.

The results that we need are summarized in the following table.

Operators	Kernels
$\mathbf{I}$	$[\Delta]$
$\mathbf{P}_t \equiv \varphi_t^*$	$P_t \equiv [\text{graph}\, \varphi_t]$
$\mathbf{K}$	K
$d \circ \mathbf{K} + \mathbf{K} \circ d$	∂K

Table 3.6

This completes the transfer of the operator equations **(A)**, **(B)**, **(C)** to the current equations (A), (B), (C). From Theorems 1.2 and 2.3 we immediately deduce the following.

Theorem 3.3. *Let* $f \in C^\infty(X)$ *be a Morse function on a compact Riemannian manifold* X *whose gradient flow* φ_t *is Morse-Stokes. Then for every differential form* $\alpha \in \mathcal{E}^k(X)$, $0 \le k \le n$, *one has*

$$\mathbf{P}(\alpha) = \lim_{t \to \infty} \varphi_t^* \alpha = \sum_{p \in \mathrm{Cr}(f)} r_p(\alpha)[S_p]$$

where the "residue" $r_p(\alpha)$ *of* α *at* p *is defined by*

$$r_p(\alpha) = \int_{U_p} \alpha$$

if $k = n - \lambda$ *and 0 otherwise. Furthermore, there is an operator* $\mathbf{T}$ *of degree -1 on* $\mathcal{E}^*(X)$ *with values in flat currents, such that*

$$\text{(3.7)} \qquad d \circ \mathbf{T} + \mathbf{T} \circ d \ = \ \mathbf{I} - \mathbf{P}$$

From Theorem 3.3, we see that $\mathbf{P} : \mathcal{E}^*(X) \longrightarrow \mathcal{D}'^*(X)$ maps onto the finite-dimensional subspace of currents

$$\mathcal{S}_f \overset{\text{def}}{=} \text{span}_{\mathbb{R}}\{[S_p]\}_{p \in \text{Cr}(f)} \tag{3.8}$$

and that

$$\mathbf{P} \circ d = d \circ \mathbf{P}. \tag{3.9}$$

This together with (3.7) leads to the following.

Corollary 3.4. *The subspace $\mathcal{S}_f$ is d-invariant and is therefore a subcomplex of $\mathcal{D}'^*(X)$. The surjective linear map of cochain complexes*

$$\widehat{\mathbf{P}} : \mathcal{E}^*(X) \longrightarrow \mathcal{S}_f$$

defined by $\mathbf{P}$ *induces an isomorphism*

$$\widehat{\mathbf{P}} : H^*_{deR}(X) \xrightarrow{\cong} H^*(\mathcal{S}_f).$$

Proof. It follows immediately from Theorem 3.3 that

$$\mathbf{P}_* = \mathbf{I}_* : H(\mathcal{E}^*(X)) \to H(\mathcal{D}'^*(X)),$$

and by [11], $\mathbf{I}_*$ is an isomorphism. Since $i_* \circ \widehat{\mathbf{P}}_* = \mathbf{P}_*$, where $i : \mathcal{S}^*_f \to \mathcal{D}'^*(X)$ denotes the inclusion, we see that $\widehat{\mathbf{P}}_*$ is injective.

However, it does not follow formally that $\widehat{\mathbf{P}}_*$ is surjective (as the example below shows). We must prove that every $S \in \mathcal{S}^*_f$ with $dS = 0$ is of the form $S = \mathbf{P}(\gamma)$ where $d\gamma = 0$.

To do this we note the following consequence of 2.1(iii). For all $p, q \in Cr(f)$ of the same index, one has that $\overline{S}_p \cap \partial U_q = \emptyset$, $\overline{S}_p \cap \overline{U}_q = \emptyset$ if $p \neq q$, and $\overline{S}_p \cap \overline{U}_q = \{p\}$. Now let $S = \sum n_p S_p$ be a k-cycle in $\mathcal{S}_f$. Then by the above there is an open neighborhood N of spt (S) such that $N \cap \partial U_q = \emptyset$ for all q of index k. Since by [11] cohomology with compact supports can be computed by either smooth forms or all currents, there exists a current σ with compact support in N such that $d\sigma = \gamma - S$ where γ is a smooth form. Now for each q of index k, U_q is a closed submanifold of N, and we can choose a family U_ϵ of smooth closed forms in N such that $\lim_{\epsilon \to 0} U_\epsilon = U_q$ in $\mathcal{D}'(N)$ and $\lim_{\epsilon \to 0}(S_p, U_\epsilon) = (S_p, U_q) = \delta_{pq}[p]$ for all $p \in Cr(f)$ of index k. This is accomplished via standard Thom form constructions for the normal bundle using canonical coordinates at q (e.g. [17]). It follows that $(\gamma - S, U_q) = \lim_{\epsilon\to 0}(\gamma - S, U_\epsilon) = \lim_{\epsilon\to 0}(d\sigma, U_\epsilon) = \lim_{\epsilon\to 0}(\sigma, dU_\epsilon) = 0$. We conclude that $\int_{U_q}\gamma = (S, U_q) = n_q$ for all q and so $\mathbf{P}(\gamma) = S$ as claimed.

Alternatively, one can choose an $\epsilon > 0$ such that $(\operatorname{spt} S)_\epsilon \cap \partial U_q = \emptyset$ for all q of index k, where $(\operatorname{spt} S)_\epsilon$ denotes an ϵ-neighborhood of spt S. Applying the Federer smoothing homotopy [13, 4.1.18] gives a current $H_\epsilon(S)$ with $\operatorname{spt} H_\epsilon(S) \subset (\operatorname{spt} S)_\epsilon$ and $dH_\epsilon(S) = \gamma - z$ where γ is a smooth closed form. Since $(\gamma - S, U_q) = (dH_\epsilon(S), U_q) = (H_\epsilon(S), \partial U_q) = 0$ we conclude as above that $\int_{U_q}\gamma = (S, U_q) = n_q$ for all q and therefore $\mathbf{P}(\gamma) = S$. q.e.d.

Example. Consider the short exact sequence of complexes

$$0 \longrightarrow K = \mathbb{R}\{a\} \longrightarrow \mathcal{E} = \mathbb{R}\{a, b\} \xrightarrow{\hat{P}} \mathcal{S} = \mathbb{R}\{b'\} \longrightarrow 0$$

where $a = db$, $db' = 0$ and $\widehat{P}(b) = b'$. Define $\mathcal{D} = \mathbb{R}\{a, b, b', c\}$ with $b' = dc$ and let $i : \mathcal{S} \to \mathcal{D}$ denote the inclusion. Define $T : \mathcal{E} \to \mathcal{D}$ by $T(a) = b$ and $T(b) = -c$, and $I : \mathcal{E} \to \mathcal{D}$ by $I(a) = a, I(b) = b$, and $P : \mathcal{E} \to \mathcal{D}$ by $P = i \circ \widehat{P}$. Then $dT + Td = I - P$, but $\widehat{P}_*$ is not surjective on homology.

4. $\mathbb{Z}$-complexes, $\mathbb{Z}/p\mathbb{Z}$-complexes

We now observe that the complex $(\mathcal{S}_f, d)$ is actually defined over the integers. Consider the lattice

$$\mathcal{S}_f^{\mathbb{Z}} \stackrel{\text{def}}{=} \operatorname{span}_{\mathbb{Z}}\{[S_p]\}_{p\in Cr(f)}$$

and note that $\mathcal{S}_f^{\mathbb{Z}}$ forms a subgroup of the integral currents $\mathcal{I}(X)$ on X.

Theorem 4.1. *The lattice $\mathcal{S}_f^{\mathbb{Z}}$ is preserved by exterior differentiation d, that is, $(\mathcal{S}_f^{\mathbb{Z}}, d)$ is a subcomplex of $(\mathcal{S}_f, d)$. Furthermore, the inclusion of complexes $(\mathcal{S}_f^{\mathbb{Z}}, d) \subset (\mathcal{I}(X), d)$ induces an isomorphism*

$$H(\mathcal{S}_f^{\mathbb{Z}}) \cong H_*(X; \mathbb{Z})$$

Proof. Corollary 3.4 implies that for any $p \in Cr(f)$ we have

$$d[S_p] = \sum_{\lambda_q = \lambda_p - 1} n_{p,q}\,[S_q] \tag{4.1}$$

for real numbers $n_{p,q}$. Furthermore, since $[S_p]$ is rectifiable, we have $n_{p,q} \in \mathbb{Z}$ for all p, q by Proposition 2.7, and the first assertion is proved.

Now the domain of the operator $\mathbf{P}$ extends to include any C^1 chain c which is transversal to the submanifolds U_p, $p \in Cr_k(f)$, while the domain of $\mathbf{T}$ extends to any C^1 chain c for which $c \times X$ is transversal to T. Standard transversality arguments show that such chain groups (over $\mathbb{Z}$) compute $H_*(X; \mathbb{Z})$. The result then follows from (3.7). q.e.d.

Corollary 4.2. *Let G be a finitely generated abelian group. Then there are natural isomorphisms*

$$H(\mathcal{S}_f^{\mathbb{Z}} \otimes_{\mathbb{Z}} G) \;\cong\; H_*(X; G).$$

Note that the proof of Theorem 4.1 can be used to give an alternate proof of Corollary 3.4.

Part one of Theorem 4.1 has an elementary proof when the flow is **Morse-Smale**, which means by definition that S_p is transversal to U_q for all $p, q \in Cr(f)$. Suppose the flow is Morse-Smale and that $p, q \in Cr(f)$ are critical points with $\lambda_q = \lambda_p - 1$. Then $U_q \cap S_p$ is the union of a finite set of flow lines from q to p which we denote $\Gamma_{p,q}$. To each $\gamma \in \Gamma_{p,q}$ we assign an index n_γ as follows. Let $B_\epsilon \subset S_p$ be a small ball centered at p in a canonical coordinate system (cf. (2.1)), and let y be the point where γ meets ∂B_ϵ. The orientation of S_p induces an orientation on $T_y(\partial B_\epsilon)$, which is identified by flowing backward along γ with $T_q(S_q)$. If this identification preserves orientations we set $n_\gamma = 1$, and if not, $n_\gamma = -1$. As in [25] Stokes' Theorem gives us the following (cf. [36]).

Proposition 4.3. *When the gradient flow of f is Morse-Smale, the coefficients in (4.1) are given by*

$$n_{p,q} \;=\; (-1)^{\lambda_p} \sum_{\gamma \in \Gamma_{p,q}} n_\gamma$$

Proof. Given a form α of degree $\lambda_p - 1$, we have

$$(-1)^{\lambda_p} d[S_p](\alpha) \;=\; \int_{S_p} d\alpha \;=\; \lim_{r \to \infty} \int_{dS_p(r)} \alpha$$

where $S_p(r) = \varphi_{-r}(S_p(\epsilon))$ as in §3. It suffices to consider forms α with support near q where $\lambda_q = \lambda_p - 1$. Near such q, the set $S_p(r)$, for large r, consists of a finite number of manifolds with boundary, transversal to U_q. There is one for each $\gamma \in \Gamma_{p,q}$. As $r \to \infty$ along one such γ, $dS_p(r)$

converges to $\pm S_q$ where the sign is determined by the agreement (or not) of the orientation of $dS_p(r)$ with the chosen orientation of S_q. q.e.d.

Remark 4.4. The integers $n_{p,q}$ have a simple definition in terms of currents. Set $S_p(r) = \varphi_{-r}(S_p(\epsilon))$ and $U_q(r) = \varphi_r(U_q(\epsilon))$ (cf. (2.3)). Then for all r sufficiently large

$$n_{p,q} = (-1)^{\lambda_p} \int_X [U_q(r)] \wedge d\,[S_p(r)] \tag{4.2}$$

where the integral denotes evaluation on the fundamental class.

5. Duality

There is a simple proof of Poincaré duality in this context. Suppose that X is compact and oriented. Given two oriented submanifolds A and B of complementary dimensions in X which meet transversally in a finite number of points, we define $A \bullet B = \int_X [A] \wedge [B]$ to be the algebraic number of intersections points (counting a point ± 1 depending on orientations as usual). Let $Cr_k(f) = \{p \in Cr(f) \ : \ \lambda_p = k\}$. Then for any k we have

$$U_q \bullet S_p = \delta_{pq} \qquad \text{for all } \ p, q \in Cr_k(f). \tag{5.1}$$

This gives a formal identification

$$\mathcal{U}_f^{\mathbb{Z}} \stackrel{\text{def}}{=} \mathbb{Z} \cdot \{[U_p]\}_{p \in Cr(f)} \cong \text{Hom}\left(\mathcal{S}_f^{\mathbb{Z}}, \mathbb{Z}\right). \tag{5.2}$$

Therefore, taking the adjoint of d gives a differential δ on $\mathcal{U}_f^{\mathbb{Z}}$ with the property that $H_{n-*}(\mathcal{U}_f^{\mathbb{Z}}, \delta) \cong H^*(X; \mathbb{Z})$. On the other hand the arguments of §§1–4 (with f replaced by $-f$) show that $\mathcal{U}_f^{\mathbb{Z}}$ is d-invariant with $H_*(\mathcal{U}_f^{\mathbb{Z}}, d) \cong H_*(X; \mathbb{Z})$. However, these two differentials on $\mathcal{U}_f^{\mathbb{Z}}$ agree up to sign as we see in the next lemma.

Lemma 5.1. *One has*

$$(dU_q) \bullet S_p = (-1)^{n-k}\, U_q \bullet (dS_p) \tag{5.3}$$

for all $p \in Cr_k(f)$ and $q \in Cr_{k-1}(f)$, and for any k.

Proof. One can see directly from the definition that the integers $n_{p,q}$ are invariant (up to a global sign) under time-reversal in the flow. However, for a simple current-theoretic proof consider the 1-dimensional current $[U_q(r)]\wedge[S_p(r)]$ consisting of a finite sum of oriented line-segments in the flow lines of $\Gamma_{p,q}$ (cf. Remark 4.4). Note that

$$d\left([U_q(r)]\wedge[S_p(r)]\right) = (d\,[U_q(r)])\wedge[S_p(r)] + (-1)^{n-k+1}[U_q(r)]\wedge(d\,[S_p(r)])$$

and apply (4.2). q.e.d.

Corollary 5.2. (Poincaré Duality)

$$H^{n-k}(X;\mathbb{Z}) \cong H_k(X;\mathbb{Z}) \qquad \textit{for all } k.$$

Note 5.3. In our operator picture the Poincaré duality isomorphism can be realized in a nice way. Let

$$T_{\text{tot}} = T^* + T = \{(y,x) : y = \varphi_t(x) \text{ for some } t \in \mathbb{R}\}.$$

Then we obtain the operator equation

$$d \circ \mathbf{T}_{\text{tot}} + \mathbf{T}_{\text{tot}} \circ d = \mathbf{P} - \check{\mathbf{P}}$$

where

$$\mathbf{P} = \sum_{p\in Cr(f)} [U_p] \times [S_p] \qquad \text{and} \qquad \check{\mathbf{P}} = \sum_{p\in Cr(f)} [S_p] \times [U_p].$$

This chain homotopy induces an isomorphism $H_*(\mathcal{U}_f^{\mathbb{Z}}) \cong H_*(\mathcal{C}_f^{\mathbb{Z}})$, which after identifying $\mathcal{U}_f^{\mathbb{Z}}$ with the cochain complex via (5.1) and (5.3), gives the duality isomorphism 5.2. When X is not oriented, a parallel analysis yields Poincaré duality with mod 2 coefficients.

6. Critical submanifolds of higher dimension

The methods introduced above apply in much greater generality. As seen in §1, one only needs the flow φ_t to be of finite volume to guarantee the existence of an operator $\mathbf{P}(\alpha) = \lim_{t\to\infty} \varphi_t^*(\alpha)$ which is chain homotopic to the identity. In this and the following sections we shall examine some important examples.

Let $f : X \to \mathbb{R}$ be a smooth function whose critical set (i.e., the set where $df = 0$) is a finite disjoint union

$$Cr(f) = \coprod_{j=1}^{\nu} F_j$$

of compact submanifolds F_j in X. We assume that Hess(f) is *non-degenerate* on the normal spaces to $Cr(f)$. Then for any f-tame gradient flow (cf. [24]) there are stable and unstable manifolds

$$S_j = \{x \in X \ : \ \lim_{t\to\infty} \varphi_t(x) \in F_j\} \text{ and } U_j = \{x \in X \ : \ \lim_{t\to -\infty} \varphi_t(x) \in F_j\}$$

with projections

$$S_j \xrightarrow{\tau_j} F_j \xleftarrow{\sigma_j} U_j. \tag{6.1}$$

where

$$\tau_j(x) = \lim_{t\to\infty} \varphi_t(x) \qquad \text{and} \qquad \sigma_j(x) = \lim_{t\to -\infty} \varphi_t(x).$$

For each j, let $n_j = \dim(F_j)$ and set $\lambda_j = \dim(S_j) - n_j$. Then $\dim(U_j) = n - \lambda_j$. For $p \in F_j$ we define $\lambda_p \equiv \lambda_j$ and $n_p \equiv n_j$.

Definition 6.1. The gradient flow φ_t of a smooth function $f \in C^\infty(X)$ on a riemannian manifold X is called a **generalized Morse-Stokes flow** if it is f-tame and:

(i) The critical set of f consists of a finite number of submanifolds $F_1, ..., F_\nu$ on the normals of which Hess(f) is non-degenerate.

(ii) The manifolds T, and T^*, and the stable and unstable manifolds S_j, U_j for $1 \le j \le \nu$ are submanifolds of finite volume. Furthermore, for each j, the fibres of the projections τ_j and σ_j are of uniformly bounded volume.

(iii) $p \prec q \quad \Rightarrow \quad \lambda_p + n_p < \lambda_q \qquad \forall p, q \in Cr(f)$.

These axioms are easily verified in a number of important cases in the algebraic and analytic category. The first main result concerning such flows is the following.

Theorem 6.2. *Suppose φ_t is a gradient flow satisfying the generalized Morse-Stokes conditions 6.1 on a compact oriented manifold X. Then there is an equation of currents*

$$\partial T \ = \ [\Delta] - P$$

on $X \times X$, where T, Δ, and P are as in Theorem 1.2, and

$$P \ = \ \sum_{j=1}^{\nu} [U_j \times_{F_j} S_j] \tag{6.2}$$

where $U_j \times_{F_j} S_j \equiv \{(y,x) \in U_j \times S_j \subset X \times X \ : \ \sigma_j(y) = \tau_j(x)\}$ *denotes the fibre product of the projections (6.1)*

Proof. The argument follows closely the proof of Theorem 2.3. Details are omitted. q.e.d.

Janko Latschev has found a Smale-type condition which yields this result in many cases where the hypothesis of 6.1 (iii) does not hold. In particular, Latschev's condition implies only that: $p \prec q \Rightarrow \lambda_p < \lambda_q$. Details appear in [24].

As in §3, this result can be translated into operator form.

Theorem 6.3. *Let* φ_t *be a gradient flow satisfying the generalized Morse-Stokes Conditions on a manifold* X *as above. Then for all smooth forms* α *on* X*, the limit*

$$\mathbf{P}(\alpha) = \lim_{t\to\infty} \varphi_t^*(\alpha)$$

exists and defines a continuous linear operator $\mathbf{P} : \mathcal{E}^*(X) \longrightarrow \mathcal{D}'^*(X)$ *with values in flat currents on* X*. This operator fits into a chain homotopy*

$$d \circ \mathbf{T} + \mathbf{T} \circ d \ = \mathbf{I} - \mathbf{P}. \tag{6.3}$$

Furthermore, $\mathbf{P}$ *is given by the formula*

$$\mathbf{P}(\alpha) \ = \ \sum_{j=1}^{\nu} Res_j(\alpha)\,[S_j] \tag{6.4}$$

where

$$Res_j(\alpha) \ \equiv \ \tau_j^*\left\{(\sigma_j)_*\left(\alpha\big|_{U_j}\right)\right\}. \tag{6.5}$$

Proof. This is a direct consequence of Theorem 6.2 except for the formulae (6.4)-(6.5). To see this consider the pull-back square

$$\begin{array}{ccc} U_j \times_{F_j} S_j & \xrightarrow{\ t_j\ } & U_j \\ {\scriptstyle s_j}\big\downarrow & & \big\downarrow{\scriptstyle \sigma_j} \\ S_j & \xrightarrow{\ \tau_j\ } & F_j \end{array} \tag{6.6}$$

where t_j and s_j are the obvious projections. One sees from the definitions (cf §3) that

$$\mathbf{P}(\alpha) \ = \ \sum_{j=1}^{\nu}(s_j)_*\left\{(t_j)^*\left(\alpha\big|_{U_j}\right)\right\}.$$

The commutativity of the diagram (6.6) allows us to rewrite these terms as in (6.5). q.e.d.

Corollary 6.4. *Suppose that $\lambda_p + n_p + 1 < \lambda_q$ for all critical points $p \prec q$. Then the homology of X is spanned by the images of the groups $H_{\lambda_j+\ell}(\overline{S_j})$ for $j = 1, ..., \nu$ and $\ell \geq 0$.*

Proof. Under this hypothesis $\partial(U_j \times_{F_j} S_j) = 0$ for all j, and so (6.2) yields a decomposition of $\mathbf{P}$ into operators that commute with d. q.e.d.

We can make this corollary more precise. Note that $\tau_j : S_j \to F_j$ can be given the structure of a vector bundle of rank λ_j. The closure $\overline{S_j} \subset X$ is a compactification of this bundle with a complicated structure at infinity. (See [10] for example.) There is nevertheless a homomorphism $\Theta_j : H_*(F_j) \longrightarrow H_{\lambda_j+*}(\overline{S_j})$ which after pushing forward to the one-point compactification of S_j, is the Thom isomorphism. This leads to the following (cf. [2]).

Theorem 6.5. *Suppose that $\lambda_p + n_p + 1 < \lambda_q$ for all critical points $p \prec q$ and that X and all F_j and S_j are oriented. Then there is an isomorphism*

$$H_*(X) \cong \bigoplus_j H_{*-\lambda_j}(F_j)$$

This result holds without the orientation assumptions if one takes homology with appropriately twisted coefficients. Much stronger versions of Theorems 6.3 and 6.5, are found in [24]. They include an extension to integral homology groups. Latschev also derives a spectral sequence associated to any *Bott-Smale* function satisfying a natural Smale-type transversality hypothesis. One virtue of this sequence is that the differentials are explicitly computable. Assuming for simplicity that everything is oriented, the E^1-term is given by

$$E^1_{p,q} = \bigoplus_{\lambda_j = p} H_q(F_j; \mathbb{Z})$$

and $E^k_{p,q} \Rightarrow H_*(X; \mathbb{Z})$

7. The relative case

In standard Morse Theory one often studies the change in the topology as one passes from $\{x : f(x) \leq a\}$ to $\{x : f(x) \leq b\}$. Our approach is easily adapted to this case.

Let $f : X \to \mathbb{R}$ be a proper Morse function, where X is not necessarily compact, and suppose that X carries a metric as in 14.3. Let $a < b$ be regular values of f and consider the compact manifold with boundary

$$Z \equiv f^{-1}([a,b]).$$

On Z we define a vector field $V = (\psi \circ f)\nabla f$ where $\psi : [a,b] \to [0,1]$ is a smooth function satisfying: (i) $\psi^{-1}(0) = \{a,b\}$, (ii) ψ is linear on $[a, a+\epsilon]$ and $[b-\epsilon, b]$, (iii) $\psi \equiv 1$ on $[a+2\epsilon, b-2\epsilon]$, (iv) $f(Cr(f)) \cap Z \subset (a+2\epsilon, b-2\epsilon)$ for some small $\epsilon > 0$.

Let $\varphi_t : Z \to Z$ be the flow of V. Note that φ_t is complete and fixes the boundary ∂Z. By 14.3, the stable and unstable manifolds of each $p \in Cr(f)$ have finite volume in Z, and so also does $T \equiv \{(y,x) : y = \varphi_t(x) \text{ for some } t,\ 0 < t < \infty\} \subset Z \times Z$.

We decompose the boundary

$$\partial Z \;=\; f^{-1}(b) - f^{-1}(a) \;\stackrel{\text{def}}{=}\; \partial_b - \partial_a.$$

In analogy with the fibre products appearing in §6 we have the following submanifolds of $Z \times Z$:

$$S(\partial_b) \;=\; \{(y,x) \in Z \times Z \;:\; y \in \partial_b \text{ and } \lim_{t\to\infty} \varphi_t(x) = y\} \qquad \text{and}$$
$$U(\partial_a) \;=\; \{(y,x) \in Z \times Z \;:\; x \in \partial_a \text{ and } \lim_{t\to-\infty} \varphi_t(y) = x\}.$$

Theorem 7.1. *On $Z \times Z$ there is an equation of integral currents*

$$\partial T \;=\; [\Delta] - \sum_{p \in Cr(f)\cap Z} [U_p] \times [S_p] - [U(\partial_a)] - [S(\partial_b)]. \tag{7.1}$$

Proof. Consider $\mathcal{T}_t \subset \mathbb{R} \times X \times X$ defined as in (1.1) and note that

$$\partial \mathcal{T}_t \;=\; \{0\} \times [\Delta_Z] - \{t\} \times [\operatorname{graph} \varphi_t] - [0,t] \times [\Delta_{\partial Z}].$$

Set $T_t = \mathrm{pr}_* \mathcal{T}_t$ and observe that since $\mathrm{pr}_*([0,t] \times [\Delta_{\partial Z}]) = 0$,

$$\partial T_t \;=\; [\Delta] - [\operatorname{graph} \varphi_t].$$

By hypothesis $T = \lim_{t\to\infty} T_t$ has finite volume, and a direct application of the arguments of §2 establishes (7.1). q.e.d.

Equation (7.1) can be translated to an operator equation. However, here we want the operator to act on the *relative* forms:

$$\mathcal{E}^*(Z, \partial_b) \;\equiv\; \left\{ \alpha \in \mathcal{E}^*(Z) \;:\; \alpha\big|_{\partial_b} = 0 \right\}. \tag{7.2}$$

We begin with the case where $f > a$ and so $\partial_a = \emptyset$.

Theorem 7.2. *Suppose $f : X \longrightarrow \mathbb{R}^+$ is a proper Morse function. Let $Z = f^{-1}(-\infty, b]$ where b is a regular value of f, and consider the operator*

$$\mathbf{P} : \mathcal{E}^*(Z, \partial Z) \longrightarrow \mathcal{D}'^*(Z)$$

defined by

$$\mathbf{P}(\alpha) \;=\; \lim_{t\to\infty} \varphi_t^*(\alpha) \tag{7.3}$$

where φ_t is the truncated gradient flow defined above. This operator is well defined and continuous. In fact, there exists a continuous operator $\mathbf{T} : \mathcal{E}^*(Z, \partial Z) \longrightarrow \mathcal{D}'^*(Z)$ of degree -1 with values in flat currents, such that

$$d \circ \mathbf{T} + \mathbf{T} \circ d \;= \mathbf{I} - \mathbf{P} \tag{7.4}$$

Furthermore, $\mathbf{P}$ is given by the formula

$$\mathbf{P}(\alpha) \;=\; \sum_{p\in Cr(f)\cap Z} r_p(\alpha)[S_p] \tag{7.5}$$

where $r_p(\alpha) = \int_{U_p} \alpha$. In particular, $\mathbf{P}$ is a continuous chain mapping onto the finite dimensional complex

$$\mathcal{S}_{f,Z} \;\overset{def}{=}\; span_{\mathbb{R}}\{[S_p]\}_{p\in Cr(f)\cap Z},$$

with differential given as in (4.1), and $\mathbf{P}$ induces an isomorphism

$$H^*_{deR}(Z, \partial Z) \;\xrightarrow{\cong}\; H(\mathcal{S}_{f,Z}).$$

Proof. This is deduced exactly as are Theorems 3.3, 4.1. We need only note that the operator given by $[S(\partial_b)]$ is zero on $\mathcal{E}^*(Z, \partial Z)$. This follows directly from the definition 3.1 and the fact that $\pi_1(S(\partial_b)) = \partial_b = \partial Z$. q.e.d.

In the more general case where $\partial_a \neq \emptyset$ we compose our operators with the projection map $\pi : \mathcal{E}^*(Z)' \longrightarrow \mathcal{E}^*(Z, \partial Z)'$, (which is adjoint to the inclusion $\mathcal{E}^*(Z, \partial Z) \subset \mathcal{E}^*(Z)$). In this case the operator corresponding to $[U(\partial_a)]$ is zero. Specifically, letting $\pi_k : X \times X \to X$ denote projection onto the kth factor, we have that

$$(\mathbf{U}(\partial_a)(\alpha), \beta) \;=\; ((\pi_1)_* \{\pi_2^*\alpha \wedge [U(\partial_a)]\},\, \beta) \;=\; 0$$

since $\pi_1(U(\partial_a)) = \partial_a$ and $\beta\big|_{\partial_a} = 0$. We conclude the following.

Theorem 7.3. *The operators*

$$\mathbf{T}, \mathbf{P} : \mathcal{E}^*(Z, \partial_b Z) \longrightarrow \mathcal{D}'^*(Z, \partial_a Z)$$

corresponding to the currents T, P from Theorem 7.2 satisfy the equation

$$d \circ \mathbf{T} + \mathbf{T} \circ d \ = \mathbf{I} - \mathbf{P}$$

where

$$\mathbf{P} \ = \sum_{p \in Cr(f) \cap Z} r_p[S_p]$$

and $r_p(\alpha) = \int_{U_p} \alpha$. Thus, $\mathbf{P}$ gives a continuous chain mapping of the relative deRham complex $\mathcal{E}^*(Z, \partial_b)$ onto the finite dimensional complex

$$\mathcal{S}_{f,Z} \ \stackrel{def}{=} \ span_{\mathbb{R}}\{[S_p]\}_{p \in Cr(f) \cap Z}$$

with differential given as in 3.5. This induces an isomorphism

$$H^*_{deR}(Z, \partial_b Z) \ \stackrel{\cong}{\longrightarrow} \ H(\mathcal{S}_{f,Z}).$$

Remark 7.4. Taking $\partial_b Z = \emptyset$ and $\partial_a Z \neq \emptyset$ in Theorem 7.3 (and interchanging the roles of a and b) gives the version of Theorem 7.2 corresponding to going backwards in time. That is, one considers

$$\mathbf{P}(\alpha) \stackrel{\text{def}}{=} \lim_{t \to -\infty} \varphi_t^* \alpha$$

and obtains a projection operator

$$\mathbf{P} : \mathcal{E}^*(Z) \longrightarrow \mathcal{S}_{f,Z} \subset \mathcal{D}'^*(Z, \partial_b)$$

where $\mathcal{S}_{f,Z}$ is a finite complex defined over $\mathbb{Z}$.

Remark 7.5. (Duality) Arguing exactly as in §5 one can retrieve the duality theorem

$$H^k(Z, \partial_b Z; \mathbb{Z}) \ \cong \ H_{n-k}(Z, \partial_a Z : \mathbb{Z}) \tag{7.6}$$

which gives, in the special case where either $\partial_a Z = 0$ or $\partial_b Z = 0$, the Lefschetz Duality Theorem.

8. Holomorphic flows and the Carrell-Lieberman-Sommese theorem

The ideas in this paper have interesting consequences in the holomorphic case. Given a $\mathbb{C}^*$-action φ_t on a compact Kähler manifold X, there is a complex graph

$$\mathcal{T} \stackrel{\text{def}}{=} \{(t, \varphi_t(x), x) \in \mathbb{C}^* \times X \times X \ : \ t \in \mathbb{C}^* \text{ and } x \in X\} \subset \mathbb{P}^1(\mathbb{C}) \times X \times X$$

similar to the graphs considered above. It is a basic result of Sommese [34] that if φ_t has fixed-points, then $\mathcal{T}$ has finite volume and its closure $\overline{\mathcal{T}}$ in $\mathbb{P}^1(\mathbb{C}) \times X \times X$ is an analytic subvariety.

$\mathbb{C}^*$-actions with fixed-points on Kähler manifolds are intimately related to Morse-Theory. The complex action φ_t can be decomposed into an "angular" S^1-action and a radial flow. Choosing an S^1-invariant Kähler metric and applying an argument of Frankel [13] gives a function $f : X \to \mathbb{R}$ of Bott-Morse type whose gradient generates the radial action. The methods of this paper can now be applied. In particular we find that $\overline{\mathcal{T}}$ defines a rational equivalence in $X \times X$ between the diagonal Δ and an analytic cycle P whose components are fibre products of stable and unstable manifolds over components of the fixed-point set.

When the fixed-points are all isolated, P becomes a sum of analytic Künneth components $P = \sum \overline{S}_p \times \overline{\overline{U}}_p$. This immediately implies that the cohomology of X is freely generated by the stable subvarieties $\{\overline{S}_p\}_{p \in \text{Zero}(\varphi)}$. Furthermore, it follows that X is algebraic and that all cohomology theories on X (eg. algebraic cycles modulo rational equivalence, algebraic cycles modulo algebraic equivalence, singular cohomology) are naturally isomorphic. (cf. [5], [12], [14]).

When the fixed-point set has positive dimension, this method yields results of Carrell-Lieberman-Sommese for $\mathbb{C}^*$-actions ([8], [9]), which assert among other things that if $\dim(X^{\mathbb{C}^*}) = k$, then $H^{p,q}(X) = 0$ for $|p - q| > k$.

9. Singularities and characteristic classes

The ideas in this paper also have an interesting application to the study of curvature and singularities. Suppose

$$\alpha : E \longrightarrow F$$

is a map between smooth vector bundles with connection over a manifold X. Let $G = G_k(E \oplus F) \to X$ denote the Grassmann bundle of k-planes in

$E \oplus F$ where $k = \text{rank}(E) \leq \text{rank}(F)$. There is a flow φ_t on G induced by the flow $\psi_t : E \oplus F \to E \oplus F$ where $\psi_t(e, f) = (te, f)$. This is a very simple generalized Morse-Stokes flow on G (written multiplicatively rather than additively). On the "affine chart" $\text{Hom}(E, F) \subset G$, one has that $\varphi_t(A) = \frac{1}{t}A$. One considers the graph T of the flow in the fibre product of G with itself, and one obtains the equation $\partial T = \Delta - P$ where Δ is the fibre diagonal and P is a sum of fibre products of stable and unstable manifolds. In the affine chart $\text{Hom}(E, F)$, the stable manifolds are exactly the universal singularity sets Σ_ℓ consisting of linear maps $A : E \to F$ of rank $k - \ell$.

Passing to the associated operators as in §3 and applying the resulting chain homotopy to Chern-Weil characteristic forms of the tautological bundle U over G leads to a "universal McPherson Formula" on $\text{Hom}(E, F)$. To pull this formula back to X we introduce the following concept.

Definition 9.1. The section α is said to be **geometrically atomic** if the submanifold

$$T(\alpha) \stackrel{\text{def}}{=} \left\{\tfrac{1}{t}\alpha(x) \in G \,:\, 0 < t \leq 1 \quad \text{and} \quad x \in X\right\}$$

has locally finite volume in G.

This hypothesis is sufficient to guarantee the existence of $\lim_{t\to 0} \alpha_t^* \phi$ where $\alpha_t \equiv \frac{1}{t}\alpha$ and where ϕ is any differential form on G. Choosing $\phi = \Phi(\Omega_U)$ where Φ is an Ad-invariant polynomial on $\mathfrak{gl}_k(\mathbb{R})$ and Ω_U is the curvature of the tautological k-plane bundle over G, one obtains the following formula for the Φ characteristic form of E:

$$\Phi(\Omega_E) \;=\; \sum_\ell \text{Res}_{\ell,\Phi}\left[\Sigma_\ell(\alpha)\right] + dT \tag{9.1}$$

where $\Sigma_\ell(\alpha)$ is a current associated to the locus where α drops rank by ℓ, $\text{Res}_{\ell,\Phi}$ is an explicit residue form defined along $\Sigma_\ell(\alpha)$, and T is an L^1_{loc}-form on X.

For bundle maps which are *normal*, i.e., transversal to the universal singularity sets in $\text{Hom}(E, F)$, this gives a local version of a basic formula of MacPherson [27], [28]. However, many important classes of bundle mappings, such as direct sum and tensor product mappings, are not normal but are generically geometrically atomic. In these cases the explicit formulas given by (9.1) are new even at the global (cohomological) level.

The critical idea is to study the operator from forms on G to currents on X whose "kernel" is given by $T(\alpha)$. The method is analogous to that of §6. Details appear in [19], [20].

When the residues in (9.1) are integers, the form T represents a Cheeger-Simons differential character canonically associated to α and the connections on E and F (cf. [22]).

10. Equivariant Morse theory

The ideas developed here carry over virtually intact to the setting of equivariant cohomology. The reason the method works directly is the simple but important fact that a closed, invariant submanifold is *equivariantly* closed (See Corollary 10.2 below). In this section we derive some consequences of the method which usefully apply to Morse functions arising in algebra and geometry (e.g., from moment map constructions). Deeper results have been obtained by J. Latschev [24]. We shall adopt the exposition of Cartan's equivariant de Rham theory found in [6].

Let G be a compact Lie group with Lie algebra $\mathfrak{g}$ acting on a compact n-manifold X. An **equivariant differential form** on X is a G-equivariant polynomial map $\alpha : \mathfrak{g} \longrightarrow \mathcal{E}^*(X)$. The set of such forms is denoted by

$$\mathcal{E}^*_G(X) \;=\; \{S^*(\mathfrak{g}^*) \otimes \mathcal{E}^*(X)\}^G$$

and is graded by defining elements of $S^p(\mathfrak{g}^*) \otimes \mathcal{E}^q(X)$ to have total degree $2p+q$. The equivariant differential $d_G : \mathcal{E}^*_G(X) \longrightarrow \mathcal{E}^{*+1}_G(X)$ is defined by setting

$$(d_G\alpha)(V) \;=\; d\alpha(V) - i_{\widetilde{V}}\alpha(V)$$

for $V \in \mathfrak{g}$ where the vector field $\widetilde{V}$ is the image of V under the natural linear map $\mathfrak{g} \longrightarrow \Gamma(TX)$, and where $i_{\widetilde{V}}$ denotes contraction with $\widetilde{V}$.

The complex $\mathcal{D}'^*_G(X)$ of **equivariant currents** is defined analogously by passing from smooth forms to generalized forms, i.e., forms with distribution coefficients.

Suppose now that $f \in C^\infty(X)$ is a G-invariant function and X is provided with a G-invariant riemannian metric. Then $g_*\nabla f = \nabla f$ for all $g \in G$ and so the gradient flow of f commutes with the action of G. Suppose this flow has finite volume and let

$$\partial T \;=\; \Delta - P \tag{10.1}$$

denote the equation discussed in §1. Let G act on $X \times X$ by the diagonal action $g \cdot (x, y) = (gx, gy)$.

Lemma 10.1. *The currents T and P are G-invariant and have the property that* $i_{\widetilde{V}}T = i_{\widetilde{V}}P = 0$ for all $V \in \mathfrak{g}$.

Proof. The current T corresponds to integration over the finite volume submanifold $\{(x, y) \in X \times X - \Delta \ : \ \exists t \in (0, \infty) \text{ s.t. } y = \varphi_t(x)\}$. Since $g\varphi_t(x) = \varphi_t(gx)$, the invariance of T is clear. The invariance of P then follows from (10.1). For the second assertion note that since V is tangent to T, $\left(i_{\tilde{V}}T\right)(\omega) = \int_T i_{\tilde{V}}\omega = 0$ for any n-form ω on $X \times X$. From the equation $d \circ i_{\tilde{V}} + i_{\tilde{V}} \circ d = \mathcal{L}_{\tilde{V}}$ (=Lie derivative) and the fact that $\mathcal{L}_{\tilde{V}}T = i_{\tilde{V}}\Delta = 0$, we conclude that $i_{\tilde{V}}P = 0$. q.e.d.

Corollary 10.2. *Consider $T \equiv 1 \otimes T \in 1 \otimes \mathcal{D}'^{n-1}(X \times X)^G$ as an equivariant current of total degree $(n-1)$ on $X \times X$. Consider Δ and P similarly as equivariant currents of degree n. Then*

$$\partial_G T = \Delta - P. \tag{10.2}$$

in the complex of equivariant currents on $X \times X$.

The dictionary given between operators and kernels in §1 carries over directly to this context. Currents in $\mathcal{D}'^{n-\ell}{}_G(X \times X)$ correspond to G-equivariant operators $\mathcal{E}^*(X) \longrightarrow \mathcal{D}'^{*+\ell}(X)$, and therefore give operators $\mathcal{E}^*_G(X) \longrightarrow \mathcal{D}'^{*+\ell}{}_G(X)$. Equation (10.2) corresponds to the operator equation

$$d_G \circ \mathbf{T} + \mathbf{T} \circ d_G = I - \mathbf{P}, \tag{10.3}$$

and the discussion of §3 proves the following.

Proposition 10.3. *Let X be a compact G-manifold and φ_t an invariant flow of finite volume on X. Then for all $\alpha \in \mathcal{E}^*_G(X)$ the limit*

$$\mathbf{P}(\alpha) = \lim_{t \to \infty} \varphi_t^* \alpha \tag{10.4}$$

exists and defines a continuous linear operator $\mathbf{P} : \mathcal{E}^*_G(X) \longrightarrow \mathcal{D}'^*_G(X)$ of degree 0, which is equivariantly chain homotopic to the identity on $\mathcal{E}^*_G(X)$.

Applying the methods of §§2–3 and the fact that

$$S^*(\mathfrak{g}^*)^G \cong H^*(BG)$$

gives the following result.

Theorem 10.4. *Let X be a compact riemannian G-manifold and $f \in C^\infty(X)$ an invariant Morse function whose gradient flow φ_t is Morse-Stokes. Then the continuous linear operator (10.4) defines a map of equivariant complexes*

$$\mathbf{P} : \mathcal{E}^*_G(X) \longrightarrow S^*(\mathfrak{g}^*)^G \otimes \mathcal{S}_f$$

where $\mathcal{S}_f = \text{span}\{[S_p]\}_{p\in Cr(f)}$ *as in (3.8) and the differential on* $S^*(\mathfrak{g}^*)^G \otimes \mathcal{S}_f$ *is* $1 \otimes \partial$. *The map* $\mathbf{P}$ *induces an isomorphism*

$$H^*_G(X) \xrightarrow{\cong} H^*(BG) \otimes H^*(X)$$

For an example of this phenomenon consider the torus acting on $\mathbb{P}^n_{\mathbb{C}}$ via the standard action on homogeneous coordinates $[z_0, ..., z_n]$, and set $f([z]) = \sum k|z_k|^2/\|z\|^2$. One sees immediately the well-known fact that $H^*_G(\mathbb{P}^n_{\mathbb{C}})$ is a free $H^*(BG)$-module with one generator in each dimension $2k$ for $k = 0, ..., n$. This extends to generalized flag manifolds and to products.

It has been pointed out by Janko Latschev that there exists an invariant Morse function for which no choice of invariant metric gives a Morse-Stokes flow.

On the other hand the method applies to much more general functions and yields results as in §§5-9. Suppose for example that f is an invariant function whose critical set consists of a finite number of non-degenerate critical orbits $\mathcal{O}_i = G/H_i$, $i = 1, ..., N$. Latschev (cf. [24], [1]) has established a spectral sequence $E^k_{*,*} \Rightarrow H^*_G(X)$ with computable differentials and (assuming for simplicity that everything is oriented)

$$E^1_{p,*} = \bigoplus_{\lambda_i = p} H^*_G(\mathcal{O}_i) = \bigoplus_{\lambda_i = p} H^*(BH_i).$$

11. Flat bundles and local coefficients

The constructions in this paper can be generalized to forms with coefficients in a flat bundle $E \to X$. The kernels of §2 become currents on $X \times X$ with coefficients in $\text{Hom}(\pi_1^* E, \pi_2^* E)$ where π_1 and π_2 are the obvious projections. More specifically suppose that φ_t is a Morse-Stokes flow on X with kernel T constructed exactly as in §2. Then given a flat bundle E we define the kernel T_E with coefficients in $\text{Hom}(\pi_1^* E, \pi_2^* E)$ by $T_E = h \otimes T$ where and $h : E_{\varphi_t(x)} \to E_x$ is parallel translation along the flow line. It is straightforward to compute that

(11.1) $$\partial T_E = \Delta_E - P_E$$

where $\Delta_E = \text{Id} \otimes \Delta$ and $P_E = \sum_p h_p \otimes ([U_p] \times [S_p])$ with $h_p : E_y \to E_x$ given by parallel translation along the broken flow line from y to x. Note that under the canonical trivializations $E\big|_{U_p} \cong U_p \times E_p$ and $E\big|_{S_p} \cong$

$S_p \times E_p$, the map h_p becomes the identity $\mathrm{Id} : E_p \to E_p$. Corresponding to (11.1) we have the operator equation

$$(11.2) \qquad d \circ \mathbf{T}_E + \mathbf{T}_E \circ d \;=\; \mathbf{I} - \mathbf{P}_E$$

where $\mathbf{P}_E$ maps onto the finite vector space

$$\mathcal{S}_E \stackrel{\mathrm{def}}{=} \bigoplus_{p \in Cr(f)} E_p \otimes [S_p]$$

by integration of forms over the unstable manifolds. By (11.2) this space is d-invariant. In fact in the Morse-Smale case the restriction of d to $\mathcal{S}_E$ is given as in 4.3 by

$$d(e \otimes [S_p]) = \sum h_{p,q}(e)[S_q]$$

where $h_{p,q} = (-1)^{\lambda_p} \sum_\gamma h_\gamma$ and $h_\gamma : E_p \to E_q$ is parallel translation along $\gamma \in \Gamma_{p,q}$. By (11.2) the complex $(\mathcal{S}_E, d)$ computes $H^*(X;\, E)$.

Reversing time in the flow shows that the complex $\mathcal{U}_E = \bigoplus_p E_p^* \otimes [U_p]$ with differential defined as above computes $H^*(X;\, E^*)$. As in §5 the obvious dual pairing of these complexes establishes the generalized Poincaré duality. Furthermore, one can extend all this to integral currents twisted by representations of $\pi_1(X)$ in $\mathrm{GL}_n(\mathbb{Z})$ or $\mathrm{GL}_n(\mathbb{Z}/p\mathbb{Z})$ and obtain duality with local coefficient systems.

12. Products

Our method has a number of interesting extensions. For example, consider the triple diagonal $\Delta_3 \subset X \times X \times X$ as the kernel of the wedge-product operator

$$\wedge : \mathcal{E}^*(X) \otimes \mathcal{E}^*(X) \to \mathcal{E}^*(X).$$

Let f and f' be functions with Morse-Stokes flows φ_t and φ'_t respectively. Assume that for all $(p, p') \in Cr(f) \times Cr(f')$ the stable manifolds S_p and $S'_{p'}$ intersect transversely in a manifold of finite volume, and similarly for the unstable manifolds U_p and $U'_{p'}$. Degenerating Δ_3 gives a kernel

$$T \;\equiv\; \{(\varphi_t(x), \varphi'_t(x), x) \in X \times X \times X \;:\; x \in X \text{ and } 0 \le t < \infty\}$$

and a corresponding operator $\mathbf{T} : \mathcal{E}^*(X) \otimes \mathcal{E}^*(X) \to \mathcal{D}'^*(X)$ of degree -1. One calculates that $\partial T \;=\; \Delta_3 - M$ where

$$M = \sum_{(p,p') \in Cr(f) \times Cr(f')} [U_p] \times [U_{p'}] \times [S_p \cap S'_{p'}].$$

The corresponding operator $\mathbf{M} : \mathcal{E}^*(X) \otimes \mathcal{E}^*(X) \to \mathcal{D}'^*(X)$ is given by

$$(12.1) \qquad \mathbf{M}(\alpha, \beta) \ = \sum_{(p,p') \in Cr(f) \times Cr(f')} \left(\int_{U_p} \alpha \right) \left(\int_{U_{p'}} \beta \right) [S_p \cap S'_{p'}].$$

The arguments of §§1-3 adapt to prove the following.

Theorem 12.1. *There is an equation of operators* $\wedge - \mathbf{M} = d \circ \mathbf{T} + \mathbf{T} \circ d$ *from* $\mathcal{E}^*(X \times X)$ *to* $\mathcal{D}'^*(X)$ *(where* $\wedge$ *denotes restriction to the diagonal). In particular for* $\alpha, \beta \in \mathcal{E}^*(X)$ *we have the chain homotopy*

$$(12.2) \quad \alpha \wedge \beta - \mathbf{M}(\alpha, \beta) \ = \ d\mathbf{T}(\alpha, \beta) + \mathbf{T}(d\alpha, \beta) + (-1)^{deg\alpha}\mathbf{T}(\alpha, d\beta)$$

between the wedge product and the operator (12.1).

Note that the operator $\mathbf{M}$ has range in the finite dimensional vector space

$$\mathcal{M} \ \stackrel{\text{def}}{=} \ \text{span}_{\mathbb{R}}\{ \ [S_p \cap S'_{p'}]\} \ _{(p,p') \in Cr(f) \times Cr(f')}.$$

It converts a pair of smooth forms α, β into a linear combination of the pairwise intersections of the stable manifolds $[S_p]$ and $[S'_{p'}]$. If $d\alpha = d\beta = 0$, then

$$(12.3) \qquad \mathbf{M}(\alpha, \beta) \ = \ \alpha \wedge \beta - d\mathbf{T}(\alpha, \beta),$$

and so $\mathbf{M}(\alpha, \beta)$ is a cycle homologous to the wedge product $\alpha \wedge \beta$. This operator also has the following properties.

Theorem 12.2. *The operator* $\mathbf{M}$ *maps onto the subspace* $\mathcal{M}$. *Furthermore, for forms* $\alpha, \beta \in \mathcal{E}^*(X)$, *it satisfies the equation*

$$(12.4) \qquad d\mathbf{M}(\alpha, \beta) \ = \ \mathbf{M}(d\alpha, \beta) + (-1)^{deg\alpha}\mathbf{M}(\alpha, d\beta)$$

Proof. To see that $\mathbf{M}$ is onto (as a linear map from $\mathcal{E}^*(X \times X)$) it suffices to see that for each non-empty intersection $S_p \cap S'_{p'}$ the current $[S_p \cap S'_{p'}]$ is in the range. However, by transversality we see that if $S_p \cap S'_{p'} \neq \emptyset$, then $\lambda_p + \lambda_{p'} \geq n$, and so $n \geq (n - \lambda_p) + (n - \lambda_{p'})$. Therefore, by transversality we have $\dim(U_q \cap U'_{q'}) \leq 0$ for all q, q'. It follows that we can find differential forms α and β such that $\int_{U_p} \alpha = \int_{U'_{p'}} \beta = 1$ and $\int_{U_q} \alpha = \int_{U'_{q'}} \beta = 0$ for all $q \neq p$ and $q' \neq p'$. Then by (12.1), $\mathbf{M}(\alpha, \beta) = [S_p \cap S'_{p'}]$ and the assertion is proved.

Equation (12.4) follows from the fact that $dM = 0$, which implies that $d \circ \mathbf{M} + \mathbf{M} \circ d = 0$, together with the standard formula for $d(\alpha \wedge \beta)$.

q.e.d.

It follows immediately that $d(\mathcal{M}) \subset \mathcal{M}$. In fact one can see from the transversality assumptions that for $(p, p') \in Cr(f) \times Cr(f')$ one has

$$(12.5) \qquad \begin{aligned} d\,[S_p \cap S'_{p'}] &= \sum_{q \in Cr(f)} n_{pq}[S_q \cap U'_{p'}] \\ &\quad + (-1)^{n-\lambda_p} \sum_{q' \in Cr(f')} n'_{p'q'}[S_p \cap U'_{q'}] \end{aligned}$$

where the n_{pq} are defined as in §4. Thus we retrieve the cup product over the integers in the Morse complex.

Example 12.3. A fundamental example of a pair satisfying our hypotheses is given by $f, -f$ where the gradient flow is Morse-Smale. In this case $U'_p = S_p$ and $S'_p = U_p$ for all $p \in Cr(f) = Cr(-f)$. Thus formula (12.1) becomes

$$\mathbf{M}(\alpha, \beta) = \sum_{p, p' \in Cr(f)} \left(\int_{U_p} \alpha \right) \left(\int_{S_{p'}} \beta \right) [S_p \cap U_{p'}].$$

In particular we have the following.

Proposition 12.4. *Suppose the gradient flow of f is Morse-Smale. Then for any cycle Y in X which is transversal to all the $S_p \cap U_{p'}$, $p, p' \in Cr(f)$, we have the formula*

$$(12.6) \qquad \int_Y \alpha \wedge \beta = \sum_{p, p' \in Cr(f)} \left(\int_{U_p} \alpha \right) \left(\int_{S_{p'}} \beta \right) [S_p \cap U_{p'} \cap Y],$$

whenever $d\alpha = d\beta = 0$.

Example 12.5. Suppose $\deg \alpha = n - \deg \beta = k$. Then (cf. [25, Prop. 12])

$$\int_X \alpha \wedge \beta = \sum_{p \in Cr_k(f)} \left(\int_{U_p} \alpha \right) \left(\int_{S_{p'}} \beta \right).$$

13. A Lefschetz theorem for the Thom-Smale complex

Let X be a compact oriented riemannian manifold, and let U_p, S_p, $p \in Cr(f)$ be the unstable and stable manifolds of a Morse-Stokes flow on X, oriented as in §2. Recall the **Lefschetz number** of a smooth mapping $F : X \to X$ defined by

$$\mathrm{Lef}(F) = \sum_i (-1)^i \mathrm{trace}\,\{F_* : H_i(X;\mathbb{R}) \to H_i(X;\mathbb{R})\}$$

Theorem 13.1. *Suppose $F : X \to X$ is a smooth mapping such that F maps S_p transversally to U_q for all $q \succ p$ (i.e., $F\big|_{S_p}$ is transversal to U_p and $F(S_p) \cap U_q = \emptyset$ for all $q \succ p$ and $q \neq p$). Let $C_p = \{x \in S_p : F(x) \in U_p\}$. Then*

$$\mathrm{Lef}(F) = \sum_{p \in Cr(f)} (-1)^{\lambda_p} \sum_{x \in C_p} \sigma_x(F)$$

where

$$\sigma_x(F) \stackrel{\mathrm{def}}{=} \begin{cases} 1 & \text{if } F_* T_x(S_p) \text{ agrees in orientation with} \\ & \text{the normal space to } U_p \text{ at } F(x) \\ -1 & \text{otherwise.} \end{cases}$$

Proof. Our transversality assumption implies that the graph $\Gamma_F = \{(F(x), x) : x \in X\}$ in $X \times X$ meets the cycle P only in its regular points $\bigcup_p U_p \times S_p$ and it is transversal there. We recall that $\mathrm{Lef}(F) = [\Delta] \bullet [\Gamma_F]$ in $X \times X$. By (2.5) and the fact that Γ_F and P meet nicely, we conclude that $\mathrm{Lef}(F) = [P] \bullet [\Gamma_F]$ which is easily computed as claimed. q.e.d.

Note that when $F = \mathrm{Id}$, the hypotheses are satisfied and we get the standard computation of the Euler characteristic from the Morse complex.

14. The genericity theorem

Let $f \in C^\infty(X)$ be a Morse function on a compact manifold X. In this section we shall prove that there exists a riemannian metric on X for which the gradient flow satisfies the Morse-Stokes conditions of §2. Some of the material in this section could be reduced by appealing to a paper of Laudenbach [25]. For completeness we have included all the details. To begin we recall the following.

Definition 14.1. The gradient flow of f for a riemannian metric on X is called **Morse-Smale** if the stable and unstable manifolds S_p and U_q intersect transversely for all $p, q \in Cr(f)$.

To simplify arguments we shall demand a little more. Recall that at each $p \in Cr(f)$ there exist **canonical local coordinate systems** $(u,v) : \mathcal{O}_p \xrightarrow{\cong} \mathcal{V}_p$ where

$$(14.1) \qquad \mathcal{V}_p = \{(u,v) \in \mathbb{R}^{\lambda_p} \times \mathbb{R}^{n-\lambda_p} : |u|^2 \leq r_p \text{ and } |v|^2 \leq r_p\}$$

such that $u(p) = v(p) = 0$ and $f(u,v) = f(p) - |u|^2 + |v|^2$.

Definition 14.2. A riemannian metric ds^2 is said to be **canonically flat near** $Cr(f)$ if $ds^2 = |du|^2 + |dv|^2$ in some canonical linear coordinate system about each $p \in Cr(f)$.

We shall prove the following.

Theorem 14.3. *Let $f \in C^\infty(X)$ be a Morse function on a compact manifold X. Suppose X is given a riemannian metric which is canonically flat near $Cr(f)$ and for which the gradient flow φ_t is Morse-Smale. Then φ_t satisfies the Morse-Stokes conditions 2.1*

Theorem 14.4. *If $f \in C^\infty(X)$ is a Morse function and ds^2 is any riemannian metric on X, then ds^2 can be modified outside some neighborhood of $Cr(f)$ so that φ_t becomes Morse-Smale. In fact this modification can be made arbitrarily small in the C^∞-topology.*

Taken together these theorems prove the following.

Theorem 14.5. *Given any Morse function f on a compact manifold X, there exists a riemannian metric on X for which the gradient flow is Morse-Stokes.*

Furthermore, this metric can be chosen to be canonically flat near $Cr(f)$.

Proof of Theorem 14.3. We first observe that if the flow of f is Morse-Smale, then

$$(14.2) \qquad p \prec q \quad \Rightarrow \quad \lambda_p < \lambda_q$$

for all $p,\ q \in Cr(f)$. To see this suppose p and q are joined by an (unbroken) flow line ℓ. Then $U_p \cap S_q \supset \ell$ and so by the transversality condition, $\dim(U_p \cap S_q) = (n - \lambda_p) + \lambda_q - n \geq 1$.

Now let $a_1 < \cdots < a_m$ be the critical values of f. For each a_k let $Cr(f, a_k) \subset Cr(f)$ be the set of critical points of f with critical value a_k. Each $p \in Cr(f, a_k)$ has a canonical local coordinate system as in (14.1) where the metric is flat. We may assume that the radius r_p is the same for all $p \in Cr(f, a_k)$. Call this radius r_k. By shrinking the neighborhoods $\mathcal{O}_p$ we may assume that these canonical coordinate systems are pairwise disjoint and that $a_{k+1} - r_{k+1} > a_k + r_k$ for all k. Furthermore, by multiplying f by some scalar $a >> 1$ and further shrinking the $\mathcal{O}_p$ we can assume that $r_p = 2$ for all $p \in Cr(f)$.

Now our manifold decomposes into "blocks":

$$X = P_0 \cup Q_0 \cup P_1 \cup Q_1 \cup P_2 \cup Q_2 \cup \ldots P_m \tag{14.3}$$

where

$$P_k = f^{-1}[a_k - 1, a_k + 1] \quad \text{and} \quad Q_k = f^{-1}[a_k + 1, a_{k+1} - 1].$$

Note that P_k and Q_k are compact manifolds with boundary. The manifolds P_k can be further decomposed. Let $\mathcal{O}'_p \subset \mathcal{O}_p$ be the subset defined by the equations $|u||v| \leq 1$ and $-1 \leq |v|^2 - |u|^2 \leq 1$. Then

$$P_k = R_k \cup \bigcup_{p \in Cr(f, a_k)} \mathcal{O}'_p$$

where R_k is the closure of $P_k - \bigcup_p \mathcal{O}'_p$.

Let ψ_s be the (incomplete) flow on $X - Cr(f)$ generated by $W \equiv \operatorname{grad} f / \|\operatorname{grad} f\|^2$, so that

$$f(\psi_s(x)) = f(x) + s$$

whenever $\psi_s(x)$ is defined. Using this vector field in the obvious way (cf. [29]) we obtain smooth product structures

$$Q_k \cong (\partial^- Q_k) \times [0,1] \tag{14.4}$$

$$R_k \cong (\partial^- R_k) \times [0,1] \tag{14.5}$$

where $\partial^- Q_k = Q_k \cap f^{-1}(a_k + 1)$ and $\partial^- R_k = R_k \cap f^{-1}(a_k - 1)$. Note that $\partial^- R_k$ is a compact manifold with non-empty boundary.

Consider now the unstable manifold U_p for some $p \in Cr(f, a_k)$ and some k. We shall show that $\text{vol}(U_p) < \infty$. To begin note that $U_p \cap \mathcal{O}'_p \cong \{(0, v) \in \mathcal{V}_p : |v| \leq 1\}$ is a smoothly embedded closed disk of dimension $\ell \stackrel{\text{def}}{=} n - \lambda_p$ and clearly has finite ℓ-volume. Its boundary $U_p \cap (\partial^- Q_k)$ is a smoothly embedded sphere, and via (14.4) we have $U_p \cap Q_k \cong (U_p \cap \partial^- Q_k) \times [0,1]$ which also has finite ℓ-volume.

We now show that $U_p \cap P_{k+1}$ has finite ℓ-volume. To begin note that via (14.5) we have a smooth product $U_p \cap R_{k+1} \cong (U_p \cap \partial^- R_{k+1}) \times [0,1]$ (which extends beyond the boundary of R_{k+1} so we needn't worry about how U_p meets this boundary). Since $U_p \cap \partial^- R_{k+1}$ is a subset of a compact $(\ell - 1)$-manifold, we see that $U_p \cap R_{k+1}$ has finite ℓ-volume.

It remains to show that $U_p \cap \mathcal{O}'_q$ has finite volume for $q \in Cr(f, a_k)$. This is equivalent to showing that $U_p \cap \mathcal{O}''_q$ has finite volume, where $\mathcal{O}''_q$ is defined by $|u| \leq 1$ and $|v| \leq 1$. (To see this push inward along the flow.) For simplicity, from here on we shall denote $\mathcal{O}''_q$ by $\mathcal{O}_q$ and $\mathcal{V}''_q$ by $\mathcal{V}_q$.

Since $q \succ p$ we know from (14.2) that

$$\dim U_p \;>\; \dim U_q.$$

In our local coordinate box $\mathcal{V}_q = \{(u, v) \in \mathbb{R}^{\lambda_q} \times \mathbb{R}^{n-\lambda_q} \; : \; |u| \leq 1, \; |v| \leq 1\}$ the flow is generated by

$$\varphi_t(u, v) \;=\; (e^{-t}u, e^t v) \tag{14.6}$$

We decompose $\partial \mathcal{V}_q$ into two pieces:

$$A = \{(u, v) \; : \; |u| = 1, \; |v| \leq 1\} \quad \text{and} \quad B = \{(u, v) \; : \; |u| \leq 1, \; |v| = 1\}.$$

There are subsets

$$\begin{aligned} A_0 &= A \cap S_q = \{(u, 0) \; : \; |u| = 1\} \cong S^{\lambda_q - 1}, \\ B_0 &= B \cap U_q = \{(0, v) \; : \; |v| = 1\} \cong S^{n-\lambda_q-1}. \end{aligned}$$

The flow determines a diffeomorphism

$$\Phi : A - A_0 \stackrel{\cong}{\longrightarrow} B - B_0$$

given by

$$\Phi(u, v) \;=\; \left(|v| u, \frac{1}{|v|} v \right)$$

Note that $\Phi(u, v)$ is the unique point in B which lies on the flow line through (u, v).

Let $\widetilde{A} \stackrel{\pi_A}{\longrightarrow} A$ be the "oriented blow-up" of A along A_0 where A_0 is replaced by the oriented normal lines to A_0. Let $\widetilde{B} \stackrel{\pi_B}{\longrightarrow} B$ be defined similarly. $\widetilde{A}$ has coordinates $(\hat{u}, \hat{v}, t)$ where $|\hat{u}| = 1$, $|\hat{v}| = 1$ and $0 \leq t \leq 1$ and $\pi_A(\hat{u}, \hat{v}, t) = (\hat{u}, t\hat{v})$. $\widetilde{B}$ has the same coordinates with $\pi_B(\hat{u}, \hat{v}, t) = (t\hat{u}, \hat{v})$.

Φ lifts to a map $\widetilde{\Phi} : \widetilde{A} \longrightarrow \widetilde{B}$ which in these coordinates is the *identity map* (and so, in particular, a diffeomorphism).

Now since U_p is transversal to S_q we know that:

$$(U_p \cap S_q) \cap \mathcal{V}_q \;\cong\; C(U_p \cap A_0) \tag{14.7}$$

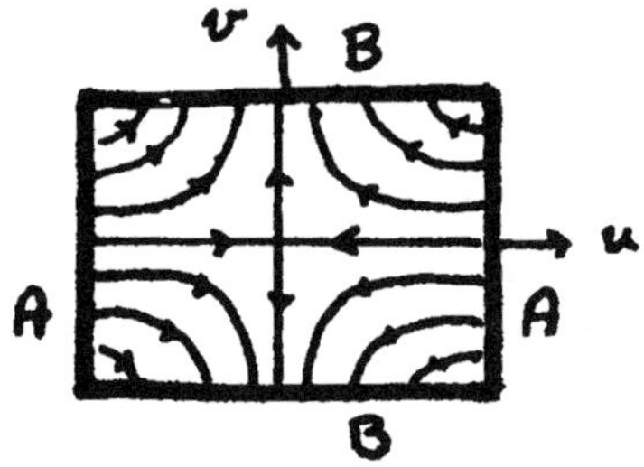

where $U_p \cap A_0$ is a compact submanifold of codimension λ_p in A_0, and where for any subset $Y \subset A_0$, $C(Y)$ is the **cone** on Y defined by

$$C(Y) = \{(tu, 0) : u \in Y \text{ and } 0 \le t \le 1\}$$

Now transversality also tells us that in a neighborhood of

$$Y \equiv U_p \cap A_0$$

the set $U_p \cap A$ is of the form $Y \times \mathbb{R}^{n-\lambda_q}$. Furthermore, $U_p \cap A$ has a "smooth proper transform" to $\widetilde{A}$, i.e., the lift of $U_p \cap (A - A_0)$ to $\widetilde{A}$ has closure which is a smooth manifold with boundary diffeomorphic to $Y \times S^{n-\lambda_q-1}$. Denote this closure by $\widetilde{U_p \cap A}$.

We can now describe the structure of $\overline{U}_p$ in our coordinate box. To begin we observe that

$$\overline{U_p \cap B} = \pi_B \widetilde{\Phi}(\widetilde{U_p \cap A})$$

and recall that $\widetilde{\Phi}$ is a diffeomorphism. It follows that $\overline{U_p \cap B}$ is a C^∞ stratified set with two strata. The top stratum is $U_p \cap B$. The singular stratum is exactly $B_0 \cong S^{n-\lambda_q-1}$. In a neighborhood of this singular stratum, $\overline{U_p \cap B}$ is diffeomorphic to $C(Y) \times S^{n-\lambda_q-1}$ where $Y = U_p \cap A_0$.

Since $\overline{U_p \cap B}$ is the image of a manifold of finite $(\ell - 1)$-volume (namely $\widetilde{U_p \cap A}$) under the smooth proper map $\pi_B \circ \widetilde{\Phi}$, it follows that $\overline{U_p \cap B}$ has finite $(\ell - 1)$-volume.

Conclusion 14.6 The closure of $U_p \cap \partial^- Q_{k+1}$ is a compact C^∞-stratified set of finite $(\ell - 1)$-volume, whose top stratum is $U_p \cap \partial^- Q_{k+1}$

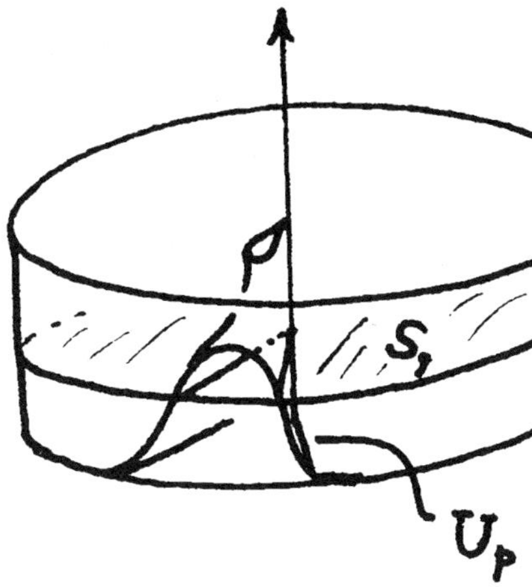

and whose singular strata are exactly the spheres $U_q \cap \partial^- Q_{k+1}$ for $q \in Cr(f, a_{k+1})$.

From the above analysis we can also conclude that U_p has finite volume in $\mathcal{V}_q$. Set

$$\widetilde{\mathcal{V}}_q = S^{\lambda_q - 1} \times S^{n-\lambda_q-1} \times [0,1] \times [0,1]$$

and consider the map $\Pi : \widetilde{\mathcal{V}}_q \to \mathcal{V}_q$ given by

$$\Pi(\hat{u}, \hat{v}, s, t) = (s\hat{u}, t\hat{v})$$

where $|\hat{u}| = |\hat{v}| = 1$. We identify $\widetilde{A}$ with the subset where $s = 1$ (and $\widetilde{B}$ with $t = 1$). Then Π restricts to be the projections π_A and π_B.

Consider the mapping

$$\Psi : \widetilde{\mathcal{V}}_q \longrightarrow \widetilde{A}$$

given by

$$\Psi(\hat{u}, \hat{v}, s, t) = (\hat{u}, \hat{v}, 1, st).$$

Then Ψ has the following two important properties. Let $M \subset \widetilde{A}$ be a closed submanifold with $\partial M \subset \partial \widetilde{A}$ which is transversal to $\partial_0 \widetilde{A} = S^{\lambda_q - 1} \times S^{n-\lambda_q-1} \times \{1\} \times \{0\}$. Then

$$\Psi \text{ is strongly transversal to } M \qquad \text{(See below)},$$

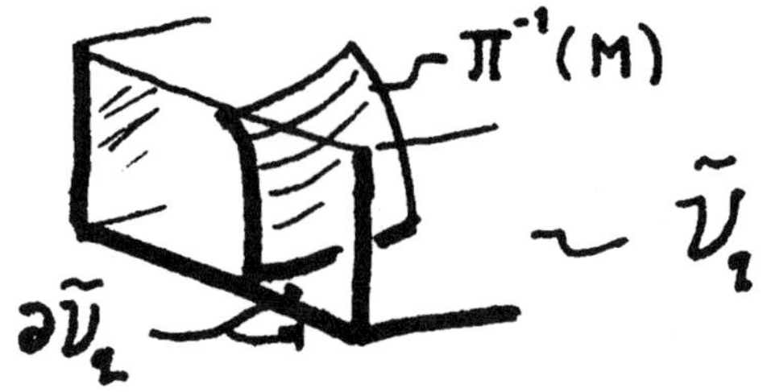

$$\Pi\left(\Psi^{-1}(M)\right) \;=\; \mathcal{V}_q \cap \bigcup_{t\geq 0} \varphi_t\left(\pi_A M\right)$$

By *strong transversality* we mean that if we extend M beyond $\partial_0\widetilde{A}$ by adding a collar, and if we extend Π to $S^{\lambda_q-1}\times S^{n-\lambda_q-1}\times(-\epsilon,1]\times(-\epsilon,1]$ using the same algebraic formula, then the extended Π is transversal to the extended M. Hence, $(\Pi^{-1}(M),\partial\Pi^{-1}(M))$ is a smooth submanifold with corners neatly embedded into $(\widetilde{\mathcal{V}}_q,\partial\widetilde{\mathcal{V}}_q)$.

In particular, if $\dim(M)=\ell-1$, then $\Psi^{-1}(M)$ has finite ℓ-volume and so does $\Pi(\Psi^{-1}(M))$

Remark 14.7. Note that the above comments also apply if M is a compact C^∞ stratified set in $\widetilde{A}$ which in a collar neighborhood $\partial_0\widetilde{A}\times[0,\epsilon)$ of $\partial_0\widetilde{A}$ is of the form $M_0\times[0,\epsilon)$.

We have now passed the first critical level beyond a_k and the closure of U_p has exited as a compact C^∞ stratified set, whose singularities are the submanifolds U_q for $q\succ p$ and $q\in Cr(f,a_{k+1})$.

We now compound the process. The product structure (14.4) shows that $U_p\cap Q_{k+1}$ has finite ℓ-volume (since it is a product of a smooth stratified set of finite $(\ell-1)$-volume with $[0,1]$). For similar reasons, $U_p\cap R_{k+2}$ has finite ℓ-volume, and it remains to examine what happens as we pass through the canonical coordinate boxes of critical points q at level $k+2$.

At points of $U_p\cap S_q$ for $q\in Cr(f,a_{k+2})$, the above analysis can be applied locally. However, to prove finite volume one must also consider points $x\in\overline{U}_p\cap S_q$ where $\overline{U}_p$ is singular. At such points x the singular set of $\overline{U}_p\cap S_q$ consists of all points of $U_{q'}$ in a neighborhood of x, for some $q'\prec q$. Now $U_{q'}$ is transversal to S_q, and $\overline{U}_p$ is locally of the form $C(Y)\times U_{q'}$ for some manifold Y. Consequently $\overline{U}_p\cap S_q$ is locally of the form $C(Y) \times \mathbb{R}^i$ and $\overline{U}_p$ is locally of the form $C(Y) \times \mathbb{R}^i \times \mathbb{R}^{n-\lambda_q}$ for some $i > 0$. In particular, $\overline{U}_p \cap A$ has a smooth proper transform to $\widetilde{A}$. This proper transform has a neat collar structure at the boundary as discussed in 14.7.

Applying the analysis above one concludes that $U_p \cap \partial^- Q_{k+2}$ is a compact C^∞ stratified set of finite ℓ-volume whose singular strata consist precisely of the sets $U_q \cap \partial^- Q_{k+2}$ for critical points $q \succ p$ of level $\leq k+2$. The same analysis also shows (using Remark 14.7) that U_q has finite ℓ-volume in a neighborhood of each critical point of level $k+2$.

One can now proceed inductively through the critical levels to prove that U_p has finite ℓ-volume.

Since S_p is the unstable manifold for -grad(f) = grad$(-f)$, we have also proved that each S_p has finite volume.

It remains to prove that the graph $\mathcal{T}$ has finite volume. For this we first observe that the arguments above apply directly to prove the following result. We say that the gradient flow of a smooth function F with non-degenerate critical manifolds is *tame* in a neighborhood of each critical point there are coordinates (x, u, v) such that $\nabla F = (0, -2u, 2v)$.

Proposition 14.8. *Let Φ_t be a tame gradient flow of a proper function $F : X \to \mathbb{R}$ bounded from below with non-degenerate critical manifolds. Let c be a non-critical value of F and suppose that $\Sigma \subset \{F \leq c\}$ is defined as the backward time image of a compact manifold $\Sigma_0 \subset \{F = c\}$. Suppose that Σ is transversal to all the unstable manifolds of the flow. Then Σ has finite volume.*

We now consider the gradient flow

$$\Phi_t(s, x, y) \;=\; \left(e^t s, \varphi_{-t}(y), x\right)$$

of the function

$$F(s, x, y) \;=\; \frac{1}{2}s^2 - f(y)$$

on $\mathbb{R} \times X \times X$ (with metric ds^2 on $\mathbb{R}$). The critical set of F consists of the non-degenerate critical manifolds

$$P_p \;=\; \{0\} \times \{p\} \times X$$

for $p \in Cr(f)$. The stable and unstable manifolds at P_p are given by

$$\mathbf{S}_p = \{0\} \times U_p \times X \qquad \text{and} \qquad \mathbf{U}_p = \mathbb{R} \times S_p \times X.$$

We consider the invariant manifold

$$\widetilde{\mathcal{T}} \;\equiv\; \left\{(e^{-s}, \varphi_s(x), x) \;:\; -\infty < s < \infty \;\; \text{and} \;\; x \in X\right\}$$

and the subdomain

$$\mathcal{T} \equiv \left\{(e^{-s}, \varphi_s(x), x) \ : \ 0 \le s < \infty \ \text{ and } \ x \in X\right\}.$$

Note that $\widetilde{\mathcal{T}}$ is merely the union of orbits passing through $\Delta = \{(1, x, x) : x \in X\}$ and

$$\mathcal{T} = \bigcup_{s \le 0} \Phi_s(\Delta)$$

is the union of the backward time orbits of Φ which begin at Δ. Recall (cf. Remark 1.6) that $\mathrm{vol}(\mathcal{T}) < \infty \Rightarrow \mathrm{vol}(T) < \infty$. Thus to complete the proof we shall prove $\mathrm{vol}(\mathcal{T}) < \infty$ by applying Proposition 14.8.

To begin, a straightforward check verifies that

(14.8) $\qquad \widetilde{\mathcal{T}}$ is transversal to $\mathbf{S}_p$ and to $\mathbf{U}_p$ for all p.

Now the intersection

$$\widetilde{\mathcal{T}}(c) \stackrel{\text{def}}{=} \widetilde{\mathcal{T}} \cap \{F = c\},$$

is always a smooth submanifold since $\widetilde{\mathcal{T}}$ is Φ_t-invariant, i.e., $\mathrm{grad}(F\big|_{\mathcal{T}}) = \mathrm{grad}(F)\big|_{\mathcal{T}} \neq 0$ on $\widetilde{\mathcal{T}}$. Furthermore, if $c > \max|f|$, then $\widetilde{\mathcal{T}}(c)$ is compact. To see this note that $\widetilde{\mathcal{T}}(c)$ is not compact iff there exists a sequence $(s_j, x_j) \in \mathbb{R} \times X$ with

$$F(e^{s_j}, x_j, \varphi_{s_j}(x_j)) = \frac{1}{2}(e^{2s_j} - f(\varphi_{-s_j}(x)) = c$$

for all j and for which there is no convergent subsequence. Clear for such a sequence we have $s_j \to -\infty$ and so $e^{2s_j} \to 0$ implying $c \le \max|f|$ as claimed. It therefore follows from Proposition 14.8 that for any $c > \max|f|$, the submanifold $\widetilde{\mathcal{T}}_{\le c} \equiv \widetilde{\mathcal{T}} \cap \{F \le c\}$ has finite volume in $\mathbb{R} \times X \times X$.

Since $\max_{\mathcal{T}} F = \frac{1}{2} + \max(f)$, we have $\mathcal{T} \subset \widetilde{\mathcal{T}}_{\le c}$ for all $c > \max|f| + \frac{1}{2}$. Thus $\mathcal{T}$ has finite volume, and the proof is complete. q.e.d.

Proof of Theorem 14.4. The following proof is inspired by arguments of Milnor given for a similar result [30, §4]. Consider the block decomposition (14.3) and the product structure 14.4 given by the flow. Proceeding in order from $k = 1$ we shall modify the metric on each subset

$$\partial^- Q_k \times (\tfrac{1}{3}, \tfrac{2}{3}) \subset \partial^- Q_k \times [0, 1]$$

so that under the new gradient flow (which agrees with the old one outside $\partial^- Q_k \times (\frac{1}{3}, \frac{2}{3})$), the unstable manifolds entering $\partial^- Q_k \times \{0\}$

become transversal to the stable manifolds at $\partial^- Q_k \times \{1\}$. By invariance under the flow this implies that each unstable manifold which meets Q_k is transversal everywhere on X to each stable manifold which meets Q_k. Modifying the metric at level k does not change the unstable manifolds below $\partial^- Q_k$. It also does not change the stable manifolds which originate below level k. From this one sees that after successively modifying the metric at level k for $k = 1, ..., m-1$, we have established the result.

We now show how to modify the metric. Fix a level k and consider the submanifolds

$$M_p = U_p \cap \partial^- Q_k \qquad \text{and} \qquad N_q = S_q \cap \partial^- Q_k$$

for $p, q \in Cr(f)$. We want to change the metric over $\partial^- Q_k \times (\frac{1}{3}, \frac{2}{3})$ so that after pushing each $M_p \times \{0\}$ forward by the new gradient flow, it becomes transversal to $N_q \times \{1\}$ for all q. We shall do this as follows. We shall construct a family of deformations of the metric, smoothly parameterized by an open subset $\mathcal{U} \subset \mathbb{R}^N$ for some N. This family will induce a smooth mapping

$$\mathcal{U} \times (\partial^- Q_k \times \{0\}) \xrightarrow{\Theta} \partial^- Q_k \times \{1\}$$

such that each $\Theta_u(\bullet) \equiv \Theta(u, \bullet)$ is the diffeomorphism induced by the new gradient flow (in the product structure of the old gradient flow).

By Sard's Theorem for Families it will suffice to prove that $\Theta\big|_{\mathcal{U} \times M_p}$ is transversal to N_q for all p, q (since then for almost all choices of $u \in \mathcal{U}$, we have that $\Theta_u\big|_{M_p}$ is transversal to N_q for each p, q). This condition will follow automatically if we show that

(14.9) $$\frac{\partial \Theta}{\partial u} \text{ is surjective at all points of } \mathcal{U} \times \partial^- Q_k$$

In fact it will suffice to show that there exists $u_0 \in \mathcal{U}$ such that

(14.10) $$\frac{\partial \Theta}{\partial u} \text{ is surjective at all points of } \{u_0\} \times \partial^- Q_k$$

for then, by the compactness of $\partial^- Q_k$, condition (14.11) will hold with $\mathcal{U}$ replaced by a small neighborhood of u_0 in $\mathcal{U}$.

We first construct such a map locally on $\partial^- Q_k$. Let $x = (x_1, ..., x_{n-1}) : \mathcal{O} \xrightarrow{\approx} \mathbb{R}^{n-1}$ be a local coordinate chart on Q_k (which maps *onto* $\mathbb{R}^{n-1}$). We then have coordinates $(x, t) \in \mathbb{R}^{n-1} \times [0, 1]$ for $\partial^- Q_k \times [0, 1]$ where $f(x, t) = t$. In these coordinates the given riemannian metric has the form

$$\langle \cdot, \cdot \rangle_{(x,t)} = (\cdot, \cdot)_{(x,t)} + dt^2$$

where $(\cdot,\cdot)$ is a family of inner products on $\mathbb{R}^{n-1}$.

Let $\mathfrak{M}$ denote the space of symmetric $n \times n$-matrices. Fix a non-trivial smooth function $\psi : [0,1] \to [0,1]$ with support in $(\frac{1}{3},\frac{2}{3})$. For each $A \in \mathfrak{M}$ with eigenvalues of absolute value < 1, we define a new metric $\langle\cdot,\cdot\rangle_A$ by

$$\langle\cdot,\cdot\rangle_A = \langle\cdot,\cdot\rangle + \psi\langle A(\cdot),\cdot\rangle. \tag{14.11}$$

Let $V = V(A)$ be the gradient of f in this new metric. Write $V = V' + V_0\partial/\partial t$ where V' is tangent to $\mathbb{R}^{n-1}$. Then for any vector field $W = W' + W_0\partial/\partial t$ we have that

$$W \cdot f = \langle V, W\rangle_A == \langle \partial/\partial t, W\rangle = \langle (I + \psi(t)A)V, W\rangle,$$

which implies that

$$\frac{\partial}{\partial t} = e_n = (I + \psi(t)A)V.$$

Our map Θ is given by taking the $\mathbb{R}^{n-1}$-component of the integral of the vector field $V = (I + \psi(t)A)^{-1}e_n$. We write the integal of V as $(\Theta_A(x,s), \tau_A(x,s))$ with respect to the decomposition $\mathbb{R}^{n-1} \times [0,1] \subset \mathbb{R}^{n-1} \times \mathbb{R}$. Since V is translation invariant in the x-variables, we have

$$\tau_A(x,s) = \tau_A(s) \qquad \text{and} \qquad \Theta_A(x,s) = \Theta_A(s) + x$$

where $\Theta_A(s) = \Theta_A(0,s)$. Note that

$$\begin{aligned}\frac{d}{ds}\tau_A(s) &= n\text{-component of } (I + \psi(\tau_A(s))A)^{-1} \cdot e_n \\ &= (n,n)\text{-component of } \sum_{k=0}^{\infty} \psi(\tau_A(s))^k A^k.\end{aligned}$$

Similarly

$$\frac{d}{ds}\Theta_A(s) = \text{pr}\left\{(I + \psi(\tau_A(s))A)^{-1} \cdot e_n\right\}$$

where $\text{pr} : \mathbb{R}^n \to \mathbb{R}^{n-1}$ is the linear projection. Hence for all A sufficiently small,

$$\begin{aligned}\Theta_A(1) &= \text{pr}\int_0^1 (I + \psi(\tau_A(s))A)^{-1} \cdot e_n\, ds \\ &= \text{pr}\int_0^1 \sum_{k=0}^{\infty} \psi(\tau_A(s))^k A^k \cdot e_n\, ds.\end{aligned}$$

Therefore, for each i, j with $1 \le i, j \le n$ we have

$$\begin{aligned}\left.\frac{\partial \Theta_A}{\partial A_{i,j}}\right|_{A=0} &= \left.\mathrm{pr}\int_0^1 \sum_{k=0}^{\infty} \frac{\partial}{\partial A_{i,j}} \psi(\tau_A(s))^k A^k \cdot e_n\, ds\right|_{A=0} \\ &= \mathrm{pr}\int_0^1 \{-\psi(\tau_0(s))E_{i,j}\} \cdot e_n\, ds \\ &= -\int_0^1 \psi(s)\, ds \begin{cases} E_{in} & \text{if } i > j = n \\ 0 & \text{otherwise.}\end{cases}\end{aligned}$$

where E_{ij} is the elementary (i,j)-matrix. This proves the surjectivity of $\partial\Theta/\partial A$ in this coordinate system.

We now modify this family by replacing A with $\xi(x)A$ where $\xi \in C_0^\infty(\mathbb{R}^{n-1})$ satisfies $\xi(x) = 1$ for $|x| \le 2$. This family of deformations now extends trivially to all of $\partial^- Q_k$ and agrees with the one above in a neighborhood of $B_1 = \{|x| \le 1\}$.

We now choose a finite family of such local coordinate systems $x^k : \mathcal{O}_k \to \mathbb{R}^{n-1}$, $k = 1, ..., \nu$ such that the open sets $(x^k)^{-1}(B_1)$ cover $\partial^- Q_k$. Let $\alpha_k = \xi_k A_k$, for $A_k \in \mathfrak{M}$, be the global section of $\mathrm{End}(T(\partial^- Q_k))$ defined above, and consider the deformations of the metric given by

$$\langle\cdot,\cdot\rangle_{A_1,...,A_\nu} = \langle\cdot,\cdot\rangle + \psi(t)\sum_{j=1}^{\nu}\langle \xi_j A_j(\cdot),\cdot\rangle$$

for $\vec{A} = (A_1, ..., A_\nu) \in \mathfrak{M}^\nu$. Then our calculation above shows that

$$\left.\frac{\partial\Theta_{\vec{A}}}{\partial\vec{A}}\right|_{\vec{A}=0} \quad \text{is surjective}$$

as desired. q.e.d.

Appendix: Currents and the kernel calculus

Let Z be a compact smooth n-dimensional manifold which is not necessarily orientable. In addition to the space $\mathcal{E}^k(Z)$, of smooth differential forms of degree k, one may consider the space $\widetilde{\mathcal{E}}^k(Z)$, of *twisted* smooth forms on Z of degree k. These are sections of the bundle $\Lambda^k T^* Z \otimes_{\mathbb{R}} \mathcal{O}_Z$ where $\mathcal{O}_Z$ is the orientation line bundle for Z. Since the transition functions for $\mathcal{O}_Z$ are $\mathbb{Z}_2 = \{-1, +1\} \subset \mathbb{R}$ valued, exterior differentiation d is naturally defined on any twisted form $\tilde{\alpha} \in \widetilde{\mathcal{E}}^k(Z)$. The resulting cohomology groups are the de Rham groups $H^k(Z; \mathcal{O}_Z)$ of Z with coefficients in the flat line bundle $\mathcal{O}_Z$.

A basic fact is that for any twisted n-form $\tilde{\alpha} \in \widetilde{\mathcal{E}}^n(Z)$ the integral $\int_Z \tilde{\alpha}$ is well defined. (This generalizes the usual definition since, if Z is oriented, then $\widetilde{\mathcal{E}}^k(Z) \cong \mathcal{E}^k(Z)$ are identified.) Following de Rham and Schwartz [33] we have the following.

Definition A.1. The space of **currents of degree k on Z** is the topological dual space

$$\mathcal{D}'^k(Z) \stackrel{\text{def}}{=} \widetilde{\mathcal{E}}^{n-k}(Z)'$$

of the space of twisted $(n-k)$-forms on Z.

Currents of degree k are a generalization of differential forms of degree k. In fact there is an embedding

$$\mathcal{E}^k(Z) \hookrightarrow \mathcal{D}'^k(Z) \tag{A.1}$$

which associates to $\alpha \in \mathcal{E}^k(Z)$, the current defined by

$$\alpha(\tilde{\beta}) \equiv \int_Z \alpha \wedge \tilde{\beta}, \qquad \text{for all } \tilde{\beta} \in \widetilde{\mathcal{E}}^{n-k}(Z). \tag{A.2}$$

Since $\int_Z d\alpha \wedge \tilde{\beta} = \int_Z d(\alpha \wedge \tilde{\beta}) - (-1)^k \int_Z \alpha \wedge d\tilde{\beta} = (-1)^{k-1} \int_Z \alpha \wedge d\tilde{\beta}$, it is natural to define the exterior derivative of a current $T \in \mathcal{D}'^*(Z)$ by:

$$(dT)(\tilde{\beta}) = (-1)^{k-1} T(d\tilde{\beta}), \qquad \text{for all } \tilde{\beta} \in \widetilde{\mathcal{E}}^{n-k}(Z). \tag{A.3}$$

Thus the de Rham complex of smooth forms $(\mathcal{E}^*(Z), d)$ is a subcomplex of the de Rham complex of currents $(\mathcal{D}'^*(Z), d)$. (In fact $\mathcal{D}'^*(Z)$ are exactly the distributional sections of $\Lambda^* T^* Z$, and d the natural extension of exterior differentiation to these sections.)

Example A.2. Let S be a codimension-k submanifold of Z which has finite volume and *oriented normal bundle.* The identification $\mathcal{O}_Z \big|_S = \mathcal{O}_S$ enables us to pull back twisted forms via the immersion $i : S \to Z$, and so S **determines a current** $[S] \in \mathcal{D}'^*(Z)$ **by integration**:

$$[S](\tilde{\beta}) = \int_S i^*(\tilde{\beta}), \qquad \text{for all } \tilde{\beta} \in \widetilde{\mathcal{E}}^*(Z). \tag{A.4}$$

It is sometiimes also useful to consider the **boundary operator** ∂ which is defined to be the dual of d on $\widetilde{\mathcal{E}}^*(Z)$, i.e.

$$\partial = (-1)^{k-1} d \qquad \text{on} \qquad \mathcal{D}'^k(Z).$$

Remark A.3. Note that in general an oriented compact submanifold does *not* define a current, but does define a **twisted current**, i.e., a linear functional on (untwisted) differential forms.

If $T \in \mathcal{D}'^p(Z)$ and $\alpha \in \mathcal{E}^q(Z)$ then the wedge product

(A.5) $$T \wedge \alpha \in \mathcal{D}'^p + q(Z) \text{ is defined by } (T \wedge \alpha)(\tilde{\beta}) = T(\alpha \wedge \tilde{\beta}).$$

More generally, for any flat bundle $E \to Z$ we have the spaces $\mathcal{E}^*(Z; E)$ of differential forms on Z with coefficients in E, and their extensions

(A.6) $$\mathcal{D}'^*(Z; E) \stackrel{\text{def}}{=} \mathcal{E}^{n-*}(Z; E^* \otimes \mathcal{O}_Z)'$$

to currents with coefficients in E.

Next we wish to represent operators on differential forms by currents on a product space. Let Y and X be manifolds with

$$\dim(Y) = n \qquad \text{and} \qquad \dim(X) = m$$

and note that $\mathcal{O}_{Y \times X} = \pi_Y^* \mathcal{O}_Y \otimes \pi_X^* \mathcal{O}_X$ where π_Y and π_X denote the projections to Y and X. Consider the space of differential forms $\mathcal{E}^*(Y \times \widetilde{X}) \stackrel{\text{def}}{=} \mathcal{E}^*(Y \times X; \pi_X^* \mathcal{O}_X)$ on $Y \times X$ twisted by the orientation line bundle $\pi_X^* \mathcal{O}_X$. For example $(\pi_Y^* \alpha) \wedge (\pi_X^* \tilde{\beta}) \in \mathcal{E}^*(Y \times \widetilde{X})$, if $\alpha \in \mathcal{E}^*(Y)$, $\tilde{\beta} \in \widetilde{\mathcal{E}}^*(X)$. The topological dual spaces

(A.7) $$\mathcal{D}'^*(\widetilde{Y} \times X) \equiv \mathcal{D}'^*(Y \times X; \pi_Y^* \mathcal{O}_Y) \equiv \mathcal{E}^{n+m-*}(Y \times \widetilde{X})',$$

is called the space of **kernels** for operators from $\mathcal{E}^*(Y)$ to $\mathcal{D}'^*(X)$. Each kernel $K \in \mathcal{D}'^n - r(\widetilde{Y} \times X)$ determines an operator $\mathbf{K} : \mathcal{E}^*(Y) \to \mathcal{D}'^* - r(X)$ by setting

(A.8) $$\mathbf{K}(\alpha)(\tilde{\beta}) = K(\pi_Y^* \alpha \wedge \pi_X^* \tilde{\beta}).$$

Since $\pi_X^* : \widetilde{\mathcal{E}}^*(X) \to \mathcal{E}^*(Y \times \widetilde{X})$ the dual map $(\pi_X)_* : \mathcal{D}'^*(\widetilde{Y} \times X) \to \mathcal{D}'^*(X)$ pushes forward kernels on $Y \times X$ to currents on X. Now the right hand side of (A.7) can be rewritten as $((\pi_X)_*(K \wedge \pi_Y^* \alpha))(\tilde{\beta}))$, so that

(A.9) $$\mathbf{K}(\alpha) = (\pi_X)_*(K \wedge \pi_Y^* \alpha)$$

provides an alternate "pull-push" definition of $\mathbf{K}$.

Proposition A.4. *Suppose $K \in \mathcal{D}'^{n-r}(\widetilde{Y} \times X)$ is a kernel whose operator* **K** *lowers degree by r. Then the kernel $\partial K \in \mathcal{D}'^{n-r+1}(\widetilde{Y} \times X)$, determines the operator*

$$\mathbf{K} \circ d + (-1)^{r-1} d \circ \mathbf{K}. \tag{A.10}$$

Proof. By definition $(\partial K)(\pi_Y^* \alpha \wedge \pi_X^* \tilde{\beta}) = K(d(\pi_Y^* \alpha \wedge \pi_X^* \tilde{\beta})$ which equals $K(\pi_Y^* d\alpha \wedge \pi_X^* \tilde{\beta}) + (-1)^{\deg \alpha} K(\pi_Y^* \alpha \wedge \pi_X^* d\tilde{\beta})$. But $(\mathbf{K}(d\alpha))(\tilde{\beta}) = K(\pi_Y^* d\alpha \wedge \pi_X^* \tilde{\beta})$, and

$$\begin{aligned}(d(\mathbf{K}(\alpha)))(\tilde{\beta}) &\equiv (-1)^{\deg \alpha - r - 1} (\mathbf{K}(\alpha))(d\tilde{\beta}) \\ &= (-1)^{\deg \alpha - r - 1} K(\pi_Y^* \alpha \wedge \pi_X^* d\tilde{\beta}).\end{aligned}$$

q.e.d.

Now we list some examples of kernels and their corresponding operators.

Example A.5. Let $\Sigma \subset Y \times X$ be a finite volume submanifold of codimension-q with a given isomorphism $\pi_X^* \mathcal{O}_X \big|_\Sigma \cong \mathcal{O}_\Sigma$. Then integration of forms over Σ defines a current $[\Sigma] \in \mathcal{D}'^q(\widetilde{Y} \times X)$.

Example A.6. (The Identity) The identity operator $\mathbf{I} : \mathcal{E}^*(X) \longrightarrow \mathcal{E}^*(X)$ on forms is represented by the kernel $I = [\Delta]$, corresponding to integration over the diagonal $\Delta \subset X \times X$, since $\alpha(\tilde{\beta}) \equiv \int_X \alpha \wedge \tilde{\beta} = \int_\Delta \pi_Y^* \alpha \wedge \pi_X^* \tilde{\beta} = [\Delta](\pi_Y^* \alpha \wedge \pi_X^* \tilde{\beta})$.

Example A.7. (The Pull Back) Suppose $\varphi : X \longrightarrow Y$. Then the operator $\mathbf{P}_\varphi \alpha \equiv \varphi^* \alpha$ is represented by the kernel $P_\varphi = [\operatorname{graph} \varphi]$; that is

$$\mathbf{P}_\varphi \;=\; (\pi_X)_*([\operatorname{graph} \varphi] \wedge \pi_Y^* \alpha).$$

Example A.8. (Projection onto ψ along $\tilde{\varphi}$). Suppose $\psi \in \mathcal{E}^*(X)$ and $\tilde{\varphi} \in \widetilde{\mathcal{E}}^*(Y)$. Then $K = \pi_Y^* \tilde{\varphi} \wedge \pi_X^* \psi \in \mathcal{D}'^*(\widetilde{Y} \times X)$ is a kernel inducing the operator

$$\mathbf{K}(\alpha) = \pm \left(\int_Y \tilde{\varphi} \wedge \alpha \right) \psi,$$

where $\pm$ equals $(-1)^{(\deg \psi)(\deg \alpha)}$. Note

$$K(\pi_Y^* \alpha \wedge \pi_X^* \tilde{\beta}) = \int_{Y \times X} \pi_Y^* \tilde{\varphi} \wedge \pi_X^* \psi \wedge \pi_Y^* \alpha \wedge \pi_X^* \tilde{\beta}.$$

Example A.9. (Projection onto $[S]$ along $[U]$). Suppose S is a submanifold with oriented normal bundle in X and $[U]$ is an oiented submanifold of Y. Let $[S]$ denote the current in X and $[U]$ the twisted current in Y determined by integration. Then $K = [U] \times [S] \in \mathcal{D}'^*(\widetilde{Y} \times X)$ is a kernel and the corresponding operator is $\mathbf{K}(\alpha) = \pm \left(\int_U \alpha\right)[S]$, where $\pm$ equals $(-1)^{(\deg S)(\deg \alpha)} = (-1)^{(\dim Y - \dim S)(\dim U)}$.

References

[1] D. M. Austin & P. J. Braam, *Morse-Bott theory and equivariant cohomology*, Topology, Geometry and Physics for Raoul Bott, Internat. Press, 1995.

[2] M. Atiyah & R. Bott, *The Yang-Mills equations over a Riemann surface*, Phil. Trans., R. Soc. London. A **308** (1982) 523–615.

[3] M. Betz, *Operad representations in Morse theory and Floer homology*, Preprint, 1996.

[4] M. Betz & R. L. Cohen, *Graph moduli spaces and cohomology operations*, Stanford Preprint, 1993.

[5] A. Bialynicki-Birula, *Some theorems on actions of algebraic groups*, Ann. of Math. **98** (1973) 480–497.

[6] N. Berline, E. Getzler & M. Vergne, *Heat kernels and the Dirac operator*, Grundl. der Math. Wiss. Band 298, Springer, Berlin-Heidelberg-New York, 1992.

[7] R. Bott & C. Taubes, *On the self-linking of knots*, J. Math. Phys. **35** (1994) 5247–5287.

[8] J. B. Carrell & D. I. Lieberman, *Holomorphic vector manifolds and compact Kähler manifolds*, Invent. Math. **21** (1973) 303–309.

[9] J. B. Carrell & A. J. Sommese, *Some topological aspects of $\mathbb{C}^*$-actions on compact Kähler manifolds*, Comm. Math. Helv. **54** (1979) 583–800.

[10] R. L. Cohen, J. D. S. Jones & G. B. Segal, *Morse theory and classifying spaces*, Stanford Preprint, 1993.

[11] G. de Rham, *Variétés différentiables*, Hermann, Paris, 1973.

[12] G. Ellingsrud & S. A. Stromme, *Toward the Chow ring of the Hilbert scheme of $\mathbb{P}^2$*, J. Reine Agnew. Math. **441** (1993) 33–44.

[13] H. Federer, *Geometric measure theory*, Springer, New York, 1969.

[14] T. T. Frankel, *Fixed points and torsion on Kähler manifolds*, Ann. of Math. **70** (1959) 1–8.

[15] W. Fulton, R. MacPherson, F. Sottile & B. Sturmfels, *Intersection theory on spherical varieties*, J. Alg. Geom. **4** (1995) 181–193.

[16] H. Gillet & C. Soulé, *Characteristic classes for algebraic vector bundles with Hermitian metrics. I, II*, Ann. of Math. **131** (1990) 163–203; 205–238.

[17] R. Harvey & H. B. Lawson, *A theory of characteristic currents associated with a singular connection*, Astèrisque, Soc. Math. de France, Montrouge, France, Vol. 213, 1993.

[18] ———, *Geometric residue theorems*, Amer. J. Math. **117** (1995) 829–874.

[19] ———, *Singularities and Chern-Weil theory. I - The local MacPherson formula*, Asian J. Math. **4** (2000) 71–96.

[20] ———, *Singularities and Chern-Weil theory. II - Geometric atomicity*, to appear.

[21] ———, *Finite volume flows and Morse Theory*, Ann. of Math. **153** (2001) 1–25.

[22] ———, *The de Rham-Federer theory of differential characters and character duality*, (to appear).

[23] R. Harvey & J. Polking, *Fundamental solutions in complex analysis, Part I*, Duke Math. J. **46** (1979) 253–300.

[24] J. Latschev, *Gradient flows of Morse-Bott functions*, Math. Ann. **318** (2000) 731–759.

[25] F. Laudenbach, *On the Thom-Smale complex*, Astérisque, An Extension of a Theorem of Cheeger and Müller, (J.-M. Bismut and W. Zhang), S.M.F., Vol. 205, 1992, 219–233.

[26] M. Kontsevich, *Feynman diagrams and low-dimensional topology*, First European Congress of Mathematics, Birhäuser, Basel, 1994, Vol. II, 97–122.

[27] R. MacPherson, *Singularities of vector bundle maps*, Proc. of Liverpool Singularities Symposium, I, Springer Lecture Notes Math. **192** (1971) 316–318.

[28] ———, *Generic vector bundle maps,* Dynamical Systems, Proc. of Symposium, University of Bahia, Salavador 1971, Academic Press, New York, 1973, 165–175.

[29] J. Milnor, *Morse theory*, Ann. of Math. Stud. 51, Princeton University Press, Princeton, N.J., 1963.

[30] ———, *Lectures on the h-cobordism theorem*, Princeton Univ. Press, Princeton, N.J., 1965.

[31] H. Rosenberg, *A generalization of Morse-Smale inequalities*, Bull. Amer. Math. Soc. **70** (1964) 422–427.

[32] S. Smale, *On gradient dynamical systems*, Ann. of Math. **74** (1961) 199–206.

[33] L. Schwartz, Théorie des Distributions, Hermann, Paris, 1966.

[34] A. J. Sommese, *Extension theorems for reductive groups actions on compact Kähler manifolds*, Math. Ann. **218** (1975) 107–116.

[35] R. Thom, *Sur une partition en cellules associées à une fonction sur une variété*, CRAS **228** (1949) 973–975.

[36] E. Witten, *Supersymmetry and Morse theory*, J. Differential Geom. **17** (1982) 661–692.

DEPARTMENT OF MATHEMATICS, RICE UNIVERSITY, HOUSTON
DEPARTMENT OF MATHEMATICS, SUNY AT STONY BROOK, NY

SURVEYS IN DIFFERENTIAL GEOMETRY, 2000
Vol. VII ©2000, INTERNATIONAL PRESS
pp. 313-326

THE ATIYAH-BOTT-SINGER FIXED POINT THEOREM AND NUMBER THEORY

F. HIRZEBRUCH

1. Introductory remarks

It was a great idea of Shing-Tung Yau to organize a meeting sponsored by the Journal of Differential Geometry and dedicated to the "Gang of Four," Atiyah, Bott, Singer, and myself. The four members of the Gang were not supposed to lecture at the meeting, they were to give dinner speeches. Perhaps the ability to lecture decreases with age, whereas the willingness to give dinner speeches increases. But nevertheless the out-of-town members of the Gang, Atiyah and I, lectured just before the opening of the meeting. I talked at MIT in the joint colloquium on May 13, 1999. My lecture had the above title and was of course dedicated to three members of the Gang. The lecture was of a rather elementary character (if one knows the ABS-theorem), but I hope ABS enjoyed it. In the dinner speech I pointed out that ABHS make up the following graph of type D_4

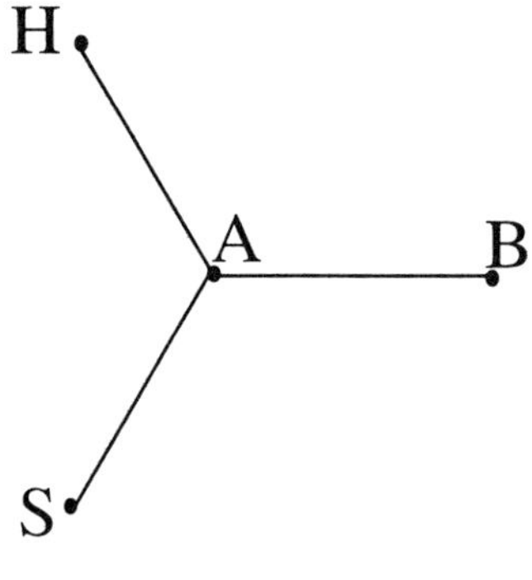

where two vertices are connected by an edge if and only if they have a joint paper. But I also said that BHS have many relations. There was much cooperation not represented by joint papers to which I look back with great pleasure and gratitude. All this began in the fifties. All four of us are good friends since more than four decades and influenced each other mathematically through all these years.

My paper is also dedicated to Michael Atiyah on the occasion of his 70th birthday. I owe him very much mathematically (we have nine joint papers) and in many other ways. It is impossible to thank him here in a proper way. I would have to write many pages. But let me mention two facts:

1. The thirty Arbeitstagungen in Bonn under my direction (1957 – 1991) were the backbone of the mathematical activity I tried to build up and to keep in Bonn (Sonderforschungsbereich Theoretische Mathematik 1969 – 1985, Max-Planck-Institut für Mathematik since 1982.) At these Arbeitstagungen Michael lectured 32 times. Very often it was the opening lecture. Everybody can see how much Bonn owes him. On July 16, 1962, Michael gave the Arbeitstagung lecture "Harmonic Spinors and Elliptic Operators". He reported on joint work with Iz Singer and on their conjecture that the $\widehat{A}$-genus equals the index of the Dirac Operator on Spin-manifolds. Here the story of the index and fixed point theorems begins. This is the origin of the line which led to the "Gang of Four"-meeting now 37 years later.

2. Michael worked for the foundation of the European Mathematical Society (EMS) through the European Mathematical Council for many years. He proposed me as the first president of the EMS when the society was finally founded in 1990.

On July 3, 1999, I received an honorary degree of the University of Konstanz. I have many connections with the mathematicians there. Konstanz was founded in 1966. It is now in a process of reform. For some time it looked as if mathematics would be reformed down to a pure service institution. In my acceptance speech of the degree I tried to make it clear that a University without mathematics hardly deserves the name University. In the Konstanz mathematical colloquium I gave a lecture in the same spirit as my MIT lecture. The manuscript was translated into English by Dr. Bruce Hunt whom I thank very much. This is the present paper. The ABS-theorem and its relation to number theory show the strength and beauty of mathematics and the unity of mathematics independently of applications and of service to other fields. But it has applications. The role of Atiyah, Bott and Singer in Mathematical Physics shows what I mean.

I thank the University of Konstanz for preparing the German TEX file and International Press for producing the English version.

Last but not least many thanks again to Shing-Tung Yau for his great energy and enthusiasm in organizing the "Gang of Four" meeting.

2. Lecture

We will first apply the Atiyah-Bott-Singer fixed point theorem ([1], page 473) to a compact connected Riemann surface X.

Let $\boldsymbol{a}$ be an automorphism of finite order of X, not equal to the identity. Then $\boldsymbol{a}$ has finitely many fixed points x and at each a rotation angle α_x with $0 < \alpha_x < 2\pi$. The automorphism $\boldsymbol{a}$ induces by lifting an action on the finite-dimensional $\mathbb{C}$-vector space $H^{1,0}(X)$ of holomorphic one-forms on X. According to ABS, one has for the trace of $\boldsymbol{a}$, denoted $\chi(\boldsymbol{a})$ (and which will also be referred to as the character), the formula

$$\chi(\boldsymbol{a}) - \overline{\chi(\boldsymbol{a})} = i \sum_{\substack{x \in X \\ ax = x}} \cot \frac{\alpha_x}{2}. \tag{1}$$

(In ABS one has a $-i$ on the right-hand side instead of i; we are using slightly different notations). According to the Lefschetz fixed point theorem of topology, we have

$$\chi(\boldsymbol{a}) + \overline{\chi(\boldsymbol{a})} = 2 - \text{ the number of fixed points,} \tag{2}$$

so that we can calculate $\chi(\boldsymbol{a})$ from these two equations.

Example. Consider the lattice $\mathbb{Z}i + \mathbb{Z}$ in $\mathbb{C}$ (with coordinate z) and the automorphism $\boldsymbol{a}$ of order 4 which is given by multiplication by i. There are two fixed points which are represented by 0 and $\frac{1}{2} + \frac{i}{2}$, both of which have rotation angle $\frac{\pi}{2}$. Hence

$$\begin{aligned} \chi(\boldsymbol{a}) - \overline{\chi(\boldsymbol{a})} &= 2i \cot \frac{\pi}{4} = 2i \\ \chi(\boldsymbol{a}) + \overline{\chi(\boldsymbol{a})} &= 0 \\ \chi(\boldsymbol{a}) &= i, \end{aligned}$$

which is the same as saying $\boldsymbol{a}^* dz = i dz$.

For the following considerations, which will lead to a theorem of HECKE (1928), compare [7] and the literature given there (E. HECKE, Mathematische Werke, page 549).

We now make the following basic assumption, without worrying at the moment whether it can be satisfied:

(∗) Let p be an odd prime number and $\boldsymbol{a}$ an automorphism of X of order p with $\frac{p-1}{2}$ fixed points, with rotation angles $2\pi\frac{k}{p}$, $1 \leq k \leq p-1$, where k a quadratic residue modulo p, i.e., $\left(\frac{k}{p}\right) = 1$.

The assumption (∗) implies

(∗∗)

$$\chi(\boldsymbol{a}) - \overline{\chi(\boldsymbol{a})} = i \sum_{\substack{1\leq k\leq p-1 \\ \left(\frac{k}{p}\right)=1}} \cot \pi \frac{k}{p}$$

$$\chi(\boldsymbol{a}) + \overline{\chi(\boldsymbol{a})} = -\frac{p-5}{2}.$$

We first consider the case $p \equiv 1 \bmod 4$. then we have $\chi(\boldsymbol{a}) - \overline{\chi(\boldsymbol{a})} = 0$, since -1 is a quadratic residue, and

$$\chi(\boldsymbol{a}) = -\frac{p-5}{4}, \qquad \text{(if (∗) holds and } p \equiv 1 \bmod 4). \tag{3}$$

For $p = 3$ we would have

$$\chi(\boldsymbol{a}) - \overline{\chi(\boldsymbol{a})} = i \cot \frac{\pi}{3} = \frac{i}{\sqrt{3}}.$$

This is impossible, as $\chi(\boldsymbol{a})$ is an algebraic integer. We now assume $p \equiv 3 \bmod 4$ and $p > 3$. Then by Gauß' theorem, we have

$$\sum_{\substack{1\leq k\leq p-1 \\ \left(\frac{k}{p}\right)=1}} \cot \pi \frac{k}{p} = \sqrt{p}\, h(-p), \tag{4}$$

where $h(-p)$ is the class number of the field $\mathbb{Q}(\sqrt{-p})$ of discriminant $-p$. Using the formula

$$\pi \cot \pi z = \sum_{n\in\mathbb{Z}} \frac{1}{z-n} \tag{5}$$

with the summation which collects the summands for n and $-n$, one can show that (4) is equivalent to the following formula

$$\sum_{k>0} \left(\frac{k}{p}\right) \frac{1}{k} = \frac{\pi}{\sqrt{p}} h(-p). \tag{6}$$

For more details compare the beautiful book [8].

From (**) and (4), we have the formula

$$\chi(\boldsymbol{a}) = \frac{1}{2}\left(-\frac{p-5}{2} + i\sqrt{p}\,h(-p)\right), \tag{7}$$

which also holds for $p \equiv 1 \bmod 4$, if one sets $h(-p) = 0$, since $-p$ is not a discriminant.

Can the fundamental assumption (*) be realized? The answer is yes for $p > 3$.

Consider the modular group $PSL_2(\mathbb{Z})$, which we will denote by Γ. This group acts on the upper half-plane $\mathbb{H}$ by means of fractional linear transformations

$$z \mapsto \frac{az+b}{cz+d}.$$

The principal congruence subgroup $\Gamma(p)$ consists of those integral unimodular matrices $\left(\begin{smallmatrix} a & b \\ c & d \end{smallmatrix}\right)$, which modulo p are equal to the identity matrix. The group $\Gamma(p)$ acts freely on $\mathbb{H}$.We obtain a non-compact Riemann surface $\Gamma(p)\backslash\mathbb{H}$, which covers $\Gamma\backslash\mathbb{H}$ finite-to-one. The Galois group of the covering is

$$PSL_2(\mathbb{F}_p) = \Gamma/\Gamma(p),$$

of order $N := \frac{1}{2}p(p^2-1)$. The Riemann surface $\Gamma\backslash\mathbb{H}$ can be identified with the complex plane $\mathbb{C}$. There are two special points, as Γ does not act freely on $\mathbb{H}$; these are representatives of the fixed points of orders 2 and 3, respectively, which one takes to be i and $\rho := \frac{1}{2}(-1+i\sqrt{3})$. The covering

$$\Gamma(p)\backslash\mathbb{H} \longrightarrow \Gamma\backslash\mathbb{H}$$

is branched at these two special points and has there $\frac{N}{2}$ and $\frac{N}{3}$ inverse images, respectively. We obtain for the Euler-Poincaré characteristic

$$e(\Gamma(p)\backslash\mathbb{H}) = -N + \frac{N}{2} + \frac{N}{3} = -\frac{N}{6} = -\frac{1}{12}p(p^2-1).$$

The Riemann surface $\Gamma\backslash\mathbb{H}$ can be compactified to the Riemannian sphere S^2 by adding one cusp. The surface $\Gamma(P)\backslash\mathbb{H}$ can be compactified in the same manner by adding $\frac{N}{p}$ cusps, yielding a compact Riemann surface $X(p)$, on which $PSL_2(\mathbb{F}_p)$ acts with quotient S^2. The isotropy group of a cusp is cyclic of order p. We denote by $i\infty$ the standard cusp whose stabilizer is generated by the transformation $\boldsymbol{a} : z \mapsto z+1$. The action of $PSL_2(\mathbb{F}_p)$ on $X(p)$ has three exceptional orbits of orders $N/2$, $N/3$ and N/p. The Euler-Poincaré characteristic of $X(p)$ is

$$e(X(p)) = -\frac{1}{12}p(p^2-1) + \frac{1}{2}(p^2-1).$$

The genus g is equal to the dimension of $H^{1,0}(X(p))$ and satisfies

$$2 - 2g = e(X(p)).$$

One has $g = 0, 3, 26, \dots$ for $p = 5, 7, 11, \dots$.

The element $\boldsymbol{a}$ mentioned above of order p generates a cyclic group U, which has $\frac{p-1}{2}$ fixed points acting on the set of $\frac{1}{2}(p^2 - 1)$ cusps, and $\frac{p-1}{2}$ orbits of p elements each (cyclic permutations). It has no further fixed points for $p > 3$. The cusps correspond to the cosets $PSL_2(\mathbb{F}_p)/U$, the fixed points to the cosets $N(U)/U$, where $N(U)$ is the normalizer of U, which consists of all maps $z \mapsto az + b$, where a is a quadratic residue modulo p. The quotient $PSL_2(\mathbb{F}_p)/N(U)$ is the projective line $\mathbb{F}_p \cup \infty$, on which $\boldsymbol{a} : z \mapsto z + 1$ acts with ∞ as sole fixed point and cyclicly permutes $\mathbb{F}_p$. Each point of the projective line represents $\frac{p-1}{2}$ cusps. Consider the case $p = 5$. Then, as is well known, $PSL_2(\mathbb{F}_5)$ is the automorphism group A_5 of an icosahedron. The 12 corners of the latter correspond to the 12 cusps of $\Gamma(5)$.

For the element $\boldsymbol{a} : z \mapsto z + 1$ the fundamental assumption $(*)$ is satisfied. It is easy to check that indeed the rotation angles at the $\frac{p-1}{2}$ fixed points are equal to $2\pi\frac{k}{p}$, with quadratic residues k modulo p.

Now we want to investigate for $p \equiv 3(4)$ the representation of $PSL_2(\mathbb{F}_p)$ on $H^{1,0}(X(p))$. Since one has the rule $(\boldsymbol{ab})^* = \boldsymbol{b}^*\boldsymbol{a}^*$ for liftings of differential forms under automorphisms $\boldsymbol{a}$ and $\boldsymbol{b}$, we may pass to the transposed representation and get a homomorphism

$$PSL_2(\mathbb{F}_p) \longrightarrow \mathrm{End} H^{1,0}(X(p)). \tag{8}$$

According to F. G. Frobenius and I. Schur the irreducible representations of $PSL_2(\mathbb{F}_p)$ for $p \equiv 3(4)$ are classified in the following manner. There is the trivial representation of degree 1, the representation of degree p which is obtained from the permutation representation on the projective line over $\mathbb{F}_p$ by splitting off the trivial representation, and there are $\frac{1}{4}(p-3)$ representations each of degrees $p-1$ and $p+1$, all of which are real, and in addition there are two conjugate representations χ^+ and χ^- of degree $\frac{1}{2}(p-1)$ with

$$\chi^+(\boldsymbol{a}) = \frac{1}{2}(-1 + i\sqrt{p}), \quad \chi^-(\boldsymbol{a}) = \frac{1}{2}(-1 - i\sqrt{p}). \tag{9}$$

These are the traces for the element

$$\boldsymbol{a} : z \mapsto z + 1, \quad \boldsymbol{a} \in PSL_2(\mathbb{F}_p)$$

mentioned above. It is interesting to recall the Gaußian sums

$$\chi^+(\boldsymbol{a}) = \sum_{\substack{1\le k\le p-1 \\ \left(\frac{k}{p}\right)=1}} \alpha^k, \quad \text{where } \alpha = e^{2\pi i/p},$$

and

$$\chi^-(\boldsymbol{a}) = \sum_{\substack{1\le k\le p-1 \\ \left(\frac{k}{p}\right)=-1}} \alpha^k = \overline{\chi^+(\boldsymbol{a})},$$

and that these relations characterize the splitting of the representations χ^+ and χ^- when these are restricted to the cyclic subgroup of order p generated by $\boldsymbol{a}$.

Let m and n be the multiplicities of χ^+ and χ^-, respectively, in the representation (8).

Theorem of Hecke. *Let $p > 3$ and $p \equiv 3(4)$. Then for the multiplicities m and n, we have*

$$m - n = h(-p). \tag{10}$$

This follows immediately from (7) and (9). By the way, one can also calculate $m + n$:

$$m + n = \frac{1}{2}\left(\frac{p-1}{6} + (-1)^{\frac{p+1}{4}}\right) + \frac{2}{3}\text{iff } p \equiv 2(3).$$

As is well-known, there are exactly 7 prime numbers $p \equiv 3(4)$ with $h(-p) = 1$, namely 3, 7, 11, 19, 43, 67, 163 (HEEGNER 1952, STARK).

p	m	n
7	1	0
11	1	0
19	1	0
43	2	1
67	3	2
163	7	6

Complex dimension two

Preliminary report (in preparation with DON ZAGIER)

We consider a compact connected Kählerian surface X, for example a complex algebraic surface. There are two fundamental topological invariants, the Euler-Poincaré characteristic $e(X)$ and the signature sign(X).

If $\boldsymbol{a}$ is an automorphism of X, then the equivariant Euler characteristic $e(X, \boldsymbol{a})$ and the equivariant signature $\operatorname{sign}(\mathrm{X}, \boldsymbol{a})$ are well-defined [1]. If $\boldsymbol{a}$ has finite order and isolated fixed points, then one has

$$\begin{aligned} & e(X,\boldsymbol{a}) = \text{number of fixed points of } \boldsymbol{a} \\ (11) \qquad & \operatorname{sign}(\mathrm{X},\boldsymbol{a}) = -\sum_{\substack{\mathrm{x}\in \mathrm{X} \\ \boldsymbol{a}\mathrm{x}=\mathrm{x}}} \cot\frac{\alpha_{\mathrm{x}}}{2}\cot\frac{\beta_{\mathrm{x}}}{2}. \end{aligned}$$

Note that (1) is the equivariant signature for the one-dimensional case (which vanishes for $\boldsymbol{a} = \mathrm{Id}$).

The first formula in (11) is the classical fixed point theorem of Lefschetz, the second is the ABS-fixed point theorem for the signature. Of course α_x, β_x denote the rotation angles of $\boldsymbol{a}$ at the fixed point x. See also [3] and [5].

As is well known, $\frac{1}{4}(e(X) + \operatorname{sign}(\mathrm{X}))$ is the arithmetic genus of the surface X. This fact can also be applied equivariantly. This leads to a formula for the character $\chi(\boldsymbol{a})$ of the action of $\boldsymbol{a}$ in the vector space $H^{2.0}(X)$ of holomorphic two-forms on X, as long as we make the following assumption.

Assumption 1. The first Betti number of X vanishes. The representation of $\boldsymbol{a}$ in $H^{2,0}$ is real, that is, equivalent to its complex conjugate representation in $\overline{H^{2.0}(X)}$.

The vanishing of the first Betti number implies that the arithmetic genus is equal to $1 + \dim H^{2.0}(X)$. Because the representation is real, we also get the relation

$$1 + \chi(\boldsymbol{a}) = \frac{1}{4}(e(X,\boldsymbol{a}) + \operatorname{sign}(\mathrm{X},\boldsymbol{a})) = \frac{1}{4}\sum_{\substack{x\in X \\ \boldsymbol{a}x=x}} \left(1 - \cot\frac{\alpha_x}{2}\cot\frac{\beta_x}{2}\right).$$

We now make an assumption which is analogous to $(*)$ in the one-dimensional case, which, however, we shall only be able to realize in very special cases.

Assumption 2. Let p be an odd prime number > 3 and $\boldsymbol{a}$ an automorphism of X of order p with $\frac{p-1}{2}$ fixed points with rotation angles $2\pi\frac{k}{p}$, $-2\pi\frac{kd}{p}$, where $1 \le k \le p-1$ and k is a quadratic residue modulo p. In this formula d denotes a given fixed coset modulo p, which is relatively prime to p.

The assumptions 1 and 2 imply

$$(12)\qquad 1+\chi(\boldsymbol{a})=\frac{1}{4}\left(\frac{p-1}{2}+\sum_{\substack{1\le k\le p-1\\ \left(\frac{k}{p}\right)=1}}\cot\frac{\pi k}{p}\cot\frac{\pi kd}{p}\right).$$

The sum of cotangents lies in the field of pth roots of unity and in fact lies in the quadratic subfield $\mathbb{Q}(\sqrt{-p})$ for $p\equiv 3(4)$ and in $\mathbb{Q}(\sqrt{p})$ for $p\equiv 1(4)$. This can be seen with the help of Galois theory: the sum is invariant upon replacing k by rk, where r is a quadratic residue modulo p.

For $p\equiv 3(4)$, -1 is a quadratic non residue. Hence the sum of cotangents is invariant under the Galois automorphisms of $\mathbb{Q}(\sqrt{-p})$ and lies in $\mathbb{Q}$, which is clear anyway since the sum is real. One has

$$(13)\qquad 1+\chi(\boldsymbol{a})=\frac{1}{8}\left(p-1+\sum_{1\le k\le p-1}\cot\frac{\pi k}{p}\cot\frac{\pi kd}{p}\right).$$

Here one of the usual Dedekind sums appears:

$$\operatorname{ded}(p,d)=\sum_{1\le k\le p-1}\cot\frac{\pi k}{p}\cot\frac{\pi kd}{p}$$

(see [3], [5]). The expression on the right-hand side of (13) is a half-integer, and an integer if and only if d is a quadratic residue modulo p ([3], formula (39), and [5]). The character must be integer-valued. This is no contradiction to our assumptions, if we assume in addition that $\left(\frac{d}{p}\right)=1$.

For $p\equiv 1(4)$, $\chi(\boldsymbol{a})$ is of the form

$$\chi(\boldsymbol{a})=\frac{u+v\sqrt{p}}{2}.$$

Because of our assumptions, $\chi(\boldsymbol{a})$ is an algebraic integer in $\mathbb{Q}(\sqrt{p})$. Hence:

$$u,\ v\in\mathbb{Z}\quad\text{and}\quad u\equiv v\text{ modulo }2.$$

The Galois automorphism of $\mathbb{Q}(\sqrt{p})$ will be denoted by $\rho\mapsto\widetilde{\rho}$ ($\rho\in\mathbb{Q}(\sqrt{p})$). Then according to (12), we have

$$(14)\qquad 2+\chi(\boldsymbol{a})+\widetilde{\chi(\boldsymbol{a})}=2+u=\frac{1}{4}(p-1+\operatorname{ded}(p,d)),$$

while at the same time

$$
\begin{aligned}
v &= \frac{\chi(\boldsymbol{a}) - \widetilde{\chi(\boldsymbol{a})}}{\sqrt{p}} \\
&= \frac{1}{4\sqrt{p}} \sum_{1 \le k \le p-1} \left(\frac{k}{p}\right) \cot \frac{\pi k}{p} \cot \frac{\pi k d}{p} \qquad (15) \\
&\stackrel{\text{DEF}}{=} f(p, d).
\end{aligned}
$$

One can show that (14) and (15) for $p > 5$ determine u, v as integers with $u \equiv v$ modulo 2. For $p > 5$ and $p \equiv 1(4)$, our assumptions again do not lead to a contradiction. Both numbers u and v are even if and only if $\left(\frac{d}{p}\right) = 1$. For $p = 5$, one has $f(p,1) = \frac{2}{5}$ and $f(p,2) = \frac{1}{5}$ and moreover $\operatorname{ded}(p,1) = -4$ and $\operatorname{ded}(p,2) = 0$.

We have been led to the introduction of the twisted Dedekind sums $f(p,d)$, which have very interesting properties, for example (for $p \equiv 1(4)$):

$$
f(p,1) = \frac{2}{5} \sum_{\substack{1 \le r < \sqrt{p} \\ r \in \mathbb{Z},\ r \text{ odd}}} \sigma_1\left(\frac{p - r^2}{4}\right)
$$

$$
\begin{aligned}
f(p,1) - 2f(p,2) &= 0 \\
f(p,1) - 3f(p,3) &= \tfrac{1}{2} h(-3p) \\
f(p,1) - 4f(p,4) &= h(-4p) \\
f(p,1) - 6f(p,6) &= 5h(-3p) \quad \text{for } p \equiv 1 \bmod 8 \\
&= -h(-3p) \quad \text{for } p \equiv 5 \bmod 8 \\
f(p,1) - 8f(p,8) &= 3h(-4p) + 2h(-8p) \quad \text{for } p \equiv 1 \bmod 8 \\
&= -h(-4p) + 2h(-8p) \quad \text{for } p \equiv 5 \bmod 8
\end{aligned}
$$

The first formula is related to the value of the Dedekind zeta function of the field $\mathbb{Q}(\sqrt{p})$ at 2 (see [4], page 192), which can be seen with the help of formula (5). The other formulas can be proved with the help of (5) and formulas of the type (6).

Naturally we would like to have examples of surfaces X with an action of $PSL_2(\mathbb{F}_p)$ so that the action of $\boldsymbol{a} \in PSL_2(\mathbb{F}_p)$ with $\boldsymbol{a} : z \mapsto z+1$ satisfies our assumptions. All representations for $p \equiv 1(4)$ are real anyhow. There are irreducible representations of degrees 1, $p, p-1$, $p+1$ and two exceptional representations χ^+, χ^- of degree $\frac{p+1}{2}$ with

$$
\chi^+(\boldsymbol{a}) = \frac{1+\sqrt{p}}{2}, \quad \chi^-(\boldsymbol{a}) = \frac{1-\sqrt{p}}{2}.
$$

Once again, the Gaußian sums are interesting. For example one has

$$\chi^+(\boldsymbol{a}) = 1 + \sum_{\substack{1\le k\le p-1 \\ \left(\frac{k}{p}\right)=1}} \alpha^k, \quad \text{where } \alpha = e^{\frac{2\pi i}{p}}.$$

All irreducible representations except for χ^+ and χ^- have characters whose values at $\boldsymbol{a}$ are in $\mathbb{Z}$. For the multiplicities m and n of χ^+ and χ^- in the representation of $PSL_2(\mathbb{F}_p)$ on $H^{2,0}(X)$, we have

$$m - n = f(p, d). \tag{16}$$

With the help of the congruence subgroups of the Hilbert modular group for real quadratic fields, we can obtain many examples, which however in general do not fulfill our simple assumptions. There is a general theory ([6] and the paper cited in that reference by H. SAITO). In [6] the twisted Dedekind sums only occur implicitly. We would like to develop the theory of these sums independently and derive some new properties of the usual Dedekind sums. For the theory of the Hilbert modular group see [4] and [2].

Examples which satisfy the assumptions 1 and 2:

Consider the field $K = \mathbb{Q}(\sqrt{5})$ and the ring of integers in K:

$$\mathcal{O} = \mathbb{Z}\cdot 1 + \mathbb{Z}\frac{1+\sqrt{5}}{2}.$$

The Hilbert modular group $\Gamma = PSL_2(\mathcal{O})$ acts on the product $\mathbb{H}\times\mathbb{H}$ of two half planes via $\left(\begin{smallmatrix} a & b \\ c & d\end{smallmatrix}\right)(z_1, z_2) = \left(\frac{az_1+b}{cz_1+d}, \frac{a'z_2+b'}{c'z_2+d'}\right)$, where the Galois automorphism of K is being denoted by $x \mapsto x'$. The surface $\Gamma\backslash\mathbb{H}^2$ has six quotient singularities, two each of the orders 2, 3 and 5. The prime numbers $p \equiv \pm 1(5)$ can be split

$$p = \pi\pi' \quad \text{with } \pi > 0,$$

where (π) is a prime ideal for which $\mathcal{O}/(\pi) = \mathbb{F}_p$. Now let $\Gamma(\pi)$ be the principal congruence subgroup of matrices in $SL_2(\mathcal{O})$, which are equivalent to the identity modulo π.

The subgroup $\Gamma(\pi)$ of Γ acts freely on $\mathbb{H}^2$. We have a Galois cover

$$\Gamma(\pi)\backslash\mathbb{H}^2 \longrightarrow \Gamma\backslash\mathbb{H}^2 \tag{17}$$

with Galois group $PSL_2(\mathbb{F}_p)$. Under the action on $\Gamma(\pi)\backslash\mathbb{H}^2$, there are only fixed points of orders 2, 3 and 5.

We must compactify the surface and consider from the start the resolution of the cusp singularities:

The surface $\Gamma\backslash\mathbb{H}^2$ is compactified by adding a rational curve with a double point. According to (17) there lie on $\Gamma(\pi)\backslash\mathbb{H}^2$ over this curve $\frac{p^2-1}{2}$ smooth rational curves, each of self-intersection number -3, which split into $p+1$ times $\frac{p-1}{2}$ curves, corresponding to the points of the projective line $\mathbb{P}^1(\mathbb{F}_p)$. Each set of $\frac{p-1}{2}$ curves forms $\frac{p-1}{2}/t$ cycles, each consisting of t curves, where t is determined as follows.

The fundamental unit $\varepsilon = \frac{1+\sqrt{5}}{2}$ as well as $\varepsilon^2 = \frac{3+\sqrt{5}}{2}$ may be viewed as cosets modulo p, as we have a $\sqrt{5}$ in $\mathbb{F}_p$ (just choose a root). The order t of the element ε^2 in $\mathbb{F}_p^*$ is a divisor of $\frac{p-1}{2}$. For $p = 11$, we have $t = 5$ ($\varepsilon^2 \equiv 5 \bmod 11$).

Hence we have 12 cycles of the form

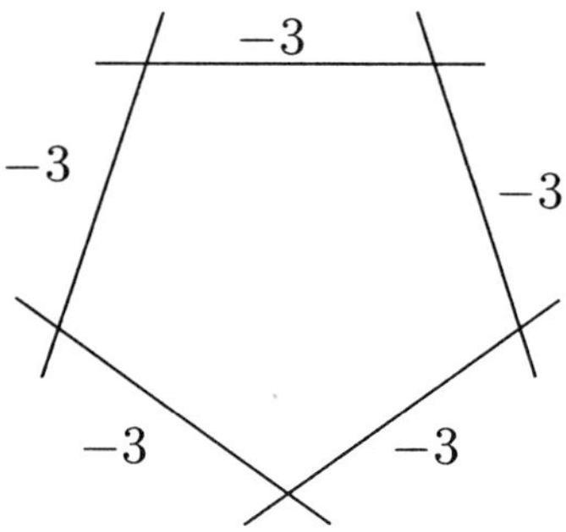

where each of the 60 curves obtained in this manner has self-intersection -3.

The element $\boldsymbol{a} : z \mapsto z+1$ of $PSL_2(\mathbb{F}_p)$ permutes p of the points of $\mathbb{P}^1(\mathbb{F}_p)$ cyclically, while fixing the point ∞. Indexed by this latter point there are $\frac{p-1}{2}$ curves, which are ordered in cycles. There are $\frac{p-1}{2}$ intersection points which are the fixed points of $\boldsymbol{a}$. Our assumptions 1 and 2 are satisfied, with $d \equiv \varepsilon^2 \bmod p$.

One can show that

$$\text{ded}(p, \varepsilon^2) = 1 - p \quad \text{for } p \equiv \pm 1(5), \tag{18}$$

for example one has $\text{ded}(11, 5) = -10$. Using this, relations (13), (14), (15) and (16) combined yield the following

Theorem. *Let p be a prime with $p \equiv \pm 1(5)$ and $X(p)$ the compact smooth Hilbert modular surface for the congruence subgroup $\Gamma(\pi)$ of the Hilbert modular group Γ for the field $\mathbb{Q}(\sqrt{5})$. Here $p = \pi\pi'$ (with $\pi > 0$) and (π) is a prime ideal of norm p. The group $PSL_2(\mathbb{F}_p)$ acts on $X(p)$ and hence on $H^{2,0}(X(p)$, the space of holomorphic two-forms on $X(p)$*

(cusp forms of weight 2 for $\Gamma(\pi)$). The character of $\boldsymbol{a} : z \mapsto z+1$ *under this action is*

$$\chi(\boldsymbol{a}) = -1 + \frac{f(p,d)}{2}\sqrt{p}, \tag{19}$$

where $d \equiv \varepsilon^2 = \frac{3+\sqrt{5}}{2}$ *modulo* p. *Here one has* $f(p,d) = 0$ *if* $p \equiv 3(4)$. *For the multiplicities* m, n *of* χ^+, χ^- *in the representation of* $PSL_2(\mathbb{F}_p)$ *we have*

$$m - n = f(p,d).$$

From [6] and the paper of H. SAITO cited in that paper we deduce that for $p \equiv \pm 1$ mod 5 and $p \equiv 1$ mod 4, $f(p,\varepsilon^2)$ vanishes for $\left(\frac{\varepsilon}{p}\right) = 1$, and for $\left(\frac{\varepsilon}{p}\right) = -1$ its value is

$$f(p,\varepsilon^2) = -2h(\mathbb{Q}(\sqrt{5},\sqrt{-\pi})).$$

Hence the expression $\frac{1}{2}|m-n|$ is the class number of a biquadratic number field. For $p = 41, 61, 109, 149, 241, 269, 281, 389$ the value of this class number is $1, 1, 1, 1, 3, 1, 3, 1$.

It follows from (19) that for $p \equiv \pm 1$ mod 5, in case $f(p,\varepsilon^2) = 0$, the representation of the cyclic subgroup of order p generated by $\boldsymbol{a}$ acting on $H^{2,0}(X(p))$ (plus the trivial representation) consists of the direct sum of $\frac{p^2-1}{120}$ cyclic permutation representations of p elements. For this, we note that for the arithmetic genus we have

$$1 + \dim H^{2,0}(X(p)) = p\frac{p^2-1}{120}.$$

(See [4]. One must multiply the Euler-volume of $\Gamma\backslash\mathbb{H}^2$ (which is $\frac{1}{15}$) with the order of $PSL_2(\mathbb{F}_p)$, which gives the Euler-Poincaré characteristic of $\Gamma(\pi)\backslash\mathbb{H}^2$, then divide by 4 to get the arithmetic genus of $X(p)$).

By the way, one can also show that for $f(p,\varepsilon^2) = 0$, the representation of $PSL_2(\mathbb{F}_p)$ on $H^{2,0}(X(p))$ (plus the trivial representation) is equivalent to the permutation representation of $PSL_2(\mathbb{F}_p)$ on the set of cosets $PSL_2(\mathbb{F}_p)/A_5$, using any embedding of A_5 in $PSL_2(\mathbb{F}_p)$.

References

[1] M. F. Atiyah & R. Bott, *A Lefschetz fixed-point formula for elliptic complexes, II. Applications,* Ann. of Math. **88** (1968) 451–491.

[2] G. van der Geer, *Hilbert modular surfaces*, Ergeb. Math., 3 Folge, Springer, Berlin, Vol. 16, 1987.

[3] F. Hirzebruch, *The signature theorem: reminiscences and recreation,* Prosp. Math., Ann. of Math. Stud. Princeton Univ. Press, Princeton, Vol. 70, 1971, 3–31.

[4] ———, *Hilbert modular surfaces*, Enseig. Math. **19** (1973) 183–281.

[5] F. Hirzebruch & D. Zagier, *The Atiyah-Singer theorem and elementary number theory*, Math. Lecture Ser. **3**, Publish or Perish, Inc., Berkeley, 1974.

[6] W. Meyer & R. Sczech, *Über eine topologische und zahlentheoretische Anwendung von Hirzebruch's Spitzenauflösung*, Math. Ann. **240** (1979) 69–96.

[7] S. H. Weintraub, $PSL_2(\mathbb{Z}_p)$ *and the Atiyah-Bott fixed-point theorem*, Houston J. Math. **6** (1980) 427–430.

[8] D. Zagier, *Zetafunktionen und quadratische Körper*, Springer, Heidelberg 1981.

MAX-PLANK-INSTITUT FÜR MATHEMATIK, BONN, GERMANY

SURVEYS IN DIFFERENTIAL GEOMETRY, 2000
Vol. VII ©2000, INTERNATIONAL PRESS
pp. 327-345

THE MODULI SPACE OF COMPLEX LAGRANGIAN SUBMANIFOLDS

N. J. HITCHIN

1. Introduction

Developments in string theory over the past few years (e.g. [13], [5]) have focussed attention on a differential geometric structure induced on the base space of an algebraically completely integrable Hamiltonian system. This has been recently formalised by D.Freed [6] as a *special Kähler structure.*

The purpose of this paper is to provide an alternative approach to the geometry of special Kähler manifolds, one that is motivated by the desire to understand a more general situation than that afforded by integrable systems. We seek the natural geometrical structure on the moduli space M of deformations of a compact, complex Lagrangian submanifold Y in a complex Kählerian symplectic manifold X. In many respects what we do parallels the approach of an earlier paper [8] which began an investigation into the geometry of the moduli space of compact special Lagrangian submanifolds in a Calabi-Yau manifold. This was motivated by a desire to understand the geometry underpinning the Strömінger-Yau-Zaslow [14] approach to mirror symmetry. We are essentially attacking here the special case where the Calabi-Yau is hyperkähler, though we shall not need the full force of the existence of a hyperkähler metric on X.

Our viewpoint, as in [8], is to pay less attention to the holomorphic structure of the situation, and more to the symplectic one. Thus a complex Lagrangian submanifold of X can be characterized as a real submanifold on which the real and imaginary parts of the holomorphic symplectic 2-form vanish. Correspondingly, the differential geometric

First published in the Asian Journal of Mathematics, 1999. Used by permission.

structure on the moduli space M is induced from a local embedding of M into $H^1(Y,\mathbf{R}) \times H^1(Y,\mathbf{R})$ which is Lagrangian with respect to two natural constant symplectic forms. We show that this "bilagrangian" condition for a submanifold of a product $V \times V$ of real symplectic vector spaces is equivalent to the structure of a special (pseudo-) Kähler metric on M. Moreover, it is easy to see from this point of view that a choice of symplectic basis of V yields the known fact that any special Kähler metric is generated by a single holomorphic function – the holomorphic prepotential.

Finally, we derive from our formalism the hyperkähler metric introduced in a string-theoretic context several years ago by Cecotti, Ferrara and Girardello [2]. It can be seen as a special case of the Legendre transform construction of Lindström and Roček [9], and yields, in the context of our moduli space, a hyperkähler metric on an open set of Markman's moduli space of Lagrangian sheaves [4].

Our approach offers a different perspective to special Kähler geometry, and in particular draws attention to a single naturally defined function ϕ which plays an important role: it is the Hamiltonian for the fundamental vector field, a potential for the Kähler metric and, with respect to one of the complex structures, a potential also for the associated hyperkähler metric.

The author wishes to thank Dan Freed for introducing the subject to him and the Institute for Advanced Study for its hospitality.

2. Complex Lagrangian submanifolds

Let X be a complex symplectic manifold of complex dimension $2n$. It has a holomorphic symplectic 2-form ω^c which we write in terms of its real and imaginary parts:

$$\omega^c = \omega_1 + i\omega_2$$

These two closed forms are real symplectic forms and define the structure of a complex symplectic manifold on X. We see this as follows. Given a closed form ω^c, we consider the distribution in $E \subset T \otimes \mathbf{C}$ defined by the complex vector fields U with $\iota(U)\omega^c = 0$. If ω^c satisfies the algebraic condition that $E \oplus \bar{E} = T \otimes \mathbf{C}$ then it defines an almost complex structure. This is integrable because if $\iota(U)\omega^c = 0$,

$$\mathcal{L}_U\omega^c = d(\iota(U)\omega^c) + \iota(U)d\omega^c = 0$$

so if U and V are sections of E, $\mathcal{L}_U\omega^c = 0$ and $\iota(V)\omega^c = 0$, so

$$\iota([U,V])\omega^c = \mathcal{L}_U(\iota(V)\omega^c) - \iota(V)(\mathcal{L}_U\omega^c) = 0$$

and hence $[U, V]$ is a section of E.

Similarly we have the following

Proposition 1. *A real $2n$-dimensional submanifold $Y \subset X$ is complex Lagrangian if and only if $\omega^c|_Y = 0$.*

Proof. If the submanifold Y is complex Lagrangian, then $\omega^c|_Y = 0$ by definition. Conversely, if $\omega^c|_Y = 0$, we need to show that Y is a complex submanifold, that its tangent spaces are complex. Now the complex structure I on X is defined algebraically by the two real symplectic forms ω_1, ω_2. Instead of the description above, we can think of each giving an isomorphism $\varphi_i : TX \cong T^*X$ and then $I = \varphi_2^{-1}\varphi_1$. Since Y is Lagrangian with respect to both symplectic forms, then both φ_1 and φ_2 map TY isomorphically to the conormal bundle $N^*Y \subset T^*X$. thus $I = \varphi_2^{-1}\varphi_1$ preserves TY. Hence Y is a complex submanifold, $\omega^c|_Y = 0$ and so Y is complex Lagrangian.

We lose nothing therefore by focussing our attention on the symplectic aspects of complex Lagrangian submanifolds. We do, however, need to know that there is a good local moduli space of deformations of $Y \subset X$. In general deformations of compact complex submanifolds can be obstructed, but it follows from the paper of Voisin [15] that this is not the case when the submanifold is Lagrangian. In fact if X has a hyperkähler metric, this is also a consequence of the differential geometric argument of McLean [11]. There thus exists a local moduli space M, which is a complex manifold, and such that there is a natural isomorphism

$$T_{[Y]}M \cong H^0(Y, N)$$

from the tangent space at the point $[Y] \in M$ representing Y and the space of holomorphic sections of the normal bundle N to $Y \subset X$. As Y is complex Lagrangian, ω^c defines a holomorphic isomorphism from N to T^*Y, so

$$T_{[Y]}M \cong H^0(Y, T^*)$$

From the holomorphic point of view this is the tangent space to the moduli of deformations of Y as a complex submanifold. The infinitesimal deformations as a complex *Lagrangian* submanifold correspond to those sections of the normal bundle for which the corresponding 1-forms in $H^0(Y, T^*)$ are *closed.* But if we assume that X has a Kähler form h so that Y is also Kähler, then any holomorphic 1-form is closed. Moreover the real dimension of M is then given by $b_1(Y) = 2\dim_{\mathbf{C}} H^0(Y, T^*)$. We make this Kählerian assumption from now on. We shall investigate next the local differential geometry of M.

3. The moduli space

Let Z be a local universal family of deformations of the complex Lagrangian submanifold $Y_0 \subset X$, so that Z is a complex manifold with a holomorphic projection $\pi : Z \to M$ and a holomorphic map $F : Z \to X$ such that

$$F(\pi^{-1}([Y])) = Y$$

Consider the 2-form $F^*\omega_1$ on Z. If $x_1, \ldots, x_{2n}, y_1, \ldots, y_{2m}$ are real local coordinates on Z with $y_1, \ldots, y_{2m}$ coordinates on M and $\pi(x_1, \ldots, y_{2m}) = (y_1, \ldots, y_{2m})$ then since each fibre Y is Lagrangian with respect to ω_1, $F^*\omega_1|_Y = 0$ and so

$$F^*\omega_1 = \sum a_{ij} dx_i \wedge dy_j + \sum b_{ij} dy_i \wedge dy_j$$

Furthermore, since $F^*\omega_1$ is closed,

$$\sum \frac{\partial a_{ij}}{\partial x_k} dx_k \wedge dx_i = 0 \tag{1}$$

We can see in concrete coordinate terms here that, for each j, the 1-form $\sum a_{ij} dx_i$ on Y is closed. More invariantly, it says that if U is a tangent vector to M at $[Y]$, then if $\tilde{U}$ is a lift to a vector field along Y, the 1-form $(\iota(\tilde{U})F^*\omega_1)|_Y$ is closed and independent of the choice of lifting.

From (1), integrating $F^*\omega_1$ over two homologous 1-cycles in a fibre of π gives the same result. Now working locally in M, we assume that M is contractible, and so we can by homotopy invariance identify the homology of each fibre. Take a homology class $A \in H_1(Z, \mathbf{Z}) \cong H_1(Y, \mathbf{Z})$ and choose a circle fibration in $\pi : Z \to M$ such that each fibre is in the class A. Integrating along the fibres, we obtain canonically a closed 1-form

$$(\pi_A)_* F^*\omega_1 \in \Omega^1(M)$$

and since M is assumed contractible, a smooth function μ_A on M, well-defined up to an additive constant, such that $d\mu_A = (\pi_A)_* F^*\omega_1$. Putting all the functions together gives a map

$$\mu : M \to H^1(Y, \mathbf{R})$$

This function by definition has the property that $d\mu(U)$ is the cohomology class of the closed form $(\iota(\tilde{U})F^*\omega_1)|_Y$.

Proposition 2. *μ is a local diffeomorphism.*

Proof. We think in terms of the holomorphic fibration $\pi : Z \to M$ and a tangent vector U at $[Y]$. The 1-form $(\iota(\tilde{U})F^*\omega^c)|_Y$ on Y is independent of the choice of lift and thus, taking local holomorphic lifts, is a well-defined global holomorphic 1-form. This is the canonical isomorphism $T_{[Y]}M \cong H^0(Y, T^*)$. Now if $d\mu(U) = 0$, by the definition of μ, the cohomology class of the real part of $\theta = (\iota(\tilde{U})F^*\omega^c)|_Y$ is zero. But Y is a Kähler manifold so we have $H^1(Y, \mathbf{C}) \cong H^{1,0} \oplus H^{0,1}$ and $\theta + \bar{\theta}$ cannot be cohomologically trivial unless $\theta = 0$. This means that $(\iota(\tilde{U})F^*\omega^c)|_Y = 0$ and $U = 0$. Thus by the inverse function theorem, μ is a local diffeomorphism.

Similarly, using the other symplectic form ω_2, we get a map

$$\nu : M \to H^1(Y, \mathbf{R})$$

and, put together, a smooth map

$$w = (\mu, \nu) : M \to H^1(Y, \mathbf{R}) \times H^1(Y, \mathbf{R})$$

Thus $w(M)$ is a smooth submanifold such that the projection onto each factor is a local diffeomorphism.

The vector space $H^1(Y, \mathbf{R})$ has a real constant symplectic form defined by the restriction of the Kähler form h on X:

$$\langle a, b \rangle = \int_Y \alpha \wedge \beta \wedge h^{n-1}$$

where α, β are representative 1-forms for a and b. This clearly only requires the cohomology class of the Kähler form h. We define two constant symplectic forms on $H^1(Y, \mathbf{R}) \times H^1(Y, \mathbf{R})$:

$$\Omega_1((a_1, a_2), (b_1, b_2)) = \langle a_1, b_2 \rangle + \langle a_2, b_1 \rangle \tag{2}$$

$$\Omega_2((a_1, a_2), (b_1, b_2)) = \langle a_1, b_1 \rangle - \langle a_2, b_2 \rangle \tag{3}$$

Take a basis for V, so that the skew form has matrix ω_{ij}, then in the corresponding linear coordinates

$$\Omega_1 = 2\sum \omega_{ij} dx_i \wedge dy_j \tag{4}$$

$$\Omega_2 = \sum \omega_{ij} dx_i \wedge dx_j - \sum \omega_{ij} dy_i \wedge dy_j \tag{5}$$

We now have

Theorem 1. *$w(M) \subset H^1(Y, \mathbf{R}) \times H^1(Y, \mathbf{R})$ is Lagrangian with respect to Ω_1 and Ω_2.*

Proof. The holomorphic symplectic form ω^c is of type $(2,0)$ so since F is holomorphic, $F^*(\omega^c)^2$ has type $(4,0)$. If U is a tangent vector on M at $[Y]$, then as we have seen, $(\iota(\tilde{U})F^*\omega^c)|_Y$ is independent of the choice of lifting $\tilde{U}$, because $\omega^c|_Y = 0$. Now

$$\iota(\tilde{U})\iota(\tilde{V})F^*(\omega^c)^2 = 2(\iota(\tilde{U})\iota(\tilde{V})F^*\omega^c)(F^*\omega^c) + 2(\iota(\tilde{U})F^*\omega^c)(\iota(\tilde{V})F^*\omega^c)$$

and restricting to Y,

$$\iota(\tilde{U})\iota(\tilde{V})F^*(\omega^c)^2|_Y = 2(\iota(\tilde{U})F^*\omega^c)|_Y \wedge (\iota(\tilde{V})F^*\omega^c)|_Y \tag{6}$$

But the left hand side is of type $(2,0)$, hence

$$\iota(\tilde{U})\iota(\tilde{V})F^*(\omega^c)^2|_Y \wedge h^{n-1}|_Y$$

is of type $(n+1, n-1)$ and so vanishes since Y is complex of dimension n. Hence from (6)

$$\int_Y \iota(\tilde{U})F^*\omega^c \wedge \iota(\tilde{V})F^*\omega^c \wedge h^{n-1} = 0$$

but this means that

$$\langle (d\mu + id\nu)(U), (d\mu + id\nu)(V)\rangle = 0$$

and so equating to zero real and imaginary parts,

$$\begin{aligned} \langle d\mu(U), d\mu(V)\rangle - \langle d\nu(U), d\nu(V)\rangle &= 0 \\ \langle d\nu(U), d\mu(V)\rangle + \langle d\mu(U), d\nu(V)\rangle &= 0 \end{aligned}$$

These two conditions are precisely the vanishing of Ω_2 and Ω_1 respectively on $w(M)$.

Remark. Theorem 1 demonstrates that the structure of the moduli space – defined as a submanifold on which two symplectic forms vanish – parallels the structure of the objects it parametrizes. This is also the philosophy behind the description in [8] of the moduli space of special Lagrangian submanifolds of a Calabi-Yau manifold.

4. Special Kähler manifolds

We shall show that, as a consequence of the "bilagrangian" property of Theorem 1, M inherits a special Kähler structure. The definition, as given in [6], is the following:

Definition 1. A special Kähler manifold is a complex manifold M with

- a Kähler metric g with Kähler form ω
- a flat torsion-free connection ∇ such that
- $\nabla\omega = 0$ and
- $d_\nabla I = 0 \in \Omega^2(M, T)$

To clarify the last property, we think of the complex structure I, an endomorphism of the tangent bundle T, as a 1-form with values in T, i.e. $I \in \Omega^1(M, T)$. The connection ∇ defines a covariant exterior derivative $d_\nabla : \Omega^p(M, T) \to \Omega^{p+1}(M, T)$ and we require $d_\nabla I = 0$. This is weaker than $\nabla I = 0$ and indeed, since ω and I determine g, the latter condition would say that ∇ is the Levi-Civita connection, and all we would be looking at is a flat Kähler manifold. The reader should be warned that special Kähler manifolds do not form a very interesting class of global Riemannian structures – it has been shown by Lu [10] that any *complete* special Kähler manifold is flat.

Now let V be a real symplectic vector space with skew form $\langle\,,\rangle$. As in (2),(3) we define two constant symplectic forms Ω_1, Ω_2 on $V \times V$. We can also define an indefinite metric g on $V \times V$ by

$$g((a,b),(a,b)) = \frac{1}{2}\langle a, b\rangle$$

Then we have the following

Theorem 2. *Let $M \subset V \times V$ be a submanifold which is Lagrangian for Ω_1 and Ω_2, and transversal to the two projections onto V. Then $g|_M$ is a special pseudo-Kähler metric. Conversely, any special pseudo-Kähler metric on a manifold M is locally induced from an embedding in $V \times V$ for some V.*

Proof. It will be convenient to use the nondegenerate form $\langle\,,\rangle$ on V to identify V with V^*. Under this identification, Ω_1 becomes essentially the canonical symplectic form on $T^*V \cong V \times V^*$, by setting $d\xi_i = \sum \omega_{ij} dy_j$ in (4),(5). We then have the following expressions for Ω_1 and Ω_2:

$$\Omega_1 = 2\sum dx_i \wedge d\xi_i \tag{7}$$

$$\Omega_2 = \sum \omega_{ij} dx_i \wedge dx_j + \sum \omega^{ij} d\xi_i \wedge d\xi_j \tag{8}$$

To begin, we use the projection onto the first factor V to locally identify M with a flat symplectic vector space. This provides us with our flat connection ∇ with $\nabla\omega = 0$. If we use the coordinates $x_1, \ldots, x_{2n}$, then covariant derivatives using ∇ are just ordinary derivatives.

Now since M is Lagrangian with respect to the canonical symplectic form Ω_1 on T^*V and transversal to the projection to V, the embedding is defined by the graph of the derivative of a function on V so in coordinates

$$\xi_j = \frac{\partial \phi}{\partial x_j}$$

for some function $\phi(x_1, \ldots, x_{2n})$. The tangent vector $\partial/\partial x_j$ of M then lies in $V \times V^*$ as the vector

$$X_j = \frac{\partial}{\partial x_j} + \sum \frac{\partial \xi_k}{\partial x_j}\frac{\partial}{\partial \xi_k} = \frac{\partial}{\partial x_j} + \sum_k \frac{\partial^2 \phi}{\partial x_k \partial x_j}\frac{\partial}{\partial \xi_k}$$

Since the metric on $V \times V^*$ is defined by

$$g((x,\xi),(x,\xi)) = \frac{1}{2}\langle x, \xi\rangle$$

the induced metric on M is

$$\sum g_{kj}dx_k dx_j = \sum g(X_k, X_j)dx_k dx_j = \sum \frac{\partial^2 \phi}{\partial x_k \partial x_j} dx_k dx_j$$

In general this metric may not be positive definite. It is nondegenerate however, for if $\sum g_{ij}a_j = 0$, then $\sum a_j X_j = \sum a_j \partial/\partial x_j$ so that projection of this tangent vector onto the second factor in $V \times V$ is zero. By the transversality assumption this means each $a_j = 0$.

Consider now the second Lagrangian condition: Ω_2 vanishes on M. From (8) this says, using $\xi_k = \partial\phi/\partial x_k$ and $g_{kj} = \partial^2\phi/\partial x_k \partial x_j$,

$$0 = \Omega_2(X_k, X_j) = \omega_{kj} + \sum g_{ka}g_{jb}\omega^{ab}$$

or, writing $I^j_k = \sum \omega^{ja}g_{ak}$, that $I^2 = -1$. This is the almost complex structure. Since $\omega_{kj} = -\omega_{jk}$, I is skew adjoint with respect to g. Now let X be the Hamiltonian vector field for the function ϕ. We have

$$X = \sum a_j \frac{\partial}{\partial x_j} = \sum \omega^{ij}\frac{\partial \phi}{\partial x_i}\frac{\partial}{\partial x_j}$$

so

$$\frac{\partial a_j}{\partial x_k} = \sum \omega^{ij} g_{ik} = I^j_k$$

Hence $I = d_\nabla X$ and so $d_\nabla I = d^2_\nabla X = 0$ since the connection is flat. Hence I satisfies the compatibility condition with the flat connection.

It remains to show that I is integrable. But consider the complex functions

$$z_j = x_j - i \sum \omega^{jk} \frac{\partial \phi}{\partial x_k} \tag{9}$$

Differentiating, we obtain the $2n$ complex 1-forms

$$\begin{aligned} dz_j &= dx_j - i \sum \omega^{jk} \frac{\partial^2 \phi}{\partial x_k \partial x_l} dx_l \\ &= dx_j - i \sum \omega^{jk} g_{kl} dx_l \\ &= dx_j - i \sum I^j_l dx_l \end{aligned}$$

and these are clearly of type $(1,0)$. We need to find n linearly independent ones. Let $E \subseteq \Lambda^{1,0}$ be the distribution spanned by $dz_1, \ldots, dz_{2n}$. Then for each j, $2dx_j = dz_j + d\bar{z}_j$, and since $dx_1, \ldots, dx_{2n}$ forms a basis, $E \oplus \bar{E} = \Lambda^1$ and the rank of E is n.

Thus the metric g and the connection ∇ satisfy all the conditions for a special pseudo-Kähler (i.e. possibly indefinite) metric.

Now consider the converse. Let M be a special pseudo-Kähler manifold, and $(x_1, \ldots, x_{2n})$ flat local coordinates, so that the covariant derivative is the ordinary derivative and the coefficients of the symplectic form are constant. Consider the 1-form

$$\alpha_k = \sum \omega_{kl} I^l_j dx_j$$

Now

$$d\alpha_k = \sum \omega_{kl} \frac{\partial I^l_j}{\partial x_m} dx_m \wedge dx_j = 0$$

since ω_{kl} is constant and $d_\nabla I = 0$. Thus locally there are functions ξ_k such that

$$\alpha_k = d\xi_k.$$

We map M to $\mathbf{R}^{2n} \times \mathbf{R}^{2n}$ by

$$(x_1, \ldots, x_{2n}) \mapsto (x_1, \ldots, x_{2n}, \xi_1, \ldots, \xi_{2n})$$

First we claim the image is Lagrangian with respect to the symplectic form $\Omega_1 = 2 \sum dx_j \wedge d\xi_j$. But restricting $\sum dx_j \wedge d\xi_j$ gives

$$dx_j \wedge \alpha_j = \sum dx_j \wedge \omega_{jl} I^l_k dx_k = \sum dx_j \wedge g_{jk} dx_k = 0$$

since the metric tensor g_{jk} is symmetric.

Next consider the symplectic form

$$\Omega_2 = \sum \omega_{ij} dx_i \wedge dx_j + \omega^{ij} d\xi_i \wedge d\xi_j.$$

Restricted to M this is

$$\sum \omega_{ij} dx_i \wedge dx_j + \omega^{ab} g_{ai} dx_i \wedge g_{bj} dx_j$$

But since $I^2 = -1$ this too is zero.

Hence taking V to be $\mathbf{R}^{2n}$ with the skew form ω_{ij}, we obtain a local embedding of M in $V \times V$, Lagrangian with respect to both forms. It is straightforward to see that the induced metric is g_{ij}.

One of the features of the above approach is the fundamental role of the function ϕ. Here is another aspect of this:

Proposition 3. *The function ϕ is a Kähler potential.*

Proof. Consider

$$d(Id\phi) = \sum \frac{\partial}{\partial x_k}(I^i_j \frac{\partial \phi}{\partial x_i}) dx_k \wedge dx_j$$

Now

$$\frac{\partial}{\partial x_k} I^i_j = \sum \frac{\partial}{\partial x_k}(\omega^{ia} g_{aj}) = \sum \omega^{ia} \frac{\partial^3 \phi}{\partial x_a \partial x_j \partial x_k}$$

so

$$d(Id\phi) = \sum \omega^{ia} \frac{\partial^3 \phi}{\partial x_a \partial x_j \partial x_k} \frac{\partial \phi}{\partial x_i} dx_k \wedge dx_j + I^i_j \frac{\partial^2 \phi}{\partial x_i \partial x_k} dx_k \wedge dx_j$$

The first term vanishes by the symmetry in j, k and the second term is

$$\sum I^i_j g_{ik} dx_k \wedge dx_j = -\sum \omega_{kj} dx_k \wedge dx_j = -\omega$$

where ω is the Kähler form.

We can now prove

Theorem 3. *The moduli space M of deformations of a complex Lagrangian submanifold of a complex Kählerian symplectic manifold X has a naturally induced special Kähler structure.*

Proof. We have to show firstly that $w(M) \subset H^1(Y,\mathbf{R}) \times H^1(Y,\mathbf{R})$ satisfies the transversality of the previous theorem, but this follows from Proposition 2 applied to μ and ν. We also need the metric to be definite. But

$$\begin{aligned} g(X,X) &= \langle d\mu(X), d\nu(X)\rangle \\ &= -\frac{i}{2}\langle (d\mu + id\nu)(X), (d\mu - id\nu)(X)\rangle \\ &= -\frac{i}{2}\int_Y \iota(X)\omega^c \wedge \iota(X)\bar{\omega}^c \wedge h^{n-1} \end{aligned}$$

which is definite. We should also show that the complex structure I coincides with the natural complex structure of the moduli space when considered as a moduli space of complex submanifolds. From (9), the functions

$$z_j = x_j - i\sum \omega^{jk}\frac{\partial \phi}{\partial x_k}$$

are antiholomorphic with respect to I. But $\xi_j = \partial\phi/\partial x_j$, so

$$z_j = x_j - i\sum \omega^{jk}\xi_k.$$

If we return from $V \times V^*$ to $V \times V$ using ω we see that this function is obtained by taking a class $A_j \in H_1(Y,\mathbf{R})$ and forming

$$z_j = \langle \mu - i\nu, A_j\rangle$$

Since $d(\mu + i\nu)(U) = \iota(\tilde{U})\omega^c$ is of type $(1,0)$, $dz_j = \langle d\mu - id\nu, A_j\rangle$ is of type $(0,1)$ with respect to the complex structure of the moduli space of compact complex submanifolds, so the two complex structures coincide.

Finally we should pass from the local to the global point of view on M, a point which is relevant in particular to the situation where M is the base space of a completely integrable system. There is an additive ambiguity in the choice of the function μ but $d\mu$ gives an isomorphism between TM and the trivial bundle $M \times H^1(Y,\mathbf{R})$, and this is the flat connection ∇ of the special Kähler structure. Globally on M, the cohomology of the fibres of $Z \to M$ defines a flat vector bundle on M (homotopy invariance of cohomology defines the Gauss-Manin connection) and so the isomorphism $d\mu$ provides a flat connection on TM. The symplectic form is preserved by the Gauss-Manin connection, and since the complex structure on M is globally defined, so is the metric g, which is defined by I and ω.

Remarks.

1. One of the well-known features of special Kähler geometry is the fact that any special Kähler metric is derived from a single holomorphic function $\mathcal{F}$ of n variables. It is known as the holomorphic prepotential on M. This fact is rather easily seen using our bilagrangian formulism. For this purpose we choose a symplectic basis on V. The corresponding coordinates $x_1, \ldots, x_{2n}$ give

$$\omega = \sum dx_j \wedge dx_{n+j}$$

and so

$$\begin{aligned}\Omega_1 &= 2\sum_1^{2n} dx_j \wedge d\xi_j \\ \Omega_2 &= \sum_1^{n} dx_j \wedge dx_{n+j} - \sum_1^{n} d\xi_j \wedge d\xi_{n+j}\end{aligned}$$

and then

$$\begin{aligned}\Omega^c &= \frac{1}{2}\Omega_1 + i\Omega_2 \\ &= \sum_i^{n} d(x_j + i\xi_{n+j}) \wedge d(\xi_j + ix_{n+j}) \\ &= \sum_1^{n} dv_j \wedge dw_j\end{aligned}$$

We see that Ω^c can be identified with the canonical complex symplectic form on $T^*\mathbf{C}^n$. From Proposition 1, a submanifold on which Ω_1 and Ω_2 vanish is the same thing as a complex Lagrangian submanifold of $T^*\mathbf{C}^n$, but this is given by the graph of the derivative of a holomorphic function

$$v_j = \frac{\partial \mathcal{F}}{\partial w_j} \tag{10}$$

From Theorem 2 this is all we need for a special Kähler manifold.

2. In the standard presentation of the prepotential, its second derivative gives a holomorphic map from the base space of an integrable system to the Siegel upper half-space – the moduli space of polarized abelian varieties, expressed as symmetric matrices with positive definite imaginary part. Such a description involves choosing a symplectic basis for $H_1(Y, \mathbf{Z})$ (the classical A and B cycles) which is what we have done to

introduce the holomorphic function $\mathcal{F}$. By contrast the real function ϕ requires no such choice. All we have chosen is projection onto the first factor in $V \times V$ to define ϕ. We postpone the discussion of the relationship between ϕ and $\mathcal{F}$ to the next section, where we study some associated hyperkähler constructions.

3. In the bilagrangian picture of $M \subset V \times V$ we get another flat torsion-free connection by projecting onto the second factor. From the second Lagrangian condition

$$0 = \Omega_2|_M = \sum \omega_{ij} dx_i \wedge dx_j - \sum \omega_{ij} dy_i \wedge dy_j$$

the pull back of the flat symplectic form under this projection is the same ω. The function ϕ is then replaced by its Legendre transform and the new flat coordinates dy_j are related to the old ones by $dy_j = I dx_j$. From the point of view of the moduli space of complex Lagrangian submanifolds, all we have done here is to replace ω_1 by ω_2, or what is essentially the same, to replace the complex symplectic form ω^c by $i\omega^c$. Clearly we can multiply ω^c by $e^{i\theta}$ and still have the same moduli space but now a family of flat connections parametrized by the circle. This is one viewpoint to the study of Higgs bundles as in Simpson's higher dimensional approach [12]. In fact any special Kähler manifold can be thought of as having a Higgs bundle structure on $T \oplus T^*$. Recall that a Higgs bundle corresponding to a local system on a Kähler manifold M consists of a holomorphic vector bundle E with a unitary connection A and a section $\Phi \in H^0(M, \operatorname{End} E \otimes T^*)$, the Higgs field, such that the connection

$$d_A + e^{i\theta}\Phi + e^{-i\theta}\Phi^*$$

is flat for all θ. In the case of a special Kähler manifold, the connection A is the Levi-Civita connection on $T \oplus T^*$ and the Higgs field has the form

$$\begin{pmatrix} 0 & 0 \\ \Theta & 0 \end{pmatrix}$$

where $\Theta \in H^0(M, Sym^3 T^*)$ is a holomorphic cubic form. Since Simpson's original approach to Higgs bundles was derived from variations of Hodge structure, this begins to take us back to the picture of a moduli space of complex manifolds which was the original motivation for this paper.

5. Hyperkähler metrics

In [2] Cecotti, Ferrara and Girardello showed how to define naturally a hyperkähler metric on a certain bundle over a special Kähler manifold (see also [6]). As we have seen a special Kähler metric can be defined via a single holomorphic function, so we have a straightforward way of constructing hyperkähler metrics. We shall show here that this construction is in fact a special case of an earlier technique called the *Legendre transform* construction of Lindström and Roček [9], [7].

Recall (see for example [1]) that a hyperkähler metric is determined by three symplectic forms $\omega_1, \omega_2, \omega_3$ satisfying some algebraic conditions, namely that if $\varphi_i : T \to T^*$ is the isomorphism determined by ω_i, and we define $J_1 = \varphi_3^{-1}\varphi_2, J_2 = \varphi_1^{-1}\varphi_3, J_3 = \varphi_2^{-1}\varphi_1$, then J_1, J_2, J_3 obey the quaternionic identities $J_1^2 = J_2^2 = J_3^2 = -1$. We work locally first and let M be a special Kähler manifold. On the product $M \times \mathbf{R}^{2k}$ take the three symplectic 2-forms $\omega_1, \omega_2, \omega_3$ defined by

$$\omega_1 = \sum \frac{\partial^2 \phi}{\partial x_j \partial x_k} dx_j \wedge dy_k \tag{11}$$

$$\omega_2 + i\omega_3 = -\frac{1}{2}\sum \omega_{jk} d(x_j + iy_j) \wedge d(x_k + iy_k) \tag{12}$$

Then

$$J_3(\frac{\partial}{\partial x_i}) = \sum \omega^{jk} g_{ik} \frac{\partial}{\partial y_j} = \sum I_i^j \frac{\partial}{\partial y_j}$$

and

$$J_3(\frac{\partial}{\partial y_i}) = \sum \omega^{jk} g_{ik} \frac{\partial}{\partial x_j} = \sum I_i^j \frac{\partial}{\partial x_j}$$

So in block matrix form

$$J_3 = \begin{pmatrix} 0 & I \\ I & 0 \end{pmatrix}$$

Similarly

$$J_2 = \begin{pmatrix} -I & 0 \\ 0 & I \end{pmatrix}$$

so that $J_3^2 = J_2^2 = -1$ and $J_2 J_3 = -J_3 J_2$. Thus J_2, J_3 generate an action of the quaternions, with $J_1 = J_2 J_3$ given by

$$J_1 = \begin{pmatrix} 0 & 1 \\ -1 & 0 \end{pmatrix}$$

and so the symplectic forms $\omega_1, \omega_2, \omega_3$ define a hyperkähler metric.

Proposition 4. *ϕ is a Kähler potential with respect to the complex structure J_1.*

Proof. From (12) and the discussion in section 2, it is clear that $z_j = x_i + iy_j$ for $1 \leq j \leq 2n$ are complex coordinates in this complex structure. But ϕ is independent of y_j so

$$\partial\bar{\partial}\phi = \sum \frac{\partial^2\phi}{\partial x_j \partial x_k} dz_j \wedge d\bar{z}_k = -2i\omega_1$$

from (11).

Remark. Note that the projection from $M \times \mathbf{R}^{2n}$ to M is holomorphic in the complex structure J_2. Using the $\partial, \bar{\partial}$ operators in that structure $\partial\bar{\partial}\phi$ gives the pull-back of the Kähler form on M, which is degenerate.

Since ω_{jk} is constant and ϕ is independent of y_j, each symplectic form is invariant by the vector field $\partial/\partial y_j$. We thus have a triholomorphic action of $\mathbf{R}^{2n}$ on $M \times \mathbf{R}^{2n}$. The Legendre transform method is a canonical construction of hyperkähler manifolds X^{4n} with a triholomorphic action of $\mathbf{R}^n$, so in our case we have far more symmetry. Strictly speaking (a point not emphasized in the literature on this method) to apply the method we need an action which also admits an equivariant hyperkähler moment map. Recall that if U is a triholomorphic vector field then for each i, $\iota(U)\omega_i = d\mu_i^U$ for some function μ_i^U and putting the moment maps μ_i^U together we get the hyperkähler moment map

$$\mu : X \to \mathbf{R}^n \otimes \mathbf{R}^3$$

which, if equivariant, represents X as the total space of a principal $\mathbf{R}^n$ bundle over an open set of $\mathbf{R}^n \otimes \mathbf{R}^3$. The Legendre transform construction reduces the hyperkähler equations in this situation to finding a real-valued function $F(\mathbf{x}_1, \ldots, \mathbf{x}_n)$ defined on $\mathbf{R}^n \otimes \mathbf{R}^3 \to \mathbf{R}$ and which satisfies the 3-dimensional Laplace equation

$$\Delta F(c_1\mathbf{x}, \ldots, c_n\mathbf{x}) = 0 \tag{13}$$

for each $(c_1, \ldots, c_n) \in \mathbf{R}^n$. Clearly if F is defined from a holomorphic function $f(z_1, \ldots, z_n)$ by

$$F(\mathbf{x}_1, \mathbf{x}_2, \ldots, \mathbf{x}_n) = Ref(u_1 + iv_1, \ldots, u_n + iv_n)$$

with $\mathbf{x}_j = (u_j, v_j, w_j)$ then it satisfies this equation. In fact this is essentially the only way to obtain a solution which is invariant under translation in one of the coordinates of each $\mathbf{x}_j \in \mathbf{R}^3$.

Proposition 5. *The hyperkähler metric defined by (11), (12) is constructed by the Legendre transform method from $F = Re\mathcal{F}$ where $\mathcal{F}$ is a holomorphic prepotential.*

Proof. Recall that the introduction of the holomorphic function $\mathcal{F}$ required the choice of a symplectic basis on V. We shall need the same to implement the Legendre transform method, because we need an equivariant moment map. To see this note that if $U_j = -\partial/\partial y_j$, then from (11), (12) the hyperkähler moment map is

$$\begin{aligned} \mu_1 &= \frac{\partial \phi}{\partial x_j} \\ \mu_2 + i\mu_3 &= i\sum \omega_{jk}(x_k + iy_k) \end{aligned}$$

and this (because of the y_k terms) is not equivariant for the full group $\mathbf{R}^{2n}$ of isometries. However, if we choose a symplectic basis so that $\omega = \sum_1^n dx_i \wedge dx_{n+i}$ and take the action of $\mathbf{R}^n$ generated by $U_1, \ldots, U_n$, we have for $1 \leq j \leq n$ the moment map for U_j

$$\mu_1 = \frac{\partial \phi}{\partial x_j} = \xi_j$$

$$\mu_2 + i\mu_3 = -i(x_{n+j} + iy_{n+j})$$

and this is equivariant since

$$U_j \cdot y_{n+k} = -\frac{\partial y_{n+k}}{\partial y_j} = 0$$

for $1 \leq j \leq n$. The hyperkähler moment map is now:

$$\mu(x) = (\xi_1, \ldots, \xi_n, -y_{n+1}, \ldots, -y_{2n}, x_{n+1}, \ldots, x_{2n}) \in \mathbf{R}^n \otimes \mathbf{R}^3$$

Since $\mathcal{F}$ is a holomorphic function of $w_j = \xi_j + ix_{n+j}$, $F = Re\mathcal{F}$ satisfies the equation (13).

To find the hyperkähler metric for such an F, we follow [7] putting

$$z_j = -y_{n+j} + ix_{n+j} = i(x_{n+j} + iy_{n+j})$$

for $1 \leq j \leq n$. These are holomorphic functions with respect to the complex structure J_1. According to [7], a Kähler potential for this complex structure is

$$K = F - \sum_1^n (u_k + \bar{u}_k)\xi_k$$

where for $1 \leq j \leq n$

$$\frac{\partial F}{\partial \xi_j} = u_j + \bar{u}_j$$

But from (10) this means that $u_j + \bar{u}_j = x_j$ and

$$K = F - \sum_1^n x_k \xi_k \tag{14}$$

To summarize, we have $F(x_{n+1}, \ldots, x_{2n}, \xi_1, \ldots, \xi_n)$ and from the complex Lagrangian submanifold structure of M we obtain from (10) for $1 \leq j \leq n$,

$$x_j = \frac{\partial F}{\partial \xi_j}, \qquad \xi_{n+j} = -\frac{\partial F}{\partial x_{n+j}} \tag{15}$$

From the Ω_1-Lagrangian description of M we also have, for $1 \leq j \leq 2n$,

$$\xi_j = \frac{\partial \phi}{\partial x_j} \tag{16}$$

so in particular we can write (14) as

$$K = F - \sum_1^n x_k \frac{\partial \phi}{\partial x_k}$$

Differentiating with respect to x_j for $1 \leq j \leq n$ we have

$$\begin{aligned}
\frac{\partial K}{\partial x_j} &= \sum_1^n \frac{\partial F}{\partial \xi_k}\frac{\partial \xi_k}{\partial x_j} - \frac{\partial \phi}{\partial x_j} - \sum_1^n x_k \frac{\partial^2 \phi}{\partial x_k \partial x_j} \\
&= \sum_1^n x_k \frac{\partial^2 \phi}{\partial x_k \partial x_j} - \frac{\partial \phi}{\partial x_j} - \sum_1^n x_k \frac{\partial^2 \phi}{\partial x_k \partial x_j} \\
&= -\frac{\partial \phi}{\partial x_j}
\end{aligned}$$

using (15) and (16) and the fact that F is independent of x_j for $1 \leq j \leq n$. Similarly for $n+1 \leq j \leq 2n$,

$$\begin{aligned}
\frac{\partial K}{\partial x_j} &= \sum_1^n \frac{\partial F}{\partial \xi_k}\frac{\partial \xi_k}{\partial x_j} + \frac{\partial F}{\partial x_j} - \sum_1^n x_k \frac{\partial^2 \phi}{\partial x_k \partial x_j} \\
&= \sum_1^n x_k \frac{\partial^2 \phi}{\partial x_k \partial x_j} - \xi_j - \sum_1^n x_k \frac{\partial^2 \phi}{\partial x_k \partial x_j} \\
&= -\frac{\partial \phi}{\partial x_j}
\end{aligned}$$

Hence $K = -\phi$ modulo an additive constant. Since ϕ is a Kähler potential from Proposition 5, we have the same metric (taking into account a difference of sign convention).

Remarks.

1. To globalise this metric presents some choice. One could, as in [6], define it on T^*M. Its local structure is, however, that of a principal bundle with structure group a translation group. As such it has no geometrically distinguished zero section. In the context of complex Lagrangian submanifolds, it can be defined on the space of pairs of a complex Lagrangian submanifold together with a line bundle of fixed Chern class over it. In this context it is defined on an open set of Markman's moduli space [4] of Lagrangian sheaves, which is itself an integrable system.

2. In [8] a Kähler metric on the moduli space of pairs consisting of a special Lagrangian submanifold of a Calabi-Yau manifold together with a flat line bundle was defined and conjectured to be itself Calabi-Yau. In the case that the Calabi-Yau is hyperkähler and the special Lagrangian submanifold is complex Lagrangian with respect to one of the complex structures, this metric is precisely the one defined above. Since it is hyperkähler it is *a fortiori* Calabi-Yau.

3. From the point of view put forward in this paper we have travelled from a complex Lagrangian submanifold of $\mathbf{C}^{2n}$ (Remark 1 of Section 4) to a hyperkähler metric. This is essentially the route followed by Cortes in [3].

References

[1] M. F. Atiyah & N. J. Hitchin, *The geometry and dynamics of magnetic monopoles*, M. B. Porter Lectures, Princeton University Press, Princeton, NJ, 1988.

[2] S. Cecotti, S. Ferrara & L. Girardello, *Geometry of type II superstrings and the moduli space of superconformal theories*, Int. J. Mod. Phys. **A4** (1989) 2475–2529.

[3] V. Cortes, *On Hyper Kähler manifolds associated to Lagrangean Kähler submanifolds of* $T^*\mathbf{C}^n$, Trans. Amer. Math. Soc. **350** (1998) 3193–3205.

[4] R. Donagi & E. Markman, *Spectral covers, algebraically completely integrable Hamiltonian systems and moduli of bundles*, Integrable Systems and Quantum Groups (Montecatini Terme, 1993), Lecture Notes in Math., Springer, Berlin, Vol. 1620, 1996, 1–119.

[5] R. Donagi, *Seiberg-Witten integrable systems*, Proc. Sympos. Pure Math., Amer. Math. Soc., Providence, RI, Vol. 62, Part 2, 1997.

[6] D. Freed, *Special Kähler manifolds*, Comm. Math. Phys. **203** (1999) 31–52.

[7] N. J. Hitchin, A. Karlhede, U. Lindström & M. Roček, *Hyperkähler metrics and supersymmetry*, Comm. Math. Phys. **108** (1987) 535–589.

[8] N. J. Hitchin, *The moduli space of Special Lagrangian submanifolds*, Dedicated to Ennio De Giorgi, Ann. Scuola Norm. Sup. Pisa Cl. Sci. **25** (1997) No. 3–4, 503–515 (1998).

[9] U .Lindström & M. Roček, *Scalar tensor duality and $N = 1, 2$ nonlinear σ models*, Nuclear Phys. **B 222** (1983) 285–308.

[10] Zhiqin Lu, *A note on special Kähler manifolds*, Math. Ann. **313** (1999) 711–713.

[11] R. C. McLean, *Deformations of calibrated submanifolds*, Comm. Anal. Geom. **6** (1997) 705–747.

[12] C. Simpson, *Higgs bundles and local systems*, Inst. Hautes Études Sci. Publ. Math. **75** (1992) 5–95.

[13] N. Seiberg & E. Witten, *Monopoles, duality and chiral symmetry breaking in $N = 2$ supersymmetric QCD*, Nuclear Phys. **B431** (1994) 484–550.

[14] A. Strominger, S.-T. Yau & E. Zaslow, *Mirror symmetry is T-duality*, Nuclear Phys. **B479** (1996) 243–259.

[15] C. Voisin, *Sur la stabilité des sous-variétés lagrangiennes des variétés symplectiques holomorphes*, Complex projective geometry (Trieste 1989/Bergen 1989), London Math. Soc. Lecture Notes, Cambridge Univ. Press, Cambridge, Vol. 179, 1992, 294–303.

Mathematical Institute, Oxford, England

Surveys in Differential Geometry, 2000
Vol. VII ©2000, International Press
pp. 347-373

WHICH SINGER IS THAT?

RICHARD V. KADISON

Most of us learn to crawl and walk; a few of us learn to run swiftly. Is Singer is one of those people — he ran swiftly and still does, as this is written. For those who doubt that, I invite them to try a set of tennis with him! Singer was visiting us (at Penn from MIT) for a few days while he gave us some lectures. Walking, with one of our graduate students, in a department hallway that had a large glass window at one end, I reached that window and looked out at some university tennis courts below. Singer was there doing what he often does as he travels: playing a set with a local pro. That student (now a famous homological algebraist) and I watched as Singer fired aces to one side of the court and the other. I asked my young, student friend, "Who is that fellow out there?" He didn't know. "That's I. M. Singer," I said. "Gosh! All that and the Atiyah–Singer Index Theorem, too!" he exclaimed. We were both impressed. As Singer runs, he also takes mathematics, and, often, physics with him along the paths he follows.

Everyone acquainted with the major developments of research mathematics in the last third of the twentieth century has had contact with the Atiyah–Singer Index Theorem. There are people who have referred to it as the "best" or "most important" theorem of the twentieth century. I have called it that, but I have heard that from other people, as well. Such declarations may not have very clear meanings; my own is based on some "absolute" feeling for depth, scope, and applicability. A good guess is that the Atiyah–Singer Index Theorem would appear on virtually every broadly educated, research mathematician's list of the ten most important theorems of the twentieth century. Each such list will contain personal favorites. Mine might include the Murray–von Neumann theorems on existence and uniqueness of the additive trace in factors of type II_1 and the uniqueness of the hyperfinite II_1 factor, as well as Tomita's Main Theorem (in the Tomita–Takesaki Theory) — to which many of the readers might respond "Huh?" My list would also include the Spectral Theorem, and in an *ecumenical spirit*, The Allendorfer, Chern–Weil versions of Werner Fenchel's original extension of the Gauss–Bonnet theorem to higher dimensions. "Huh?" would probably not be heard in connection with these latter choices, and certainly not with the Atiyah–Singer Index Theorem. The Atiyah–Singer Index Theorem is at the core of a vast body of work, created by Atiyah and Singer, which has touched and influenced most of current mathematics and much of theoretical physics.

Since pure mathematics is, in my view, the poetry of basic science, it's not surprising that its results and advances are less accessible to the general public than those of, say, biology and chemistry. Nevertheless, if we were seeking a result in those areas with an effect analogous to the discovery of the Atiyah–Singer Index, we might point to the Crick–Watson discovery of the double helix nature of DNA. Where the Crick–Watson work discloses something of the fundamental biology of life, the Atiyah–Singer Index Theorem reveals something as fundamental as the interplay among the topological, geometric and analytic patterns in the fabric of our universe. There is also something analogous in the process by which both results were discovered. Two superbly talented scientists, with a very clear view of where they are headed, a sure knowledge of what they want to achieve, and a firm grasp of the techniques they need, organize the large scientific enterprise needed to arrive at their goal and orchestrate the results and methods used from the many component subdisciplines into a powerful solution of their major problems.

The Atiyah–Singer Index Theorem extended the algebraic Riemann–Roch Theorem to complex manifolds using analytic techniques in the tradition of Hodge, Weyl, and Kodaira. It produced new topological invariants that topologists are still challenged to describe by traditional methods, thirty years later. Atiyah and Singer provided a unified treatment of the Riemann–Roch, Hirzebruch Signature, and Gauss–Bonnet Theorems by their use of the Dirac operator. The Index Theorem removed many barriers between algebraic geometry, differential geometry, topology, and analysis.

They gave two proofs. The first proof involved a cobordism argument that required the solution to an elliptic boundary value problem. The methods they discovered remain valuable today. Their second proof was an axiomatic treatment of the topological and analytic index. This approach lends itself to a natural extension to operator algebras and noncommutative geometry. A third proof, by P. Gilkey, is based on the heat equation approach to index theory, initiated by H. McKean and Singer. There is a masterful presentation by Atiyah, Bott, and Patodi. The Feynman-Kac formula expresses the heat kernel as a path integral. Using supersymmetric path integrals, Witten derived the index formula for the Dirac operator in a simple, elegant way .

Because the Index Theorem, its statement and proofs, encompasses so much mathematics, it has had a great impact on virtually every area of modern mathematics. Among the topics to which it has had crucial application, one can list: invariants of actions of groups on manifolds — the fixed point formula, families of elliptic operators and the determinant line bundle, the value of Hecke L-series at 0 in number theory, spectral flow and the theory of anomalies in physics, K-homology in operator theory, gauge theories as applied to three- and four-dimensional topology, the non-existence of spaces of positive scalar curvature in differential geometry.

The last paragraph of a book review (N. Hitchin, BAMS, Vol. 15, 1986, pp. 243-245) sums up, very nicely, what many of us feel on the subject of the Atiyah–Singer Index Theorem. "Like Stonehenge, the theorem stands there as an immovable edifice, with each generation giving its own interpretation. For one it is a computational device, for another a more mystical representation of supersymmetry. Either way, it has created a bridge between mathematics and physics and has given mathematicians and physicists a deeper, or at least more sympathetic, understanding of each other's work. The Dirac operator will never be reinvented a third time!"

Given the monumental nature of the work just mentioned, it is surprising to realize that Singer has run swiftly in other directions. He is recognized as one of the great geometers of our time. Few of the younger geometers are aware that the geometry of higher-dimensional manifolds was anything but the smoothly functioning apparatus that they know today. Of course, E. Cartan was the great initial developer and set the theory on paper in a form that contained most of the important beginning ideas; but that account had a noticeably different form from what one sees today. Chern understood, in the deepest sense, what Cartan was saying and taught it to us, along with his own deep contributions, in our graduate student days (1949) at the University of Chicago. It was a wonderful experience. Cartan's largely descriptive account (collected works) now found itself in a working mathematical form, though still far from the precise style available to us today. It must be remembered that Norman Steenrod was in the process of developing and writing his celebrated book on fiber bundles (Princeton University Press 1951), a theory implicit in Elie Cartan's treatment. We must also recall that there were no copying machines available, so preliminary copies of vital manuscripts did not "cover" the mathematical landscape as they do today (in both paper and electronic form). In that environment, Singer and his more senior colleague at MIT, Warren Ambrose, armed with the notes from the Chern course, undertook to make precise mathematical sense of connections, holonomy, parallel translation and all the other key concepts of differential geometry that were cloaked in mystery for all but a handful of "initiates." Independently, Ch. Ehresmann and they produced the form of the subject substantially as it is today. From my own personal observation of that process, it was an heroic effort. Their paper, "A Theorem on Holonomy" (TAMS Vol. 75, 1953, pp. 428–443), remains a classic account of that theory. At the same time, their Ph. D. students have written several "best sellers" based on the lecture notes of his courses (among them, Bishop-Crittenden, Hicks, and Warner).

Singer's current research in string theory and other differential geometric aspects of quantum physics is well known and highly respected in a large segment of the mathematics and theoretical physics communities. His work in that area earned him the Wigner Medal (1988) for contributions to theoretical physics. It is unusual for a pure mathematician to have students (*e.g.*, D. Freed, D. Friedan, and J. Lott) who have contributed significantly to high energy theoretical physics.

Even less known is Singer's powerful influence on the mid-century development of functional analysis. In the early fifties, during his association with UCLA, he teamed with Richard Arens to usher in a new era in the study of commutative Banach Algebras. This study broadens the scope of the theory of several complex variables and recasts it in the framework of functional analysis. Some of Singer's Ph. D. students and postdocs, notably Hugo Rossi and Ken Hoffman, became leading researchers in this area of analysis.

The background for the title of this article is an incident that occurred when Jacques Dixmier was revising his von Neumann algebra book. I was visiting Paris and spent an afternoon with Dixmier discussing some of the additions he wanted to make. In particular, I told him about Singer's early contribution to the subject of derivations of operator algebras (more about that, shortly). He interrupted me during that description to ask, "Which Singer is that?" I was puzzled and asked him what he meant. On an earlier occasion, I had mentioned to him that I often "wrote up" our joint papers for publication "since Singer had trouble writing them." He replied that there were the Singers who did differential geometry, commutative Banach algebras, operator algebras (and others). I don't know how, but I managed to display no more than a smile, and responded, "They are all the same Singer." Dixmier mused for a moment and said, "No wonder he has trouble writing up papers!"

One evening, at a 1953 conference, Irving Kaplansky asked Is Singer what he thought the derivations of $C(X)$ (the algebra of all continuous, complex-valued, functions on the compact Hausdorff space X under pointwise operations and supremum norm) were. The next day, Is showed us a sweet little argument that each such derivation is 0. I can't resist giving it here! To recall, a linear mapping δ of $C(X)$ into itself satisfying the Leibnitz rule (for differentiation of products), $\delta(fg) = \delta(f)g + f\delta(g)$, is called a *derivation* (of $C(X)$ into itself). Since each f in $C(X)$ is the sum of a "real" and "purely imaginary" function in $C(X)$ it suffices to show that $\delta(f) = 0$ for each real f in $C(X)$. Of course, $\delta(1) = \delta(1^2) = 2\delta(1)$, where '1' denotes the function whose value is 1 at each point of X. Thus $\delta(1) = 0$; by linearity, $\delta(a) = 0$ for each constant function a. Given a point x in X, $\delta(f - f(x)) = \delta(f)$. Let h be $f - f(x)$, h_+ be $\frac{1}{2}(|h| + h)$, and h_- be $\frac{1}{2}(|h| - h)$. Then h_+ and h_- are positive functions in $C(X)$, $h_+ - h_- = h$, $h_+h_- = 0$, and $h_+(x) = h_-(x) = 0$. Now, $h_+ = g^2$ for some (positive) g in $C(X)$. Then $g(x) = 0$, and $\delta(h_+)(x) = 2g(x)\delta(g)(x) = 0$. Similarly, $\delta(h_-)(x) = 0$. Thus $\delta(f)(x) = \delta(h)(x) = 0$. Since x was arbitrarily chosen in X, $\delta(f) = 0$. Hence $\delta = 0$. Kaplansky went on from there to write his famous paper [11] showing (among other things) that all derivations of type I von Neumann algebras are inner. Singer and Wermer [19] brought Singer's argument into a commutative Banach algebra context and extended it. A veritable army of researchers took the theory of derivations of operator algebras to dizzying heights — producing a theory of cohomology of operator algebras as well as much information about automorphisms of operator algebras. It all started with Kaplansky's thoughts and Singer's argument nearly fifty years ago.

Along with their efforts to put global differential geometry on a firm foundation and make it broadly accessible, Ambrose and Singer concentrated on understanding the basic structure of the Murray–von Neumann factors of type II_1. They were trying to display such a factor as a "matrix algebra" relative to an appropriate "orthonormal basis." In this instance there are complex entries at each "point" of the matrix — the rows and columns are thought of as the unit interval $[0, 1]$ and each has an appropriate measure on it. Multiplication of elements in the factor becomes matrix multiplication with the row-colunm product being integrated rather than summed. The mathematical problems they encountered were daunting; one of their most baffling questions was answered just a few years ago — more than 45 years after it was posed. What remains in print is a single-sentence abstract, "L_2-matrices are studied," and some reference to the project in [18].

It is Singer's role in the early development of the theory of operator algebras that is the primary focus of this article; it is a role that is too little known outside the field of operator algebras. I had better advise the reader that, at this point, the account turns into what has become known as a "research-expository" article. Let's pause, first, to establish some notation and background information. With $\mathcal{H}$ a Hilbert space over the complex numbers $\mathbb{C}$, we denote by '$\mathcal{B}(\mathcal{H})$' the family of all linear transformations of $\mathcal{H}$ into itself ("operators") continuous relative to the metric topology on $\mathcal{H}$ induced by the norm that assigns to each x in $\mathcal{H}$ its length ($\|x\| =$) $< x, x >^{\frac{1}{2}}$, where $< u, v >$ is the inner product of u and v in $\mathcal{H}$. If T is in $\mathcal{B}(\mathcal{H})$, $\sup\{\|Tx\| : \|x\| \leq 1\}$ ($= \|T\|$) is finite (T is "bounded" with "bound" or "norm" $\|T\|$). The metric topology on $\mathcal{B}(\mathcal{H})$ associated with the norm $T \to \|T\|$ is called the *norm* or *uniform* topology. Equipped with this norm, $\mathcal{B}(\mathcal{H})$ is a complete normed space, a Banach space.

The usual operations of addition, multiplication by a scalar, and multiplication (iteration) of linear transformations of a vector space into itself provide $\mathcal{B}(\mathcal{H})$ with the structure of an associative algebra. It has a unit element I, the *identity* operator, that assigns x to each x is $\mathcal{H}$. Each T in $\mathcal{B}(\mathcal{H})$ has associated with it a unique operator T^*, called its *adjoint*, characterized by the equality $< Tx, y > = < x, T^*y >$ for all x and y in $\mathcal{H}$.

The properties, $(aA + B)^* = \bar{a}A^* + B^*$, $(AB)^* = B^*A^*$, $(A^*)^* = A$, $\|T\| = \|T^*\|$, and $\|T^*T\| = \|T\|^2$ are established by simple computations. If $\mathcal{F}$ is a subset of $\mathcal{B}(\mathcal{H})$, we denote by '$\mathcal{F}^*$' the family $\{T^* : T \in \mathcal{F}\}$ and say that $\mathcal{F}$ is *self-adjoint* when $\mathcal{F} = \mathcal{F}^*$. In particular, a self-adjoint subalgebra $\mathfrak{A}$ of $\mathcal{B}(\mathcal{H})$ (called a *self-adjoint operator algebra*) is a *C^*-algebra* when it is norm closed in $\mathcal{B}(\mathcal{H})$. The topology on $\mathcal{B}(\mathcal{H})$ corresponding to *strong-operator convergence* ($A_n \to A$ when $A_n x \to Ax$, in the metric of $\mathcal{H}$, for each x in $\mathcal{H}$) is *the strong-operator topology* on $\mathcal{B}(\mathcal{H})$.

The *von Neumann algebras* are the self-adjoint operator algebras on a Hilbert space, containing I, that are strong-operator closed. Those whose centers consist of just the scalar multiples of I are called *factors*. The von Neumann algebras were introduced in [16] (as "rings of operators"), where it is proved that for each such $\mathcal{R}$, $\mathcal{R} = \mathcal{R}''$. (The *commutant* $\mathcal{F}'$ of a subset $\mathcal{F}$ of $\mathcal{B}(\mathcal{H})$ is $\{T : T \in \mathcal{B}(\mathcal{H}),\ TA = AT \text{ for all } A \text{ in } \mathcal{F}\}$.)

The factors were studied, intensively, in a series of papers [12], [13], [14] and [15] published between 1936 and 1943. Of course, $\mathcal{B}(\mathcal{H})$ itself is an example of a self-adjoint operator algebra, C*-algebra, von Neumann algebra, and factor. The operators that "project" a vector in $\mathcal{H}$, orthogonally, onto a given (closed) subspace of $\mathcal{H}$ are called *projections*. Each projection E lies in $\mathcal{B}(\mathcal{H})$, $\|E\| = 1$ (0 when E is 0), $E = E^*$, and $E^2 = E$. The last two properties characterize the projections in $\mathcal{B}(\mathcal{H})$. An operator A in $\mathcal{B}(\mathcal{H})$ *commutes* with a projection E if and only if A and A^* leave the space on which E projects (its range) invariant. Each von Neumann algebra is the norm closure of the linear span of its projections. In particular, there are many projections in a von Neumann algebra, while a C*-algebra may have no projections other than 0 and I. The projections are ordered by the size of their ranges: $E \le F$ when $E(\mathcal{H}) \subseteq F(\mathcal{H})$. This is equivalent to the equality, $EF = E$ (and agrees with their ordering as self-adjoint operators). If E lies in a von Neumann algebra $\mathcal{R}$, is non-zero, and no smaller projection in $\mathcal{R}$ distinct from it is non-zero, we call E a *minimal projection* in $\mathcal{R}$. One of the earliest results proved by Murray and von Neumann classifies factors that have a minimal projection.

Theorem. *A factor that has a minimal projection is isomorphic to $\mathcal{B}(\mathcal{H})$ for some Hilbert space $\mathcal{H}$.*

Murray and von Neumann found examples of factors without minimal projections. The first construction of such examples [12] employ a countable group G, with unit e, of one-to-one, measurability-and-measure-zero-preserving transformations of a (countably separated, σ-finite) measure space $\mathcal{S}$ (with measure μ). The action is *free* (each transformation, other than e, has a fixed-point set of measure 0). Using the Radon–Nikodým derivative of μ, transformed by a group element g of G, relative to μ, we can associate with g a unitary operator U_g on the Hilbert space $L_2(\mathcal{S}, \mu)$ $(= \mathcal{H})$. If the transformations are measure preserving, the Radon–Nikodým derivatives are (1 and) not needed.

Let $\mathcal{K}$ be the linear space of functions ϕ from G to $\mathcal{H}$, under pointwise addition and multiplication by scalars, for which $\sum_{g\in G} \|\phi(g)\|^2 < \infty$. Provided with the inner product $<\phi, \psi> = \sum_{g\in G} <\phi(g), \psi(g)>$, $\mathcal{K}$ becomes a Hilbert space. Of course, $\mathcal{K}$ may be identified with the direct sum of copies $\mathcal{H}_g$ of $\mathcal{H}$, g in G. Thus operators in $\mathcal{B}(\mathcal{K})$ have representations as matrices with rows and columns indexed by the elements of G and entries from $\mathcal{B}(\mathcal{H})$. With T in $\mathcal{B}(\mathcal{H})$, let $\Phi(T)$ be the operator on $\mathcal{K}$ whose matrix has T at each diagonal entry and 0 at all other entries (so $(\Phi(T)(\phi))(g) = T(\phi(g))$ $(\phi \in \mathcal{K},\ g \in G)$).

With f a bounded measurable function on $\mathcal{S}$, we denote by 'M_f' the (*multiplication*) operator that assigns $f \cdot h$ to h in $\mathcal{H}$. The family $\mathcal{A}_0$ of all such multiplication operators is an abelian von Neumann subalgebra of $\mathcal{B}(\mathcal{H})$. Each T in $\mathcal{B}(\mathcal{H})$ commuting with all the elements of $\mathcal{A}_0$ is a multiplication operator, whence $\mathcal{A}_0$ is a *maximal abelian, self-adjoint subalgebra* of $\mathcal{B}(\mathcal{H})$ (a *masa*). With f a function on $\mathcal{S}$ and g in G, let $f_g(s)$ be $f(g^{-1}(s))$, where '$g(s)$' denotes the result of g acting on s in $\mathcal{S}$. Assuming that g acts by measure-preserving transformations, we note that

$$\begin{aligned}(U_g M_f U_g^{-1} h)(s) &= (M_f U_{g^{-1}} h)(g^{-1}(s)) = f(g^{-1}(s))(U_{g^{-1}} h)(g^{-1}(s)) \\ &= f_g(s) h(g(g^{-1}(s))) = (M_{f_g} h)(s),\end{aligned}$$

for each g in G, h in $\mathcal{H}$, s in $\mathcal{S}$, and bounded measurable f on $\mathcal{S}$. Thus $U_g M_f U_g^{-1} = M_{f_g}$. It follows that G acts by automorphisms ($M_f \to U_g M_f U_g^{-1} = M_{f_g}$) on $\mathcal{A}_0$.

Let $(V_g \phi)(g')$ be $U_g \phi(g^{-1} g')$ for all g and g' in G and ϕ in $\mathcal{K}$. The mapping $g \to V_g$ is a unitary representation of G on $\mathcal{K}$. Moreover,

$$\begin{aligned}[(V_g \Phi(M_f) V_g^{-1})\phi](g') &= U_g([\Phi(M_f) V_{g^{-1}}(\phi)](g^{-1} g')) \\ &= U_g[M_f(V_{g^{-1}}(\phi)(g^{-1} g'))] \\ &= U_g[M_f(U_{g^{-1}} \phi(g g^{-1} g'))] \\ &= M_{f_g} \phi(g') = (\Phi(M_{f_g})\phi)(g')\end{aligned}$$

for all g and g' in G, ϕ in $\mathcal{K}$, and bounded measurable f on $\mathcal{S}$. Thus $V_g \Phi(M_f) V_g^{-1} = \Phi(M_{f_g})$, and G acts by automorphisms ($\Phi(M_f) \to V_g \Phi(M_f) V_g^{-1} = \Phi(M_{f_g})$) of the abelian von Neumann algebra $\Phi(\mathcal{A}_0)$ on $\mathcal{K}$. The von Neumann algebra $\mathcal{R}$ generated by $\Phi(\mathcal{A}_0)$ on $\mathcal{K}$ and $\{V_g : g \in G\}$ provides us with the example we want. The von Neumann algebra $\mathcal{R}$ is a factor if and only if G acts "ergodically" on $\mathcal{S}$ (that is, $\cup_{g \in G} g(\mathcal{S}_0)$ or $\mathcal{S} \setminus \cup_{g \in G} g(\mathcal{S}_0)$ has measure 0 for each measurable set $\mathcal{S}_0$). When $\mathcal{S}$ has no atoms (no sets of positive measure without subsets of smaller positive measure), the factor has no minimal projections. If $\mu(\mathcal{S}) < \infty$, the factor is one of type II_1 — the factors that most of us studied in the early days of the subject. A specific example is given by the group of rotations of the circle, with Lebesgue measure, generated by a single rotation through an irrational multiple of π. A description of $\mathcal{R}$ in matrix terms follows. Let $(T_{p,q})$ be the matrix of T in $\mathcal{B}(\mathcal{K})$.

Theorem. *An operator T in $\mathcal{B}(\mathcal{K})$ lies in $\mathcal{R}$ if and only if there is a mapping $g \to A(g)$ from G into $\mathcal{A}_0$ such that $T_{p,q} = U_{pq^{-1}} A(pq^{-1})$. An* *operator T' lies in $\mathcal{R}'$ if and only if there is a mapping $g \to A'(g)$ of G into $\mathcal{A}_0$ such that $T'_{p,q} = U_p A'(q^{-1}p) U_{p^{-1}}$.*

Note that the diagonal entries $T_{p,p}$ of each T in $\mathcal{R}$ are equal to a single element $A(e)$ of $\mathcal{A}_0$, while those $T'_{p,p}$ of T' are the transforms $U_p A'(e) U_{p^{-1}}$ of a single element $A'(e)$ of $\mathcal{A}_0$. If we use the mapping $g \to A(g)$ that assigns 0 to each g other than e, the resulting operator T in $\mathcal{R}$ is $\Phi(A(e))$, the diagonal operator with $A(e)$ at each diagonal entry. It is not hard to show that the abelian von Neumann subalgebra $\Phi(\mathcal{A}_0)$ of $\mathcal{R}$ is a masa in $\mathcal{R}$, by using the assumption of free action of G on $\mathcal{S}$.

Let us suppose that G acts ergodically, $\mathcal{S}$ has no atoms, and $\mu(\mathcal{S}) = 1$. The algebra $\mathcal{M}$ we construct is a factor of type II_1 in this case. If u is the element of $\mathcal{K}$ that assigns the constant function 1 on $\mathcal{S}$ to e and 0 to each other g in G, the linear functional τ on $\mathcal{M}$ that takes the value $< Tu, u >$ ($= \int f \, d\mu$, where $A(e)$ in $\mathcal{A}_0$ is $\mathcal{M}_f$ in our preceding notation) has special properties. To begin with, it is a *state* of $\mathcal{M}$. A *state* on a C*-algebra is a linear functional on the algebra that takes non-negative, real values on positive operators and is 1 at I. States arising from vectors (as τ does from u) are called *vector states*. Using the facts that $U_g(1) = 1$, $U_g \mathcal{A}_0 U_g^* = \mathcal{A}_0$, and $\mathcal{A}_0$ is abelian, we also have that $\tau(AB) = \tau(BA)$. In this case, we call τ a *tracial state* of $\mathcal{M}$. In the explicit example of factors of type II_1, just described, we exhibit a tracial state in terms of the special construction. It is a deep fact (proved in [13]) that such a state exists for each factor of type II_1 and is unique. For present purposes, we may define the factors of type II_1 as those factors that have no minimal projections and satisfy the condition that $VV^* = I$ when $V^*V = I$ for some V in the factor. As innocent and insignificant as this condition may seem, it is a simple expression of the property that leads to a stunningly rich structure. The factors of type II_1 are at least as natural a replacement of the finite-dimensional matrix algebra $M_n(\mathbb{C})$, in the infinite dimensional case, as $\mathcal{B}(\mathcal{H})$ is, with $\mathcal{H}$ infinite dimensional. For one thing, the factors of type II_1 are simple algebras. For another, they have the all-important tracial state. If we divide the trace of a matrix in $M_n(\mathbb{C})$ by n, we arrive at the unique tracial state τ_n on $M_n(\mathbb{C})$. It assigns to projections in $M_n(\mathbb{C})$ one of the values $0, \frac{1}{n}, \frac{2}{n}, \ldots, 1$ and all these values are assumed. For a factor of type II_1, the tracial state assigns all values in $[0, 1]$ to the projections in the factor. "Discrete" dimensionality has been replaced by "continuous" dimensionality in the infinite-dimensional case; it makes serious sense to speak of projections with dimensions $\sqrt{2}/2$ and $1/e$ in factors of type II_1.

As remarked, earlier, Ambrose and Singer undertook to represent a factor of type II_1 as a "matrix algebra" with complex entries at each point of a square and measures associated with each column and row. To go from transformations to matrices requires the choice of a basis — an orthonormal basis, when the adjoint operation is part of the structure considered. In view of "non-atomicity" of factors of type II_1, we cannot allow ourselves the luxury of an orthonormal basis of vectors. An orthonormal basis $e_1, e_2, \ldots$, for a (separable) Hilbert space $\mathcal{H}$ is determined, up to a phase factor (that is, a c in $\mathbb{C}$ of modulus 1), by the family of one-dimensional projections $E_1, E_2, \ldots$, there the range $E_j(\mathcal{H})$ of E_j is spanned by e_j. The family $\{E_j\}$ generates an abelian von Neumann algebra in which $\{E_j\}$ is precisely the family of minimal projections. Relative to the basis $\{e_j\}$, the operators in the algebra are precisely those with diagonal matrices. That algebra is a masa in $\mathcal{B}(\mathcal{H})$. In this sense, we may speak of each masa $\mathcal{A}$ in $\mathcal{B}(\mathcal{H})$ as a "generalized orthonormal basis." In this instance, some (or all) of the "vectors" in the basis may correspond to "Dirac delta functions." In the same way, an "orthonormal basis," where the factor $\mathcal{M}$ of type II_1 replaces $\mathcal{B}(\mathcal{H})$, is a masa $\mathcal{A}$ in $\mathcal{M}$.

Applying a special process (the "GNS construction") to the tracial state τ of a factor $\mathcal{M}$ of type II_1, we "represent" $\mathcal{M}$ as a factor of type II_1 acting on a Hilbert space $\mathcal{H}$ in such a way that τ is the vector state of $\mathcal{M}$ corresponding to a (unit) vector u whose transforms under the elements of $\mathcal{M}$ form a dense submanifold of $\mathcal{H}$. (We say that u is a *cyclic* or *generating* vector for $\mathcal{M}$.) In this situation, $\mathcal{M}'$ is also a factor of type II_1 and the tracial state of $\mathcal{M}'$ is the vector state corresponding to u. We say that $\mathcal{M}$ is in *standard form* in this case. For each T in $\mathcal{M}$, there is a unique T' in $\mathcal{M}'$ such that $Tu = T'u$. The mapping $T \to T'$ is an adjoint-preserving, anti-isomorphism of $\mathcal{M}$ onto $\mathcal{M}'$. In particular, the image $\mathcal{B}$ of a masa $\mathcal{A}$ in $\mathcal{M}$ under this mapping is a masa in $\mathcal{M}'$. Of course, the von Neumann algebra generated by $\mathcal{A}$ and $\mathcal{B}$ is abelian. Is it a masa in $\mathcal{B}(\mathcal{H})$? In the parallel situation of $\mathcal{M}_n(\mathbb{C})$ acting on $\mathcal{H}$ in such a way that its commutant is also (isomorphic to) $\mathcal{M}_n(\mathbb{C})$, there is a unit vector giving rise to the tracial states on each of the algebras and an adjoint-preserving, anti-isomorphism of the algebra onto its commutant. The abelian algebra generated by a masa and its image is a masa in $\mathcal{B}(\mathcal{H})$, in this case. As we shall show, shortly, the algebra generated algebraically by $\mathcal{A}$ and $\mathcal{B}$ is maximal abelian in the algebra generated algebraically by $\mathcal{M}$ and $\mathcal{M}'$. We shall also give an example showing that the von Neumann algebra generated by $\mathcal{A}$ and $\mathcal{B}$ need not be a masa in $\mathcal{B}(\mathcal{H})$ (the von Neumann algebra generated by the factors $\mathcal{M}$ and $\mathcal{M}'$). As Ambrose and Singer discovered, when $\mathcal{A}$ and $\mathcal{B}$ generate a masa in $\mathcal{B}(\mathcal{H})$, $\mathcal{A}$ serves as a particularly useful "orthonormal basis" for their matrix representation. (It could be effected in terms of "single-sheeted," rather than "multiple-sheeted" matrices.) They called such a masa in $\mathcal{M}$ *simple* and knew that the masa $\Phi(\mathcal{A}_0)$ of the measure-theoretic construction we discussed is simple. Returning to that example, we note that if T in $\mathcal{M}$ is positive, then $T_{e,e}$ $(= A(e))$ is positive (in $\mathcal{B}(\mathcal{H})$). If $< Tu, u >= 0$, and $A(e) = M_f$, then $f(s) \geq 0$ for almost all s in $\mathcal{S}$ and $\int f\, d\mu = 0$. Hence f is 0 almost everywhere on $\mathcal{S}$, and $A(e) = 0$. The diagonal entries $T_{p,p}$ of T are all 0 and $T = 0$ (since T is assumed positive). It follows that S in $\mathcal{M}$ is 0 when $Su = 0$. We say that u is a *separating vector* for $\mathcal{M}$, in this case. This implies that $\mathcal{M}'u$ is dense in $\mathcal{K}$, that is u is generating for $\mathcal{M}'$. Suppose $< T'u, u >= 0$ for some positive T' in $\mathcal{M}'$. If $T'_{p,q} = U_pA'(q^{-1}p)U_{p^{-1}}$ and $A'(e) = M_h$, then $T'_{e,e} = A'(e) = M_h \geq 0$ and $h(s) \geq 0$ for almost all s in $\mathcal{S}$. Moreover, $0 =< T'u, u >= \int h\, d\mu$, and h is 0 almost everywhere. Thus $0 = A'(e) = U_pA'(e)U_{p^{-1}} = T'_{p,p}$ and $T' = 0$. Hence u is separating for $\mathcal{M}'$. It follows that u is generating for $\mathcal{M}$ $(= \mathcal{M}'')$. When a *trace vector* u is generating for a II_1 factor $\mathcal{M}$, u is also a generating trace vector for $\mathcal{M}'$.

The mapping $g \to A'(g)$ that assigns 0 to each g other than e, produces the operator T' in $\mathcal{M}'$ with matrix whose diagonal entry $T'_{p,p}$ is $U_pA'(e)U_{p^{-1}}$ and all off-diagonal entries are 0. If $T = \Phi(A(e))$, $A(e) = M_f = A'(e)$, then Tu and $T'u$ are the vector in $\mathcal{K}$ that assigns f to e and 0 to each other g in G. In particular, $Tu = T'u$. The adjoint-preserving, anti-isomorphism of $\mathcal{M}$ onto $\mathcal{M}'$ corresponding to u maps T to T' and transforms $\Phi(\mathcal{A}_0)$ onto the algebra $\mathcal{B}$ of diagonal matrices in $\mathcal{M}'$. Thus $\mathcal{B}$ is a masa in $\mathcal{M}'$. We note, finally, that $\Phi(\mathcal{A}_0)$ and $\mathcal{B}$ generate a masa in $\mathcal{B}(\mathcal{K})$. If S, with matrix $(S_{p,q})$, commutes with $\Phi(\mathcal{A}_0)$, then each $S_{p,q}$ is a multiplication operator since $\mathcal{A}_0$, the algebra of multiplication operators on $L_2(\mathcal{S}, \mu)$, is maximal abelian in $\mathcal{B}(\mathcal{H})$. If, in addition, S commutes with all diagonal matrices in $\mathcal{M}'$, then $S_{p,q}U_pEU_p^* = S_{p,q}U_qEU_q^*$ for each projection E in $\mathcal{A}_0$. Replacing E by $U_p^*EU_p$, we have that $S_{p,q}E = S_{p,q}U_gEU_g^*$, where $g = qp^{-1}$. If $p \neq q$, then $g \neq e$, and there is a non-zero subprojection E_0 of E in $\mathcal{A}_0$ such that $E_0U_gE_0U_g^* = 0$ (from the freeness of the action of G on $\mathcal{S}$ [8, Lemma 8.6.5]). With E_0 in place of E, $S_{p,q}E_0 = S_{p,q}U_gE_0U_g^*$, whence $S_{p,q}E_0 = S_{p,q}E_0^2 = S_{p,q}U_gE_0U_g^*E_0 = 0$. Each non-zero projection E in $\mathcal{A}_0$ has a non-zero subprojection E_0 such that $S_{p,q}E_0 = 0$ when $p \neq q$. By using Zorn's lemma, we find a maximal orthogonal family $\{E_a\}$ of projections in $\mathcal{A}_0$ such that $S_{p,q}E_a = 0$ for each a. If $I - \sum_a E_a$ were not 0, there would be a non-zero subprojection E_0 of it in $\mathcal{A}_0$ such that $S_{p,q}E_0 = 0$, contradicting the maximality of $\{E_a\}$. It follows that $S_{p,q} = S_{p,q}\sum_a E_a = \sum_a S_{p,q}E_a = 0$ (with convergence in the strong-operator topology), when $p \neq q$. Thus the algebra $\tilde{\mathcal{A}}'$ of operators in $\mathcal{B}(\mathcal{K})$ commuting with both $\Phi(\mathcal{A}_0)$ and $\mathcal{B}$ (hence, with the von Neumann algebra $\tilde{\mathcal{A}}$ they generate) is a subalgebra of the abelian algebra $\mathcal{D}$ of diagonal matrices with diagonal entries from $\mathcal{A}_0$. As $\tilde{\mathcal{A}}$ and $\tilde{\mathcal{A}}'$ are abelian, $\tilde{\mathcal{A}} \subseteq \tilde{\mathcal{A}}' \subseteq \tilde{\mathcal{A}}'' = \tilde{\mathcal{A}}$. Thus $\tilde{\mathcal{A}} = \tilde{\mathcal{A}}'$, $\tilde{\mathcal{A}}$ is maximal abelian in $\mathcal{B}(\mathcal{K})$, and $\Phi(\mathcal{A}_0)$ is a simple masa in $\mathcal{M}$. For our example of a masa that is not simple, we turn to a construction of factors of type II_1 introduced in [14]. We start with a countably infinite, discrete group and construct an operator-algebra group algebra. Let G be a countable (discrete) group and $\mathcal{H}$ be the separable Hilbert space $l_2(G)$, that is

$$\{\phi : \sum_{g \in G} |\phi(g)|^2 < \infty\}, \qquad < \phi, \psi >= \sum_{g \in G} \phi(g)\overline{\psi(g)}),$$

Let $(L_g\phi)(g')$ be $\phi(g^{-1}g')$ and $(R_g\phi)(g')$ be $\phi(g'g)$ for ϕ in $\mathcal{H}$. Then L_g and R_g are unitary operators. Let $\mathcal{L}_G$ and R_G (the *left and right-von Neumann-group algebras of* G) be the von Neumann algebras generated by $\{L_g\}$ and $\{R_g\}$, respectively.

Theorem. *$\mathcal{L}_G$ and $\mathcal{R}_G$ are factors iff all conjugacy classes in G but $\{e\}$ are infinite. In this case, $\mathcal{L}_G$ and $\mathcal{R}_G$ are factors of type* II_1, *$\mathcal{L}_G = \mathcal{R}'_G$, and $\mathcal{R}_G = \mathcal{L}'_G$.*

The free (non-abelian) group $\mathcal{F}_n$ on $n(> 1)$ generators and Π, the group of "finite" permutations of the integers, are examples of these *i.c.c groups.*

Murray and von Neumann took a crucial step, proving [14] that there are factors of type II_1 acting on separable Hilbert spaces that are not isomorphic. We now have examples of uncountably many non-isomorphic factors of type II_1 (as was to be expected after the Murray–Von Neumann result that follows).

Theorem. *$\mathcal{L}_{\mathcal{F}_n}$ is not isomorphic to $\mathcal{L}_\Pi$.*

As this is written, we do not know if $\mathcal{L}_{\mathcal{F}_2}$ is isomorphic to $\mathcal{L}_{\mathcal{F}_3}$, but deep work of Voiculescu, laying the foundations of a non-commutative, free probability theory, has given us such results as:

Theorem(Voiculescu). *The* II_1 *factors $\mathcal{L}_{\mathcal{F}_2}$ and $M_2(\mathcal{L}_{\mathcal{F}_5})$, the algebra of 2×2 matrices with entries from $\mathcal{L}_{\mathcal{F}_5}$, are isomorphic.*

This same work of Voiculescu provides the technical basis for a brilliant proof [6] that the factors $\mathcal{L}_{\mathcal{F}_n}$, among others, do not have simple masas. (This came after the proof of a difficult intermediate result by Voiculescu [20] that such factors do not possess a "Cartan subalgebra.")

Let $x_g(h)$ be 1 when $h = g$ and 0 otherwise. Then $\{x_g : g \in G\}$ is an orthornormal basis for $\mathcal{H}$ and each x_g is a (unit) trace vector for $\mathcal{L}_G$ and for $\mathcal{R}_G$ (that is, $< ABx_g, x_g >=< BAx_g, x_g >$ when $A, B \in \mathcal{L}_G$ or $A, B \in \mathcal{R}_G$). In general, each element of $\mathcal{L}_G$ ($\mathcal{R}_G$) is uniquely representable as $\sum_{g\in G} \eta_g L_g$ ($\sum_{g\in G} \eta_g R_g$), where the sum converges in the strong-operator topology over the net of finite subsums. Defining $\eta(g)$ to be η_g, $\eta \in l_2(G)$, but not each η in $l_2(G)$ appears in this way. Since $L_g x_e = x_g = R_{g^{-1}} x_e$, the anti-isomorphism $A \to A'$ of $\mathcal{L}_G$ onto $\mathcal{R}_G$ (*reflection about the trace vector* x_e) maps $\sum_{g\in G} \eta_g L_g$ onto $\sum_{g\in G} \eta'_g R_g$, where $\eta'_g = \eta_{g^{-1}}$

To complete our construction of a masa that is not simple, we choose $\mathcal{L}_{\mathcal{F}_2}$ for G. Let a and b be (free) generators of $\mathcal{F}_2$. We show that the algebra $\mathcal{A}$ generated by L_a in $\mathcal{L}_{\mathcal{F}_2}$ is a masa in $\mathcal{L}(\mathcal{F}_2)$. In any case, it consists of elements representable as $\sum \eta_g L_g$, where $\eta_g = 0$ unless $g = a^m$ for some integer m. Suppose $A = \sum \eta_g L_g$ and $L_a A = AL_a$. Then $\sum \eta_g L_{ag} = \sum \eta_g L_{ga}$. Thus $\eta_g = \eta_{aga^{-1}}$ for each g in G. So $\eta_g = \eta_{aga^{-1}} = \eta_{a^2 g a^{-2}} = \ldots = \eta_{a^n g a^{-n}}$ for each integer n. If $g \notin \{a^m : m \in \mathbb{Z}\}$, then $\{a^n g a^{-n} : n \in \mathbb{Z}\}$ is an infinite subset of $\mathcal{F}_2$. Since $\eta \in l_2(G)$, $\eta(g) = \eta_g = 0$, in this case. Thus $AL_a = L_a A$ if and only if $\eta_g = 0$ unless $g = a^m$ for some integer m. It follows that $\mathcal{A}$ is a masa in $\mathcal{L}_{\mathcal{F}_2}$. Of course, this argument and conclusion applies to the von Neumann subalgebra generated by any one of the (free) generators of $\mathcal{L}_{\mathcal{F}_n}$.

Theorem. *The masa generated by L_a in $\mathcal{L}_{\mathcal{F}_n}$ is not simple.*

Proof. With A in $\mathcal{B}(l_2(G))$, if $Ax_d = \sum_c \alpha_{c,d} x_c$, then $\alpha_{c,d}$ is the entry in row c and column d of the matrix for A relative to $\{x_g\}$. Since $L_a x_c = x_{ac}$, the matrix for L_a has a 1 in row ac at column c and 0 at all other entries of that column, for each c in $\mathcal{F}_2$. Similarly, $R_a x_c = x_{ca^{-1}}$, and the matrix for R_a has a 1 in row ca^{-1} at column c and 0 at all other entries of that column. Hence, with A as before, if $L_a A = AL_a$, then $\alpha_{c,d} = \alpha_{ac,ad}$ for all c, d in G. If $R_a A = AR_a$, then $\alpha_{c,d} = \alpha_{ca,da}$. Conversely, these conditions on the matrix of A imply commutativity with L_a and R_a.

Let $\mathcal{A}$ and $\mathcal{B}$ be the von Neumann subalgebras of $\mathcal{L}(\mathcal{F}_2)$ and $\mathcal{R}(\mathcal{F}_2)$ generated by L_a and R_a, respectively. Since $L_a x_e = x_a = R_{a^{-1}} x_e$, $\mathcal{B}$ is the reflection of $\mathcal{A}$ about the trace vector x_e. Let $\sum_{n,m=-k}^{k} \alpha_{n,m} L_{a^n} R_{a^m}$ be C. Such sums C form a weak-operator dense subalgebra of $\mathcal{A}_0$, the (abelian) von Neumann algebra generated by $\mathcal{A}$ and $\mathcal{B}$. Moreover, $< Cx_b, x_{b^2} >= 0$ for each such sum C. Thus $< Tx_b, x_{b^2} >= 0$ for each T in $\mathcal{A}_0$. Let A be the linear operator that maps $x_{a^n b a^m}$ to $x_{a^n b^2 a^m}$, for $n, m = 0, \pm 1, \pm 2, \ldots$ and x_c to 0 for each other c in $\mathcal{F}_2$. Then A is the product of a "permutation unitary" (relative to the basis $\{x_g\}$) and the projection onto the subspace generated by $\{x_{a^nba^m} : n, m \in \mathbb{Z}\}$. Thus $A \in \mathcal{B}(l_2(\mathcal{F}_2))$.

The matrix for A satisfies $\alpha_{c,d} = 1$ if $c = a^n b^2 a^m$ and $d = a^n b a^m$ for some n and m in $\mathbb{Z}$, otherwise, $\alpha_{c,d} = 0$. If $\alpha_{c,d} = 1$, then $\alpha_{ac,ad} = 1$. If $\alpha_{c,d} = 0$, then $\alpha_{ac,ad} = 0$. Similarly, $\alpha_{c,d} = \alpha_{ca,da}$ for all c and d in $\mathcal{F}_2$. Thus $AL_a = L_a A$, $AR_a = R_a A$, and $A \in \mathcal{A}_0'$. But $< Ax_b, x_{b^2} >= \alpha_{b^2,b} = 1$. Since $< Tx_b, x_{b^2} >= 0$ for all T in $\mathcal{A}_0$, $A \notin \mathcal{A}_0$. It follows that $\mathcal{A}_0$ is not maximal abelian in $\mathcal{B}(l_2(G))$ and $\mathcal{A}$ is not a simple masa in $\mathcal{L}_{\mathcal{F}_2}$. q.e.d.

This same argument applies to the abelian von Neumann subalgebra generated by L_a for each (free) generator of $\mathcal{F}_n$; each is a masa in $\mathcal{L}_{\mathcal{F}_n}$ but none is simple. To what extent does the finite-dimensional situation (where a masa in a factor and its reflection about a trace vector generate a masa in the algebra of all linear transformations on the finite-dimensional space) carry over to infinite dimensions? The theorem that follows shows that it does transfer in the algebraic sense. It is proved in fairly general terms. For the case we have been discussing, $\mathcal{R}$ and $\mathcal{S}$ should be taken to be the same factor of type II_1, and $\mathcal{R}'$ and $\mathcal{T}$ to be its commutant.

Theorem. *Let $\mathcal{R}$ be a von Neumann algebra, with center $\mathcal{Z}$, acting on a Hilbert space $\mathcal{H}$, $\mathcal{S}$ and $\mathcal{T}$ be von Neumann subalgebras, containing $\mathcal{Z}$, of $\mathcal{R}$ and $\mathcal{R}'$, respectively, and $\mathcal{A}$ and $\mathcal{B}$ be masas in $\mathcal{S}$ and $\mathcal{T}$, respectively. Then the algebra $\mathcal{C}$ generated by $\mathcal{A}$ and $\mathcal{B}$ is maximal abelian in the algebra $\mathcal{D}$ generated by $\mathcal{S}$ and $\mathcal{T}$.*

Proof. Let D be an element of $\mathcal{D}$ commuting with $\mathcal{C}$. Then $D = S_1T_1 + \cdots + S_nT_n$, for some $S_1, \ldots, S_n$ in $\mathcal{S}$ and some $T_1, \ldots, T_n$ in $\mathcal{T}$. Let $\tilde{S}$ be the $n \times n$ matrix whose first row is $\{S_1, \ldots, S_n\}$ and all of whose other entries are 0. Let $\tilde{\mathcal{H}}$ be the n-fold direct sum of $\mathcal{H}$ with itself and a be the norm of $\tilde{S}$ acting on $\tilde{\mathcal{H}}$. We wish to show that $D \in \mathcal{C}$ and, thence, that $\mathcal{C}$ is maximal abelian in $\mathcal{D}$. If $a = 0$, then $D = 0$, and $D \in \mathcal{C}$. We may assume that $a > 0$ and that $\|\tilde{S}\| = 1$, after multiplying each S_j by a^{-1} and T_j by a.

Let $\{A_1, \ldots, A_m\}$ be a finite subset of $\mathcal{A}$. Since $\mathcal{A} \subseteq \mathcal{C}$ and D commutes with $\mathcal{C}$, we have that

$$0 = A_1D - DA_1 = (A_1S_1 - S_1A_1)T_1 + \cdots + (A_1S_n - S_nA_1)T_n.$$

(Note, too, for this that each $A_j \in \mathcal{A} \subseteq \mathcal{S} \subseteq \mathcal{R}$, and each $T_j \in \mathcal{R}'$.) From [7, Theorem 5.5.4], there are operators C_{jk} in $\mathcal{Z}$ $(j, k \in \{1, \ldots, n\})$

such that the $n \times n$ matrix $\tilde{C}$, acting on $\tilde{\mathcal{H}}$, with C_{jk} as j, k entry, is an orthogonal projection and

$$\sum_{j=1}^{n}(A_1S_j - S_jA_1)C_{jk} = 0 \quad (k \in \{1, \dots, n\}),$$

$$\sum_{k=1}^{n} C_{jk}T_k = T_j \quad (j \in \{1, \dots, n\}).$$

Hence

$$A_1\left(\sum_{j=1}^{n} S_jC_{jk}\right) = \left(\sum_{j=1}^{n} S_jC_{jk}\right)A_1 \qquad (k = 1, \dots, n)$$

and

$$\sum_{k=1}^{n}\left(\sum_{j=1}^{n} S_jC_{jk}\right)T_k = \sum_{j=1}^{n} S_j\left(\sum_{k=1}^{n} C_{jk}T_k\right) = \sum_{j=1}^{n} S_jT_j.$$

Thus $\sum_{j=1}^{n} S_jT_j = \sum_{k=1}^{n} S_{k1}T_k$, where $S_{k1} = \sum_{j=1}^{n} S_jC_{jk}$ for k in $\{1, \dots, n\}$, and each $S_{k1} \in \mathcal{S}$ (since $\mathcal{Z} \subseteq \mathcal{S}$).

The matrix $\tilde{S}_1$ with first row $\{S_{11}, \dots, S_{n1}\}$ and all other entries 0 is $\tilde{S}\tilde{C}$. Thus $\|\tilde{S}\| \le 1$. With A_2 in place of A_1 and S_{k1} in place of S_k, proceeding as before, we find operators C'_{jk} in $\mathcal{Z}$ such that the $n \times n$ matrix with j, k entry C'_{jk} is an orthogonal projection on $\tilde{\mathcal{H}}$, $\sum_{j=1}^{n} S_{j1}C'_{jk}$ $(= S_{k2})$ commutes with A_2 and lies in $\mathcal{S}$, and the $n \times n$ matrix $\tilde{S}_2$ with first row $\{S_{12}, \dots, S_{n2}\}$ and all other entries 0 has norm not exceeding 1. In addition, $D = \sum_{k=1}^{n} S_{k2}T_k$ and each S_{k2} commutes with A_1, as well as A_2, since each S_{j1} and each C'_{jk} commute with A_1.

Continuing in this way, we construct operators $\{S_{1m}, \dots, S_{nm}\}$ in $\mathcal{S}$ such that each S_{jm} commutes with all of $A_1, \dots, A_m$, $D = \sum_{k=1}^{n} S_{km}T_k$, and the $n \times n$ matrix $\tilde{S}_m$ with $\{S_{1m} \dots, S_{nm}\}$ as first row and all other entries 0 has norm 1 or less. In general, if '$\mathcal{F}$' denotes the finite subset $\{A_1, \dots, A_m\}$ of $\mathcal{A}$, we write '$S_{j\mathcal{F}}$' in place of 'S_{jm}' and '$\tilde{S}_{\mathcal{F}}$' in place of '$\tilde{S}_m$.' Since $\|\tilde{S}_{\mathcal{F}}\| \le 1$ for each finite subset $\mathcal{F}$ of $\mathcal{A}$, each $S_{j\mathcal{F}}$ lies in $(\mathcal{S})_1$, the closed unit ball of $\mathcal{S}$.

The net $\{S_{j\mathcal{F}}\}_{\mathcal{F}\in\mathbb{A}}$, indexed by the family $\mathbb{A}$ of finite subsets of $\mathcal{A}$ ordered by inclusion has a weak-operator convergent cofinal subnet since $(\mathcal{S})_1$ is weak-operator compact. Starting with a convergent subnet $\{S_{1\mathcal{F}(1)}\}$ of $\{S_{1\mathcal{F}}\}$, passing to a convergent cofinal subnet $\{S_{2\mathcal{F}(2)}\}$ of $\{S_{2\mathcal{F}(1)}\}$ and, successively, to a convergent, cofinal subnet $\{S_{m\mathcal{F}(m)}\}$, we have that each cofinal subnet $\{S_{j\mathcal{F}(m)}\}$ of $\{S_{j\mathcal{F}}\}$ converges, in the weak-operator toplogy, to some A'_j in $\mathcal{S}$. We shall show that each $A'_j \in \mathcal{A}$ and that $D = A'_1T_1 + \cdots + A'_nT_n$.

If $A \in \mathcal{A}$, the terms of $\{S_{j\mathcal{F}(m)}\}$ such that $A \in \mathcal{F}(m)$ form a cofinal subnet of it, and each of these terms commutes with A, by construction of $S_{j\mathcal{F}(m)}$ (with A in $\mathcal{F}(m)$). Hence the weak-operator limit A'_j of this cofinal subnet commutes with A. Thus each A'_j commutes with $\mathcal{A}$. Since $A'_j \in \mathcal{S}$ and $\mathcal{A}$ is maximal abelian in $\mathcal{S}$, each $A'_j \in \mathcal{A}$.

For each finite subset $\mathcal{F}$ of $\mathcal{A}$, we have, by construction, that $\sum_{j=1}^{n} S_{j\mathcal{F}}T_j = D$. Thus

$$\sum_{j=1}^{n} < S_{j\mathcal{F}}T_j x, y >=< Dx, y > \qquad (x, y \in \mathcal{H}).$$

Passing to weak-operator limits over the appropriate subnet, we conclude that

$$\left\langle \left(\sum_{j=1}^{n} A'_j T_j \right) x, y \right\rangle =< Dx, y > \qquad (x, y \in \mathcal{H}).$$

Thus $D = A'_1 T_1 + \cdots + A'_n T_n$.

Applying what we have just proved, with $\mathcal{S}$ and $\mathcal{T}$ interchanged, $\mathcal{A}$ and $\mathcal{B}$ interchanged, and A'_j in place of S_j, we see that there are operators $B'_1, \ldots, B'_n$ in $\mathcal{B}$ with the property that $D = A'_1 B'_1 + \cdots + A'_n B'_n \in \mathcal{C}$. Hence $\mathcal{C}$ is maximal abelian in $\mathcal{D}$. q.e.d.

If we limit the scope of the preceding theorem by assuming that $\mathcal{H}$ is separable, then $\mathcal{A}$ is generated (as a von Neumann algebra) by a single self-adjoint operator A. With A in place of A_1, we conclude that $\sum_{j=1}^{n} S_j C_{jk} \in \mathcal{A}$, for each k in $\{1, \ldots, n\}$. Letting A'_k be $\sum_{j=1}^{n} S_j C_{jk}$, we arrive at the equality $\sum_{k=1}^{n} A'_k T_k = D$ without the need to introduce nets.

Is $\mathcal{C}^=$, the norm closure of $\mathcal{C}$, (the C*-algebra generated by $\mathcal{A}$ and $\mathcal{B}$) maximal abelian in $\mathcal{D}^=$, the C*-algebra generated by $\mathcal{S}$ and $\mathcal{T}$?

While almost nothing of the Ambrose–Singer project for representing a II_1 factor as a matrix algebra appeared in print, it still had an important influence on the development of the theory of operator algebras. In one way or another, word of it reached the ears of capable people over the years. Among other routes, I included the question of whether all factors of type II_1 possess a simple masa in my Baton Rouge list of problems (from the 1967 conference at LSU in honor of Jacques Dixmier).

A paper [18] of Singer's, that appeared in 1955, makes reference to the Ambrose-Singer project in a footnote on p. 121. The talk that Singer gave at the 1953 conference (mentioned earlier in connection with his derivation result) was based on the results in [18]. In that paper, Singer analyzes special automorphisms of a factor $\mathcal{M}$ of type II_1 arising from a countable group G acting as measure-preserving transformations of a measure space $(\mathcal{S}, \mu)$ $(\mu(\mathcal{S}) = 1)$ that we discussed before. We use the notation of that discussion. Singer studies the group $\mathrm{Aut}_1(\mathcal{M})$ of automorphisms α of $\mathcal{M}$ that map the masa $\Phi(\mathcal{A}_0)$ onto itself. Each such automorphism α gives rise to a measure-preserving transformation α' of $\mathcal{S}$ onto itself. He characterizes the elements of $\mathrm{Aut}_1(\mathcal{M})$ in terms of the action of α' on $\mathcal{S}$.

Theorem. *A measure-preserving transformation α' of $\mathcal{S}$ is induced by an automorphism α in $\mathrm{Aut}_1(\mathcal{M})$ if and only if there are measurable sets X_h^g in $\mathcal{S}$ $(g, h \in G)$ such that*

(*i*) $\mu(X_h^g \cap X_k^g) = 0$ *when* $h \neq k$;

(*ii*) $\mu(\cup_{h \in G} X_h^g) = 1$;

(*iii*) $(\alpha'^{-1} h^{-1} \alpha')(x) = g^{-1}(x)$ *for almost every* x *in* $\alpha'(X_h^g)$.

Ambrose [1] developed a framework for studying groups of measure-preserving transformations, his H-systems, that is roughly equivalent to the Murray-von Neumann, group-measure-space construction. In [18], Singer passes freely between both formulations, using the one he found better suited to a particular situation. This probably led to the article [18] not receiving as much attention as it deserved. In section 6 of [18], the last section, consisting of two brief paragraphs, Singer notes that the Murray-von Neumann construction (in our terminology) could be effected without assuming ergodicity of G on $\mathcal{S}$. The resulting von Neumann algebra would not, then, be a factor. He remarks, that the resulting operator algebra can be studied in terms of factors through the then-recently-published "direct integral theory" [17]. He notes that that is not his main interest. He was concerned, primarily, with the factor case.

In the second paragraph of that section, he notes that the Murray-von Neumann construction really occurs *algebraically* in terms of the multiplication algebra $\mathcal{A}_0$ and G acting by automorphisms of $\mathcal{A}_0$. He suggests that this construction can be carried out with another algebra in place of $\mathcal{A}_0$, and notes that it would probably lead to different and interesting examples of factors. Of course, Singer is anticipating the "crossed product" construction in this comment (compare [8, Chapter 13]). It has, indeed, become one of the basic constructions of the subject of operator algebras, leading to new and vital aspects of the theory.

Singer and I have several joint articles. The question of what an orthonormal basis is has been a dominant theme in most of that research. At first glance, every trained mathematician will think that the construction and properties of such bases form one of the less strenuous and most completely understood chapters in twentieth century mathematics! Is there really anything left to say? Certainly, the question of the existence of a simple masa in a factor of type II_1, needed as a "preferred basis" for the Ambrose-Singer project of assigning a "matrix" to each of the elements of that factor, is one aspect of that question. It led us on a merry chase for nearly fifty years!

A good way to start thinking of the meaning of orthonormal bases is to consider the uses to which we put these bases. In one instance, if we are given an especially interesting basis for the topic we are studying, we may want to expand all or some of the elements of $\mathcal{H}$ in terms of that basis. We recognize the L_2-theory of Fourier series as one aspect of that use of orthonormal bases.

We can turn that process around — instead of having an interesting basis given to us, we may want to *find* a particularly appropriate basis for some purpose, say, one that diagonalizes a self-adjoint operator on $\mathcal{H}$ or a commuting family of such operators. Let's phrase this example in a more physical way. Given a compatible family of observables, we want to find a complete set of simultaneous eigenstates for them. Dirac speaks of finding a "representation" and even presents an agenda for this. The following is quoted from pp. 74-75 of the Third Edition of his famous "Quantum Mechanics." Oxford University Press, London 1947

"To introduce a representation in practice

(i) We look for observables which we would like to have diagonal either because we are interested in their probabilities or for reasons of mathematical simplicity;

(ii) We must see that they all commute — a necessary condition since diagonal matrices always commute;

(iii) We then see that they form a complete commuting set, and if not we add some more commuting observables to them to make them into a complete commuting set;

(iv) We set up an orthogonal representation with this complete commuting set diagonal."

The representation is then completely determined except for arbitrary phase factors. For most purposes the arbitrary phase factors are unimportant and trivial, so that we may count the representation as being completely determined by the observables that are diagonal ... "

The emphasis, above, is mine. What would that say if it were put down in precise mathematical form? For one thing, Dirac talks about finding a basis that diagonalizes a self-adjoint operator, and while that is always possible when $\mathcal{H}$ is finite dimensional, there are perfectly respectable self-adjoint operators on infinite-dimensional Hilbert space that do not have a single eigenvector, in the strict sense. Still, we do have a "spectral resolution" of such operators. Again, Dirac's way of going at that problem is inspiring. On pp. 57-58, he writes:

> We have not yet considered the lengths of the basic vectors. With an orthogonal representation, the natural thing to do is to normalize the basic vectors, rather than leave their lengths arbitrary, and so introduce a further stage of simplification into the representation. However, it is possible to normalize them only if the parameters which label them all take on discrete values. If any of these parameters are continuous variables that can take on all values in a range, the basic vectors are eigenvectors of some observable belonging to eigenvalues in a range and are of infinite length..."

Dirac's "ranges" are "intervals" and his "continuous variables" are points in the interval. At this stage, Dirac introduces his δ-functions and develops their formalism. But without eigenstates that are vectors in $\mathcal{H}$, there are problems with what we mean by a "diagonalizing orthonormal basis" — especially, if we are "representing" *families* of compatible observables.

Let's see what this means in the case of a classical basis $\{e_1, e_2, \ldots\}$. If $\mathcal{A}_d$ is the family of all bounded operators on $\mathcal{H}$ that are diagonal relative to that basis, then $\mathcal{A}_d$ is abelian, as Dirac notes, and it is "complete" in his sense — that is "maximal abelian" in $\mathcal{B}(\mathcal{H})$. We have noted that $\mathcal{A}_d$ is a "masa" in $\mathcal{B}(\mathcal{H})$. Of course, there is no difficulty, here, in identifying the "simultaneous eigenstates" for our "complete commuting" family of observables; they are the vectors e_n of our basis. But what are they when our observables have "ranges" in their spectra. Dirac has his δ-functions, his vectors of "infinite length." This is a bit cumbersome, from the rigorous mathematical point of view. What we want to do is to replace the vectors e_n by some acceptable mathematical construct that is effectively the same as the vector, when there is one, and gives us something precise and usable when there is only a δ-function. Something that works very well is the vector state ω_{e_n} corresponding to e_n ($\omega_{e_n}(T) =< Te_n, e_n >$ for each T in $\mathcal{B}(\mathcal{H})$). With this notation, ω_{e_n} is "the expectation functional" of the state, in physical terminology, corresponding to the vector being replaced. The value $\omega_{e_n}(T)$, the expectation value of T in the state corresponding to e_n, is what is measured in the laboratory. If the observable corresponding to T is measured many times with the physical system in the state corresponding to e_n and those measurements are averaged, the resulting number is (close to) $\omega_{e_n}(T)$.

Of course, ω_{e_n} is a state of $\mathcal{B}(\mathcal{H})$. We are not there as yet; the states of $\mathcal{B}(\mathcal{H})$ are not quite the "replacement" for the (unit) vectors of $\mathcal{H}$. The states ω_x of $\mathcal{B}(\mathcal{H})$ corresponding to unit vectors x in $\mathcal{H}$ have another crucial property; they are "pure." A state ω is a *pure state* when $\omega = \frac{1}{2}(\omega_1+\omega_2)$ only if $\omega = \omega_1 = \omega_2$. In physical language, ω is pure when it is not a proper mixture of other states. The pure states of $\mathcal{B}(\mathcal{H})$ are the "generalized unit vectors in $\mathcal{H}$," the smoothly functioning replacement for the δ-function in this quantum-measurement context.

We can certainly speak of states of operator algebras other than $\mathcal{B}(\mathcal{H})$ — and pure states of those algebras — states that are not proper mixtures of other states of the algebra. As luck would have it, the pure states of $\mathcal{A}_{\mathrm{d}}$ are precisely the (non-zero) multiplicative linear functionals on $\mathcal{A}_{\mathrm{d}}$. More generally, the pure states of each abelian operator algebra are the (non-zero) multiplicative functionals on the algebra.

For each unit vector x in $\mathcal{H}$, ω_x is a pure state of $\mathcal{B}(\mathcal{H})$. But there are others — many! If there weren't, we wouldn't have succeeded at including all the δ-functions, the "eigenstates" of observables with "ranges" in their spectra. Even in the case of the classical basis $\{e_n\}$, there are "simultaneous eigenstates" of $\mathcal{A}_{\mathrm{d}}$ other than the states ω_{e_n} — again, many! When we try to deal with the non-vector eigenstates of a system in a rigorous mathematical fashion, we open a large Pandora's Box. But it's one that we must open, as we shall soon note.

When we speak of an "orthonormal basis," or as Dirac does, "a representation," shall we talk about all the pure states of the masa $\mathcal{A}$ or just those that correspond to unit vectors in $\mathcal{H}$? As remarked, $\mathcal{A}_{\mathrm{d}}$ has many other pure states. The vector states are the only ones that are "normal" (that is, strong-operator continuous on the unit ball of $\mathcal{B}(\mathcal{H})$). If we want to deal with the system (masa) $\mathcal{A}_{\mathrm{c}}$ generated by an observable whose spectrum is the "range" $[0, 1]$, for example, the position observable of a particle oscillating back and forth on the unit interval, there are no normal eigenstates, and we want to talk about eigenstates of that masa. We can say that the "generalized orthonormal basis" "representing" a masa $\mathcal{A}$ is the set of all simultaneous eigenstates of $\mathcal{A}$, and wind up with a "few" more eigenstates than we need in the case of $\mathcal{A}_{\mathrm{d}}$. If we insist on *normal* eigenstates in the case of $\mathcal{A}_{\mathrm{c}}$, we wind up with nothing — there are no normal pure states. In the end, the best approach is to say that $\mathcal{A}$, itself, is the (generalized) orthonormal basis.

Definition. A generalized orthonormal basis for $\mathcal{H}$ is a masa on $\mathcal{B}(\mathcal{H})$.

We do know all these generalized bases.

Theorem. *Each masa on a separable Hilbert space is unitarily equivalent to one of* $\mathcal{A}_d$*, where the underlying Hilbert space can have any finite dimension or* $\aleph_0$*, to* $\mathcal{A}_c$*, or to* $\mathcal{A}_d \oplus \mathcal{A}_c$*.*

There are, however, a number of basic things about generalized orthonormal bases that we do not know. Of course, each unit vector x in $\mathcal{H}$ is contained in an orthonormal basis — so, ω_x is multiplicative on some masa. Is each generalized unit vector "contained" in a masa? That is, if ω is a pure state of $\mathcal{B}(\mathcal{H})$, is it multiplicative on some masa $\mathcal{A}$? That question has been with us for more than fifty years. There's still no answer. In [4] it is proved that, for a countably generated C*-algebra, each pure state is multiplicative (pure) on some masa. In [3], it is proved that the restriction of that pure state to the masa has unique state extension.

What becomes of Dirac's statement in this framework: "so that we may count the representation as being completely determined by the observables that are diagonal ..."? First, we must interpret it in our rigorous language. If two generalized unit vectors (pure states of $\mathcal{B}(\mathcal{H})$) ω_1 and ω_2 give rise to the same eigenstate (pure state) of a masa $\mathcal{A}$, are ω_1 and ω_2 equal? Put in another way, can a pure state (multiplicative linear functional) of $\mathcal{A}$ have distinct pure state extensions to $\mathcal{B}(\mathcal{H})$? This is the problem of "uniqueness of pure state extension" (from a masa to $\mathcal{B}(\mathcal{H})$).

In [9], Singer and I showed that answer is "No!" in general in the case of pure states of $\mathcal{A}_c$. We proved something stronger. Using a technique of von Neumann [16], we defined and produced a "diagonalization process" for $\mathcal{B}(\mathcal{H})$ relative to a masa $\mathcal{A}$ of $\mathcal{B}(\mathcal{H})$. This "process" is a module mapping Φ of $\mathcal{B}(\mathcal{H})$ onto $\mathcal{A}$, where $\mathcal{B}(\mathcal{H})$ is a two-sided module over $\mathcal{A}$ (under left and right multiplication by elements of $\mathcal{A}$) that takes positive operators to positive operators and I to I. (It is a "conditional expectation" of $\mathcal{B}(\mathcal{H})$ onto $\mathcal{A}$, in present day terminology.) If ρ is a state of $\mathcal{A}$, then $\rho \circ \Phi$ is a state of $\mathcal{B}(\mathcal{H})$. We proved that there are distinct diagonalization processes for $\mathcal{A}_c$. If Φ_1 and Φ_2 are two such and T is an operator in $\mathcal{B}(\mathcal{H})$ such that $\Phi_1(T) \neq \Phi_2(T)$ then there is a pure state ρ of $\mathcal{A}$ such that $\rho(\Phi_1(T)) \neq \rho(\Phi_2(T))$ (the pure states of $\mathcal{A}$ "separate" the elements of $\mathcal{A}$). Let ρ_1 be $\rho \circ \Phi_1$ and ρ_2 be $\rho \circ \Phi_2$. With A in $\mathcal{A}$, $\Phi_1(A) = A\Phi_1(I) = A = \Phi_2(A)$. Thus $\rho_1(A) = \rho(A) = \rho_2(A)$, and ρ_1, ρ_2 are distinct state extensions of ρ from $\mathcal{A}$ to $\mathcal{B}(\mathcal{H})$. The set of all extensions of ρ from $\mathcal{A}$ to $\mathcal{B}(\mathcal{H})$ is convex and compact in a special "weak" topology, whence, it is the closed convex hull of its extreme points (from the Krein-Milman theorem). Each of these extreme points extends ρ and is a pure state of $\mathcal{B}(\mathcal{H})$, since ρ is a pure state of $\mathcal{A}$. Since the set of state extensions of ρ from $\mathcal{A}$ to $\mathcal{B}(\mathcal{H})$ does not consist of a single element, there are distinct *pure* state extensions of ρ from $\mathcal{A}$ to $\mathcal{B}(\mathcal{H})$.

We showed that each ω_{e_n} has a unique (pure) state extension from $\mathcal{A}_d$ to $\mathcal{B}(\mathcal{H})$. We raised the question of whether or not the other pure states of $\mathcal{A}_d$ have unique extension. The techniques we developed in proving the non-uniqueness of conditional expectations from $\mathcal{B}(\mathcal{H})$ onto $\mathcal{A}$ make it possible to reduce this problem to inequalities with matrices. Some of these matrix problems have arisen in other contexts. While much work has been done on this set of (equivalent) problems, they remain open.

The discussion of orthonormal bases, to this point, has focussed on their "general meaning" and the nature of the "vectors" in those bases. There is another aspect of an orthonormal basis, inherent in the way we usually use such bases, that is less recognized. That aspect is an *ordering* of the basis. Typically, we are dealing with a separable Hilbert space $\mathcal{H}$ and we choose our orthonormal basis as $e_1, e_2, \ldots$. In terms of this basis, it is easy to describe the "one-way-shift" operator V that maps each e_n onto e_{n+1}. The operator V is a non-unitary isometry of $\mathcal{H}$ into itself with spectrum the closed unit disk in $\mathbb{C}$. If we want to describe the "two-way-shift," a unitary operator U on $\mathcal{H}$ with spectrum all complex numbers of modulus 1, it's convenient to choose our orthonormal basis labeled by all the integers $\{e_n\}_{n\in\mathbb{Z}}$. With this basis, U is the unitary operator that maps e_n to e_{n+1}. Of course, we are using bases labeled by a linearly ordered set: the ordering type of the positive integers, with a smallest element but no largest element, in the first case, and the ordering type of all integers with no smallest element and no largest element, in the second case. There are other ordered sets that will serve as labels for an orthonormal basis, for example, the set $\mathbb{Q}_1$ of rationals in the interval $[0,1]$. The basis so labeled can be used to confound "the unsuspecting." Recalling an earlier quote of Dirac, "However, it is possible to normalize them only if the parameters which label them all take on discrete values. If any of these parameters are continuous variables that can take on all values in a range, the basic vectors are eigenvectors of some observable belonging to eigenvalues in a range and are of infinite length...," we can form the bounded self-adjoint operator A that assigns re_r to e_r, for each basis element e_r of an orthonormal basis labeled by $\mathbb{Q}_1$, for a separable Hilbert space $\mathcal{H}$. Then A is diagonalized by the basis $\{e_r\}$, each e_r is an eigenvector for A ("normalized" to have length 1) corresponding to the eigenvalue r, and $\|A\| = 1$. Since the spectrum of A is a closed subset of $[0,1]$ containing $\mathbb{Q}_1$, that spectrum is $[0,1]$. Each point of $\mathbb{Q}_1$ lies in the "range" $[0,1]$ and is an eigenvalue corresponding to an eigenvector of finite length 1.

Ordered bases serve many purposes; it is well worth understanding what an ordered basis is. Singer and I studied that question in a paper [10] that appeared in 1960. Work on that paper began while I was visiting MIT during the academic year 1956-57. My permanent job was at Columbia University, at that time. On occasion, I shared Is Singer's office with him at MIT. A large part of our joint work was done sitting and talking together, in the office, at home, and while driving; we traded ideas, thought about them, and then commented to one another about them. Of course, a good deal of work was done privately — trying to make computations and lemmas "go." At first, the guiding question was what it meant to put an operator on a Hilbert space in "triangular form" — that is, to view it as part of the algebra of, say, upper triangular matrices. So, we tried to isolate what it should mean to say that an algebra of bounded operators on a Hilbert space is the algebra of all upper triangular matrices. Of course, we thought first of the algebra of upper triangular matrices of finite order. We see this algebra as upper triangular matrices only after we have chosen an appropriate basis *and* put that basis in an appropriate order. We knew that we didn't want to be too literal in our interpretation of "basis" when dealing with infinite-dimensional Hilbert space $\mathcal{H}$, and we knew what a generalized orthonormal basis should be in that case, namely, a masa on $\mathcal{H}$. From the algebra of finite matrices of a given order, a good working definition seemed to be: $\mathcal{T}$ is the algebra of all triangular matrices when $\mathcal{T} \cap \mathcal{T}^*$ is a given masa $\mathcal{A}$ and $\mathcal{T}$ is maximal with respect to that property. So, we tried that in infinite dimensions. Zorn's lemma gave us maximal algebras $\mathcal{T}$ for a given $\mathcal{A}$. We called these algebras *maximal triangular* and $\mathcal{A}$ the *diagonal* of the algebras. The important question at the earliest stage of our work was whether there is a family of projections in $\mathcal{A}$, totally ordered, generating $\mathcal{A}$ as a von Neumann algebra, each member of the family invariant under the operators of $\mathcal{T}$. We called such a projection a *hull* and the von Neumann algebra $\mathcal{C}$ generated by these projections, the hulls, the *core* of $\mathcal{T}$. If $\mathcal{S}$ is a set of vectors in $\mathcal{H}$, the closure of the linear span of $\{Tx : x \in \mathcal{S}, T \in \mathcal{T}\}$ is invariant under each operator in $\mathcal{T}$, in particular, under the operators in $\mathcal{A}$. Thus the projection E with this closure as range commutes with $\mathcal{A}$. Since $\mathcal{A}$ is maximal abelian, $E \in \mathcal{A}$. Since the range of E is invariant under each operator in $\mathcal{T}$, E is a hull in $\mathcal{T}$. With F a projection in $\mathcal{T}$, we denote by '$h(F)$' the projection constructed in this way when the range of F is taken for $\mathcal{S}$. We call $h(F)$ the *hull of* F. There was no difficulty in showing that $\mathcal{C}$ is contained in $\mathcal{A}$. With some effort, we showed that the set of hulls of $\mathcal{T}$

is totally ordered (by the usual ordering on self-adjoint operators). At that point, we knew that we had a theory before us, and there was no turning back.

Is it the case that the core is always $\mathcal{A}$? That was the next question that we tackled. In a short while, we knew that the algebra generated by $\mathcal{A}$ and a unitary operator U that induces an ergodic automorphism of $\mathcal{A}$ (no projection E in $\mathcal{A}$ such that $UEU^* = E$ other than 0 and I) is triangular; Zorn's lemma then gives us maximal triangular algebras $\mathcal{T}$ containing it. Of course, the core of such a $\mathcal{T}$ is just the scalars. We called those triangular algebras (with core the scalars) *irreducible*. A specific example of an irreducible maximal triangular algebra is obtained by choosing the multiplication algebra of the unit circle in $\mathbb{C}$ with Lebesgue measure for $\mathcal{A}$ and the unitary operator induced by a rotation of that circle through an irrational multiple of π radians for U. The maximal triangular algebras whose core is the diagonal we called *hyperreducible*. We proved several general results about the hyperreducible maximal triangular algebras and then classified them completely algebraically and with respect to their action on the underlying Hilbert space. We did not get much further than establishing the existence of the irreducible maximal triangular algebras. The main problem was that the final passage to the full algebra through the use of Zorn's lemma did not give us much of a handle on the elements in the final algebra. Although we had found examples of such algebras, we had not constructed examples in which we had any control over the general element in the algebra. In the case of a von Neumann algebra, our examples were usually arrived at as the strong-operator closures of a self-adjoint algebra whose operators could be easily described — we could approach the general operator in the algebra with nets or sequences of the operators in that self-adjoint algebra. That gave us a handle, though not necessarily an easy path to a proof. There certainly are (uncountably many) non-isomorphic irreducible maximal triangular algebras but that hasn't been proved as this is written.

Theorem. *If $\{E_a\}$ is a totally-ordered family of projections that generates a maximal abelian algebra $\mathcal{A}$, then $\mathcal{T}$, the set of all bounded operators that leave each E_a invariant, is maximal triangular with core and diagonal $\mathcal{A}$. Each hyperreducible algebra arises in this way.*

Theorem. *If $\mathcal{T}$ is a maximal triangular algebra with diagonal $\mathcal{A}$ generated by its family $\{E_a\}$ of minimal projections, then $\mathcal{T}$ is hyperreducible. If we order $\{E_a\}$ by the relation $\precsim$, where $E_a \precsim E_b$ precisely when $h(E_a) \le h(E_b)$ then $\precsim$ is a total ordering. Two maximal triangular* *algebras with totally-atomic diagonals are unitarily equivalent if and only if their sets of minimal projections are order isomorphic. Corresponding to each total-ordering type there is a maximal triangular algebra with a totally-atomic diagonal whose set of minimal projections has this order type.*

Theorem. *If $\mathcal{T}$ is hyperreducible, its diagonal $\mathcal{A}$ has no minimal projections, and $\mathcal{H}$ is separable, then $\mathcal{T}$ is unitarily equivalent to the algebra of all bounded operators on $L_2([0,1],\mu)$, where μ is Lebesgue measure, leaving each F_λ invariant, where F_λ is the multiplication operator corresponding to the characteristic function of $[0,\lambda]$.*

Singer and I felt that our maximal triangular algebras played roughly the role for the theory of non-self-adjoint operator algebras that von Neumann algebras played in the self-adjoint theory. In any event, the theory of non-self-adjoint operator algebras was initiated by [10]. It has developed into a flourishing subject with a large number of very talented research workers. Some of the original questions that we asked are still open as this is written.

As we began to develop an intuition for the subject, we felt that the irreducible case corresponds to factors and the hyperreducible case corresponds to maximal abelian von Neumann algebras. Of course, we understood that a masa is a generalized orthonormal basis — and we realized that we should add "unordered orthonormal basis" to that understanding. It was at a fairly early stage, certainly during that academic year, 1956-1957, that we knew that the hypperreducible maximal triangular algebra was precisely what should be meant by a generalized ordered basis. The ordering of the hulls corresponds to the ordering of the basis and the maximal abelian algebra that serves as the diagonal is the unordered basis. We called these hyperreducible algebras (*generalized*) *ordered bases.*

After that initial development, the main thrust of our paper was classifying the ordered bases — the hyperreducible case — roughly, the equivalent of handling the abelian case in the self-adjoint theory. We were able to do that completely. Each ordered basis corresponds to a closed subset of $[0,1]$ containing 0 and 1 up to what we called Lebesgue order isomorphism — that is a homeomorphism of $[0,1]$ onto itself preserving orientation and Lebesgue null sets. Given such an equivalence class of closed sets, there is a canonically constructed ordered basis that corresponds to it. Two ordered bases are unitarily equivalent if and only if they correspond to the same equivalence class of closed sets.

The most difficult technical lemma we had to prove in connection with this classification is the following. If $\mathfrak{X}$ and $\mathfrak{Y}$ are two dense denumerable subsets of $[0,1]$ containing 0 and 1 and m, M are two numbers such that $0 < m < M < 1$, then there is a homeomorphism f of $[0,1]$ onto itself such that $f(0) = 0$, $f(1) = 1$, f maps $\mathfrak{X}$ onto $\mathfrak{Y}$, and $m(x-y) \leq f(x) - f(y) \leq M(x-y)$ when $y \leq x$ and x, y are in $[0,1]$.

The work leading to [9], described before, grew out of the project with triangular operator algebras. A year after that MIT work, Singer was visiting me at Columbia. We were sitting together trading ideas on some of the problems we still had with triangular operator algebras. Singer suggested something. I thought about it and said, "To carry that out, we would have to settle the question of uniqueness of pure state extension from maximal abelian algebras." That was a problem that Is and I had discussed on occasion over the nine preceding years. At that point, Singer said, "OK, let's settle it!" Two to three weeks later we had settled it. You may ask, with some justice, "And how about the ones that got away?" There were plenty of those — but that's another story!

Toward the end of my year at MIT, Singer and I were sitting in his office — at about 1 AM — each reading material that the other had written on our project. We were at desks against opposite walls with our backs to one another. Suddenly, Singer began to laugh uncontrollably. I turned around, smiling, and began to laugh, as well — it was catching, and we were both slightly giddy after a long day of work. Singer asked, "Dick, are you trying to become the William Faulkner of mathematics?" He had just been reading some particularly complex prose I had written — the syntax was correct, but required an oscilloscope for its analysis. Well, the years have gone by; I can't say anything about my becoming the William Faulkner of mathematics, but I know who has become the Pavarotti–Sinatra! Those two gentlemen have a duet on the popular hit, "My Way." Singer could teach them each something on that topic, and he'd have his usual standing-room-only audience while doing it.

References

[1] W. Ambrose, *The L_2 system of a unimodular group. I*, Trans. Amer. Math. Soc. **65** (1949) 27–48.

[2] W. Ambrose & I. Singer, *A theorem on holonomy*, Trans. Amer. Math. Soc. **75** (1953) 428–443.

[3] C. Akemann, *Approximate units and maximal abelian C^*-subalgebras*, Pacific J. Math. **33** (1970) 543–550.

[4] J. Aarnes & R. Kadison, *Pure states and approximate identities*, Proc. Amer. Math. Soc. **21** (1969) 749–752.

[5] P. Dirac, *The Principles of Quantum Mechanics,* Third Edition, Oxford University Press, London, 1930.

[6] L. Ge, *Applications of free entropy to finite von Neumann algebras. II*, Ann. of Math. **147** (1998) 143–157.

[7] R. Kadison & J. Ringrose, *Fundamentals of the theory of operator algebras. I,* Academic Press, Orlando, 1983.

[8] R. Kadison & J. Ringrose, *Fundamentals of the theory of operator algebras. II,* Academic Press, Orlando, 1986.

[9] R. Kadison & I. Singer, *Extensions of pure states*, Amer. J. Math. **81** (1959), 383–400.

[10] ———, *Triangular operator algebras*, Amer. J. Math. **82** (1960) 227–259.

[11] I. Kaplansky, *Modules over operator algebras*, Amer. J. Math. **75** (1953) 839–858.

[12] F. Murray & J. von Neumann, *On rings of operators*, Ann. of Math. **37** (1936) 116–229.

[13] ———, *On rings of operators. II*, Trans. Amer. Math. Soc. **41** (1937) 208–248.

[14] ———, *On rings of operators. IV*, Ann. of Math. **44** (1943) 716–808.

[15] J. von Neumann, *On rings of operators. III*, Ann. of Math. **41** (1940) 94–161.

[16] ———, *Zur Algebra der Funktionaloperationen und Theorie der normalen Operatoren*, Math. Ann. **102** (1930) 370–427.

[17] ———, *On rings of operators. Reduction theory*, Ann. of Math. **50** (1949) 401–485.

[18] I. Singer, *Automorphisms of finite factors*, Amer. J. Math. **77** (1955) 117–133.

[19] I. Singer & J. Wermer, *Derivations on commutative normed algebras*, Math. Ann. **129** (1955) 260–264.

[20] D. Voiculescu, *The analogues of entropy and of Fisher's information measure in free probability theory. III*, Geom. Funct. Anal. **6** (1996) 172–199.

DEPARTMENT OF MATHEMATICS UNIVERSITY OF PENNSYLVANIA, PHILADELPHIA

SURVEYS IN DIFFERENTIAL GEOMETRY, 2000
Vol. VII ©2000, INTERNATIONAL PRESS
pp. 375-432

CURVATURE AND FUNCTION THEORY ON RIEMANNIAN MANIFOLDS

PETER LI

Function theory on Euclidean domains in relation to potential theory, partial differential equations, probability, and harmonic analysis has been the target of investigation for decades. There is a wealth of classical literature in the subject. Geometers began to study function theory with the primary reason to prove a uniformization type theorem in higher dimensions. It was first proposed by Greene-Wu and Yau to study the existence of bounded harmonic functions on a complete manifold with negative curvature. While uniformization in dimension greater than 2 still remains an open problem, the subject of function theory on complete manifolds takes on life of its own. The seminal work of Yau [107] provided a fundamental technique in handling analysis on noncompact, complete manifolds. It also opens up many interesting problems which are essential for the understanding of analysis on complete manifolds. Since Yau's paper in 1975, there are many developments in this subject. The aim of this article is to give a rough outline of the history of a specific point of view in this area, namely, the interplay between the geometry – primarily the curvature – and the function theory. Throughout this article, unless otherwise stated, we will assume that M^n is an n-dimensional, complete, non-compact, Riemannian manifold without boundary. In this case, we will simply say that M is a complete manifold.

One of the goal of this survey is to demonstrate, by way of known theorems, the two major steps which are common in many geometric analysis programs. First, we will show how one can use assumptions on the curvature to conclude function theoretic properties of the manifold M. Secondly, we will showed that function theoretic properties can in turn be used to conclude geometrical and topological statements about

Research partially supported by NSF grant #DMS-9626310.

the manifold. In many incidents, combining the two steps will result in a theorem which hypothesizes on the curvature and concludes on either the topological, geometrical, or complex structure of the manifold.

The references will not be comprehensive due to the vast literature in the subject. It is merely an indication of the flavor of the field for the purpose of whetting one's appetite. As examples of areas not being discussed in this note are harmonic analysis (function theory) on symmetric spaces, Lie groups, and discrete groups. The contributors to this subject are Furstenberg, Varopoulos, Coulhon, Saloff-Coste, and etc. Another point of view which was systematically taken up by Lyons-Sullivan, and later by Varopoulos, is to relate the group theoretic property of the covering group to the function theory of a covering space.

1. Curvature assumptions and notations

In this paper, we will impose different curvature assumptions on various occasions. The two primary notions of curvature we will use are the sectional curvature and the Ricci curvature. For a given point $x \in M$ and a 2-plane section $\sigma \subset T_xM$, we denote its sectional curvature by $K_M(\sigma)$. The notation $K_M(x)$ means the sectional curvature functional defined on all 2-plane sections at the point x. The Ricci curvature will be denoted by $Ric_M(x)$, which is a symmetric 2-tensor at the point $x \in M$. In the first half of this paper, there are primarily four different types of curvature assumptions that are related to one another.

(1) Non-negative Ricci curvature: We assume that M has non-negative Ricci curvature at every point, i.e.,

$$Ric_M(x) \geq 0$$

for all $x \in M$.

(2) Non-negative Ricci curvature near infinity: There exists a compact subset $D \subset M$, such that

$$Ric_M(x) \geq 0$$

for all $x \in M \setminus D$.

(3) Asymptotically non-negative Ricci curvature: There exists a monotonically non-increasing function $\alpha(r) > 0$ satisfying

$$\int_0^\infty r^{n-1}\alpha(r)\,dr < \infty,$$

such that,

$$Ric_M(x) \geq -\alpha(\rho(x)),$$

where $\rho(x)$ is the distance function from a fixed point $p \in M$.

(4) Almost non-negative Ricci curvature: There exists a sufficiently small $\epsilon > 0$, such that,

$$Ric_M(x) \geq -\epsilon\, \rho^{-2}(x)$$

for all $x \in M$.

One easily verifies that the above assumptions satisfy the following monotonically decreasing ordering:

$$(1) \Rightarrow (2) \Rightarrow (3) \Rightarrow (4).$$

We would also like to take this opportunity to point out that assumptions on the Ricci curvature yield much less information on the manifold as similar assumptions on the sectional curvature. For instance, the soul theorem of Cheeger-Gromoll asserts that:

Theorem 1.1 (Cheeger-Gromoll [22]). *If M has non-negative sectional curvature, then there exists a compact totally geodesic submanifold $N \subset M$ such that M is diffeomorphic to the normal bundle of N.*

The sectional curvature assumption places stringent topological restriction on a manifold. In particular, M must have the topological type of a compact manifold. In a similar spirit, Abresch took their argument a step further.

Theorem 1.2 (Abresch [1], [2]). *Suppose M has asymptotically non-negative sectional curvature, i.e., there exists a positive, monotonically non-increasing function $\gamma(r)$ satisfying*

$$\int_0^\infty r\,\gamma(r)\,dr < \infty,$$

such that, $K_M(x) \geq -\gamma(\rho(x))$ for all $x \in M$. Then M must have bounded topological type. Moreover, the number of ends of M and the total Betti number of M can be estimated in terms of n and γ.

Contrary to the rigid topological restriction imposed on a manifold with the sectional curvature assumptions in the last two theorems, Sha-Yang [97] showed that there are manifolds with positive Ricci curvature which has infinite topological type. In fact, their example is diffeomorphic to $\mathbb{R}^4$ connected sum with k copies of $\mathbb{CP}^2$, for any $k = 1, 2, \ldots, \infty$. Notice that the notions of asymptotically non-negative sectional curvature and asymptotically non-negative Ricci curvature differs by a factor of r^{n-1} in the integrand. This factor seems to arise more naturally for

Ricci curvature than sectional curvature. However, there are no concrete examples which indicate that this factor is not a mere technical assumption.

Definition 1.3. Let $D \subset M$ be a compact subset of M. An *end* E of M with respect to D is a connected unbounded component of $M \setminus D$. When we say that E is an end, it is implicitly assumed that E is an end with respect to some compact subset $D \subset M$.

From the definition, it is clear that if D_1 and D_2 are compact subsets with $D_1 \subset D_2$, then the number of ends with respect to D_1 is at most the number of ends with respect to D_2. This monotonicity property allows us to define the number of ends of a manifold.

Definition 1.4. M is said to have *finitely many ends* if there exists $0 < k < \infty$, such that, for any $D \subset M$, the number of ends with respect to D is at most k.

In this case, we denote $\pi_0^\infty(M)$ to be the smallest such k. Obviously, $\pi_0^\infty(M)$ must be an integer. Also, one readily concludes that there exists $D_0 \subset M$, such that, the number of ends with respect to D_0 is precisely $\pi_0^\infty(M)$. If M has infinitely ends, we will still use $\pi_0^\infty(M) = \infty$ to denote the number of ends.

2. Function theory

Definition 2.1. A *Green's function* $G(x, y)$ is a function defined on $(M \times M) \setminus \{(x, x)\}$ satisfying the following properties:

- $G(x, y) = G(y, x)$, and
- $\Delta_y G(x, y) = -\delta_x(y)$,

for all $x \neq y$.

It was proved by Malgrange [84] that every manifold admits a Green's function. Recently, Li-Tam [69] gave a constructive argument for the existence of $G(x, y)$. As in the difference between $\mathbb{R}^2$ and $\mathbb{R}^n$ for $n \geq 3$, some manifolds admit Green's functions which are positive and others may not. This special property distinguishes the function theory of complete manifolds into two classes.

Definition 2.2. A complete manifold M is said to be *non-parabolic* if it admits a positive Green's function. Otherwise, M is said to be *parabolic.*

For the sake of future reference, we will outline the construction procedure in [69] for $G(x,y)$. Let $p \in M$ be a fixed point and $\{\Omega_i\}$ be a compact exhaustion of M satisfying

$$\{p\} \subset \Omega_1 \subset \cdots \subset \Omega_i \subset \cdots \subset M$$

and

$$\cup_i \Omega_i = M.$$

Let $G_i(p,\cdot)$ be the positive Dirichlet Green's function on Ω_i with pole at p. The fact that

$$\Omega_i \subset \Omega_j$$

for $i \leq j$ and the maximum principle implies that

$$G_i(p,\cdot) \leq G_j(p,\cdot).$$

In particular, if $G_i(p,\cdot)$ monotonically converges to some function $G(p,\cdot)$, then G is a positive Green's function, and hence M is non-parabolic. In this case, one checks readily that G is the minimal positive Green's function. The minimality property determines G uniquely.

In the event that $G_i(p,\cdot) \nearrow \infty$, by defining

$$a_i = \sup_{\partial B_p(1)} G_i(p,\cdot),$$

one can show that $G_i(p,\cdot) - a_i$ converges to some function $G(p,\cdot)$. This function will indeed be a Green's function which changes sign and, in this case, M is parabolic. From this construction, one sees that

$$G(p,\cdot) \leq 0$$

on $M \setminus B_p(1)$. Note that G is not unique and may depend on the choice of the compact exhaustion.

Let us now examine the situation when $G_i(p,\cdot)$ converges to a positive Green's function. It was shown [69] that this occurs if and only if there exists a harmonic function h defined on $M \setminus B_p(1)$ with the property that

$$h = 1 \text{ on } \partial B_p(1)$$

and

$$\inf_{M \setminus B_p(1)} h = 0.$$

To understand the existence of h, we consider the corresponding problem on annuli of the form $A_p(1,r) = B_p(r) \setminus B_p(1)$. For each $r > 1$, let h_r be the harmonic function defined on $A_p(1,r)$ with the properties that

$$h_r = 1 \text{ on } \partial B_p(1)$$

and

$$h_r = 0 \text{ on } \partial B_p(r).$$

Clearly, h_r is the minimizer for the Dirichlet integral

$$\int_{A_p(1,r)} |\nabla f|^2$$

among all functions in the space

$$H_r = \{f \in H_{1,2}(A_p(1,r)) \,|\, f = 1 \text{ on } \partial B_p(1), \; f = 0 \text{ on } \partial B_p(r)\}.$$

If we define

$$E(r) = \inf_{H_r} \int_{A_p(1,r)} |\nabla f|^2 = \int_{A_p(1,r)} |\nabla h_r|^2$$

then clearly $H_r \subset H_R$ for $r \leq R$. Hence $E(r)$ is a monotonically non-increasing function of r. Due to the boundary conditions, the sequence h_r satisfies $h_r \leq h_R$ for $r \leq R$. The fact that $h_r \leq 1$ because of the maximum principle implies that the sequence $\{h_r\}$ converges uniformly on compact subsets to a harmonic function h_∞. Moreover, h_∞ has the property that

$$h_\infty = 1 \text{ on } \partial B_p(1).$$

Clearly, unless h_∞ is identically constant 1, the function

$$h = \frac{h_\infty - \inf h_\infty}{1 - \inf h_\infty}$$

will be the desired harmonic function we wish to construct.

We now claim that h_∞ is the constant function 1 if and only if $E(r) \searrow 0$. Indeed, using the fact that h_r is harmonic and the boundary conditions, we can rewrite the integral

$$\begin{aligned} E(r) &= \int_{A_p(1,r)} |\nabla h_r|^2 \\ &= \int_{\partial B_p(r)} h_r \frac{\partial h_r}{\partial \nu} - \int_{\partial B_p(1)} h_r \frac{\partial h_r}{\partial \nu} \\ &= - \int_{\partial B_p(1)} \frac{\partial h_r}{\partial \nu}. \end{aligned}$$

Hence the strong maximum principle asserts that h_∞ is identically constant if and only if $E(r) \searrow 0$. In particular, this implies that $E(r) \searrow 0$ if and only if M is parabolic. The quantity

$$\lim_{r \to \infty} E(r)$$

is sometimes called the capacity of M at infinity. With this equivalent condition for parabolicity, Royden's theorem [92] follows immediately.

Definition 2.3. A manifold M is said to be *quasi-isometric* to another manifold N if there exists a diffeomorphism $\phi : M \to N$ and a constant $C > 0$, such that,

$$C^{-1}\, ds_M^2 \leq \phi^*(ds_N^2) \leq C\, ds_M^2.$$

Theorem 2.4 (Royden [92]). *Let M be quasi-isometric to N. Then M is parabolic if and only if N is parabolic.*

Definition 2.5. An end E is said to be a *non-parabolic end* if it admits a positive Green's function with Neumann boundary condition on ∂E. Otherwise, it is said to be a *parabolic end.* We will denote $\Pi_0^\infty(M)$ to be the number of parabolic ends of M.

From the construction of [69] outlined above, one verifies that M is non-parabolic if and only if it has a non-parabolic end. Indeed, if E is a non-parabolic end, then it admits a Neumann Green's function $G(x,y)$. For a fixed $x \in E$, the strong maximum principle asserts that $G(x,\cdot)$ must be positive on ∂E. If we define

$$g = \min\{G(x,\cdot), a\}$$

for some sufficiently large constant $a > 0$, then g is a positive superharmonic function define on E with

$$\inf_E g = 0$$

and

$$\inf_{\partial E} g = b > 0$$

for some constant b. Clearly, the function $b^{-1}g$ can be used as a barrier to solve for a positive harmonic function on E with

$$h = 1 \text{ on } \partial E$$

and

$$\inf_E h = 0.$$

The existence of h implies M is non-parabolic as indicated above.

Conversely, if M admits a positive Green's function then the minimal positive Green's function will have the property that

$$\inf_M G(x,\cdot) = 0.$$

Let E be an end with respect to some compact set containing x such that

$$\inf_E G(x,\cdot) = 0.$$

Clearly, the above construction together with $G(x,\cdot)|_E$ can be used to construct a positive Neumann Green's function on E.

It is useful to point out that Nakai [90] (also see [91]) showed at if M is parabolic then there exists a Green's function $G(p,\cdot)$ with the property that

$$G(p,x) \to -\infty \text{ as } x \to \infty.$$

3. Geometric criteria for parabolicity

Though the definition of parabolicity is purely analytical, in some incidents, there are geometric description of parabolicity. It was first pointed out by Cheng and Yau [26] that if the volume growth of M satisfies

$$V_p(r^2) \leq C\, r^2$$

for some constant $C > 0$, then M must be parabolic. The sharp condition was proved by Ahlfors for dimension 2, and later independently by Grigor'yan [43], [44] and Varopoulos [103] for all dimensions, that a necessary condition for a manifold to be non-parabolic is that there exists $p \in M$, such that, the volume $V_p(t)$ of geodesic ball centered at p of radius t satisfies the growth condition

$$\int_1^\infty \frac{t\,dt}{V_p(t)} < \infty. \tag{3.1}$$

Observe that this property holds at one point if and only if it holds at all points of M. Moreover, this condition is clearly invariant under quasi-isometry. The obvious question is to determine if this condition is also sufficient. Unfortunately, the following example of Greene (see [103]) indicated that this is not true in general.

Example. Let M be $\mathbb{R}^2$ endowed with the metric of the form

$$ds^2 = \begin{cases} y^{-2}(dx^2+dy^2) & \text{for} \quad y \geq 2 \\ f(y)(dx^2+dy^2) & \text{for} \quad 0 \leq y \leq 2 \\ dx^2+dy^2 & \text{for} \quad y \leq 0, \end{cases}$$

where f is any smooth function satisfying $f(0) = 1$ and $f(2) = 1/4$. This manifold is obviously parabolic because it is conformally equivalent to the standard flat metric on $\mathbb{R}^2$. However, direct computation shows that (3.1) holds.

An interesting phenomenon is that for manifolds with non-negative Ricci curvature, condition (3.1) is also sufficient for non-parabolicity.

Theorem 3.1 (Varopoulos [102]). *If M has non-negative Ricci curvature, then M is non-parabolic if and only if*

$$\int_1^\infty \frac{t\,dt}{V_p(t)} < \infty$$

for some $p \in M$, where $V_p(t)$ is the volume of geodesic ball centered at p of radius t.

In fact, in the case of non-negative Ricci curvature, one can estimate the Green's function by the volume growth.

Theorem 3.2 (Li-Yau [79]). *If M has non-negative Ricci curvature, then there exists positive constants C_1 and C_2, such that, the minimal positive Green's function satisfies*

$$C_1 \int_{\rho(x,y)}^\infty \frac{t\,dt}{V_x(t)} \le G(x,y) \le C_2 \int_{\rho(x,y)}^\infty \frac{t\,dt}{V_x(t)},$$

where $\rho(x,y)$ denotes the geodesic distance between x and y.

In 1995, Li-Tam managed to prove that the volume growth condition is sufficient for non-parabolicity for a larger class of manifolds.

Theorem 3.3 (Li-Tam [73]). *Let us assume that there is a constant $C_1 > 0$ such that the Ricci curvature of M satisfies*

$$Ric_M(x) \ge -C_1\,\rho^{-2}(x)$$

for all $x \in M$. Assume that there exists $p \in M$ and $C_2 > 0$, such that, the volume comparison condition

$$V_p(R) \le C_2\, V_x(R/2),$$

is satisfied for all $x \in \partial B_p(R)$, then M is non-parabolic if and only if

$$\int_1^\infty \frac{t\,dt}{V_p(t)} < \infty.$$

Corollary 3.4. *If M has non-negative Ricci curvature near infinity and finite first Betti number, then M is non-parabolic if and only if*

$$\int_1^\infty \frac{t\,dt}{V_p(t)} < \infty$$

for some $p \in M$.

Corollary 3.5. *If M is quasi-isometric to a manifold satisfying the assumption of Theorem 3.3, then M is non-parabolic if and only if*

$$\int_1^\infty \frac{t\,dt}{V_p(t)} < \infty$$

for some $p \in M$.

In [73], the authors obtained estimates for the Green's function on manifolds satisfying the hypothesis of Theorem 3.3. However, the estimates are not as clean as those of Theorem 3.4. Recently, Colding-Minicozzi [31] showed that if M^n with $n \geq 3$ has non-negative Ricci curvature and maximal volume growth then the Green's function has an asymptotic limit. In a joint work [74] of Tam, Wang, and the author, they gave a short proof of the asymptotic limit and also gave sharp upper and lower bounds for G. In this case, maximal volume growth means that there exists $p \in M$ such that

$$\liminf_{r\to\infty} r^{-n}\, V_p(r) > 0.$$

Bishop comparison theorem implies that, in fact,

$$\theta_p(r) = r^{-n}\, V_p(r) > 0$$

is a monotonically non-increasing function of r. Also, it is easy to see that if

$$\theta = \lim_{r\to\infty} \theta_p(r)$$

then θ is independent of p.

Theorem 3.6 (Li-Tam-Wang [74]). *Let M be a complete manifold with non-negative Ricci curvature of dimension at least 3. Assume that M has maximal volume growth, and let ρ be the distance function to the point $p \in M$. For any $\delta > 0$, there exists a constant $C > 0$ depending only on n and θ, so that the minimal positive Green's function on M satisfies*

$$\begin{aligned}(1+9\delta)^{1-\frac{n}{2}} \quad & \frac{\rho^{2-n}(x)}{n(n-2)\theta_p(\delta\,\rho(x))} \\ & \leq G(p,x) \\ & \leq (1+C(\delta+\beta))\,(1-\delta)^{1-\frac{n}{2}}\,\frac{\rho^{2-n}(x)}{n(n-2)\theta},\end{aligned}$$

where

$$\beta = \delta^{-2n} \max_{r \geq (1-\delta)\rho(x)} \left\{ 1 - \frac{\theta_p(r)}{\theta_p(\delta^{2n+1} r)} \right\}.$$

In particular,

$$\lim_{x \to \infty} \rho^{n-2}(x)\, G(p,x) = \frac{1}{n(n-2)\theta}.$$

Let us consider the special case when M is a complete manifold with a rotationally symmetric metric with respect to a point $p \in M$. If $A_p(t)$ denotes the area of $\partial B_p(t)$, then let us assume that

$$\int_1^\infty \frac{dt}{A_p(t)} < \infty. \tag{3.2}$$

In this case, M is non-parabolic and the minimal positive Green's function with the pole at p is given by

$$G(p,x) = \int_{\rho(p,x)}^\infty \frac{dt}{A_p(t)}. \tag{3.3}$$

Indeed, using the fact that $A_p(t)$ is asymptotically

$$A_p(t) \sim n\omega_n\, t^{n-1}$$

as $t \to 0$, where ω_n denotes the volume of the Euclidean unit n-ball, we verify that

$$\int_{\rho(p,x)}^\infty \frac{dt}{A_p(t)} \sim \frac{1}{n(n-2)\omega_n}\, \rho^{2-n}(p,x)$$

as $x \to p$. Also, since the metric is rotationally symmetric, the Laplacian in terms of polar coordinates can be written as

$$\Delta = \frac{\partial^2}{\partial r^2} + \frac{A_p'}{A_p} \frac{\partial}{\partial r},$$

hence

$$\Delta \left(\int_{\rho(p,x)}^\infty \frac{dt}{A_p(t)} \right) = 0$$

for $x \neq p$, and (3.3) is verified. In fact, a similar computation will confirm that the function

$$\int_1^{\rho(p,x)} \frac{dt}{A_p(t)} \tag{3.4}$$

is a Green's function on a rotationally symmetric manifold regardless of parabolicity. In case (3.2) holds, then (3.4) differs from (3.3) by a additive constant. If (3.2) is not valid, then (3.4) is still a Green's function and M is parabolic. Notice that if M satisfies some non-negativity assumption on the Ricci curvature, then one can show that $tA_p(t)$ is equivalent to $V_p(t)$, which explains the validity of Theorem 3.3.

In a recent works of Holopainen [50] and Holopainen-Koskela [51], the authors gave a criteria upon which the condition (3.1) is equivalent to non-parabolicity. In particular, one criterion has the property that it is localized on a cone neighborhood of a geodesic ray.

Theorem 3.7 (Holopainen-Koskela [51]). *Let M be a complete manifold. Suppose there exists a geodesic ray $\gamma : [0, \infty) \to M$ satisfying the following two properties:*

- *There exists a constant $C_1 > 0$, such that, for all $t > 0$ and for all geodesic ball $B_x(2r) \subset B_{\gamma(t)}(\frac{t}{2})$ the volume doubling condition*

$$C_1 \, V_x(r) \geq V_x(2r)$$

  is satisfied.

- *There exists a constant $C_2 > 0$, such that, for all $t > 0$ and for all $B_x(2r) \subset B_{\gamma(t)}(\frac{t}{2})$ the Poincare inequality*

$$C_2 \, r \left(\int_{B_x(2r)} |\nabla f|^2 \right)^{frac12} \geq \inf_{k \in \mathbb{R}} \int_{B_x(r)} |f - \bar{f}|.$$

  is satisfied for all $f \in H_{1,2}(B_x(r))$ with $\bar{f} = V_x(r)^{-1} \int_{B_x(r)} f$.

The manifold M is non-parabolic if and only if

$$\int_1^\infty \frac{t \, dt}{V_p(t)} < \infty$$

for some $p \in M$.

We would like to remark that the authors actually proved a more general version of this theorem which holds for the p-Laplacian.

4. A basic theorem on harmonic functions

In this section, we will indicate that various spaces of harmonic functions will play certain roles in reflecting the topology of the underlying manifold.

Definition 4.1. Define $\mathcal{H}_D^\infty(M)$ to be the space of bounded harmonic functions with finite Dirichlet integral on M.

Definition 4.2. Define $\mathcal{H}^\infty(M)$ to be the space of bounded harmonic functions on M.

Definition 4.3. Define $\mathcal{H}^+(M)$ to be the space spanned by the set of positive harmonic functions on M.

Definition 4.4. Define $\mathcal{H}'(M)$ to be the space spanned by the set of harmonic functions which are bounded on one side at each end of M. More precisely, a harmonic function, f, is bounded on one side at each end if there exists a compact set $D \subset M$ such that f is either bounded from above or from below when restricted to each end with respect to D.

It follows directly from the definitions that these spaces satisfy the monotonic relations

$$\{\text{constants}\} \subset \mathcal{H}_D^\infty(M) \subset \mathcal{H}^\infty(M) \subset \mathcal{H}^+(M) \subset \mathcal{H}'(M).$$

In particular, their respective dimensions satisfy

$$1 \leq \dim \mathcal{H}_D^\infty(M) \leq \dim \mathcal{H}^\infty(M) \leq \dim \mathcal{H}^+(M) \leq \dim \mathcal{H}'(M).$$

Observe that if M has only one end, then $\mathcal{H}^+(M) = \mathcal{H}'(M)$.

Definition 4.5. A manifold is said to have the *strong Liouville property* if it does not admit any non-constant positive harmonic function, i.e., $\dim \mathcal{H}^+(M) = 1$.

Definition 4.6. A manifold is said to have the *Liouville property* if it does not admit any non-constant bounded harmonic function, i.e., $\dim \mathcal{H}^\infty(M) = 1$.

An interesting, but unrelated fact concerning the space $\mathcal{H}_D^\infty(M)$ is a theorem of Sario-Schiffer-Glasner [95]. It asserts that if M admits a non-constant harmonic function with finite Dirichlet integral, then it must also admits a non-constant bounded harmonic function with finite Dirichlet integral. We are now ready to state the theorem which relates the dimension of these spaces of harmonic functions to $\pi_0^\infty(M)$ and $\Pi_0^\infty(M)$.

Theorem 4.7 (Li-Tam [72]). *Let M be any complete Riemannian manifold without boundary. The the number of ends, $\pi_0^\infty(M)$, of M satisfies the upper bound*

$$\pi_0^\infty(M) \leq \dim \mathcal{H}'(M).$$

If M is non-parabolic, then we have the improved estimate

$$\pi_0^\infty(M) \leq \dim \mathcal{H}^+(M),$$

and the number of non-parabolic ends, $\Pi_0^\infty(M)$, satisfies the bound

$$\Pi_0^\infty(M) \leq \dim \mathcal{H}_D^\infty(M).$$

We should point out that the last case of Theorem 4.7 also follows from the work of Grigor'yan [46], where he related $\dim \mathcal{H}_D^\infty(M)$ to the number disjoint of D-massive sets. At this point, perhaps it is useful to consider a few examples so we have a better understanding of this theorem.

Example 1. Let $M = \sharp\, k\, (\mathbb{R}^n)$ be the connected sum of k copies of $\mathbb{R}^n$. If $n = 2$, then M is parabolic. In this case,

$$\pi_0^\infty(M) = k = \dim \mathcal{H}'(M).$$

If $n \geq 3$, then M is non-parabolic, and

$$\pi_0^\infty(M) = \Pi_0^\infty(M) = k.$$

Moreover,

$$\dim \mathcal{H}'(M) = \dim \mathcal{H}^+(M) = \dim \mathcal{H}_D^\infty(M) = k.$$

In any event, Theorem 4.7 is sharp.

Example 2. Let M be the hyperbolic plane $\mathbb{H}^2$. In this case,

$$\dim \mathcal{H}'(M) = \dim \mathcal{H}^+(M) = \dim \mathcal{H}_D^\infty(M) = \infty.$$

However, M is non-parabolic and has only 1 end.

Example 3. Recall that a well-known theorem of Yau asserts that:

Theorem 4.8 (Yau [107]). *If M has non-negative Ricci curvature, then M has the strong Liouville property.*

On the other hand, the splitting theorem of Cheeger-Gromoll implies that:

Theorem 4.9 (Cheeger-Gromoll [21]). *If M has non-negative Ricci curvature, then either $M = N \times \mathbb{R}$, for some compact manifold N with non-negative Ricci curvature, or M has only 1 end.*

Using these two theorems, we will analyze the situation when M has non-negative Ricci curvature. First, let us consider the case when $M = N \times \mathbb{R}$. In this case, clearly M is parabolic according to the volume growth condition (3.1) for non-parabolicity. One also checks easily that the space $\mathcal{H}'(M)$ is spanned by the constant function 1 and the function $t \in \mathbb{R}$. Hence

$$\dim \mathcal{H}'(M) = 2 = \pi_0^\infty(M).$$

The remaining case is when M has only 1 end. This implies that

$$\mathcal{H}'(M) = \mathcal{H}^+(M)$$

from the definition of $\mathcal{H}'(M)$. In this case, Yau's result implies that

$$\dim \mathcal{H}'(M) = \dim \mathcal{H}^+(M) = 1.$$

Hence whether M is parabolic or not, Theorem 4.7 is sharp for manifolds with non-negative Ricci curvature.

To give a more quantitative description of Theorem 4.7, let us first consider the case when M is parabolic. For any compact subset $D \subset M$, let $\{e_i\}_{i=1}^s$ be the set of all ends. For each $i \neq 1$, there exists a harmonic function f_i satisfying

$$f_i(x) \to -\infty \qquad \text{as} \qquad x \to e_1(\infty),$$

$$f_i(x) \to \infty \qquad \text{as} \qquad x \to e_i(\infty),$$

and f_i is bounded on e_j for all $j \neq 1, i$. The notation $x \to e_i(\infty)$ means that $x \to \infty$ and $x \in e_i$. One checks readily that the set $\{f_i\}_{i=2}^s$ together with the constant function form a linearly independent set. Hence, $s \leq \dim \mathcal{H}'(M)$ and the first assertion of Theorem 4.7 follows because D is arbitrary.

When M is non-parabolic, for any compact subset D, let $\{e_i\}_{i=1}^s$ be the set of parabolic ends with respect to D and $\{E_\alpha\}_{\alpha=1}^\ell$ be the set of non-parabolic ends with respect to D. If $s > 0$, then for each $1 \leq i \leq s$, there exists a positive harmonic function g_i satisfying

$$\inf_{E_\alpha} g_i(x) = 0 \qquad \text{for all} \qquad \alpha,$$

$$g_i(x) \to \infty \qquad \text{as} \qquad x \to e_i(\infty),$$

and g_i is bounded on e_k for all $k \neq i$. Also, for any $1 \leq \alpha \leq \ell$, there exists a bounded harmonic function h_α with the properties that

$$\sup_{E_\alpha} h_\alpha = 1,$$

$$\inf_{E_\beta} h_\alpha = 0 \qquad \text{for} \qquad \beta \neq \alpha,$$

and h_α is bounded on e_i for all $1 \leq i \leq s$. Similarly to the parabolic case, the existence of these harmonic functions implies the inequality asserted in Theorem 4.7.

5. Historical background

Before we proceed to discuss various applications of Theorem 4.7, perhaps it is useful to point out some historical aspects that lead to the development of the theorem. In his fundamental paper [107], Yau introduced the method of gradient estimate to prove Theorem 4.8. Right after this work, Cheng-Yau [26] provided a local argument for the gradient estimate which will become a basic technique and a starting point for the theory of harmonic functions in years to come. The version of the gradient estimate that is related to the content of this article is as follows:

Theorem 5.1 (Cheng-Yau [26]). *Let M be a manifold with boundary, ∂M. Suppose $p \in M$ and $r > 0$ such that the geodesic ball $B_p(r)$ centered at p of radius r satisfies $B_p(r) \cap \partial M = \emptyset$. If f is a positive harmonic function defined on $B_p(r)$, then for any $0 < \alpha < 1$, there exists a constant $C > 0$ depending only on $n = \dim M$ and α, such that,*

$$|\nabla f|(x) \leq C\, r^{-1}\, f(x)$$

for all $x \in B_p(\alpha r)$. In particular,

$$f(x) \leq C\, f(y)$$

for all x, $y \in B_p(\alpha r)$.

In 1987, in an attempt to understand and generalize Yau's strong Liouville theorem to a larger class of manifold, the author and Luen-fai Tam considered manifolds which behave like a manifold with non-negative Ricci curvature. The most elementary situation which we considered is the case when M is a connected sum of k copies of $\mathbb{R}^n$ as given by Example 1 of the previous section. The example indicated that the validity of Yau's theorem hinges on Theorem 4.9, even though Yau's proof is completely independent of Theorem 4.9. Also, the fact that the manifold is Euclidean at each end allows us to use the explicit form of the Green's function as barriers. In fact, modeling on Example 1, Li-Tam [68] successfully determined the spaces of bounded and positive harmonic functions on a manifold with non-negative sectional curvature near infinity.

Theorem 5.2 (Li-Tam [68]). *Let M be a complete manifold with non-negative sectional curvature near infinity. It follows that M must have finitely many ends. Hence there exists $p \in M$ and $r > 0$ such that the number of ends with respect to $B_p(r)$ is precisely $\pi_0^\infty(M)$. An end E is non-parabolic if and only if*

$$\int_r^\infty \frac{t\,dt}{V_E(t)} < \infty.$$

Suppose $\{e_i\}_{i=1}^s$ as the set of ends satisfying the volume growth condition

$$\int_r^\infty \frac{t\,dt}{V_{e_i}(t)} = \infty.$$

where $V_{e_i}(t)$ denotes the volume of the set $B_p(t) \cap e_i$. Also, let $\{E_\alpha\}_{\alpha=1}^\ell$ be the set of ends satisfying the volume growth condition

$$\int_r^\infty \frac{t\,dt}{V_{E_\alpha}(t)} < \infty.$$

In particular, we have

$$\Pi_0^\infty(M) = \ell \qquad \textit{and} \qquad \pi_0^\infty(M) = s + \ell.$$

If $\Pi_0^\infty(M) = 0$ then $\dim \mathcal{H}^+(M) = 1$. If $\Pi_0^\infty(M) > 0$, then for each e_i there exists a positive harmonic function g_i satisfying

$$g_i(x) \to 0 \qquad \textit{as} \qquad x \to E_\alpha(\infty) \qquad \textit{for all} \qquad \alpha,$$

$$g_i(x) \to \infty \qquad \textit{as} \qquad x \to e_i(\infty),$$

and g_i is bounded on e_k for all $k \neq i$. Also, for any $1 \leq \alpha \leq \ell$, there exists a bounded harmonic function h_α with the properties that

$$h_\alpha(x) \to 1 \qquad \textit{as} \qquad x \to E_\alpha(\infty),$$

$$h_\alpha(x) \to 0 \qquad \textit{as} \qquad x \to E_\beta(\infty)$$

for all $\beta \neq \alpha$, and h_α is bounded on e_i for all $1 \leq i \leq s$. Moreover, the set $\{h_\alpha\}_{\alpha=1}^\ell$ spans the space of bounded harmonic functions denoted by $\mathcal{H}^\infty(M)$, and the set $\{h_\alpha\}_{\alpha=1}^\ell \cup \{g_i\}_{i=1}^s$ spans the space $H^+(M)$. In particular,

$$\Pi_0^\infty(M) = \mathcal{H}^\infty(M) \qquad \textit{and} \qquad \pi_0^\infty(M) = \mathcal{H}^+(M).$$

The reason that sectional curvature was assumed is because some of the arguments used in proving Theorem 1.1 can be used to restrict the topology and geometry at infinity of these manifolds. In particular, the fact that these manifolds have finitely many ends, with each end homeomorphic to a product space $N \times [0, \infty)$, is extensively used in the proof. In the paper [68], the authors raised the question that if we replace the sectional curvature assumption in Theorem 5.2 by the Ricci curvature, to what extend will the consequences of the theorem still remain valid. The first obstacle in proving this is to determine if manifolds with non-negative Ricci curvature near infinity has only finitely many ends.

Around the same time, Donnelly [37] proved that the space of bounded harmonic functions, $\mathcal{H}^\infty(M)$, on a manifold with non-negative Ricci curvature near infinity must be finite. Later, in an unpublished work, Cheng showed that if M has non-negative Ricci curvature outside a set D with diameter a, and if the Ricci curvature is bounded from below by $-K$ on D for some $K > 0$, then there exists a constant $C(n, a\sqrt{K}) > 0$ depending only on n and $a\sqrt{K}$ such that

$$\dim \mathcal{H}^\infty(M) \leq C(n, a\sqrt{K}).$$

He also proved that $\mathcal{H}^+(M)$ must be finite dimensional. In view of these developments, if Theorem 5.2 holds for manifolds with non-negative Ricci curvature near infinity, then it will imply that M has finitely many ends if M is non-parabolic. In fact, this provides the motivation behind Theorem 4.7.

6. Applications to Riemannian geometry

Theorem 4.7 allows us to estimate the number of ends, by estimating $\dim \mathcal{H}'(M)$.

Theorem 6.1 (Li-Tam [72]). *Assume that M has asymptotically non-negative Ricci curvature as defined in §1. Then there exists a constant $C(\alpha, n) > 0$ depending only on α and $n = \dim M$ such that*

$$\pi_0^\infty(M) \leq C(\alpha, n).$$

For the special case when M has non-negative Ricci curvature outside some compact set D, then the estimate on $\pi_0^\infty(M)$ takes the form

$$\pi_0^\infty(M) \leq C_1 \exp(C_2\, a\sqrt{K}) + 1$$

where a is the diameter of D, $-K < 0$ is the lower bound of the Ricci curvature on D, and C_1 and C_2 are constants depending only on n.

We would like to point out that independently Cai [17] used a Riemannian geometric method to prove a slightly weaker estimate for the case when M has non-negative Ricci curvature near infinity. Later, Cai-Colding-Yang [18] refined Cai's argument and showed that if $a\sqrt{K}$ is sufficiently small, then M has at most 2 ends. This can be viewed as a generalization of the consequence of the splitting theorem (Theorem 4.9). Using some of the argument of Cai, Liu [82] also proved a ball covering property for these manifolds.

Theorem 6.2 (Liu [82]). *Let M be a complete manifold with non-negative Ricci curvature outside a compact set $D \subset B_p(a)$. Let $-K < 0$ be the lower bound of the Ricci curvature on D. For any $\mu > 0$, there exists a constant $C(n, a\sqrt{K}, \mu) > 0$, such that, for any $r > 0$ there exists a set of points $\{p_1, \ldots, p_k\} \subset \overline{B_p(r)}$ with $k \le C(n, a\sqrt{K}, \mu)$ satisfying*

$$\overline{B_p(r)} \subset \cup_{i=1}^k B_{p_i}(\mu r).$$

Observe that the ball covering theorem implies that

$$\pi_0^\infty(M) \le C(n, a\sqrt{K}, 1/2).$$

It is interesting to point out that it is still not known if the ball covering property holds for manifolds with asymptotically non-negative Ricci curvature. Note that for a non-parabolic manifold, in order to prove that the inequality

$$\dim \mathcal{H}_D^\infty(M) \ge \Pi_0^\infty(M)$$

is indeed an equality, it is necessary to show that any bounded harmonic function must have a unique infinity behavior up to a scalar multiple at each non-parabolic end. For example, for the case when M has non-negative sectional curvature near infinity, the authors [68] showed that a bounded harmonic function must be asymptotically constant at infinity of each non-parabolic end. One way to show this is to develop a spherical Harnack inequality, which asserts that there is a constant depending only on M such that, if f is a positive harmonic function defined on E then

$$f(x) \le C\, f(y)$$

for all x, $y \in \partial B_p(r) \cap E$. This type of inequality allows us to conclude that if

$$\liminf_{x \to E(\infty)} f = 0,$$

then

$$\lim_{x \to E(\infty)} f = 0.$$

If M has non-negative Ricci curvature on E then using Theorem 5.1, we conclude that

$$f(x) \leq C\, f(y) \tag{6.1}$$

for $x \in \partial B_p(r)$ and $y \in B_p(r/2)$. Hence, if we know that $\partial B_p(r) \cap E$ is connected, the ball covering property implies that one can iterate the inequality at most $C(n, a\sqrt{K}, 1/2)$ times and obtain the spherical Harnack inequality. It turns out that if we assume M has finite first Betti number then one can show the basic connectedness of $\partial B_p(r) \cap E$. This line of argument yield the following theorem:

Theorem 6.3 (Li-Tam [73]). *Let M be a complete manifold with non-negative Ricci curvature near infinity. Suppose the first Betti number of M is finite, then all the inequalities of Theorem 4.7 become equalities. In particular,*

$$\pi_0^\infty(M) = \dim \mathcal{H}'(M),$$

and if M is non-parabolic then

$$\pi_0^\infty(M) = \dim \mathcal{H}^+(M) = \dim \mathcal{H}'(M)$$

and

$$\Pi_0^\infty(M) = \dim \mathcal{H}_D^\infty(M).$$

We do not know of a complete manifold with non-negative Ricci curvature near infinity, but have infinite first Betti number. It is plausible that the finiteness of $b_1(M)$ is a consequence of the curvature assumption.

We should also point out that, in proving Theorem 3.3, the authors [73] proved that the ball covering property holds on a manifold satisfying the hypothesis of Theorem 3.3. However, it is not known that the volume comparison condition asserted in Theorem 3.3 holds even on manifolds with non-negative Ricci curvature near infinity and has only 1 end.

Theorem 4.7 can also be applied to study stable minimal hypersurfaces. In 1976, in their study of stable minimal hypersurfaces, Schoen and Yau [97] showed that a complete, oriented, stable minimal hypersurface M^n in a manifold of non-negative Ricci curvature must have

$$\dim H_D^\infty(M) = 1.$$

Exploiting this fact, Cao, Shen, and Zhu proved that such a manifold must have only 1 end.

Theorem 6.4 (Cao-Shen-Zhu [19]) *If M^n $(n \geq 3)$be a complete, oriented, stable minimal hypersurface in $\mathbb{R}^{n+1}$, then*

$$\pi_0^\infty(M) = 1.$$

Their argument used the Sobolev inequality of Michael-Simon [85] to conclude that each end of M must be non-parabolic. Hence one can apply the estimate to conclude that

$$\pi_0^\infty(M) = \Pi_0^\infty(M).$$

An upshot of their argument is the following general fact on complete manifolds. If a complete manifold M satisfies a Sobolev inequality of the form

$$\left(\int_{B_p(r)} |f|^{2p}\right)^{\frac{1}{p}} \leq C \int_{B_p(r)} |\nabla f|^2 \tag{6.2}$$

for some constants $C > 0$, $p \geq 1$ and for all $f \in H^c_{1,2}(B_p(r)$, then each end E of M must either have finite volume or be non-parabolic. In particular, using the necessary criteria (3.1) for non-parabolicity, one concludes that either the volume of E is finite or

$$\int_1^\infty \frac{t\,dt}{V_E(t)} < \infty.$$

This constitutes a gap phenomenon for the volume growth on manifolds satisfying (6.2). Note that a finite volume end is possible. This can be seen by taking a complete metric on a 2-dimensional annulus which has constant -1 curvature. One can arrange the metric to have finite volume on one end, but infinite volume on the other end. In this case, one verifies easily that (6.2) holds for $p = 1$. Also, the manifold is non-parabolic, due to the existence of one non-parabolic end.

7. Function theory under quasi-isometries

Recall that Theorem 2.4 asserts that parabolicity is a quasi-isometric invariant. On the other hand, any topological data is certainly invariant under quasi-isometries. Therefore, it is interesting to ask if the dimensions of the spaces $\mathcal{H}'$, $\mathcal{H}^+$, and $\mathcal{H}_D^\infty$ are quasi-isometric invariants.

An example of Lyons [83] shows that there are manifolds M and N which are quasi-isometric but

$$\dim \mathcal{H}^+(M) = 1 \qquad \text{and} \qquad \dim \mathcal{H}^\infty(N) > 1.$$

On the other hand, Grigor'yan [45], [46] proved that the dimension of the space $\mathcal{H}_D^\infty$ is invariant under quasi-isometry. This leads us to the question that perhaps there are spaces of harmonic functions $\mathcal{H}_1$ and $\mathcal{H}_2$ which play the same roles as $\mathcal{H}^+$ and $\mathcal{H}'$ in Theorem 4.7, but their dimensions are quasi-isometric invariants.

Recall that the De Giorgi-Nash-Moser theory implies that if a manifold M is quasi-isometric to $\mathbb{R}^n$, then it must have the strong Liouville property, namely,

$$\dim \mathcal{H}^+(M) = 1.$$

In view of this Yau conjectured that if a manifold M is quasi-isometric to a manifold with non-negative Ricci curvature then

$$\dim \mathcal{H}^+(M) = 1.$$

In fact, this was verified by Grigor'yan and Saloff-Coste independently.

Theorem 7.1 (Grigor'yan [47] and Saloff-Coste [93]). *Let M be a complete manifold satisfying the following two properties:*

- *Volume doubling property which asserts that there exists a constant $\eta > 0$ depending only on M such that*

$$2^\eta V_p(r) \geq V_p(2r) \tag{7.1}$$

  for all $p \in M$ and $r > 0$; and

- *Weak Poincaré inequality which asserts that there exists a constant $C > 0$ depending only on M such that*

$$\int_{B_p(2r)} |\nabla f|^2 \geq C\, r^{-2} \inf_k \int_{B_p(r)} (f-k)^2 \tag{7.2}$$

  for all functions $f \in H_{1,2}(B_p(2r))$.

Then

$$\dim \mathcal{H}^+(M) = 1.$$

Since both the volume doubling property and the weak Poincaré inequality are invariants under quasi-isometries, and they both hold for manifolds with non-negative Ricci curvature, this implies Yau's conjecture. Along the same direction, Sung pushed this one step further.

Theorem 7.2 (Sung [99]). *Let M be quasi-isometric to a manifold N with non-negative Ricci curvature near infinity. If M has finite first Betti number, then all the inequalities in Theorem 4.7 become equalities. In particular*

$$\dim \mathcal{H}'(M) = \dim \mathcal{H}'(N) = \pi_0^\infty(N) = \pi_0^\infty(N).$$

Moreover if M is non-parabolic then

$$\dim \mathcal{H}^+(M) = \dim \mathcal{H}^+(N) = \pi_0^\infty(N) = \pi_0^\infty(M),$$

and

$$\dim \mathcal{H}_D^\infty(M) = \dim \mathcal{H}_D^\infty(N) = \Pi_0^\infty(N) = \Pi_0^\infty(M).$$

A weaker version of isometry was defined by Kanai [55].

Definition 7.3. A map $f : X \to Y$ between two metric spaces X and Y is a *rough isometry* if there exists constants $k \geq 1$, $b > 0$, and $c > 0$, such that, for all $y \in Y$ there exists $x \in X$ with the properties that

$$d_Y(y, f(x)) \leq c,$$

and for any x_1, $x_2 \in X$

$$k^{-1}\, d_X(x_1, x_2) - b \leq d_Y(f(x_1), f(x_2)) \leq k\, d_X(x_1, x_2) + b.$$

He studied the effect of function theory under rough isometries for a special class of manifolds.

Definition 7.4. A complete manifold is said to have *bounded Ricci geometry* if its Ricci curvature is bounded from below and its injectivity radius is strictly positive.

In [55], Kanai showed that if M has bounded Ricci geometry and it is roughly isometric to $\mathbb{R}^m$ then M satisfies the strong Liouville property. He [56] also showed that if M is roughly isometric to N and both manifolds have bounded Ricci geometry, then M is parabolic if and only if N is parabolic. In 1993, Holopainen [49] generalized these theorems for the p-Laplacian on manifolds with bounded Ricci geometry and finitely many ends, all of which are roughly isometric to Euclidean space. Recently Coulhon and Saloff-Coste generalized Kanai's theorem.

Theorem 7.4 (Coulhon-Saloff-Coste [36]). *Suppose $f : M \to N$ is a rough isometry. Assume that there exists a constant $C > 0$ such that f satisfies*

$$C^{-1} V_x(1) \leq V_{f(x)}(1) \leq C\, V_x(1) \tag{7.3}$$

for all $x \in M$. Also, assume that both manifolds have Ricci curvatures bounded from below, then M is parabolic if and only if N is parabolic. Moreover, if M has non-negative Ricci curvature then N satisfies the strong Liouville property.

In fact, the Ricci curvature lower bound can be replaced by the assumption that both manifolds satisfy a local parabolic Harnack inequality. Also the non-negative Ricci curvature assumption can be replaced by a global parabolic Harnack inequality.

In [36], the authors define a rough isometry satisfying (7.3) as an isometry at infinity. An upshot of their analysis is that if a manifold M has Ricci curvature bounded from below, and it is isometric at infinity to a manifold with non-negative Ricci curvature, then M must satisfy (7.1) and (7.2). This fact will revisit in the discussion in §11.

8. Applications to Kähler geometry

In this section, we will discuss various applications of harmonic function theory to Kähler geometry.

Theorem 8.1 (Napier-Ramachandran [91]). *Let M be a complete Kähler manifold. Assume that M has bounded geometry, or that it admits a pluri-subharmonic exhaustion function, then the following statements hold:*

(a) *If $\pi_0^\infty(M) \geq 2$ then $b_1(M) > 0$;*

(b) *If $\pi_0^\infty(M) \geq 3$ then there exists a complete Riemann surface Σ and a proper, surjective, holomorphic map $h : M \to \Sigma$ with compact fibers.*

Theorem 8.2 (Li-Ramachandran [66]). *Let M be a complete Kähler manifold. Suppose $R(x)$ is function defined on M which is a lower bound of the Ricci curvature satisfying*

$$Ric_M(x) \geq R(x)$$

for all $x \in M$. *Let*

$$R_-(x) = \max\{-R(x), 0\}$$

be the negative part of the function R *which is assumed to be integrable, i.e.,*

$$\int_M R_- < \infty.$$

Also assume that

$$R(x) \geq -\epsilon\, \rho^{-2}(x),$$

for some sufficiently small $\epsilon > 0$. *Then the following statements hold:*

(a) *If* $\pi_0^\infty(M) \geq 2$ *then* M *must be parabolic and* $b_1(M) > 0$;

(b) *If* $\pi_0^\infty(M) \geq 3$ *then there exists a complete parabolic Riemann surface* Σ *and a proper, surjective, holomorphic map* $h : M \to \Sigma$ *with compact fibers.*

Let us remark that the curvature assumption in Theorem 8.2 is sharp. In fact, let us consider $M = \mathbb{C}^2 \setminus \{p_1, \ldots, p_k\}$, where $\{p_i\}$ are k well-spaced points in $\mathbb{C}^2$. For any $\delta > 0$, there exists [66] a complete Kähler metric on M such that the Ricci curvature satisfies

$$Ric_M(x) \geq -(1+\delta)\, \rho^{-2}(x).$$

Obviously the conclusion of Theorem 8.2 is invalid. In particular this indicates that ϵ has to be less than 1 in the assumption of Theorem 8.2. In this example, we can also take the number of points k to be infinite.

Theorem 8.3 (Li [61], [62]). *Let* M *be a complete Kähler manifold with non-negative sectional curvature near infinity. Then the conclusion of Theorem 8.2 holds. Moreover, if* $\pi_o^\infty(M) \geq 3$, *then for each end* E *of* M, *the fibration*

$$h : E \to h(E) \subset \Sigma$$

is a Riemannian fibration with fiber given by a compact Kähler manifold, N, *with non-negative sectional curvature. Locally* E *is a Riemannian product of* N *and open subsets* $U \subset \Sigma$. *Also,* Σ *is a parabolic surface with non-negative curvature near infinity.*

Corollary 8.4. *If* M *be a complete Kähler manifold with positive sectional curvature near infinity, then* M *has most 2 ends.*

Using a vanishing theorem of Li-Yau [80], one can prove a rather general theorem which put a restriction on the number of non-parabolic ends for a Kähler manifold.

Theorem 8.5 (Li-Tam [72]). *Let M be a complete Kähler manifold of complex dimension m. Suppose $R(x)$ is function defined on M which is a lower bound of the Ricci curvature satisfying*

$$Ric_M(x) \geq R(x)$$

for all $x \in M$. Let

$$R_-(x) = \max\{-R(x), 0\}$$

be the negative part of the function R. If

$$\int_M R_- < \infty,$$

and the L^q-norm of R_- over the geodesic ball of radius r centered at some fixed point $p \in M$ satisfies

$$\int_{B_p(r)} R_-^q = o(r^{\beta(q-1)})$$

for some $q > m$ and $\beta < 2/(m-2)$, then

$$\Pi_0^\infty(M) \leq 1.$$

9. Harmonic functions of polynomial growth

In 1980, Cheng [24] observed that the localized version of the Yau's gradient estimate (Theorem 5.1) can be used to show that a manifold with non-negative Ricci curvature does not admit any non-constant sublinear growth harmonic functions.

Theorem 9.1 (Cheng [24]). *Let M be a complete manifold with non-negative Ricci curvature. There are no non-constant harmonic functions defined on M which is of sublinear growth, i.e.,*

$$|f(x)| \leq o(\rho(x))$$

as $x \to \infty$, where $\rho(x)$ denotes the distance function to some fixed point $p \in M$.

In fact, in the same paper, Cheng proved that a similar statement is true for harmonic maps into a Cartan-Hadamard manifold. Note that on the n-dimensional Euclidean space, $\mathbb{R}^n$, the set of harmonic polynomials generate all the polynomial growth harmonic functions. In particular, for each $d \in \mathbb{Z}^+$, the space of harmonic polynomials $\mathcal{H}_d(\mathbb{R}^n)$ of degree at most d is of dimension

$$\begin{aligned}\dim \mathcal{H}_d(\mathbb{R}^n) &= \binom{n+d-1}{d} + \binom{n+d-2}{d-1} \\ &\sim \frac{2}{(n-1)!} d^{n-1}.\end{aligned}$$

Cheng's theorem asserts that manifolds with non-negative Ricci curvature is quite similar to $\mathbb{R}^n$ for harmonic functions which grow sublinearly. In view of this result, and the fact that all polynomial growth harmonic functions in $\mathbb{R}^n$ are generated by harmonic polynomials, Yau conjectured that the space of harmonic functions on a manifold with non-negative Ricci curvature of at most polynomial growth at a fixed degree must be of finite dimensional. To state this more precisely, let us define the following spaces of harmonic functions.

Definition 9.2. Let $\mathcal{H}_d(M)$ be the space of harmonic functions f defined on a complete manifold M satisfying the growth condition

$$|f(x)| = O(\rho^d(x)).$$

Note that in this notation, $\mathcal{H}_0(M) = \mathcal{H}^\infty(M)$.

Conjecture 9.3 (Yau [109]). *Let M be a complete manifold with non-negative Ricci curvature. The dimension of $\mathcal{H}_d(M)$ is finite for all $d \in \mathbb{R}^+$.*

In fact, Yau also raised the question if

$$\dim \mathcal{H}_d(M) \leq \dim \mathcal{H}_d(\mathbb{R}^n)$$

for manifolds with non-negative Ricci curvature. In 1989, the author and L. F. Tam [70] considered the case when $d = 1$.

Theorem 9.4 (Li-Tam [70]). *Let M be a complete manifold with non-negative Ricci curvature. Suppose the volume growth of M satisfies*

$$V_p(r) = O(r^k)$$

for some constant $k > 0$. Then

$$\dim \mathcal{H}_1(M) \leq \dim \mathcal{H}_1(\mathbb{R}^k) = k + 1.$$

Observe that the assumption on the Ricci curvature and the Bishop comparison theorem assert that

$$V_p(r) \leq \omega_n \, r^n,$$

where ω_n is the volume of the unit ball in $\mathbb{R}^n$. On the other hand, a theorem of Yau [108] (also see [23]) asserts that $V_p(r)$ must grow at least linearly. Hence the constant k in Theorem 9.3 must exist and satisfy $1 \leq k \leq n$.

Corollary 9.5. *Let M be a complete manifold with non-negative Ricci curvature. Then*

$$\dim \mathcal{H}_1(M) \leq n + 1.$$

This theorem leads us to consider two obvious questions.

Question 9.6 (Li-Tam [70]). Let M be a complete manifold with non-negative Ricci curvature. Suppose the volume growth of M satisfies

$$V_p(r) = O(r^k)$$

for some constant $k > 0$. Is it true that

$$\dim \mathcal{H}_d(M) \leq \dim \mathcal{H}_d(\mathbb{R}^k) = \binom{k+d-1}{d} + \binom{k+d-2}{d-1} \; ?$$

The answer to this question was affirmatively verified by Kasue [57] and Li-Tam [71], independently, for the case when M is of dimension 2. In fact, they considered surfaces satisfying a much weaker curvature condition. We will defer the discussion of this until the next section.

Question 9.7. What can we say about the manifold on which equality is achieved in the upper bound given by Corollary 9.5, or even Theorem 9.4?

The first result in this direction, was due to the author, where he assumed, in addition to non-negative Ricci curvature, that the manifold is Kähler.

Theorem 9.8 (Li [64]). *Let M be a complete Kähler manifold with non-negative Ricci curvature. If*

$$\dim \mathcal{H}_1(M) = 2m + 1$$

where $m = \dim_{\mathbb{C}}(M)$*, then* M *must be isometrically biholomorphic to* C^m*.*

Later, Cheeger-Colding-Minicozzi proved this theorem without the Kähler assumption. In fact, they proved a splitting type theorem for the tangent cone at infinity.

Theorem 9.8 (Cheeger-Colding-Minicozzi [20]). *Let* M *be a complete manifold with non-negative Ricci curvature. If*

$$\dim \mathcal{H}_1(M) = k + 1$$

then any tangent cone $C(M)$ *at infinity of* M *must spit into* $\mathbb{R}^k \times N$ *where* N *is a (possibly singular) metric cone. In particular, if*

$$\dim \mathcal{H}_1(M) = n + 1$$

then M *must be isometric to* $\mathbb{R}^n$*.*

In a recent paper, Wang [104] estimated $\dim \mathcal{H}_1(M)$ for manifolds with non-negative Ricci curvature outside a compact set and have finite first Betti number.

Theorem 9.10 (Wang [104]). *Let* M *be a complete manifold with non-negative Ricci curvature outside the geodesic ball* $B_p(a)$ *centered at* $p \in M$ *of radius* $a > 0$*. Assume that the first Betti number of* M *is finite. Suppose that the Ricci curvature on* $B_p(a)$ *has a lower bound given by*

$$Ric_M \geq -K$$

for some constant $K > 0$*. There exist a constant* $C(n, a, K) > 0$ *depending only on* n*,* a*, and* K *such that*

$$\dim \mathcal{H}_1(M) \leq C(n, a, K).$$

10. Surfaces of finite total curvature

Definition 10.1. A complete surface M is said to have *finite total curvature* if the negative part of its Gaussian curvature is integrable. More precisely, if $K(x)$ denotes the Gaussian curvature on M and its negative part is defined by

$$K_-(x) = \max\{-K(x), 0\},$$

then M has finite total curvature if

$$\int_M K_- < \infty.$$

These kind of surfaces were first studied by Cohn-Vossen [28] in connection to generalizing a Gaussian-Bonnet formula for complete surfaces. He showed that if for any compact exhaustion Ω_i of a complete surface M, the sequence

$$\int_{\Omega_i} K \to \int_M K$$

converges to a possibly infinite limit denoted by $\int_M K$, then the inequality

$$\int_M K \leq 2\pi\chi(M) \tag{10.1}$$

holds, where $\chi(M)$ is the Euler characteristic of M. This inequality is referred to as the Cohn-Vossen inequality. Later, Huber [53] showed that if

$$\int_M K_- < \infty$$

then M must be conformally equivalent to a compact Riemann surface with finite punctures. Moreover, the Cohn-Vossen inequality is valid. Note that since

$$K = K_+ - K_-$$

for

$$K_+(x) = \max\{K(x), 0\}$$

being the positive part of K, the Cohn-Vossen inequality implies that

$$\int_M K_+ \leq \int_M K_- + 2\pi\chi(M).$$

On the other hand, Huber's theorem asserts that the right hand side is finite, hence

$$\int_M K_+ < \infty$$

follows as a consequence. An upshot of this is that

$$\int_M K_- < \infty$$

implies

$$\int_M |K| < \infty.$$

This justifies the term total curvature in Definition 10.1. After Huber, there were much work done [38], [48], [39] in understanding this class of surfaces. In particular, Hartman – though had not explicitly stated in his paper – showed that the correction term in Cohn-Vossen inequality can be computed in terms of the volume growth of each end. Specifically, since M is conformally equivalent to a compact Riemann surface with finite punctures, M has finitely many ends given by $\{e_i\}_{i=1}^k$. Moreover, each end e_i is conformally equivalent to a punctured disk, hence must be parabolic. The finite total curvature assumption implies that the volume growth of M is at most quadratic. For each end e_i, we can define

$$\alpha_i = 1 - \lim_{r\to\infty} \frac{V_{e_i}(r)}{\pi r^2},$$

with $\alpha_i \leq 1$. Hartman showed that

$$2\pi\chi(M) - \int_M K = 2\pi \sum_{i=1}^k \lim_{r\to\infty} (1 - \alpha_i).$$

The next theorem indicates that these constants α_i also play an important role in the function theory of M.

Theorem 10.2 (Li-Tam [71]). *Let M be a complete surface with finite total curvature. Then*

$$\dim \mathcal{H}_d(M) \leq \sum_{i=1}^k \dim \mathcal{H}_{d(1-\alpha_i)}(\mathbb{R}^2),$$

and if M has quadratic area growth, i.e., $\alpha_i < 1$ for some i, then for any $\epsilon > 0$

$$\dim \mathcal{H}_d(M) \geq \sum_{i=1}^k \dim \mathcal{H}_{d(1-\alpha_i)-\epsilon}(\mathbb{R}^2) - k'$$

where k' is the number of ends with $\alpha_i = 1$. Here we are taking the convention that

$$\dim \mathcal{H}_d(\mathbb{R}^2) = 0$$

for $d < 0$.

This estimate can be sharpen, when the manifold has non-negative curvature near infinity.

Theorem 10.3 (Li-Tam [71]). *Let M be a complete manifold with non-negative Gaussian curvature near infinity, then*

$$\dim \mathcal{H}_d(M) = \sum_{i=1}^{k} \dim \mathcal{H}_{d(1-\alpha_i)}(\mathbb{R}^2)$$

for all $d \geq 1$.

Note that when we restrict ourselves to manifolds with non-negative Gaussian curvature, then either M is a cylinder $\mathbb{S}^1 \times \mathbb{R}$ or M has only one end. In the first case when $M = \mathbb{S}^1 \times \mathbb{R}$, the polynomial growth harmonic functions on M are generated by the constant function and the linear function t which parameterizes $\mathbb{R}$. Hence

$$\dim \mathcal{H}_d(M) = \begin{cases} 1 & \text{if} \quad d < 1 \\ 2 & \text{if} \quad d \geq 1. \end{cases}$$

When M has only one end, then according to Theorem 10.3,

$$\dim \mathcal{H}_d(M) = \dim \mathcal{H}_{d(1-\alpha)}(\mathbb{R}^2)$$

where

$$1 - \alpha = \lim_{r \to \infty} \frac{V(r)}{\pi\, r^2}.$$

Hence if M has linear volume growth, then

$$\dim \mathcal{H}_d(M) = 1$$

for all $d \geq 1$. On the other hand, if M has quadratic volume growth, then the curvature assumption implies that $\alpha \geq 0$ and

$$\dim \mathcal{H}_d(M) \leq \dim \mathcal{H}_d(\mathbb{R}^2).$$

In either case, Question 9.6 is answered affirmatively for surfaces. We should point out that Kasue [57] independently proved the upper bound in Theorem 10.2.

11. High dimensions

Before we discuss the higher dimensional development of Yau's conjecture, we would like to point out different points of view of this type of problems. The first is to consider polynomial growth solutions for elliptic operators in $\mathbb{R}^n$. Let

$$L = \frac{\partial}{\partial x^i}\left(a_{ij}\frac{\partial}{\partial x^j}\right) \tag{11.1}$$

be an elliptic operator defined on $\mathbb{R}^n$ with measurable coefficients (a_{ij}) satisfying the uniformly bounded conditions,

$$\lambda\,(\delta_{ij}) \le (a_{ij}) \le \Lambda\,(\delta_{ij}) \tag{11.2}$$

for some constants $\lambda, \Lambda > 0$. The Harnack inequality of De Giorgi-Nash-Moser implies that L has no non-constant bounded solutions. In fact, if we define

$$\mathcal{H}_d(L) = \{f \in H^{1,2}(\mathbb{R}^n) \mid L(f) = 0, |f|(x) = O(\rho^d)\}$$

then the De Giorgi-Nash-Moser theory implies that there exists $d_0 > 0$ depending on Λ/λ, such that,

$$\dim \mathcal{H}_d(L) = 1$$

for all $d < d_0$. For general d, Avelleneda and Lin [6] first considered the special case when the coefficients (a_{ij}) are periodic, Lipschitz continuous functions in all the variables. They showed that there is a linear isomorphism between polynomial growth solutions of

$$L(f) = 0$$

to harmonic polynomials in $\mathbb{R}^n$, hence gave a precise estimate on $\dim \mathcal{H}_d(L)$. The Lipschitz condition was later dropped in a paper of Moser-Struwe [88]. In a recent work of Lin [81], he considered elliptic operators satisfying both (11.1) and (11.2) plus an asymptotically conic condition (see Definition 2.1 of [81]). The condition roughly says that the operator is asymptotic to a unique conic operator. With this extra condition, Lin proved that

$$\dim \mathcal{H}_d(M) < \infty$$

for all $d > 0$. Moreover, the dimension of each $\mathcal{H}_d(M)$ can be estimated explicitly using information on L.

Recently, Zhang [110] proved a similar dimension estimate for $\dim \mathcal{H}_d(L)$ for a class of uniformly elliptic operators of divergence form that is more general than those in [81]. He considered those operators which are not necessarily asymptotic to a unique conic operator, but those who are asymptotic to a periodic family of conic operators. In this case, he proved that

$$\dim \mathcal{H}_d(L) \le C d^{n-1}.$$

Another class of elliptic operators which have some baring to this problem are uniformly elliptic operators of non-divergence form. Let

$$L = a_{ij}\frac{\partial}{\partial x^i}\frac{\partial}{\partial x^j} \tag{11.3}$$

be an elliptic operator defined on $\mathbb{R}^n$ with coefficients (a_{ij}) satisfying (11.2). Then the Harnack inequality of Krylov-Safonov (see [41]) implies that there exists $d_0 > 0$ depending on Λ/λ, such that,

$$\dim \mathcal{H}_d(L) = 1.$$

In yet another direction, Bombieri-Giusti [13] proved a Harnack inequality for uniformly elliptic operators on area minimizing hypersurfaces M in $\mathbb{R}^n$. Hence in the same spirit as above,

$$\dim \mathcal{H}_d(M, L) = 1$$

for d sufficiently small, where L is a uniformly elliptic operator on M

Recently there has been substantial developments on Yau's conjecture in higher dimensions. We will take this opportunity to document various contributions and give the historical account in this direction. The first partial result was indirectly given by Bando-Kasue-Nakajima [10]. They proved that if the sectional curvature, K_M of a complete n-dimensional manifold satisfies

$$|K_M(x)| \leq C\,\rho^{-(2+\epsilon)}(x)$$

for some constants $C, \epsilon > 0$ and if the volume growth for each end E satisfies

$$V_E(r) \geq C\,r^n,$$

then M is asymptotically locally Euclidean. This fact is sufficient [11], [57], [58] to imply that

$$\dim \mathcal{H}_d(M) < \infty$$

for all d. In a series of papers, Colding-Minicozzi [29], [34] proved a number of theorems which eventually lead to and went beyond Yau's conjecture. First, they proved the case when M has non-negative Ricci curvature and has maximal volume growth. Eventually, they improved their argument to give a dimension estimate for $\dim \mathcal{H}_d(M)$ for manifolds satisfying the volume doubling property (7.1) and the Poincaré inequality.

In the context of this section, we will say that a manifold satisfies the Poincaré inequality if there exists a constant $\alpha > 0$, such that, the first Neumann eigenvalue for the Laplacian on $B_x(r)$ satisfies

$$\lambda_1(B_x(r)) \geq \alpha\, r^{-2} \tag{11.4}$$

for all $x \in M$ and $r > 0$. It is worth pointing out that though (11.4) is stronger than the weak Poincaré inequality (7.2), a covering argument of Jerison [54] asserts that the volume doubling property (7.1) together with the weak Poincaré inequality (7.2), in fact, imply that Poincaré inequality. In [34], they also considered a volume growth property, which asserts that there exists a constant $\nu > 0$ such that

$$\left(\frac{r'}{r}\right)^{\nu} V_x(r) \geq V_x(r') \tag{11.5}$$

for all $x \in M$ and $0 < r < r'$. Using these conditions, the main result which they proved can be stated as follows:

Theorem 11.1 (Colding-Minicozzi [32], [34]). *Let M be a complete manifold satisfying the Poincaré inequality (11.4).*

1) Suppose M also satisfies the volume doubling property (7.1), then there exists a constant $C > 0$ depending only on n and α, such that,

$$\dim \mathcal{H}_d(M) \leq C\, d^{\eta}$$

for all $d > 1$.

2) Suppose M also satisfies the volume growth property (11.5), then there exists a constant $C > 0$ depending only on n an α, such that,

$$\dim \mathcal{H}_d(M) \leq C\, d^{\nu-1}.$$

for all $d > 1$.

In particular, this confirms Yau's conjecture since manifolds with non-negative Ricci curvature satisfy both the Poincaré inequality [16] and the volume growth property [12]. In this case, $\nu = n$. This gives a sharp growth rate as $d \to \infty$, as indicated by the case when $M = \mathbb{R}^n$. We would also like to point out that the first estimate of Colding-Minicozzi using the volume doubling property is not sharp in the power of d. The sharp power should be $\eta - 1$, since $\eta = n$ if $M = \mathbb{R}^n$. Also, the volume doubling property and the volume growth property are related. It is clear that, the volume growth property implies the volume doubling property with $\eta = \nu$. Moreover, one can easily argue that the volume doubling property (7.1) implies

$$\left(\frac{2r'}{r}\right)^{\eta} V_p(r) \geq V_p(r')$$

for $r' \geq r$. In this sense, the volume doubling property is weaker than the volume growth property. In view of the relationship between the volume doubling property and the volume growth property, it is convenient to define the weak volume growth property which encapsulate both properties. A manifold is said to have the weak volume growth property if there exists constants $C_1 > 0$ and $\eta > 0$ such that

$$C_1 \frac{(r')^{\eta} - r^{\eta}}{r^{\eta}} V_p(r) \geq V_p(r') - V_p(r) \tag{11.6}$$

for all $p \in M$ and $0 < r \leq r'$.

Due to the work of Grigor'yan [47] and Saloff-Coste [93], [94] it is known that the volume doubling property (7.1) together with the Poincaré inequality (11.4) is equivalent to the parabolic Harnack inequality which implies the elliptic Harnack inequality. Grigor'yan has informed me, by way of an example, that the parabolic Harnack inequality is stronger than the elliptic Harnack inequality. In any case, (7.1) and (11.4) imply a mean value inequality of the form

$$V_x(r)\, f(x) \leq C_2 \int_{B_x(r)} f \tag{11.7}$$

for some constant C_2, and for any non-negative function f defined on $B_x(r)$ satisfying

$$\Delta f \geq 0.$$

In this case, C_2 will depend only on η, α, and n. Indeed, it was argued in [94] that (7.1) and (11.4) imply a Sobolev inequality of the form

$$\begin{aligned}\left(\int_{B_x(r)} f^{\frac{2\mu}{\mu-2}}\right)^{\frac{\mu-2}{\mu}} \\ \leq C_3\, V_x(r)^{-2/\mu} \left(r^2 \int_{B_x(r)} |\nabla f|^2 + \int_{B_x(r)} f^2\right)\end{aligned} \tag{11.8}$$

for any compactly supported function $f \in H^c_{1,2}(B_x(r))$, where $C_3 > 0$ and $\mu > 2$ are some fixed constants, and $x \in M$ and $r > 0$ are arbitrary. It is now clear that by running the Moser iteration argument [87] using

(7.1) and (11.8), one obtains (11.7). In fact, Moser's argument actually implies the mean value inequality not only for non-negative subharmonic functions, but for non-negative functions satisfying

$$\Delta f \geq -g\, f \tag{11.9}$$

where g is a non-negative function satisfying some appropriate decay condition (see [63]). In this case, C_2 in (11.7) will depend on μ, C_3 and g. In particular, a special case of this situation is when g has compact support.

Colding-Minicozzi circulated an announcement [32] of Theorem 11.1 in June 1996 together with a number of applications using Theorem 11.1. In [33] and [34], they proved many of the announced theorems, including Theorem 11.1. Shortly after the circulation of [34], the author [65] came up with a simple argument using a weaker assumption.

Theorem 11.2 (Li [65]). *Let M be a complete manifold satisfying the weak volume growth property (11.6). Let K be a linear space of sections of a rank-q vector bundle E over M. Suppose each $u \in K$ satisfies the growth condition*

$$|u|(x) = O(\rho^d(x))$$

as the distance ρ to some fixed point $p \in M$ goes to infinity for some constant $d \geq 1$, and the mean value inequality

$$C_2 \int_{B_x(r)} |u|^2 \geq V_x(r)\, |u|^2(x)$$

for all $x \in M$ and $r > 0$. Then there exists a constant $C > 0$ depending only on η and C_1, such that

$$\dim K \leq q\, C\, C_2\, d^{\eta - 1}.$$

In their announcement [32], the authors also announced, without indication of the proof, that $\mathcal{H}_d(M)$ is finite dimensional if M is a minimal submanifold in Euclidean space with Euclidean volume growth. In the same note, they also announced a finite dimensionality result for polynomial growth harmonic sections of at most degree d on a Hermitian vector bundle with nonnegative curvature over a manifold with non-negative Ricci curvature. Shortly after the circulation of [65], Colding-Minicozzi circulated a new preprint [35] providing the proofs for the minimal submanifold and the harmonic sections cases. In this paper, they also used a form of mean value inequality similar to (11.7). However their argument did not provide the sharp power in d.

As indicated in the above discussion, as long as $|u|^2$ satisfies an inequality of the form (11.9) for some compactly supported g, and M also satisfies the Sobolev inequality (11.8), then the mean value inequality follows as a consequence. In particular, the following corollaries can be deduced from Theorem 11.2.

Corollary 11.3. *Let M^n be a complete manifold satisfying conditions (11.6) and (11.7) for non-negative subharmonic functions. Then*

$$\dim \mathcal{H}_d(M) \leq C\, C_2\, d^{\eta-1}$$

for all $d \geq 1$. In particular, if M is quasi-isometric to a manifold with non-negative Ricci curvature, then

$$\dim \mathcal{H}_d(M) \leq C\, d^{n-1}$$

for all $d \geq 1$.

In view of the discussion after Theorem 7.4, the work of Coulhon and Saloff-Coste [36] together with Theorem 11.2 implies that one can also deal with the case when M is roughly isometric to a manifold with non-negative Ricci curvature.

Corollary 11.4. *Let M^n be a complete manifold with Ricci curvature bounded from below. Suppose M is isometric at infinity to a manifold with non-negative Ricci curvature, then*

$$\dim \mathcal{H}_d(M) \leq C\, d^{n-1}$$

for all $d \geq 1$.

Using the fact that the mean value inequality holds for functions satisfying (11.9), Theorem 11.2 also implies the next corollary.

Corollary 11.5. *Let M be a manifold whose metric ds^2 is obtained by a compact perturbation of another metric ds_0^2 which has non-negative Ricci curvature. Suppose*

$$\mathcal{H}^p_d(M) = \{u \in \Lambda^p(M) \,|\, \delta u = 0,\ |u|(x) = O(\rho^d(x)) \text{ as } \rho \to \infty\}$$

denotes the space of harmonic p-forms of at most polynomial growth of degree $d \geq 1$. Then

$$\dim \mathcal{H}^1_d(M) = \dim \mathcal{H}^{n-1}_d(M) \leq n\, C\, d^{n-1}.$$

If we further assume that ds_0^2 has non-negative curvature operator, then

$$\dim \mathcal{H}_d^p(M) \le \binom{n}{p} C\, d^{n-1}.$$

Corollary 11.6. *Let M be a complete Ricci flat manifold. Suppose $\mathcal{K}_d(M)$ is the space of Killing vector fields on M which has polynomial growth of at most degree $d \ge 1$. Then*

$$\dim \mathcal{K}_d(M) \le C\, d^{n-1}.$$

Corollary 11.7. *Let M^m be a complete Kähler manifold of complex dimension m. Assume that M satisfies conditions (11.6) and (11.7) for non-negative subharmonic functions. Suppose E is a rank-q Hermitian vector bundle over M and that the mean curvature (in the sense defined in [59]) of E is non-positive. Let $\bar{\mathcal{H}}_d(M, E)$ be the space of holomorphic sections which is polynomial growth of at most degree $d \ge 1$. Then*

$$\dim \bar{\mathcal{H}}_d(M, E) \le q\, C\, \lambda\, d^{2m-1}.$$

In particular the space of polynomial growth holomorphic functions of at most degree $d \ge 1$ is bounded by

$$\dim \bar{\mathcal{H}}_d(M) \le C\, \lambda\, d^{2m-1}.$$

Complex and algebraic geometers have been interested in estimating the dimension of $\bar{\mathcal{H}}_d(M, E)$ for many years. We would like to refer to the survey article of Mok [86] for a more detail history and reference in this direction. Another interesting result was due to Wu, Tam and the author, where they considered Kähler manifolds with at most quadratic volume growth. In this case, no additional assumption on the manifold is necessary.

Theorem 11.8 (Wu [105], [106]). *Let M be a complete Kähler manifold. If M has subquadratic volume growth*

$$V_p(r) = o(r^2),$$

then M does not admit any non-constant polynomial growth holomorphic functions. If M has quadratic volume growth

$$V_p(r) = O(r^2),$$

then there exists constants $C(m,d) > 0$ *depending only on* m *and* d *such that*

$$\dim \bar{\mathcal{H}}_d(M) \leq C(m,d).$$

Theorem 11.2 also applies to uniformly elliptic operators on $\mathbb{R}^n$. The following corollaries are consequences of the theorem.

Corollary 11.9. *Let*

$$L = \frac{\partial}{\partial x^i}\left(a_{ij}\frac{\partial}{\partial x^j}\right)$$

be an elliptic operator of divergence form defined on $\mathbb{R}^n$ *with uniformly bounded coefficients satisfying (11.2). Let*

$$\mathcal{H}_d(L) = \{u \in H^{loc}_{1,2}(\mathbb{R}^n) \,|\, L(u) = 0,\ |u|(x) = O(\rho^d(x)) \text{ as } \rho \to \infty\}$$

be the space of L*-harmonic functions that has polynomial growth of degree at most* $d \geq 1$. *Then*

$$\dim \mathcal{H}_d(L) \leq C\, d^{n-1}.$$

Corollary 11.10. *Let*

$$L = a_{ij}\frac{\partial^2}{\partial x^i \partial x^j}$$

be an elliptic operator of non-divergence form defined on $\mathbb{R}^n$ *with uniformly bounded coefficients satisfying (11.2). Let*

$$\mathcal{H}_d(L) = \{u \in H^{loc}_{2,n}(\mathbb{R}^n) \,|\, L(u) = 0,\ |u|(x) = O(\rho^d(x)) \text{ as } \rho \to \infty\}$$

be the space of L*-harmonic functions that has polynomial growth of degree at most* $d \geq 1$. *Then*

$$\dim \mathcal{H}_d(L) \leq C\, d^{n-1}.$$

Corollary 11.11. *Let* M^n *be a complete minimal surface in* $\mathbb{R}^N$. *Suppose* ρ_0 *is the distance function of* $\mathbb{R}^N$ *with respect to some fixed point* $p \in M$. *Assume the volume growth of* M *satisfies*

$$V(B_0(r) \cap M) \leq C r^n$$

where $B_0(r) \subset \mathbb{R}^N$ is the Euclidean ball center at p of radius r. Let L be a uniformly elliptic operator defined on M. Suppose $\mathcal{H}_d(M, L)$ is the space of L-harmonic functions f on M satisfying the growth condition

$$|f|(x) = O(\rho_0^d(x))$$

for some $d \geq 1$. Then

$$\dim \mathcal{H}_d(M, L) \leq C\, d^{n-1}$$

for some constant C depending on M and the ellipticity constants of L.

We would like to comment that, the mean value inequality (11.7) is weaker than the Poincaré inequality (11.5). An interesting fact is that Theorem 11.2 allows one to prove that

$$\dim \mathcal{H}_0(M) < \infty$$

without implying

$$\dim \mathcal{H}_0(M) = 1.$$

On the other hand, Theorem 7.1 asserts that the Poincaré inequality and the volume doubling property imply that

$$\dim \mathcal{H}_0(M) = 1.$$

An example of a manifold satisfying the hypothesis of Theorem 11.2 but

$$\dim \mathcal{H}_0(M) > 1$$

is $\mathbb{R}^n \sharp \mathbb{R}^n$ for $n \geq 3$.

In [35], the authors verified that a complete manifold satisfying the assumptions of Theorem 3.3 has the volume doubling property and the mean value inequality. Hence, Theorem 11.2 applies to this case. In a recent preprint of Tam [101], he relaxed the volume comparison condition of Theorem 3.3. Instead of assuming that the volume comparison condition holds on the whole manifold, he only assumed that it hold for each end individually. Note that since the volume growth of each end may be different, this covers a more general situation. In the same article, he also considered harmonic forms on surfaces of finite total curvature and on manifolds with asymptotically non-negative curvature operator .

Surprisingly, as it turned out, Wang and the author observed that the conditions to ensure the validity of Yau's original conjecture, namely the finite dimensionality of $\mathcal{H}_d(M)$, can be weaken. If one does not aim for the sharp order estimate as in Theorem 11.2, the weak volume growth property (11.6) can be replaced by a polynomial volume growth assumption. Sometimes it is also convenient to replace the mean value inequality (11.7) by the weak mean value inequality of the form

$$V_x(r)\, f(x) \leq C_4 \int_{B_x(br)} f \tag{11.10}$$

for some constants $C_4 > 0$, $b \geq 1$, and for any non-negative subharmonic function f defined on $B_x(\beta r)$. Note that if the manifold satisfy the weak volume growth property (11.6), then the weak mean value inequality is equivalent to the mean value inequality. On the other hand, without the weak volume growth property, the weak mean value property is, in general, easier to obtain. For example, if a manifold satisfies the Sobolev inequality (11.8), then the Moser iteration argument yields the weak mean value inequality.

Theorem 11.12 (Li-Wang [76]). *Let M be a complete manifold whose volume growth satisfies*

$$V_x(r) = O(r^\nu)$$

as $r \to \infty$ for some $x \in M$ and $\nu > 0$. Assume that M also satisfies the weak mean value inequality (11.10). Then

$$\dim \mathcal{H}_d(M) \leq C_4(2b+1)^{(2d+\nu)}.$$

As we pointed out, the Sobolev inequality (11.8) implies the weak mean value inequality. If we choose $f \in H^c_{1,2}(B_x(r))$ to be the non-negative function satisfying $f = 1$ on $B_x(1)$, and $f = 0$ on $M \setminus B_x(2)$, then after applying to (11.8), we conclude that $V_x(r) = O(r^\mu)$. Hence, Theorem 11.12 can be stated with only the assumption of (11.8). However, as indicated by a recent paper of Li-Wang [77], one can actually do much better on the estimate if we assume (11.8).

Theorem 11.13 (Li-Wang [77]). *Let M be a complete manifold satisfying the Sobolev inequality (11.8). Then*

$$\dim \mathcal{H}_d(M) \leq C\, d^\mu$$

for some constant $C > 0$.

It is also worthwhile to point out that Theorem 11.12 can be applied to harmonic sections of vector bundles. In particular, a weaker estimate as in Corollary 11.7 holds for manifolds satisfying the conditions of Theorem 11.12. The argument of [65] and [76] also can be applied to study d-massive sets [75], which yields interesting applications to the image structure of harmonic maps.

Recently, Sung-Tam-Wang [100] considered the effect of $\dim \mathcal{H}_d(M)$ under connected sums. They proved a formula for $\dim \mathcal{H}_d(M_1 \sharp M_2)$ in terms of $\dim \mathcal{H}_d(M_1)$ and $\dim \mathcal{H}_d(M_2)$.

In an attempt to give an affirmative answer to Question 9.6, Li-Wang [77] recently proved a sharp asymptotic estimate for $\dim \mathcal{H}_d(M)$ when M has non-negative sectional curvature.

Theorem 11.14 (Li-Wang [77]). *Let M^n be a complete manifold with non-negative sectional curvature. Let us define α by*

$$\alpha = \lim_{r\to\infty} r^{-n}\, V_p(r).$$

The Bishop comparison theorem asserts that $0 \le \alpha \le omega_n$ where ω_n is the volume of the unit Euclidean ball. Then the truncated sum of $\dim \mathcal{H}_d(M)$ must satisfy

$$\limsup_{d\to infty} d^{-n} \sum_{i=1}^{d} \dim \mathcal{H}_i(M) \le \frac{2\alpha}{n!\,\omega_n}.$$

Moreover, the equality

$$\limsup_{d\to infty} d^{-n} \sum_{i=1}^{d} \dim \mathcal{H}_i(M) = \frac{2}{n!}$$

holds if and only if $M = \mathbb{R}^n$.

In another recent paper of Li-Wang [78], they also proved a parallel version of this theorem for uniformly elliptic operators of divergence form. The estimate depends on the ratio of the ellipticity bounds at infinity. In particular, if

$$L = \frac{\partial}{\partial x^i}\left(a_{ij}\frac{\partial}{\partial x^j}\right)$$

is an elliptic operator of divergence form defined on $\mathbb{R}^n$ with uniformly bounded measurable coefficients satisfying (11.2). We define the ellipticity bounds λ_r and Λ_r on the complement of the Euclidean ball of radius r centered at the origin so that they satisfy

$$\lambda_r\,(\delta_{ij}) \le (a_{ij}(x)) \le \Lambda_r\,(\delta_{ij})$$

for all $x \in \mathbb{R}^n \setminus B_0(r)$. Obviously, both λ_r and Λ_r are bounded, monotonic functions, hence we can define

$$\lambda_\infty = \lim_{r \to \infty} \lambda_r$$

and

$$\Lambda_\infty = \lim_{r \to \infty} \Lambda_r.$$

Theorem 11.15 (Li-Wang [78]). *Let L be a uniformly elliptic operator of divergence form defined on $\mathbb{R}^n$. Let*

$$\mathcal{H}_d(L) = \{u \in H^{loc}_{1,2}(\mathbb{R}^n) \mid L(u) = 0, \; |u|(x) = O(\rho^d(x)) \text{ as } \rho \to \infty\}$$

be the space of L-harmonic functions that has polynomial growth of degree at most $d \geq 1$. Then

$$\sum_{i=1}^{d} \dim \mathcal{H}_i(L) \leq \left(\frac{\Lambda_\infty}{\lambda_\infty}\right)^{n-1} \frac{2}{n!} (d + 2n)^n.$$

12. L^p conditions

Another type of growth conditions which appear naturally in geometric problems are integrability conditions. For example, a natural question to ask is whether a manifold possesses any non-trivial L^2 harmonic functions. This was first answered by Yau in 1976. Notice that since the absolute value of a harmonic function is subharmonic, we may generalize this discussion to non-negative subharmonic functions which satisfy some integrability conditions.

Theorem 12.1 (Yau [108]). *Let u be a non-negative subharmonic function defined on a complete manifold M. If $u \in L^p(M)$ for some $p > 1$, then u must be identically constant. This constant must be zero if M has infinite volume. In particular, a complete manifold does not admit any non-constant L^p harmonic functions for $p > 1$.*

It turns out that for $p \leq 1$, the situation is not as definitive, but geometrically more interesting. In a joint work of the author and Schoen [67], they studied these cases and found out that the curvature of M plays a role. In fact, the case $p = 1$ is also different from the remaining cases $p < 1$.

Theorem 12.2 (Li-Schoen [67]). *Let M be a complete manifold. Suppose $p \in M$ is a fixed point and ρ is the distance function to p. If* *there exists constants $C > 0$ and $\alpha > 0$ such that the Ricci curvature of M satisfies*

$$Ric_M(x) \geq -C\,(1+\rho^2(x))(\log(1+\rho^2(x))^{-\alpha},$$

then any non-negative L^1 subharmonic must be identically constant. Moreover, this constant must be zero if M has infinite volume.

Theorem 12.3 (Li-Schoen [67]). *Let M be a complete manifold. Suppose $p \in M$ is a fixed point and ρ is the distance function to p. There exists a constant $\delta(n) > 0$ depending only on n, such that, if the Ricci curvature satisfies M satisfies*

$$Ric_M(x) \geq -\delta(n)\,\rho^{-2}(x),$$

as $x \to \infty$, then any non-negative L^p subharmonic must be identically constant for $p \in (0,1)$. Moreover, this constant must be zero if M has infinite volume.

In the same paper, Li and Schoen also produced examples of manifolds which possess non-constant L^p harmonic functions. They showed that for any $\epsilon > 0$, there are manifolds with sectional curvature decay at the order of

$$K_M \sim -C\,\rho^{2+\epsilon}$$

as $\rho \to \infty$, which admit non-constant L^1 harmonic functions. Also, for any $p < 1$,, there exists manifolds with sectional curvature behave like

$$K_M \sim -C\,\rho^{-2}$$

as $\rho \to \infty$ which admit non-constant L^p harmonic functions. These examples show that the curvature condition in Theorem 12.3 is sharp and the condition in Theorem 12.2 is almost sharp. In fact, a sharp curvature condition was later found by the author for the case $p = 1$.

Theorem 12.4 (Li [60]). *Let M be a complete manifold. Suppose $p \in M$ is a fixed point and ρ is the distance function to p. If there exists a constant $C > 0$ such that the Ricci curvature of M satisfies*

$$Ric_M(x) \geq -C\,(1+\rho^2(x)),$$

then any non-negative L^1 subharmonic function must be identically constant. Moreover, this constant must be zero if M has infinite volume.

Other than lower bounds on the Ricci curvature, there are also other conditions which will imply the non-existence of L^p harmonic functions.

Theorem 12.5 (Li-Schoen [67]). *Let M be a complete manifold satisfying one of the following conditions:*

- *M is a Cartan-Hadamard manifold.*
- *M has Ricci curvature bounded from below and the volume of every unit geodesic ball is uniformly bounded from below.*

Then for all $p \in (0, 1]$, any non-negative L^p subharmonic function must be constant.

For the sake of application, Yau's theorem can be relaxed to the following form:

Proposition 12.6 (Yau [108]). *Let M be a complete manifold. Suppose u is a non-negative subharmonic function whose L^p-norm satisfies the growth condition*

$$\int_{B_x(r)} u^p = o(r^2)$$

as $r \to \infty$ for some fixed point $x \in M$. Then u must be identically constant. Moreover, this constant must be zero if the volume growth of M satisfies

$$\limsup_{r\to\infty} r^{-2}\, V_x(r) > 0$$

as $r \to \infty$.

The interested reader should also refer to the work of Nadirashvili [89] for a different type of integrability condition for the Liouville theorem.

13. Cartan-Hadamard manifolds

The function theory on a hyperbolic disk is quite different from the Euclidean plane. Our previous discussion, in many ways models on the Euclidean case. In this section, we will discuss the higher dimensional analog of the hyperbolic case. With the intend of proving a uniformization type theorem for higher dimensional Kähler manifolds, Greene-Wu and Yau asked if a complete, simply connected, Kähler manifold with sectional curvature bounded from above by -1 is biholomorphic to a bounded domain in $\mathbb{C}^n$. Clearly, to prove such a statement, one needs to produce many bounded holomorphic functions to be used as embedding functions. The first step is to study the real analog of this statement and see if one can produce enough bounded harmonic functions. In fact, Greene-Wu [42] posted the following conjecture:

Conjecture 13.1 (Greene-Wu [42]). *Let M be a Cartan-Hadamard manifold whose sectional curvature satisfies the upper bound*

$$K_M(x) \leq -C\,\rho^{-2}(x)$$

for some constant $C > 0$, where ρ is the distance function to a fixed point. Then M must admit a non-constant bounded harmonic function.

In 1983, Sullivan [98] proved that there are abundance of bounded harmonic functions on a strongly negatively curved Cartan-Hadamard manifold. Anderson [4] later used an argument of Choi [27] gave another proof of the same statement. To describe the space of bounded harmonic functions on Cartan-Hadamard manifold with strongly negative curvature, we need to define the geometric boundary.

Definition 13.2. Let M be a Cartan-Hadamard manifold. We define the *geometric boundary* $M(\infty)$ of M to be the set of equivalent classes of geodesic rays defined by the equivalence relation that two geodesic rays $\gamma_1(t)$ and $\gamma_2(t)$ are equivalent if $\rho(\gamma_1(t), \gamma_2(t))$ is a bounded function to $t \in [0, \infty)$.

The geometric boundary $M(\infty)$ together with M form a compactification of M, and $M \cup M(\infty)$ has a natural topology inherited from M, namely the cone topology. The cone $C_p(v, \delta)$ about a tangent vector $v \in T_pM$ of angle δ is defined by

$$C_p(v, \delta) = \{x \in M \mid \text{the geodesic } \gamma \text{ joining } p \text{ to } x \text{ satisfies } \langle \gamma', v\rangle < \delta\}.$$

The open sets of the cone topology is generated by the sets of all truncated cones $C_p(v, \delta) \setminus B_p(r)$ and geodesic balls $B_q(r)$, for $p, q \in M$, $v \in T_pM$, $\delta > 0$, and $r > 0$. Using the Toponogov comparison theorem, one checks [5] that if the sectional curvature of M is strongly negative, i.e.,

$$-a \geq K_M \geq -b$$

for some constants $0 < a \leq b < \infty$, then $M(\infty)$ has a natural $C^{a/b}$ structure.

Theorem 13.3 (Sullivan [98], Anderson [4]). *Let M be a complete, simply connected manifold. Assume that the sectional curvature of M satisfy the bound*

$$-a \geq K_M \geq -b$$

for some constants $0 < a \leq b < \infty$. Then for any continuous function ϕ defined on the geometric boundary $M(\infty)$ of M, there exists a function f defined on $M \cup M(\infty)$ which is harmonic on M, and

$$f = \phi$$

on $M(\infty)$.

Shortly after this theorem, Anderson and Schoen considered the existence of positive harmonic functions on the same class of manifolds. In fact, they showed that the Martin boundary is homeomorphic to the geometric boundary. The Martin boundary is defined on non-parabolic manifolds. Let $p \in M$ be a fixed point and $x, y \in M$. Suppose G is the minimal positive Green's function defined on M, then we defined the normalized Green's function with pole at y by

$$h_y(x) = \frac{G(y,x)}{G(y,p)}.$$

Clearly, the normalization yields $h_y(p) = 1$. Let y_i be a non-convergent sequence of points in M, then the sequence $\{h_i(x) = h_{y_i}(x)\}$ of harmonic functions are uniformly bounded on compact subsets of M. The Harnack inequality implies that there exists a subsequence $\{h_{i_j}\}$ which converges uniformly on compact subsets to a positive harmonic function $\bar{h}$ defined on M with the property that $\bar{h}(p) = 1$. The corresponding subsequence of points $\{y_{i_j}\}$ is denoted to be a fundamental sequence. We say that two fundamental sequences are equivalent if the corresponding limiting harmonic functions are the same.

Definition 13.4. The *Martin boundary* $\mathcal{M}$ of a manifold M consists of the equivalent classes of fundamental sequences $\bar{y} = [y_i]$. To each $\bar{y} \in \mathcal{M}$ there associates a positive harmonic function $h_{\bar{y}}$ from the above construction.

The Martin boundary together with M form a compactification for M. The topology on $M \cup \mathcal{M}$ can be defined by the distance function $\bar{\rho}$ given by

$$\bar{\rho}(y, y') = \sup_{x \in B_p(1)} |h_y(x) - h_{y'}(x)|.$$

One checks readily that this structure is independent of the choice of p, and this topology coincides with the topology induced by the Riemannian structure of M.

When M is a complete manifold with non-negative Ricci curvature near infinity and if M has finite first Betti number and it is non-parabolic, then one can show that the Martin boundary consists of $\pi_0^\infty(M)$ points. The compactification $M \cup \mathcal{M}$ is simply a 1-point compactification at each end of M.

Theorem 13.5 (Anderson-Schoen [5]). *Let M be a complete, simply connected manifold. Assume that the sectional curvature of M satisfies the bound*

$$-a \geq K_M \geq -b$$

for some constants $0 < a \leq b < \infty$. *Then the Martin boundary* $\mathcal{M}$ *is homeomorphic to the geometric boundary and the homeomorphism is* C^α. *In particular, there exist a Poisson kernel* $K(x, \bar{y})$ *defined on* $M \times M(\infty)$ *which is* C^α *in the variable* $\bar{y}$ *such that for any positive harmonic function* f, *there exists a unique, finite, positive, Borel measure* $d\mu$ *defined on* $M(\infty)$ *such that*

$$f(x) = \int_{M(\infty)} K(x, \bar{y})\, d\mu(\bar{y}).$$

Two years later, Ancona [3] generalized this theorem to a larger class of second order elliptic operator. In particular, a special case of Ancona's theorem asserts that the conclusion of Theorem 13.2 holds for manifolds which are quasi-isometric to a strongly negatively curved Cartan-Hadamard manifold.

Theorem 13.6 (Ancona [3]). *Let* M *be a complete, simply connected manifold. Assume that* M *is quasi-isometric to a manifold* N *satisfying the curvature bound*

$$-a \geq K_N \geq -b$$

for some constants $0 < a \leq b < \infty$. *Then the Martin boundary* $\mathcal{M}$ *of* M *is homeomorphic to* N_∞.

Other progress has been made to relax the curvature assumption of these theorems. For example, in [52], Hsu and Marsh relaxed the bounds on the curvature assumption. They generalized Theorem 13.3 to Cartan-Hadamard manifolds whose section curvature satisfies the estimate

$$-C\,\rho^{-2} \geq K_M \geq -b$$

for some constants $b > 0$ and $C > 2$. In 1992, Borbély [14] relaxed the lower bound by assuming that the sectional curvature satisfies

$$-a \geq K_M \geq -b\, e^{\lambda r}$$

for some constants $0 < a \leq b < \infty$ and $\lambda < 1/3$. In this case, he proved that the Dirichlet problem at infinity can be solved as in Theorem 13.3. Recently, Cheng proved the existence of non-constant bounded harmonic functions by assuming a pointwise curvature pinching condition.

Theorem 13.7 (Cheng [25]). *Let* M *be a Cartan-Hadamard manifold. Assume that the lower bound of the spectrum* $\lambda_1(M)$ *for the Laplacian on* M *is positive. Suppose there exists* $p \in M$ *and a constant* $C > 0$ *such that the sectional curvatures* $K_M(\sigma)$ *and* $K_M(\sigma')$ *satisfy*

$$|K_M(\sigma)| \leq C\, |K_M(\sigma')|$$

for any pair of 2-plane sections σ and σ' at x containing the tangent vector of the geodesic joining x to p. Then for any continuous function ϕ defined on the geometric boundary $M(\infty)$ of M, there exists a function f defined on $M \cup M(\infty)$ which is harmonic on M, and

$$f = \phi$$

on $M(\infty)$.

Note that unlike the previous theorems in this section, Cheng's theorem allows points where the curvature of M may vanish. The following theorem of Ballmann also allows this possibility, but rather than a pinching condition he assumed that the manifold is of rank one.

Definition 13.8. A Cartan-Hadamard manifold is said to have *rank one* if it admits a geodesic σ with no parallel Jacobi field along σ perpendicular to σ'.

If M is a Cartan-Hadamard manifold which is irreducible and admits a discrete, co-compact, isometry group, then it is known [7], [15] that either M has rank one or M is a symmetric space of noncompact type of rank at least 2.

Theorem 13.9 (Ballmann [8]). *Let M be an irreducible, Cartan-Hadamard manifold which admits a discrete, co-compact isometry group. If M has rank one, then for any continuous function ϕ defined on the geometric boundary $M(\infty)$ of M, there exists a function f defined on $M \cup M(\infty)$ which is harmonic on M, and*

$$f = \phi$$

on $M(\infty)$.

In a subsequent joint paper of Ballmann and Ledrappier [9], they showed that, in fact, one can represent any bounded harmonic function on M by a Poisson representation formula.

Theorem 13.10 (Ballmann-Ledrappier [9]). *Let M be an irreducible, Cartan-Hadamard manifold which admits a discrete, co-compact isometry group. If M has rank one, then there exists an equivalent class of harmonic measures $d\nu_p$ defined on $M(\infty)$ for each $p \in M$, such that, for any bounded measurable function ϕ the bounded function defined by*

$$f(x) = \int_{M(\infty)} \phi(\bar{x}) d\nu_x(\bar{x})$$

is a harmonic extension of ϕ *to* $M \cup M(\infty)$. *Conversely, any bounded harmonic function* f *can be such represented by some bounded measurable function* ϕ *on* $M(\infty)$.

In view of the theorems of Cheng, Ballmann, and Ballmann-Ledrappier, the natural questions to ask is whether it is true that the Martin boundaries of these manifolds are the same as their geometric boundaries? Also, is there a Poisson representation formula similar to the case of strongly negatively curved Cartan-Hadamard manifolds? What can one say about the Martin boundary for manifolds which are quasi-isometric to these manifolds?

Obviously, the set of positive harmonic functions on a complete manifold does not form a vector space. However, if f and g are positive harmonic functions then linear combinations of the form

$$a\, f(x) + b\, g(x),$$

where $a,\, b > 0$, is again a positive harmonic function. Hence the set of positive harmonic functions form a convex positive cone in a vector space. The boundary points of this convex cone determines the cone itself. A positive harmonic function f which is a boundary point of this cone has the property that if g is another positive harmonic function satisfying

$$g(x) \leq f(x),$$

then

$$g(x) = a\, f(x)$$

for some constant $0 < a \leq 1$. This property is called minimal. The set of positive harmonic functions are given by the positive span of minimal positive harmonic functions. In his paper [40], Freire considered the Martin boundary for the product of Riemannian manifolds by studying the set of minimal positive harmonic functions.

Theorem 13.11 (Freire [40]). *Let* $M = M_1 \times M_2$ *be a product to two complete Riemannian manifolds whose Ricci curvatures are bounded from below. If* f *is a minimal positive harmonic function defined on* M, *then* $f(x) = f(x_1, x_2)$ *can be written as a product*

$$f(x) = f_1(x_1)\, f_2(x_2)$$

of positive functions defined on each factor. Moreover, the functions $f_i > 0$ *satisfies*

$$\Delta_i f_i(x_i) = \lambda_i\, f_i(x_i)$$

on M_i with constants λ_i for $i = 1, 2$ such that $\lambda_1 + \lambda_2 = 0$. Also, each f_i is a minimal positive $(\Delta_i - \lambda_i)$-harmonic function. Conversely, the product of two minimal positive $(\Delta_i - \lambda_i)$-harmonic functions as above yields a minimal positive harmonic function on M.

References

[1] U. Abresch, *Lower curvature bounds, Toponogov's theorem, and bounded topology*, Ann. Sci. École Norm. Sup. **18** (1985) 651–670.

[2] ———, *Lower curvature bounds, Toponogov's theorem, and bounded topology. II*, Ann. Sci. École Norm. Sup. **20** (1987) 475–502.

[3] A. Ancona, *Negatively curved manifolds, elliptic operators, and the Martin boundary*, Ann. of Math. **125** (1987) 495–536.

[4] M. Anderson, *The Dirichlet problem at infinity for manifolds of negative curvature*, J. Differential Geom. **18** (1983) 701–721.

[5] M. Anderson & R. Schoen, *Positive harmonic functions on complete manifolds of negative curvature*, Ann. of Math. **121** (1985) 429–461.

[6] M. Avellaneda & F. H. Lin, *Une thèorème de Liouville pour des èquations elliptique à coefficients pèriodiques*, Compt. Rendus Acad. Sci. Paris **309** (1989) 245–250.

[7] W. Ballmann, *Nonpositively curved manifolds of higher rank*, Ann. of Math. **122** (1982) 131–144.

[8] ———, *On the Dirichlet Problem at infinity for manifolds of nonpositive curvature*, Forum Math. **1** (1989) 201–213.

[9] W. Ballmann & F. Ledrappier, *Poisson boundary for rank one manifolds and their cocompact lattices*, Forum Math. **6** (1994) 301–313.

[10] S. Bando, A. Kasue & H. Nakajima, *On a construction of coordinates at infinity on manifolds with fast curvature decay and maximal volume growth*, Invent. Math. **97** (1989) 313–349.

[11] R. Bartnik, *The mass of an asymptotically flat manifold*, Comm. Pure Appl. Math. **39** (1986) 661–693.

[12] R. Bishop & R. Crittenden, *Geometry of manifolds*, Academic Press, New York and London, 1964.

[13] E. Bombieri & E. Giusti, *Harnack's inequality for elliptic differential equations on minimal surfaces*, Invent. Math. **15** (1972) 24–46.

[14] A. Borbély, *A note on the Dirichlet problem at infinity for manifolds of negative curvature*, Proc. Amer. Math. Soc. **114** (1992) 865–872.

[15] K. Burns & R. Spatzier, *Manifolds of nonpositive curvature and their buildings*, Inst. Hautes Études Sci. Publ. Math. **65** (1987) 35–59.

[16] P. Buser, *On Cheeger's inequality* $\lambda_1 \geq h^2/4$ *in Geometry of the Laplace operator*, Proc. Symp. Pure Math., Amer. Math. Soc. **36** (1980) 29–77.

[17] M. Cai, *Ends of Riemannian manifolds with nonnegative Ricci curvature outside a compact set*, Bull. Amer. Math. Soc. **24** (1991) 371–377.

[18] M. Cai, T. Colding & D. Yang, *A gap theorem for ends of complete manifolds*, Proc. Amer. Math. Soc. **123** (1995) 247–250.

[19] H. Cao, Y. Shen & S. Zhu, *The structure of stable minimal hypersurfaces in* $\mathbb{R}^{n+1}$, Preprint.

[20] J. Cheeger, T. Colding & W. Minicozzi, *Linear growth harmonic functions on complete manifolds with non-negative Ricci curvature*, Geom. Func. Anal. **5** (1995) 948–954.

[21] J. Cheeger & D. Gromoll, *The splitting theorem for manifolds of nonnegative Ricci curvature*, J. Differential Geom. **6** (1971) 119–128.

[22] ———, *On the structure of complete manifolds of nonnegative curvature*, Ann. of Math. **92** (1972) 413–443.

[23] J. Cheeger, M. Gromov & M. Taylor, *Finite propagation speed, kernel estimates for functions of the Laplace operator, and the geometry of complete Riemannian manifolds*, J. Differential Geom. **17** (1983) 15–53.

[24] S. Y. Cheng, *Liouville theorem for harmonic maps*, Geometry of the Laplace operator, Proc. Symp. Pure Math. Amer. Math. Soc. **36** (1980) 147–151.

[25] ———, *The Dirichlet problem at infinity for nonpositively curved manifolds*, Comm. Anal. Geom. **1** (1993) 101–112.

[26] S. Y. Cheng & S. T. Yau, *Differential equations on Riemannian manifolds and their geometric applications*, Comm. Pure Appl. Math. **28** (1975) 333–354.

[27] H. I. Choi, *Asymptotic Dirichlet problems for harmonic functions on Riemannian manifolds*, Trans. Amer. Math. Soc. **281** (1984) 691–716.

[28] S. Cohn-Vossen, *Kürzeste Wege and Totalkrümmung auf Flächen*, Compositio Math. **2** (1935) 69–133.

[29] T. Colding & W. Minicozzi, *On function theory on spaces with a lower Ricci curvature bound,* , Math. Res. Lett. **3** (1996) 241–246.

[30] ———, *Harmonic functions with polynomial growth*, Preprint.

[31] ———, *Large scale behavior of kernels of Schrödinger operators*, Preprint.

[32] ———, *Generalized Liouville properties for manifolds*, Preprint.

[33] ———, *Harmonic functions on manifolds*, Preprint.

[34] ———, *Weyl type bounds for harmonic functions*, Preprint.

[35] ———, *Liouville theorems for harmonic sections and applications manifolds*, Preprint.

[36] Th. Coulhon & L. Saloff-Coste, *Variétés riemanniennes isométriques à l'infini*, Revista Mat. Iber. **11** (1995) 687–726.

[37] H. Donnelly, *Bounded harmonic functions and positive Ricci curvature*, Math. Z. **191** (1986) 559–565.

[38] F. Fiala, *Le problème des isopérimètres sur les surfaces ouvertes à courbure positive*, Comm. Math. Helv. **13** (1940) 293–346.

[39] R. Finn, *On a class of conformal metrics with application to differential geometry in the large*, Comm. Math. Helv. **40** (1965) 1–30.

[40] A. Freire, *On the Martin boundary of Riemannian products*, J. Differential Geom. **33** (1991) 215–232.

[41] D. Gilbarg & N. S. Trudinger, *Elliptic partial differential equations of second order*, 2nd Ed. Springer, Berlin, 1983.

[42] R. E. Greene & H. Wu, *Function theory on manifolds which possess a pole*, Lect. Notes in Math., Springer, Berlin, Vol. 699.

[43] A. Grigor'yan, *On the existence of a Green function on a manifold*, Uspechi Matem. Nauk **38** (1983) 161–162; Engl transl: Russian Math. Surveys **38** (1983) 190–191.

[44] ———, *On the existence of positive fundamental solution of the Laplace equation on Riemannian manifolds*, Mat. Sbornik **128** (1985) 354–363; Engl transl: Math. USSR Sbornik **56** (1987) 349–358.

[45] ———, *On Liouville theorems for harmonic functions with finite Dirichlet integral*, Math. USSR Sbornik **60** (1988) 485–504.

[46] ———, *On the dimension of spaces of harmonic functions*, Math. Notes **48** (1990) 1114–1118.

[47] ———, *The heat equation on noncompact Riemannian manifolds*, Math. USSR Sbornik **72** (1992) 47–77.

[48] P. Hartman, *Geodesic parallel coordinates in the large*, Amer. J. Math. **86** (1964) 705–727.

[49] I. Holopainen, *Solutions of elliptic equations on manifolds with roughly Euclidean ends*, Math. Z. **217** (1994) 459–477.

[50] ———, *Volume growth, Green's functions, and parabolicity of ends*, Duke Math. J. **97** (1999) 319–346.

[51] I. Holopainen & P. Koskela, *Volume growth and parabolicity*, Preprint.

[52] P. Hsu & P. Marsh, *The limiting angle of certain Riemannian Brownian motions*, Comm. Pure Appl. Math. **38** (1985) 755–768.

[53] A. Huber, *On subharmonic functions and differential geometry in the large*, Comm. Math. Helv. **32** (1957) 13–72.

[54] D. Jerison, *The Poincaré inequality for vector fields satisfying Hörmander's condition*, Duke Math. J. **53** (1986) 503–523.

[55] M. Kanai, *Rough isometries, and combinatorial approximations of geometries of non-compact Riemannian manifolds*, J. Math. Soc. Japan **37** (1985) 391–413.

[56] ———, *Rough isometries and the parabolicity of Riemannian manifolds*, J. Math. Soc. Japan **38** (1986) 227–238.

[57] A. Kasue, *Harmonic functions with growth conditions on a manifold of asymptotically non-negative curvature. II*, Recent Topics in Differential and Analytic Geometry Adv. Stud. Pure Math., North-Holland, 1989, Vol. 18.

[58] J. L. Kazdan, *Parabolicity and the Liouville property on complete Riemannian manifolds*, Seminar on New Results in Nonlinear Partial Differential Equations, Max-Planck-Inst. Bonn, 1987, 153–166.

[59] S. Kobayashi, *Differential geometry of complex vector bundles*, Princeton Univ. Press, Princeton, 1987.

[60] P. Li, *Uniqueness of L^1 solutions for the Laplace equation and the heat equation on Riemannian manifolds*, J. Differential Geom. **20** (1984) 447–457.

[61] ———, *On the structure of complete Kähler manifolds with nonnegative curvature near infinity*, Invent. Math. **99** (1990) 579–600.

[62] ———, *Erratum "On the structure of complete Kähler manifolds with nonnegative curvature near infinity,"* Invent. Math. **104** (1991) 447-448.

[63] ———, *Lecture notes on geometric analysis*, Lecture Notes Series No. 6, Research Institute of Mathematics and Global Analysis Research Center, Seoul, Seoul National University, 1993.

[64] ———, *Harmonic functions of linear growth on Kähler manifolds with non-negative Ricci curvature*, Math. Res. Lett. **2** (1995) 79–94.

[65] ———, *Harmonic sections of polynomial growth*, Math. Res. Lett. **4** (1997) 35–44.

[66] P. Li & M. Ramachandran, *Kähler manifolds with almost non-negative Ricci curvature*, Amer. J. Math. **118** (1996) 341–353.

[67] P. Li & R. Schoen, *L^p and mean value properties of subharmonic functions on Riemannian manifolds*, Acta Math. **153** (1984) 279–301.

[68] P. Li & L. F. Tam, *Positive harmonic functions on complete manifolds with non-negative curvature outside a compact set*, Ann. of Math. **125** (1987) 171–207.

[69] ———, *Symmetric Green's functions on complete manifolds*, Amer. J. Math. **109** (1987) 1129–1154.

[70] ———, *Linear growth harmonic functions on a complete manifold*, J. Differential Geom. **29** (1989) 421–425.

[71] ———, *Complete surfaces with finite total curvature*, J. Differential Geom. **33** (1991) 139–168.

[72] ———, *Harmonic functions and the structure of complete manifolds*, J. Differential Geom. **35** (1992) 359–383.

[73] ———, *Green's functions, harmonic functions, and volume comparison*, J. Differential Geom. **41** (1995) 277–318.

[74] P. Li, L. F. Tam & J. Wang, *Some sharp bounds for the Green's function and the heat kernel*, Math. Res. Lett. **4** (1997) 589–602.

[75] P. Li & J. P. Wang, *Convex hull properties of harmonic maps*, J. Differential Geom. **48** (1998) 497–530.

[76] ———, *Mean value inequalities*, Indiana Math. J. **48** (1999) 1257–1283.

[77] ———, *Counting massive sets and dimensions of harmonic functions*, J. Differential Geom. **53** (1999) 237–278.

[78] ———, *Counting dimensions of L-harmonic functions*, Ann. of Math. **152** (2000) 645–658.

[79] P. Li & S. T. Yau, *On the parabolic kernel of the Schrödinger operator*, Acta Math. **156** (1986) 153–201.

[80] ———, *Curvature and holomorphic mappings of complete Kähler manifolds*, Compositio Math. **73** (1990) 125–144.

[81] F. H. Lin, *Asymptotically conic elliptic operators and Liouville type theorems*, Volume dedicated to S. Hildebrandt on the occasion of his 60th birthday, (ed. J. Jost), to appear.

[82] Z. Liu, *Ball covering property and nonnegative Ricci curvature outside a compact set*, Differential Geometry: Riemannian Geometry, Proc. Sypm. Pure Math. Amer. Math. Soc. **54 (3)** (1993) 459–464.

[83] T. J. Lyons, *Instability of the Liouville property for quasi-isometric Riemannian manifolds and reversible Markov chains*, J. Differential Geom. **26** (1987) 33–66.

[84] B. Malgrange, *Existence et approximation des solutions der équations aux dérivées partielles et des équations de convolution*, Ann. Inst. Fourier **6** (1955) 271–355.

[85] J. H. Michael & L. Simon, *Sobolev and mean-value inequalities on generalized submanifolds of* $\mathbb{R}^n$, Comm. Pure Appl. Math. **26** (1973) 361–379.

[86] N. Mok, *A survey on complete noncompact Kähler manifolds of positive curvature*, Complex analysis of several variables, Proc. Symp. Pure Math. Amer. Math. Soc. **41** (1984) 151–162.

[87] J. Moser, *On Harnack's theorem for elliptic differential equations*, Comm. Pure Appl. Math. **14** (1961) 577–591.

[88] J. Moser & M. Struwe, *On a Liouville-type theorem for linear and nonlinear elliptic differential equations on a torus*, Bol. Soc. Brasil Mat. **23** (1992) 1–20.

[89] N. Nadirashvili, *A theorem of Liouville type on a Riemannian manifold*, Russian Math. Surveys **40** (1986) 235–236.

[90] M. Nakai, *On Evans potential*, Proc. Japan Acad. **38** (1962) 624–629.

[91] T. Napier & M. Ramachandran, *Structure theorems for complete Kähler manifolds and applications to Lefschetz type theorems*, Geom. Funct. Anal. **5** (1995) 809–851.

[92] H. Royden, *Harmonic functions on open Riemann surfaces*, Trans. Amer. Math. Soc. **73** (1952) 40–94.

[93] L. Saloff-Coste, *Uniformly elliptic operators on Riemannian manifolds*, J. Differential Geom. **36** (1992) 417–450.

[94] ______, *A note on Poincaré, Sobolev, and Harnack inequalities*, Internat. Math. Res. Notices **2** (1992) 27–38.

[95] L. Sario, M. Schiffer & M. Glasner, *The span and principal functions in Riemannian spaces*, J. Anal. Math. **15** (1965) 115–134.

[96] J. P. Sha & D. G. Yang, *Examples of manifolds of positive Ricci curvature*, J. Differential Geom. **29** (1989) 95–103.

[97] R. Schoen & S. T. Yau, *Harmonic maps and the topology of stable hypersurfaces and manifolds of nonnegative Ricci curvature*, Comm. Math. Helv. **39** (1976) 333–341.

[98] D. Sullivan, *The Dirichlet problem at infinity for a negatively curved manifold*, J. Differential Geom. **18** (1983) 723–732.

[99] C. J. Sung, *Harmonic functions under quasi-isometry*, J. Geom. Anal. **8** (1998) 143–161

[100] C. J. Sung, L. F. Tam & J. P. Wang, *Spaces of harmonic functions*, J. London Math. Soc., to appear.

[101] L. F. Tam, *A note on harmonic forms on complete manifolds*, Preprint.

[102] N. Varopoulos, *The Poisson kernel on positively curved manifolds*, J. Funct. Anal. **44** (1981) 359–380.

[103] ______, *Potential theory and diffusion on Riemannian manifolds,* Conference on harmonic analysis in honor of Antoni Zygmund, Vol I, II Belmont, Calif., Wadsworth Math. Ser., Wadsworth, 1983, 821–837.

[104] J. Wang, *Linear growth harmonic functions on complete manifolds*, Comm. Anal. Geom. **4** (1995) 683–698.

[105] H. Wu, *Polynomial functions on complete Kähler manifolds*, Several complex variables and complex geometry, Proc. Symp Pure Math, Amer. Math. Soc. **52** (1989) 601–610.

[106] ———, *Subharmonic functions and the volume of a noncompact manifold*, Differential Geometry, Pitman Monographs Surveys Pure Appl. Math. **52** (1991) 351–368.

[107] S. T. Yau, *Harmonic functions on complete Riemannian manifolds*, Comm. Pure Appl. Math. **28** (1975) 201–228.

[108] ———, *Some function-theoretic properties of complete Riemannian manifolds and their applications to geometry*, Indiana Math. J. **25** (1976) 659–670.

[109] ———, *Nonlinear analysis in geometry*, L'Enseignement Mathématique, Série des Conférences de l'Union Mathématique Internationale, No. 8, Genève, SRO-KUNDIG, 1986.

[110] L. Zhang, *On generic eigenvalue flow of a family of metrics and its application in dimension estimates of polynomial growth harmonic functions,* Comm. Anal. Geom. **7** (1999) 259–278.

UNIVERSITY OF CALIFORNIA, IRVINE

SURVEYS IN DIFFERENTIAL GEOMETRY, 2000
Vol. VII ©2000, INTERNATIONAL PRESS
pp. 433-474

MIRROR PRINCIPLE. III

BONG H. LIAN, KEFENG LIU & SHING-TUNG YAU

Abstract

We generalize the theorems in *Mirror Principle I* and *II* to the case of general projective manifolds without the convexity assumption. We also apply the results to balloon manifolds, and generalize to higher genus.

1. Introduction

The present paper is a sequel to *Mirror Principle I* and *II* [29] [30]. For motivations and the main ideas of mirror principle, we refer the reader to the introductions of these two papers.

Let X be a projective manifold, and $d \in A_1(X)$. Let $M_{0,k}(d,X)$ denote the moduli space of k-pointed, genus 0, degree d, stable maps $(C, f, x_1, .., x_k)$ with target X [26]. *Note that our notation is without the bar.* By the construction of [27] (see also [6], [14]), each nonempty $M_{0,k}(d,X)$ admits a homology class $LT_{0,k}(d,X)$ of dimension $dim\ X + \langle c_1(X), d\rangle + k - 3$. This class plays the role of the fundamental class in topology, hence $LT_{0,k}(d,X)$ is called the virtual fundamental class. For background on this, we recommend [28].

Let V be a convex vector bundle on X. (i.e., $H^1(\mathbf{P}^1, f^*V) = 0$ for every holomorphic map $f : \mathbf{P}^1 \to X$.) Then V induces on each $M_{0,k}(d,X)$ a vector bundle V_d, with fiber at $(C, f, x_1, .., x_k)$ given by the section space $H^0(C, f^*V)$. Let b be any multiplicative characteristic class [20]. (i.e., if $0 \to E' \to E \to E'' \to 0$ is an exact sequence of vector bundles, then $b(E) = b(E')b(E'')$.) The problem we study here is to understand the intersection numbers

$$K_d := \int_{LT_{0,0}(d,X)} b(V_d)$$

First printed in Asian Journal of Mathematics, 1999. Used by permission.

and their generating function:

$$\Phi(t) := \sum K_d \, e^{d \cdot t}.$$

There is a similar and equally important problem if one starts from a concave vector bundle V [29]. (i.e., $H^0(\mathbf{P}^1, f^*V) = 0$ for every holomorphic map $f : \mathbf{P}^1 \to X$.) More generally, V can be a direct sum of a convex and a concave bundle. Important progress made on these problems has come from mirror symmetry. All of it seems to point toward the following general phenomenon [9], which we call *the Mirror Principle.* Roughly, it says that there are functional identities which can be used to either constrain or to compute the K_d often in terms of certain explicit special functions, loosely called generalized hypergeometric functions. In this paper, we generalize this principle to include all projective manifolds. We apply this theory to compute the multiplicative classes $b(V_d)$ for vector bundles on balloon manifolds. The answer is in terms of certain universal virtual classes which are independent of V, b.

When X is a toric manifold, b is the Euler class, and V is a sum of line bundles, there is a general formula derived in [21], [23] based on mirror symmetry, giving $\Phi(t)$ in terms of generalized hypergeometric functions [15]. Similar functions were studied [16] in equivariant quantum cohomology theory based on a series of axioms. For further background, see introduction of [29].

Acknowledgements. We thank Y. Hu, C.H. Liu, and G. Tian, for numerous helpful discussions with us during the course of this project. We owe special thanks to J. Li for patiently explaining to us his joint work with Tian, for tirelessly providing a lot of technical assistance, and for proofreading a substantial part of this manuscript. B.H.L.'s research is supported by NSF grant DMS-9619884. K.L.'s research is supported by NSF grant DMS-9803234 and the Terman fellowship. S.T.Y.'s research is supported by DOE grant DE-FG02-88ER25065 and NSF grant DMS-9803347.

1.1 Outline.

In Section 2, we do the necessary preparation to set up the version of localization theorem we need. This is a (functorial localization) formula which translates a commutative square diagram into a relation between localizations on two T spaces related by an equivariant map.

We do basically three things in Section 3. After we introduced the necessary notations, first we apply functorial localization to stable map moduli spaces. Second, we prove one of the main results of this paper: Theorem 3.6, which translate structure of fixed points on stable map moduli into an algebraic identity on the homology of a projective manifold (with or without T action). This motivates the notion of Euler data and Euler series. These are essentially solutions to the algebraic identity just mentioned. Third, we prove the main Theorems 3.12-3.13 which relate the generating functions $\Phi(t)$ with an Euler series $A(t)$ arising from induced bundles on stable map moduli.

In Section 4, we specialize results in Section 3 to balloon manifolds, and introduce the notion of linking. The main theorems here are 4.5 and 4.7. The first of these gives a description of an essential polar term of $A(t)$ upon localizing at a fixed point in X. The second theorem gives a sufficient condition for computing $A(t)$ in terms of certain universal virtual classes on stable map moduli. We then specialize this to the case when b_T is the Euler class or the Chern polynomial.

In Section 5, we explain some other ways to compute $A(t)$, first by relaxing those sufficient conditions, then by finding an explicit closed formula for those universal virtual classes above by using an equivariant short exact sequence for the tangent bundle. This includes toric manifolds as a special case. We then formulate an inductive method for computing $A(t)$ in full generality for any balloon manifold. Next, we discuss a method in which functorial localization is used to study $A(t)$ via a resolution of the image of the collapsing map. In certain cases, this resolution can be described quite explicitly. Finally, we discuss a generalization of mirror principle to higher genus.

2. Set-up.

Basic references: on intersection theory on algebraic schemes and stacks, we use [13], [40]; on the virtual classes, we follow [27]; on their equivariant counterparts, see [1], [2], [7], [25], [12], [17], [41].

T denotes an algebraic torus. T-equivariant Chow groups (homology) with complex coefficients are denoted by $A_*^T(\cdot)$. T-equivariant operational Chow groups (cohomology) with complex coefficients are denoted by $A_T^*(\cdot)$. For $c \in A_T^p(X)$, and $\beta \in A_q^T(X)$, we denote by $c \cap \beta = \beta \cap c$ the image of $c \otimes \beta$ under the canonical homomorphisms

$$A_T^p(X) \otimes A_q^T(X) \to A_{q-p}^T(X).$$

The product on $A_T^*(X)$ is denoted by $a \cdot b$. The homomorphisms $\cap$ define an $A_T^*(X)$-module structure on the homology $A_*^T(X)$. When X is nonsingular, there is a compatible intersection product on $A_*^T(X)$ which we denote by $\beta \cdot \gamma$.

Given a T-equivariant (proper or flat) map $f : X \to Y$, we denote by

$$f_* : A_*^T(X) \to A_*^T(Y), \qquad f^* : A_*^T(Y) \to A_*^T(X)$$

the equivariant (proper) pushforward and (flat) pullback; the notations f^* and f_* are also used for pullback and (flat) pushforward on cohomology. All maps used here will be assumed proper. A formula often used is the projection formula:

$$f_*(f^*c \cap \beta) = c \cap f_*(\beta)$$

for cohomology class c on Y and homology class β on X. Note that both $A_*^T(X)$ and $A_T^*(X)$ are modules over the algebra $A_T^*(pt) = \mathbf{C}[\mathcal{T}^*]$, where $\mathcal{T}^*$ is the dual of the Lie algebra of T, and the homomorphisms f_*, f^* are module homomorphisms. We often extend these homomorphisms over the field $\mathbf{C}(\mathcal{T}^*)$ without explicitly saying so. Finally, suppose we have a fiber square

$$\begin{array}{ccc} F & \xrightarrow{i'} & M \\ p\downarrow & & \downarrow q \\ X & \xrightarrow{i} & Y \end{array}$$

where i is a regular embedding of codimension d, then we have

$$p_* i^! \beta = i^* q_* \beta$$

for any homology class β on M. Here $i^! : A_*^T(M) \to A_{*-d}^T(F)$ is the refined Gysin homomorphism.

2.1 Functorial localization.

Let X be an algebraic stack with a T action and equipped with a suitable perfect obstruction theory (see [27], [17]). Let F_r denote the fixed point components in X. Let $[X]^{vir}, [F_r]^{vir}$ be the equivariant virtual classes of X and the F_r. Then by [17],

$$[X]^{vir} = \sum_r i_{r*} \frac{[F_r]^{vir}}{e_T(F_r/X)}$$

where $i_r : F_r \to X$ are the inclusions, and $e_T(F_r/X)$ the equivariant Euler class of the virtual normal bundle of $F_r \subset X$. Then for any cohomology class c on X, we have

$$c \cap [X]^{vir} = \sum_r i_{r*} \frac{i_r^* c \cap [F_r]^{vir}}{e_T(F_r/X)}. \tag{2.1}$$

Throughout this subsection, let

$$f: \quad X \to Y$$

be an equivariant map with Y smooth. Let E be a fixed point component in Y, and let F be the fixed points in $f^{-1}(E)$. Let g be the restriction of f to F, and $j_E : E \to Y$, $i_F : F \to X$ be the inclusion maps. Thus we have the commutative diagram:

$$\begin{array}{ccc} F & \xrightarrow{i_F} & X \\ g \downarrow & & \downarrow f \\ E & \xrightarrow{j_E} & Y. \end{array} \tag{2.2}$$

Then we have the following *functorial localization formula.*

lemma 2.1. *Given a cohomology class $\omega \in A_T^*(X)$, we have the equality on E:*

$$\frac{j_E^* f_*(\omega \cap [X]^{vir})}{e_T(E/Y)} = g_* \left(\frac{i_F^* \omega \cap [F]^{vir}}{e_T(F/X)} \right).$$

Proof. Applying (5.5) to the class $c = \omega \cdot f^* j_{E*} 1$ on X, we get

$$\omega \cdot f^* j_{E*} 1 \cap [X]^{vir} = i_{F*} \left(\frac{i_F^*(\omega \cdot f^* j_{E*} 1) \cap [F]^{vir}}{e_T(F/X)} \right).$$

Note that the contributions from fixed components other than F vanish. Applying f_* to both sides, we get

$$f_*(\omega \cap [X]^{vir}) \cap j_{E*} 1 = f_* i_{F*} \left(\frac{i_F^*(\omega \cdot f^* j_{E*} 1) \cap [F]^{vir}}{e_T(F/X)} \right).$$

Now $f \circ i_F = j_E \circ g$ which, implies

$$f_* i_{F*} = j_{E*} g_*, \quad i_F^* f^* = g^* j_E^*.$$

Thus we get

$$f_*(\omega \cap [X]^{vir}) \cap j_{E*} 1 = j_{E*} g_* \left(\frac{i_F^* \omega \cdot g^* e_T(E/Y) \cap [F]^{vir}}{e_T(F/X)} \right).$$

Applying j_E^* to both sides here, we get

$$\begin{aligned} & j_E^* f_*(\omega \cap [X]^{vir}) \cap e_T(E/Y) \\ & = e_T(E/Y) \cap g_* \left(\frac{i_F^* \omega \cdot g^* e_T(E/Y) \cap [F]^{vir}}{e_T(F/X)} \right) \\ & = e_T(E/Y)^2 \cap g_* \left(\frac{i_F^* \omega \cap [F]^{vir}}{e_T(F/X)} \right). \end{aligned}$$

Since $e_T(E/Y)$ is invertible, our assertion follows. q.e.d.

Note that if F has more than one component, then the right hand side of the formula above becomes a sum over those components in an obvious way.

Corollary 2.2. *Let Y' be a T-invariant submanifold of Y, $f' : X' = f^{-1}(Y') \to Y'$ be the restriction of $f : X \to Y$ to the substack X', and $j : Y' \to Y$, $i : X' \to X$ be the inclusions. Then for any $\omega \in A_T^*(X)$, we have*

$$\frac{j^* f_*(\omega \cap [X]^{vir})}{e_T(Y'/Y)} = f'_* \left(\frac{i^*\omega \cap [X']^{vir}}{e_T(X'/X)} \right).$$

Proof. Let E be any fixed point component of Y contained in Y', and F be the fixed points in $f^{-1}(E)$, as in the preceding lemma. Then we have the commutative diagram

$$(2.3)\qquad \begin{array}{ccccc} F & \xrightarrow{i'_F} & X' & \xrightarrow{i} & X \\ g \downarrow & & f' \downarrow & & f \downarrow \\ E & \xrightarrow{j'_E} & Y' & \xrightarrow{j} & Y. \end{array}$$

We will show that

$$(*) \quad {j'_E}^* \left(\frac{j^* f_*(\omega \cap [X]^{vir})}{e_T(Y'/Y)} \right) = {j'_E}^* f'_* \left(\frac{i_F^*\omega \cap [X']^{vir}}{e_T(X'/X)} \right).$$

Then our assertion follows from the localization theorem.

Put $j_E := j \circ j'_E$, $i_F := i \circ i'_F$. The left hand side of (*) is

$$\frac{{j'_E}^* j^* f_*(\omega \cap [X]^{vir})}{{j'_E}^* e_T(Y'/Y)} = e_T(E/Y') \cap \frac{j_E^* f_*(\omega \cap [X]^{vir})}{e_T(E/Y)}$$
$$= e_T(E/Y') \cap g_* \left(\frac{i_F^*\omega \cap [F]^{vir}}{e_T(F/X)} \right)$$
$$\text{(preceding lemma)}.$$

Now apply the left hand square in (5.5) and the preceding lemma again to the class $\frac{i^*\omega}{e_T(X'/X)}$ on X'. Then the right hand side of (*) becomes

$${j'_E}^* f'_* \left(\frac{i_F^*\omega}{e_T(X'/X)} \cap [X']^{vir} \right) = e_T(E/Y') \cap g_* \left(\frac{{i'_F}^* \frac{i^*\omega}{e_T(X'/X)} \cap [F]^{vir}}{e_T(F/X')} \right)$$
$$= e_T(E/Y') \cap g_* \left(\frac{i_F^*\omega \cap [F]^{vir}}{e_T(F/X)} \right).$$

This proves (*). q.e.d.

3. General projective T-manifolds

Let X be a projective T-manifold. Let $M_d(X)$ be the degree $(1,d)$, arithmetic genus zero, 0-pointed, stable map moduli stack with target $\mathbf{P}^1 \times X$. The standard $\mathbf{C}^\times$ action on $\mathbf{P}^1$ together with the T action on X induces a $G = \mathbf{C}^\times \times T$ action on $M_d(X)$. Let $LT_d(X) \in A_*^G(M_d(X))$ be the virtual class of this moduli stack. This is an equivariant homology class of dimension $\langle c_1(X), d\rangle + dim\ X$.

The $\mathbf{C}^\times$ fixed point components F_r, labelled by $0 \preceq r \preceq d$, in $M_d(X)$ can be described as follows (see [30]). Let F_r be the substack

$$F_r := M_{0,1}(r, X) \times_X M_{0,1}(d-r, X)$$

obtained from gluing the two one pointed moduli stacks. More precisely, consider the map

$$e_r^X \times e_{d-r}^X : M_{0,1}(r, X) \times M_{0,1}(d-r, X) \to X \times X$$

given by evaluations at the corresponding marked points; and

$$\Delta :\ X \to X \times X$$

the diagonal map. Then we have

$$F_r = (e_r^X \times e_{d-r}^X)^{-1} \Delta(X).$$

Note that $F_d = M_{0,1}(d, X) = F_0$ by convention, but F_0 and F_d will be embedded into $M_d(X)$ in two different ways. The F_r can be identified with a $\mathbf{C}^\times$ fixed point component of $M_d(X)$ as follows. Consider the case $r \neq 0, d$ first. Given a pair $(C_1, f_1, x_1) \times (C_2, f_2, x_2)$ in F_r, we get a new curve C by gluing C_1, C_2 to $\mathbf{P}^1$ with x_1, x_2 glued to $0, \infty \in \mathbf{P}^1$ respectively. The new curve C is mapped into $\mathbf{P}^1 \times X$ as follows. Map $\mathbf{P}^1 \subset C$ identically onto $\mathbf{P}^1$, and collapse C_1, C_2 to $0, \infty$ respectively; then map C_1, C_2 into X with f_1, f_2 respectively, and collapse the $\mathbf{P}^1$ to $f(x_1) = f(x_2)$. This defines a stable map (C, f) in $M_d(X)$. For $r = d$, we glue (C_1, f_1, x_1) to $\mathbf{P}^1$ at x_1 and 0. For $r = 0$, we glue (C_2, f_2, x_2) to $\mathbf{P}^1$ at x_2 and ∞.

Notations.

(i) We identify F_r as a substack of $M_d(X)$ as above, and let

$$i_r : F_r \to M_d(X)$$

denote the inclusion map.

(ii) We have evaluation maps

$$e^X : F_r \to X,$$

which sends a pair in F_r to the value at the common marked point. While the notation e^X doesn't reflect the dependence on r, the domain F_r that e^X operates on will be clear.

(iii) We have the obvious inclusion

$$\Delta' : F_r \subset M_{0,1}(r,X) \times M_{0,1}(d-r,X),$$

and projections

$$p_0 : F_r \to M_{0,1}(r,X), \qquad p_\infty : F_r \to M_{0,1}(d-r,X).$$

(iv) Let L_r denote the universal line bundle on $M_{0,1}(r,X)$.

(v) We have the natural forgetting, evaluation, and projection maps:

$$\begin{aligned} &\rho : M_{0,1}(d,X) \to M_{0,0}(d,X) \\ &e_d^X : M_{0,1}(d,X) \to X \\ &\pi : M_d(X) \to M_{0,0}(d,X). \end{aligned}$$

We also have the obvious commutative diagrams

$$\begin{array}{ccc} M_d(X) & & \\ \pi \downarrow & \nwarrow i_0 & \\ M_{0,0}(d,X) & \stackrel{\rho}{\longleftarrow} & M_{0,1}(d,X) \end{array}$$

$$\begin{array}{ccc} F_r & \stackrel{\Delta'}{\longrightarrow} & M_{0,1}(r,X) \times M_{0,1}(d-r,X) \\ e^X \downarrow & & \downarrow e_r^X \times e_{d-r}^X \\ X & \stackrel{\Delta}{\longrightarrow} & X \times X \end{array} \tag{3.1}$$

where Δ is the diagonal map. Note that we have a diagram similar to (3.1) but with X replaced by Y in the bottom row. From the fiber square (3.1), we have a refined Gysin homomorphism

$$\Delta^! : A_*^T(M_{0,1}(r,X) \times M_{0,1}(d-r,X)) \to A_{*-dim\ X}^T(F_r).$$

We refer the reader to Section 6 [27] for the following

Lemma 3.1. ([27]). *For $r \neq 0, d$,*

$$[F_r]^{vir} = \Delta^!(LT_{0,1}(r,X) \times LT_{0,1}(d-r,X)).$$

(vi) Let α be the weight of the standard $\mathbf{C}^\times$ action on $\mathbf{P}^1$. We denote by $A_*^T(X)(\alpha)$ the algebra obtained from $A_*^T(X)[\alpha]$ by inverting the classes ω such that $(i_F^*\omega)^{-1}$ is well-defined in $A_*(F) \otimes \mathbf{C}(\mathcal{T}^*)(\alpha)$, for every fixed point component F. If β is an element in $A_*^T(X)(\alpha)$, we let $\overline{\beta}$ be the class obtained from β by replacing $\alpha \to -\alpha$. We also introduce formal variables $\zeta = (\zeta_1, ..., \zeta_m)$ such that $\bar{\zeta}_a = -\zeta_a$. Denote $\mathcal{R} = \mathbf{C}(\mathcal{T}^*)[\alpha]$. When a multiplicative class b_T, such as the Chern polynomial $c_T = x^r + xc_1 + \cdots + c_r$, is considered, we must replace the ground field $\mathbf{C}$ by $\mathbf{C}(x)$, so that c_T takes value in Chow groups with appropriate coefficients. This change of ground field will be implicit whenever necessary.

(vii) For each d, let $\varphi : M_d(X) \to W_d$ be a G-equivariant map into smooth manifold (or orbifold) W_d with the property *that the* $\mathbf{C}^\times$ *fixed point components in* W_d *are* G*-invariant submanifolds* Y_r *such that* $\varphi^{-1}(Y_r) = F_r$.

The spaces W_d exist but are not unique. Two specific kinds will be used here. First, choose an equivariant projective embedding

$$\tau : X \to Y = \mathbf{P}^{n_1} \times \cdots \times \mathbf{P}^{n_m}$$

which induces an isomorphism $A^1(X) \cong A^1(Y)$. Then we have a G-equivariant embedding

$$M_d(X) \to M_d(Y).$$

There is a G-equivariant map (see [29] and references there)

$$M_d(Y) \to W_d := N_{d_1} \times \cdots \times N_{d_m}$$

where the $N_{d_a} := \mathbf{P}H^0(\mathbf{P}^1, \mathcal{O}(d_a))^{n_a+1} \cong \mathbf{P}^{(n_a+1)d_a+n_a}$, which are the linear sigma model for the $\mathbf{P}^{n_i}$. Thus composing the two maps above, we get a G-equivariant map $\varphi : M_d(X) \to W_d$. It is also easy to check that the $\mathbf{C}^\times$ fixed point components in W_d have the desired property. Second, if X is a toric variety, then there exist toric varieties W_d [31] where Y_r are submanifolds of X. We postpone the discussion of this till Section 5 when we discuss the case of toric manifolds. From now on, unless specified otherwise, W_d will be the first kind as defined above.

(viii) We denote the equivariant hyperplane classes on W_d by κ_a (which are pullbacked from the each of the N_{d_a} to W_d). We denote the equivariant hyperplane classes on Y by H_a (which are pullbacked from each of the $\mathbf{P}^{n_a}$ to Y). We use the same notations for their restrictions to X. We write $\kappa \cdot \zeta = \sum_a \kappa_a \zeta_a$, $H \cdot t = \sum_a H_a t_a$, $d \cdot t = \sum_a d_a t_a$, where the t and ζ are formal variables.

3.1 Localization on stable map moduli.

Clearly we have the commutative diagram:

$$\begin{array}{ccc} F_r & \xrightarrow{i_r} & M_d(X) \\ e^Y \downarrow & & \downarrow \varphi \\ Y_r & \xrightarrow{j_r} & W_d. \end{array} \tag{3.2}$$

Let $\varphi : M_d(X) \to W_d$, $e^Y : F_r \to Y_r$ play the respective roles of $f : X \to Y$, $f' : X' \to Y'$ in functorial localization. Then it follows that

Lemma 3.2. *Given a cohomology class ω on $M_d(X)$, we have the following equality on $Y_r \cong Y$ for $0 \preceq r \preceq d$:*

$$\frac{j_r^* \varphi_*(\omega \cap LT_d(X))}{e_G(Y_r/W_d)} = e_*^Y \left(\frac{i_r^* \omega \cap [F_r]^{vir}}{e_G(F_r/M_d(X))} \right).$$

Following [29], one can easily compute the Euler classes $e_G(Y_r/W_d)$, and they are given as follows. For $d = (d_1, .., d_m)$, $r = (r_1, ..., r_m) \preceq d$, we have

$$e_G(Y_r/W_d) = \prod_{a=1}^{m} \prod_{i=0}^{n_a} \prod_{k=0 k \neq r_a}^{d_a} (H_a - \lambda_{a,i} - (k - r_a)\alpha)$$

where the $\lambda_{a,i}$ are the T weights of $\mathbf{P}^{n_a}$. Note that e^Y is the composition of $e^X : F_r \to X$ with $\tau : X \to Y = Y_r$. Thus

$$e_*^Y = \tau_* e_*^X.$$

It follows that

Lemma 3.3. *Given a cohomology class ω on $M_d(X)$, we have the following equality on X for $0 \preceq r \preceq d$:*

$$\tau^* \left(\frac{j_r^* \varphi_*(\omega \cap LT_d(X))}{e_G(Y_r/W_d)} \right) = e_T(X/Y) \cap e_*^X \left(\frac{i_r^* \omega \cap [F_r]^{vir}}{e_G(F_r/M_d(X))} \right).$$

Now if ψ is a cohomology class on $M_{0,0}(d, X)$, then for $\omega = \pi^* \psi$, we get $i_0^* \omega = i_0^* \pi^* \psi = \rho^* \psi$. It follows that

Lemma 3.4. *Given a cohomology class ψ on $M_{0,0}(d,X)$, we have the following equality on X:*

$$\tau^*\left(\frac{j_0^*\varphi_*(\pi^*\psi\cap LT_d(X))}{e_G(Y_0/W_d)}\right)=e_T(X/Y)\cap e_*^X\left(\frac{\rho^*\psi\cap LT_{0,1}(d,X)}{e_G(F_0/M_d(X))}\right).$$

Lemma 3.5. *For $r\neq 0,d$,*

$$e_G(F_r/M_d(X))=\alpha(\alpha+p_0^*c_1(L_r))\cdot\alpha(\alpha-p_\infty^*c_1(L_{d-r})).$$

For $r=0,d$,

$$e_G(F_0/M_d(X))=\alpha(\alpha-c_1(L_d)),\qquad e_G(F_d/M_d(X))=\alpha(\alpha+c_1(L_d)).$$

The computation done in Section 2.3 of [29] and in Section 3 of [30] (see also references there), for the normal bundles $N_{F_r/M_d(X)}$, makes no use of the convexity assumption on TX. Therefore it carries over here with essentially no change.

3.2 From gluing identity to Euler data.

Fix a T-equivariant multiplicative class b_T. Fix a T-equivariant bundle of the form $V=V^+\oplus V^-$, where $V^\pm$ are respectively the convex/concave bundles on X. We assume that

$$\Omega:=\frac{b_T(V^+)}{b_T(V^-)}$$

is a well-defined invertible class on X. By convention, if $V=V^\pm$ is purely convex/concave, then $\Omega=b_T(V^\pm)^{\pm1}$. Recall that the bundle $V\to X$ induces the bundles

$$V_d\to M_{0,0}(d,X),\quad U_d\to M_{0,1}(d,X),\quad \mathcal{U}_d\to M_d(X).$$

Moreover, they are related by $U_d=\rho^*V_d$, $\mathcal{U}_d=\pi^*V_d$, Define linear maps

$$i_r^{vir}:A_G^*(M_d(X))\to A_*^T(X)(\alpha),\qquad i_r^{vir}\omega:=e_*^X\left(\frac{i_r^*\omega\cap[F_r]^{vir}}{e_G(F_r/M_d(X))}\right).$$

Theorem 3.6. *For $0\preceq r\preceq d$, we have the following identity in $A_*^T(X)(\alpha)$:*

$$\Omega\cap i_r^{vir}\pi^*b_T(V_d)=\overline{i_0^{vir}\pi^*b_T(V_r)}\cdot i_0^{vir}\pi^*b_T(V_{d-r}).$$

Proof. For simplicity, let's consider the case $V = V^+$. The general case is entirely analogous. The proof here is the one in [29], [30], but slightly modified to take into account the new ingredient coming from the virtual class.

Recall that a point (f, C) in $F_r \subset M_d$ comes from gluing together a pair of stable maps $(f_1, C_1, x_1), (f_2, C_2, x_2)$ with $f_1(x_1) = f_2(x_2) = p \in X$. From this, we get an exact sequence over C:

$$0 \to f^*V \to f_1^*V \oplus f_2^*V \to V|_p \to 0.$$

Passing to cohomology, we have

$$0 \to H^0(C, f^*V) \to H^0(C_1, f_1^*V) \oplus H^0(C_2, f_2^*V) \to V|_p \to 0.$$

Hence we obtain an exact sequence of bundles on F_r:

$$0 \to i_r^*\mathcal{U}_d \to U_r' \oplus U_{d-r}' \to e^{X*}V \to 0.$$

Here $i_r^*\mathcal{U}_d$ is the restriction to F_r of the bundle $\mathcal{U}_d \to M_d(X)$. And U_r' is the pullback of the bundle $U_r \to M_{0,1}(d, X)$, and similarly for U_{d-r}'. Taking the multiplicative class b_T, we get the identity on F_r:

$$e^{X*}b_T(V) \cdot b_T(i_r^*\mathcal{U}_d) = b_T(U_r') \cdot b_T(U_{d-r}').$$

We refer to this as the *gluing identity*.

Now put

$$\omega = \frac{b_T(U_r)}{e_G(F_r/M_r(X))} \times \frac{b_T(U_{d-r})}{e_G(F_0/M_{d-r}(X))} \cap LT_{0,1}(r, X) \times LT_{0,1}(d-r, X)$$

From the commutative diagram (3.1), we have the identity:

$$e_*^X \Delta^!(\omega) = \Delta^*(e_r^X \times e_{d-r}^X)_*(\omega).$$

On the one hand is

$$\begin{aligned}
\Delta^*&(e_r^X \times e_{d-r}^X)_*(\omega) \\
&= (e_r^X)_* \frac{b_T(U_r) \cap LT_{0,1}(r, X)}{e_G(F_r/M_r(X))} \\
&\qquad \cdot (e_{d-r}^X)_* \frac{b_T(U_{d-r}) \cap LT_{0,1}(d-r, X)}{e_G(F_0/M_{d-r}(X))} \\
&= (e_r^X)_* \frac{\rho^* b_T(V_r) \cap LT_{0,1}(r, X)}{e_G(F_r/M_r(X))} \\
&\qquad \cdot (e_{d-r}^X)_* \frac{\rho^* b_T(V_{d-r}) \cap LT_{0,1}(d-r, X)}{e_G(F_0/M_{d-r}(X))} \\
&= \overline{i_0^{vir}\pi^* b_T(V_r) \cdot i_0^{vir}\pi^* b_T(V_{d-r})}.
\end{aligned}$$

On the other hand, applying the gluing identity, we have

$$\begin{aligned}
e_*^X \Delta^!(\omega) &= e_*^X \left(\frac{b_T(U_r')}{\alpha(\alpha + p_0^* c_1(L_r))} \cdot \frac{b_T(U_{d-r}')}{\alpha(\alpha - p_\infty^* c_1(L_{d-r}))} \cap [F_r]^{vir} \right) \\
&= e_*^X \left(\frac{e^{X*} b_T(V) \cdot i_r^* b_T(\mathcal{U}_d) \cap [F_r]^{vir}}{e_G(F_r/M_d(X))} \right) \\
&= b_T(V) \cap e_*^X \left(\frac{i_r^* b_T(\mathcal{U}_d) \cap [F_r]^{vir}}{e_G(F_r/M_d(X))} \right) \\
&= b_T(V) \cap i_r^{vir} \pi^* b_T(V_d).
\end{aligned}$$

This proves our assertion. q.e.d.

Specializing the theorem to $b_T \equiv 1$, we get

Corollary 3.7. *$i_r^{vir} 1_d = \overline{i_0^{vir} 1_r} \cdot i_0^{vir} 1_{d-r}$ where 1_d is the identity class in on $M_d(X)$.*

For a given convex/concave bundle V on X, and multiplicative class b_T, we put

$$A^{V,b_T}(t) = A(t) := e^{-H \cdot t/\alpha} \sum_d A_d \, e^{d \cdot t}$$

$$A_d := i_0^{vir} \pi^* b_T(V) = e_*^X \left(\frac{\rho^* b_T(V_d) \cap LT_{0,1}(d, X)}{e_G(F_0/M_d(X))} \right).$$

Here we will use the convention that $A_0 = \Omega$, and the sum is over all $d = (d_1, ..., d_m) \in \mathbf{Z}_+^m$. When the reference to V, b_T is clear, we'll drop them from the notations. The special case in the corollary will play an important role. So we introduce the notation:

$$\mathbb{1}(t) := e^{-H \cdot t/\alpha} \sum_d \mathbb{1}_d e^{d \cdot t}, \qquad \mathbb{1}_d = i_0^{vir} 1_d.$$

By the preceding theorem and Lemma 3.2, it follows immediately that for $\omega = \varphi_*(\pi^* b_T(V_d) \cap LT_d(X))$, we have

$$\begin{aligned}
\int_{W_d} \omega \cap e^{\kappa \cdot \zeta} &= \sum_{0 \preceq r \preceq d} \int_{Y_r} \frac{j_r^* \omega}{e_G(Y_r/W_d)} \, e^{(H + r\alpha) \cdot \zeta} \\
&= \sum_r \int_{Y_r} \tau_* i_r^{vir} \pi^* b_T(V_d) \, e^{(H + r\alpha) \cdot \zeta} \\
&= \sum_r \int_X i_r^{vir} \pi^* b_T(V_d) \, e^{(H + r\alpha) \cdot \zeta} \\
&= \sum_r \int_X \Omega^{-1} \cap \bar{A}_r \cdot A_{d-r} \, e^{(H + r\alpha) \cdot \zeta} \qquad \text{(Theorem 3.6).}
\end{aligned}$$

Since $\omega \in A_*^G(W_d)$, hence

$$\int_{W_d} \omega \cap c \in A_*^G(pt) = \mathbf{C}[\mathcal{T}^*, \alpha]$$

for all $c \in A_G^*(W_d)$, it follows that both sides of the eqn. above lie in $\mathcal{R}[[\zeta]]$. This motivates the following (cf. [16])

Definition 3.8. Let $\Omega \in A_T^*(X)$, invertible. We call a power series of the form

$$B(t) := e^{-H\cdot t/\alpha} \sum_d B_d \, e^{d\cdot t}, \qquad B_d \in A_*^T(X)(\alpha)$$

an Ω-Euler series if

$$\sum_{0 \preceq r \preceq d} \int_X \Omega^{-1} \cap \bar{B}_r \cdot B_{d-r} \; e^{(H+r\alpha)\cdot\zeta} \in \mathcal{R}[[\zeta]]$$

for all d.

Thus we have seen above that an elementary consequence of the gluing identity in Theorem 3.6 is that

Corollary 3.9. *$A^{V,b_T}(t) = e^{-H\cdot t/\alpha} \sum_d i_0^{vir} \pi^* b_T(V_d) \; e^{d\cdot t}$ is an Euler series.*

Definition 3.10. ([29]). Let $\Lambda \in A_T^*(Y)$. We call a sequence $P: \; P_d \in A_G^*(W_d)$ an Λ-Euler data if

$$\Lambda \cdot j_r^* P_d = \overline{j_0^* P_r} \cdot j_0^* P_{d-r}, \quad 0 \preceq r \preceq d.$$

Let P be an Λ-Euler data such that $\tau^*\Lambda$ is invertible. Then we have

$$(3.3) \qquad \tau^*\Lambda \cdot \tau^* j_r^* P_d \cap i_r^{vir} 1_d = \tau^* \overline{j_0^* P_r} \cdot \tau^* j_0^* P_{d-r} \cap \overline{i_0^{vir} 1_r} \cdot i_0^{vir} 1_{d-r}.$$

By Lemma 3.3,

$$\begin{aligned} \tau^* j_r^* P_d \cap i_r^{vir} 1_d &= \tau^* j_r^* P_d \cdot e_T(X/Y)^{-1} \cap \tau^* \left(\frac{j_r^* \varphi_* LT_d(X)}{e_G(Y_r/W_d)} \right) \\ &= e_T(X/Y)^{-1} \cap \tau^* \left(\frac{j_r^* \varphi_* (\varphi^* P_d \cap LT_d(X))}{e_G(Y_r/W_d)} \right) \\ &= i_r^{vir} \varphi^* P_d. \end{aligned}$$

Thus (5.5) becomes

$$\tau^*\Lambda \cap i_r^{vir} \varphi^* P_d = \overline{i_0^{vir} \varphi^* P_r} \cdot i_0^{vir} \varphi^* P_{d-r}.$$

(cf. Theorem 3.6.) From this we get, as before,

$$\int_{W_d} \varphi_* LT_d(X) \cap P_d \cdot e^{\kappa\cdot\zeta} = \int_X \tau^*\Lambda^{-1} \cap \overline{i_0^{vir}\varphi^* P_r} \cdot i_0^{vir}\varphi^* P_{d-r}\ e^{(H+r\alpha)\cdot\zeta} \in \mathcal{R}[[\zeta]].$$

Therefore, that

$$B(t) = e^{-H\cdot t/\alpha} \sum_d i_0^{vir}\varphi^* P_d\ e^{d\cdot t}$$

is an Euler series, is just an elementary consequence of the Euler data identity. More generally, we have

Theorem 3.11. *Let P be an Λ-Euler data as before, and let $\mathbf{O}(t) = e^{-H\cdot t/\alpha}\sum_d O_d\ e^{d\cdot t}$ be any Ω-Euler series. Then*

$$B(t) = e^{-H\cdot t/\alpha} \sum_d \tau^* j_0^* P_d \cap O_d\ e^{d\cdot t}$$

is an $\Omega \cdot \tau^*\Lambda$-Euler series.

Proof. Define P'_d on W_d by setting

$$j_r^* P'_d := \tau_*(\Omega^{-1}\bar{O}_r \cdot O_{d-r}) \cap e_G(Y_r/W_d).$$

By the localization theorem, this defines a class on W_d. Moreover, we have

$$\int_{W_d} P'_d \cap e^{\kappa\cdot\zeta} = \sum_r \int_X \Omega^{-1}\bar{O}_r \cdot O_{d-r}\ e^{(H+r\alpha)\cdot\zeta} \in \mathcal{R}.$$

It follows that $P'_d \in A_*^G(W_d) \otimes \mathcal{R}$ (see proof of Lemma 2.15 [29]). Now

$$\int_{W_d} P_d \cap P'_d\ e^{\kappa\cdot\zeta} = \sum_r \int_X \Omega^{-1}\tau^*\Lambda^{-1}\ \overline{(\tau^* j_0^* P_r \cap O_r)} \cdot (\tau^* j_0^* P_{d-r} \cap O_{d-r})\ e^{(H+r\alpha)\cdot\zeta},$$

which lies in $\mathcal{R}$ because $P_d \cap P'_d$ lies in $A_*^G(W_d) \otimes \mathcal{R}$. q.e.d.

Note that if $O_d = \mathbb{1}_d$, then P'_d in the proof above is just $\varphi_* LT_d(X)$. For explicit examples of Euler data, see [29], [30].

3.3 From Euler data to intersection numbers.

Again, fix the data V, b_T as before. From now on we write e^X simply as e. We recall the notations

$$A^{V,b_T}(t) = A(t) = e^{-H\cdot t/\alpha} \sum_d A_d \, e^{d\cdot t},$$

$$A_d = i_0^{vir}\pi^* b_T(V_d) = e_*^X \left(\frac{\rho^* b_T(V_d) \cap LT_{0,1}(d,X)}{e_G(F_0/M_d(X))} \right)$$

$$K_d^{V,b} = K_d = \int_{LT_{0,0}(d,X)} b(V_d)$$

$$\Phi^{V,b} = \Phi = \sum K_d \, e^{d\cdot t}.$$

Theorem 3.12. *(i)* $deg_\alpha A_d \le -2$.

(ii) If for each d the class $b_T(V_d)$ has homogeneous degree the same as the dimension of $LT_{0,0}(d,X)$, then in the nonequivariant limit we have

$$\int_X e^{-H\cdot t/\alpha} A_d = \alpha^{-3}(2 - d\cdot t) K_d$$

$$\int_X \left(A(t) - e^{-H\cdot t/\alpha}\Omega \right) = \alpha^{-3}(2\Phi - \sum t_i \frac{\partial \Phi}{\partial t_i}).$$

Proof. By definition,

$$A_d = e_* \left(\frac{\rho^* b(V_d) \cap LT_{0,1}(d,X)}{e_G(F_0/M_d(X))} \right).$$

So assertion (i) follows immediately from this formula Lemma 3.5.

The second equality in assertion (ii) follows from the first equality in (ii). Now consider

$$\begin{aligned} I &:= \int_X e^{-H\cdot t/\alpha} A_d \\ &= \int_{LT_{0,1}(d,X)} e^{-e^* H\cdot t/\alpha} \frac{\rho^* b(V_d)}{e_{\mathbf{C}^\times}(F_0/M_d(X))} \\ &= \int_{LT_{0,0}(d,X)} b(V_d) \; \rho_* \left(\frac{e^{-e^* H\cdot t/\alpha}}{e_{\mathbf{C}^\times}(F_0/M_d(X))} \right). \end{aligned}$$

Now $b(V_d)$ has homogeneous degree the same as the dimension of $LT_{0,0}(d,X)$. The second factor in the last integrand contributes a scalar factor given by integration over a fiber E of ρ. By Lemma 3.5, the degree 1 term in the second factor is $\frac{-e^*H\cdot t}{\alpha^3}+\frac{c}{\alpha^3}$ where $c=c_1(L_d)$.

Now the line bundle L_d on $M_{0,1}(d,X)$ is the restriction of the universal bundle L'_d on $M_{0,1}(d,Y)$ ($Y=\mathbf{P}^{n_1}\times\cdots\times\mathbf{P}^{n_m}$), and the map $\rho: M_{0,1}(d,X)\to M_{0,0}(d,X)$, is the restriction of the forgetting map $\rho': M_{0,1}(d,Y)\to M_{0,0}(d,Y)$. For the latter, the general fiber of ρ' is smooth $E'\cong\mathbf{P}^1$ so that

$$\int_{E'} i^*_{E'}c_1(L'_d)=\int_{E'}c_1(TE')=2.$$

Since ρ' is flat,

$$\int_{E} i^*_{E}c_1(L_d)=\int_{E'}i^*_{E'}c_1(L'_d)=2.$$

Restricting to a fiber E say over $(C,f)\in M_{0,0}(d,X)$, the evaluation map e is equal to f, which is a degree d map $E\to X$. It follows that

$$\int_E e^*H=d.$$

So we have

$$I=(-\frac{d\cdot t}{\alpha^3}+\frac{2}{\alpha^3})K_d. \qquad q.e.d.$$

Theorem 3.13. *More generally suppose b_T is an equivariant multiplicative class of the form*

$$b_T(V)=x^r+x^{r-1}b_1(V)+\cdots+b_r(V),\quad rk\ V=r$$

where x is a formal variable, b_i is a class of degree i. Suppose

$$s:=rk\ V_d-exp.\ dim\ M_{0,0}(d,X)\geq 0$$

is independent of $d\succ 0$. Then in the nonequivariant limit,

$$\frac{1}{s!}\left(\frac{d}{dx}\right)^s|_{x=0}\int_X e^{-H\cdot t/\alpha}A_d=\alpha^{-3}x^{-s}(2-d\cdot t)K_d$$

$$\frac{1}{s!}\left(\frac{d}{dx}\right)^s|_{x=0}\int_X\left(A(t)-e^{-H\cdot t/\alpha}\Omega\right)=\alpha^{-3}x^{-s}(2\Phi-\sum t_i\frac{\partial\Phi}{\partial t_i}).$$

Proof. The proof is entirely analogous to (ii) above. q.e.d.

In the case of $b_T(V)=1$, one can improve the α degree estimates for $A_d=\mathbf{1}_d$ given by Theorem 3.12 (i).

Lemma 3.14. *For all d,*

$$deg_\alpha\ \mathbb{1}_d \leq min(-2, -\langle c_1(X), d\rangle).$$

Proof. If $\langle c_1(X), d\rangle \leq 2$, then the assertion is a special case Theorem 3.12 (i). So suppose that $\langle c_1(X), d\rangle > 2$. The class $LT_{0,1}(d,X)$ is of dimension

$$s = exp.dim\ M_{0,1}(d,X) = \langle c_1(X), d\rangle + dim\ X - 2.$$

Let $c = c_1(L_d)$. Then $c^k \cap LT_{0,1}(d,X)$ is of dimension $s-k$, and so $e_*(c^k \cap LT_{0,1}(d,X))$ lies in the group $A^T_{s-k}(X)$. But this group is zero unless $s-k \leq dim\ X$ or $k \geq s - dim\ X = \langle c_1(X), d\rangle - 2$. Now by Lemma 3.5, it follows that

$$\mathbb{1}_d = i_0^{vir} 1_d = \sum_{k \geq \langle c_1(X), d\rangle - 2} \frac{1}{\alpha^{k+2}}\ e_*(c^k \cap LT_{0,1}(d,X)).$$

This completes the proof. q.e.d.

Remark 3.15. The entire theory discussed in this section obviously specializes to the case $T = 1$, hence applies to any projective manifold X.

4. Linking

Definition 4.1. A projective T-manifold X is called a balloon manifold if X^T is finite, and if for $p \in X^T$, the weights of the isotropic representation T_pX are pairwise linearly independent.

The second condition in the definition is known as the GKM condition [18]. We will assume that our balloon manifold has the property that if $p, q \in X^T$ such that $c(p) = c(q)$ for all $c \in A^1_T(X)$, then $p = q$. *From now on, unless stated otherwise, X will be a balloon manifold with this property.* If two fixed points p, q in X are connected by a T-invariant 2-sphere, then we call that 2-sphere a balloon and denote it by pq. For examples and the basic facts we need to use about these manifolds, see [30] and references there. All the results in Sections 5-6 in [30] are proved for balloon manifolds without any convexity assumption, and are therefore also applicable here. We will quote the ones we need here without proof, but with only slight change in notations and terminology.

Definition 4.2. Two Euler series A, B are linked if for every balloon pq in X and every $d = \delta[pq] \succ 0$, the function $(A_d - B_d)|_p \in \mathbf{C}(\mathcal{T}^*)(\alpha)$ is regular at $\alpha = \frac{\lambda}{\delta}$ where λ is the weight on the tangent line $T_p(pq) \subset T_pX$.

Theorem 4.3 (Theorem 5.4 [30]). *Suppose A, B are linked Euler series satisfying the following properties: for $d \succ 0$,*

(i) For $p \in X^T$, every possible pole of $(A_d - B_d)|_p$ is a scalar multiple of a weight on T_pX.

(ii) $deg_\alpha(A_d - B_d) \leq -2$.

Then we have $A = B$.

Theorem 4.4 (Theorem 6.6 [30]). *Suppose that A, B are two linked Euler series having property (i) of the preceding theorem. Suppose that $deg_\alpha A_d \leq -2$ for all $d \succ 0$, and that there exists power series $f \in \mathcal{R}[[e^{t_1}, .., e^{t_m}]]$, $g = (g_1, .., g_m)$, $g_j \in \mathcal{R}[[e^{t_1}, .., e^{t_m}]]$, without constant terms, such that*

$$e^{f/\alpha}B(t) = \Omega - \Omega\frac{H \cdot (t+g)}{\alpha} + O(\alpha^{-2}) \tag{4.1}$$

when expanded in powers of α^{-1}. Then

$$A(t+g) = e^{f/\alpha}\ B(t).$$

The change of variables effected by f, g above is an abstraction of what's known as mirror transformations [9].

Theorem 4.5. *Let $p \in X^T$, $\omega \in A_T^*\ (M_{0,1}\ (d, X))\ [\alpha]$, and consider $i_p^* e_* \left(\frac{\omega \cap LT_{0,1}(d,X)}{e_G(F_0/M_d(X))}\right) \in \mathbf{C}(\mathcal{T}^*)(\alpha)$ as a function of α. Then*

(i) Every possible pole of the function is a scalar multiple of a weight on T_pX.

(ii) Let pq be a balloon in X, and λ be the weight on the tangent line $T_p(pq)$. If $d = \delta[pq] \succ 0$, then the pole of the function at $\alpha = \lambda/\delta$ is of the form

$$e_T(p/X)\ \frac{1}{\delta}\frac{1}{\alpha(\alpha - \lambda/\delta)}\frac{i_F^*\omega}{e_T(F/M_{0,1}(d,X))}$$

where F is the (isolated) fixed point $(\mathbf{P}^1, f_\delta, 0) \in M_{0,1}(d, X)$ with $f_\delta(0) = p$, and $f_\delta : \mathbf{P}^1 \to X$ maps by a δ-fold cover of pq.

Proof. Consider the commutative diagram

$$\begin{array}{ccc} \{F\} & \xrightarrow{i_F} & M_{0,1}(d,X) \\ e' \downarrow & & e \downarrow \\ p & \xrightarrow{i_p} & X \end{array}$$

where e is the evaluation map, $\{F\}$ are the fixed point components in $e^{-1}(p)$, e' is the restriction of e to $\{F\}$, and i_F, i_p are the usual inclusions. By functorial localization we have, for any $\beta \in A_T^*(M_{0,1}(d,X))(\alpha)$,

$$\begin{aligned} i_p^* e_*(\beta \cap LT_{0,1}(d,X)) &= e_T(p/X) \sum_F e'_* \left(\frac{i_F^* \beta \cap [F]^{vir}}{e_T(F/M_{0,1}(d,X))} \right) \\ &= e_T(p/X) \sum_F \int_{[F]^{vir}} \frac{i_F^* \beta}{e_T(F/M_{0,1}(d,X))}. \end{aligned} \tag{4.2}$$

We apply this to the class

$$\beta = \frac{\omega}{e_G(F_0/M_d(X))} = \frac{\omega}{\alpha(\alpha - c)}$$

where $c = c_1(L_d)$ (cf. Lemma 3.5). For (i), we will show that a pole of the sum (5.5) is at either $\alpha = 0$ or $\alpha = \lambda'/\delta'$ for some tangent weight λ' on T_pX. For (ii), we will show that only one F in the sum (5.5) contributes to the pole at $\alpha = \lambda/\delta$, that the contributing F is the isolated fixed point $(\mathbf{P}^1, f_\delta, 0)$ as asserted in (ii), and that the contribution has the desired form.

A fixed point (C, f, x) in $e^{-1}(p)$ is such that $f(x) = p$, and that the image curve $f(C)$ lies in the union of the T-invariant balloons in X. The restriction of the first Chern class c to an F must be of the form

$$i_F^* c = c_F + w_F$$

where $c_F \in A^1(F)$, and $w_F \in \mathcal{T}^*$ is the weight of the representation on the line T_xC induced by the linear map $df_x : T_xC \to T_pX$ (cf. [26]). The image of df_x is either 0 or a tangent line $T_p(pr)$ of a balloon pr. Thus w_F is either zero (when the branch $C_1 \subset C$ containing x is contracted), or $w_F = \lambda'/\delta'$ (when $C_1 \xrightarrow{f} X$ maps by a δ'-fold cover of a balloon pr with tangent weight λ'). The class $e_T(F/M_{0,1}(d,X))$ is obviously independent of α. Since c_F is nilpotent, a pole of the sum (5.5) is either at $\alpha = 0$ or $\alpha = w_F$ for some F. This proves (i).

Now, an F in the sum (5.5) contributes to the pole at $\alpha = \lambda/\delta$ only if $w_F = \lambda/\delta$. Since the weights on T_pX are pairwise linearly independent, that $\lambda/\delta = \lambda'/\delta'$ implies that $\lambda = \lambda'$ and $\delta = \delta'$. Since $d = \delta[pq]$, it follows that the only fixed point contributing to the pole at $\alpha = \lambda/\delta$ is (C, f, x) where $C \xrightarrow{f} X$ maps by a δ-fold cover of the balloon pq with $C \cong \mathbf{P}^1$ and $f(x) = 0$. This is an isolated fixed point, which we denote by $F = (\mathbf{P}^1, f_\delta, 0)$. It contributes to the sum (5.5) the term

$$\int_F \frac{i_F^*\beta}{e_T(F/M_{0,1}(d,X))} = \frac{1}{\delta}\frac{i_F^*\omega}{\alpha(\alpha - \lambda/\delta)}\frac{1}{e_T(F/M_{0,1}(d,X))}.$$

Here F is an orbifold point of order δ, and hence the integration contributes the factor $1/\delta$. This proves (ii). q.e.d.

Fix the data V, b_T and a Λ-Euler data $P:\ P_d$ such that

$$\tau^*\Lambda = \Omega := b_T(V^+)/b_T(V^-).$$

We now discuss the interplay between four Euler series: $A^{V,b_T}(t)$, $\mathbb{1}(t)$, and two others

$$\mathbf{O}(t) := e^{-H\cdot t/\alpha}\sum O_d\ e^{d\cdot t}$$
$$B(t) := e^{-H\cdot t/\alpha}\sum_d \tau^* j_0^* P_d \cap O_d\ e^{d\cdot t}$$

where $\mathbf{O}(t)$ *denote some unspecified Euler series linked to* $\mathbb{1}(t)$. (In particular $\mathbf{O}(t)$ may be specialized to $\mathbb{1}(t)$ itself.) That $B(t)$ is an Euler series follows from Theorem 3.11.

Corollary 4.6. *Suppose that at* $\alpha = \lambda/\delta$ *and* $F = (\mathbf{P}^1, f_\delta, 0)$, *we have* $i_p^* j_0^* P_d = i_F^* \rho^* b_T(V_d)$ *for all* $d = \delta[pq]$. *Then* $B(t)$ *is linked to* $A^{V,b_T}(t)$.

Proof. Since $\mathbf{O}(t)$ is linked to $\mathbb{1}(t)$ by assumption, it follows trivially that

$$B(t) = e^{-H\cdot t/\alpha}\sum_d \tau^* j_0^* P_d \cap O_d\ e^{d\cdot t}$$
$$C(t) = e^{-H\cdot t/\alpha}\sum_d \tau^* j_0^* P_d \cap \mathbb{1}_d\ e^{d\cdot t}$$

are linked. So it suffices to show that $A(t)$ and $C(t)$ are linked. Denote their respective Fourier coefficients by A_d, C_d. Then

$$\begin{aligned} i_p^* C_d - i_p^* A_d =& i_p^* j_0^* P_d \cdot i_p^* e_* \left(\frac{LT_{0,1}(d,X)}{e_G(F_0/M_d(X))}\right) \\ &- i_p^* e_* \left(\frac{\rho^* b_T(V_d) \cap LT_{0,1}(d,X)}{e_G(F_0/M_d(X))}\right). \end{aligned} \tag{4.3}$$

By Theorem 4.5 (ii), this difference is regular because the zero of the function $i_p^* j_0^* P_d - i_F^* \rho^* b_T(V_d)$ cancels the *simple* pole of each term in (5.5) at $\alpha = \lambda/\delta$. q.e.d.

We now formulate one of the main theorems of this paper. It'll also give a more directly applicable form of Theorem 4.4. Given the data $V, b_T, \mathbf{O}(t), P$, and

$$B(t) := e^{-H\cdot t/\alpha} \sum_d \tau^* j_0^* P_d \cap O_d \; e^{d\cdot t},$$

assume that the preceding corollary holds. Suppose in addition, that

(*) For each d, we have the form

$$\tau^* j_0^* P_d = \Omega \alpha^{\langle c_1(X), d\rangle} \left(a + (a' + a'' \cdot H)\alpha^{-1} + \cdots \right),$$

for some $a, a', a_i'' \in \mathbf{C}(\mathcal{T}^*)$ (depending on d).

(**) For each d, we have the form (written in cohomology $A_{\mathcal{T}}^*(X)$):

$$O_d = \alpha^{-\langle c_1(X), d\rangle} \left(b + (b' + b'' \cdot H)\alpha^{-1} + \cdots \right),$$

for some $b, b', b_i'' \in \mathbf{C}(\mathcal{T}^*)$ (depending on d).

Theorem 4.7. Suppose that $A^{V,b_T}(t), B(t)$ are as in the preceding corollary. Under the assumptions (*)-(**), there exist power series $f \in \mathcal{R}[[e^{t_1}, .., e^{t_m}]]$, $g = (g_1, .., g_m)$, $g_j \in \mathcal{R}[[e^{t_1}, .., e^{t_m}]]$, without constant terms, such that

$$A^{V,b_T}(t+g) = e^{f/\alpha} B(t).$$

Proof. Recall that

$$B(t) := e^{-H\cdot t/\alpha} \sum_d \tau^* j_0^* P_d \cap O_d \; e^{d\cdot t}.$$

By the preceding corollary, $B(t)$ is linked to $A(t)$. We will use the asymptotic forms (*)-(**) to explicitly construct f, g satisfying the condition (4.1). Our assertion then follows from Theorem 4.4.

By (*)-(**), the Fourier coefficient B_d, $d \succ 0$, of $B(t)$ has the form

$$B_d = \Omega \left(ab + (ab' + a'b)\alpha^{-1} + (ab'' + a''b) \cdot H\alpha^{-1} + \cdots \right)$$

(and $B_0 = \Omega$). Multiplying this by $e^{-H\cdot t/\alpha} = 1 - H \cdot t\alpha^{-1} + \cdots$, and $e^{d\cdot t}$, and then sum over d, we get the form

$$B(t) = \Omega \left(C + (C' + C'' \cdot H - C\ H \cdot t)\alpha^{-1} + \cdots \right)$$

where $C, C', C''_i \in \mathbf{C}(\mathcal{T}^*)[[e^{t_1}, .., e^{t_m}]]$ having constant terms $1, 0, 0$ respectively. It follows that

$$\frac{e^{-C'/C\alpha}}{C} B(t) = \Omega \left(1 - (t - \frac{C''}{C}) \cdot H\alpha^{-1} + \cdots \right)$$

So putting $f = -\alpha log\ C - \frac{C'}{C}$ and $g = -\frac{C''}{C}$ yields the eqn. (4.1). This completes the proof. q.e.d.

Corollary 4.8. *The preceding theorem holds if we specialize the choice of $O(t)$ to $\mathbb{1}(t)$, i.e.,*

$$B(t) = e^{-H\cdot t/\alpha} \sum_d \tau^* j_0^* P_d \cap \mathbb{1}_d\ e^{d\cdot t}.$$

Proof. The preceding theorem holds for any Euler series $\mathbf{O}(t)$ satisfying the condition (**) linked to $\mathbb{1}(t)$. Now by Lemma 3.14, $\mathbb{1}(t)$ satisfies condition (**); and obviously it is also linked to itself. q.e.d.

4.1 Linking values.

In this subsection, we continue using the notations $V, b_T, \mathbb{1}(t), O(t), A(t)$, introduced above, where $O(t)$ is linked to $\mathbb{1}(t)$. We will apply Theorem 4.7 to the case when b_T is the Euler class or the Chern polynomial.

For simplicity, we will assume that V has the following property: that there exist nontrivial T-equivariant line bundles $L_1^+, .., L_{N^+}^+; L_1^-, .., L_{N^-}^-$ on X with $c_1(L_i^+) \geq 0$ and $c_1(L_j^-) < 0$, such that for any balloon $pq \cong \mathbf{P}^1$ in X we have

$$V^{\pm}|_{pq} = \oplus_{i=1}^{N^{\pm}} L_i^{\pm}|_{pq}.$$

Note that $N^{\pm} = rk\ V^{\pm}$. We also require that

$$\text{(4.4)} \qquad \Omega := b_T(V^+)/b_T(V^-) = \prod_i b_T(L_i^+)/\prod_j b_T(L_j^-).$$

In this case we call the list $(L_1^+, .., L_{N^+}^+; L_1^-, .., L_{N^-}^-)$ the splitting type of V. Note that V is not assumed to split over X.

Theorem 4.9. *Let $b_T = e_T$ be the equivariant Euler class. Let pq be a balloon, $d = \delta[pq] \succ 0$, and λ be the weight on the tangent line $T_q(pq)$. Let $F = (\mathbf{P}^1, f_\delta, 0)$ be the fixed point, as in Theorem 4.5 (ii). Then*

$$
(4.5)\qquad
\begin{aligned}
i_F^*\rho^* b_T(V_d) = \prod_i \prod_{k=0}^{\langle c_1(L_i^+),d\rangle} \left(c_1(L_i^+)|_p - k\lambda/\delta\right) \\
\times \prod_j \prod_{k=1}^{-\langle c_1(L_j^-),d\rangle - 1} \left(c_1(L_j^-)|_p + k\lambda/\delta\right).
\end{aligned}
$$

In particular, $A^{V,e_T}(t)$ is linked to the Euler series

$$
B(t) = e^{-H\cdot t/\alpha} \sum_d B_d \, e^{d\cdot t}
$$

where

$$
B_d = O_d \cap \prod_i \prod_{k=0}^{\langle c_1(L_i^+),d\rangle} (c_1(L_i^+) - k\alpha) \times \prod_j \prod_{k=1}^{-\langle c_1(L_j^-),d\rangle-1} (c_1(L_j^-) + k\alpha).
$$

Proof. Define $P: \ P_d \in A_G^*(W_d)$ by

$$
P_d := \prod_i \prod_{k=0}^{\langle c_1(L_i^+),d\rangle} (\hat{L}_i^+ - k\alpha) \times \prod_j \prod_{k=1}^{-\langle c_1(L_j^-),d\rangle-1} (\hat{L}_j^- + k\alpha),
$$

where $\hat{L}_i^{\pm} \in A_G^*(W_d)$ denotes the canonical lifting of $c_1(L_i^{\pm}) \in A_T^*(Y)$. Then P is an Ω-Euler data (see Section 2.2 [29]). By Theorem 3.11, it follows that $B_d = \tau^* j_0^* P_d \cap O_d$ is an Euler series. By (corollary to) Theorem 4.5, $A(t)$ is linked to $B(t)$, provided that eqn. (5.5) holds. We now prove eqn. (5.5).

We first consider a single convex line bundle $V = L$. As before, the fixed point $F = (\mathbf{P}^1, f_\delta, 0)$ in $M_{0,1}(d, X)$ is a δ-fold cover of the balloon $pq \cong \mathbf{P}^1$. We can write it as

$$
f_\delta : \mathbf{P}^1 \to pq \cong \mathbf{P}^1, \quad [w_0, w_1] \mapsto [w_0^\delta, w_1^\delta].
$$

Note that the T-action on X induces the standard rotation on $pq \cong \mathbf{P}^1$ with weight λ. Clearly, we have

$$
i_F^*\rho^* e_T(V_d) = i_{\rho(F)}^* e_T(V_d) = e_T(i_{\rho(F)}^* V_d).
$$

The right hand side is the product the weights of the T representation on the vector space

$$i_{\rho(F)}^* V_d = H^0(\mathbf{P}^1, f_\delta^* L) = H^0(\mathbf{P}^1, f_\delta^* \mathcal{O}(l))$$

where $l = \langle c_1(L), [pq] \rangle$. Thus we get (cf. Section 2.4 [29])

$$e_T(i_{\rho(F)}^* V_d) = \prod_{k=0}^{l\delta} (c_1(L)|_p - k\frac{\lambda}{\delta}).$$

This proves (5.5) for a single convex line bundle.

Similarly for a concave line bundle $V = L$, if its restriction to the balloon pq is $\mathcal{O}(-l)$ with $-l = \langle c_1(L), [pq] \rangle$, then

$$e_T(i_{\rho(F)}^* V_d) = \prod_{k=1}^{l\delta-1} (c_1(L)|_p + k\frac{\lambda}{\delta}).$$

This is (5.5) for a single concave line bundle. The general case can clearly be obtained by taking products. q.e.d.

A parallel argument for $b_T =$ the Chern polynomial yields

Theorem 4.10. *Let $b_T = c_T$ be the equivariant Chern polynomial, with the rest of the notations as in the preceding theorem. Then*

$$i_F^* \rho^* c_T(V) = \prod_i \prod_{k=0}^{\langle c_1(L_i^+),d\rangle} (x + c_1(L_i^+)|_p - k\lambda/\delta)$$
$$\times \prod_j \prod_{k=1}^{-\langle c_1(L_j^-),d\rangle-1} \left(x + c_1(L_j^-)|_p + k\lambda/\delta\right).$$

In particular, $A^{V,e_T}(t)$ is linked to the Euler series

$$B(t) = e^{-H\cdot t/\alpha} \sum_d B_d \, e^{d\cdot t}$$

where

$$B_d = O_d \cap \prod_i \prod_{k=0}^{\langle c_1(L_i^+),d\rangle} (x + c_1(L_i^+) - k\alpha) \times \prod_j \prod_{k=1}^{-\langle c_1(L_j^-),d\rangle-1} (x - c_1(L_j^-) + k\alpha).$$

By Theorem 4.7, we can therefore compute $A(t) = A^{V,b_T}(t)$ in terms of the Euler series $B(t)$ given above, provided that the Euler data P and the Euler series $O(t)$ both have the appropriate asymptotic forms (*)-(**) required by Theorem 4.7.

Corollary 4.11. *Let b_T be either e_T or c_T. Suppose that*

$$c_1(V^+) - c_1(V^-) \leq c_1(X). \tag{4.6}$$

Then the condition (*) holds for the Euler data P in the two preceding theorems. In this case, if $O(t)$ is any Euler series linked to $\mathbb{1}(t)$ and satisfies condition (**), then Theorem 4.7 applies to compute $A^{V,b_T}(t)$ in terms of $O(t)$ and P.

Proof. The Euler data P in either of the preceding theorems has the form: for each $d \succ 0$,

$$\tau^* j_0^* P_d = \Omega \alpha^{\langle c_1(V^+)-c_1(V^-),d\rangle - N_-} \left(a + (a' + a'' \cdot H)\alpha^{-1} + \cdots\right),$$

for some $a, a', a''_i \in \mathbf{C}(cT^*)$ (depending on d). By assumption,

$$\langle c_1(V^+) - c_1(V^-), d\rangle \leq \langle c_1(X), d\rangle.$$

This implies that P satisfies the condition (*). q.e.d.

This result shows that if $\Omega = b_T(V^+)/b_T(V^-)$ has a certain factorized form (4.4), and if there is a suitable bound (4.6) on first Chern classes, then $A(t) = A^{V,b_T}(t)$ is computable in terms of the $\mathbb{1}(t)$ (or a suitable Euler series $O(t)$ linked to it). Note that even though $\mathbb{1}(t)$ is not known explicitly in closed form in general, it is *universal* in the sense that it is natural and is independent of any choice of V or b_T. Its Fourier coefficients also happen to be related to the universal line bundle on $M_{0,1}(d, X)$. In the next section, we specialize $O(t)$ to something quite explicit. We also discuss some other ways to compute $A(t)$. We consider situations in which the first Chern class bound and the factorization condition on Ω can be removed.

5. Applications and generalizations.

Throughout this section, we continue to use the same notations: $V, b_T, \Omega, A(t), \ldots$.

5.1 Inverting $\mathbb{1}_d$.

Suppose $\mathbb{1}_d$ is invertible for all d. Then obviously, there exist unique $B_d \in A_T^*(X)(\alpha)$ such that

$$A(t) = e^{-H\cdot t/\alpha} \sum B_d \cap \mathbb{1}_d \, e^{d\cdot t}.$$

In particular this says that for $d = \delta[pq]$, $F = (\mathbf{P}^1, f_\delta, 0)$, we must have

$$i_p^* B_d = i_F^* \rho^* b_T(V_d) \tag{5.1}$$

at $\alpha = \lambda/\delta$. By Theorem 4.3, the B_d are the unique classes in $A_T^*(X)(\alpha)$ such that

(i) eqn. (5.5 holds.

(ii) $deg_\alpha\ B_d \cap \mathbb{1}_d \leq -2$.

(iii) $e^{-H\cdot t/\alpha} \sum B_d \cap \mathbb{1}_d\ e^{d\cdot t}$ is an Ω-Euler series.

In other words these algebraic conditions completely determine the B_d. Thus in principle the B_d can be computed in terms of the classes $\mathbb{1}_d$ and the linking values (5.5). The point is that *this is true whether or not the bound (4.6) or the factorized form Ω (4.4) holds.* Here are a few examples.

Example 1. $X = Y$ is a product of projective space with the maximal torus action. In this case,

$$\mathbb{1}_d = \frac{1}{e_G(Y_0/W_d)}$$

which is given explicitly in Section 2. We also have

$$B_d = j_0^* \varphi_*(\pi^* b_T(V_d) \cap LT_d(X)) \in A_T^*(X)[\alpha]$$

(cf. Lemma 3.3). Finding the B_d explicitly amounts to finding polynomials in H_a, α with the prescribed values (5.5), and the degree bound (ii). This is a linear problem! This approach is particularly useful for computing $b_T(V_d)$ for nonsplit bundles V (e.g. $V = TX$), or for bundles where the bound (4.6) fails (e.g. $\mathcal{O}(k)$ on $\mathbf{P}^n$ with $k > n+1$).

Example 2. Suppose X is a balloon manifold such that every balloon pq generates the integral classes in $A_1(X)$. Then every integral class $d \in A_1(X)$ is of the form $\delta[pq]$ (e.g. Grassmannians). We claim that, in this case, $\mathbb{1}_d$ is invertible for all d. It suffices to show that $i_p^* \mathbb{1}_d$ is nonzero for every fixed point p in X. Given p, we know that there are $n = dim\ X$ other fixed points q joint to p by balloons pq. Pick such a q. Then $d = \delta[pq]$ for some δ. It follows from Theorem 4.5 that the function $i_p^* \mathbb{1}_d$ has a *nontrivial* simple pole at $\alpha = \lambda/\delta$ where λ is the weight on the tangent line $T_p(pq)$. This completes the proof.

Obviously, we can take product of these examples and still get invertible $\mathbb{1}_d$ for the product manifold.

5.2 Toric manifolds.

Let X be a toric manifold of dimension n. Denote by D_a, $a = 1,..,N$, the T-invariant divisors in X. We denote by the same notations the equivariant homology classes they represent. Recall that [3], [11], [32] X can be represented as an orbit space

$$X = (\Gamma - Z)/K$$

where K is an algebraic torus of dimension $N - n$, $\Gamma = \mathbf{C}^N$ is a linear representation of K, and Z is a K-invariant monomial variety of $\mathbf{C}^N$, all determined by the fan of X. The T action on the orbit space is induced by $(\mathbf{C}^\times)^N$ acting on Γ by the usual scaling. Define

$$O(t) = e^{-H\cdot t/\alpha} \sum_d O_d \; e^{d\cdot t},$$

$$O_d := \frac{\prod_{\langle D_a,d\rangle<0} \prod_{k=0}^{-\langle D_a,d\rangle-1} (D_a + k\alpha)}{\prod_{\langle D_a,d\rangle\geq 0} \prod_{k=1}^{\langle D_a,d\rangle} (D_a - k\alpha)}. \tag{5.2}$$

We will prove that $O(t)$ is a 1-Euler series.

First we recall a construction in [31], [42]. Given an integral class $d \in A_1(X)$, let $\Gamma_d = \oplus_a H^0(\mathbf{P}^1, \mathcal{O}(\langle D_a, d\rangle))$. Let K act on Γ_d by $\phi_a \mapsto t^{\lambda_a}\phi_a$ where the λ_a are the same weights with which K acts on Γ. Let

$$Z_d = \{\phi \in \Gamma_d | \phi(z,w) \in Z, \; \forall (z,w) \in \mathbf{C}^2\}.$$

(Note that ϕ here is viewed as a polynomial map $\mathbf{C}^2 \to \mathbf{C}^N$.) It is obvious that Z_d is K-invariant. Define the orbit space

$$\mathcal{W}_d := (\Gamma_d - Z_d)/K.$$

(i) If not empty, $\mathcal{W}_d$ is a toric manifold of dimension

$$dim \; \mathcal{W}_d = \sum_a{}' (\langle D_a, d\rangle + 1) - dim \; K$$

where $\sum_a'$ means summing only those terms which are positive.

(ii) T acts on $\mathcal{W}_d$ in an obvious way. There is also a $\mathbf{C}^\times$ action on $\mathcal{W}_d$ induced by the standard action on $\mathbf{P}^1$ with weight α. Each $\mathbf{C}^\times$ fixed point component in $\mathcal{W}_d$ is (consisting of K-orbits of)

$$\mathcal{Y}_r = \{\phi = (x_1 w_0^{\langle D_1,r\rangle} w_1^{\langle D_1,d-r\rangle}, ..., x_N w_0^{\langle D_N,r\rangle} w_1^{\langle D_N,d-r\rangle}) | (x_1,..,x_N) \in \mathbf{C}^N,$$
$$x_b = 0 \; if \; the \; corresp. \; monomial \; has \; negative \; exponent\}.$$

Let $j_r : \mathcal{Y}_r \to \mathcal{W}_d$ be the inclusion maps. If nonempty, $\mathcal{Y}_r$ is canonically isomorphic to a T-invariant submanifold in X given by intersecting those divisors $x_b = 0$ corresponding to negative exponents above. Denote the canonical inclusions by $\tau_r : \mathcal{Y}_r \to X$. Then $\tau_{r*}(1) = \prod_{\langle D_a, r\rangle<0 \text{ or } \langle D_a, d-r\rangle<0} D_a$. We will denote the class of $D_a \cap \mathcal{Y}_r$ in $\mathcal{Y}_r$ simply by D_a.

(iii) The $G = \mathbf{C}^\times \times T$-equivariant Euler class of the normal bundle of $\mathcal{Y}_r$ in $\mathcal{W}_d$ is

$$e_G(\mathcal{Y}_r/\mathcal{W}_d) = \prod_{\langle D_a,d\rangle \geq 0} \prod_{k=0}^{\langle D_a,d\rangle}{}_{k\neq\langle D_a,r\rangle} (D_a + \langle D_a, r\rangle\alpha - k\alpha).$$

(iv) Corresponding canonically to every T-divisor class D_a on X is a G-divisor class $\hat{D}_a$ on $\mathcal{W}_d$. It is determined by

$$j_r^* \hat{D}_a = D_a + \langle D_a, r\rangle\alpha.$$

Similarly, every linear combination D of the D_a corresponds to some $\hat{D}$ on $\mathcal{W}_d$.

Lemma 5.1. *$O(t)$ introduced above is a 1-Euler series.*

Proof. Let

$$\omega_d = \prod_{\langle D_a,d\rangle<0} \prod_{k=1}^{-\langle D_a,d\rangle-1} (\hat{D}_a + k\alpha) \in A_*^G(\mathcal{W}_d). \tag{5.3}$$

By the localization theorem,

$$\int_{\mathcal{W}_d} \omega_d \cdot e^{\hat{H}\cdot\zeta} = \sum_r \int_{\mathcal{Y}_r} \frac{j_r^*\omega_d}{e_G(\mathcal{Y}_r/\mathcal{W}_d)} e^{(H+r\alpha)\cdot\zeta}.$$

Obviously, the left hand side lies in $A_G^*(pt)[[\zeta]] \subset \mathcal{R}[[\zeta]]$. Now observe that the right hand side is

$$\sum_r \int_{\mathcal{Y}_r} \frac{\prod_{\langle D_a,d\rangle<0}\prod_{k=1}^{-\langle D_a,d\rangle-1}(D_a+\langle D_a,r\rangle\alpha+k\alpha)}{\prod_{\langle D_a,d\rangle\geq 0}\prod_{k=0}^{\langle D_a,d\rangle}{}_{k\neq\langle D_a,r\rangle}(D_a+\langle D_a,r\rangle\alpha-k\alpha)} e^{(H+r\alpha)\cdot\zeta}$$

$$= \sum_r \int_X \frac{\prod_{\langle D_a,r\rangle<0}\prod_{k=0}^{-\langle D_a,r\rangle-1}(D_a-k\alpha)}{\prod_{\langle D_a,r\rangle\geq 0}\prod_{k=1}^{\langle D_a,r\rangle}(D_a+k\alpha)}$$

$$\times \frac{\prod_{\langle D_a,d-r\rangle<0}\prod_{k=0}^{-\langle D_a,d-r\rangle-1}(D_a+k\alpha)}{\prod_{\langle D_a,d-r\rangle\geq 0}\prod_{k=1}^{\langle D_a,d-r\rangle}(D_a-k\alpha)} \cdot e^{(H+r\alpha)\cdot\zeta}$$

$$= \sum_r \int_X \bar{O}_r \cdot O_{d-r}\, e^{(H+r\alpha)\cdot\zeta}.$$

This shows that $O(t)$ is a 1-Euler series. q.e.d.

Remark 5.2. One can define the notion of Euler data on the basis of $\mathcal{W}_d$ in a way similar to Definition 3.10. The classes (5.5) in fact give an example of Euler data for $\mathcal{W}_d$. One can also construct the whole parallel theory of mirror principle for toric manifolds using $\mathcal{W}_d$.

Lemma 5.3. *The two Euler series $O(t)$ and $\mathbb{1}(t)$ are linked.*

Proof. Let $p \in X^T$, pq be a balloon in X, $d = \delta[pq] \succ 0$, and λ be the weight on the tangent line on $T_p(pq)$. Let $F = (\mathbf{P}^1, f_\delta, 0)$ be the fixed point in $M_{0,1}(d, X)$, as given in Theorem 4.5, which says that the function $i_p^*\mathbb{1}_d$ has the polar term, at $\alpha = \lambda/\delta$,

$$e_T(p/X)\,\frac{1}{\lambda}\frac{1}{\alpha-\lambda/\delta}\frac{1}{e_T(F/M_{0,1}(d,X))}. \tag{5.4}$$

We now compute the contribution from $e_T(F/M_{0,1}(d,X))$ for a toric manifold X. The virtual normal bundle of the point $F = (C = \mathbf{P}^1, f_\delta, 0)$ in $M_{0,1}(d,X)$ is

$$N_{F/M_{0,1}(d,X)} = [H^0(C, f_\delta^*TX)] - [H^1(C, f_\delta^*TX)] - A_C$$

(notation as in Section 2.3 [29]). From the Euler sequence of X [24], we get an equivariant exact sequence for every balloon pq in X,

$$0 \to \mathcal{O}^{N-n} \to \oplus_a \mathcal{O}(D_a)|_{pq} \to TX|_{pq} \to 0$$

where $\mathcal{O}$ is the trivial line bundle. At p, there are exactly n nonzero $D_a(p) := i_p^* D_a$ giving the weights for the isotropic representation T_pX, and $N-n$ zero $D_a(p)$ corresponding to the trivial representation $\mathcal{O}^{N-n}|_p$.

As usual we ignore the zero weights below, which must drop out at the end.

Let $\lambda = D_b(p)$. Note that $\langle D_b, d\rangle = 1$ (Section 2.3 [30]). The bundle $\mathcal{O}(D_b)$ contributes to $e_T([H^0(C, f_\delta^* TX)])$ the term

$$\prod_{k=0}^{\delta-1}(D_b(p) - k\lambda/\delta).$$

For each $a \neq b$ with $\langle D_a, d\rangle \geq 0$, the bundle $\mathcal{O}(D_a)$ contributes to $e_T([H^0(C, f_\delta^* TX)])$ the term

$$\prod_{k=0}^{\langle D_a,d\rangle} (D_a(p) - k\lambda/\delta) \ \ if\ D_a(p) \neq 0,$$

$$\prod_{k=1}^{\langle D_a,d\rangle} (D_a(p) - k\lambda/\delta) \ \ if\ D_a(p) = 0.$$

For each a with $\langle D_a, d\rangle < 0$, the bundle $\mathcal{O}(D_a)$ contributes to $e_T([H^1(C, f_\delta^* TX)])$ the term

$$\prod_{k=1}^{-\langle D_a,d\rangle-1} (D_a(p) + k\lambda/\delta).$$

The automorphism group A_C contributes $e_T(A_C) = -\lambda/\delta$. Finally, we have

$$e_T(p/X) = \prod_{D_a(p)\neq 0} D_a(p).$$

Combining all the contributions, we see that (5.5) becomes $\frac{1}{\alpha-\lambda/\delta}$ times

$$\frac{-1}{\delta}\frac{\prod_{\langle D_a,d\rangle<0}\prod_{k=0}^{-\langle D_a,d\rangle-1}(D_a(p)+k\lambda/\delta)}{\prod_{\langle D_a,d\rangle\geq 0_{a\neq b}}\prod_{k=1}^{\langle D_a,d\rangle}(D_a(p)-k\lambda/\delta)\times\prod_{k=1}^{\delta-1}(D_b(p)-k\lambda/\delta)}.$$

But this coincides with

$$lim_{\alpha\to\lambda/\delta}\ (\alpha - \lambda/\delta) i_p^* O_d.$$

This shows that $i_p^* O_d - i_p^* \mathbb{1}_d$ is regular at $\alpha = \lambda/\delta$. q.e.d.

Note that $O_d = \alpha^{\langle c_1(X),d\rangle} +$ lower order terms, because $\sum D_a = c_1(X)$. Thus $O(t)$ is an Euler series linked to $\mathbb{1}(t)$ and meets the condition (**) of Theorem 4.7. In particular to apply to the case $b_T = e_T$ or

c_T, all we need is the form (4.4) for Ω and the bound (4.6). For then Corollary 4.11 holds.

Example. Take $b_T = c_T$. Take V to be any direct sum of convex equivariant line bundles L_i, so that (4.6) holds. Note that in this case (4.4) holds automatically. Then Theorem 4.7 yields an explicit formula for $A^{V,b_T}(t)$ in terms of the O_d (5.2) and the P_d in Theorem 4.10. For $b_T = e_T$, and V a direct sum of convex line bundles L_i with $\sum_i c_1(L_i) = c_1(X)$, we get a similar explicit formula for $A(t)$. Plugging this formula into Theorem 3.12 in the nonequivariant limit, we get

Corollary 5.4. *Let*

$$B(t) = e^{-H\cdot t/\alpha} \sum_d \prod_i \prod_{k=0}^{\langle c_1(L_i),d\rangle} (c_1(L_i) - k\alpha) \cap \frac{\prod_{\langle D_a,d\rangle<0} \prod_{k=0}^{-\langle D_a,d\rangle-1} (D_a + k\alpha)}{\prod_{\langle D_a,d\rangle\geq 0} \prod_{k=1}^{\langle D_a,d\rangle} (D_a - k\alpha)} e^{d\cdot t},$$

as in Theorem 4.7. Then we have

$$\int_X \left(e^{f/\alpha} B(t) - e^{-H\cdot T/\alpha} \Omega \right) = \alpha^{-3} (2\Phi - \sum T_i \frac{\partial \Phi}{\partial T_i})$$

where $T = t + g$, and f, g are the power series computed in Theorem 4.7.

This is the general mirror formula in [21], [22] (see also references there), formulated in the context of mirror symmetry and reflexive polytopes [4], [5].

5.3 A generalization.

We have now seen several ways to compute $A(t) = A^{V,b_T}(t)$ under various assumptions on either V, b_T, or TX, or the classes $\mathbb{1}_d$, or some combinations of these assumptions. We now combine these approaches to formulate an algorithm for computing $A(t)$ in full generality on any balloon manifold X, for arbitrary V, b_T. The result will be in terms of certain (computable) T representations.

(i) By Lemma 3.4, the A_d is of the form

$$A_d = \frac{\tau^* \phi_d}{e_G(X/W_d)},$$

where $\phi_d \in A_*^T(Y)[\alpha]$, hence represented by a polynomial $\mathbf{C}[\mathcal{T}^*][H_1, ..., H_m, \alpha]$. Note that the denominator of A_d is

$$e_G(X/W_d) = e_T(X/Y) \cdot \tau^* e_G(Y_0/W_d).$$

Thus the goal is to compute the class $\tau^*\phi_d$ for all d. We shall set up a (over-determined) system of linear equations with a solution (unique up to $ker\ \tau^*$) given by the ϕ_d.

(ii) By Theorem 3.12, the degree of ϕ_d is bounded according to

$$deg_\alpha A_d \leq -2.$$

(iii) By Theorem 4.5 (i), at any fixed point p, the function $i_p^* A_d$ is regular away from $\alpha = 0$ or λ/δ, where λ is a weight on T_pX. In other words,

$$Res_{\alpha=\gamma}(\alpha - \gamma)^k\ i_p^* A_d = 0$$

for all $\gamma \neq 0, \lambda/\delta$ and $k \geq 0$. Note that these are all linear conditions on ϕ_d.

(iv) By Theorem 4.5 (ii) (see notations there), for any balloon pq in X and $d = \delta[pq] \succ 0$, we have

$$\begin{aligned} & lim_{\alpha\to\lambda/\delta}(\alpha - \lambda/\delta)\ i_p^* A_d \\ (5.5) \qquad & = \frac{e_T(p/X)}{\lambda\ e_T(F/M_{0,1}(d, X))} i_F^* \rho^* b_T(V_d) \\ & = \frac{-e_T(p/X)}{\delta} \frac{e_T[H^1(\mathbf{P}^1, f_\delta^* TX)]'}{e_T[H^0(\mathbf{P}^1, f_\delta^* TX)]'}\ b_T(i_{\rho(F)}^* V_d). \end{aligned}$$

Here we have used the fact that

$$N_{F/M_{0,1}(d,X)} = [H^0(\mathbf{P}^1, f_\delta^* TX)] - [H^1(\mathbf{P}^1, f_\delta^* TX)] - A_C$$

(cf. Section 5.2). The prime in the Euler classes above means that we drop the zero weights in the T representations $[H^i(\mathbf{P}^1, f_\delta^* TX)]$. Now if $V = V^+ \oplus V^-$ is a convex/concave bundle on X, then we have the T representation

$$i_{\rho(F)}^* V_d = H^0(\mathbf{P}^1, f_\delta^* V^+) \oplus H^1(\mathbf{P}^1, f_\delta^* V^-).$$

Thus

$$b_T(i_{\rho(F)}^* V_d)) = b_T[H^0(\mathbf{P}^1, f_\delta^* V^+)]\ b_T[H^1(\mathbf{P}^1, f_\delta^* V^-)],$$

which is just the value of b_T for a trivial bundle over a point. Note that if U is any T representation with weight decomposition $U = \oplus_i \mathbf{C}_{\nu_i}$, then $b_T(U) = \prod_i b_T(\mathbf{C}_{\nu_i})$ by the multiplicativity of b_T. Hence once the T representations $H^i(\mathbf{P}^1, f_\delta^* V^\pm)$ are given, eqn. (5.5) becomes a linear condition on the ϕ_d, where the right hand side is some known element in $\mathbf{C}(\mathcal{T}^*)$.

(v) Finally, we know that $A(t)$ is an Ω-Euler series. This is (inductively) a linear condition on the ϕ_d.

(vi) By Theorem 4.3, any solution to the linear conditions in (ii)-(v) necessarily represents the class $\tau^*\phi_d$ we seek.

Of course, this algorithm relies on knowing the T representations $[H^i(\mathbf{P}^1, f_\delta^* TX)]$, $[H^i(\mathbf{P}^1, f_\delta^* V)]$ induced by the T-equivariant bundles $TX|_{pq}$ and $V|_{pq}$ on each balloon $pq \cong \mathbf{P}^1$. But describing them for any given X and V is clearly a classical question. We have seen that these representations are easily computable in many cases. We now discuss a general situation in which these representations can also be computed similarly.

Let V be any T-equivariant vector bundle on X and let

$$0 \to V_N \to \cdots \to V_1 \to V \to 0$$

be an equivariant resolution. Then by the Euler-Poincare Principle, we have

$$\begin{aligned}[H^0(\mathbf{P}^1, f_\delta^* V)] &- [H^1(\mathbf{P}^1, f_\delta^* V)] \\ &= \sum_a (-1)^{a+1}([H^0(\mathbf{P}^1, f_\delta^* V_a)] - [H^1(\mathbf{P}^1, f_\delta^* V_a)]).\end{aligned}$$

Note that there is a similar equality of representations whenever V is a term in any given exact sequence

$$0 \to V_N \to \cdots \to V_i \to V \to V_{i-1} \to \cdots \to V_1 \to 0.$$

Now suppose that each V_a is a direct sum of T-equivariant line bundles. Then each summand L will contribute to $[H^0(\mathbf{P}^1, f_\delta^* V_a)] - [H^1(\mathbf{P}^1, f_\delta^* V_a)]$ the representations

$$c_1(L)|_p - k\lambda/\delta, \quad k = 0, 1, ..., l\delta,$$

if $l = \langle c_1(L), [pq]\rangle \geq 0$; and

$$c_1(L)|_p + k\lambda/\delta, \quad k = 1, ..., l\delta - 1,$$

if $-l = \langle c_1(L), [pq]\rangle < 0$ (cf. proof of Theorem 4.9). In this case, $[H^0(\mathbf{P}^1, f_\delta^* V)] - [H^1(\mathbf{P}^1, f_\delta^* V)]$ are then determined completely. Thus whenever a T-equivariant resolution by line bundles is known for TX and the convex/concave bundle $V^\pm$, the right hand side of eqn. (5.5) becomes known.

Example. Consider the case $X = \mathbf{P}^n$, $V = TX$, and b_T the Chern polynomial. This will be an example where V has no splitting type, but $A(t)$ can be computed via a T-equivariant resolution nevertheless. Recall the T equivariant Euler sequence

$$0 \to \mathcal{O} \to \oplus_{i=0}^n \mathcal{O}(H - \lambda_i) \to TX \to 0.$$

For $F = (\mathbf{P}^1, f_\delta, 0)$, where f_δ is the δ-fold cover of the balloon pq, this gives

$$b_T(i_{\rho(F)}^* V_d) = \frac{1}{x} \prod_i \prod_{k=0}^d (x + \lambda_j - \lambda_i - k\lambda/\delta).$$

Here p, q are the jth and the lth fixed points in $\mathbf{P}^n$, so that $\lambda = \lambda_j - \lambda_l$. We can use this to set up a linear system to solve for $A(t)$ inductively. However, there is an easier way to compute $A(t)$ in this case. Observe that $\Omega = b_T(V) = \frac{1}{x}\prod_i (x + H - \lambda_i)$, and that

$$P: \quad P_d := \frac{1}{x} \prod_i \prod_{k=0}^d (x + \kappa - \lambda_i - k\alpha)$$

defines an Ω-Euler data (see Section 2.2 [29]). Since $j_0^* \kappa = H$ and $i_p^* H = \lambda_j$, it follows that

$$b_T(i_{\rho(F)}^* V_d) = i_p^* j_0^* P_d$$

at $\alpha = \lambda/\delta$. By the corollary to Theorem 4.5, the Euler series

$$B(t) := e^{-H \cdot t/\alpha} \sum j_0^* P_d \cap \mathbb{1}_d$$

is linked to $A(t)$. Obviously, we have $deg_\alpha j_0^* P_d = (n+1)d$, hence P meets condition (*) of Theorem 4.7 (τ is the identity map here). For $O_d = \mathbb{1}_d$, condition (**) there is also automatic. It follows that

$$A(t + g) = e^{f/\alpha} B(t)$$

where f, g are explicitly computable functions from Theorem 4.7. Note that $rank\ V_d = (n+1)d + n$, and so Theorem 3.13 yields immediately the codimension $s = 3$ Chern class of V_d.

5.4 Blowing up the image.

In this section, we discuss another approach to compute $A(t)$. For clarity, we restrict to the case of a convex T-manifold X (T may be trivial), and $b_T \equiv 1$, so that $A(t) = \mathbb{1}(t)$. Thus we will study the classes

$$\mathbf{1}_d = e_*^X \left(\frac{LT_{0,1}(d,X)}{e_G(F_0/M_d(X))} \right).$$

We will actually be interested in the integrals $\int_X \tau^* e^{H\cdot\zeta} \cap \mathbf{1}_d$, where $\tau : X \to Y$ is a given projective embedding. For the purpose of studying the intersection numbers in Section 3.3, this is adequate. Since X is assumed convex, $LT_{0,1}(d,X)$ is represented by $M_{0,1}(d,X)$. Likewise for $M_d(X)$.

Suppose that we have a commutative diagram

$$(5.6) \qquad \begin{array}{ccccc} F_0 & \xrightarrow{e^Y} & Y_0 & \xleftarrow{g} & E_0 \\ \downarrow i & & \downarrow j & & \downarrow k \\ M_d & \xrightarrow{\varphi} & W_d & \xleftarrow{\psi} & \mathcal{Q}_d. \end{array}$$

Here the left hand square is as in (3.2) ($i_0, j_0, M_d(X)$ there are written as i, j, M_d here for clarity). We assume that $\mathcal{Q}_d$ is a G-manifold, that $\psi : \mathcal{Q}_d \to W_d$ is a G-equivariant resolution of singularities of $\varphi(M_d)$, and that E_0 is the fixed points in $\psi^{-1}(Y_0)$. Here g denotes the restriction of ψ, and k the inclusion. Recall that φ is an isomorphism into its image away from the singular locus of M_d. The singularities in $\varphi(M_d)$ is the image of the compactifying divisor in M_d, which has codimension at least 2. Then we have the equality in $A_*^G(W_d)$:

$$\varphi_*[M_d] = \psi_*[\mathcal{Q}_d].$$

Applying functorial localization to the left hand square in (5.6) as in Section 3.1, we get

$$\frac{j^*\varphi_*[M_d]}{e_G(Y_0/W_d)} = e_*^Y \left(\frac{[F_0]}{e_G(F_0/M_d)} \right).$$

Doing the same for the right hand square, we get

$$\frac{j^*\psi_*[\mathcal{Q}_d]}{e_G(Y_0/W_d)} = g_* \left(\frac{[E_0]}{e_G(E_0/\mathcal{Q}_d)} \right).$$

It follows that

Lemma 5.5. *In $A_*^T(Y)$, we have the equality*

$$\tau_* \mathbb{1}_d = e_*^Y\left(\frac{[F_0]}{e_G(F_0/M_d)}\right) = g_*\left(\frac{[E_0]}{e_G(E_0/\mathcal{Q}_d)}\right).$$

It follows that

$$\begin{aligned}\int_X \tau^* e^{H\cdot\zeta} \cap \mathbb{1}_d &= \int_{Y_0} e^{H\cdot\zeta} \cap g_*\left(\frac{[E_0]}{e_G(E_0/\mathcal{Q}_d)}\right) \\ &= \int_{E_0} \frac{g^* e^{H\cdot\zeta}}{e_G(E_0/\mathcal{Q}_d)} \\ &= \int_{E_0} \frac{g^* j^* e^{\kappa\cdot\zeta}}{e_G(E_0/\mathcal{Q}_d)} \\ &= \int_{E_0} \frac{k^* \psi^* e^{\kappa\cdot\zeta}}{e_G(E_0/\mathcal{Q}_d)}.\end{aligned}$$

In many cases, the spaces $\mathcal{Q}_d$ can be explicitly described, and the classes $\psi^*\kappa$ on $\mathcal{Q}_d$ can be expressed in terms of certain universal classes. For example, when X is a flag variety, then the $\mathcal{Q}_d$ can be chosen to be the Grothendieck Quot scheme (cf. [10], [8]). Integration on the Quot scheme can be done by explicit localization (cf. [39]). When X is a Grassmannian and $\tau : X \to Y = \mathbf{P}^N$ is the Plucker embedding, then $\psi^*\kappa = -c_1(S)$, where S is a universal subbundle on the Quot scheme. In this case, the image $\psi(Q_d)$ has been studied extensively in [37], [38].

When X is not convex, a similar method still works if we can find an explicit cycles Z_d in $\mathcal{Q}$ such that

$$\varphi_* LT_d(X) = \psi_*[Z_d]$$

in $A_*^G(W_d)$. This approach deserves further investigation.

5.5 Higher genus.

In this section, we discuss a generalization of mirror principle to higher genus. More details will appear elsewhere. As before X will be a projective T-manifold, and $\tau : X \to Y$ a given T-equivariant projective embedding. (Again, T may be the trivial group.)

Let $M_{g,k}(d, X)$ denote the k-pointed, arithmetic genus g, degree d, stable map moduli stack of X. Let M_d^g denote $M_{g,0}((d,1), X \times \mathbf{P}^1)$. Note that for each stable map $(C, f) \in M_d^g$ there is a unique branch $C_0 \cong \mathbf{P}^1$ in C such that f composed with the projection $X \times \mathbf{P}^1 \to \mathbf{P}^1$

maps $C_0 \to \mathbf{P}^1$ isomorphically. Moreover, C is a union of C_0 with some disjoint curves $C_1, .., C_N$, where each C_i intersects C_0 at a point $x_i \in C_0$. The map f composed with $X \times \mathbf{P}^1 \to \mathbf{P}^1$ collapses all $C_1, .., C_N$.

The standard $\mathbf{C}^\times$ action on $\mathbf{P}^1$ induces an action on M_d^g. The fixed point components are labelled by $F_{d_1,d_2}^{g_1,g_2}$ with $d_1 + d_2 = d$, $g_1 + g_2 = g$. As in the genus zero case, a stable map (C, f) in this component is given by gluing two 1-pointed stable maps

$$(f_1, C_1, x_1) \in M_{g_1,1}(d_1, X), \ (f_2, C_2, x_2) \in M_{g_2,1}(d_2, X)$$

with $f_1(x_1) = f_2(x_2)$, to a $\mathbf{P}^1$ at 0 and ∞ at the marked points (cf. Section 3). We can therefore identify $F_{d_1,d_2}^{g_1,g_2}$ with

$$M_{g_1,1}(d_1, X) \times_X M_{g_2,1}(d_2, X).$$

We denote by

$$F_{d_1,d_2}^g := \coprod_{g_1+g_2=g} F_{d_1,d_2}^{g_1,g_2}, \quad i_{d_1,d_2} : F_{d_1,d_2}^g \to M_d^g,$$

the disjoint union and inclusions. There are two obvious projection maps

$$p_0 : F_{d_1,d_2}^g \to \coprod_{g_1=0}^{g} M_{g_1,1}(d_1, X), \quad p_\infty : F_{d_1,d_2}^g \to \coprod_{g_2=0}^{g} M_{g_2,1}(d_2, X).$$

The map p_0 strips away the stable maps (f_2, C_2, x_2) glued to ∞ and forgets the $\mathbf{P}^1$; p_∞ strips away the stable map (f_1, C_1, x_1) glued to 0 and forgets the $\mathbf{P}^1$. We also have the usual evaluation maps, and the forgetting map:

$$e_{d_1,d_2} : F_{d_1,d_2}^g \to X, \quad e_d : M_{g,1}(d, X) \to X, \quad \rho : M_{g,1}(d, X) \to M_{g,0}(d, X).$$

Relating and summarizing the natural maps above is the following diagram:
(5.7)

$$\begin{array}{ccccccc}
X & \xleftarrow{e_{d_1,d_2}} & F_{d_1,d_2}^{g_1,g_2} & \xrightarrow{i_{d_1,d_2}} & M_d^g & \xrightarrow{\pi} & M_{g,0}(d, X) \\
e_{d_1} \uparrow & p_0 \swarrow & & \searrow p_\infty & & & \\
M_{g_1,1}(d_1, X) & & & & M_{g_2,1}(d_2, X) & & \\
\rho \downarrow & & & & \downarrow \rho & & \\
M_{g_1,0}(d_1, X) & & & & M_{g_2,0}(d_2, X) & &
\end{array}$$

Fix a class $\Omega \in A_T^*(X)$. We call a list $b_d^g \in A_T^*(M_{g,0}(d, X))$ an Ω-gluing sequence if we have the identities on the F_{d_1,d_2}^g:

$$e_{d_1,d_2}^* \Omega \cdot i_{d_1,d_2}^* \pi^* b_d^g = \sum_{g_1+g_2=g} p_0^* \rho^* b_{d_1}^{g_1} \cdot p_\infty^* \rho^* b_{d_2}^{g_2}.$$

It is easy to verify that $b_d^g \equiv 1$ is an example of a 1-gluing sequence. Restricted to $g = 0$, the identity above is precisely the gluing identity in Section 3.2. There we have found that the gluing identity results in an Euler series. It turns out that a gluing sequence too leads to an Euler series. For $\omega \in A_G^*(M_d^g)$ and $d = d_1 + d_2$, define (cf. Section 3.2)

$$i_{d_1,d_2}^{vir}\omega := e_{d_1,d_2*}\left(\frac{i_{d_1,d_2}^*\omega \cap [F_{d_1,d_2}^g]^{vir}}{e_G(F_{d_1,d_2}^g/M_d^g)}\right) \in A_*^T(X)(\alpha).$$

Then for a given gluing sequence $b_d^g \in A_T^*(M_{g,0}(d,X))$, we have the identities

$$\Omega \cap i_{d_1,d_2}^{vir}\pi^* b_d^g = \sum_{g_1+g_2=g} \overline{i_{0,d_1}^{vir}\pi^* b_{d_1}^{g_1}} \cdot i_{0,d_2}^{vir}\pi^* b_{d_2}^{g_2}.$$

Again, putting $g = 0$, we get the identities in Theorem 3.6. The argument in the higher genus case is essentially the same as the genus zero case. Here, one chases through a fiber diagram analogous to (3.1) using the associated refined Gysin homomorphism, together with the diagram (5.7).

Now given a gluing sequence, we put

$$A_d^g := i_{0,d}^{vir}\pi^* b_d^g, \quad A_d := \sum_g A_d^g\, \mu^g, \quad A(t) := e^{-H\cdot t/\alpha}\sum_d A_d\, e^{d\cdot t}.$$

Here μ is a formal variable. Then $A(t)$ is an Euler series. (We must, of course, replace the ring $\mathcal{R}$ by $\mathcal{R}[[\mu]]$.) The argument is also similar to the genus zero case: one applies functorial localization to the diagram

$$\begin{array}{ccc} F_{d_1,d_2}^g & \stackrel{i_{d_1,d_2}}{\longrightarrow} & M_d^g \\ e_{d_1,d_2}\downarrow & & \downarrow\varphi, \\ X \subset Y & \stackrel{j_{d_1,d_2}}{\longrightarrow} & W_d \end{array}$$

the same way we have done to diagram (3.2) in Section 3.2.

We can proceed further in a way parallel to the genus zero case. Namely, to find further constraints to a gluing sequence, we should compute the linking values of the Euler series $A(t)$. For this, let's assume that X is a balloon manifold, as in Sections 4.1 and 5.3. In genus zero, the linking values of an Euler series, say coming from $b_T(V_d)$, are determined by the restrictions $i_F^* b_T(\rho^* V_d)$ to the isolated fixed point $F = (\mathbf{P}^1, f_\delta, 0) \in M_{0,1}(d,X)$, which is a δ-fold cover of a balloon pq in X (see Theorem 4.5). In higher genus, this is replaced by a component in

$M_{g,1}(d, X)$ consisting of the following stable maps (C, f, x). Here C is a union of two curves C_1 and $C_0 \cong \mathbf{P}^1$ such that $C_0 \xrightarrow{f} pq$ is a δ-fold cover with $f(x) = p$, and $f(C_1) = q$. Therefore this fixed point component can be identified canonically with $\bar{M}_{g,1}$, the moduli space of genus g, 1-pointed, stable curves. Let's abbreviate it F. The linking values of $A(t)$ for this component is then a power series summing over integrals on $\bar{M}_{g,1}$ of classes given in terms of $i_F^* \rho^* b_d^g$ and $e_T(F/M_{g,1}(d, X))$ (cf. Theorem 4.5).

The entire discussion in this section can be generalized to the case of multiple marked points, i.e., $M_{g,k}(d, X)$. The details will appear elsewhere.

References

[1] C. Allday & V. Puppe, *Cohomology methods in transformation groups*, Cambridge University Press, 1993.

[2] M. Atiyah & R. Bott, *The moment map and equivariant cohomology*, Topology **23** (1984) 1–28.

[3] M. Audin, *The topology of torus actions on symplectic manifolds*, Prog. in Math. Birkhäuser, Vol. 93, 1991.

[4] V. Batyrev, *Dual polyhedra and mirror symmetry for Calabi–Yau hypersurfaces in toric varieties*, J. Algeb. Geom. **3** (1994) 493–535.

[5] V. Batyrev & L. Borisov, *On Calabi-Yau complete intersections in toric varieties*, alg-geom/9412017.

[6] K. Behrend & B. Fantechi, *The intrinsic normal cone*, Invent. Math. **128** (1997) 45–88.

[7] N. Berline & M. Vergne, *Classes caracteéristiques equivariantes, Formula de localisation en cohomologie equivariante*, 1982.

[8] A. Bertram, *Towards a Schubert calculus for maps from Riemann surface to Grassmannian*, Inter. J. Math. **5:6** (1994) 811–825.

[9] P. Candelas, X. de la Ossa, P. Green & L. Parkes, *A pair of Calabi-Yau manifolds as an exactly soluble superconformal theory*, Nucl. Phys. **B359** (1991) 21–74.

[10] I. Ciocan-Fontanine, *Quantum cohomology of flag manifolds*, IMRN, **6** (1995) 263–277.

[11] D. Cox, *Homogeneous coordinate ring of a toric variety*, J. Algeb. Geom. **4** (1995) 17–50.

[12] D. Edidin & W. Graham, *Equivariant intersection theory*, alg-geom/9609018.

[13] W. Fulton, *Intersection theory*, 2nd Ed., Springer, 1998.

[14] K. Fukaya & K. Ono, *Arnold conjecture and Gromov-Witten invariants,* Preprint, 1996.

[15] I. Gel'fand, M. Kapranov & A. Zelevinsky, *Hypergeometric functions and toral manifolds*, Funct. Anal. Appl. **23** (1989) 94–106.

[16] A. Givental, *A mirror theorem for toric complete intersections*, alg-geom/9701016.

[17] T. Graber & R. Pandharipande, *Localization of virtual classes*, alg-geom/9708001.

[18] M. Goresky, R. Kottwitz, & R. MacPherson, *Equivariant cohomology, Koszul duality and the localization theorem*, Invent. Math. **131** (1998) 25–83.

[19] V. Guillemin & C. Zara, *Equivariant de Rham Theory and Graphs*, math. DG/9808135.

[20] F. Hirzebruch, *Topological methods in algebraic geometry*, 3rd Ed., Springer, Berlin, 1995.

[21] S. Hosono, A. Klemm, S. Theisen & S. T. Yau, *Mirror symmetry, mirror map and applications to complete intersection Calabi-Yau spaces*, hep-th/9406055, Nucl. Phys. **B433** (1995) 501–554.

[22] S. Hosono, B. H. Lian & S. T. Yau, *GKZ-generalized hypergeometric systems and mirror symmetry of Calabi-Yau hypersurfaces*, alg-geom/9511001, Commun. Math. Phys. **182** (1996) 535–578.

[23] ———, *Maximal Degeneracy Points of GKZ Systems*, alg-geom/9603014, J. Amer. Math. Soc. **10:2** (1997) 427–443.

[24] K. Jaczewski, *Generalized Euler sequence and toric varieties*, Contemporary Math. **162** (1994) 227–247.

[25] A. Kresch, *Cycle groups for Artin Stacks,* math. AG/9810166.

[26] M. Kontsevich, *Enumeration of rational curves via torus actions,* The Moduli Space of Curves, (R. Dijkgraaf, C. Faber, G. van der Geer, eds.), Progr. Math. Birkhäuser, Vol. 129, 1995, 335–368.

[27] J. Li & G. Tian, *Virtual moduli cycle and Gromov-Witten invariants of algebraic varieties,* J. of Amer. math. Soc. **11:1** (1998) 119–174.

[28] ———, *A brief tour of GW invariants*, to appear in Surveys in Differential Geom. **6** (1999).

[29] B. Lian, K. Liu, & S. T. Yau, *Mirror Principle. I*, Asian J. Math. **1** (1997) 729–763.

[30] ———, *Mirror Principle. II*, Asian J. Math. **3** (1999) 109–146.

[31] D. Morrison & R. Plesser, *Summing the instantons: quantum cohomology and mirror symmetry in toric varieties*, alg-geom/9412236.

[32] I. Musson, *Differential operators on toric varieties*, JPAA **95** (1994) 303–315.

[33] T. Oda, *Convex Bodies and Algebraic Geometry*, Series of Modern Surveys in Math., Springer, Vol. 15, 1985.

[34] Y. B. Ruan, *Virtual neighborhoods and pseudo-holomorphic curves*, math. AG/9611021.

[35] Y. B. Ruan & G. Tian, *A mathematical theory of quantum cohomology*, J. Differential Geom. **42** (1995) 259–367.

[36] B. Siebert, *Gromov-Witten invariants for general symplectic manifolds,* alg-geom/ 9608005.

[37] F. Sottile, *Real rational curves in Grassmannians*, Math.AG/9904167.

[38] F. Sottile & B. Sturmfels, *A sagbi basis for the quantum Grassmannian*, Math. AG/9908016.

[39] S. A. Stromme, *On parametrized rational curves in Grassmann varieties*, Space curves, (Ghione et al., eds.), Lecture Notes Math. Springer, Vol. 1266, 1987.

[40] A. Vistoli, *Intersection theory on algebraic stacks and their moduli,* Invent. Math. **97** (1989) 613–670.

[41] A. Vistoli, *Equivariant Grothendieck groups and equivariant Chow groups in Classification of irregular varieties*, Lecture Notes Math. **1515** 1992 112–133.

[42] E. Witten, *Phases of N=2 theories in two dimension*, hep-th/9301042.

Brandeis University
Stanford University
Harvard University, Chinese University of Hong Kong

Surveys in Differential Geometry, 2000
Vol. VII ©2000, International Press
pp. 475-496

MIRROR PRINCIPLE. IV

BONG H. LIAN, KEFENG LIU & SHING-TUNG YAU

In honor of Professors M. Atiyah, R. Bott, F. Hirzebruch, and I. Singer.

Abstract

This is a continuation of *Mirror Principle. III*, [14].

1. Some background

This paper is a sequel to [12], [13], [14]. In this series of papers we develop *mirror principle* in increasing generality and breadth. Given a projective manifold X, mirror principle is a theory that yields relationships for and often computes the intersection numbers of cohomology classes of the form $b(V_D)$ on stable moduli spaces $\bar{M}_{g,k}(d,X)$. Here V_D is a certain induced vector bundles on $\bar{M}_{g,k}(d,X)$ and b is any given multiplicative cohomology class. In the first paper [12], we consider this problem in the genus zero $g=0$ case when $X=\mathbf{P}^n$ and V_D is a bundle induced by any convex and/or concave bundle V on $\mathbf{P}^n$. As a consequence, we have proved a mirror formula which computes the intersection numbers via a generating function. When $X=\mathbf{P}^n$, V is a direct sum of positive line bundles on $\mathbf{P}^n$, and b is the Euler class, a second proof of this special case has been given in [15], [2] following an approach proposed in [6]. Other proofs in this case has also been given in [1], and when V includes negative line bundles, in [3]. In [13], we develop mirror principle when X is a projective manifold with TX convex. In [14], we consider the $g=0$ case when X is an arbitrary projective manifold. Here emphasis has been put on a class of T-manifolds (which we call balloon manifolds) because in this case mirror principle yields a (linear!) reconstruction algorithm which computes in principle all the intersection numbers above for *any* convex/concave equivariant bundle

V on X and *any* equivariant multiplicative class b. Moreover, specializing this theory to the case of line bundles on toric manifolds and b to Euler class, we give a proof of the mirror formula for toric manifolds. In [14], we have also begun to develop mirror principle for higher genus. In this paper, we complete the theory for hight genus. We also extend the theory to include the intersection numbers for cohomology classes of the form $ev^*(\phi)b(V_D)$. Here $ev : \bar{M}_{g,k}(d,X) \to X^k$ is the usual evaluation map into the product X^k of k copies of X, and ϕ is any cohomology class on X^k.

For motivations and the main ideas of mirror principle, we refer the reader to the introduction of [12], [13].

Acknowledgement. B.H.L. wishes to thank the organizers for inviting him to lecture in the Conference on Geometry and Topology in honor of M. Atiyah, R. Bott, F. Hirzebruch, and I. Singer. Once again, we owe our special thanks to J. Li who has been patiently lending his help to us throughout this project. B.H.L.'s research is supported by NSF grant DMS-0072158. K.L.'s research is supported by NSF grant DMS-9803234 and the Terman fellowship and the Sloane fellowship. S.T.Y.'s research is supported by DOE grant DE-FG02-88ER25065 and NSF grant DMS-9803347.

2. Higher genus

We assume that the reader is familiar with [14]. We follow the notations introduced there. Most results proved here have been summarized in Section 5.5 there.

In the first subsection, we give some examples of gluing sequences, a notion introduced in [14]. We also prove a quadratic identity, which is a generalization of Theorem 3.6 in [14] to higher genus plus multiple marked points. In the second subsection, we give a reconstruction algorithm which allows us to reconstruct the Euler series $A(t)$ associated to a gluing sequence in terms of Hodge integrals and some leading terms of $A(t)$.

1. Throughout this note, we abbreviate the data $(g, k; d)$ as D and write

$$M_D = M_{g,k}((d,1), X \times \mathbf{P}^1).$$

We denote by $LT_D(X), LT_{g,k}(d,X)$ the Li-Tian class of M_D and $M_{g,k}(d,X)$ respectively.

2. In the last subsection, we prove the regularity of the collapsing

map

$$\varphi : M_{g,0}((d,1), \mathbf{P}^n \times \mathbf{P}^1) \to N_d$$

generalizing Lemma 2.6 in [12]. Replacing $\mathbf{P}^n$ by the product

$$Y = \mathbf{P}^{n_1} \times \cdots \times \mathbf{P}^{n_m},$$

and N_d by

$$W_d = N_{d_1} \times \cdots \times N_{d_m},$$

we get the map

$$M_{g,0}((d,1), Y \times \mathbf{P}^1) \to W_d.$$

Given an equivariant projective embedding $\tau : X \to Y$ with $A^1(X) \cong A^1(Y)$ (see [14]), we have an induced map

$$M_D = M_{g,k}((d,1), X \times \mathbf{P}^1) \to M_{g,0}((d,1), Y \times \mathbf{P}^1).$$

Composing this with $M_{g,0}((d,1), Y \times \mathbf{P}^1) \to W_d$, we get a G-equivariant map

$$M_D \to W_d$$

which we also denoted by φ. This map will be used substantially to do functorial localization in the first subsection.

3. The standard $\mathbf{C}^\times$ action on $\mathbf{P}^1$ induces an action on each M_D (see section 5.5 [14]). The fixed point components are labelled by F_{D_1,D_2} with $D_1 = (g_1, k_1; d_1)$, $D_2 = (g_2, k_2; d_2)$, $g_1 + g_2 = g$, $k_1 + k_2 = k$, $d_1 + d_2 = d$. As in the genus zero case, when $d_1, d_2 \neq 0$, a stable map $(C, f, y_1, ..., y_k)$ in this component is given by gluing two pointed stable maps $(f_1, C_1, y_1, .., y_{k_1}, x_1) \in M_{g_1,k_1+1}(d_1, X)$, $(f_2, C_2, y_{k_1+1}, ..., y_k, x_2) \in M_{g_2,k_2+1}(d_2, X)$ with $f_1(x_1) = f_2(x_2)$, to $C_0 = \mathbf{P}^1$ at 0 and ∞ at the marked points (cf. Section 3). We can therefore identify F_{D_1,D_2} with $M_{g_1,k_1+1}(d_1, X) \times_X M_{g_2,k_2+1}(d_2, X)$. We also have a special component $F_{D,O}$ which is obtained by gluing a $k+1$ pointed stable map to $\mathbf{P}^1$ at either 0, as described above. Likewise for $F_{O,D}$. We denote by

$$i : F_{D_1,D_2} \to M_D,$$

the inclusions.

4. There are two obvious projection maps

$$p_0 : F_{D_1,D_2} \to M_{g_1,k_1+1}(d_1, X), \quad p_\infty : F_{D_1,D_2} \to M_{g_2,k_2+1}(d_2, X).$$

The map p_0 strips away (with the notations above) the stable maps $(f_2, C_2, y_{k_1+1}, ..., y_k, x_2)$ glued to the $\mathbf{P}^1$ at ∞, and forgets the $\mathbf{P}^1$; p_∞

strips away the stable map $(f_1, C_1, y_1, ..., y_{k_1}, x_1)$ glued to the $\mathbf{P}^1$ at 0 and forgets the $\mathbf{P}^1$.

5. We also have the following evaluation maps, and the forgetting map:

$$e : F_{D_1,D_2} \to X, \quad e_D : M_{g,k+1}(d,X) \to X,$$
$$\rho : M_{g,k+1}(d,X) \to M_{g,k}(d,X).$$

Here e evaluates a stable map in F_{D_1,D_2} at the gluing point, e_D evaluates a $k+1$ pointed stable map at the last marked point, and ρ forgets the last marked point. Relating and summarizing the natural maps above is the following diagram:

(1)

$$\begin{array}{ccccccc} X & \xleftarrow{e} & F_{D_1,D_2} & \xrightarrow{i} & M_D & \xrightarrow{\pi} & M_{g,k}(d,X) \\ e_{D_1}\uparrow & p_0 \swarrow & & \searrow p_\infty & & & \\ M_{g_1,k_1+1}(d_1,X) & & & & M_{g_2,k_2+1}(d_2,X) & & \\ \rho\downarrow & & & & \downarrow\rho & & \\ M_{g_1,k_1}(d_1,X) & & & & M_{g_2,k_2}(d_2,X) & & \end{array}$$

Here π is natural morphism which maps $(C, f, y_1, .., y_k) \in M_D$ to the stabilization of $(C, \pi_1 \circ f, y_1, .., y_k)$, where $\pi_1 : X \times \mathbf{P}^1 \to X$ is the projection. Note that we can identify $M_{g,k+1}(d,X)$ with $F_{O,D}$ via p_∞. When $D_1 = O := (0,0;0)$, the right part of the diagram above completes to a commutative triangle, i.e., $\pi \circ i = \rho$.

6. Let $\bar{M}_{g,k}$ be the Deligne-Mumford moduli space of k-pointed, genus g stable curves. Recall the map

$$M_{g,k}(d,X) \to \bar{M}_{g,k}$$

which sends $(C, f, y_1, .., y_k)$ to the stabilization of $(C, y_1, ..., y_k)$. Let $\mathcal{L}$ and $\mathcal{H}$ be respectively the universal line bundle and the Hodge bundle on $\bar{M}_{g,k+1}$. Thus $\mathcal{L}, \mathcal{H}$ have fibers at $(C, y_1, ..., y_k, x)$ given by $T_x C$ and $H^0(C, K_C) \cong H^1(C, \mathcal{O})^*$ respectively. We denote

$$\Lambda_g(\xi) = \sum_{i=0}^{g} \xi^i c_{g-i}(\mathcal{H})$$

for any formal variable ξ (cf. [4]). We denote by the same notation the pullback of $\Lambda_g(\xi)$ to $M_{g,k+1}(d,X)$. We denote by L_D the universal line bundle on $M_{g,k+1}(d,X)$. corresponding to the last marked point.

7. **Notations.** In all formulas below involving d, d_1, d_2, and g, g_1, g_2, it is always assumed that

$$\begin{aligned} D &= (g, k; d) \\ D_1 &= (g_1, k_1; d_1) \\ D_2 &= (g_2, k_2; d_2) \\ g &= g_1 + g_2 \\ k &= k_1 + k_2 \\ d &= d_1 + d_2. \end{aligned}$$

Lemma 2.1 (cf. Lemma 3.5 [14].) *Let $g = g_1 + g_2$, $k_1 + k_2 = k$, $d = d_1 + d_2$. For $d_1, d_2 \neq 0$, the equivariant Euler class of the virtual normal bundle $N_{F_{D_1,D_2}/M_D}$ is*

$$e_G(F_{D_1,D_2}/M_D) \\ = p_0^* \left(\alpha(\alpha + c_1(L_{D_1}))\ \Lambda_{g_1}(\alpha)^{-1}\right) p_\infty^* \left(\alpha(\alpha - c_1(L_{D_2}))\ \Lambda_{g_2}(-\alpha)^{-1}\right).$$

For $d_1 = 0,\ g_1 = 0,\ k_1 = 0$,

$$e_G(F_{O,D}/M_D) = -\alpha(-\alpha + c_1(L_D))\ \Lambda_g(-\alpha)^{-1}.$$

For $d_2 = 0,\ g_2 = 0,\ k_2 = 0$,

$$e_G(F_{D,O}/M_D) = \alpha(\alpha + c_1(L_D))\ \Lambda_g(\alpha)^{-1}.$$

Proof. We consider the first equality, the other two being similar. We will compute the virtual normal bundle $N_{F_{D_1,D_2}/M_D}$ of F_{D_1,D_2} in M_D following the methods in [10], [7], [12], using the description of F_{D_1,D_2} given above. We must identify the terms appearing in the tangent obstruction sequence of M_D. (See [7] Section 4.)

Consider the bundle $V := \pi_1^* TX \oplus \pi_2^* T\mathbf{P}^1$ on $X \times \mathbf{P}^1$, where π_1 and π_2 are the projections from $X \times \mathbf{P}^1$ to X and $\mathbf{P}^1$ respectively.

According to the description above, for each stable map $(C, f, x_1, ..., x_k)$ in F_{D_1,D_2}, we have an exact sequence over C:

$$0 \to f^*V \to f_1^*V \oplus f_0^*V \oplus f_2^*V \to V_{x_1} \oplus V_{x_2} \to 0. \tag{2}$$

Here f_0 is the restriction of f to C_0, and V_{x_1}, V_{x_2} denote respectively the bundles

$$\pi_1^* T_x X \oplus \pi_2^* T_0 \mathbf{P}^1, \quad \pi_1^* T_x X \oplus \pi_2^* T_\infty \mathbf{P}^1$$

where $x = f_0(C_0) \in X$. From the long exact sequence associated to (8), we get the equality in the K-group:

$$\begin{aligned} &H^0(C, f^*V) - H^1(C, f^*V) \\ &\quad = \sum_{j=0}^{2} [H^0(C_j, f_j^*V) - H^1(C_j, f_j^*V)] - (V_{x_1} + V_{x_2}). \end{aligned}$$

The tangent complex of M_D restricted to F_{D_1,D_2} is

$$\begin{aligned} &H^0(C, f^*V) - H^1(C, f^*V) + T_{x_1}C_1 \otimes T_0\mathbf{P}^1 \\ &\qquad\qquad + T_{x_2}C_2 \otimes T_\infty\mathbf{P}^1 - A_C \\ &\quad = \sum_{j=0}^{2} [H^0(C_j, f_j^*V) - H^1(C_j, f_j^*V)] - (V_{x_1} + V_{x_2}) \\ &\qquad + T_{x_1}C_1 \otimes T_0\mathbf{P}^1 + T_{x_2}C_2 \otimes T_\infty\mathbf{P}^1 - A_C. \end{aligned} \tag{3}$$

where $T_{x_1}C_1 \otimes T_0\mathbf{P}^1$ and $T_{x_2}C_2 \otimes T_\infty\mathbf{P}^1$ are contributions from the deformation of the nodes at x_1, x_2 of C, and A_C is the contribution from the infinitesimal automorphisms of C. To get the moving parts of this, we subtract from it the fixed parts corresponding to $F_{D_1,D_2} = M_{g_1,k_1+1}(d_1, X) \times_X M_{g_2,k_2+1}(d_2, X)$ (see description above). This is given by

$$\sum_{j=1}^{2} [H^0(C_j, f_j^*\pi_1^*TX) - H^1(C_j, f_j^*\pi_1^*TX)] - \pi_1^*T_xX - A_{C'}, \tag{4}$$

where C' is the curve obtained from C by contracting the component C_0. Note that here we have ignored the contributation coming from the deformation of C_1, C_2 in both (3) and (4), because the same contributation appear in both and hence this contributation cancels out in the difference.

We now compute and compare the terms in both (3) and (4) above. Since f_0 maps C_0 to a point x, we have

$$\begin{aligned} H^1(C_0, f_0^*\pi_1^*TX) &\simeq H^1(C_0, O) \otimes \pi_1^*TX = 0 \\ H^1(C_0, f_0^*\pi_2^*T\mathbf{P}^1) &\simeq H^1(C_0, TC_0) = 0 \\ H^0(C_0, f_0^*\pi_1^*TX) &= \pi_1^*T_xX. \end{aligned}$$

Similarly for C_1 and f_1, we have

$$\begin{aligned} H^1(C_1, f_1^*V) &= H^1(C_1, f_1^*\pi_1^*TX) + H^1(C_1, f_1^*\pi_2^*T\mathbf{P}^1) \\ H^1(C_1, f_1^*\pi_2^*T\mathbf{P}^1) &\simeq H^1(C_1, O) \otimes \pi_2^*T_0\mathbf{P}^1 \\ H^0(C_1, f_1^*V) &= H^0(C_1, f_1^*\pi_1^*TX) + H^0(C_1, f_1^*\pi_2^*T\mathbf{P}^1) \\ H^0(C_1, f_1^*\pi_2^*T\mathbf{P}^1) &\simeq \pi_2^*T_0\mathbf{P}^1. \end{aligned}$$

Likewise, we have similar relations with $C_1, f_1, T_0\mathbf{P}^1$ replaced by $C_2, f_2, T_\infty\mathbf{P}^1$ everywhere.

Putting these formulas together, we get

$$\begin{aligned} N_{F_{D_1,D_2}/M_D} = {} & H^0(C_0, f_0^*\pi_2^*T\mathbf{P}^1) + T_{x_1}C_1 \otimes T_0\mathbf{P}^1 \\ & + T_{x_2}C_2 \otimes T_\infty\mathbf{P}^1 - H^1(C_1, O) \otimes \pi_2^*T_0\mathbf{P}^1 \\ & - H^1(C_2, O) \otimes \pi_2^*T_\infty\mathbf{P}^1 - A_{C_0}. \end{aligned}$$

By taking equivariant Euler classes, we get the required formula. Here the terms $-H^1(C_1, O) \otimes \pi_2^*T_0\mathbf{P}^1$ and $-H^1(C_2, O) \otimes \pi_2^*T_\infty\mathbf{P}^1$ contribute $p_0^*\Lambda_{g_1}(\alpha)^{-1}$ and $p_\infty^*\Lambda_{g_2}(-\alpha)^{-1}$ respectively. The terms $T_{x_1}C_1 \otimes T_0\mathbf{P}^1$ and $T_{x_2}C_2 \otimes T_\infty\mathbf{P}^1$ contribute $\alpha + p_0^*c_1(L_{D_1})$ and $-\alpha + p_\infty^*c_1(L_{D_2})$ respectively, and the term $H^0(C_0, f_0^*\pi_2^*T\mathbf{P}^1) - A_{C_0}$ contributes $-\alpha^2$ (see [12, Section 2.3]).

Similarly, when $d_1 = 0, g_1 = 0, k_1 = 0$, we have

$$\begin{aligned} N_{F_{O,D}/M_D} = {} & H^0(C_0, f_0^*\pi_2^*T\mathbf{P}^1) + T_{x_2}C_2 \otimes T_\infty\mathbf{P}^1 \\ & - H^1(C_2, O) \otimes \pi_2^*T_\infty\mathbf{P}^1 - A_{C_0}. \end{aligned}$$

The term $-A_{C_0}$ contributes a factor $-\alpha$. Similarly for $d_2 = 0$ and $g_2 = 0$, we have

$$\begin{aligned} N_{F_{D,O}/M_D} = {} & H^0(C_0, f_0^*\pi_2^*T\mathbf{P}^1) + T_{x_1}C_1 \otimes T_0\mathbf{P}^1 \\ & - H^1(C_1, O) \otimes \pi_2^*T_0\mathbf{P}^1 - A_{C_0}, \end{aligned}$$

and now the term $-A_{C_0}$ contributes a factor α. q.e.d.

2.1 Gluing sequences

Fix a class $\Omega \in A_T^*(X)$, such that Ω^{-1} is well-defined. We call the list of classes

$$b_D \in A_T^*(M_{g,k}(d, X)).$$

an Ω-gluing sequence if we have the (gluing) identities on the F_{D_1,D_2}:

$$e^*\Omega \cdot i^*\pi^*b_D = p_0^*\rho^*b_{D_1} \cdot p_\infty^*\rho^*b_{D_2}.$$

(This generalizes the definition in [14] to include the cases with multiple marked points.) Gluing sequences have the following obvious multiplicative property: that if b_D and b'_D form two gluing sequences with respective to, say Ω and Ω', then the product $b_D b'_D$ form a gluing sequence with respective to $\Omega\Omega'$.

Let V be a T-equivariant bundle on X. Suppose that $H^0(C, f^*V) = 0$ for every positive degree map $f : C \to X$ where C a nonsingular genus g curve. Then V induces on each $M_{g,k}(d, X)$ a vector bundle V_D whose fiber at $(C, f, y_1, .., y_k)$ is $H^1(C, f^*V)$. We call such a V a *concave bundle* on X.

Example 1. $X = \mathbf{P}^n$, and $V = \mathcal{O}(-k)$, $k < 0$.

Example 2. If X is a projective manifold with $V = K_X < 0$, then V induces the bundles V_D. This is the situation in *local mirror symmetry* [12], [8].

Let b_T be a T-equivariant multiplicative class. such that $\Omega = b_T(V)^{-1}$ is well-defined. Consider the classes $b_T(V_D) \in A_T^*(M_{g,k}(d, X))$.

Lemma 2.2. *The cohomology classes $b_T(V_D)$ form an Ω-gluing sequence.*

Proof. The proof is essentially the same as the argument for the genus zero gluing identity for a concave bundle V. See the first half of the proof of Theorem 3.6 [14]. q.e.d.

We now discuss a second example of a gluing sequence. Recall that a point (f, C) in F_{D_1,D_2} comes from gluing together a pair of stable maps $(f_1, C_1, y_1, .., y_{k_1}, x_1), (f_2, C_2, y_{k_1+1}, .., y_k, x_2)$ with $f_1(x_1) = f_2(x_2)$, to $C_0 = \mathbf{P}^1$ at 0 and ∞, so that we have a long exact sequence

$$\begin{aligned}0 &\to H^0(C, \mathcal{O}) \to H^0(C_1, \mathcal{O}) \oplus H^0(C_2, \mathcal{O}) \to H^0(C_0, \mathcal{O}) \\ &\to H^1(C, \mathcal{O}) \to H^1(C_1, \mathcal{O}) \oplus H^1(C_2, \mathcal{O}) \to H^1(C_0, \mathcal{O}) \to 0.\end{aligned}$$

Thus we have a natural isomorphism

$$H^1(C, \mathcal{O}) \cong H^1(C_1, \mathcal{O}) \oplus H^1(C_2, \mathcal{O}).$$

This implies the isomorphism

$$i^*\pi^*\mathcal{H}_D = p_0^*\rho^*\mathcal{H}_{D_1} \oplus p_\infty^*\rho^*\mathcal{H}_{D_2}$$

of bundles on F_{D_1,D_2}. Here $\mathcal{H}_D$ denotes the bundle on $M_{g,k}(d, X)$ with fiber $H^1(C, \mathcal{O})$. (Note that for $g \geq 2$, then $\mathcal{H}_D$ is the pullback of the dual of the Hodge bundle $\mathcal{H}$ via the natural map $M_{g,k}(d,X) \to \bar{M}_{g,k}$.) Thus if b is a multiplicative class, and $b_D := b(\mathcal{H}_D)$, then the isomorphism above yields the gluing identity

$$i^*\pi^* b_D = p_0^*\rho^* b_{D_1} \cdot p_\infty^* \rho^* b_{D_2}$$

with $\Omega = 1$. To summarize, we have

Lemma 2.3. *The cohomology classes $b_D := b(\mathcal{H}_D)$ above form a 1-gluing sequence.*

Note that in both examples above, each class b_D is naturally the pullback, via the forgetful map, of a class $b_d^g \in A_T^*(M_{g,0}(d,X))$. We will call *a list of classes $b_d^g \in A_T^*(M_{g,0}(d,X))$ an Ω-gluing sequence, if their pullbacks to $M_{g,k}(d,X)$ form a gluing sequence in the sense introduced above.*

We now discuss a third construction. Fix a set of generators ϕ_i of $A_T^*(X)$, as a free module over $A_T^*(pt)$. Let

$$\phi = \sum_i s_i \phi_i$$

where s_i are formal variables. Let

$$\phi_D := \prod_{j=1}^k ev_j^* \phi \in A_T^*(M_{g,k}(d,X)).$$

Here the map ev_j evaluates at the jth marked point of a stable map. Since the forgetting map ρ commutes with evaluations, we have the identity

$$i^*\pi^*\phi_D = p_0^*\rho^*\phi_{D_1} \cdot p_\infty^*\rho^*\phi_{D_2}.$$

Thus we have

Lemma 2.4. *The cohomology classes ϕ_D above form a 1-gluing sequence.*

Combining with the multiplicative property of gluing sequences, as explained above, this construction allows us to consider the intersection numbers of classes of the form $ev^*(\phi)\ b(V_D)$ on stable map moduli. In particular, this yields the GW-invariants twisted by a multiplicative class of the form $b(V_D)$. Here ev is the evaluation map $M_{g,k}(d,X) \to X^k$ into the product of k copies of X. The results below are easily generalized to the cases involving the additional factor $ev^*(\phi)$.

For $\omega \in A_G^*(M_D)$, introduce the notation (cf. Section 3.2 [14])

$$J_{D_1,D_2}\omega := e_*\left(\frac{i^*\omega \cap [F_{D_1,D_2}]^{vir}}{e_G(F_{D_1,D_2}/M_D)}\right) \in A_*^T(X)(\alpha).$$

Theorem 2.5 (cf. Theorem 3.6 [14]). *Given a gluing sequence $b_D \in A_T^*(M_{g,k}(d,X))$, we have the following identities in $A_*^T(X)(\alpha)$:*

$$\Omega \cap J_{D_1,D_2}\pi^* b_D = \overline{J_{O,D_1}\pi^* b_{D_1}} \cdot J_{O,D_2}\pi^* b_{D_2}.$$

Proof. Consider the fiber square

$$\begin{array}{ccc} F_{D_1,D_2} & \xrightarrow{\Delta'} & M_{g_1,k_1+1}(d_1,X) \times M_{g_2,k_2+1}(d_2,X) \\ e\downarrow & & \downarrow e_{D_1} \times e_{D_2} \\ X & \xrightarrow{\Delta} & X \times X \end{array} \tag{5}$$

where Δ is the diagonal inclusion. Recall also that (see Section 6 in [11])

$$[F_{D_1,D_2}]^{vir} = \Delta^!(LT_{g_1,k_1+1}(d_1,X) \times LT_{g_2,k_2+1}(d_2,X)).$$

Put

$$\omega = \frac{\rho^* b_{D_1}}{e_G(F_{D_1,O}/M_{D_1})} \times \frac{\rho^* b_{D_2}}{e_G(F_{O,D_2}/M_{D_2})} \cap LT_{g_1,k_1+1}(d_1,X) \times LT_{g_2,k_2+1}(d_2,X).$$

From the fiber square (5), we have

$$e_*\Delta^!(\omega) = \Delta^*(e_{D_1} \times e_{D_2})_*(\omega).$$

On the one hand is

$$\begin{aligned} \Delta^*(e_{D_1} \times e_{D_2})_*(\omega) &= e_{D_1 *}\frac{\rho^* b_{D_1} \cap LT_{g_1,k_1+1}(d_1,X)}{e_G(F_{D_1,O}/M_{D_1})} \\ &\quad \cdot e_{D_2 *}\frac{\rho^* b_{D_2} \cap LT_{g_2,k_2+1}(d_2,X)}{e_G(F_{O,D_2}/M_{D_2})} \\ &= \overline{J_{O,D_1}\pi^* b_{D_1}} \cdot J_{O,D_2}\pi^* b_{D_2}. \end{aligned}$$

Here we have use the fact that $\pi \circ i = \rho$. On the other hand, applying the gluing identity and Lemma (2.1), we have

$$\begin{aligned} e_*\Delta^!(\omega) &= e_*\left(p_0^*\frac{\rho^* b_{D_1}}{e_G(F_{D_1,O}/M_{D_1})} \cdot p_\infty^*\frac{\rho^* b_{D_2}}{e_G(F_{O,D_2}/M_{D_2})} \cap [F_{D_1,D_2}]^{vir}\right) \\ &= e_*\left(\frac{e^*\Omega \cdot i^*\pi^* b_D \cap [F_{D_1,D_2}]^{vir}}{e_G(F_{D_1,D_2}/M_D)}\right) \\ &= \Omega \cap J_{D_1,D_2}\pi^* b_D. \end{aligned}$$

This proves our assertion. q.e.d.

Lemma 2.6 (cf. Lemma 3.2 [14]). *Given a cohomology class ω on M_D, we have the following identities on the $\mathbf{C}^\times$ fixed point component $Y_{d_1,d_2} \cong Y$ in W_d:*

$$\frac{j^*\varphi_*(\omega \cap LT_D(X))}{e_G(Y_{d_1,d_2}/W_d)} = \sum_{g_1+g_2=g, k_1+k_2=k} \tau_* e_* \left(\frac{i^*\omega \cap [F_{D_1,D_2}]^{vir}}{e_G(F_{D_1,D_2}/M_D)} \right).$$

Proof. This follows from applying functorial localization to the diagram

$$\begin{array}{ccc} \{F_{D_1,D_2}\} & \stackrel{i}{\longrightarrow} & M_D \\ \tau \circ e \downarrow & & \downarrow \varphi \\ Y_{d_1,d_2} & \stackrel{j}{\longrightarrow} & W_d. \end{array} \tag{6}$$

q.e.d.

Now given a gluing sequence b_D, we put

$$A_D := J_{O,D}\pi^* b_D, \quad A_d := \sum_{g,k} A_D\ \nu^g \mu^k, \quad A(t) := e^{-H\cdot t/\alpha} \sum_d A_d\ e^{d\cdot t}.$$

Here ν, μ are formal variables. Consider the class $\beta = \varphi_*(\pi^* b_D \cap LT_D(X))$. We have

$$\begin{aligned} \int_{W_d} \beta \cap e^{\kappa\cdot\zeta} &= \sum_{d_1+d_2=d} \int_{Y_{d_1,d_2}} \frac{j^*\beta}{e_G(Y_{d_1,d_2}/W_d)}\ e^{(H+d_1\alpha)\cdot\zeta} \\ &= \sum_{D_1+D_2=D} \int_{Y_{d_1,d_2}} \tau_* J_{D_1,D_2}\pi^* b_D\ e^{(H+d_1\alpha)\cdot\zeta} \quad \text{(Lemma 2.6)} \\ &= \sum_{D_1+D_2=D} \int_X J_{D_1,D_2}\pi^* b_D\ e^{(H+d_1\alpha)\cdot\zeta} \\ &= \sum_{D_1+D_2=D} \int_X \Omega^{-1} \cap \overline{A_{D_1}} \cdot A_{D_2}\ e^{(H+d_1\alpha)\cdot\zeta} \quad \text{(Theorem 2.5)}. \end{aligned}$$

Since $\beta \in A_*^G(W_d)$, hence $\int_{W_d} \beta \cap c \in A_*^G(pt) = \mathbf{C}[\mathcal{T}^*, \alpha]$ for all $c \in A_G^*(W_d)$, it follows that both sides of the eqn. above lie in $\mathcal{R}[[\zeta]]$. Thus we get

Corollary 2.7. *$A(t)$ is an Euler series.*

We call V be a D-critical concave bundle if the homogeneous degree of the class $b(V_D)$ is the same as the expected dimension of $M_{g,k}(d,X)$. We denote

$$K_D = \int_{LT_{g,k}(d,X)} b(V_D).$$

Lemma 2.8. *If V is a D-critical concave bundle, then in the T-nonequivariant limit we have the following formula*

$$\int_X e^{-H\cdot t/\alpha} J_{O,D}\pi^* b(V_D) = (-1)^g \alpha^{g-3}(2-2g-k-d\cdot t)K_D.$$

Proof. The integral above is equal to

$$\int_{LT_{g,k+1}(d,X)} e^{-e^* H\cdot t/\alpha} \frac{\rho^* b(V_D)\Lambda_g(-\alpha)}{\alpha(\alpha-c)}$$
$$= \int_{LT_{g,k}(d,X)} b(V_D)\Lambda_g(-\alpha)\rho_* \left(\frac{e^{-e^* H\cdot t/\alpha}}{\alpha(\alpha-c)}\right).$$

Here $\rho : M_{g,k+1}(d,X) \to M_{g,k}(d,X)$ forgets the last marked point.

Since the fiber of ρ is of dimension 1, we take the degree 1 term in the fiber integral. The integration along the fiber E is done in essentially the same way as in the genus zero case (see Theorem 3.12(ii) of [14]). It yields

$$\int_E e^* H = d$$

and

$$\int_E c = 2-2g-k$$

by Gauss-Bonnet formula. Since the degree of $b(V_D)$ coincides with the dimension of $LT_{g,k}(d,X)$, it follows that only the α^0 term of $\Lambda_g(-\alpha)$ contributes to the integral above. q.e.d.

2.2 Reconstruction

From now on, we assume that X is a balloon manifold, as in Sections 4.1 and 5.3 of [14]. We will find further constraints on a gluing sequence by computing the linking values of the Euler series $A(t)$. Recall that in genus zero, the linking values of an Euler series, say coming from $b_T(V_d)$, are determined by the restrictions $i_F^* b_T(\rho^* V_d)$ to the isolated fixed point $F = (\mathbf{P}^1, f_\delta, 0) \in M_{0,1}(d,X)$, which is a δ-fold cover of a balloon pq in X. In higher genus with multiple marked points, this will be replaced by a component in $M_{g,k+1}(d,X)$ consisting of the following stable maps $(C, f, y_1, .., y_k, x)$. Here C is a union of two curves C_1 and $C_0 \cong \mathbf{P}^1$ such that $y_1, .., y_k \in C_1$ and that $C_0 \xrightarrow{f} pq$ is a δ-fold cover with $f(x) = p$, $f(C_1) = q$. Therefore this fixed point component can be identified canonically with $\bar{M}_{g,k+1}$. For clarity, we will restrict the following discussion to the case of $k = 0$. We'll denote the component by F^g. By convention, F^0 is the isolated fixed point $(\mathbf{P}^1, f_\delta, 0)$. Recall that (section 5.3 [14])

$$e_G(F^0/M_{0,1}(d,X))^{-1} = \frac{-\lambda}{\delta} \frac{e_T[H^1(\mathbf{P}^1, f_\delta^* TX)]'}{e_T[H^0(\mathbf{P}^1, f_\delta^* TX)]'}. \tag{7}$$

Theorem 2.9. *Suppose $g > 0$. Let $p \in X^T$, $\omega \in A_T^*(M_{g,1}(d,X))[\alpha]$, and consider $i_p^* e_* \left(\frac{\omega \cap LT_{g,1}(d,X)}{e_G(F_{O,D}/M_D)}\right) \in \mathbf{C}(\mathcal{T}^*)(\alpha)$ as a function of α. Then*

(i) Every possible pole of the function is a scalar multiple of a weight on T_pX.

(ii) Let pq be a balloon in X, and λ be the weight on the tangent line $T_p(pq)$. If $d = \delta[pq] \succ 0$, then the pole of the function at $\alpha = \lambda/\delta$ is of the form

$$e_T(p/X) \frac{1}{\alpha(\alpha - \lambda/\delta)} \frac{1}{e_T(F^0/M_{0,1}(d,X))} \cdot \int_{[F^g]^{vir}} \frac{i_{F^g}^* \omega \; \Lambda_g(\alpha) \; e_T(\mathcal{H}^* \otimes T_qX)}{(-\frac{\lambda}{\delta} + c_1(\mathcal{L}))}.$$

Proof. The proof here is a slight modification of the genus zero case. We repeat the details here for the readers' convenience. Consider the commutative diagram

$$\begin{array}{ccc} \{F\} & \xrightarrow{i_F} & M_{g,1}(d,X) \\ e' \downarrow & & e \downarrow \\ p & \xrightarrow{i_p} & X \end{array}$$

where e is the evaluation map, $\{F\}$ are the fixed point components in $e^{-1}(p)$, e' is the restriction of e to $\{F\}$, and i_F, i_p are the usual inclusions. By functorial localization we have, for any $\beta \in A_T^*(M_{g,1}(d,X))(\alpha)$,

$$\begin{aligned} i_p^* e_*(\beta \cap LT_{g,1}(d,X)) &= e_T(p/X) \sum_F e'_* \left(\frac{i_F^* \beta \cap [F]^{vir}}{e_T(F/M_{g,1}(d,X))} \right) \\ &= e_T(p/X) \sum_F \int_{[F]^{vir}} \frac{i_F^* \beta}{e_T(F/M_{g,1}(d,X))}. \end{aligned} \tag{8}$$

We apply this to the class

$$\beta = \frac{\omega}{e_G(F_{O,D}/M_D)} = \frac{\omega \; \Lambda_g(-\alpha)}{\alpha(\alpha - c)}$$

where $c = c_1(L_D)$ (cf. Lemma (2.1)). For (i), we will show that a pole of the sum (8) is at either $\alpha = 0$ or $\alpha = \lambda'/\delta'$ for some tangent weight λ' on T_pX. For (ii), we will show that only one component F in the sum (8) contributes to the pole at $\alpha = \lambda/\delta$, that the contributing component is F^g, and that the contribution has the desired form.

A fixed point (C, f, x) in $e^{-1}(p)$ is such that $f(x) = p$, and that the image curve $f(C)$ lies in the union of the T-invariant balloons in X. The restriction of the first Chern class c to an F must be of the form

$$i_F^* c = c_F + w_F$$

where $c_F \in A^1(F)$, and $w_F \in \mathcal{T}^*$ is the weight of the representation on the line T_xC induced by the linear map $df_x : T_xC \to T_pX$. The image of df_x is either 0 or a tangent line $T_p(pr)$ of a balloon pr. Thus w_F is either zero (when the branch $C_1 \subset C$ containing x is contracted), or $w_F = \lambda'/\delta'$ (when $C_1 \xrightarrow{f} X$ maps by a δ'-fold cover of a balloon pr with tangent weight λ'). The class $e_T(F/M_{g,1}(d, X))$ is obviously independent of α. Since c_F is nilpotent, a pole of the sum (8) is either at $\alpha = 0$ or $\alpha = w_F$ for some F. This proves (i).

Now, an F in the sum (8) contributes to the pole at $\alpha = \lambda/\delta$ only if $w_F = \lambda/\delta$. Since the weights on T_pX are pairwise linearly independent, that $\lambda/\delta = \lambda'/\delta'$ implies that $\lambda = \lambda'$ and $\delta = \delta'$. Since $d = \delta[pq]$, a fixed point (C, f, x) contributing to the pole at $\alpha = \lambda/\delta$ must have the following form: that there is a branch $C_0 \cong \mathbf{P}^1$ in C such that $f|_{C_0} : C_0 \to pq$ is a δ-fold cover with $f(x) = p$. Let $y \in C_0$ be the preimage of q under this covering. Then the curve C is a union of C_0 and a genus g curve C_1 meeting C_0 at y, and $f(C_1) = q$. In other words, the fixed point component F contributing to the pole at $\alpha = \lambda/\delta$ is $F^g \cong \bar{M}_{g,1}$. It contributes to the sum (8) the term

$$\int_{[F^g]^{vir}} \frac{i_{F^g}^* \beta}{e_T(F^g/M_{0,1}(d, X))} = \int_{F^g} \frac{i_{F^g}^* \omega \; \Lambda_g(-\alpha)}{\alpha(\alpha - \lambda/\delta - c_{F^g})} \frac{1}{e_T(F^g/M_{g,1}(d, X))}.$$

Here $c_{F^g} \in A^1(F^g)$ is zero because the universal line bundle L_d^g restricted to F^g is trivial (the line T_xC is located at the marked point x).

We now compute the virtual normal bundle $N_{F^g/M_{0,1}(d,X)}$. A point (C, f, x) in F^g can be viewed as gluing two stable maps $(C_0, f_0, x, y) \in M_{0,2}(d, X)$, $(C_1, f_1, x_1) \in M_{g,1}(0, X)$, by identifying $x_1 \equiv y$. Here $f_0 : C_0 \to pq$ is a δ-fold cover with $f_0(x) = p$, $f_0(y) = q$, and $f_1(C_1) = q$. As before, we have

$$\begin{aligned} N_{F^g/M_{g,1}(d,X)} &= [H^0(C, f^*TX)] - [H^1(C, f^*TX)] \\ &\quad + [T_yC_1 \otimes T_yC_0] + A_{C_0} \\ &= \left([H^0(C_0, f_0^*TX)] - H^1(C_0, f_0^*TX)] - A_{C_0}\right) \\ &\quad - [H^1(C_1, f_1^*TX)] + [T_yC_1 \otimes T_yC_0]. \end{aligned}$$

Note that the first three terms collected in the parentheses is the virtual normal bundle of F^0 in $M_{0,1}(d, X)$. The Euler class of this is constant on F^g, as given in eqn. (7). Since $f_1 : C_1 \to X$ maps to the point q, it follows that $[H^1(C_1, f_1^*TX)] = \mathcal{H}^* \otimes T_qX$ where $\mathcal{H}$ is the Hodge bundle on $\bar{M}_{g,1}$. Clearly $[T_yC_1 \otimes T_yC_0] = \mathcal{L} \otimes [\frac{-\lambda}{\delta}]$, where $\mathcal{L}$ is the universal line bundle on $\bar{M}_{g,1}$, and $[\frac{-\lambda}{\delta}]$ is a 1 dimensional representation of that given weight. Therefore, we get

$$e_T(F^g/M_{g,1}(d,X)) = e_T(F^0/M_{0,1}(d,X))\,(-\frac{\lambda}{\delta} + c_1(\mathcal{L}))\; e_T(\mathcal{H}^* \otimes T_qX)^{-1}.$$

Hence the contribution of the sum (8) to the pole at $\alpha = \lambda/\delta$ is

$$\begin{aligned} e_T(p/X) &\int_{[F^g]^{vir}} \frac{i^*_{F^g}\beta}{e_T(F^g/M_{0,1}(d,X))} \\ &= e_T(p/X)\, \frac{1}{\alpha(\alpha - \lambda/\delta)} \frac{1}{e_T(F^0/M_{0,1}(d,X))} \\ &\quad \cdot \int_{[F^g]^{vir}} \frac{i^*_{F^g}\omega\; \Lambda_g(-\alpha)\; e_T(\mathcal{H}^* \otimes T_qX)}{(-\frac{\lambda}{\delta} + c_1(\mathcal{L}))}. \end{aligned}$$

This proves (ii). q.e.d.

Let V be a concave bundle on X, and b_T a choice of multiplicative class as before. Define the genus $g \geq 0$, degree $d = \delta[pq]$, *linking values at the balloon* pq:

$$Lk_g := e_T(p/X) \int_{[F^g]^{vir}} \frac{i^*_{F^g}\beta}{e_T(F^g/M_{g,1}(d,X))}, \qquad \beta := \frac{\rho^* b_T(V_D)}{e_G(F_{O,D}/M_D)}.$$

Corollary 2.10. *For $g > 0$,*

$$Lk_g = Lk_0 \;\times \int_{\bar{M}_{g,1}} \frac{b_T(\mathcal{H}^* \otimes V|_q)\; \Lambda_g(-\alpha)\; e_T(\mathcal{H}^* \otimes T_qX)}{(-\frac{\lambda}{\delta} + c_1(\mathcal{L}))}.$$

Proof. Restricting the bundle $\rho^* V_D$ on $M_{g,1}(d, X)$ to the component F^g, we get

$$i^*_{F^g}\rho^* V^g_d = [H^1(C, f^*V)] = [H^1(C_0, f_0^*V)] \oplus [H^1(C_1, f_1^*V)].$$

So, we have

$$i_{F^g}^* \rho^* b_T(V_D) = i_{F^0}^* \rho^* b_T(V_{0,0;d})\ b_T(\mathcal{H}^* \otimes V|_q).$$

Again, the first factor is constant on F^g. Note that F^g consists of orbifold points of order δ. Thus the integral $\int_{[F^g]^{vir}}$ can be written as $\frac{1}{\delta}\int_{\bar{M}_{g,1}}$. Now applying the preceding theorem with $\omega = \rho^* b_T(V_D)$ yields the desired result. q.e.d.

In the special case $b_T = e_T$, the linking values become

$$Lk_g = Lk_0 \ \times \int_{\bar{M}_{g,1}} \frac{\prod_i \Lambda_g(-\xi_i)\ \prod_j \Lambda_g(-\lambda_j)\ \Lambda_g(-\alpha)}{(-\frac{\lambda}{\delta} + c_1(\mathcal{L}))}$$

where the ξ_i and λ_j are the weights on the isotropic representations $V|_q$ and T_qX respectively. These integral are nothing but Hodge integrals on $\bar{M}_{g,1}$. Their values have been fully determined in [4].

Fix a concave bundle V and multiplicative class b_T. Consider the associated Euler series $A(t) = e^{-H\cdot t/\alpha} \sum A_D\ \nu^g\ e^{d\cdot t}$ with coefficients

$$A_D = e_* \left(\frac{\rho^* b_T(V_D) \cap LT_{g,1}(d,X)}{e_G(F_{O,D}/M_D)} \right).$$

By Lemma (2.1), we see that

$$deg_\alpha\ A_D \leq -2 + g.$$

Theorem 2.11. *Consider the gluing sequence $b_D = \Lambda_g(\alpha)$ and suppose that $c_1(X) > 0$. Then for a given g, the A_D can be reconstructed from the linking values Lk_g and from finitely many degrees d.*

Proof. Recall that the homology class $LT_{g,1}(d,X)$ has dimension $s = exp.dim\ M_{g,1}(d,X) = (1-g)(dim\ X-3)+\langle c_1(X), d\rangle+1$. Let $c = c_1(L_D)$. Then $c^k \cap LT_{g,1}(d,X)$ is of dimension $s-k$, and so $e_*(c^k \cap LT_{g,1}(d,X))$ lies in the group $A^T_{s-k}(X)$. But this group is zero unless $s - k \leq dim\ X$. The last condition means that

$$-k + 2g \leq -\langle c_1(X), d\rangle + g(dim\ X - 1) + 2.$$

For given g, the right hand side is negative for all but finitely many d. Suppose that A_D are known for those finitely many d. Now

$$\begin{aligned} A_D &= \sum_{k\geq 0} \alpha^{-k-2} e_* \left(\Lambda(\alpha)\Lambda(-\alpha) c^k \cap LT_{g,1}(d,X)\right) \\ &= \sum_{k\geq 0} (-1)^g \alpha^{-k-2+2g} e_*(c^k \cap LT_{g,1}(d,X)). \end{aligned}$$

So each of the unknown A_D has order α^{-2+p}. where $p < 0$. By Theorem 4.3 [14], these A_D can be reconstructed from the linking values. q.e.d.

The same argument shows that if $\{b_D\}$ is a given gluing sequence with the property that for a given g, the number

$$-\langle c_1(X), d\rangle + g(dim\ X - 2) + 2 + deg\ b_D$$

is negative for all but finitely many d, then the theorem above holds for this gluing sequence.

2.3 The collapsing lemma

Let $X = \mathbf{P}^1 \times \mathbf{P}^n$ and let p_1, p_2 be the first and the second projection of $\mathbf{P}^1 \times \mathbf{P}^n$. We let $M_g(d, X)$ be the moduli space (stack) of stable morphisms from genus g curves to X of bi-degree $(1, d)$. The case $g = 0$ was treated in [12]. Here we will prove a similar lemma in case $g \geq 1$. Note that there are no degree 1 maps from positive genus smooth curves to $\mathbf{P}^1$. Thus the domain of any stable morphism $f : C \to X$ in $M_g(d, X)$ must have a distinguished irreducible component $C_0 \cong \mathbf{P}^1$ with

$$p_1 \circ f|_{C_0} : C_0 \xrightarrow{\sim} \mathbf{P}^1$$

and all other components mapping to points via $p_1 \circ f$. Let d_0 be the degree of $p_1 \circ f|_{C_0}$. Use the collapsing map $M_0(d_0, X) \to N_{d_0}$, which depend on a choice of basis of $H^0(\mathcal{O}_{\mathbf{P}^n}(1))$, we obtain $(n+1)$ sections

$$[\phi_0, \cdots, \phi_n] \in H^0(\mathcal{O}_{\mathbf{P}^1}(d_0))^{\oplus(n+1)}/\mathbf{C}^\times.$$

Let $C_1, \cdots, C_k$ be other irreducible components of C and let $z_i \in \mathbf{P}^1$ be $f(C_i)$ and d_i be the degree of $f^* p_2^* \mathcal{O}_{\mathbf{P}^n}(1)$ over C_i. Note $d = \sum_{i=0}^k d_i$. Then using imbedding of sheaves

$$\mathcal{O}_{\mathbf{P}^1}(d_0) \to \mathcal{O}_{\mathbf{P}^1}(\sum_{i=0}^k d_i z_i) \cong \mathcal{O}_{\mathbf{P}^1}(d)$$

we obtain $(n+1)$-tuple of sections

$$[\tilde{\phi}_0, \cdots, \tilde{\phi}_n] \in H^0(\mathcal{O}_{\mathbf{P}^n}(1))^{\oplus(n+1)}/\mathbf{C}^\times,$$

which will be a point in N_d. This defines a correspondence

$$\tilde{\varphi} : M_g(d, X) \longrightarrow N_d.$$

Lemma 2.12. *The correspondence $\tilde{\varphi}$ is induced by a morphism $\varphi : M_g(d, X) \to N_d$. Moreover φ is equivariant with respect to the induced action of $\mathbf{C}^\times \times T$.*

Proof. The following proof is given by J. Li. Let $\mathcal{S}$ be the category of all schemes of finite type (over $\mathbf{C}$) and let $\mathcal{F} : \mathcal{S} \longrightarrow$ (Set) be the the contra-variant functor associating to each $S \in \mathcal{S}$ the set of families of stable morphisms

$$F : \mathcal{X} \longrightarrow \mathbf{P}^1 \times \mathbf{P}^n \times S$$

over S of bi-degree $(1, d)$ of arithmetic genus g, modulo the obvious equivalence relation. Note that $\mathcal{F}$ is represented by the moduli stack $M_g(d, X)$. Hence to define φ it suffices to define a transformation

$$\Psi : \mathcal{F} \longrightarrow \mathrm{Mor}\,(-, N_d).$$

We now define such a transformation. Let $S \in \mathcal{S}$ and let $\xi \in \mathcal{F}(S)$ be represented by $(\mathcal{X}, F)$. We let p_i be the composite of F with the i-th projection of $\mathbf{P}^1 \times \mathbf{P}^n \times S$ and let p_{ij} be the composite of F with the projection from $\mathbf{P}^1 \times \mathbf{P}^n \times S$ to the product of its i-th and j-th components. We consider the locally free sheaves $p_2^*\mathcal{O}_{\mathbf{P}^n}(k)$, $k = 0$ or 1, of $\mathcal{O}_{\mathcal{X}}$-modules and its direct image complex

$$\mathcal{L}_\xi(k) = R^\bullet p_{13*} p_2^* \mathcal{O}_{\mathbf{P}^n}(k).$$

We claim that $\mathcal{L}_\xi(k)$, which is a complex of $\mathcal{O}_{\mathbf{P}^1 \times S}$-modules, is quasi-isomorphic to a perfect complex. Since this is a local problem, we can assume S is affine. We pick a sufficiently relative-ample line bundle H on $\mathcal{X}/S$ so that $p_2^*\mathcal{O}_{\mathbf{P}^n}(-k)$ is a quotient sheaf of

$$\mathcal{V}_1 = p_3^* p_{3*}(\mathcal{O}(H) \otimes p_2^*\mathcal{O}_{\mathbf{P}^n}(-k)) \otimes \mathcal{O}(-H).$$

Let $\mathcal{V}_2$ be the kernel of $\mathcal{V}_1 \to p_2^*\mathcal{O}_{\mathbf{P}^n}(-k)$. (Here $\mathcal{V}_i$ are implicitly depending on k.) Since H is sufficiently ample, both $\mathcal{V}_1$ and $\mathcal{V}_2$ are locally free. Hence we have a short exact sequence of locally free sheaves of $\mathcal{O}_{\mathcal{X}}$-modules

$$0 \longrightarrow p_2^*\mathcal{O}_{\mathbf{P}^n}(k) \longrightarrow \mathcal{V}_1^\vee \longrightarrow \mathcal{V}_2^\vee \longrightarrow 0.$$

We then apply $R^\bullet p_{13*}$ to this exact sequence,

$$0 \longrightarrow p_{13*}p_2^*\mathcal{O}_{\mathbf{P}^n}(k) \longrightarrow p_{13*}\mathcal{V}_1^\vee \longrightarrow p_{13*}\mathcal{V}_2^\vee \longrightarrow R^1 p_{13*}p_2^*\mathcal{O}_{\mathbf{P}^n}(k) \longrightarrow 0.$$

Here all other terms vanish because H is sufficiently ample and fibers of p_{13} have dimension at most one. As argued in [12] both $p_{13*}\mathcal{V}_1^\vee$ and $p_{13*}\mathcal{V}_2^\vee$ are flat over S. Because $\mathbf{P}^1 \times S \to S$ is smooth,

$$p_{13*}\mathcal{V}_1^\vee \longrightarrow p_{13*}\mathcal{V}_2^\vee, \tag{9}$$

and hence $\mathcal{L}_\xi(k)$, is quasi-isomorphic to a perfect complex.

The complex $\mathcal{L}_\xi(k)$ satisfies the following base change property: let $\rho : T \to S$ be any base change and let $\rho^*(\xi) \in \mathcal{F}(T)$ be the pull back of ξ. Then there is a canonical isomorphism of complex of sheaves of $\mathcal{O}_T$-modules

$$\mathcal{L}_{\rho^*(\xi)}(k) \cong (\mathbf{1}_{\mathbf{P}^1} \times \rho)^* \mathcal{L}_\xi(k).$$

Since $\mathcal{L}_\xi(k)$ is quasi-isomorphic to a perfect complex, we can define the determinant line bundle[1] of $\mathcal{L}_\xi(k)$, denoted by $\det \mathcal{L}_\xi(k)$. It is an invertible sheaf of $\mathcal{O}_{\mathbf{P}^1\times S}$-modules. Using the Riemann-Roch theorem, one computes that the degree of $\det(\mathcal{L}_\xi(k))$ along fibers of $\mathbf{P}^1 \times S \to S$ are $kd - g$. Further, because $\mathcal{L}_\xi(k)$ has rank one, there is a canonical homomorphism

$$\mathcal{L}_\xi(k) \longrightarrow \det \mathcal{L}_\xi(k)$$

defined away from the support of the torsions subsheaves of $p_{13*}p_2^*\mathcal{O}_{\mathbf{P}^n}(k)$ and $R^1p_{13*}p_2^*\mathcal{O}_{\mathbf{P}^n}(k)$. Now let w be any element in $H^0(\mathbf{P}^n, \mathcal{O}_{\mathbf{P}^n}(1))$. Its pull back provides a canonical meromorphic section

$$\sigma_{\xi,w} \in H^0\left(\mathbf{P}^1 \times S, \mathbf{M}(\det \mathcal{L}_\xi(1))\right).$$

For similar reason, the section $1 \in H^0(\mathcal{O}_{\mathbf{P}^n})$ provides a canonical meromorphic section δ of $\det \mathcal{L}_\xi(0)$. Combined, they provide a canonical meromorphic section

$$\eta_{\xi,w} = \sigma_{\xi,w} \cdot \delta^{-1} \in H^0\left(\mathbf{P}^1 \times S, \mathbf{M}(\det \mathcal{L}_\xi(1) \otimes \det \mathcal{L}_\xi(0)^{-1})\right).$$

We now show that $\eta_{\xi,w}$ extends to a regular section. Let $s \in S$ be any closed point. We first assume that there are no irreducible components of $\mathcal{X}_s$ that are mapped entirely to $\mathbf{P}^1 \times w^{-1}(0) \subset \mathbf{P}^1 \times \mathbf{P}^n$ under F_s. Here $F_s : \mathcal{X}_s \to \mathbf{P}^1 \times \mathbf{P}^n$ is the restriction of F to the fiber over $t \in S$. By shrinking S if necessary, we can assume all $F_t : \mathcal{X}_t \to \mathbf{P}^1 \times \mathbf{P}^n$, $t \in S$, have this property. Then the section w induces a short exact sequence

$$0 \longrightarrow \mathcal{O}_{\mathcal{X}} \longrightarrow p_2^*\mathcal{O}_{\mathbf{P}^n}(1) \longrightarrow \mathcal{R} \longrightarrow 0$$

[1] All materials concerning determinant line bundle are taken from [9].

and a long exact sequence

$$\longrightarrow R^\bullet p_{13*}\mathcal{O}_\mathcal{X} \longrightarrow R^\bullet p_{13*}p_2^*\mathcal{O}_{\mathbf{P}^n}(1) \longrightarrow R^\bullet p_{13*}\mathcal{R} \longrightarrow R^{\bullet+1}p_{13*}\mathcal{O}_\mathcal{X} \longrightarrow .$$

By our assumption on w, $R^1p_{13*}\mathcal{R} = 0$. Next, we let $\mathcal{E}_1 \to \mathcal{E}_2$ (resp. $\mathcal{F}_1 \to \mathcal{F}_2$) be the complex (9) associated to $k = 1$ (resp. $k = 0$). Clearly, we have canonical commutative diagram

$$\begin{array}{ccccccccccc} 0 & \longrightarrow & p_{13*}\mathcal{O}_\mathcal{X} & \longrightarrow & \mathcal{F}_1 & \longrightarrow & \mathcal{F}_2 & \longrightarrow & R^1p_{13*}\mathcal{O}_\mathcal{X} & \longrightarrow & 0 \\ & & \downarrow & & \downarrow & & \downarrow & & \downarrow & & \\ 0 & \longrightarrow & p_{13*}p_2^*\mathcal{O}_{\mathbf{P}^n}(1) & \longrightarrow & \mathcal{E}_1 & \longrightarrow & \mathcal{E}_2 & \longrightarrow & R^1p_{13*}p_2^*\mathcal{O}_{\mathbf{P}^n}(1) & \longrightarrow & 0 \end{array}$$

of short exact sequences. Let

$$\mathcal{K}_1 : \quad \mathcal{F}_1 \longrightarrow \mathcal{E}_1 \oplus \mathcal{F}_2 \longrightarrow \mathcal{E}_2$$

be the induced complex. Note the last arrow is surjective. Let $\mathcal{A}_1$ be $\mathcal{F}_1$ and $\mathcal{A}_2$ be the kernel of the last arrow of the above complex. Hence $\mathcal{K}_1$ is quasi-isomorphic to the complex

$$\mathcal{K}_2 : \quad \mathcal{A}_1 \longrightarrow \mathcal{A}_2. \tag{10}$$

Note that both $\mathcal{A}_1$ and $\mathcal{A}_2$ are $\mathcal{O}_S$-flat. Hence we can define the determinant line bundle $\det \mathcal{K}_2$. We then have canonical isomorphisms

$$\det \mathcal{K}_2 \cong \det \mathcal{K}_1 \cong \det \mathcal{L}_\xi(1)^{-1} \otimes \det \mathcal{L}_\xi(0). \tag{11}$$

Now let $t \in S$ be any closed point. $t \in S$. It is clear that the restriction of (10) to general points of $\mathbf{P}^1_t$ is an isomorphism. Hence $\det \mathcal{K}_2^{-1}$ has a canonical section over $\mathbf{P}^1 \times S$ [5]. It is direct to check that under the isomorphism (11) this section is the extension of $\eta_{\xi,w}$. Since $\mathbf{P}^1 \times S \to S$ is proper, such extension is unique.

It remains to show that $\eta_{\xi,w}$ can be extended even the assumption on $w^{-1}(0)$ does not hold. Note that in this case, we can find two sections w_1 and w_2 in $H^0(\mathcal{O}_{\mathbf{P}^n}(1))$ so that $w = w_1 + w_2$ and that both w_1 and w_2 satisfies the condition about $w_1^{-1}(0)$ and $w_2^{-1}(0)$. Here we might need to shrink S if necessary. Then η_{ξ,w_1} and η_{ξ,w_2} both can be extended to regular sections in

$$H^0\left(\mathbf{P}^1 \times S, \det \mathcal{L}_\xi(1) \otimes \det \mathcal{L}_\xi(0)^{-1}\right).$$

Further, over the open subset $Z \subset \mathbf{P}^1 \times S$ where all $R^ip_{13*}p_2^*\mathcal{O}_{\mathbf{P}^n}(k)$, $i, k = 0, 1$, are torsion free, we obviously have

$$\eta_{\xi,w} = \eta_{\xi,w_1} + \eta_{\xi,w_2}.$$

Since $Z \cap \mathbf{P}^1 \times \{t\} \neq \emptyset$ for all $t \in S$, the right hand side of the above identity provides an extension of $\eta_{\xi,w}$. This proves that for any $w \in H^0(\mathcal{O}_{\mathbf{P}^n}(1))$ the meromorphic sections $\eta_{\xi,w}$ extends to a regular section

$$\eta^{\mathrm{ex}}_{\xi,w} \in H^0\left(\mathbf{P}^1 \times S, \det \mathcal{L}_\xi(1) \otimes \det \mathcal{L}_\xi(0)^{-1}\right).$$

Again since $\mathbf{P}^1 \times S \to S$ is smooth and proper, the extension is unique.

Now we define the morphism $S \to N_d$. Let $\{w_0, \cdots, w_n\}$ be a basis of $H^0(\mathcal{O}_{\mathbf{P}^n}(1))$. Then we obtain $(n+1)$ canonical regular sections

$$\eta^{\mathrm{ex}}_{\xi,w_0}, \cdots, \eta^{\mathrm{ex}}_{\xi,w_n} \in H^0\left(\mathbf{P}^1 \times S, \det \mathcal{L}_\xi(1) \otimes \det \mathcal{L}_\xi(0)^{-1}\right).$$

Hence, after fixing an isomorphism

$$\det \mathcal{L}_\xi(1) \otimes \det \mathcal{L}_\xi(0)^{-1} \cong \pi_S^* \mathcal{M} \otimes \pi_{\mathbf{P}^1}^* \mathcal{O}_{\mathbf{P}^1}(d)$$

for some invertible sheaf $\mathcal{M}$ of $\mathcal{O}_S$-modules, we obtain $(n+1)$ canonical sections of

$$H^0(\mathcal{O}_{\mathbf{P}^1}(d)) \otimes_{\mathbf{C}} \mathcal{M}$$

which defines a morphism

$$S \longrightarrow H^0(\mathcal{O}_{\mathbf{P}^n}(1))^{\oplus(n+1)}/\mathbf{C}^*.$$

Since the image is always away from 0, it defines a morphism

$$S \longrightarrow N_d.$$

It is routine to check that this construction satisfies the base change property, and hence defines a morphism $M_g(d, X) \to N_d$, as desired.

To check that this morphism gives rise to the correspondence mentioned before, it suffices to check the case where S is a closed point. In this case one sees immediately that the complex $\mathcal{L}_\xi(1) - \mathcal{L}_\xi(0)$ has locally free part isomorphic to $\mathcal{O}_{\mathbf{P}^1}(d_0) - \mathcal{O}_{\mathbf{P}^1}$, and has torsion part supported at z_i of length d_i. Further, a direct check shows that the sections $\sigma_{\xi,w_i} \cdot \delta^{-1}$ are a non-zero constant multiple of the sections $\tilde{\phi}_i$ mentioned in the definition of the correspondence. This shows that the morphism defines the correspondence constructed. The equivariant property of this morphism again follows from the base change property of this construction. This completes the proof of the Collapsing Lemma. q.e.d.

References

[1] A. Bertram, *Another way to enumerate rational curves with torus action*, math. AG/9905159.

[2] G. Bini, C. De Concini, M. Polito & C. Procesi, *Givental's work relative to mirror symmetry*, math.AG/9805097.

[3] A. Elezi, *Mirror symmetry for concavex bundles on projective spaces*, math. AG/0004157.

[4] C. Faber & R. Pandharipande, *Hodge integrals and Gromov-Witten theory*, math. AG/9810173.

[5] R. Friedman, J. W. Morgan, *Smooth four-manifolds and complex surfaces,* Ergebnisse der Mathematik und ihrer Grenzgebiete (3), Results in Mathematics and Related Areas (3), Springer, Berlin, Vol. 27, 1994.

[6] A. Givental, *Equivariant Gromov-Witten invariants*, alg-geom/9603021.

[7] T. Graber & R. Pandharipande, *Localization of virtual classes*, alg-geom/9708001.

[8] S. Katz, A. Klemm & C. Vafa, *Geometric engineering of quantum field theories*, Nucl. Phys. **B497** (1997) 173–195.

[9] F. F. Knudsen & D. Mumford, *The projectivity of the moduli space of stable curves. I. Preliminaries on "det" and "Div"*, Math. Scand. **39** (1976) 19–55.

[10] M. Kontsevich, *Enumeration of rational curves via torus actions*, The Moduli Space of Curves, (eds. R. Dijkgraaf, C. Faber, G. van der Geer), Progr. Math., Birkhäuser, Vol. 129, 1995, 335–368.

[11] J. Li & G. Tian, *Virtual moduli cycle and Gromov-Witten invariants of algebraic varieties*, J. Amer. Math. Soc. **11** (1998) 119–174.

[12] B. Lian, K. Liu & S.T. Yau, *Mirror principle. I*, Asian J. Math. **1** (1997) 729–763.

[13] ———, *Mirror principle. II*, Asian J. Math. **3** (1999) ???–???.

[14] ———, *Mirror principle. III*, math.AG/9912038.

[15] R. Pandharipande, *Rational curves on hypersurfaces (after givental)*, math. AG/9806133.

BRANDEIS UNIVERSITY
STANFORD UNIVERSITY
HARVARD UNIVERSITY

SURVEYS IN DIFFERENTIAL GEOMETRY, 2000
Vol. VII ©2000, INTERNATIONAL PRESS
pp. 497-554

THREE CONSTRUCTIONS OF FROBENIUS MANIFOLDS: A COMPARATIVE STUDY

YU. I. MANIN

Abstract

The paper studies three classes of Frobenius manifolds: Quantum Cohomology (topological sigma–models), unfolding spaces of singularities (K. Saito's theory, Landau–Ginzburg models), and the recent Barannikov–Kontsevich construction starting with the Dolbeault complex of a Calabi–Yau manifold and conjecturally producing the B–side of the Mirror Conjecture in arbitrary dimension. Each known construction provides the relevant Frobenius manifold with an extra structure which can be thought of as a version of "non–linear cohomology". The comparison of these structures sheds some light on the general Mirror Problem: establishing isomorphisms between Frobenius manifolds of different classes. Another theme is the study of tensor products of Frobenius manifolds, corresponding respectively to the Künneth formula in Quantum Cohomology, direct sum of singularities in Saito's theory, and presumably, the tensor product of the differential Gerstenhaber–Batalin–Vilkovyski algebras. We extend the initial Gepner's construction of mirrors to the context of Frobenius manifolds and formulate the relevant mathematical conjecture.

0. Introduction

0.1. Frobenius manifolds. Frobenius manifolds were introduced and investigated by B. Dubrovin as the axiomatization of a part of the rich mathematical structure of the Topological Field Theory (TFT): cf. [9].

According to [9] and [26], *a Frobenius manifold* is a quadruple $(M, \mathcal{T}_M^f, g, A)$. Here M is a supermanifold in one of the standard categories (C^∞, analytic, algebraic, formal, ...), $\mathcal{T}_M^f$ is the sheaf of flat vector fields tangent to an affine structure, g is a flat Riemannian metric

First published in The Asian Journal of Mathematics, 1999. Used by permission.

(non–degenerate even symmetric quadratic tensor) such that $\mathcal{T}_M^f$ consists of g–flat tangent fields. Finally, A is an even symmetric tensor $A : S^3(\mathcal{T}_M) \to \mathcal{O}_M$. All these data must satisfy the following conditions;

a) *Potentiality of* A. Everywhere locally there exists a function Φ such that $A(X, Y, Z) = (XYZ)\Phi$ for any flat vector fields X, Y, Z.

b) *Associativity.* A and g together define a unique symmetric multiplication $\circ : \mathcal{T}_M \otimes \mathcal{T}_M \to \mathcal{T}_M$ such that

$$A(X, Y, Z) = g(X \circ Y, Z) = g(X, Y \circ Z).$$

This multiplication must be associative.

In other words, in flat coordinates the tensor of the third derivatives $\Phi_{ab}{}^c$ must constitute the set of structure constants of an associative algebra.

If one excludes the trivial case when Φ is a cubic form with constant coefficients in flat coordinates, the first large class of Frobenius manifolds was discovered by Kyoji Saito even before Dubrovin's axiomatization (see [31], [32] and [30]):

(i) Moduli spaces of unfolding (germs of) isolated singularities of hypersurfaces carry natural structures of Frobenius manifold.

Each such structure is determined by a choice of Saito's *good primitive form.*

In [9] a more global variation of this construction is described (Hurwitz's spaces).

Physicists call the relevant TFT the topological sector of the Landau–Ginzburg theory: cf. [7].

The second large class of Frobenius manifolds was discovered by physicists (Witten, Dijkgraaf, Vafa) and is called Quantum Cohomology. For an axiomatic treatment, see [20]. The correlators of this theory are called Gromov–Witten (GW) invariants. Their actual construction in the algebraic–geometric framework was carried out in [3] following [5] and [4].

(ii) The formal completion at zero of the cohomology (super)space of any smooth projective or compact symplectic manifold carries a natural structure of formal Frobenius manifold.

The third large class of Frobenius manifolds was recently constructed by S. Barannikov and M. Kontsevich ([2]).

(iii) The formal moduli spaces of solutions to the Maurer–Cartan equations modulo gauge equivalence, related to a class of the differential Gerstenhaber–Batalin–Vilkovyski (dGBV) algebras, carry a natural structure of formal Frobenius manifold.

As their main application, Barannikov and Kontsevich construct the dGBV–algebra starting with the Dolbeault complex of an arbitrary Calabi–Yau manifold, and conjecture that the resulting formal Frobenius manifold (B–model) can be identified with the quantum cohomology of the mirror dual Calabi–Yau manifold (A–model). S. Merkulov ([29]) recently invented a similar construction applicable to any symplectic manifold satisfying the strong Lefschetz condition (cf. below, subsections 5.9, 5.10, 6.5 and 6.6). Yet another possible source of dGBV–algebras (or rather their homotopy version) is provided by the BRST cohomology of certain chiral algebras: cf. [25].

The Mirror Conjecture is a part of the gradually emerging considerably more general pattern. Within the Calabi–Yau domain, it should be a consequence of the Kontsevich's conjecture about the equivalence of the Fukaya triangulated category associated to one member of the mirror pair and the derived category of sheaves on the other member. Furthermore, one expects the extension of the mirror picture to other classes of varieties, non–necessarily smooth, compact or having trivial canonical class. For some exciting recent results on mirrors, cf. [14] and [24].

Isomorphisms of Frobenius manifolds of different classes remain the most direct expression, although by no means the final one, of various mirror phenomena.

From this vantage point, the three classes of examples considered above should be compared at least in two ways.

First, one looks for isomorphisms between Frobenius manifolds (and their submanifolds) constructed by different methods.

Second, one tries to generalize to other classes of Frobenius manifolds additional structures peculiar to each of the known classes.

Consider, for example, Quantum Cohomology. Physically, quantum cohomology of a manifold V is only the tree level small phase space part of the topological sigma model with target space V. In particular, the correlators of this theory, which are essentially the coefficients of the formal Frobenius potential, can be mathematically defined in terms of the intersection indices on moduli spaces of stable maps of curves of genus zero to V. This set of the correlators of Quantum Cohomology of V can be extended to a much vaster structure involving, first, curves of arbitrary genus, and second, the so called gravitational descendants, mathematically expressible via Chern classes of certain tautological bundles on the moduli spaces of stable maps (cf. [21] for precise statements).

This leads to two natural questions.

What is the differential geometric meaning of the arbitrary genus correlators with descendants in the framework of general geometry of Frobenius manifolds?

Which Frobenius manifolds admit extensions of this type?

Of course, Frobenius manifolds admitting such an extension include those that are isomorphic to Quantum Cohomology. Therefore a better understanding of this problem could shed some light on the Mirror Conjecture as well. In [21] it is shown that at least the correlators with gravitational descendants in any genus g can be reconstructed from the additional data consisting of two different parts.

One part is the genus $\leq g$ Cohomological Field Theory in the sense of [20] whose correlators take values in the cohomology of the moduli spaces of stable curves with marked points. At the moment it is unclear which abstract Frobenius manifolds can be extended to such genus $\leq g$ geometry and how it can be done.

Another part of the data concerns only genus zero correlators and therefore in principle can be formulated in terms of arbitrary Frobenius manifold. Its existence, however, poses non–trivial restrictions on the manifold which are axiomatized below in the notion of qc–type.

This whole setup can be illuminated by comparison with the motivic philosophy. In principle, any natural structure on the cohomology of an algebraic manifold can be considered as a realization of its motive, and the question which abstract structures of a given type arise from cohomology ("are motivated") is a typical question of the theory of motives.

Quantum Cohomology is a highly nonlinear realization of the motives of smooth projective manifolds. It is functorial, at least in the naive sense, only with respect to isomorphisms. Hence it cannot be extended to the category of the Grothendieck motives in an obvious way. Nevertheless, the natural monoidal structure of motives extends to Frobenius manifolds. Their tensor product in the formal context furnishing the Künneth formula for Quantum Cohomology was constructed in [20], and [22]. R. Kaufmann (cf. [18] and paper in preparation) has shown that the tensor product of convergent potentials converges, and the resulting Frobenius manifold in a sense does not depend on the choice of the base points. This adds some flexibility to the motivic perspective. For example, Frobenius manifolds provide a context in which one can meaningfully speak about cohomology of fractional weight: cf. e.g. the treatment of A_n–manifolds in 2.3..1 and 3.5 below. It should be also compared with S. Cecotti's suggestion that the TFT's of Landau–Ginzburg type naturally give rise to mixed Hodge structures (see [7]).

0.2. Plan of the paper. This paper is dedicated to the study of the Frobenius manifolds of the three classes (i), (ii) and (iii) from the perspective described above.

In §1 we start with reminding the formalism of Gromov–Witten invariants and the structure of the potential of the Quantum Cohomology. In the treatment of [20] its terms of degree ≤ 2 were not fixed because of absence of stable curves of genus zero with ≤ 3 marked points. However, the Divisor Axiom allows an unambiguous definition of these terms (Proposition 1.3.1). This simple remark is essential for the definition of ≤ 2 point correlators for abstract formal Frobenius manifolds. We introduce the notion of the manifold of qc–type, which embodies a version of the Divisor Axiom, and show that it allows us to define for such manifolds the correlators with gravitational descendants. This is an elaboration of the picture sketched in §3 of [21].

In §2 we introduce the notion of *the (strong) Saito framework.* It axiomatizes those properties of the spaces of miniversal deformations of isolated singularities of functions which directly lead to the Frobenius structure. The most difficult and deep aspects of Saito's theory are thereby neatly avoided and become "existence theorems". (This illustrates the advantages of theft in comparison with honest work, as was justly remarked about the axiomatic method in the beginning of this century).

This part is taken from my notes to a lecture course and is included here on suggestion of A. Givental.

The main result of this section is summarized in the formulas (2.12), (2.13) which in the context of Saito's theory refer to the unfolding space of the direct sum of singularities, and in our axiomatic treatment are stated in terms of abstract direct sum diagrams. Another proof of these formulas using oscillating integrals was shown to me by A. Givental.

In §3 we show first of all, using (2.12) and (2.13), that the direct sum of Saito's frameworks corresponds to the tensor product of the associated Frobenius manifolds.

Looking then at the tensor products M of the Frobenius manifolds A_n (deformation space of the singularity z^{n+1} at zero) and more general manifolds with rational spectra we find out that the integral part of their spectra define Frobenius submanifolds HM which look like quantum cohomology of a manifold with trivial canonical class, at least on the level of discrete invariants (cf. below). This argument exactly corresponds to the well known idea of D. Gepner ([11], [16], [7]) of building Calabi–Yau sigma models from the tensor products of minimal models. The numerology is also similar.

Here is an example of our results.

0.2.1. Claim. The manifold $A_n^{\otimes n+1}$ contains a (canonically defined) pointed Frobenius submanifold $HA_n^{\otimes n+1}$ whose spectrum looks formally like that of even–dimensional part of quantum cohomology of an $(n-1)$–dimensional algebraic (or symplectic) manifold V.

More precisely, V must have Betti numbers

$$h^{2m}(V) := \text{the number of } (i_1, \dots, i_{n+1}) \in \mathbf{Z}_{\geq 0}^{n+1}, \text{ satisfying}$$

$$\frac{1}{m}\sum_{k=1}^{n+1} i_k = m(n+1),\ 0 \leq i_k \leq n-1 \tag{0.1}$$

and vanishing (modulo torsion) $c_1(V)$.

For example, even Betti numbers must be (1,19,1) for $n = 3$, and (1,101,101,1) for $n = 4$.

The Poincaré symmetry of them is generally established by the involution $(i_k) \mapsto (n-1-i_k)$, $m \mapsto n-1-m$.

0.2.2. Problem. Is there actually a manifold V_n whose Quantum Cohomology contains $HA_n^{\otimes n+1}$? Is it at least true that $HA_n^{\otimes n+1}$ is Frobenius manifold of qc–type? (Notice that A_n itself is not of qc–type).

As was explained above, $A_n^{\otimes n+1}$ is the unfolding space at zero of the singularity of $x_1^{n+1} + \dots + x_{n+1}^{n+1}$. An argument which I learned from [7] and (in a different version) from A. Givental then shows that $HA_n^{\otimes n+1}$ carries the variation of Hodge structure corresponding to the middle cohomology of the hypersurface $x_1^{n+1} + \dots + x_{n+1}^{n+1} = 0$. More precisely, the volume form periods constitute the horizontal sections of one of the structure connections of the Frobenius manifold in question.

Thus Problem 0.2.2 has the flavor of Mirror Conjecture, and of course Gepner's idea was a precursor of the modern studies of the mirrors. Hence at least the case $n = 4$ of the Problem 0.2.2 might be reducible to the Givental's treatment of the toric CY threefold (see [14] and the subsequent developments due to B.–H. Lian, K. Liu, S.-T. Yau in [24]).

The Barannikov–Kontsevich construction [2] conjecturally provides another, and quite general, class of Frobenius manifolds of Calabi–Yau B–type. To describe it succinctly, notice that the space of vector fields on a Frobenius manifold is simultaneously a Lie (super)algebra and a (super)commutative algebra. The classical example of such combination is the algebra of functions on a Poisson (super)manifold. So it would appear that Frobenius manifolds could come from Poisson structures. However, such a relation cannot be straightforward, because the Poisson identity between the bracket and the multiplication does not hold on Frobenius manifolds. The ingenious twist in [2] consists in considering instead *odd Poisson algebras with two differentials.* Under appropriate conditions, the Frobenius manifold structure is then induced on the homology space of such Poisson algebra.

The sections 4–6 of this paper constitute a completely self–contained account of the theory of [2] in the axiomatic context of dGBV–algebras. Specifically, in §3 we supply direct elementary proofs of all results related to the formality and to the structure of Maurer–Cartan moduli spaces at their non–obstructed points. In §4 we collect a list of basic general properties of dGBV–algebras. Finally, in §6 we define the relevant formal Frobenius manifolds. We would like to stress the similarity of the formulas defining $\circ$–multiplication in the Saito's and Barannikov–Kontsevich constructions: compare (2.3) and (6.1). This supports the expectation that both construction might be special cases of a more general picture.

As a comment to the title of Cecotti's paper [7], it is instructive to compare the extensions of the Calabi–Yau variations of Hodge structure (VHS) given by Saito's theory to the construction of [2] and the equivariant theory of Givental ([14]).

Barannikov and Kontsevich embed any Calabi–Yau VHS into a Frobenius supermanifold that has all discrete invariants perfectly matching those of the mirror dual quantum cohomology. Thus it has a good chance to be the correct B–model in the classical Mirror Conjecture picture. A drawback of this embedding is that the relevant Frobenius manifolds are not semisimple, and so the identification of them hardly can be achieved by formal calculations.

To the contrary, whenever methods of Gepner, Saito and Givental are applicable (quasi–homogeneous singularities, anticanonical hypersurfaces in toric compactifications), they embed (parts of) Calabi–Yau VHS into Frobenius manifolds that are generically semisimple or into families of manifolds with generically semisimple general fiber. This makes more accessible the direct check of mirror isomorphisms (cf. §3 below). But these techniques do not give the full dimension spectrum (e.g. odd–dimensional cohomology is skipped) and they are not directly applicable to those Calabi–Yau manifolds which are not toric anticanonical hypersurfaces (or complete intersections).

Acknowledgements. The first draft of this paper was written after stimulating discussions of Gepner's work with V. Schechtman. It was read and extensively commented by A. Givental who corrected some mistakes and supplied illuminating explanations of the Saito theory. In particular, the whole §2 owes its existence to Givental's suggestion. In writing the last three sections, I have greatly benefited from M. Kontsevich's lectures at the MPI and his handwritten notes.

1. Quantum cohomology and Frobenius manifolds

1.1. Gromov–Witten (GW) invariants. We start with reminding some basic notation and facts from [20], [5], [3].

Let V be a smooth projective algebraic manifold over an algebraically closed field of characteristic zero, $B = B(V)$ the semigroup of effective one–dimensional algebraic cycles modulo numerical equivalence. For any $\beta \in B(V)$, $g, n \geq 0$ we can define the Deligne–Mumford stack $\overline{M}_{g,n}(V, \beta)$ parametrizing stable maps of curves of genus g with n labelled points, landing in class β. This stack comes equipped with virtual fundamental class in the homological Chow group with rational coefficients $J_{g,n}(V, \beta) \in A_s(\overline{M}_{0,n}(V, \beta))$ where

$$s = (1 - g)\,(\dim V - 3) + (c_1(V), \beta) + n.$$

Moreover, there are canonical morphisms $ev : \overline{M}_{g,n}(V, \beta) \to V^n$ sending a stable map to the image of the family of labelled points. In the stable range, that is when $2g - 2 + n > 0$, there is also a map $st : \overline{M}_{g,n}(V, \beta) \to \overline{M}_{g,n}$ forgetting V and stabilizing the curve. They can be used in order to define *the Gromov–Witten correspondences* in the Chow rings

$$I_{g,n}(V, \beta) := (ev, st)_*(J_{g,n}(V, \beta)) \in A_s(V^n \times \overline{M}_{g,n}).$$

This family of Chow correspondences is the most manageable embodiment of motivic quantum cohomology forgetting just the right amount of geometric information encoded in the rather uncontrollable stack of stable maps. For genus zero, the situation further simplifies. Since V^n and $\overline{M}_{0,n}$ are smooth, we can identify A_* with A^*. As S. Keel proved, $A^*(V^n \times \overline{M}_{0,n}) = A^*(V^n) \otimes A^*(\overline{M}_{0,n})$, and $A^*(\overline{M}_{0,n})$ is a finite–dimensional self–dual linear space. Hence one can identify $I_{0,n}(V, \beta)$ with the induced map $A^*(\overline{M}_{0,n}) \to A^*(V^n)$. The space $A^*(\overline{M}_{0,n})$ is spanned by the dual classes of the boundary strata $\overline{M}(\tau)$ indexed by n–trees. So calculating $I_{0,n}(V,\beta)$ amounts to calculating a finite family of elements in $A^*(V^n)$ indexed by these trees.

1.2. Frobenius manifolds. All our examples of Frobenius manifolds (see 0.1) will come equipped with two additional structures: a flat vector field e which is *identity* with respect to $\circ$, and *an Euler vector field* E expressing the scaling invariance of the Frobenius manifold M. More precisely, we must have $\mathrm{Lie}_E(g) = Dg$ for some constant D, and $\mathrm{Lie}_E(\circ) = d_0\circ$ for another constant d_0 (in the context of Frobenius manifolds g means the metric, not the genus, which is zero for the relevant GW–invariants). If the first condition is satisfied, the second one is equivalent to

$$E\Phi = (d_0 + D)\Phi + \text{a polynomial in flat coordinates of degree } \leq 2.$$

For any Euler field E we have $[E, \mathcal{T}_M^f] \subset \mathcal{T}_M^f$. Assume for simplicity that the spectrum $\{d_a\}$ of $-\operatorname{ad} E$ on flat vector fields belongs to the base field. We understand $\{d_a\}$ as a family of constants with multiplicities. The constant d_0 introduced earlier is in addition the eigenvalue corresponding to e. The family $D, \{d_a\}$ is called the spectrum of M. Since any multiple of E is an Euler field together with E, in the case $d_0 \neq 0$ we can normalize E by the condition $d_0 = 1$.

We also put $d = 2 - D$ and $q_a = 1 - d_a$ and call the family $d, \{q_a\}$ *the d–spectrum* of M.

If M is the formal spectrum of the ring of formal series in flat coordinates, Φ is a formal solution of the Associativity Equations, we call M *formal Frobenius manifold.* Formal Frobenius manifolds can be tensor multiplied. The underlying metric space of flat fields of the tensor product is the usual tensor product of the respective spaces of factors. The potential of the tensor product is defined in a much subtler way: see [20] and [26]. If the factors are additionally endowed with flat identities and normalized Euler fields, they can be used to produce a canonical flat identity and Euler field on the tensor product: see [18].

1.3. From genus zero GW invariants to Frobenius manifolds. In the situation of 1.1, we can construct a formal Frobenius manifold $H^*_{\mathrm{quant}}(V)$ whose underlying linear supermanifold is the completion of $H^*(V,\Lambda)$ at zero, with obvious flat structure and Poincaré form as metric. Here Λ is a $\mathbf{Q}$–algebra endowed with the universal character $B(V) \to \Lambda : \beta \mapsto q^\beta$, with values in the Novikov ring Λ which is the completed semigroup ring of $B(V)$ eventually localized with respect to the multiplicative system q^β. It is topologically spanned by the monomials $q^\beta = q_1^{\beta_1} \dots q_m^{b_m}$ where $\beta = (b_1, \dots, b_m)$ in a basis of the numerical class group of 1–cycles, and $(q_1, \dots, q_m)$ are independent formal variables. This is needed to make convergent the formal sums

$$I_{g,n}(V) := \sum_{\beta \in B} q^\beta I_{g,n}(V, \beta) \in A_\Lambda^*(V^n \times \overline{M}_{g,n}) \tag{1.1}$$

For H^* we can take any cohomology theory functorial with respect to Chow correspondences. The construction of the potential requires only the top degree terms of the genus zero GW invariants. To be concrete, choose a homogeneous basis $\{\Delta_a \,|\, a = 0, \dots r\}$ of $H^*(V, \mathbf{Q})$. Denote by $\{x_a\}$ the dual coordinates and by $\Gamma = \sum_a x_a \Delta_a$ the generic even element of the cohomology superspace. This means that $\mathbf{Z}_2$–parity of x_a equals 0 (resp. 1), if Δ_a is even–dimensional (resp. odd–dimensional). Put for $\gamma_i \in H^*(V)$

$$\langle I_{g,n}(V) \rangle (\gamma_1 \otimes \dots \otimes \gamma_n) = \langle \gamma_1 \dots \gamma_n \rangle_{g,n} := \int_{I_{g,n}(V)} pr^*(\gamma_1 \otimes \dots \otimes \gamma_n) \tag{1.2}$$

where $pr : V^n \times \overline{M}_{g,n} \to V^n$ is the projection. Then the quantum cohomology potential is

$$\Phi(x) = \langle e^\Gamma \rangle_0 := \sum_{n, (a_1, \dots, a_n)} \epsilon(a) \frac{x_{a_1} \dots x_{a_n}}{n!} \langle \Delta_{a_1} \dots \Delta_{a_n} \rangle_{0,n} \tag{1.3}$$

where $\epsilon(a)$ is the sign resulting from rewriting $\prod x_i \Delta_{a_i}$ as $\epsilon(a) \prod x_i \prod \Delta_{a_i}$. Assume that the dual fundamental class Δ_0 of V is the part of our basis. Then the flat identity is $\partial_0 = \partial / \partial x_0$. Moreover, the Euler field is

$$E = \sum_a \left(1 - \frac{|\Delta_a|}{2}\right) x^a \partial_a + \sum_{b:\, |\Delta_b| = 2} r^b \partial_b, \tag{1.4}$$

where $\Delta_a \in H^{|\Delta_a|}(V)$, and r^b are defined by

$$c_1(\mathcal{T}_V) = -K_V = \sum_{b:\, |\Delta_b| = 2} r^b \Delta_b. \tag{1.5}$$

Clearly, $-\operatorname{ad} E$ is semisimple on flat vector fields. Let $H(d_a)$ be the eigenspace correspondng to d_a. We have $H(d_a) = H^{2q_a}(V)$. Hence the total spectrum is

$$\begin{aligned} &D = 2 - \dim V, \; d_0 = 1, \\ &d_a = 1 - \frac{|\Delta_a|}{2} \text{ of multiplicity } \dim H^{|\Delta_a|}(V) \end{aligned} \tag{1.6}$$

and the d–spectrum is

(1.7) $$d = \dim V,\ \{q \text{ of multiplicity } h^{2q}(V)\}.$$

Quantum cohomology of the direct product of manifolds is the tensor product of the respective Frobenius manifolds. So the d–spectrum behaves as is expected.

E itself is not flat, but it has the flat projection $E(0)$ to $H(D) = H^2$ which is just the anticanonical class, if E is normalized as above by $d_0 = 1$ and $H^*(V)$ is identified with the space of flat vector fields. This is evident from (1.3) and (1.4).

The spaces $H(d_a), H(d_b)$ are orthogonal unless $d_a + d_b = D$, or equivalently $q_a + q_b = d$, and dual in the latter case. They also all have integral structure compatible with metric. If we work with a coefficient ring Λ, these subspaces are direct (free) sumbodules. Identity belongs to $H(1) = H^0(V)$. Especially important are $H^2(V) = H(D)$, $H_2(V) = H(0)$, and the semigroup $B \subset H_{2,\mathbf{Z}}$ of effective algebraic classes, in which every element is finitely decomposable, and zero is indecomposable. They are never trivial for projective smooth V of positive dimension.

Returning to the potential Φ, we see that since $\overline{M}_{0,n}$ is empty for $n \leq 2$, the definition (1.3) specifies only its terms of degree ≥ 3 in x_a. The validity of the Associativity Equations is not sensitive to this indeterminacy. However, the missing terms can be uniquely normalized either geometrically, by integrating over $J_{0,n}(V,\beta)$ in $\overline{M}_{0,n}(V,\beta)$ for $n \leq 2$, or formally, by using the Divisor Axiom of [20]. Since this normalization is important for the future use, we describe it explicitly. Denote by $\delta := \sum_{a:\, d_a = 0} x_a \Delta_a$ the generic even element of H^2.

1.3.1. Proposition. *There exists a unique formal function Φ differing from (1.3) only by terms of degree ≤ 2 which is representable as a formal Fourier series in $q^\beta e^{(\beta,\delta)}$, $\beta \in B$, with coefficients which are formal series of the remaining coordinates, having the following properties. Put $\Phi = \Psi + c$ where c is the constant ($\beta = 0$) term of the Fourier series. Then, assuming $d_0 = 1$ and denoting by $E(0)$ the anticanonical class summand of E, we have:*

a) $E\Psi = (D+1)\Psi$.

b) c is a cubic form with $(E - E(0))\,c = (D+1)c$, the classical cubic self–intersection index divided by 6.

In fact, if one puts formally $q^\beta = 0$ for $\beta \neq 0$ in the structure constants of the quantum multiplication, one gets the classical cup multiplication on $H^*(V)$ which together with metric determines c starting even with non necessarily normalized potential.

Proof. We use the properties of $I_{0,n}(V,\beta)$ stated axiomatically in [20] and proved in [3].

The initial potential Φ is expressed via Gromov–Witten invariants

$$\Phi(\Gamma) := \sum_{n\geq 3}\sum_{\beta} \frac{q^\beta}{n!}\langle I_{0,n}(V,\beta)\rangle(\Gamma^{\otimes n})$$

where $\langle I_{0,n}(V,\beta)\rangle$ is defined by the same formula as (1.2) only with integral taken over $I_{0,n}(V,\beta)$. The part of Φ corresponding to $\beta=0$ is exactly $c=\frac{1}{6}(\Gamma^3)$.

The maps $\langle I_{0,n}(V,\beta)\rangle : H^*(V)^{\otimes n} \to \Lambda$ are defined for $n\geq 3$ and satisfy

$$\langle I_{0,n}(V,\beta)\rangle(\alpha\otimes\delta) = (\beta,\delta)\,\langle I_{0,n-1}(V,\beta)\rangle(\alpha)$$

for $\delta\in H^2(V)$ (we write (β,δ) for $g(\beta,\delta)$). This follows from the Divisor Axiom. It is easy to check that there exists a unique polylinear extension of $\langle I_{0,n}(V,\beta)\rangle$ to all $n\geq 0$ satisfying this identity. In fact, it suffices to put

$$\langle I_{0,n}(V,\beta)\rangle(\alpha) = (\beta,\delta)^{-m}\langle I_{0,n+m}(V,\beta)\rangle(\alpha\otimes\delta^{\otimes m})$$

for any (m,δ) with $m+n\geq 3$ and invertible (β,δ).

Now put

$$\Psi(\Gamma) := \sum_{n\geq 0}\sum_{\beta\neq 0} \frac{q^\beta}{n!}\langle I_{0,n}(V,\beta)\rangle(\Gamma^{\otimes n})$$

Clearly, $\Psi+c$ differs from the initial Φ by terms of degree ≤ 2. Moreover, c is a linear combination of $x^a x^b x^c$ with $|\Delta_a|+|\Delta_b|+|\Delta_c| = 2\dim V$ so that $(E-E(0))\,c=(D+1)c$. As for $E\Psi$, we have for $\Gamma=\gamma_0+\delta$

$$\begin{aligned}\Psi(\gamma_0+\delta) &= \sum_{i,k\geq 0}\sum_{\beta\neq 0}\frac{q^\beta}{i!\,k!}\langle I_{0,n}(V,\beta)\rangle(\gamma_0^{\otimes i}\otimes\delta^{\otimes k})\\ &= \sum_{i\geq 0}\sum_{\beta\neq 0}\frac{q^\beta e^{(\beta,\delta)}}{i!}\langle I_{0,n}(V,\beta)\rangle(\gamma_0^{\otimes i}).\end{aligned}\tag{1.8}$$

Let us apply now E to any summand in (1.8). The $E(0)$ part acts only upon $e^{(\beta,\delta)}$ and multiplies it by $(c_1(V),\beta)$. The $E-E(0)$ part multiplies any monomial $x_{a_1}\dots x_{a_n}$ in non–divisorial coordinates by $\sum_i(1-$

$|\Delta_{a_i}|/2)$. From (1.2) and (1.3) we see that β can furnish a non–zero contribution to such term only if

$$\dim I_{g,n}(V,\beta) = \dim V - 3 + (c_1(V),\beta) + n = \sum_{i=1}^{n} \frac{|\Delta_{a_i}|}{2}.$$

Hence every non–vanishing term of (1.8) is an eigenvector of E with eigenvalue $D+1 = 3 - \dim V$.

This proves the Proposition.

Notice in conclusion that $q^\beta e^{(\beta,\delta)}$ is the universal character of B together with q^β. We have introduced q^β only to achieve the formal convergence. If it holds without q^β, we can forget about it. Moreover, if the formal Fourier series actually converges for δ lying somewhere in the complexified ample cone, $\Psi(x)$ has a free abelian symmetry group: translations by an appropriate discrete subgroup in the space $H^2(V, i\mathbf{R})$. Conversely, in the analytic category this condition is necessary for the existence of the appropriate Fourier series.

1.4. Potentials of qc–type. Based upon the analysis above, we will introduce the following definition. Its first goal is to axiomatize a part of the structures of sec. 1.3 which suffices for the construction of the coupling of a formal Frobenius manifold with gravity in the sense of [21]. As we will recall below, this construction is divided into two steps: the construction of the modified gravitational descendants which can be done for any formal Frobenius manifold, and the construction of a linear operator T on the big phase space which requires additional structures.

The second goal is to provide an intermediate step in the problem of checking whether a given formal Frobenius manifold is quantum cohomology. We must be able at least to detect the following structures.

1.4.1. Definition. Let $(M = \mathrm{Spf}\ k[[H^t]], g, \Phi_0)$ be a formal Frobenius manifold over a $\mathbf{Q}$–algebra k with flat identity, Euler field E, and spectrum $D, \{d_a\}$ in k as above. Here H is a free $\mathbf{Z}_2$–graded k–module of flat vector fields, and H^t is the dual module of flat coordinates vanishing at the origin. Put $H^2 = H(0), H_2 = H(D)$.

Assume that there exists a semigroup $B \subset H_2$ with finite decomposition and indecomposable zero, and the cubic form c on H, such that by eventually changing terms of degree ≤ 2 in Φ_0 we can obtain the potential of the form

$$\Phi = \Psi + c,\ E\Psi = (D+1)\Psi,\ (E - E(0))c = (D+1)c,$$

$$\Psi(\gamma_0+\delta)=\sum_{i\geq 0}\sum_{\beta\in B\setminus\{0\}}\frac{e^{(\beta,\delta)}}{i!}\,I_\beta(\gamma_0^{\otimes i}) \tag{1.9}$$

such that all summands in the last sum are eigenvectors of E with eigenvalue $D+1$. Here γ is a generic even element of H, δ its "divisorial" H^2–part, $\gamma_0=\gamma-\delta$. The coefficient $I_\beta(\gamma_0^{\otimes i})$ is a form in non–divisorial coordinates.

A formal Frobenius manifold satisfying these conditions will be called of qc–type.

A flat identity e in this language is an element $e\in H$ which considered as a derivation satisfies

$$e\,\Psi=0,\ e\,c=g$$

(see [26], p. 29 for the same expressions in coordinates).

1.4.2. Correlators of qc–manifolds. Let M be a formal Frobenius manifold of qc–type. Recall that $\Phi_{ab}{}^c$ are the structure constants of the quantum multiplicarion. On qc–manifolds there are two useful specializations of this structure.

a) The "small quantum multiplication" obtained by restricting $\Phi_{ab}{}^c$ to $\gamma_0=0$. We will denote this multiplication by dot.

b) The cup multiplication $\cup$ obtained by putting formally $e^{(\beta,\gamma)}=0$ for all $\beta\neq 0$ ("large volume limit"). In other words, this is the multiplication, for which c can be written as

$$c(\gamma)=\frac{1}{6}\,g(\gamma,\gamma\cup\gamma).$$

We now define correlators $\langle\dots\rangle : H^{\otimes n}\to k$ as S_n–invariant polylinear functions whose values are derivatives of Φ at zero. In other words, for a basis $\{\Delta_a\}$ of H and dual coordinates $\{x_a\}$ as above, we have

$$\Phi(x)=\sum_{n,a_1,\dots,a_n}\epsilon(a)\,\frac{x_{a_1}\dots x_{a_n}}{n!}\,\langle\Delta_{a_1}\dots\Delta_{a_n}\rangle$$

In the qc–case we can write

$$\langle\Delta_{a_1}\dots\Delta_{a_n}\rangle=\sum_{\beta\in B\setminus\{0\}}\langle\Delta_{a_1}\dots\Delta_{a_n}\rangle_\beta+\langle\Delta_{a_1}\dots\Delta_{a_n}\rangle_0$$

where the first sum comes from Ψ and the second, nonvanishing only for triple arguments, from c.

Looking at (1.9) one sees, that small quantum multiplication depends only on the triple correlators of non–divisorial elements of the basis.

1.4.3. Claim. The correlators of the Frobenius manifolds of qc–type satisfy the following Divisor Identity: if $\delta \in H^2$, $\beta \neq 0$,

$$\langle \delta\gamma_1 \dots \gamma_n \rangle_\beta = (\delta, \beta)\langle \gamma_1 \dots \gamma_n \rangle_\beta .$$

Reading backwards the proof of (1.8), one sees that this property follows from (1.9).

This formula allows us to extend the definition of the correlators to $n \leq 2$ arguments.

1.5. Gravitational descendants for the Frobenius manifolds of qc–type. Let now M be a formal Frobenius manifold as in 1.4.1 whose space we identify with the linear superspace H, At first we do not assume that it is of qc–type. Following [21], we can define its modified correlators with gravitational descendants. They are polylinear functions on *the big phase space* $\oplus_{d\geq 0} H[d]$ where $H[d]$ are copies of the space H identified with the help of the shift operator $\tau : H[d] \to H[d+1]$. To define them explicitly, we recall that any formal Frobenius manifold gives rise to the genus zero Cohomological Field Theory. Namely, there exists a unique sequence of linear maps $I_n^M : H^{\otimes n} \to H^*(\overline{M}_{0,n}, k)$, $n \geq 3$, satisfying the folowing properties.

a) I_n^M are S_n–invariant and compatible with restriction to the boundary divisors (cf. [20] or [26], p. 101).

b) The top degree term of I_n^M capped with the fundamental class of $\overline{M}_{0,n}$ is the correlator of M with n arguments.

Moreover, in the quantum cohomology case

$$I_n^M = \sum_\beta q^\beta I^V_{0,n,\beta}$$

where $I^V_{0,n,\beta}$ are the genus zero Gromov–Witten invariants.

Now let $C \to \overline{M}_{0,n}$ be the universal curve, $s_i : \overline{M}_{0,n} \to C$, $i = 1, \dots, n$ its structure sections, ω_C the relative dualizing sheaf, $\psi_i := c_1(s_i^*(\omega_c)) \in A^1(\overline{M}_{0,n}$. The *the modified correlators with gravitational descendants* for M are defined by the formula

$$(1.10) \quad \langle \tau^{d_1}\Delta_{a_1} \dots \tau^{d_n}\Delta_{a_n} \rangle := \int_{\overline{M}_{0,n}} I_n^M(\Delta_{a_1} \otimes \dots \otimes \Delta_{a_n})\psi_1^{d_1} \dots \psi_n^{d_n}$$

and the generator function for them, *the modified potential*, by the formula

$$(1.11) \quad G^M(x) = \sum_{n\geq 3, (a_i, d_i)} \epsilon(a) \frac{x_{d_1,a_1} \dots x_{d_n,a_n}}{n!} \langle \tau^{d_1}\Delta_{a_1} \dots \tau^{d_n}\Delta_{a_n} \rangle$$

where $(x_{d,a})$ are coordinates dual to $\tau^d\Delta_a$.

If we assume in addition that M is of qc–type, we can define also the (unmodified) two argument correlators $\langle\tau_d\gamma_1\,\gamma_2\rangle$ by the inductive formula (25) from [21]:

$$\langle\tau_d\gamma_1\,\gamma_2\rangle = \sum_{j=1}^{d}(-1)^{j+1}\partial_\delta^{-j}\langle\tau_{d-j}\gamma_1\,\delta\cdot(\delta^{j-1}\cup\gamma_2)\rangle$$
$$+(-1)^d\partial_{\gamma_0}^{-(d+1)}[\langle\delta\,\gamma_1\,\delta^d\cup\gamma_2\rangle-\langle\delta\,\gamma_1\,\delta^d\cup\gamma_2\rangle_0]. \tag{1.12}$$

Here $\delta\in H^2$ is an arbitrary (say, generic) element such that $(\delta,\beta)\neq 0$ for all $\beta\in B\setminus\{0\}$ and the operator ∂_δ^{-1} divides $\langle\dots\rangle_\beta$ by (δ,β).

Furthermore, put

$$y_{c,b}=x_{c,b}+\sum_{(a,d),g\geq c+1}x_{d,a}\langle\tau_{d-c-1}\Delta_a\,\Delta^b\rangle.$$

Then the big phase space potential of M is, by definition, $F^M(x):=G^M(y)$, and the unmodified correlators with gravitational descendants of M are defined as coefficients of F:

$$F^M(x)=\sum_{n\geq 3,(a_i,d_i)}\epsilon(a)\,\frac{x_{d_1,a_1}\dots x_{d_n,a_n}}{n!}\,\langle\tau_{d_1}\Delta_{a_1}\dots\tau_{d_n}\Delta_{a_n}\rangle \tag{1.13}$$

The main result of [21] is that if M is the quantum cohomology of V, this prescription provides the correlators with descendants of the topological sigma model with target space V. The latter are defined by the formula similar to (1.10) but with $\overline{M}_{0,n}$ replaced by $\overline{M}_{0,n}(V)$, I replaced by J, and the respective change in the meaning of ψ_i.

In conclusion notice that the sigma model correlators satisfy, partly demonstrably, partly conjecturally, some additional identities, of which the most interesting are probably the Virasoro constraints. I do not know which of these identities might be valid for the more general qc–type manifolds. In any case, it would be interesting to determine differential equations at least for the modified potential with descendants for general formal Frobenius manifolds or particular examples like the manifolds A_n (see 2.3. below).

2. K. Saito's frameworks

2.1. Setup. Let $p: N\to M$ be a submersion of complex analytic or algebraic manifolds, generally non–compact, F a holomorphic function on N. We consider F as a family of functions on the fibers of p parametrized by points of M. In local coordinates $z = (z_a), t = (t_b)$ where t_b are constant along the fibers of p we write $F = F(z, t)$.

Let $d_p : \mathcal{O}_N \to \Omega^1_{N/M}$ be the relative differential. Denote by C the closed analytic subspace (or subscheme) of the critical points of the restrictions of F to the fibers given by the equation $d_pF = 0$. Its ideal J_F is locally generated by the partial derivatives XF where X are vertical vector fields on N. Derivatives $\partial F/\partial z_a$ of course suffice. Let $i_C : C \to N$ be the natural embedding, p_C the restriction of p to C.

Denote by $\Omega^{max}_{N/M}$ the invertible sheaf of holomorphic vertical volume forms on N, $L := i_C^*(\Omega^{max}_{N/M})$. The Hessian Hess $(F) \in \Gamma(C, L^2)$ is a well defined section of L^2 which in local coordinates as above can be written as

$$\operatorname{Hess}(F) = i_C^* \left[\det \left(\frac{\partial^2 F}{\partial z_a \partial z_b} \right) (dz_1 \wedge \cdots \wedge dz_n)^2 \right].$$

We denote by $G_C \subset C$ the subspace Hess $(F) = 0$. Let $\mathcal{T}_M$ be the tangent sheaf of M. Finally, let ω be a nowhere vanishing global section of $\Omega^{max}_{N/M}$.

2.1.1. Definition. The family $(p : N \to M, F, \omega)$ is called Saito's framework, if the following conditions are satisfied:

a) Let the map $s : \mathcal{T}_M \to p_{C*}(\mathcal{O}_C)$ be defined by $X \mapsto \overline{X}F \bmod J_F$, where $\overline{X}$ is any local (in N) lift of X. Then s is an isomorphism of $\mathcal{O}_M$–modules. In particular, C is finite and flat over M. Assume moreover that G_C is a divisor, and $p_C : C \to M$ is étale on the complement to the divisor $G = i_{C*}(G_C) \subset M$.

b) Define the following 1–form ϵ on $M \setminus G$. Its value on the vector field $X = s^{-1}(f)$ corresponding to the local section f of $\pi_*(\mathcal{O}_C)$ equals

$$i_X(\epsilon) := \operatorname{Tr}_{C/M} \left(\frac{f\, i_C^*(\omega^2)}{\operatorname{Hess}(F)} \right) = \sum_{i=1}^{\mu} \frac{f(\rho_i)}{\det((\partial^2 F/\partial z_a \partial z_b)(\rho_i))} \tag{2.1}$$

where ρ_i are the local branches of the critical locus C over M, (z_a) is any vertical local coordinate system unimodular with respect to ω.

Then the scalar product $g : S^2(\mathcal{T}_{M\setminus G}) \to \mathcal{O}_{M \setminus G}$ defined by

$$g(X, Y) := i_{X \circ Y}(\epsilon) \tag{2.2}$$

is a flat metric. Both ϵ and g (as flat metric) extend regularly to M.

2.2. The (pre–)Frobenius structure associated to the Saito framework. Let $(p : N \to M, F, \omega)$ be a Saito framework.

Denote by $\circ$ the multiplication in $\mathcal{T}_M$ induced by the one in $p_*(\mathcal{O}_C)$:

$$X \circ Y := s^{-1}(\overline{X}F \cdot \overline{Y}F) \bmod J_F$$

or equivalently

$$\overline{X \circ Y}F \equiv \overline{X}F \cdot \overline{Y}F \bmod J_F. \tag{2.3}$$

Clearly, the vector field $e := s^{-1}(1 \bmod J_F)$ is the identity for $\circ$. Let $\mathcal{T}_M^f$ be the sheaf of vector fields flat with respect to g. Finally, put

$$A(X,Y,Z) = g(X \circ Y, Z) = g(X, Y \circ Z).$$

The last equality follows from (2.2) and the associativity of $\circ$. The tensor A is symmetric because $\circ$ is commutative. Therefore we have:

2.2.1. Claim. The data $(\mathcal{T}_M^f, g, A)$ define on M the structure of pre–Frobenius manifold in the sense of [26], Ch. I, Def. 1.1.1.

One can say more about this structure restricted to $M \setminus G$. Call a connected open subset U in $M \setminus G$ *small* if $p_C^{-1}(U)$ is the disjoint union of $\mu = \dim M$ connected components U_i canonically isomorphic to U. For concreteness, we will arbitrarily number them by $\{1, \dots, \mu\}$ as in (2.1). Then we have natural ring isomorphisms

$$\Gamma(U, p_{C*}(\mathcal{O}_C)) = \oplus_{i=1}^{\mu} \Gamma(U_i, \mathcal{O}_C) = \Gamma(U, \mathcal{O}_M)^{\mu}.$$

This $\Gamma(U, \mathcal{O}_M)$–algebra has a basis of idempotents $f_i := \delta_{ij}$ on U_j. Defining $e_i \in \Gamma(U, \mathcal{T}_M)$ by $s(e_i) = f_i$, that is, $\overline{e}_i F \bmod J_F = f_i$, we get a local $\mathcal{O}_M$–basis of $\mathcal{T}_M$ satisfying $e_i \circ e_j = \delta_{ij} e_j$ and $e = \sum_i e_i$. Denote by $u^i \in \Gamma(U, \mathcal{O}_M)$ the restriction of F to U_i pushed down to U that is, put

$$u^i = F(\rho_i)$$

in the notation of (2.1). Small subsets cover $M \setminus G$ so that the structure group of $\mathcal{T}_M$ is reduced to S_μ. Summarizing, we have:

2.2.2. Proposition. *The data $(\mathcal{T}_M^f, g, A)$ define on $M \setminus G$ the structure of semisimple pre–Frobenius manifold in the sense of [26], Ch. I, Def. 3.1 and 3.2. Moreover, we have $e_i u^j = \delta_{ij}$ so that (u^j) form a local coordinate system (Dubrovin's canonical coordinates) and $[e_i, e_j] = 0$ because $e_i = \partial/\partial u^i$.*

Proof. Only the last statement might need some argumentation. We have $p^*(e_i u^j) = \overline{e}_i p^*(u^j)$ for any lift $\overline{e}_i$ of e_i. To calculate the right hand side we can restrict it to any local section of p since it is constant along the fibers. We choose $\overline{e}_i$ tangent to U_j and restrict the right hand side to U_j where $p^*(u^j)$ coincides with F. The result is δ_{ij} by the definition of e_i.

For the future use, we can reformulate this as follows. Dualizing s we get the isomorphism $s^t : \Omega^1_M \to \mathcal{H}om_{\mathcal{O}_M}(p_{C*}(\mathcal{O}_C), \mathcal{O}_M)$. Then $s^t(du^i) : p_{C*}(\mathcal{O}_C) \to \mathcal{O}_M$ is the map which annihilates j–components for $j \neq i$ and coincides with the pushforward on the i–th component.

2.2.3. Theorem. *The structure $(M, \mathcal{T}^f_M, g, A)$ associated to the Saito framework is Frobenius iff $d\epsilon = 0$.*

Proof. To check the Frobenius property on $M \setminus G$ we appeal to [26], Ch. I, Th. 3.3 (Dubrovin's criterium), both conditions of which, $[e_i, e_j] = 0$ and $d\epsilon = 0$ are satisfied. To pass from $M \setminus G$ to M one can use a continuity argument, e. g. in the following form, again due to Dubrovin. Let ∇_0 be the Levi–Civita connection of g, and ∇_λ the pencil of connections on $\mathcal{T}_M$ determined by its covariant derivatives $\nabla_{\lambda,X}(Y) := \nabla_{0,X}(Y) + X \circ Y$. Then M if Frobenius iff ∇_λ is flat for some $\lambda \neq 0$, and so automatically for all λ. Clearly, this is the closed property.

We will now discuss when e is flat.

On a small U, we can define functions η_j by $\eta_j = i_{e_j}(\epsilon) = g(e_j, e_j)$. When $e_j = \partial/\partial u^j$, the closedness of $\epsilon = \sum_i \eta_i du^i$ means that $\eta_j = e_j \eta$ for a local function η well defined up to addition of a constant, or else $\epsilon = d\eta$. In the notations (2.1)

$$\eta_i = \frac{1}{\det\left((\partial^2 F/\partial z_a \partial z_b)(\rho_i)\right)}. \tag{2.4}$$

2.2.4. Theorem. *Assume that the conditions of the Theorem 2.2.3. are satisfied.*

The identity e is flat, iff for all i, $e\eta_i = 0$, or equivalently, $e\eta = g(e,e) = \mathrm{const}$. This holds automatically in the presence of an Euler field E with $D \neq 2d_0$ (see 1.2 above and 2.2.5 below).

This is Prop. 3.5 from [26], Ch. I.

One important remark about the identity is in order. Namely, in all examples I know of there exists a lift $\overline{e}$ of e to N such that $\overline{e}F = 1$ identically, so that in the appropriate coordinate system we have $F = F_0 + t_0$ where F_0 does not depend on t_0 which is lifted from M, and $\overline{e} = \partial/\partial t_0$.

It remains to clarify what Euler fields this structure can have.

2.2.5. Theorem. *Assume that the conditions of the Theorem 2. 2.3. hold. Let E be a vector field on a small subset U in M.*

a) We have $\mathrm{Lie}_E(\circ) = d_0 \circ$ *iff*

$$E = d_0 \sum_i (u^i + c^i) e_i \tag{2.5}$$

for some constants c^i, where (du^i) are 1–forms dual to (e_j).

In particular, for non–zero E we have $d_0 \neq 0$ so that we may normalize E by $d_0 = 1$. Furthermore, if the monodromy representation of the fundamental group of $M \setminus G$ on H_0 of the fibers of $C \to M$ has only one-dimensional trivial subrepresentation, the global vector field E of this form with fixed d_0 is defined uniquely up to addition of a multiple of e.

b) For a field E of the form (2.5) and a constant D, we have $\mathrm{Lie}_E(g) = Dg$ *iff*

$$E\eta = (D - d_0)\eta + \mathrm{const}\,. \tag{2.6}$$

In particular, if e is flat, adding a multiple of e does not change the validity of this property.

This follows from [26], Ch. I, Th. 3.6.

When M comes from the Saito framework, we have a natural candidate for the global Euler field with $d_0 = 1$ suggested by our identification of local coordinates u^i. Namely, put on any small U

$$E_F := \sum_{i=1}^{\mu} F(\rho_i) e_i = \sum_{i=1}^{\mu} u^i e_i. \tag{2.7}$$

Assume that it is in fact an Euler field and that we are in the conditions when it is defined uniquely up to a shift by a multiple of e. Assume furthermore that there exists a point 0 in M to which E_F extends and at which it vanishes (0 may lie in G, and in the theory of singularities it does so). Since e cannot vanish, the choice of such 0 fixes E_F completely.

2.2.6. Definition. Saito's framework $(p : N \to M, F, \omega)$ is called the strong Saito framework, if the structure $(M, \mathcal{T}_M^f, g, A)$ described above is Frobenius, with flat identity e and Euler field E_F.

2.2.7. Remark. Since the definitions of the pre–Frobenius and Frobenius structures, and also of the identity and Euler fields, are local, we can lift all these structures from $M \setminus G$ to $C \setminus G_C$.

2.3. Unfolding singularities. K. Saito's theory (cf. [31], [32], [30] and the references therein) produces (a germ of) a strong Saito's framework starting with a germ of holomorphic function $f(z_1, \dots, z_n)$ with isolated singularity at zero.

Namely, one can choose holomorphic germs $\phi_0 = 1, \phi_1, \dots, \phi_{\mu-1}$ whose classes constitute a basis of the Milnor ring $\mathbf{C}\{\{z\}\}/(\partial f/\partial z_a)$ in such a way that $F := f + \sum t_i \phi_i$ is the miniversal unfolding of f. Then $N = N_f$, resp. $M = M_f$ is a neighborhood of zero in the (z, t)–, resp. (t)–space, and F is defined above.

The crucial piece of the structure is the choice of ω encoded in the Saito notion of a good primitive form. Generally its existence is established in an indirect way. For the singularities of ADE–type one can take $\omega = dz_1 \wedge \dots \wedge dz_n$. Generally, if f is a quasi–homogeneous polynomial, most of the data constituting the Saito framework are algebraic varieties, rational maps and rational differential forms so that the whole setup has considerably more global character.

In order to help the reader to compare notation, we notice that Saito's S is our M, and our function η is denoted τ on p. 630 of [30]. Starting with the germ of z^{n+1} at zero, one obtains in this way the following Frobenius manifold.

2.3.1. Example: manifolds A_n. Denote by N , resp. M, the affine space with coordinates $(z; a_1, \dots, a_n)$, resp. $a_1, \dots, a_n)$, and by p the obvious projection. Put

$$F = F(z; a_1, \dots, a_n) = z^{n+1} + a_1 z^{n-1} + \dots + a_n.$$

Then C iz given by the equation $F'(z) = 0$. We choose $\omega = d_p z$.

Making explicit the basic structures described above we get the following description of A_n. Consider the global covering of M whose points consist of total orderings of the roots $\rho_1, \dots, \rho_n$ of $F'(z)$. On the semisimple part of it where $F'(z)$ has no multiple roots and $u^i := F(\rho_i)$ are local coordinates we have the flat metric

$$g := \sum_{i=1}^{n} \frac{(du^i)^2}{F''(\rho_i)}$$

with metric potential

$$\eta = \frac{a_1}{n+1} = \frac{1}{n-1} \sum_{i<j} \rho_i \rho_j = -\frac{1}{2(n-1)} \sum \rho_i^2.$$

Furthermore, e, E and flat coordinates $x_1, \dots, x_n$ can be calculated through $(a_1, \dots, a_n)$ (which are generically local coordinates as well):

$$e = \partial/\partial a_n, \text{ i. e., } ea_n = 1, ea_i = 0 \text{ for } i < n.$$

$$E = \frac{1}{n+1} \sum_{i=1}^{n} (i+1) a_i \frac{\partial}{\partial a_i},$$

x_i are the first Laurent coefficients of the inversion of

$$w = \sqrt[n+1]{p(z)} = z + O(1/z)$$

near $z = \infty$:

$$z = w + \frac{x_1}{w} + \frac{x_2}{w^2} + \dots + \frac{x_n}{w^n} + O(w^{-n-1}).$$

For the direct proof of these statements, see e.g. [26], Ch. I, 4.5.

The spectrum of A_n is

$$D = \frac{n+3}{n+1}, \; d^{(i)} = \frac{i+1}{n+1}, \quad 1 \le i \le n,$$

more precisely, $Ex_i = \dfrac{i+1}{n+1} x_i$. Hence the d–spectrum is

$$d^{(n)} = \frac{n-1}{n+1}, \; q_i^{(n)} = \frac{i}{n+1}, \; i = 0, \dots, n-1.$$

Now, Φ is analytic in x_a and the spectrum of $-\operatorname{ad} E$ is strictly positive. Therefore Φ must be a polynomial in flat coordinates. One can check that its degree is precisely $n+2$. Hence for $n \ge 2$ it cannot be of qc–type, and by the method described in §1 we can define for A_n only modified correlators with gravitational descendants.

Comparing the spectrum of A_n with that of the quantum cohomology of projective spaces, one can somewhat imaginatively say that A_n represents "projective space of dimension $\dfrac{n-1}{n+1}$, with rank one cohomology in each dimension $\dfrac{i}{n+1}$, $0 \le i \le n-1$ and with vanishing canonical class".

2.3.2. Example: Gepner's manifolds $V_{n,k}$. Let $n \ge 2, k \ge 1, h = n+k$. We will call *Gepner's Frobenius manifold* $V_{n,k}$ the manifold which is produced from the Saito's framework obtained by unfolding the polynomial

$$f_{n,k}(z_1, \dots, z_{n-1}) := \frac{1}{h} \sum_{i=1}^{n-1} y_i^h$$

where y and z are related by

$$\prod_{j=1}^{n-1}(1+y_jT)=1+\sum_{l=1}^{n-1} z_lT^l.$$

In particular, if one assigns to z_l weight l, $f_{n,k}$ becomes quasi-homogeneous of weight h. Its unfolding space is spanned by the classes of appropriate monomials, and a Zariski open dense subset $V_{n,k}$ of this space carries the structure of the Frobenius manifold as above. This subspace contains the point m corresponding to the *fusion potential*

$$g_{n,k}(z_1,\dots,z_{n-1}) := f_{n+1,k-1}(z_1,\dots,z_{n-1},1).$$

As D. Gepner ([12]) proved, the tangent space $\mathcal{T}_m V_{n,k}$ with $\circ$-multiplication, that is, the Milnor algebra of $g_{n,k}$, is isomorphic to the Verlinde algebra (fusion ring) of the $su(n)_k$ WZW model of the conformal field theory. Zuber in [34] conjectured, and Varchenko and Gusein–Zade in [17] proved, that the lattice of the Verlinde algebra and the respective bilinear form can be interpreted in terms of vanishing cycles of $f_{n,k}$.

The total Frobenius manifold $V_{n,k}$ is thus a deformation of this fusion ring, in much the same way as quantum cohomology is the deformation of the usual cohomology ring.

2.4. Direct sum diagram. We will consider now three Saito's frameworks $(p: N\to M, F,\omega)$ and $(p_i: N_i\to M_i, F_i,\omega_i)$, $i=1,2$. We will call *the direct sum diagram* any cartesian square

$$\begin{array}{ccc} N_1\times N_2 & \xrightarrow{\ \nu\ } & N \\ {\scriptstyle (p_1,p_2)}\downarrow & & \downarrow{\scriptstyle p} \\ M_1\times M_2 & \xrightarrow{\ \nu_M\ } & M \end{array} \tag{2.8}$$

with the following properties:

(i) $\nu^*(F)=F_1\boxplus F_2$.

(ii) $\nu^*(\omega)=\omega_1\boxtimes\omega_2$.

Thus in a neighborhood of any point of N lying over the image of ν_M there exist local coordinates $(z_a^{(1)},z_b^{(2)};t_e)$ such that t_e are lifted from M, and (i) can be written as

$$F(z_a^{(1)},z_b^{(2)};\nu_M^*(t_e))=F_1(z_a^{(1)};t_c^{(1)})+F_2(z_b^{(2)};t_d^{(2)}) \tag{2.9}$$

and similarly (ii) can be written as

$$\omega(z_a^{(1)},z_b^{(2)};\nu_M^*(t_e))=\omega_1(z_a^{(1)};t_c^{(1)})\wedge\omega_2(z_b^{(2)};t_d^{(2)}). \tag{2.10}$$

2.4.1. Properties of the direct sum diagrams. Clearly, $\nu^{-1}(C)$ is defined by the equations $d_{p_1}(F_1) \boxplus d_{p_2}(F_2) = 0$. Both summands then must vanish so that $\nu^{-1}(C) = C_1 \times C_2$. Denote by $\nu_C : C_1 \times C_2 \to C$ the restriction of ν. From (2.9) one then sees that

$$\nu_C^*(\mathrm{Hess}\,(F)) = \mathrm{Hess}\,(F_1) \boxtimes \mathrm{Hess}\,(F_2) \tag{2.11}$$

and hence $\nu_M^{-1}(G) = G_1 \times M_2 \cup M_1 \times G_2$. Let now $m = \nu_M(m_1, m_2)$, $m_i \in M_i$. Choose small neighborhoods $m \in U$ in M, $m_i \in U_i$ in M_i such that $\nu_M(U_1 \times U_2) \subset U$. Number the connected components $U_i^{(1)}$ of $p_{C_1}^{-1}(U_1)$, resp. $U_j^{(2)}$ of $p_{C_2}^{-1}(U_2)$, by some indices i, resp. j, as in (2.1). Then the connected components of $p_C^{-1}(U)$ are naturally numbered by the ordered pairs $I = (ij)$ in such a way that

$$\nu_C(U_i^{(1)} \times U_j^{(2)}) \subset U_I.$$

From now on we will assume that all the frameworks we are considering are strong ones. Then one can define e_I, u^I, η_I etc as above, and from (2.3.), (2.4), (2.9)–(2.11) one immediately sees that

$$u^I(m) = u_1^i(m_1) + u_2^j(m_2),\ \ \eta_I(m) = \eta_i^{(1)}(m_1)\eta_j^{(2)}(m_2) \tag{2.12}$$

where in the right hand side we have the respective local functions on M_1, M_2.

The following slightly less evident restriction formula will be also needed in the next section.

2.4.2 Proposition. *Let $I = (ij), K = (kl), \eta_{IK} = e_I\eta_K = e_K\eta_I$, and similarly $\eta_{ik}^{(1)} = e_i\eta_k^{(1)}$ etc. Then we have in the same notations as in (2.12):*

$$\eta_{IK}(m) = \delta_{jl}\eta_{ik}^{(1)}(m_1)\,\eta_l^{(2)}(m_2) + \delta_{ik}\eta_k^{(1)}(m_1)\,\eta_{jl}^{(2)}(m_2). \tag{2.13}$$

Proof. Calculate $\nu_M^*(d\eta_I)$ in two ways. On the one hand, we have

$$\nu_M^*(d\eta_I) = \sum_K \nu_M^*(\eta_{IK})\nu_M^*(du^K). \tag{2.14}$$

As at the end of the proof of Prop. 2.2.2, we can identify du^K with a map from $p_{C*}(\mathcal{O}_C)$ to $\mathcal{O}_M$ vanishing on all components except for the K–th one where it is the canonical pushforward. After restriction to

$M_1 \times M_2$ it may therefore be non–vanishing only on $U_k^{(1)} \times U_l^{(2)}$ so that we can calculate $\nu_M^*(\eta_{IK})$ by restricting $\nu_M^*(d\eta_I)$ to this product.

On the other hand, in view of (2.12),

$$\begin{aligned} \nu_M^*(d\eta_I) &= d\nu_M^*(\eta_I) = d(\eta_i^{(1)} \boxtimes \eta_j^{(2)}) \\ &= d\eta_i^{(1)} \boxtimes \eta_j^{(2)} + \eta_i^{(1)} \boxtimes d\eta_j^{(2)} \\ &= \sum_r \eta_{ir}^{(1)} du_1^r \boxtimes \eta_j^{(2)} + \sum_s \eta_i^{(1)} \boxtimes \eta_{js}^{(2)} du_2^s. \end{aligned} \tag{2.15}$$

Only the k–th summand in the first sum restricted to $U_k^{(1)} \times U_l^{(2)}$ may be non–vanishing and considered as a map (cf. above) it equals $\delta_{jl}\eta_{ik}^{(1)} \boxtimes \eta_j^{(2)}$ times the pushforward map. We have the similar expression for the l–th summand of the second sum. Comparison with (2.14) furnishes (2.13) because $\nu_M^*(du^K) = du_1^k \boxplus du_2^l$.

2.5. Direct sums of singularities. In the theory of singularities, we can compare the miniversal unfolding spaces M_f, M_g, M_{f+g} of the germs f, g and $f+g$. It so happens that they fit into the direct sum diagram (2.8) (the only choice that remains is that of the volume form ω on the space of $f+g$ which is natural to take decomposable as in 2.4 above).

By iteration, we can consider arbitrary number of summands. In particular, the Frobenius manifold $A_{n_1,\dots,n_k}$ which is obtained by unfolding the quasi–homogeneous singularity at zero $f(z) := z_1^{n_1} + \dots + z_k^{n_k}$ is related to the summands A_{n_i} in the way described above. We will show in the next section, that the formulas (2.12) and (2.13) imply a much neater description: $M_{f+g} = M_f \otimes M_g$, and in particular $A_{n_1,\dots,n_k}$ is the tensor product of A_{n_i} in the sense of [22] (in the context of formal Frobenius manifolds) and [18] (in the global context).

3. Tensor products and their submanifolds

3.1. Tensor product of formal Frobenius manifolds. Let us first of all recall the general construction of the tensor product of formal Frobenius manifolds over a common coefficient ring k. Instead of $(M = \text{Spf } k[[H^t]], g, \Phi)$ as in the Definition 1.4.1 we will be writing (H, g, Φ). We will not assume that our manifolds are of qc–type.

Let $(H^{(i)}, g^{(i)}, \Phi^{(i)})$, $i = 1, 2$, be two formal Frobenius manifolds. Then

$$(H^{(1)}, g^{(1)}, \Phi^{(1)}) \otimes (H^{(2)}, g^{(2)}, \Phi^{(2)}) = (H^{(1)} \otimes H^{(2)}, g^{(1)} \otimes g^{(2)}, \Phi)$$

where the terms of Φ of degree n are defined in the following way: reconstruct the Cohomological Field Theories $I_n^{(i)}$ for the two factors as in 1.5, put $I_n := I_n^{(1)} \cup I_n^{(2)}$ (cup product in $H^*(\overline{M}_{0,n})$) and cap I_n with the fundamental class of $\overline{M}_{0,n}$.

If the factors are endowed with flat identities and Euler fields, with $d_0^{(i)} = 1$ for both of them, one can define in a canonical way the flat identity and the Euler field for the product ([18]). Moreover, the d–spectrum of the product is

$$d = d^{(1)} + d^{(2)},\ \{q_A\} = \{q_a + q_b\}$$

as a sum of families with multiplicities. Notice that if ad E is semisimple for both factors, it is semisimple for the product, and for $A = (a, b)$ one can identify (∂_A, q_A) with $(\partial_a \otimes \partial_b, q_a + q_b)$.

Since the tensor product potential is defined coefficientwise and involves multiplication in all cohomology algebras $H^*(\overline{M}_{0,n})$, $n \geq 3$, it is practically impossible to calculate it directly. The problem becomes much more manageable if we deal with (germs of) semisimple analytic manifolds.

The reason for this is that generally a germ of Frobenius manifold of a given dimension depends on functional parameters, even in the presence of a flat identity and Euler field. The semisimplicity condition puts sufficiently strong constraints in order to leave undetermined only a finite number of constants, and then it is reasonable to rxpect that the tensor product is calculable in terms of these constants. Below we review the relevant results following [26] and [27].

3.2. Moduli space and tensor product of germs of semisimple Frobenius manifolds. Consider a pointed germ M of analytic Frobenius manifold over $\mathbf{C}$, (or a formal manifold with zero as the base point), as usual, with flat e, Euler E and $d_0 = 1$, and having pure even dimension. It will be called *tame semisimple* if the operator $E\circ$ has simple spectrum $(u_0^1, \dots, u_0^n)$ on the tangent space to the base point. We have the following general facts already partly invoked in the specific situation of Saito's framework in 2.2 above:

a) In a neighborhood of the base point, eigenvalues $(u^1, \dots, u^n)$ of $E\circ$ on $\mathcal{T}_M$ form a local coordinate system (Dubrovin's canonical coordinates), taking the values $(u_0^1, \dots, u_0^n)$ at the base point. The potential Φ is an analytic function of these coordinates. If the initial manifold was only assumed to be formal, from tame semisimplicity it follows that it is in fact the completion of a pointed analytic germ.

b) Put $e_i = \partial/\partial u^i$. Then $e_i \circ e_j = \delta_{ij}$. In particular, $e = \sum_i e_i$. It follows that the $\circ$ multiplication on the tangent spaces is semisimple.

c) We have $g(e_i, e_j) = 0$ for $i \neq j$. Furthermore, there exists a function η defined up to addition of a constant such that $g(e_i, e_i) = e_i \eta := \eta_i$. Moreover, we have $eg = \text{const}$, $Eg = (D-1)\eta + \text{const}$. Finally, $E = \sum_i u^i e_i$.

A very important feature of canonical coordinates is that a given tame semisimple germ can be uniquely extended to the Frobenius structure on the universal covering of the total (u^i)–space with deleted partial diagonals. This follows from the Painlevé property of the solutions of Schlesinger's equations: cf. [26], Ch. II, sec. 1–3. We will call this extension *the maximal tame continuation* of the initial germ. The qualification "tame" is essential. It may well happen that a further extension containing non–tame semisimple points or even points with non–semisimple multiplication on the tangent space is possible: e.g. points in A_n where $F'(z)$ has multiple roots have the latter property.

3.2.1. Definition. Special coordinates of a tame semisimple pointed germ of Frobenius manifold consist of the values at the base point of the following functions:

$$(u^i,\, \eta_j, v_{ij} := \frac{1}{2}\,(u^j - u^i)\,\frac{\eta_{ij}}{\eta_j}\,) \tag{3.1}$$

Here $\eta_{ij} := e_i e_j \eta$.

To avoid any misunderstanding, let us stress that the canonical coordinates are functions on a germ, whereas special coordinates are functions on the moduli space of germs.

For a description of the necessary and generically sufficient conditions for a system of mumbers to form special coordinates of a Frobenius germ, see [27], 2.7, pp. 26–27, and 2.6, p. 23, where some inaccuracies of [26] are corrected. The following Theorem summarizes the properties of special coordinates that we will use.

3.2.2. Theorem. *(i) Any tame semisimple pointed germ with labelled spectrum of $E\circ$ is uniquely (up to isomorphism) defined by its special coordinates.*

(ii) Let $(u'^i,\, \eta'_j, v'_{ij})$ for $i \in S$ and $(u''^i,\, \eta''_j, v''_{ij})$ for $j \in T$ be special coordinates of two pointed germs. If the family $u'^i + u''^j$ consists of pairwise distinct elements, then the tensor product of the two germs defined through their completions is again a tame semisimple pointed germ whose

canonical coordinates are naturally labelled by the pairs $I \in S \times T$ and have the following form: for $I = (i, j)$, $K = (k, l)$,

$$u^I = u'^i + u''^j,\ \eta_I = \eta'_i\eta''_j,\ v_{IK} = \delta_{jl}v'_{ik} + \delta_{ik}v''_{jl}. \tag{3.2}$$

(iii) Let another two germs be obtained from the initial ones by analytic continuation and subsequent shifts of base points. Then their tensor product can be obtained from the initial tensor product by analytic continuation and the appropriate shift of the base point. In this sense, the tensor product does not depend on the choice of base points.

The first statement is proved in [26] and [27]. The second and the third ones are due to R. Kaufmann ([18]). Actually, the third statement is proved in [18] in the considerably more general context: Kaufmann uses flat coordinates and does not assume semisimplicity or absence of odd coordinates. The fact that the tensor product of two convergent germs is again convergent is proved in his paper in preparation, without semisimplicity assumption as well.

Kaufmann remarks that in order to prove (3.2) it suffices to control the relevant potentials only to the fourth order in flat coordinates, and the necessary calculation can then be done directly.

We can now deduce from (3.2) the following corollary.

3.2.3. Theorem. *Assume that we have the direct sum diagram of Saito's frameworks as in 2.4 above. Then the Frobenius manifold M is (canonically isomorphic to) the tensor product of the Frobenius manifolds $M_1 \otimes M_2$.*

Proof. In the notation of (2.12) we may assume that m_1, m_2 and m are tame semisimple, because tameness is the open property. Then (2.12) coincides with the first two formulas of (3.2). The third one follows directly from (2.13) and the definition of v_{ij} in (3.1).

We will now prove that the integral part of the spectrum corresponds to a Frobenius submanifold.

3.3. Proposition. *Assume that we have an analytic or formal Frobenius manifold M with an Euler field E, $d_0 = 1$, $D \in \mathbf{Z}$, and flat identity. Let $-\operatorname{ad} E$ be semisimple on flat vector fields with spectrum d_a, (x_a) a flat coordinate system with*

$$E = \sum_{a:\, d_a \neq 0} d_a x^a \partial_a + \sum_{b:\, d_b = 0} r_b \partial_b.$$

and $e = \partial_0$. *Define the submanifold* $HM \subset M$ *by the equations*

$$x_c = 0 \text{ for all } c \text{ such that } d_c \notin \mathbf{Z}.$$

Finally, assume that at least one of the following conditions is satisfied:

(i) $r_b = 0$ *for all* b *with* $|\Delta_b| = 2$.

(ii) M *is of qc–type, and* $\sum_{b:\, d_b=0} r_b \Delta_b$ *takes only integral values on* B.

Then HM *with induced metric,* $\circ$*–multiplication,* E *and* e *is a Frobenius manifold.*

Remark. From the proof it will be clear that one can replace integers in this statement by any arithmetic progression containing 0 to which D and d_0 belong.

Proof of the Proposition 3.3. If d_c is not integral, the functions $Ex_c = d_c x_c$, $ex_c = 0$ vanish on HM. Hence E and e are tangent to HM and can be restricted to it. From the equation $(d_a + d_b - D)g_{ab} = 0$ ([26], p. 32, (2.17)) one sees that if $d_a, D \in \mathbf{Z}$, $d_b \notin \mathbf{Z}$, we have $g_{ab} = 0$. Therefore the restriction of g to HM is non–degenerate (it is obviously flat), and x_a for $d_a \in \mathbf{Z}$ restrict to a flat coordinate system on HM. The $\circ$–product of two vector fields tangent to HM at the points of M does not contain the transverse components. In fact, we have $E\Phi_{ab}{}^c = (d_0 - d_a - d_b + d_c)\Phi_{ab}{}^c$ ([26], p. 32, (2.18)). Hence if $d_a, d_b \in \mathbf{Z}$, $d_c \notin \mathbf{Z}$, then in the case (i) every monomial in the series Φ^c_{ab} must be an eigenvector of E with non–integral eigenvalue, and therefore it must contain some x_e with $d_e \notin \mathbf{Z}$ so that it vanishes on HM. In the case (ii) we apply the same reasoning separately to the generalized (involving exponentials) monomials contributing to the third derivatives of Ψ in (1.9) and to the third derivatives of c.

The same reasoning shows that the induced multiplication of vector fields on $\mathcal{T}_{HM}$ is defined by the third derivatives of the induced potential.

3.4. Special coordinates of A_n. We return to the notation of 2.3.1.

3.4.1. Proposition. *Consider the points of* A_n *where* $a_1, \dots, a_{n-2} = 0, a_{n-1}$, a_n *arbitrary. Choose a primitive root* ζ *of* $\zeta^n = 1$ *and a root* b *of* $b^n = -\dfrac{a_{n-1}}{n+1}$. *At these points we have:*

$$u^i = a_n + \frac{n}{n+1}\zeta^i a_{n-1} b, \tag{3.3}$$

(3.4)
$$\eta_i = \frac{\zeta^i}{n(n+1)\,b^{n-1}},$$

(3.5)
$$v_{jk} = \frac{1}{(n+1)\,(1-\zeta^{k-j})}.$$

Remark. It is suggestive to compare these coordinates with those for the quantum cohomology of $\mathbf{P}^{n-1}$ ([26], p. 71) on the plane spanned by the identity (coordinate x_0) and the dual hyperplane section (coordinate x_1):

(3.3*a*)
$$u^i = x_0 + n\,\zeta^i e^{\frac{x_1}{n}},$$

(3.4*a*)
$$\eta_i = \frac{\zeta^i}{n}\,e^{-x_1\,\frac{n-1}{n}},$$

(3.5*a*)
$$v_{jk} = \frac{1}{1-\zeta^{k-j}}.$$

Proof. At our subspace $F(z) = z^{n+1} + a_{n-1}z + a_n$. Hence
$$F'(z) = (n+1)(z^n + \frac{a_{n-1}}{n+1})$$
has roots $\rho_i = \zeta^i b$. But for A_n–manifolds we have universally $u^i = F(\rho_i)$, $\eta_i = \dfrac{1}{F''(\rho_i)}$. This furnishes (3.3) and (3.4).

The proof of (3.5) is longer. We have to calculate the values of functions

(3.6)
$$\frac{1}{2}\,(u^k - u^j)\,\frac{\eta_{jk}}{\eta_k},$$

restricted to the plane of our base points.

At a generic point of A_n, we can calculate η_{jk} in the following three–step way:

(3.7)
$$\eta_{jk} = \frac{\partial \eta_j}{\partial u^k} = \sum_{l,m=1}^{n} \frac{\partial \eta_j}{\partial \rho_m}\,\frac{\partial \rho_m}{\partial a_l}\,\frac{\partial a_l}{\partial u^k}.$$

Since
$$\eta_j = \frac{1}{F''(z)} = \frac{1}{(n+1)\prod_{i:\,i\neq j}(\rho_i - \rho_j)},$$

we have

$$\frac{\partial \eta_j}{\partial \rho_m} = -\frac{\eta_j}{\rho_m - \rho_j}, \tag{3.8}$$

if $m \neq j$, and

$$\frac{\partial \eta_j}{\partial \rho_j} = \eta_j \sum_{i:\, i\neq j} \frac{1}{\rho_i - \rho_j}. \tag{3.9}$$

Moreover,

$$\frac{\partial \rho_m}{\partial a_l} = -(n-l)\, \rho_m^{n-l-1} \eta_m. \tag{3.10}$$

This can be checked by derivating the identity $F'(\rho_m) = 0$.

Finally, according to [26], p. 47, (4.24), we have

$$\sum_{l=1}^{n} \frac{\partial a_l}{\partial u^k}\, \rho_i^{n-l} = \delta_{ik}. \tag{3.11}$$

We will now restrict (3.8)–(3.11) to our plane.

Using (3.3) and (3.4), we get consecutively:

$$\frac{\partial \eta_j}{\partial \rho_m} = \frac{1}{n\, a_{n-1}} \frac{1}{\zeta^{m-j} - 1} \tag{3.12}$$

if $m \neq j$, and

$$\frac{\partial \eta_j}{\partial \rho_j} = \frac{n-1}{2n\, a_{n-1}}, \tag{3.13}$$

$$\frac{\partial \rho_m}{\partial a_l} = -\frac{n-l}{n(n+1)}\, b^{-l} \zeta^{-ml}. \tag{3.14}$$

Solving (3.11) for partial derivatives, we also find

$$\frac{\partial a_l}{\partial u^k} = \frac{1}{n}\, b^{l-n} \zeta^{kl}. \tag{3.15}$$

It remains to substitute (3.12)–(3.14) into (2.7) to get after some calculation

$$\eta_{jk} = \frac{2\zeta^{k-j}}{(\zeta^{k-j} - 1)^2} \frac{1}{n\, a_{n-1}^2}. \tag{3.16}$$

Finally, substituting (3.3), (3.4) and (3.16) in (3.6), we obtain (3.5).

3.5. Tensor products of A_n's. We want to describe $(n_1, \dots, n_N)$ with non-trivial $H(A_{n_1} \otimes \cdots \otimes A_{n_N})$. We can assume $n_i \geq 2$ because A_1 is identity with respect to the tensor multiplication. The first necessary condition, following from (1.20) is

$$\mathbf{d} := \sum_{i=1}^{N} \frac{n_i - 1}{n_i + 1} \in \mathbf{Z}. \tag{3.17}$$

If it is satisfied, the full d-spectrum of the tensor product consists of certain rational points between 0 and $\mathbf{d}$. Multiplicity of 0 and $\mathbf{d}$ is one. Generally, the multiplicity of some $m \leq \mathbf{d}$ is

$$h^{2m}(H(\otimes_i A_{n_i})) := \text{the number of } (i_1, \dots, i_N) \in \mathbf{Z}^N_{\geq 0} \text{ satisfying}$$

$$\sum_{k=1}^{N} \frac{i_k}{n_k + 1} = m, \ 0 \leq i_k \leq n_k - 1. \tag{3.18}$$

The d-spectrum of $H(\otimes A_{n_k})$ consists of the part of (3.18) for all integer m. Clearly, (0.1) is a particular case of (3.18).

The flat part of E in the total tensor product and in the H-part of it vanishes because it vanishes on all factors (cf. [18], Theorem 6.3).

Let us show that $\otimes_k A_{n_k}$ admits a tame semisimple base point which is the "sum" of the points the special coordinates of which we have calculated. Choose base points on all A_{n_k} as in Prop. 3.4.1. For our purpose, we may even assume that $a_{n_k} = 0$ on each A_{n_k}. Therefore, slightly changing notation of (3.3), we will assume that canonical coordinates of the base point of A_{n_k} are of the form $u^i = \zeta_k^i c_k$, where ζ_k is the primitive root of unit of degree n_k and $0 \leq i \leq n_k - 1$. Then in view of the Theorem 3.2.2 for $I = (i_1, \dots, i_N)$, $J = (j_1, \dots, j_N)$ we have a pair of canonical coordinates u^I, u^J on the full tensor product whose values at the base point of this product are

$$u^I = \sum_{k=1}^{N} \zeta_k^{i_k} c_k, \ u^J = \sum_{k=1}^{N} \zeta_k^{j_k} c_k.$$

One easily sees that with generic choice of c_k these coordinates are distinct for all $I \neq J$.

3.6. Involutive pairs of Gepner's manifolds. In the notations of 2.3..2, consider a pair of Frobenius manifolds $V_{n+1,k}$ and $V_{k+1,n}$. They contain respectively points producing Verlinde's algebras of $su(n+1)_k$ and $su(k+1)_n$ which are isomorphic according to the reasoning of [17] (the level–rank duality). This reasoning runs as follows. Assume for concreteness that $n \geq k+1$. Consider the function

$$\widetilde{f}_{n+1,k}(z_1,\dots,z_n) := f_{k+1,n}(z_1,\dots,z_k) + \sum_{i=k+1}^{n} z_i z_{n+k-i+1}. \tag{3.19}$$

Then $f_{n+1,k}$ and $\widetilde{f}_{n+1,k}$ are quasi–homogeneous polynomials of the same degree, depending on the same set of weighted variables, and having an isolated critical point at the origin. Hence they belong to a connected family of polynomials with the same property, and whatever structures can be derived from their lattices of vanishing cycles, they can be identified. On the other hand, $\widetilde{f}_{n+1,k}$ is obtained from $f_{k+1,n}(z_1,\dots,z_k)$ by adding a sum of squares which again does not change the structure of vanishing cycles, except for that of the intersection form which changes in a controlled way.

In fact, adding a sum of squares does not change the respective Frobenius manifolds: this agrees with the fact that A_1 is the tensor identity. Hence $V_{n+1,k}$, or at least its germ at the origin, is deformable to (the germ of) $V_{k+1,n}$.

Perhaps, these Frobenius manifolds, or at least their appropriate coverings, are themselves isomorphic. To check this, it would suffice to identify their special coordinates at an appropriate pair of tame semisimple points. The Gepner–Verlinde points $m \in V_{n,k}$ (cf. 2.3..2 above) are certainly not tame because the potential has only n, not $n+k$ different critical values at m.

4. Maurer–Cartan and master equations

4.1. Maurer–Cartan equations. Fix a supercommutative $\mathbf{Q}$–algebra k. All our structures are $\mathbf{Z}_2$–graded, notation like $\widetilde{x}$ means the parity of a homogeneous element x. Let $g = g_0 \oplus g_1$ be a Lie superalgebra over k, supplied with an odd differential d satisfying $d[a,b] = [da,b] + (-1)^{\widetilde{a}}[a,db]$.

Put $Z = Z(g,d) := \operatorname{Ker} d, B = B(g,d) := \operatorname{Im} d, H := H(g,d) = Z/B$. Clearly, Z is a Lie subalgebra, and B its ideal, so that H with induced bracket product is a Lie superalgebra.

The differential d can be shifted.

For $\gamma \in g_1$, put $d_\gamma(a) := da + [\gamma, a]$. Clearly,

$$d_\gamma[a,b] = [d_\gamma a, b] + (-1)^{\tilde{a}}[a, d_\gamma b].$$

4.1.1. Claim. a) We have $d_\gamma^2 = 0$ if

$$d\gamma + \frac{1}{2}[\gamma,\gamma] = 0. \tag{4.1}$$

b) Let $\gamma' = \gamma + \epsilon\beta$, $\beta \in g$, ϵ an even or odd constant with $\epsilon^2 = 0$ such that $\epsilon\beta$ is odd. Assume that γ satisfies (4.1). Then γ' satisfies (4.1) as well iff

$$d_\gamma(\beta) = d\beta + [\gamma,\beta] = 0. \tag{4.2}$$

This is straightforward.

If K is another supercommutative k–algebra, we define $g_K = K \otimes_k g, d_K = 1 \otimes d$. We will always work with K flat over k so that $Z_K :=$ Ker $d_K = K \otimes_k Z$, and similarly for B and H. Claim 4.4.1 is of course applicable to (g_K, d_K) as well.

We want to produce from this setting a non–linear version of the homology $H(g,d)$ or rather of the diagram $g \supset Z \to H$.

The most straightforward is the case when, say, g is free of finite rank over k. We then replace g by the linear superspace $\mathcal{G} :=$ Spec $k[\Pi g^t]$, where Π is the parity inversion functor, Z by the closed subspace $\mathcal{Z} \subset \mathcal{G}$ defined by the equations (4.1). In order to understand what should be the non–linear version of B, we interpret the Claim 4.1.1 (ii) as saying that d_γ–cycles form the Zariski tangent space to the point γ of the Maurer–Cartan space (4.1). It then contains the subspace of d_γ–boundaries, and we can construct the distribution $\mathcal{B}$ generated by the boundaries. If the quotient space $\mathcal{H} = \mathcal{Z}/\mathcal{B}$ in some sense exists, it can be regarded as the non–linear cohomology of (g,d).

In more down–to–earth terms, choose a (homogeneous) basis $\{\gamma_i\}$ of g and a family of independent (super)commuting variables t^i such that $\widetilde{t^i} = \widetilde{\gamma}_i + 1$. Then $\Gamma := \sum_i t^i\gamma_i$ is a generic odd element of g (or rather of $k[t^i] \otimes g$), and the equation $d\Gamma + \frac{1}{2}[\Gamma,\Gamma] = 0$ is equivalent to the system of equations

$$\forall k: \quad \sum_i (-1)^i t^i D_i^k + \frac{1}{2}\sum_{i,j} t^i t^j (-1)^{i(j+1)} L_{ij}{}^k = 0. \tag{4.3}$$

Here we define the structure constants by $d\gamma_i = \sum_k D_i^k \gamma_k$ and $[\gamma_i, \gamma_j] = \sum_k L_{ij}{}^k \gamma_k$ and use the following shorthand for the signs: $(-1)^{i(j+1)}$ means $(-1)^{\widetilde{\gamma}_i(\widetilde{\gamma}_j+1)}$ etc.

These equations define the coordinate ring R of the affine scheme which we called $\mathcal{Z}$. Obviously, $\mathcal{Z}$ represents the following functor on the category of supecommutative k–algebras K:

$$K \mapsto \{solutions\ to\ (4.1)\ in\ (K \otimes g)_1\}. \tag{4.4}$$

Similarly, if we have any odd d_Γ–cycle $\epsilon\beta = \sum_a \epsilon s^a \gamma_a$ with coefficients in $K \otimes R$, the statement 4.1.1 b) means that the map $X_\beta : t^a \to s^a$ descends to the derivation of $K \otimes R$ over K that is, to a vector field on $\mathcal{G}_K$ of parity $\widetilde{\epsilon}$. Of course, the adequate functorial language for derivations is that of the first order infinitesimal deformations of points, because generally the vector fields implied by 4.1.1 b) are defined only in the infinitesimal neighborhood of γ.

We will stop now discussing the case of finite rank g because in most interesting examples this does not hold, and only $H(g,d)$ is of finite rank.

So we step back and try to produce a formal section of $\mathcal{Z}$ passing through $\gamma = 0$ and transversal to the distribution $\mathcal{B}$. We want it to be of the same size as H, or rather ΠH, and we will assume henceforth that H is free of finite rank. From now on in this section, we denote $K := k[[\Pi H^t]] = k[[x_i]]$ where x_i are coordinate functions on ΠH dual to a basis of ΠH. Any element $\Gamma \in g_K$ can be uniquely written as $\sum_{n\geq 0} \Gamma_n$ where Γ_n is homogeneous of degree n in x_i. Such an element can be naturally called a formal section of $\mathcal{Z}$, or *a generic (formal) solution to (4.1)*, if it has the following properties:

a) $\Gamma \in (g_K)_1, \Gamma_0 = 0, \Gamma_1 = \sum_i x_i c_i$ where $dc_i = 0$ and classes of c_i form a basis of H odd dual to $\{x_i\}$.

b) $d_K \Gamma + \dfrac{1}{2}[\Gamma, \Gamma] = 0$.

The necessary condition for the existence of Γ is the identical vanishing of the Lie bracket induced on $H(g,d)$. In fact, the equation $d\Gamma + \dfrac{1}{2}[\Gamma,\Gamma] = 0$ implies (assuming a) above) $d\Gamma_2 + \dfrac{1}{2}[\Gamma_1, \Gamma_1] = 0$. Hence $[c_i, c_j] \in B$. However, generally it is not sufficient. In fact, the next equation reads $d\Gamma_3 + [\Gamma_1, \Gamma_2] = 0$, but since Γ_2 may be non–closed, we cannot conclude that $[\Gamma_1, \Gamma_2]$ is a boundary. The manageable sufficient condition is stronger: (g,d) must be quasi–isomorphic to the differential Lie algebra $H(g,d)$ with zero bracket and zero differential. For a considerably more general treatment see [15]. Our direct and elementary approach is self–contained and produces slightly more detailed information in the cases essential for the theory of Frobenius manifolds.

4.2. Theorem. *(i) Assume that there exists a surjective morphism of differential Lie superalgebras* $\phi : (g, [,], d) \to (H, 0, 0)$ *inducing isomorphism on the homology. Then there exists a generic formal solution* Γ *to (4.1).*

Moreover, Γ *can be chosen in such a way that for any* $n \geq 2$, $\Gamma_n \in K \otimes \operatorname{Ker} \phi$. *In other words,* $(\mathrm{id} \otimes \phi)(\Gamma) = \sum_i x_i[c_i]$. *Such a solution will be called normalized.*

(ii) If (i) is satisfied, then for any generic solution Γ, *non necessarily normalized, the map* $\phi_K = \mathrm{id} \otimes \phi : g_K \to H_K$ *is the surjective morphism of differential Lie superalgebras* $(g_K, [,]_K, d_{K,\Gamma}) \to (H_K, 0, 0)$ *inducing isomorphism on the homology.*

Proof. (i) Let $n \geq 1$. Assuming that Γ_i for $i \leq n$ are already constructed, and writing d instead of d_K we must find Γ_{n+1} from the equation

$$d\Gamma_{n+1} = -\frac{1}{2} \sum_{i,j:\, i+j=n+1} [\Gamma_i, \Gamma_j]. \tag{4.5}$$

First of all we check that the right hand side of (4.5) is closed in g_K. In fact, since the components $\Gamma_1, \dots, \Gamma_n$ satisfy the similar equations by the inductive assumption, the differential of the rhs equals

$$\frac{1}{2} \sum_{i+j+k=n+1} [[\Gamma_i, \Gamma_j], \Gamma_k].$$

This expression vanishes because the Jacobi identity for odd elements reads

$$[[\Gamma_i, \Gamma_j], \Gamma_k] + [[\Gamma_k, \Gamma_i], \Gamma_j] + [[\Gamma_j, \Gamma_k], \Gamma_i] = 0.$$

Hence the coefficients of the rhs of (4.5) (as polynomials in x_i) belong to $Z \cap [g, g]$. But $[g, g] \in \operatorname{Ker} \phi$ and $Z \cap \operatorname{Ker} \phi = B$ because ϕ is a quasi–isomorphism. Thus we can solve (4.5).

We can add to any solution elements of Z_K of degree $n + 1$. But $Z + \operatorname{Ker} \phi = g$ because ϕ induces surjection on homology. Hence we can normalize Γ_{n+1} by the requirement $\Gamma_{n+1} \in K \otimes \operatorname{Ker} \phi$.

(ii) Now fix Γ satisfying (4.5) for all n. We will write d_Γ instead of $d_{K,\Gamma}$ and put $Z_\Gamma := \operatorname{Ker} d_\Gamma \subset g_K$, $B_\Gamma = d_\Gamma(g_K)$. We have $B \subset \operatorname{Ker} \phi_K$ and $[g_K, g_K] \subset \operatorname{Ker} \phi_K$, hence $B_\Gamma \subset \operatorname{Ker} \phi_K$. Therefore, ϕ_K is compatible with zero bracket and zero differential on H_K. The natural inclusion

$Z_\Gamma + \mathrm{Ker}\ \phi_K \to g_K$ becomes surjection after the reduction modulo the ideal (x_i) of K, because ϕ is surjective. Hence this inclusion is surjective, and ϕ_K is surjective as well. It remains to show that ϕ_K induces injection on homology, that is,

$$Z_\Gamma \cap \mathrm{Ker}\ \phi_K \subset B_\Gamma. \tag{4.6}$$

Let $c = \sum_{n\geq 0} c_n \in Z_\Gamma$. This means that $dc_0 = 0$ and in general

$$dc_n = - \sum_{i+j=n} [\Gamma_i, c_j]. \tag{4.7}$$

(we keep writing d for d_K). Assuming that $\phi_K(c) = 0$ we want to deduce the existence of homogeneous elements a_n of degree n in g_K such that

$$c_{n+1} = da_{n+1} + \sum_{i+j=n+1} [\Gamma_i, a_j]. \tag{4.8}$$

We have $dc_0 = 0$ and $\phi(c_0) = 0$, hence c_0 is a boundary because ϕ is the quasi–isomorphism. Assuming that $a_0, \ldots, a_n$ are found, we will establish the existence of a_{n+1} satisfying (4.8), if we manage to prove that $c_{n+1} - \sum_{i+j=n+1}[\Gamma_i, a_j]$ is d–closed. In fact, this element also belongs to Ker ϕ_K and so must be a boundary.

The differential of this element is

$$dc_{n+1} + \sum_{i+j=n+1} [\Gamma_i, da_j] - \sum_{i+j=n+1} [d\Gamma_i, a_j]. \tag{4.9}$$

Replace in the first sum da_j by $c_j - \sum_{k+l=j}[\Gamma_k, a_l]$ for $j \leq n$ (this holds by induction). Replace in the second sum $d\Gamma_i$ by the sum of commutators from (4.5). The terms containing c_j will cancel thanks to (4.7). The remaining terms can be written as

$$- \sum_{i+j+k=n+1} [\Gamma_i, [\Gamma_j, a_k]] + \frac{1}{2} \sum_{i+j+k=n+1} [[\Gamma_i, \Gamma_j], a_k].$$

This expression vanishes because of Jacobi identity.

4.2.1. Corollary. *Define the map $\psi = \psi_\Gamma : H_K \to g_K$ as the K–linear extension of*

$$H \to g_K : \ X \mapsto \overline{X}\Gamma$$

where $\overline{X}$ acts on $K \otimes g$ as the right g–linear extension of the derivation on K acting as $(\Pi X, *)$ on ΠH^t.

Then ψ is a section of ϕ_K if Γ is normalized.

Proof. First of all, we have $\overline{X}(d\Gamma + \frac{1}{2}[\Gamma,\Gamma]) = 0$ from which it follows that $d_\Gamma(\overline{X}\Gamma) = 0$, that is, $\overline{X}\Gamma$ is a d_Γ–cycle. Its image in H_K is $(\overline{X}\Gamma_1 + \sum_{n\geq 2}\overline{X}\Gamma_n)$ mod B_Γ. The first term is clearly X. The remaining ones are in $\overline{K} \otimes \mathrm{Ker}\ \phi_K$, if Γ is normalized.

4.3. Odd Lie (super)algebras. As in 4.1, let now $g = g_0 \oplus g_1$ be a k–module endowed with a bilinear operation *odd bracket* $(a,b) \mapsto [a \bullet b]$ which satisfies the following conditions:

a) *parity of $[a \bullet b]$ equals $\widetilde{a} + \widetilde{b} + 1$,*

b) *odd anticommutativity:*

$$[a \bullet b] = -(-1)^{(\widetilde{a}+1)(\widetilde{b}+1)}[b \bullet a], \tag{4.10}$$

c) *odd Jacobi identity:*

$$[a \bullet [b \bullet c]] = [[a \bullet b] \bullet c] + (-1)^{(\widetilde{a}+1)(\widetilde{b}+1)}[b \bullet [a \bullet c]]. \tag{4.11}$$

Such a structure will be called an odd Lie (super)algebra. We consider such algebras endowed with an odd differential satisfying

$$d[a \bullet b] = [da \bullet b] + (-1)^{\widetilde{a}+1}[a \bullet db]. \tag{4.12}$$

Physicists sometimes denote such multiplication $\{,\}$ (see e.g. [25]). Our choice of notation allows one to use consistently the standard sign mnemonics of superalgebra, if $\bullet$ counts as an element of parity one.

If (g,d) is the usual differential Lie superalgebra, the parity change functor $g \mapsto \Pi g$ turns the usual bracket product $[\,,]$ into the odd bracket product, and defines an equivalence of the two categories (the differential changes sign). It seems therefore that there is not much point in considering odd brackets. However, in the context of GBV–algebras they come together with usual supercommutative multiplication, and parity change then turns this multiplication into odd one (see the next section). This is, of course, a particular case of the general operadic formalism over the category of superspaces, where any operation can be inherently even or odd.

In the next section we choose to work with even multiplication and odd bracket product. But we will use the results of this section, with appropriately modified parities and signs, for odd Lie superalgebras. In particular, the odd Maurer–Cartan equation in the physical literature is called *the master equation*:

$$d\Gamma + \frac{1}{2}[\Gamma \bullet \Gamma] = 0. \tag{4.13}$$

The Theorem 4.2 provides conditions of its solvability in $k[[H^t]]\otimes g$ rather than $k[[\Pi H^t]] \otimes g$. Notice also that Γ in (4.13) must be even.

5. Gerstenhaber–Batalin–Vilkovyski algebras

5.1. Gerstenhaber–Batalin–Vilkovyski algebras. Let $\mathcal{A}$ be a supercommutative algebra with identity over another supercommutative algebra k (constants). Consider an odd k–linear operator $\Delta : \mathcal{A} \to \mathcal{A}, \Delta(1) = 0$, with the following property:

$$\forall a \in \mathcal{A}, \qquad \partial_a := (-1)^{\widetilde{a}}([\Delta, l_a] - l_{\Delta a})$$

$$\text{is the derivation of parity } \widetilde{a} + 1 \text{ over } k. \tag{5.1}$$

Here l_a denotes the operator of left multiplication by a, and brackets denote the supercommutator. Explicitly,

$$\partial_a b = (-1)^{\widetilde{a}}\Delta(ab) - (-1)^{\widetilde{a}}(\Delta a)\, b - a\, \Delta b.$$

The sign ensures the identity $\partial_{ca} = c\partial_a$ for any constant c. By definition of derivation,

$$[\partial_a, l_b] = l_{\partial_a b}. \tag{5.2}$$

The pair $(\mathcal{A}, \Delta)$ is called a *GBV–algebra* if, in addition, $\Delta^2 = 0$. There is an obvious operation of scalar extension.

5.1.1. Lemma. *In any GBV–algebra we have*

$$[\Delta, \partial_a] = \partial_{\Delta a}, \tag{5.3}$$

$$[\partial_a, \partial_b] = \partial_{\partial_a b} \tag{5.4}$$

Proof. From (5.1) we have

$$[\Delta, \partial_a] = (-1)^{\widetilde{a}}([\Delta, [\Delta, l_a]] - [\Delta, l_{\Delta a}]).$$

From the Jacobi identity for operators and $[\Delta, \Delta] = 0$ we find $[\Delta, [\Delta, l_a]] = 0$ because

$$[\Delta, [\Delta, l_a]] = [[\Delta, \Delta], l_a] - [\Delta, [\Delta, l_a]])$$

From (5.1) with Δa replacing a we have $[\Delta, l_{\Delta a}] = (-1)^{\tilde{a}+1}\partial_{\Delta a}$. Hence

$$[\Delta, \partial_a] = (-1)^{\tilde{a}+1}[\Delta, l_{\Delta a}] = \partial_{\Delta a}.$$

To prove (5.4), we notice that since $[\partial_a, \partial_b]$ must be a derivation, in the intermediate calculations we are allowed not to register all the summands which are left multiplications: they will cancel anyway. So we have, denoting such summands by dots and using consecutively (5.1), (5.2), Jacobi and (5.3), and again (5.1) with $\partial_a b$ replacing a:

$$\begin{aligned}[\partial_a, \partial_b] &= (-1)^{\tilde{b}}[\partial_a, [\Delta, l_b] - l_{\Delta b}] = (-1)^{\tilde{b}}[\partial_a, [\Delta, l_b]] + \dots \\ &= (-1)^{\tilde{b}+\tilde{a}+1}[\Delta, [\partial_a, l_b]] + \dots = (-1)^{\tilde{b}+\tilde{a}+1}[\Delta, l_{\partial_a b}] + \dots = \partial_{\partial_a b}.\end{aligned}$$

Define now the odd bracket operation on $\mathcal{A}$ by the formula

$$[a \bullet b] := \partial_a b. \tag{5.5}$$

5.1.2. Proposition. *The pair of bilinear operations (multiplication and odd bracket) defines on $\mathcal{A}$ the structure of the odd Poisson algebra in the following sense:*

(i) The odd bracket satisfies the odd anticommutativity, the odd Jacobi and the odd Poisson identities:

$$\begin{aligned}[a \bullet b] &= -(-1)^{(\tilde{a}+1)(\tilde{b}+1)}[b \bullet a], \\ [a \bullet [b \bullet c]] &= [[a \bullet b] \bullet c] + (-1)^{(\tilde{a}+1)(\tilde{b}+1)}[b \bullet [a \bullet c]], \\ [a \bullet bc] &= [a \bullet b]c + (-1)^{\tilde{b}(\tilde{a}+1)}b\,[a \bullet c].\end{aligned} \tag{5.6}$$

(ii) Δ is the derivation with respect to the odd brackets so that $(\mathcal{A}, \bullet, \Delta)$ is the differential odd Lie algebra.

Proof. The anticommutativity can be checked directly. The Jacobi identity follows from (5.4) written as $[\partial_a, \partial_b] = \partial_{[a\bullet b]}$. The Poisson identity means that ∂_a is a derivation. The last statement follows from (5.1).

Notice that with respect to the usual multiplication Δ is the differential operator of order ≤ 2 and not necessarily derivation.

5.2. Additional differential. Assume now that we have an additional k–linear odd map $\delta : \mathcal{A} \to \mathcal{A}$ which is the derivation with respect to the multiplicative structure of $\mathcal{A}$ satisfying

$$\delta^2 = [\delta, \Delta] = \delta\Delta + \Delta\delta = 0. \tag{5.7}$$

We will say that $(\mathcal{A}, \Delta, \delta)$ is *a differential GBV–algebra (dGBV).*

5.2.1. Lemma. *We have*

$$[\,\delta, \partial_a] = \partial_{\delta a}. \tag{5.8}$$

Therefore δ is the derivation with respect to the odd bracket as well.

Proof. Since $[\,\delta, \partial_a]$ is a derivation of $\mathcal{A}$, we can calculate omitting the multiplication operators as above:

$$\begin{aligned}[\,\delta, \partial_a] &= (-1)^{\tilde{a}}[\delta, [\Delta, l_a]] + \dots \\ &= (-1)^{\tilde{a}}([[\,\delta, \Delta], l_a] - [\Delta, [\,\delta, l_a]]) + \dots \\ &= \,-\,(-1)^{\tilde{a}}[\Delta, l_{\delta a}] + \dots = \partial_{\delta a}.\end{aligned}$$

Furthermore,

$$\begin{aligned}\delta\,[a \bullet b] &= \delta\,\partial_a b = [\delta, \partial_a]b + (-1)^{\tilde{a}+1}\partial_a \delta b \\ &= \partial_{\delta a} b + (-1)^{\tilde{a}+1}\partial_a \delta b \\ &= [\,\delta a \bullet b] + (-1)^{\tilde{a}+1}[a \bullet \delta b].\end{aligned} \tag{5.9}$$

5.2.2. Shifted differential. Let δ be a differential satisfying (5.7). For an even $a \in \mathcal{A}$ put

$$\delta_a := \delta + \partial_a, \;\; \delta_a(b) = \delta b + [a \bullet b]. \tag{5.10}$$

Then we have $\delta_a^2 = 0$ if the odd Maurer–Cartan equation is satisfied:

$$\delta a + \frac{1}{2}[a \bullet a] = 0. \tag{5.11}$$

Furthermore,

$$[\delta_a, \Delta] = 0 \quad \text{if} \quad \Delta a = 0. \tag{5.12}$$

Therefore, from (5.11), (5.12) it follows that $(\mathcal{A}, \Delta, \delta_a)$ is a differential GBV–algebra (dGBV). In particular,

$$[\delta_a, \partial_b] = \partial_{\delta_a b}. \tag{5.13}$$

We can in the same way shift Δ. The essential difference is that, as Δ itself, the shifted differential generally will not be the derivation with respect to the associative multiplication.

5.3. Homology of $(\mathcal{A}, \delta)$. Since δ is the derivation with respect to both multiplications in $\mathcal{A}$ (associative one and the bracket), Ker δ is the subalgebra with respect to both of them, and Im δ is the ideal in this subalgebra with respect to both structures. Therefore the homology group $H(\mathcal{A}, \delta)$ inherits both multiplications, satisfying the identities (5.6) and (5.7).

This reasoning holds for $H(\mathcal{A}, \delta_a)$ as well, if a satisfies the Maurer–Cartan equation (5.11).

5.4. Homology of $(\mathcal{A}, \Delta)$. The same reasoning furnishes only the structure of odd Lie algebra on $H(\mathcal{A}, \Delta)$, because Δ is not a derivation with respect to the associative multiplication. However, if δ and Δ satisfy conditions (A) and (B) below, we will have the natural isomorphism $H(\mathcal{A}, \Delta) = H(\mathcal{A}, \delta)$.

The Lemma below is well known, see e.g. [15].

5.4.1. Lemma. *Let $\mathcal{A}$ be an additive group supplied with two endomorphisms δ and Δ satisfying $\delta^2 = \Delta^2 = 0$ and $\delta\Delta = \alpha\Delta\delta$ where α is an automorphism of $\mathcal{A}$ such that $\alpha(\mathrm{Im}\ \Delta\delta) = Im\,\Delta\delta$. Then clearly, $\mathrm{Im}\ \delta\Delta = \mathrm{Im}\ \Delta\delta \subset \mathrm{Im}\ \delta \cap \mathrm{Ker}\ \Delta$ and similarly with δ and Δ permuted.*

The following statements are equivalent:

(i) The inclusions of the differential subgroups $i : (\mathrm{Ker}\ \Delta, \delta) \subset (\mathcal{A}, \delta)$ and $j : (\mathrm{Ker}\ \delta, \Delta) \subset (\mathcal{A}, \Delta)$ are quasi-isomorphisms (that is, induce isomorphisms of homology).

(ii) We have actually equalities:

$$(A) \qquad \mathrm{Im}\ \delta\Delta = \mathrm{Im}\ \Delta\delta = \mathrm{Im}\ \delta \cap \mathrm{Ker}\ \Delta,$$

$$(B) \qquad \mathrm{Im}\ \delta\Delta = \mathrm{Im}\ \Delta\delta = \mathrm{Im}\ \Delta \cap \mathrm{Ker}\ \delta.$$

Assume that these conditions are satisfied. Then the both homology groups in (i) are naturally isomorphic to

$$(\mathrm{Ker}\ \Delta \cap \mathrm{Ker}\ \delta)/\,\mathrm{Im}\ \delta\Delta.$$

Moreover, the natural map $\mathrm{Ker}\ \Delta \to H(\mathcal{A}, \Delta)$ induces the surjection of the differential groups $(\mathrm{Ker}\ \Delta, \delta) \to (H(\mathcal{A}, \Delta), 0)$ which is a quasi-isomorphism, and similarly with δ and Δ interchanged. Hence the both differential groups $(\mathcal{A}, \Delta)$ and $(\mathcal{A}, \delta)$ are formal.

Proof. We have:

$$H(i) \text{ is injective } \iff \text{Ker } \Delta \cap \text{Im } \delta = \delta\,(\text{Ker } \Delta). \tag{5.14}$$

$$\begin{aligned} H(i) \text{ is surjective } &\iff \text{Ker } \delta \subset \text{Ker } \Delta + \text{Im } \delta \\ &\Rightarrow \Delta\,(\text{Ker } \delta) = \text{Im } \delta\Delta. \end{aligned} \tag{5.15}$$

Here and below all kernel and images are taken in $\mathcal{A}$. In the right hand side of (5.14), the inclusion $\supset$ is evident, and the injectivity of $H(i)$ supplies the reverse inclusion. The last arrow in (5.15) is obtained by applying Δ to the previous inclusion: this gives $\Delta\,(\text{Ker } \delta) \subset \Delta\,(\text{Im } \delta) = \text{Im } \delta\Delta$ whereas the reverse inclusion is obvious.

Interchanging δ and Δ we find

$$H(j) \text{ is injective } \iff \text{Ker } \delta \cap \text{Im } \Delta = \Delta\,(\text{Ker } \delta). \tag{5.16}$$

$$\begin{aligned} H(j) \text{ is surjective } &\iff \text{Ker } \Delta \subset \text{Ker } \delta + \text{Im } \Delta \\ &\Rightarrow \delta\,(\text{Ker } \Delta) = \text{Im } \delta\Delta. \end{aligned} \tag{5.17}$$

Taken together, (5.14) and (5.17) prove (A), and (5.15) and (5.16) prove (B), so that we have established the implication (i) $\Rightarrow$ (ii).

Conversely, assume that (A) and (B) hold.

Then $H(i)$ induces surjection on the homology, because if $\delta a = 0$, we have $\Delta a \in \text{Ker } \delta \cap \text{Im } \Delta$ so that by (B), $\Delta a = \Delta\delta b$, and then $a - \delta b \in \text{Ker } \Delta$ represents the same homology class as a.

Moreover, $H(i)$ induces injection on the homology, because if $a \in \text{Ker } \Delta$, $a = \delta b$ for some $b \in \mathcal{A}$, then $a \in \text{Ker } \Delta \cap \text{Im } \delta$ so that by (A), $a = \delta c$ for some $c \in \text{Im } \Delta \subset \text{Ker } \Delta$.

By symmetry, the same holds for $H(j)$.

The cycle subgroup for both differential groups $(\text{Ker } \delta, \Delta)$ and $(\text{Ker } \Delta, \delta)$ is $(\text{Ker } \delta \cap \text{Ker } \Delta)$, and if (i) and (ii) hold, the boundaries can be identified with $\text{Im } \delta\Delta$, cf. (5.15) and (5.17). It remains to deduce formality, say, from (A) and (B).

The natural map $\text{Ker } \Delta \to H(\mathcal{A}, \Delta)$ is compatible with differentials, because if $a \in \text{Ker } \Delta$, then $\delta a \in \text{Im } \delta \cap \text{Ker } \Delta$ so that by (A), $\delta a = \Delta\delta b$ for some b, and hence the map is compatible with the zero differential on $H(\mathcal{A}, \Delta)$.

This map is surjective on the homology. In fact, consider the class of a, $\Delta a = 0$ in $H(\mathcal{A}, \Delta)$. Then $\delta a \in \text{Im } \delta \cap \text{Ker } \Delta$ so that in view of (A), $\delta a = \delta\Delta b$, and the δ–cycle $a - \Delta b$ represents the same class as a.

Finally, the map is injective on the homology. In fact, if $a \in \text{Im}\,\Delta$ and $\delta a = 0$, then in view of (B), $a \in \text{Im}\,\delta\Delta \subset \delta\,(\text{Ker}\,\Delta)$.

Thus we established the two–step quasi–isomorphism of $(\mathcal{A}, \delta)$ with $(H(\mathcal{A}, \Delta), 0)$ and by symmetry of $(\mathcal{A}, \Delta)$ with $(H(\mathcal{A}, \delta), 0)$. But the first two groups are also naturally quasi–isomorphic. So they are formal.

5.4.2. Remarks. In the context of dGBV–algebras, we will apply this identification to $(\mathcal{A}, \Delta, \delta_a)$ with variable or formal generic a. Then we will be able to interpret the "constant" space $H = H(\mathcal{A}, \Delta)$ as the flat structure on the family of algebras $H(\mathcal{A}, \delta_a)$ parametrized by the points of the generic formal section of the Maurer–Cartan manifold. The important technical problem will be then deriving the conditions (A) and (B) for the variable a.

Notice that taken together, (A) and (B) are equivalent to

$$(C) \qquad \text{Im}\,\delta\Delta = \text{Im}\,\Delta\delta = (\text{Ker}\,\delta \cap \text{Ker}\,\Delta) \cap (\text{Im}\,\delta + \text{Im}\,\Delta).$$

To deduce, say, (A) from (C), one omits the last term in (C) and gets $\text{Im}\,\delta\Delta \supset \text{Im}\,\delta \cap \text{Ker}\,\Delta$ whereas the inverse inclusion is obvious. Similarly, (C) follows from (A) and (B) together.

Assume that $\mathcal{A}$ is finite dimensional over a field and δ varies in a family, say $\{\delta_a\}$. After a generalization, dimension of $\text{Im}\,\delta\Delta$ can only jump, and that of $\text{Ker}\,\delta$ only drop. Hence if (B) holds at a point, it holds in an open neighborhood of it. In the case of the Dolbeault complex (cf [2]), only the cohomology will be finite–dimensional. The validity of (C) for a particular $\delta = \delta_0$ follows from the Kähler formalism. The argument of the previous section (Theorem 4.2 (ii)) furnishes the same result for the generic formal deformation.

5.5. Integral. Let $(\mathcal{A}, \Delta, \delta)$ be a dGBV–algebra. An even k–linear functional $\int : \mathcal{A} \to k$ is called *an integral* if the following two conditions are satisfied:

$$(5.18) \qquad \forall a, b \in \mathcal{A}, \quad \int (\delta a) b = (-1)^{\tilde{a}+1} \int a \delta b,$$

$$(5.19) \qquad \forall a, b \in \mathcal{A}, \quad \int (\Delta a) b = (-1)^{\tilde{a}} \int a \Delta b.$$

Notice that (5.18) is equivalent to $\forall a \in \mathcal{A}, \int \delta a = 0$ because δ is a k–derivation. Applying (5.19) to $b = 1$, we see that $\forall a \in \mathcal{A}, \int \Delta a = 0$ as well.

5.5.1. Proposition. *Let $\int$ be an integral for $(\mathcal{A}, \Delta, \delta)$.*

(i) If a or b belongs to Ker Δ, *we have*

$$\int \partial_a b = \int [a \bullet b] = 0. \tag{5.20}$$

Hence if a satisfies (5.11) and (5.12), $\int$ is an integral for $(\mathcal{A}, \Delta, \delta_a)$ as well.

(ii) $\int$ induces a linear functional on $H(\mathcal{A}, \Delta)$ and $H(\mathcal{A}, \delta_a)$ for all a as above. These functionals are compatible with the identifications following from the condition (C).

Proof. If, say, $\Delta a = 0$, we have

$$\begin{aligned}\int \partial_a b &= \int ((-1)^{\tilde{a}}\Delta(ab) - (-1)^{\tilde{a}}(\Delta a)\, b - a\,\Delta b) \\ &= -\int a\,\Delta b = -(-1)^{\tilde{a}} \int \Delta a\, b = 0.\end{aligned}$$

The rest is straightforward.

5.6. Metric. If $\int$ is an integral on $(\mathcal{A}, \Delta, \delta)$, we can define the scalar products on $H(\mathcal{A}, \delta_a)$ induced by the symmetric scalar product $(a, b) \mapsto \int ab$ on $\mathcal{A}$. For the construction of Frobenius manifolds, it is necessary to ensure that these scalar products are non–degenerate.

Integral and metric are compatible with base extensions.

5.7. Additional grading. Assume now that $\mathcal{A}$ as commutative k–superalgebra is graded by an additive subgroup of k. Thus $\mathcal{A} = \oplus_n \mathcal{A}^n$, $k \in \mathcal{A}^0$, $\mathcal{A}^m\mathcal{A}^n \subset \mathcal{A}^{m+n}$, and each $\mathcal{A}^i$ is graded by parity. We write $|a| = i$ if $a \in \mathcal{A}^i$. Various induced gradings and degrees of homogeneous operations are denoted in the same way. (In the main example of [2], $\mathcal{A}$ is $\mathbf{Z}$–graded, and each $\mathcal{A}^i$ is either even, or odd, but this plays no role in general).

All base extensions then must be furnished by the similar grading or its topological completion.

We will assume also that $|\Delta| = -1$. It follows that $|[a\bullet b]| = |a|+|b|-1$ which we interpret as $|\bullet| = -1$. Moreover, we postulate that $|\delta| = 1$. This means that the shifted differential δ_γ can be homogeneous only for $|\gamma| = 2$, and similarly for extended base.

Homology space H in all its incarnations (cf. Lemma 5.4.1) inherits the grading from $\mathcal{A}$. The dual space H^t is graded in such a way that the pairing $H^t \otimes H \to k$ has degree zero. This induces the additional grading (or more precisely, the notion of homogeneity) on $K = k[[H^t]]$ (which might be the product rather than the sum of its homogeneous components).

Integral is supposed to have a definite degree, not necessarily zero (and usually non–zero).

5.8. Tensor product of GBV–algebras. Let $(\mathcal{A}_i, \Delta_i)$, $i = 1, 2$, be two GBV–algebras over k. Put

$$\mathcal{A} := \mathcal{A}_1 \otimes \mathcal{A}_2, \Delta := \Delta_1 \otimes 1 + 1 \otimes \Delta_2 : \mathcal{A} \to \mathcal{A}.$$

5.8.1. Proposition. *$(\mathcal{A}, \Delta)$ is a GBV–algebra. We have for $a_i, b_i \in \mathcal{A}_i$*

$$\partial_{a_1 \otimes a_2} = \partial_{a_1} \otimes (-1)^{\tilde{a}_2} l_{a_2} + l_{a_1} \otimes \partial_{a_2}, \tag{5.21}$$

or equivalently

$$\begin{aligned}[a_1 \otimes a_2 \bullet b_1 \otimes b_2] = (-1)^{\tilde{a}_2(\tilde{b}_1+1)}[a_1 \bullet b_1] \otimes a_2 b_2 \\ + (-1)^{\tilde{b}_1(\tilde{a}_2+1)} a_1 b_1 \otimes [a_2 \bullet b_2].\end{aligned} \tag{5.22}$$

Proof. (5.21) is established by a straightforward calculation which we omit. From (5.21) it follows that $\partial_{a_1 \otimes a_2}$ are derivations. Hence ∂_a are derivations for all $a \in \mathcal{A}$ so that $(\mathcal{A}, \Delta)$ is a GBV–algebra. (5.22) is a rewriting of (5.21).

Clearly, tensor product is commutative and associative with respect to the standard isomorphisms.

If $\delta_i : \mathcal{A}_i \to \mathcal{A}_i$ are odd derivations of $(\mathcal{A}_i, \Delta_i)$ satisfying (5.7), then $\delta := \delta_1 \otimes 1 + 1 \otimes \delta_2$ is an odd derivation of $\mathcal{A}_1 \otimes \mathcal{A}_2$ satisfying (5.7).

If $\mathcal{A}_i$ are furnished with additional gradings having the properties postulated above, then the total grading on $\mathcal{A}_1 \otimes \mathcal{A}_2$ satisfies the same conditions.

5.8.2. Decomposable solutions to the Maurer–Cartan equation. In the notation of the previous subsection, let $(\mathcal{A}, \Delta, \delta)$ be the tensor product of $(\mathcal{A}_i, \Delta_i, \delta_i), i = 1, 2$. Assume that $a_i \in \mathcal{A}_i$ satisfy the Maurer–Cartan equation (5.11). Then from (5.22) it follows that $a := a_1 \otimes 1 + 1 \otimes a_2$ satisfies (5.11) as well. Moreover, if $\Delta_i a_i = 0$, then $\Delta a = 0$, so that $(\mathcal{A}, \Delta, \delta_a)$ is the differential GBV–algebra. Such structures will be called decomposable ones.

5.9. Example: dGBV algebras related to the Calabi–Yau manifolds ([2]). Let W be a compact complex Kähler manifold with the property $\Omega_W^{\max} \cong \mathcal{O}_W$ ("weak Calabi–Yau"). Choose once for all a nonzero holomorphic volume form Ω on W. Consider the $\mathbf{C}$–algebra

$$\mathcal{A}_W := \bigoplus_{p,q\geq 0} \Gamma_{C^\infty}(W, \wedge^q(\overline{T}_M^*) \otimes \wedge^p(T_W)) \tag{5.23}$$

with $\mathbf{Z}_2$–grading $(p+q)$ mod 2.

The map $\gamma \mapsto \gamma \vdash \Omega$ identifies $\mathcal{A}_W$ with the complexified de Rham complex of W. Let Δ correspond to ∂ with respect to this identification. One can directly check that Δ satisfies conditions of the first paragraph of 5.1 (Tian–Todorov lemma), hence claims 5.1.1 and 5.1.2 as well. Furthermore, since $\wedge^*(T_W)$ is a holomorphic vector bundle, $\mathcal{A}_W$ can be endowed with the differential $\overline{\partial}$ which we identify with δ. Again, (5.7) can be checked directly so that $(\mathcal{A}_W, \Delta, \delta)$ is a dGBV.

The key property is the validity of Lemma 5.4.1: this is essentially the $\partial\overline{\partial}$–lemma from [8]. Notice that only the existence of Kähler structure on W is needed for its validity, concrete choice does not matter.

The homology space is

$$H(\mathcal{A}_W, \delta) = H^*(W, \wedge^*(T_W)). \tag{5.24}$$

Clearly, using Ω, one can identify it with $H^*(W, \Omega_W^*)$ as well.

Define the integral by

$$\int \gamma := \int_W (\gamma \vdash \Omega) \wedge \Omega. \tag{5.25}$$

It does not vanish only on the component $q = p = \dim W$. Properties (5.18) and (5.19) follow from the Stokes formula.

The algebra $\mathcal{A}_W$ possesses the additional $\mathbf{Z}$–grading by $q+p$ satisfying all the conditions of sec. 5.7.

5.10. Example: dGBV algebras related to the symplectic manifolds satisfying the strong Lefschetz condition ([29]). Let now (U, ω) be a real manifold of dimension $2m$ endowed with a symplectic form ω. Denote by $\langle\,,\rangle$ the pairing on $\Omega^*(U)$ induced by the symplectic form. Put

$$(\mathcal{B}_U, \Delta, \delta) := (\Omega^*(U), (-1)^{*+1} \star d\star, d) \tag{5.26}$$

where $\star : \Omega^k(U) \to \Omega^{2m-k}(U)$ is the symplectic star operator defined by

$$\beta \wedge (\star\alpha) = \langle \beta, \alpha \rangle \frac{\omega^m}{m!}. \tag{5.27}$$

$\mathbf{Z}_2$–grading is the degree of the diferential form mod 2. Calculating Δ in local coordinates, one sees that it is the differential operator of second order satisfying (5.1), whereas (5.7) follows from (5.27). Thus $(\mathcal{B}_U, \Delta, \delta)$ is a dGBV–algebra.

5.10.1. Proposition. *Assume that (U, ω) satisfies the strong Lefschetz condition, that is, the cup product*

$$[\omega^k] \cup : H^{m-k}(U) \to H^{m+k}(U)$$

is an isomorphism for each $k \leq m$. Then $(\mathcal{B}_U, \Delta, \delta)$ satisfies Lemma 5.4.1.

S. Merkulov [29] proves this, completing some earlier results from [23], [6] and [28].

From now on, we will assume that the strong Lefschetz condition holds, so that U is compact. Then we can define the integral on $\mathcal{B}_U$:

$$\int \gamma := \int_U \gamma. \tag{5.28}$$

Properties (5.18) and (5.19) follow from the Stokes formula combined with the identities $\star(\star\alpha) = \alpha$ and $\beta \wedge (\star\alpha) = (\star\beta) \wedge \alpha$.

The standard $\mathbf{Z}$–grading of $\Omega^*(W)$ then satisfies all conditions of sec. 5.7.

6. From dGBV–algebras to Frobenius manifolds

6.1. Normalized formal solution to the master equation. In this section, we fix a dGBV k–algebra $(\mathcal{A}, \Delta, \delta)$ and the derived odd bracket $[\bullet]$ on it. We will assume that this algebra satisfies a series of assumptions which will be introduced and numbered consecutively.

Assumption. 1. $(\mathcal{A}, \Delta, \delta)$ satisfies conditions of the Lemma 5.4.1. Moreover, the homology group $H = H(\mathcal{A}, \delta)$ (and any group naturally isomorphic to it) is a free k–module of finite rank.

Choosing an indexed basis $[c_i], c_i \in \mathcal{A}$ of H and the dual basis (x_i) of H^t we will always assume that $c_0 = 1$. As in 4.1, but now conserving parity, we put $K := k[[H^t]] = k[[x_i]]$. We will denote by $X_i = \partial/\partial x_i$ the respective partial derivatives acting on K and on $K \otimes \mathcal{A}, K \otimes H$ etc via the first factor.

6.1.1. Proposition. *If $(\mathcal{A}, \Delta, \delta)$ satisfies Assumption 1 above, then there exists a generic even formal solution $\Gamma = \sum_i \Gamma_i \in K \otimes \operatorname{Ker} \Delta$*

to the master equation

$$\delta\Gamma + \frac{1}{2}[\Gamma \bullet \Gamma] = 0$$

with the following properties:

(i) $\Gamma_0 = 0, \Gamma_1 = \sum x_i c_i, \Gamma_n \in K \otimes \mathrm{Im}\,\Delta$ *for all* $n \geq 2$. *Here* $c_i \in \mathrm{Ker}\,\Delta \cap \mathrm{Ker}\,\delta$, *and* Γ_n *is the homogeneous component of* Γ *of degree* n *in* (x_i).

(ii) Moreover, this Γ *can be chosen in such a way that* $X_0\Gamma = 1$.

Such a solution will be called normalized.

Proof. The first statement follows from the Theorem 4.2 (i) applied to the odd differential Lie superalgebra $(\mathrm{Ker}\,\Delta, [\bullet], \delta)$.

We must only check that the conditions of the applicability of this theorem are satisfied. To facilitate the bookkeeping for the reader, we register the correspondences between the old and the new notation: Ker d becomes $\mathrm{Ker}\,\Delta \cap \mathrm{Ker}\,\delta$, Ker ϕ turns into Im Δ, Im d corresponds to Im $\delta\Delta$. All of this forms a part of Lemma 5.4.1. From the second formula in 5.1 it follows that Ker Δ is closed with respect to $[\bullet] : [a \bullet b] = (-1)^{\tilde{a}}\Delta(ab)$. This formula shows as well that $[\bullet]$ induces zero operation on $H(\mathrm{Ker}\,\Delta, \delta)$: if $a, b \in \mathrm{Ker}\,\Delta \cap \mathrm{Ker}\,\delta$, then $[a \bullet b] \in \mathrm{Im}\,\Delta \cap \mathrm{Ker}\,\delta = \mathrm{Im}\,\delta\Delta$.

It remains to check the assertion (ii). Clearly, our choice $c_0 = 1$ assures that $X_0\Gamma_1 = 1$. Assume by induction that $\Gamma_2, \ldots, \Gamma_n$ do not depend on x_0. Clearly, $[c_0 \bullet a]$=0 for any a, so that in the (odd version of the) equation (4.5) the right hand side is independent of x_0 as well. Since the argument showing the existence of Γ_{n+1} in the proof of 4.2 can be applied to each coefficient of the monomials in x_i separately, we may find Γ_{n+1} independent of x_0. The final normalization argument can be also applied coefficientwise.

6.2. The (pre)–Frobenius manifold associated to $(\mathcal{A}, \Delta, \delta)$. Consider the formal manifold M, the formal spectrum of K over k. The flat coordinates will be by definition (x_i) so that the space of flat vector fields can be canonically identified with H. We fix a normalized Γ as above.

6.2.1. Lemma. *The bi–differential group* $(\mathcal{A}_K, \Delta_K, \delta_\Gamma)$ *satisfies the conditions and conclusions of Lemma 5.4.1.*

Proof. Clearly, $\delta_\Gamma^2 = \Delta_K^2 = [\Delta_K, \delta_\Gamma] = 0$ (the latter follows from (5.3)). From the Assumption 1 and the proof of 6.6.1 above we see that we can apply the Theorem 4.2 (ii) to $(K \otimes \text{Ker } \Delta, \delta_\Gamma)$ instead of $(g_K, d_{K,\Gamma})$. The inclusion (4.6) reads in this context:

$$\text{Im } \Delta_K \cap \text{Ker } \delta_\Gamma \subset \text{Im } \delta_\Gamma \Delta_K$$

which implies the condition (B).

To check (A), consider the inclusion map

$$\text{Im } \delta_\Gamma \Delta_K \to \text{Im } \delta_\Gamma \cap \text{Ker } \Delta_K.$$

It becomes an isomorphism after reduction modulo (x_i) in view of the Assumption 1. Hence it is an isomorphism.

6.2.2. $\circ$–multiplication on tangent fields. We will define now the K–linear $\circ$–multiplication on the K–module of all vector fields $\mathcal{T}_M = K \otimes H = H_K$.

To this end we first apply Theorem 4.2 (ii) to the odd differential Lie algebra $(\text{Ker } \Delta, \delta_\Gamma)$. It shows that the homology of this algebra is naturally identified with H_K.

In view of Lemmas 5.4.1 and 6.2.1 we know that the injection $(\text{Ker } \Delta_K, \delta_\Gamma) \to (\mathcal{A}_K, \delta_\Gamma)$ induces isomorphism of homology $H_K = \text{Ker } \delta_\Gamma / \text{Im } \delta_\Gamma$. But Ker δ_Γ is a commutative K–subalgebra of A_K and Im δ_Γ is an ideal in it. Hence H_K inherits the multiplication which we denote $\circ$. We record the following "explicit" formula for it. Interpreting any $X \in H_K$ as the derivation $\overline{X}$ of $K \otimes \mathcal{A}$ acting through the first factor (cf. Corollary 4.2.1), we have:

$$\overline{X \circ Y}\Gamma \equiv \overline{X}\Gamma \cdot \overline{Y}\Gamma \text{ mod Im } \delta_\Gamma \tag{6.1}$$

(dot here means the associative multiplication in $\mathcal{A}_K$). This follows directly from the Corollary 4.2.1 applied to our situation. Notice that whereas $\overline{X}\Gamma$ and $\overline{Y}\Gamma$ lie in Ker $\Delta_K \cap$ Ker δ_Γ, their product generally lies only in the larger group Ker δ_Γ.

Directly from the initial definition one sees that $e := X_0$ is the flat identity for $\circ$.

In order to complete the description of the pre–Frobenius structure, it remains to choose a flat metric on M.

Assumption 2. There exists an integral $\int$ for $(\mathcal{A}, \Delta, \delta)$ such that the bilinear form on $H = H(A, \delta)$ induced by $(X, Y) \mapsto \int \overline{X}\Gamma \cdot \overline{Y}\Gamma$ is non–degenerate.

Denoting this form g we clearly have the invariance property defining the symmetric multiplication tensor A:

$$g(X, Y \circ Z) = g(X \circ Y, Z) := A(X, Y, Z). \tag{6.2}$$

We will check now that this structure is actually Frobenius. Since the $\circ$–multiplication is associative, we have only to establish its potentiality.

To this end we will check Dubrovin's criterium: the structure connection ∇_λ on $\mathcal{T}_M$ is flat (cf. [26], Ch. I, Theorem 1.5).

To be more precise, let ∇_0 be the flat connection on $\mathcal{T}_M$ whose horizontal sections are H. Clearly, $\nabla_{0,Y}(Z) = \overline{Y}(Z)$ where this time $\overline{Y}$ means Y acting on $K \otimes H$ via K. By definition,

$$\nabla_{\lambda,Y}(Z) = \overline{Y}(Z) + \lambda Y \circ Z \tag{6.3}$$

where λ is an even parameter.

We have the canonical surjection $\mathrm{Ker}\ \Delta_K \to H_K$ and the two lifts of X both denoted by $\overline{X}$ are compatible with this surjection, and also with embedding $\mathrm{Ker}\ \Delta_K \subset \mathcal{A}_K$. Therefore the section of $\mathrm{Ker}\ \Delta_K \to H_K$ denoted ψ in the Corollary 4.2.1 sends $\overline{Y}(Z)$ to $\overline{Y}(\overline{Z}\Gamma)$, and $\nabla_{\lambda,Y}(Z)$ lifts to $\overline{Y}(\overline{Z}\Gamma) + \lambda\overline{Y}\Gamma \cdot \overline{Z}\Gamma$ in view of (6.1) and (6.3).

Our preparations being now completed, we can prove

6.2.3. Theorem. *The connection ∇_λ is flat. Hence the pre–Frobenius structure defined above is potential.*

Proof. Applying (6.3) twice, we find

$$\nabla_{\lambda,X}\nabla_{\lambda,Y}(Z) = \overline{X}(\overline{Y}(Z)) + \lambda\overline{X}(Y \circ Z) + \lambda X \circ \overline{Y}(Z) + \lambda^2 X \circ Y \circ Z. \tag{6.4}$$

We may and will consider only the case when X, Y supercommute (e.g. $X, Y \in H$). In order to establish flatness, it suffices to check that

$$\overline{X}(Y \circ Z) + X \circ \overline{Y}(Z) = (-1)^{\tilde{X}\tilde{Y}}(\overline{Y}(X \circ Z) + Y \circ \overline{X}(Z)). \tag{6.5}$$

We will see that already the ψ–lifts of both sides of (6.5) coincide up to $\mathrm{Im}\ \delta_\Gamma$. In fact, $\overline{X}(Y \circ Z)$ lifts to $\overline{X}(\overline{Y}\Gamma \cdot \overline{Z}\Gamma)$, $X \circ \overline{Y}(Z)$ lifts to $\overline{X}\Gamma \cdot \overline{Y}(\overline{Z}\Gamma)$ so that (6.5) becomes

$$\overline{X}(\overline{Y}\Gamma) \cdot \overline{Z}\Gamma + (-1)^{\tilde{X}\tilde{Y}}\overline{Y}\Gamma \cdot \overline{X}(\overline{Z}\Gamma) + \overline{X}\Gamma \cdot \overline{Y}(\overline{Z}\Gamma)$$
$$= (-1)^{\tilde{X}\tilde{Y}}\overline{Y}(\overline{X}\Gamma) \cdot \overline{Z}\Gamma + \overline{X}\Gamma \cdot \overline{Y}(\overline{Z}\Gamma) + (-1)^{\tilde{X}\tilde{Y}}\overline{Y}\Gamma \cdot \overline{X}(\overline{Z}\Gamma).$$

This finishes the proof.

6.3. Euler field. Assume now that $\mathcal{A}$ is endowed with a grading satisfying the conditions of 5.7. All the previous discussion makes sense, and the results hold true, if we add appropriate grading conditions at certain places, the most important of which is $|\Gamma| = 2$ implying $|\circ| = 2$ in view of (6.1).

Denote by E the derivation of K defined by the following Euler condition:

$$\forall f \in K,\ Ef = \frac{1}{2}|f|f, \tag{6.6}$$

where $|\ |$ is the grading induced on H^t from $\mathcal{A}$ via H. In cooordinates as in 6.1 we have

$$E = \frac{1}{2}\sum_i |x_i|\, x_i X_i. \tag{6.7}$$

Assumption 3. Assume that the integral is homogeneous and denote its degree by $2D - 4$.

For the general discussion of spectrum cf. [26], Chapter I, §2, cf. also 1.3 above for spectrum of quantum cohomology.

6.3.1. Proposition. *E is an Euler field on the formal Frobenius manifold described in 6.2. Its spectrum is (D; d_i with multiplicity* dim H^{-2d_i}*), and $d_0 = 1$.*

Proof. Comparing (6.7) with the notation of [26], we see that the spectrum of $-\operatorname{ad} E$ on $H = \mathcal{T}_M^f$ is d_i with multiplicity dim H^{-2d_i} where

$$d_i = \frac{1}{2}|x_i| = -\frac{1}{2}|X_i|. \tag{6.8}$$

Since $X_0 \circ X = X$ and $|\circ| = 2$, we have $d_0 = 1$.
We must now check the formula

$$E(g(X,Y)) - g([E,X],Y) - g(X,[E,Y]) = Dg(X,Y). \tag{6.9}$$

It suffices to do this for the case when X, Y are flat vector fields having definite degrees. Then $[E,X] = \dfrac{|X|}{2}X$. Since $g(X,Y) \in k$, (6.9) becomes

$$(|X| + |Y| + 2D)g(X,Y) = 0.$$

But $g(X,Y) = \int \overline{X\Gamma} \cdot \overline{Y\Gamma}$ vanishes unless $2D - 4 + |X| + 2 + |Y| + 2 = 0$ which proves (6.9).

Furthermore, from (6.1) we infer that $|X \circ Y| = |X| + |Y| + 2$. Hence if $X_i \circ X_j = \sum_k A_{ij}{}^k X_k$, we have

$$E\, A_{ij}{}^k = \frac{1}{2}\, |A_{ij}{}^k| \cdot A_{ij}{}^k = \frac{1}{2}\, (|X_i| + |X_j| - |X_k| + 2)\, A_{ij}{}^k.$$

Comparing this with the formula (2.18) of [26], Chapter I and taking into account (6.8), we see that E satisfies the Definition 2.2.1 of [26], loc. cit. This finishes the proof.

Notice that the Euler field (6.7) contains no flat summand: X_i with $d_i = 0$ do not contribute. Hence if this construction furnishes a Frobenius manifold which is quantum cohomology of some V, then $c_1(V)$ must vanish (modulo torsion).

6.3.2. Remark. Comparing the Frobenius manifold produced from a dGBV–algebra with grading as above to a quantum cohomology Frobenius manifold, one must first shift the dGBV–grading by two. Then X_0 and $\circ$ acquire the degree zero.

6.4. Explicit potential. The direct way to establish potentiality is to find an even series $\Phi \in K$ such that for all $X, Y, Z \in H$ we have $A(X, Y, Z) = XYZ\Phi$ (from now on, we write X instead of $\overline{X}$ in order to denote derivations on various K–modules acting through K). Moreover, it suffices to check this for $X = Y = Z$. We will give here the beautiful formula of Chern–Simons type for Φ discovered in [2].

Extend the integral to the K–linear map $\int : \mathcal{A}_K \to K$. For a fixed normalized Γ put $\Gamma = \Gamma_1 + \Delta B$ where $B_0 = B_1 = 0$ and Δ means Δ_K.

6.4.1. Theorem. *The formal function*

$$\Phi := \int \left(\frac{1}{6}\Gamma^3 - \frac{1}{2}\,\delta B\, \Delta B \right). \tag{6.10}$$

is a potential for the Frobenius manifold defined above.

Proof. We have to prove that for any $X \in H$

$$A(X, X, X) = \int (X\Gamma)^3 = X^3\Phi. \tag{6.11}$$

We supply below the detailed calculation consisting of the series of elementary steps, each being an application of one of the identities (5.18), (5.19), Leibniz rule for (super)derivations and the fact that δ, Δ, X pairwise supercommute. Moreover, we use the master equation in the form $\Delta\Gamma^2 = -2\,\delta\Gamma$ following from $\Delta\Gamma = 0$. Finally, $\delta\Gamma_1 = X^n\Gamma_1 = 0$ for $n \geq 2$ so that $\delta\Gamma = \delta\Delta B, X^n\Gamma = X^n\Delta B$.

We start with treating the first summand of the right hand side of (6.10). The derivation X is interchangeable with integration, so we have by the Leibniz rule

$$X^3\left(\frac{1}{6}\int\Gamma^3\right) = \int (X\Gamma)^3 + \int\left((2+(-1)^{\tilde{X}})\Gamma\cdot X\Gamma\cdot X^2\Gamma + \frac{1}{2}\Gamma^2\cdot X^3\Gamma\right). \tag{6.12}$$

The second summand of (6.10) is added in order to cancel the extra terms in (6.12).

First, we rewrite it:

$$\frac{1}{2}\int \delta B\,\Delta B = \frac{1}{2}\int B\,\delta\Delta B = \frac{1}{2}\int B\,\delta\Gamma = -\frac{1}{4}\int B\,\Delta(\Gamma^2) = \frac{1}{4}\int \Delta B\cdot\Gamma^2. \tag{6.13}$$

(We could have chosen the last expression in (6.13) from the start).

Now, again by Leibniz rule,

$$\frac{1}{4}X^3\int\Delta B\cdot\Gamma^2 = \frac{1}{4}\int\Big(X^3(\Delta B)\,\Gamma^2 + (2+(-1)^{\tilde{X}})X^2(\Delta B)\cdot X(\Gamma^2) + (2+(-1)^{\tilde{X}})X(\Delta B)\cdot X^2(\Gamma^2) + \Delta B\cdot X^3(\Gamma^2)\Big). \tag{6.14}$$

The first two summands in (6.14) can be directly rewritten in the same form as in (6.12):

$$X^3(\Delta B)\cdot\Gamma^2 + (2+(-1)^{\tilde{X}})X^2(\Delta B)\cdot X(\Gamma^2) = X^3\Gamma\cdot\Gamma^2 + (2+(-1)^{\tilde{X}})X^2\Gamma\cdot X(\Gamma^2). \tag{6.15}$$

The third summand takes somewhat more work:

$$\begin{aligned}\int X(\Delta B)\cdot X^2(\Gamma^2) &= -\int XB\cdot X^2(\Delta\Gamma^2) = 2\int XB\cdot X^2(\delta\Gamma)\\ &= 2\int XB\cdot\delta(X^2\Gamma) = 2\int XB\cdot\delta\Delta(X^2B)\\ &= -2\int \delta\Delta XB\cdot X^2B = -2\int X\delta\Gamma\cdot X^2B\\ &= \int X\Delta(\Gamma^2)\cdot X^2B = \int X(\Gamma^2)\cdot X^2\Delta B\\ &= \int X(\Gamma^2)\cdot X^2\Gamma.\end{aligned}\tag{6.16}$$

Finally, the fourth summand is calculated similarly, but in two steps. We start with an expression of the second order in X:

$$\begin{aligned}\int \Delta B\cdot X^2(\Gamma^2) &= -\int B\cdot X^2(\Delta\Gamma^2) = 2\int B\cdot X^2(\delta\Gamma)\\ &= 2\int B\cdot\delta(X^2\Gamma) = 2\int B\cdot\delta\Delta(X^2B)\\ &= -2\int \delta\Delta B\cdot X^2B = -2\int \delta\Gamma\cdot X^2B\\ &= \int \Delta(\Gamma^2)\cdot X^2B = \int \Gamma^2\cdot X^2\Delta B\\ &= \int \Gamma^2\cdot X^2\Gamma.\end{aligned}\tag{6.17}$$

Apply now X to the first and the last expressions of (6.17). We get

$$\begin{aligned}&\int X(\Delta B)\cdot X^2(\Gamma^2) + \int \Delta B\cdot X^3(\Gamma^2)\\ &\qquad = \int X(\Gamma^2)\cdot X^2\Gamma + \int \Gamma^2\cdot X^3\Gamma.\end{aligned}\tag{6.18}$$

Comparing this with (6.16), one gets

$$\int \Delta B\cdot X^3(\Gamma^2) = \int \Gamma^2\cdot X^3\Gamma.\tag{6.19}$$

Putting all of this together, one obtains finally (6.11).

6.5. Example: Frobenius manifolds of B–type, related to the Calabi–Yau manifolds. Returning now to the examples of 5.9, one sees that all assumptions of this section hold so that we get a class of Frobenius manifolds, which we may call BK–models of Calabi–Yau manifolds W. In particular, we can easily calculate the d–spectrum which is:

$$(6.20)\qquad (w; d \text{ with multiplicity } \sum_{q+p=2d} h^{p,w-q}(W)),\ w := \dim_{\mathbf{C}}(W).$$

6.6. Example: Frobenius manifolds related to the symplectic manifolds satisfying the strong Lefschetz condition. Similarly, in the situation of 5.10 we obtain the Frobenius manifold with the d–spectrum

$$(6.21)\qquad (m; d \text{ with multiplicity } \dim H^{2d}(W)).$$

Notice that in this case as well the anticanonical component of the Euler field vanishes.

It would be interesting to establish isomorphisms between these examples and to understand when they furnish Frobenius manifolds of qc–type. Notice that if W, U are mirror dual Calabi–Yau manifolds, then the spectra of $\mathcal{A}_W$ and $\mathcal{B}_U$ coincide.

References

[1] V. Arnold, S. Gusein-Zade & A. Varchenko, *Singularities of differentiable maps*, Vols. I, II, Birkhäuser, Boston, 1985 and 1988.

[2] S. Barannikov & M. Kontsevich, *Frobenius manifolds and formality of Lie algebras of polyvector fields*, Preprint, alg–geom/97010072.

[3] K. Behrend, *Gromov–Witten invariants in algebraic geometry*, Invent. Math. **127** (1997) 601–617.

[4] K. Behrend & B. Fantechi, *The intrinsic normal cone*, Invent. Math. **128** (1997) 45–88.

[5] K. Behrend & Yu. Manin, *Stacks of stable maps and Gromov–Witten invariants*, Duke Math. J. **85** (1996) 1–60.

[6] J.-L. Brylinski, *A differential complex for Poisson manifolds*, J. Differential Geom. **28** (1988) 93–114.

[7] S. Cecotti, $N = 2$ *Landau–Ginzburg vs. Calabi–Yau* σ*–models: non-perturbative aspects*, Int. J. Mod. Phys. A **6:10** (1991) 1749–1813.

[8] P. Deligne, Ph. Griffiths, J. Morgan & D. Sullivan, *Real homotopy theory of Kähler manifolds*, Invent. Math. **29** (1975) 245–274.

[9] B. Dubrovin, *Geometry of 2D topological field theories*, Springer Lecture Note Math. **1620** (1996) 120–348.

[10] B. Dubrovin & Youjin Zhang, *Extended affine Weyl groups and Frobenius manifolds*, Preprint SISSA 67/96/FM.

[11] D. Gepner, *On the spectrum of 2D conformal field theory*, Nucl. Phys. **B287** (1987) 111–126.

[12] ———, *Fusion rings and geometry*, Comm. Math. Phys. **141** (1991) 381–411.

[13] M. Gerstenhaber, *The cohomology structure of an associative ring*, Ann. of Math. **78** (1963) 267–288.

[14] A. Givental, *Equivariant Gromov-Witten invariants*, Int. Math. Res. Notes **13** (1996) 613–663.

[15] W. Goldman & J. Millson, *The deformation theory of representations of fundamental groups of compact Kähler manifolds*, Inst. Hautes. Études Sci. Publ. Math. **86** (1988) 43–96.

[16] B. Green, *Constructing mirror manifolds*, Mirror Symmetry II, (B. Greene and S. T. Yau, eds.), Amer. Math. Soc.–Internat. Press, 1996, 29–69.

[17] S. Gusein-Zade & A. Varchenko, *Verlinde algebras and the intersection form on vanishing cycles*, Selecta Math., New. Ser. **3** (1997) 79–97.

[18] R. Kaufmann, *The geometry of moduli spaces of pointed curves, the tensor product in the theory of Frobenius manifolds, and the explicit Künneth formula in quantum cohomology*, Ph. D. thesis, MPI für Mathematik, Bonn, 1997.

[19] M. Kontsevich, *Enumeration of rational curves via torus actions*, The Moduli Space of Curves, (R. Dijkgraaf, C. Faber, G. van der Geer, eds.), Progr. Math., Birkhäuser. Vol. 129, 1995, 335–368.

[20] M. Kontsevich & Yu. Manin, *Gromov-Witten classes, quantum cohomology, and enumerative geometry*, Comm. Math. Phys. **164** (1994) 525–562.

[21] ———, *Relations between the correlators of the topological sigma–model coupled to gravity*, Preprint, alg–geom/970824.

[22] ———, *Quantum cohomology of a product (with Appendix by R. Kaufmann)*, Invent. Math. **124** (1996) 313–339.

[23] J.-L. Koszul, *Crochet de Schouten–Nijenhuis et cohomologir*, Elie Cartan et les Mathématiques D'aujourd'huis, Astérisque, 1985, 251–271.

[24] B. H. Lian, K. Liu & S.-T. Yau, *Mirror Principle. I*, alg-geom/9712011.

[25] B. H. Lian & G. Zuckerman, *New perspectives on the BRST–algebraic structure of string theory*, Comm. Math. Phys. **154** (1993) 613–646; hep–th/9211072.

[26] Yu. Manin, *Frobenius manifolds, quantum cohomology, and moduli spaces*, (Chapters I, II, III) Preprint MPI, 1996, 96–113.

[27] Yu. Manin & S. Merkulov, *Semisimple Frobenius (super)manifolds and quantum cohomology of* $\mathbf{P}^r$, Topological Methods in Nonlinear Analysis **9** (1997) 107–161; alg–geom/9702014.

[28] O. Mathieu, *Harmonic cohomology classes of symplectic manifolds*, Comm. Math. Helv. **70** (1995) 1–9.

[29] S. Merkulov, *Formality of canonical symplectic complexes and Frobenius manifolds*, Preprint, math/9805072.

[30] T. Oda, *K. Saito's period map for holomorphic functions with isolated critical points*, Adv. Stud. Pure Math. **10** (1987), Algebraic Geometry, Sendai, 1985, 591–648.

[31] K. Saito, *Primitive forms for a universal unfolding of a function with an isolated critical point*, J. Fac. Sci. Univ. Tokyo, Sec. IA **28** (1982) 775–792.

[32] K. Saito, *Period mapping associated to a primitive form*, Publ. RIMS, Kyoto Univ. **19** (1983) 1231–1264.

[33] J. Stasheff, *Deformation theory and the Batalin–Vilkovisky master equation*, Deformation Theory and Symplectic Geometry, (D. Sternheimer, et. al., eds.), Kluwer, 1997, 271–284.

[34] J.-B. Zuber, *Graphs and reflection groups*, Comm. Math. Phys. **179** (1996) 265–294.

MAX PLANCK INSTITUTE FOR MATHEMATICS, GERMANY

SURVEYS IN DIFFERENTIAL GEOMETRY, 2000
Vol. VII ©2000, INTERNATIONAL PRESS
pp. 555-564

ON RICCI-FLAT TWISTOR THEORY

ROGER PENROSE

1. Background

In the autumn of 1952, I had the honour to be taken on as a research student at the University of Cambridge, to work in algebraic geometry under the supervision of the renowned mathematician William V.D. Hodge. As I recall it, there were four of us, starting under Hodge at the same time. The research that then interested him was broadly divided into that which was centred on algebraic geometry and a more topological line arising from his work on harmonic integrals. I had specifically started on the algebraic geometry side, but I was finding things rather too strictly "algebraic", for my tastes, with not much of a realization of this algebra into what I thought of as "geometry". Noticing that I was not entirely happy with spending my time dealing with questions in ideal theory, local rings, and so on, Hodge suggested that I might like to sit in on a supervision session, the supervisee being the only one of the four of us who was working on the harmonic integrals side of things. The idea intrigued me because that work seemed to be rather more geometrical in nature than the problems that I had been looking at, so with considerable expectations I turned up. The student was a "Mr Attia"—or, at least, that is how Hodge used to refer to him—and I remember being totally snowed under by Mr Attia's breadth of knowledge and comprehension; indeed, I recall not understanding a single word of what was going on. Of course "Attia" was really "Atiyah"–and one of the difficulties about being a research student, especially at a place like Cambridge, is that one never knows who one's co-research students really are (or will be)! "Not

First printed in Asian Journal of Mathematics, 1999. Used by permission.

understanding a single word" may perhaps be nothing to be ashamed of under such circumstances.

Over my research-student period, there was much interest in what was then referred to as "the theory of stacks". I remember trying to struggle with stacks, for a little, but I then made life easier for myself by deciding that my interests lay largely elsewhere, so I spent a good deal of my time learning about general relativity, quantum mechanics, mathematical logic, and various other matters purely mathematical. After I left Cambridge, in 1959, my interests had moved more and more in the direction of theoretical physics, mainly general relativity, but also quantum mechanics. Later I developed my interest in what I referred to as the theory of "twistors", which took advantage of many algebraic/geometrical notions that I had learned about during my student days—most particularly the Klein representation of lines in projective 3-space as points of a 4-quadric.

The basic idea of twistor theory (for flat Minkowski space-time M) was, in effect, to take the Klein representation "in reverse", where the conformally compactified space-time $M^{\#}$ is, roughly speaking, taken as the "Klein Quadric" of another space PN (Penrose 1967). More precisely, we regard the natural complexification $CM^{\#}$, of $M^{\#}$, as the Klein representation of complex straight lines in a certain CP^3 called projective twistor space PT. The space PN is a real 5-submanifold (given by the vanishing of a Hermitian quadratic form of signature $++--$) of the real 6-manifold PT. The projective lines which lie in PN are "Klein-represented" by the points of of the real 4-manifold $M^{\#}$. Then, using this description, the basic physical notions of space-time, particles, fields, etc. would be interpreted in terms of the projective geometry of PT or PN, or of the geometry/analysis of the underlying vector space T, simply called *twistor space.* It later turned out (Penrose 1969) that *massless* fields, in particular, find an elegant description in terms of contour integrals in twistor space. In particular, linearized gravitational fields (massless fields of spin 2) can be neatly accommodated within this scheme. (See Penrose 1987, for an account of the curious history of all this.)

Yet, this approach did not directly cope with the space-time curvature which would be needed in order that the gravitational field proper could be incorporated into twistor theory, in accordance with Einstein's general relativity. However, through a roundabout route, originating with an idea due to E.T. Newman (Newman 1976, cf. Penrose 1992 for the relevant history), I had come to the conclusion that "half" of the gravitational field—the "left-handed" half that is described by an anti-self-dual (ASD) Weyl curvature—can indeed be incorporated into twistor geometry, where the notion of twistor space has to be generalized away from the *flat* twistor model PT (or T) to a "curved" one $P\mathcal{T}$ (or $\mathcal{T}$), this "curvature" not being anything that shows up at the *local* level (T and $\mathcal{T}$ being locally identical), but arising from the *global* structure of $P\mathcal{T}$. I had realized that I needed to understand how to describe *defomations* of complex manifolds (particularly non-compact ones) and that this could indeed accommodate genuine ASD Weyl curvature into the (complex) "space-time". Moreover, the condition of *Ricci-flatness* for such ASD complex-Riemannian 4-manifolds can be easily incorporated. I consulted a few people about how to describe such deformations and under what circumstances the needed *4-parameter* family of "Klein" lines would persist in $P\mathcal{T}$, but it was not until Michael Atiyah explained Kodaira's various theorems on this question to me—and more importantly, how to *use* these theorems in the context that I needed, that I began to see what all those "stacks"—now called *sheaves*—had really been about, all the time. This provided the necessary background for the construction that I referred to as "the non-linear graviton" (Penrose 1976) in which complex ASD Ricci-flat 4-manifolds can be described in terms of a kind of "Klein representation" of lines in appropriate complex 3-manifolds.

By then I had been in Oxford for several years, where Michael now was, and he made a special point of providing me and my research group with illuminating expository sessions, in which he explained to us, in his characteristically revealing way, the beauty, the essential simplicity, and the relevant uses of sheaf cohomology. One point, in particular, that I found valuable was Michael's deliberate use of *Čech* cohomology in his expositions, rather than the more frequently used Dolbeault approach. In my opinion, the Čech approach provided a much greater clarity, in the context of the problems of relevance to us, and it was certainly sufficient for our immediate needs. It soon emerged, on the basis of Michael's encouraging insights, that the contour integral expressions that I had previously adopted for the description of (linear) massless fields really were themselves expressions of (Čech) sheaf cohomology. Accordingly, an (analytic) massless field of helicity $n/2$ in M would be interpreted as an element of $H^1(Q, \mathcal{O}(-n-2))$, where Q is some suitable open subregion of PT, related to the domain of definition (assumed appropriate) of the massless field in M, and where $\mathcal{O}(-n-2)$ is the sheaf of twisted holomorphic functions on PT, locally given by holomorphic functions on T of homogeneity degree $-n-2$ (cf. Eastwood, Penrose, and Wells 1981; here n is an integer, the *spin* of the field being $|n|/2$, where the *sign* of n tells us the "handedness" of the field). This clarified numerous points of previous confusion.

These insights also led to a direct interpretation of the linear massless fields of helicity-2 (linearized ASD gravity) as providing infinitesimal deformations of (regions of) projective twistor space PT (cf. Penrose and Rindler 1986), this being a *weak-field* version of the above "non-linear graviton". A point to note is that we are here concerned with transition functions that are constructed from holomorphic twistor functions $f_{ij}(Z^\alpha)$ that are homogeneous of degree $+2$ (corresponding to helicity $n/2 = -2$). (Here, I am beginning to use the standard 2-spinor/twistor index-notation of Penrose and Rindler 1986. The *twistor* Z^α is an element of flat twistor space T, "α" being a 4-dimensional *abstract index*.) Thus, the family of Čech representative functions $\{f_{ij}\}$, for the $H^1(Q, \mathcal{O}(-n-2))$ element, defined on the overlaps $U_i \cap U_j$ of a suitable Čech cover $\{U_i\}$ of Q (with $f_{ij} = -f_{ji}$, and $f_{ij} - f_{ik} + f_{jk} = 0$ on triple overlaps), directly provides the family of infinitesimal transition functions for piecing together the infinitesimally curved twistor space $\mathcal{T}$. These infinitesimal transition functions are provided by "sliding infinitesimally along" the vector field

$$\varepsilon^{AB} \frac{\partial f_{ij}}{\partial \omega^A} \frac{\partial}{\partial \omega^B},$$

where I now adopt the 2-spinor/twistor index-notation $(\omega^A, \pi_{A'})$ for the spinor parts of the twistor Z^α, taken with respect to some origin O in M. Note that the homogeneity degree $+2$ of f_{ij} exactly balances the two $\partial/\partial\omega$ contributions, each of degree -1.

2. The Googly problem

Although all this was remarkably satisfying, a definite problem began to loom large. For if twistor theory is to be taken to be a *physical* theory, the gravitational field as it is *actually* understood, must be described by a (Weyl) curvature for a space-time which possesses *both* an SD (self-dual) and an ASD part. In the case of *weak-field* gravity, regarded as a massless field of spin 2, this is neatly accommodated because the $\mathcal{O}(-6)$ Čech cohomology handles the right-handed (SD) part of the gravitational field in a closely analogous way to the $\mathcal{O}(+2)$ Čech cohomology description of the left-handed (ASD) part of the gravitational field. Moreover, if we regard these as referring to the non-projective twistor space T rather than to the projective PT, then there is an easy way of expressing the *sum* of the SD and ASD parts, to obtain a twistor-cohomological description of *full* (neither SD nor ASD) weak-field gravity. Yet, for this to provide an actual deformation of twistor space, we need an active role for the $\mathcal{O}(-6)$-cohomology, analogous, in some appropriate way, to the way in which the

$\mathcal{O}(+2)$-cohomology infinitesimally *deforms* twistor space, thus leading to the "non-linear graviton" construction referred to above. The problem of introducing SD Weyl curvature into the geometry of twistor space has been referred to as the (gravitational) *googly* problem of twistor theory—in reference to the cricketing term "googly" for a ball that spins in a right-handed sense even though the bowling action suggests a left-handed spin. Taking the cricketing analogy further, I now refer to the original "non-linear graviton" (mentioned above; as given in Penrose 1976) as the *leg-break construction.*

Somewhat over a year ago, a new approach to the relevant googly geometry has come about (see Penrose 1999), in which the googly (SD) information is encoded in the way that the twistor space $\mathcal{T}$ sits above its "projective" version $P\mathcal{T}$, where the leg-break (ASD) information resides in the structure of $P\mathcal{T}$, essentially just as before. In 1978 Michael and his colleagues showed (Atiyah, Hitchin, and Singer 1978) how my original leg-break construction could be adapted to the case of an ordinary (positive-definite) ASD Riemannian Ricci-flat 4-space (the ASD condition being non-trivial in the positive-definite case, unlike the situation with the Lorentzian signature of general relativity). The purpose of this article is to point out that there is also a Riemannian version of the new googly geometry, although I have not worked out all the requirements for this. It is my hope that these ideas will be taken up seriously by someone, and that there may be some interesting new things to say about general (neither SD nor ASD) Ricci-flat Riemannian 4-spaces in accordance with these *twistorial* ideas.

I can give only a very brief account of the new googly geometry here; otherwise there is danger of things getting unhelpfully bogged down in the notation. In any case, it is probable that any Riemannian geometric approach would rely upon some different concepts which might be better expressed in ways other than those that naturally suit Lorentzian space-time geometry. It should be made clear, also, that there are still major unresolved issues with regard to the googly geometry, parts of the programme being still in a conjectural state. Moreover, there are some aspects of the construction that rely upon conditions of asymptotic flatness that are appropriate in the Lorentzian case, whereas I do not know to what extent these Lorentzian ideas can be taken over to the case of a Riemannian Ricci-flat 4-space.

The Riemannian case does have one clear advantage over the Lorentzian case, in relation to the ideas of twistor theory. Since the condition of Ricci-flatness becomes a set of *elliptic* equations, we must expect that the solutions are analytic in the interior regions. Indeed, this is the case (see

Kazdan 1983). Thus, for any Riemannian Ricci-flat 4-manifold $\mathcal{M}$, there exists a (local) complexification $C\mathcal{M}$, which need be merely a "thickening" of the real 4-manifold $\mathcal{M}$ into a (non-compact) real 8-manifold which is a complex 4-manifold $C\mathcal{M}$ of topology $\mathcal{M} \times R^4$.

The first step in the proposed construction of a "twistor space" $\mathcal{T} = \mathcal{T}(\mathcal{M})$, for $\mathcal{M}$ is to produce the *relative* twistor space $\mathcal{T}_p$, where p is any point of $\mathcal{M}$. This is a perfectly rigorous procedure, which I shall outline shortly. The second step would be to attempt to provide a local identification between what I shall call a "comprehensive" (open) region of $\mathcal{T}_p$, and an analogous comprehensive region of $\mathcal{T}_q$, for different points $p, q \in \mathcal{M}$ for which p and q are close enough to each other for this to be achieved. I shall describe the idea behind the notion of "comprehensive" in a moment. In the absence of a more satisfactory procedure, this identification could be via some "ideal" twistor space $\mathcal{T}_\infty$, which we try to think of as being defined as a limit of $\mathcal{T}_p$, as $p \to \infty$, there being identifications of compehensive regions of each of $\mathcal{T}_p$ and $\mathcal{T}_q$ with one and the same compehensive region of $\mathcal{T}_\infty$.

The idea behind this "compehensive" notion is that such a comprehensive region contains the essential global structure that is to be carried from $\mathcal{T}_p$ to $\mathcal{T}_q$ (perhaps via $\mathcal{T}_\infty$). This is to be analogous to what happens in the procedure of *analytic continuation*, as applied to $C\mathcal{M}$. In fact, something of this very nature is already part of the original leg-break construction, although this point does not seem to have been particularly emphasized before. In that case (now taking $C\mathcal{M}$ to be ASD), we can construct the standard leg-break twistor spaces $\mathcal{T}_a$, $\mathcal{T}_b$ of intersecting open neighbourhoods of points $a, b \in C\mathcal{M}$. The twistor space of the *intersection* of these neighbourhoods, provides an identification between open regions of $\mathcal{T}_a$ and $\mathcal{T}_b$ that is sufficiently "comprehensive" that the essential analytic geometry of $C\mathcal{M}$ is carried from $\mathcal{T}_a$ to $\mathcal{T}_b$ via this region. In simple enough ASD situations, it is possible to "glue" all the $\mathcal{T}_x$-spaces together so as to obtain *one* all-inclusive (Hausdorff) twistor space $\mathcal{T}$, but there are other situations when this is not possible, at least if one requires a Hausdorff geometry. When $C\mathcal{M}$ is not ASD, the situation appears to be like this, but essentially more complicated, and some appropriate attitude towards this geometry (not yet fully formulated) seems to be required.

It is not yet clear to me how all this is to work, for general Ricci-flat Riemannian 4-spaces, but there is a "generic" family of Lorentzian space-times for which it can indeed be carried out. These are the space-times that I refer to as "strongly asymptotically flat" radiative analytic vacuums. Think of a sourceless (analytic) gravitational wave that comes in from infinity and then finally disperses out to infinity again, leaving no remnant in the form of a black hole or ay other kind of undispersed localized curvature. In fact, it is only the final dispersing of the wave out to infinity that is needed here, and the work of Friedrich (1986, 1998) is sufficient to establish the "generic" nature of solutions of the Einstein vaccum equations satisfying the needed conditions. What is required is an analytic future-null conformal infinity $\mathcal{I}^+$, with a regular future vertex i^+ (see Penrose and Rindler 1986, Chapter 9). In this case, the required twistor space "$\mathcal{T}^\infty$" actually does exist, this being the space $\mathcal{T}_{i^+}$, and for points $a, b, \cdots$ of $C\mathcal{M}$, "close enough" to i^+, there will indeed be comprehensive regions of $\mathcal{T}_a, \mathcal{T}_b, \cdots$ that can be identified with comprehensive regions of this $\mathcal{T}^\infty$. It is probably not appropriate to go into the details, here, of why this appears to work in the Lorentzian case, but in any case I do not see any reason to expect that this should directly carry over to the Reimannian situation.

Let me leave this issue aside as largely unresolved. However, I should try to explain, briefly, how the relative twistor spaces $\mathcal{T}_a$, are to be constructed. Here, there is no real difference between the Lorentzian and Riemannian cases. In $C\mathcal{M}$, each point $a \in C\mathcal{M}$ has its light cone $\mathcal{C}_a$, consisting of all the points of $C\mathcal{M}$ that lie on null geodesics through a. On $\mathcal{C}_a$, there are curves known as α-lines, which are the curves that "appear intrinsically" to be the intersections of $\mathcal{C}_a$ with α-planes in $C\mathcal{M}$ (SD totally null complex 2-sufaces), even though there may be no actual α-planes in $C\mathcal{M}$. The equation of an α-line, with tangent vector $o^A\pi^{A'}$ can be expressed as

$$\pi^{B'}\nabla_{0B'}\pi_{A'} \;\propto\; \pi_{A'}$$

on $\mathcal{C}_a$. Here *suffixes* 0 and $0'$ are to denote components obtained by contraction with spinors o^A and with $\tilde{o}^{A'}$, respectively, where the tangents to the null geodesics through a (i.e. generators of $\mathcal{C}_a$) are the null vectors $o^A\tilde{o}^{A'}$. (When $\mathcal{M}$ is SD the twistor lines are null geodesics on $\mathcal{C}_a$, but in the general case they are not.) It should be remarked that the definition of a twistor line is *conformally invariant.*

The *points of the projective* relative twistor space $P\mathcal{T}_a$ are just the α-lines on $\mathcal{C}_a$. We define the *non*-projective relative twistor space $\mathcal{T}_a$ by fixing the proportionality scale in the above equation according to the conformally invariant equation

$$\pi^{B'}\nabla_{0B'}\pi_{A'} \;=\; K\,\pi_{A'} \times (\pi_{0'})^{-5}\text{Þ}_c\tilde{\psi}_{0'0'0'0'}\;,$$

along the α-lines on $\mathcal{C}_a$. Here Þ_c is the conformally invariant "thorn" operator defined in *Spinors and Space-Time*, Vol. 1 (Penrose and Rindler

1984) p. 395, which is a modified version of the covariant derivative operator $\nabla_{00'}$, and $\tilde{\psi}_{A'B'C'D'}$ is the (conformally invariant) helicity +2 massless field related to the SD Weyl spinor $\tilde{\Psi}_{A'B'C'D'}$ by

$$\tilde{\psi}_{A'B'C'D'} = \Omega^{-1}\tilde{\Psi}_{A'B'C'D'} ,$$

where Ω is a *conformal factor* which is needed when we go to a new metric $\Omega^2 g$ which is regular on $\mathcal{I}^+$, where g is the given metric of $\mathcal{M}$. We shall require this for $\mathcal{T}^\infty$, though for $\mathcal{T}_a$ we can take $\Omega = 1$. The quantity K is a particular numerical constant whose value has not yet been determined, at the time of writing.

For the detailed meaning of all these quantities, see Penrose and Rindler (1986), Penrose (1999). Apart from the precise (as yet undetermined) value of K, the form of this equation is dictated by requirements of conformal invariance. The space $\mathcal{T}^\infty$ (and hence, each $\mathcal{T}_a$) has a structure determined from a 1-form ι and a 3-form θ (just given up to proportionality), subject to

$$\iota \wedge d\iota = 0 \ , \ \iota \wedge \theta = 0$$

and a further condition that can be given as

$$d\theta \otimes \iota = -2\theta \oslash d\iota$$

where the bilinear operator $\oslash$, acting between an n-form and a 2-form, is defined by

$$\eta \oslash (dp \wedge dq) = \eta \wedge dp \otimes dq - \eta \wedge dq \otimes dp \ .$$

In the original leg-break construction, the forms θ and ι provide the essential local structure of $\mathcal{T}$. In flat space we have

$$\iota = \varepsilon_{A'B'}\pi^{A'} d\pi^{B'} ,$$
$$\theta = 1/6\, \varepsilon_{\alpha\beta\gamma\delta}\, Z^\alpha \wedge dZ^\beta \wedge dZ^\gamma \wedge dZ^\delta \ .$$

Here, we merely have

$$\Pi = d\theta \otimes \iota \text{ and } \Sigma = d\theta \otimes d\theta \otimes \theta$$

(or something equivalent) as being specified as local structure assigned to $\mathcal{T}^+$. We also retain the condition $d\theta \otimes \iota = -2\theta \oslash d\iota$. For any particular choice of ι and θ, consistent with these relations, we can provide a definition of the "Euler vector field" $\Upsilon = \theta \div \phi$, and the projective space

$P\mathcal{T}^+$ is the factor space of $\mathcal{T}^+$ by the integral curves of Υ. For further details, see Penrose (1999).

ACKNOWLEDGEMENT. I am grateful to the Institute for Theoretical Physics, Santa Barbara for support under contract PHY94-07194; while this paper was being prepared, and also to Penn State University, for support under PHY93-96246.

References

[1] M. F. Atiyah & R. S. Ward, *Instantons and algebraic geometry geometry*, Comm. Math. Phys. **55** (1977) 111–124.

[2] M. F. Atiyah, N. J. Hitchin & I. M. Singer, *Self-duality in four-dimensional Riemannian geometry*, Proc. Roy. Soc. (London), Vol. A362, 1978, 425–461.

[3] M. G. Eastwood, R. Penrose & Jr. R. O. Wells, *Cohomology and massless fields*, Comm. Math. Phys. **78** (1981) 305–351.

[4] H. Friedrich, *On the existence of n-geodesically complete or future complete solutions of Einstein's field equations with smooth asymptotic structure*, Commun. Math. Phys. **107** (1986) 587–609.

[5] ______, *Einstein's equation and conformal structure*, The Geometric Universe; Science, Geometry, and the Work of Roger Penrose, (S. A. Huggett, L. J. Mason, K. P. Tod, S. T. Tsou, and N. M. J. Woodhouse, eds.), Oxford Univ. Press, Oxford, 1998, 81–98.

[6] J. L. Kazdan, *Some applications of partial differential equations to problems in geometry*, Surveys in Geom. Ser., Tokyo Univ., 1983 (revised notes 1993).

[7] E. T. Newman, *Heaven and its properties*, Gen. Rel. Grav. **7** (1976) 107–111.

[8] R. Penrose, *Twistor algebra*, J. Math. Phys. **8** (1967) 345–366.

[9] ______, *Solutions of the zero rest-mass equations*, J. Math. Phys. **10** (1969) 38–39.

[10] ______, *Non-linear gravitons and curved twistor theory*, Gen. Rel. Grav. **7** (1976) 31–52.

[11] ______, *On the origins of twistor theory*, Gravitation and Geometry: a volume in honour of I. Robinson, (W. Rindler and A. Trautman, eds.), Bibliopolis, Naples, 1987, 341–361.

[12] ______, *$\mathcal{H}$-space and twistors*, Recent Advances in General Relativity, (Einstein Studies, Vol. 4), (Allen I. Janis and John R. Porter, eds.), Birkhäuser, Boston, 1992, 6–25.

[13] ______, *The central programme of twistor theory*, Chaos, Solitons & Fractals **10** (1999) 581–611.

[14] R. Penrose & W. Rindler, *Spinors and space-time. Vol. 1: Two-Spinor Calculus and Relativistic Fields*, Cambridge University Press, Cambridge, 1984; *Spinors and Space-Time. Vol. 2: Spinor and Twistor Methods in Space-Time Geometry*, Cambridge University Press, Cambridge, 1986.

Mathematical Institute, University of Oxford, England

SURVEYS IN DIFFERENTIAL GEOMETRY, 2000
Vol. VII ©2000, INTERNATIONAL PRESS
pp. 565-623

ON THE GEOMETRY OF NILPOTENT ORBITS

WILFRIED SCHMID & KARI VILONEN

1. Introduction

In this paper we describe certain geometric features of nilpotent orbits in a real semisimple Lie algebra $\mathfrak{g}_{\mathbb{R}}$. Our tools are Ness' moment map [12] and the proof of the Hodge-theoretic SL_2-orbit theorem [13], [1]; our aim is a better understanding of the Kostant-Sekiguchi correspondence [16].

Let us recall the nature of the correspondence. We choose a Cartan decomposition $\mathfrak{g}_{\mathbb{R}} = \mathfrak{k}_{\mathbb{R}} \oplus \mathfrak{p}_{\mathbb{R}}$, which we complexify to $\mathfrak{g} = \mathfrak{k} \oplus \mathfrak{p}$. Four groups will be of interest: the automorphism group $G = \mathrm{Aut}(\mathfrak{g})^0$, the real form $G_{\mathbb{R}} = \mathrm{Aut}(\mathfrak{g}_{\mathbb{R}})^0$, the connected subgroup K with Lie algebra $\mathfrak{k}$, and $K_{\mathbb{R}} = G_{\mathbb{R}} \cap K$, which is maximal compact in both $G_{\mathbb{R}}$ and K. Sekiguchi [16] and Kostant (unpublished) establish a bijection between the set of nilpotent $G_{\mathbb{R}}$-orbits in $\mathfrak{g}_{\mathbb{R}}$ on the one hand and, on the other hand, the set of nilpotent K-orbits in $\mathfrak{p}$ – this is the Kostant-Sekiguchi correspondence.

Our proof [15] of a representation theoretic conjecture of Barbasch and Vogan depends on a particular geometric description of the correspondence. In very rough terms, our version of the correspondence amounts to an explicit (but subtle) deformation of any nilpotent K-orbit in $\mathfrak{p}$ into the $G_{\mathbb{R}}$-orbit that it corresponds to. Earlier [15] we had reduced this result – Theorem 7.22 below – to certain geometric statements about nilpotent orbits. These statements – Lemmas 8.5 and 8.10

First published in The Asian Journal of Mathematics, 1999. Used by permission.
The first author was partially supported by NSF, and the second author by NSA, NSF, the Guggenheim Foundation, and MPI Bonn.

– are proved in the final section of this paper. Along the way, we obtain several results on nilpotent orbits that look interesting in their own right. What we do has implications for Kronheimer's instanton flow [9]: the flow is real analytic at infinity, with a power series expansion that we describe recursively.

To give some idea of our methods, we consider a nilpotent $G_{\mathbb{R}}$-orbit $\mathcal{O}$ in $\mathfrak{g}_{\mathbb{R}} - \{0\}$. Ness' moment map [12] is a real analytic, $K_{\mathbb{R}}$-invariant map $m : \mathbb{S}(\mathcal{O}) \to \mathfrak{p}_{\mathbb{R}}$; here $\mathbb{S}(\mathcal{O}) \simeq \mathbb{R}^+ \backslash \mathcal{O}$ denotes the set of unit vectors in $\mathcal{O}$. The square norm $\|m\|^2$ assumes its minimum value exactly along a $K_{\mathbb{R}}$-orbit in $\mathbb{S}(\mathcal{O})$, which we call the *core* of $\mathcal{O}$, and denote by $C(\mathcal{O})$. Each point of the core determines, and is determined by, an embedding of $\mathfrak{sl}(2,\mathbb{R}) \hookrightarrow \mathfrak{g}_{\mathbb{R}}$, compatibly with the Cartan involutions. This fact – in effect, a refined version of the Jacobson-Morozov theorem – is a crucial ingredient of Sekiguchi's description of his correspondence. The core contains much information about the orbit; for example, $\mathcal{O}$ is $K_{\mathbb{R}}$-equivariantly and real analytically isomorphic to $T_{C(\mathcal{O})}\mathcal{O}$, the normal bundle of the core.

The properties of nilpotent $G_{\mathbb{R}}$ orbits we mentioned so far all carry over to nilpotent orbits attached to involutions: if $H_{\mathbb{R}} \subset G_{\mathbb{R}}$ is the fixed point group of an involutive automorphism $\sigma : G_{\mathbb{R}} \to G_{\mathbb{R}}$, then $H_{\mathbb{R}}$ acts on the nilpotents in the (-1)-eigenspace of σ on $\mathfrak{g}_{\mathbb{R}}$. Orbits of this type have cores, which again can be characterized as the set of minima of $\|m\|^2$, and orbits in this setting are again isomorphic to the normal bundles along their cores. Since K is the group of fixed points of the Cartan involution, this discussion applies to nilpotent K-orbits in $\mathfrak{p}$. The core of any such orbit $\mathcal{O}_{\mathfrak{p}}$ corresponds to a $K_{\mathbb{R}}$-orbit of Cartan-compatible embeddings of $\mathfrak{sl}(2,\mathbb{R})$ into $\mathfrak{g}_{\mathbb{R}}$, just as in the case of a nilpotent $G_{\mathbb{R}}$-orbit. Orbits of the two types are Sekiguchi-related precisely when their cores coincide via the description of cores in terms of embeddings of $\mathfrak{sl}(2,\mathbb{R})$ into $\mathfrak{g}_{\mathbb{R}}$. This shows, in particular, that the cores of any two Sekiguchi-related orbits are $K_{\mathbb{R}}$-equivariantly, real analytically isomorphic.

Not only are the cores of Sekiguchi-related orbits isomorphic, but also their normal bundles. We show this by giving a description, inspired by the nilpotent orbit theorem [1], [13], of the fibers of the normal bundles, in terms of Cartan-compatible linear maps $\mathfrak{sl}(2,\mathbb{R}) \to \mathfrak{g}_{\mathbb{R}}$. Since the orbits are isomorphic to the normal bundles of the cores, we thus get $K_{\mathbb{R}}$-equivariant, real analytic isomorphisms between related orbits. The existence of isomorphisms of this type had been deduced earlier from Kronheimer's results [9] by Vergne [17].

The description of the normal bundles, in conjunction with arguments in [1], [13], leads to our refinements of Kronheimer's results. We recall those results in §3, and state and prove the refinements in §5. Neither the logic nor the exposition of the proof of our version of the Kostant-Sekiguchi correspondence depends on these two sections.

We wish to thank David Vogan for informative discussions. In particular, he alerted us to the fact that the isomorphism between a nilpotent orbit and the normal bundle of its core is a particular instance of a general property of homogeneous spaces of reductive Lie groups.

2. Nilpotent orbits and the moment map

We consider a real semisimple Lie algebra $\mathfrak{g}_{\mathbb{R}}$, and let $G_{\mathbb{R}}$ denote the identity component of $\mathrm{Aut}(\mathfrak{g}_{\mathbb{R}})$. Further notation: $K_{\mathbb{R}} \subset G_{\mathbb{R}}$ is a maximal compact subgroup,

$$\mathfrak{g}_{\mathbb{R}} = \mathfrak{k}_{\mathbb{R}} \oplus \mathfrak{p}_{\mathbb{R}} \tag{2.1}$$

is the Cartan decomposition, and $\theta : \mathfrak{g}_{\mathbb{R}} \to \mathfrak{g}_{\mathbb{R}}$ the Cartan involution. We define the inner product

$$(\zeta_1, \zeta_2) = -B(\zeta_1, \theta\zeta_2) \qquad (\, \zeta_1, \zeta_2 \in \mathfrak{g}_{\mathbb{R}} \,) \tag{2.2}$$

in terms of the Killing form B. It is positive definite and $K_{\mathbb{R}}$-invariant. We use the term "Killing form" loosely: a $G_{\mathbb{R}}$-invariant symmetric bilinear form which is negative definite on $\mathfrak{k}_{\mathbb{R}}$.

Ness [12] has defined a moment map for linear group actions. In our situation, it is a $K_{\mathbb{R}}$-invariant, real algebraic map

$$m : \mathfrak{g}_{\mathbb{R}} - \{0\} \longrightarrow \mathfrak{p}_{\mathbb{R}}, \tag{2.3}$$

described implicitly by the equation

$$(m(\zeta), \eta) = \frac{1}{2\,\|\zeta\|^2} \left(\frac{d}{dt} \,\| \operatorname{Ad} \exp(t\eta)\zeta\|^2 \right) |_{t=0}. \tag{2.4}$$

As η runs over $\mathfrak{g}_{\mathbb{R}}$ in this equation, $m(\zeta)$ becomes determined as vector in $\mathfrak{g}_{\mathbb{R}}$. But the inner product is $K_{\mathbb{R}}$-invariant, hence $m(\zeta)$ does lie in $\mathfrak{p}_{\mathbb{R}}$. The $K_{\mathbb{R}}$-invariance also implies

$$m(\operatorname{Ad}(k)\zeta) = \operatorname{Ad}(k)(m(\zeta)) \qquad (\, k \in K_{\mathbb{R}} \,), \tag{2.5}$$

i.e., the map m is $K_{\mathbb{R}}$-equivariant. To get an explicit formula for $m(\zeta)$, we calculate:

$$\begin{aligned}
\frac{1}{2}\left(\frac{d}{dt}\|\operatorname{Ad}\exp(t\eta)\zeta\|^2\right)|_{t=0} &= ([\eta,\zeta],\zeta)\\
&= -\ B([\eta,\zeta],\theta\zeta)\\
&= -\ B(\eta,[\zeta,\theta\zeta])\\
&= B(\eta,\theta[\zeta,\theta\zeta])\\
&= -\ (\eta,[\zeta,\theta\zeta])\\
&= -\ ([\zeta,\theta\zeta],\eta),
\end{aligned} \tag{2.6}$$

for every test vector $\eta \in \mathfrak{g}_{\mathbb{R}}$, hence

$$m(\zeta) \ = \ -\frac{[\zeta,\theta\zeta]}{\|\zeta\|^2}. \tag{2.7}$$

The moment map is invariant under scaling, hence descends to the projectivized Lie algebra $\mathbb{P}(\mathfrak{g}_{\mathbb{R}})$. For our purposes, it is preferable to work on

$$\mathbb{S}(\mathfrak{g}_{\mathbb{R}}) \ = \ \mathbb{R}^+\backslash(\mathfrak{g}_{\mathbb{R}} - \{0\}), \tag{2.8a}$$

the universal (two-fold) cover of $\mathbb{P}(\mathfrak{g}_{\mathbb{R}}) = \mathbb{R}^*\backslash(\mathfrak{g}_{\mathbb{R}} - \{0\})$. Note that

$$\mathbb{S}(\mathfrak{g}_{\mathbb{R}}) \ \cong \ \{\,\zeta \in \mathfrak{g}_{\mathbb{R}} \mid \|\zeta\|^2 = 1\,\}; \tag{2.8b}$$

however, to see the action of $G_{\mathbb{R}}$, one must think in terms of the description (2.8a) of $\mathbb{S}(\mathfrak{g}_{\mathbb{R}})$.

For our next statement, we fix a particular nilpotent $G_{\mathbb{R}}$-orbit $\mathcal{O} \subset \mathfrak{g}_{\mathbb{R}} - \{0\}$. By Jacobson-Morozov, any $\zeta \in \mathcal{O}$ can be embedded in an essentially unique $\mathfrak{sl}_2$-triple. In other words, there exist τ, ζ_- in $\mathfrak{g}_{\mathbb{R}}$ such that

$$[\tau,\zeta] = 2\zeta, \quad [\tau,\zeta_-] = -2\zeta_-, \quad [\zeta,\zeta_-] = \tau, \tag{2.9}$$

τ is unique up to conjugacy by the centralizer of ζ in $G_{\mathbb{R}}$, and ζ_- becomes unique once τ has been chosen. In particular, the orbit $\mathcal{O}$ determines τ up to $G_{\mathbb{R}}$-conjugacy. Thus, when we re-scale B by requiring

$$B(\tau,\tau) \ = \ 2, \tag{2.10}$$

the normalization depends on the orbit $\mathcal{O}$, not on the particular choice of ζ. By construction, the re-scaled B restricts to the linear span of ζ, ζ_-, τ as the trace form of $\mathfrak{sl}(2,\mathbb{R})$, to which this linear span is isomorphic. The one parameter subgroup of $G_{\mathbb{R}}$ generated by τ normalizes ζ and acts on it via $\mathbb{R}^+$. This establishes the well-known fact that nilpotent orbits are invariant under scaling by positive numbers. The action of $K_{\mathbb{R}}$ on the nilpotent orbit $\mathcal{O}$ commutes with scaling, so the product group $K_{\mathbb{R}} \times \mathbb{R}^+$ acts on $\mathcal{O}$.

2.11. Lemma. *A point $\zeta \in \mathcal{O}$ is a critical point of the function $\zeta \mapsto \|m(\zeta)\|^2$ if and only if there exists a real number a, $a < 0$, such that*

$$[[\zeta, \theta\zeta], \zeta] \;=\; a\,\zeta \quad \textit{and} \quad [[\zeta, \theta\zeta], \theta\zeta] \;=\; -a\,\theta\zeta.$$

The set of critical points is non-empty and consists of a single $K_{\mathbb{R}} \times \mathbb{R}^+$-orbit. The function $\|m\|^2$ on $\mathcal{O}$ assumes its minimum value exactly on the critical set.

Proof. Most of the assertions of the lemma follow readily from an adaptation of [12, Theorems 6.1, 6.2] to the case of real group actions [10]. It is also possible to argue directly in our particular situation, as follows. To begin with, ζ is a critical point if and only if $\operatorname{ad}(m)(\zeta)$ normalizes the line $\mathbb{R}\zeta$; this comes down to a short calculation, as in the proof of [12, Theorem 6.1]. Hence ζ is a critical point if and only if

$$[[\zeta, \theta\zeta], \zeta] \;=\; a\,\zeta \tag{2.12a}$$

for some $a \in \mathbb{R}$. Applying θ to both sides, we find

$$[[\zeta, \theta\zeta], \theta\zeta] \;=\; -a\,\theta\zeta. \tag{2.12b}$$

Next we argue that (2.12a), plus the nilpotency of ζ, forces $a < 0$. Indeed, $[\zeta, \theta\zeta]$ lies in the (-1)-eigenspace of θ, i.e., in $\mathfrak{p}_{\mathbb{R}}$, on which B is positive definite. Thus

$$a\|\zeta\|^2 \;=\; -B([[\zeta, \theta\zeta], \zeta], \theta\zeta) \;=\; -B([\zeta, \theta\zeta], [\zeta, \theta\zeta]) \;=\; -\|[\zeta, \theta\zeta]\|^2 \;\leq 0\,.$$

Equality cannot hold: write $\zeta = \zeta_1 + \zeta_2$ with $\zeta_1 \in \mathfrak{k}_{\mathbb{R}}$, $\zeta_2 \in \mathfrak{p}_{\mathbb{R}}$; $[\zeta, \theta\zeta] = 0$ implies $[\zeta_1, \zeta_2] = 0$; both summands are semisimple, making ζ simultaneously semisimple and nilpotent – impossible, since $\mathcal{O} \neq \{0\}$ by assumption. This gives the first assertion of the lemma. Continuing with the assumption that ζ is a critical point, we rescale ζ by a positive multiple to make $a = -2$. Then, if we set $\tau = -[\zeta, \theta\zeta]$ and $\zeta_- = -\theta\zeta$, the triple ζ, τ, ζ_- is a strictly normal S-triple in the sense of Sekiguchi [16]. The set of all $\zeta \in \mathcal{O}$ which can be embedded into a strictly normal S-triple consists of exactly one $K_{\mathbb{R}}$-orbit [16]. Thus, as claimed, the critical set in $\mathcal{O}$ is non-empty, and $K_{\mathbb{R}} \times \mathbb{R}^+$ acts transitively on it. The moment map (2.4) extends naturally to the complexification $\mathcal{O}_{\mathbb{C}}$ of $\mathcal{O}$ – i.e., the orbit of $\mathrm{Aut}(\mathfrak{g})^0$ in $\mathfrak{g} = \mathbb{C} \otimes_{\mathbb{R}} \mathfrak{g}_{\mathbb{R}}$ passing through $\mathcal{O}$. Any critical point of $\|m\|^2 : \mathcal{O} \to \mathbb{R}_{>0}$ remains critical for the function $\|m\|^2$ on $\mathcal{O}_{\mathbb{C}}$. According to [12, Theorem 6.2], the set of critical points of $\|m\|^2 : \mathcal{O}_{\mathbb{C}} \to \mathbb{R}_{>0}$ coincides with the set of minima of $\|m\|^2$ on $\mathcal{O}_{\mathbb{C}}$. We conclude that all critical points of $\|m\|^2$ on $\mathcal{O}$ are minima, as asserted by the lemma.

q.e.d.

Let us rephrase the lemma in slightly different terms. Since $\mathbb{R}^+$ acts on the nilpotent orbit $\mathcal{O}$, we can define

$$\mathbb{S}(\mathcal{O}) \;=\; \mathbb{R}^+ \backslash \mathcal{O} \;\cong\; \{\, \zeta \in \mathcal{O} \mid \|\zeta\|^2 = 1 \,\} \tag{2.13}$$

in analogy to (2.8). We shall call

$$C(\mathcal{O}) \;=\; \{\, \zeta \in \mathbb{S}(\mathcal{O}) \mid \zeta \text{ is a critical point for } \|m\|^2 \,\} \tag{2.14}$$

the *core* of $\mathcal{O}$. The core becomes a submanifold of $\mathcal{O}$ when we identify $\mathbb{S}(\mathcal{O})$ with the set of unit vectors in $\mathcal{O}$: in analogy to (2.8b),

(2.15) $C(\mathcal{O})$ is the set of all critical points in $\mathcal{O}$ of unit length.

According to Lemma 2.11,

(2.16)
a) $C(\mathcal{O})$ is non-empty,
b) $K_{\mathbb{R}}$ acts transitively on $C(\mathcal{O})$, and
c) $\mathbb{R}^+ \cdot C(\mathcal{O})$ is the critical set in $\mathcal{O}$.

The simplest example of a pair $(\mathfrak{g}_{\mathbb{R}}, \mathfrak{k}_{\mathbb{R}})$ satisfying our hypotheses is $(\mathfrak{sl}(2,\mathbb{R}), \mathfrak{so}(2))$. To simplify the notation, we set

$$\begin{aligned} \mathfrak{s}_{\mathbb{R}} \;&=\; \mathfrak{sl}(2,\mathbb{R}), \quad \text{with Cartan involution} \\ &\theta_{\mathfrak{s}} : \mathfrak{s} \to \mathfrak{s}\,, \quad \theta_{\mathfrak{s}}(\zeta) \;=\; -\,{}^t\zeta. \end{aligned} \tag{2.17a}$$

The three elements

$$e \;=\; \begin{pmatrix} 0 & 1 \\ 0 & 0 \end{pmatrix}, \quad f \;=\; \begin{pmatrix} 0 & 0 \\ 1 & 0 \end{pmatrix}, \quad h \;=\; \begin{pmatrix} 1 & 0 \\ 0 & -1 \end{pmatrix} \tag{2.17b}$$

constitute a basis of $\mathfrak{s}_{\mathbb{R}}$ and satisfy the relations

$$\begin{aligned} [h,e] \;&=\; 2e, \quad [h,f] \;=\; -2f, \quad [e,f] \;=\; h, \\ \theta_{\mathfrak{s}}(e) \;&=\; -f\,, \quad \theta_{\mathfrak{s}}(h) \;=\; -h. \end{aligned} \tag{2.17c}$$

Although we are interested primarily in real Lie algebras, it is useful for certain purposes to complexify. We write $\mathrm{Mor}(\mathfrak{s}, \mathfrak{g})$ for the set of non-zero

Lie algebra homomorphisms from $\mathfrak{s} = \mathfrak{sl}(2,\mathbb{C})$ to the complexification $\mathfrak{g} = \mathbb{C} \otimes_{\mathbb{R}} \mathfrak{g}_{\mathbb{R}}$ of $\mathfrak{g}_{\mathbb{R}}$, and define

$$\begin{aligned} \operatorname{Mor}^{\mathbb{R}}(\mathfrak{s},\mathfrak{g}) &= \{\, \Phi \in \operatorname{Mor}(\mathfrak{s},\mathfrak{g}) \mid \Phi \text{ is defined over } \mathbb{R} \,\}, \\ \operatorname{Mor}^{\theta}(\mathfrak{s},\mathfrak{g}) &= \{\, \Phi \in \operatorname{Mor}(\mathfrak{s},\mathfrak{g}) \mid \theta \circ \Phi = \Phi \circ \theta_{\mathfrak{s}} \,\}, \\ \operatorname{Mor}^{\mathbb{R},\theta}(\mathfrak{s},\mathfrak{g}) &= \operatorname{Mor}^{\mathbb{R}}(\mathfrak{s},\mathfrak{g}) \cap \operatorname{Mor}^{\theta}(\mathfrak{s},\mathfrak{g}). \end{aligned} \tag{2.18}$$

Note that $\operatorname{Mor}^{\mathbb{R}}(\mathfrak{s},\mathfrak{g})$ is naturally isomorphic to $\operatorname{Mor}(\mathfrak{s}_{\mathbb{R}},\mathfrak{g}_{\mathbb{R}})$, the set of non-trivial morphisms between the real Lie algebras $\mathfrak{s}_{\mathbb{R}}$, $\mathfrak{g}_{\mathbb{R}}$. The group $K_{\mathbb{R}}$ acts on $\operatorname{Mor}^{\mathbb{R}}(\mathfrak{s},\mathfrak{g})$ through the adjoint action on $\mathfrak{g}_{\mathbb{R}}$: $(k\,\Phi)(\zeta) =_{\text{def}} \operatorname{Ad} k(\Phi(\zeta))$.

2.19. Lemma. *The map $\Phi \mapsto \Phi(e)$ establishes a $K_{\mathbb{R}}$-equivariant isomorphism*

$$\{\, \Phi \in \operatorname{Mor}^{\mathbb{R},\theta}(\mathfrak{s},\mathfrak{g}) \mid \Phi(e) \in \mathcal{O} \,\} \;\cong\; C(\mathcal{O})\,.$$

Proof. Note that any $\Phi \in \operatorname{Mor}^{\mathbb{R},\theta}(\mathfrak{s},\mathfrak{g})$ is uniquely determined by its value on e – cf. (2.17c). If $\zeta = \Phi(e)$ lies in the orbit $\mathcal{O}$, it is a critical point, as follows from Lemma 2.11, coupled with the relations (2.17c); any such ζ has unit length since the normalization (2.10) of the Killing form makes Φ an isometry, relative to the trace form on $\mathfrak{s}_{\mathbb{R}}$. This makes the map $\Phi \mapsto \Phi(e)$ well defined and injective. It is surjective because $a = -2$ in the proof of Lemma (2.11) if and only if $\|\zeta\| = 1$; in that case, the triple ζ, ζ_-, τ defined in that proof satisfy the same relations (2.17c) as e, f, h. The equivariance, finally, is obvious from the definition of the action. q.e.d.

Lemma 2.19, together with 2.16, formally implies a statement that appears, in different language, in [16]: the set of nilpotent $G_{\mathbb{R}}$-orbit in $\mathfrak{g}_{\mathbb{R}} - \{0\}$ corresponds bijectively to the set of $K_{\mathbb{R}}$-orbits in $\operatorname{Mor}^{\mathbb{R},\theta}(\mathfrak{s},\mathfrak{g})$.

The inner product (2.2), normalized as in (2.10), determines a $K_{\mathbb{R}}$-invariant Riemannian metric on $\mathbb{S}(\mathfrak{g}_{\mathbb{R}})$. We use this metric to give meaning to the gradient vector field $\nabla\|m\|^2$ on $\mathbb{S}(\mathfrak{g}_{\mathbb{R}})$. Note that $\mathfrak{g}_{\mathbb{R}}$ acts on $\mathbb{S}(\mathfrak{g}_{\mathbb{R}})$ by infinitesimal translation. For $\eta \in \mathfrak{g}_{\mathbb{R}}$, $\ell(\eta)$ shall denote the vector field corresponding to η. A simple calculation shows

$$(\nabla\|m\|^2)|_{\zeta} \;=\; 2\ell(m(\zeta)) \tag{2.20}$$

[12]. In particular, the gradient vector field – both on $\mathcal{O}$ and on $\mathbb{S}(\mathcal{O})$ – is tangential to $G_{\mathbb{R}}$-orbits.

2.21. Proposition. *The function $\|m\|^2 : \mathbb{S}(\mathcal{O}) \to \mathbb{R}$ is Bott-Morse. It assumes its minimum value on the core $C(\mathcal{O})$, and has no other critical points. Its gradient flow establishes a natural $K_{\mathbb{R}}$-equivariant real analytic map from $\mathbb{S}(\mathcal{O})$ to the core $C(\mathcal{O})$ which exhibits $C(\mathcal{O})$ as a strong deformation retract of $\mathbb{S}(\mathcal{O})$.*

The normalization (2.10) specifies the value of $\|m\|^2$ on $C(\mathcal{O})$ as 2. Thus we can conclude:

2.22. Corollary. *The family of open sets*

$$\{\eta \in \mathbb{S}(\mathcal{O}) \mid \|m\|^2(\eta) < 2 + \epsilon\}\,, \quad \epsilon > 0\,,$$

forms a neighborhood basis of $C(\mathcal{O})$.

Since $\mathbb{S}(\mathcal{O}) = \mathbb{R}^*\backslash\mathcal{O}$, we can combine the retraction $\mathbb{S}(\mathcal{O}) \to C(\mathcal{O})$ with $\mathbb{R}^* \to \{1\}$ to construct a retraction of $\mathcal{O}$:

2.23. Corollary. *There exists a $K_{\mathbb{R}}$-equivariant, real analytic, strong deformation retraction $\mathcal{O} \to C(\mathcal{O})$.*

Proof of Proposition 2.21. Recall the notion of a Bott-Morse function: the critical set is a compact manifold, and the Hessian descends to a non-degenerate bilinear form on the normal bundle. Lemma 2.11 implies that $\|m\|^2$ assumes its minimum along $C(\mathcal{O})$ and has no critical points outside of $C(\mathcal{O})$, which is surely smooth and compact. Ness [12, Theorem 6.2] points out that the non-degeneracy is a general property of moment maps attached to linear actions of semisimple groups. This establishes all but the final assertion. For the last assertion, let us consider the unstable set of the gradient flow of $\|m\|^2$ associated to $C(\mathcal{O})$, i.e., the union of the integral curves of $\nabla\|m\|^2$ emanating from $C(\mathcal{O})$. Because the function $\|m\|^2$ is Bott-Morse, this set is a manifold. We shall show:

(2.24) the unstable set of the gradient flow associated to the critical set $C(\mathcal{O})$ consists of all of $\mathbb{S}(\mathcal{O})$.

The existence of a retraction from $\mathbb{S}(\mathcal{O})$ to $C(\mathcal{O})$ will then follow. To establish (2.24), we may work on $\mathbb{P}(\mathcal{O}) = \{\pm 1\}\backslash\mathbb{S}(\mathcal{O})$. This allows us to complexify the situation, replacing $\mathbb{P}(\mathfrak{g}_{\mathbb{R}})$ by $\mathbb{P}(\mathfrak{g})$, the projectivization of the complexified Lie algebra $\mathfrak{g}$, and correspondingly $\mathbb{P}(\mathcal{O})$ by $\mathbb{P}(\mathcal{O}_{\mathbb{C}})$, the appropriate orbit of $G = \mathrm{Aut}^0(\mathfrak{g})$. The inner product on $\mathfrak{g}_{\mathbb{R}}$ extends to a hermitian inner product on $\mathfrak{g}$, which is preserved by $U_{\mathbb{R}}$, the unique maximal compact subgroup of G which contains $K_{\mathbb{R}}$. The Definition (2.4) of the moment map carries over to the complexified setting, where it agrees with the usual (symplectic) moment map associated to the action of $U_{\mathbb{R}}$ on $\mathbb{P}(\mathfrak{g})$ [12]. According to Kirwan [6, Theorem 6.18], the stratification defined by the gradient flow on $\mathbb{P}(\mathfrak{g})$ is G-invariant. In particular, it is $G_{\mathbb{R}}$-invariant. Since the gradient flow on $\mathbb{P}(\mathfrak{g}_{\mathbb{R}})$ is tangential to the $G_{\mathbb{R}}$-orbits, we can deduce that the stratification of $\mathbb{P}(\mathfrak{g}_{\mathbb{R}})$ defined by the gradient flow is $G_{\mathbb{R}}$-invariant. q.e.d.

The assertion of Corollary 2.23 can be strengthened considerably: the orbit $\mathcal{O}$ is isomorphic to the normal bundle of its core $C(\mathcal{O})$. David Vogan pointed out to us that this is a particular instance of a general fact about homogeneous spaces of semisimple Lie groups. Mostow [11, Theorem 5] proved that any quotient of a semisimple group by a semisimple subgroup fibers equivariantly over an orbit of a maximal compact subgroup, with Euclidean fibers; the fibers are then necessarily the fibers of the normal bundle. The analogous statement in general case, i.e., for the quotient of a semisimple group by a closed subgroup, can be reduced to Mostow's theorem. Below we shall sketch the argument for nilpotent orbits, since we know of no statement in the literature that would imply it.

2.25. Proposition. *There exists a $K_{\mathbb{R}}$-equivariant, real analytic isomorphism $\mathcal{O} \cong T_{C(\mathcal{O})}\mathcal{O}$.*

Proof. We fix a point $\Phi(e) \in C(\mathcal{O})$ and use $\Phi \in \operatorname{Mor}^{\mathbb{R},\theta}(\mathfrak{s},\mathfrak{g})$ to identify $\mathfrak{s}$ with a subalgebra of $\mathfrak{g}$. In particular, e, f, h now all lie in $\mathfrak{g}_{\mathbb{R}}$, the Cartan involution maps e to $-f$ and h to $-h$, and e lies in $C(\mathcal{O})$. We define

$$
(2.26)\qquad
\begin{aligned}
\mathfrak{m}_{\mathbb{R}} &= \text{centralizer of } h \text{ in } \mathfrak{g}_{\mathbb{R}},\\
\mathfrak{n}_{\mathbb{R}} &= \text{direct sum of all eigenspaces of } \operatorname{ad} h \text{ in } \mathfrak{g}_{\mathbb{R}}\\
&\qquad \text{corresponding to strictly positive eigenvalues},\\
M_{\mathbb{R}} &= \text{centralizer of } h \text{ in } G_{\mathbb{R}},\\
N_{\mathbb{R}} &= \exp \mathfrak{n}_{\mathbb{R}}.
\end{aligned}
$$

Then $\mathfrak{m}_{\mathbb{R}} \oplus \mathfrak{n}_{\mathbb{R}} \subset \mathfrak{g}_{\mathbb{R}}$ is a parabolic subalgebra and $M_{\mathbb{R}} \cdot N_{\mathbb{R}}$ (semi-direct product) the corresponding parabolic subgroup of $G_{\mathbb{R}}$. Since $h \in \mathfrak{p}_{\mathbb{R}}$, the Cartan involution fixes $\mathfrak{m}_{\mathbb{R}}$, $K_{\mathbb{R}} \cap M_{\mathbb{R}}$ is maximal compact in $M_{\mathbb{R}}$, and

$$
(2.27)\qquad G_{\mathbb{R}} \simeq K_{\mathbb{R}} \times_{K_{\mathbb{R}} \cap M_{\mathbb{R}}} (M_{\mathbb{R}} \cdot N_{\mathbb{R}}) \qquad \text{(fiber product)}.
$$

The symbol $(G_{\mathbb{R}})_e$ shall denote the centralizer of e in $G_{\mathbb{R}}$, with the analogous convention applying also to subgroups of $G_{\mathbb{R}}$ and subalgebras of

$\mathfrak{g}_{\mathbb{R}}$. We claim:

$$(2.28)\qquad \begin{aligned} &\text{a)}\quad (G_{\mathbb{R}})_e \;=\; (M_{\mathbb{R}})_e \cdot (N_{\mathbb{R}})_e, \quad \text{and} \\ &\text{b)}\quad (N_{\mathbb{R}})_e \;=\; \exp((\mathfrak{n}_{\mathbb{R}})_e). \end{aligned}$$

To see this, we suppose that $\operatorname{Ad} g(e) = e$, and express g using the decomposition (2.27) of $G_{\mathbb{R}}$ and the Cartan decomposition of $M_{\mathbb{R}}$:

$$(2.29)\ \operatorname{Ad}(k \exp\xi \, \exp\eta)\, e \;=\; e, \quad \text{with } k \in K_{\mathbb{R}}\,,\ \xi \in \mathfrak{m}_{\mathbb{R}} \cap \mathfrak{p}_{\mathbb{R}}\,,\ \eta \in \mathfrak{n}_{\mathbb{R}}.$$

Then $\tilde{e} =_{\text{def}} \operatorname{Ad}(\exp\xi \, \exp\eta)\, e \;=\; \operatorname{Ad}(k^{-1})e$ lies in $C(\mathcal{O})$. Because of (2.19), the triple $\tilde{e}$, $\tilde{f} = -\theta\tilde{e}$, $\tilde{h} = [\tilde{e}, \tilde{f}]$ satisfies the same commutation relations as the triple e, f, h. In particular, $[[\tilde{e}, \tilde{f}], \tilde{f}] = -2\tilde{f}$. Conjugating by the inverse of $\theta(\exp\xi \, \exp\eta) = \exp(-\xi) \, \exp(\theta\eta)$, we find

$$(2.30)\qquad \begin{aligned} [\,[\operatorname{Ad}(\exp(-\theta\eta) \, \exp(2\xi) \, \exp\eta)e\,,\, f\,]\,,\, f\,] \;&=\; -\,2\,f \\ &= [\,[\,e\,,\, f\,]\,,\, f\,]\,. \end{aligned}$$

From the definition of $\mathfrak{n}_{\mathbb{R}}$, one finds that $\operatorname{Ad}(\exp\eta) - 1_{\mathfrak{g}_{\mathbb{R}}}$ raises h-weights. Similarly, $\operatorname{Ad}(-\theta \exp\eta) - 1_{\mathfrak{g}_{\mathbb{R}}}$ lowers weights, and $\operatorname{Ad}(\exp\xi)$ acts semisimply with strictly positive eigenvalues, while preserving weights. We conclude: either $\exp\eta$ commutes with e, or else

$$\operatorname{Ad}(\exp(-\theta\eta) \, \exp(2\xi) \, \exp\eta)e \;=\; \textstyle\sum_\ell \zeta_\ell$$

is a linear combination of weight vectors ζ_ℓ, with $\zeta_k \neq 0$ for at least one weight $k > 2$. This latter possibility is incompatible with the identity (2.30): in any finite dimensional representation of $\mathfrak{s}$, f^2 lowers weights exactly by four and is injective on all weight spaces corresponding to weights $k \geq 2$. Conclusion: $\exp\eta \in (N_{\mathbb{R}})_e$. Arguing analogously, we find that $\operatorname{Ad}\exp(2\xi)e = e$, and even $\operatorname{Ad}(\exp\xi)e = e$ because of the nature of the action of $\operatorname{Ad}(\exp\xi)$. Now, in view of (2.29), k must also commute with e. Any element of $K_{\mathbb{R}}$ that commutes with e must commute with $f = -\theta e$, hence with $h = [e, f]$. This puts $k \exp\xi$ into $(M_{\mathbb{R}})_e$, as asserted by (2.28a). Finally, if $\exp\eta$, with $\eta \in \mathfrak{n}_{\mathbb{R}}$, centralizes e, then so do $\eta = \log(\exp\eta)$ and the one parameter group generated by η. This implies (2.28b).

The centralizers of e in $K_{\mathbb{R}}$ and $M_{\mathbb{R}}$ commute with all of $\mathfrak{s}$. In the case of $K_{\mathbb{R}}$, we just gave the argument; for $M_{\mathbb{R}}$ it follows from the observation that any two members of an $\mathfrak{sl}_2$-triple – in our case, e and h – determine the third. For emphasis,

$$(2.31)\qquad (K_{\mathbb{R}})_e \;=\; (K_{\mathbb{R}})_{\mathfrak{s}}, \quad (M_{\mathbb{R}})_e \;=\; (M_{\mathbb{R}})_{\mathfrak{s}}.$$

In particular, $(K_{\mathbb{R}})_e$ and $(M_{\mathbb{R}})_e$ normalize both $\mathfrak{n}_{\mathbb{R}}$ and $(\mathfrak{n}_{\mathbb{R}})_e$. We can choose a linear complement $\mathfrak{c}_{\mathbb{R}}$ to $(\mathfrak{n}_{\mathbb{R}})_e$ in $\mathfrak{n}_{\mathbb{R}}$, which is $(K_{\mathbb{R}})_e$-invariant and $(M_{\mathbb{R}})_e$-invariant: we decompose $\mathfrak{g}_{\mathbb{R}}$ $\mathfrak{s}_{\mathbb{R}}$-isotypically; in the isotypic subspace of highest weight r, we take the sum of all eigenspaces corresponding to eigenvalues strictly between 0 and r; then $\mathfrak{c}_{\mathbb{R}}$, the sum of all of these spaces for $r > 0$, has the required properties. Since $\mathfrak{n}_{\mathbb{R}} = \mathfrak{c}_{\mathbb{R}} \oplus (\mathfrak{n}_{\mathbb{R}})_e$ (direct sum of vector spaces),

$$\mathfrak{c}_{\mathbb{R}} \times (\mathfrak{n}_{\mathbb{R}})_e \xrightarrow{\sim} N_{\mathbb{R}}, \quad (\zeta, \eta) \mapsto \exp\zeta \cdot \exp\eta, \tag{2.32}$$

is a $(K_{\mathbb{R}})_e$-invariant, $(M_{\mathbb{R}})_e$-invariant, real analytic isomorphism. Indeed, the diffeomorphism statement can be reduced to an assertion about nilpotent matrix groups, which can be verified using Engel's theorem; the invariance properties are a consequence of the particular choice of $\mathfrak{c}_{\mathbb{R}}$. Because of (2.27-28) and (2.31-32),

$$\begin{aligned} G_{\mathbb{R}} &\simeq K_{\mathbb{R}} \times_{K_{\mathbb{R}} \cap M_{\mathbb{R}}} (M_{\mathbb{R}} \cdot N_{\mathbb{R}}) \\ &\simeq K_{\mathbb{R}} \times_{K_{\mathbb{R}} \cap M_{\mathbb{R}}} M_{\mathbb{R}} \times_{(M_{\mathbb{R}})_{\mathfrak{s}}} ((M_{\mathbb{R}})_{\mathfrak{s}} \times \mathfrak{c}_{\mathbb{R}} \times (N_{\mathbb{R}})_e) \\ &\simeq K_{\mathbb{R}} \times_{K_{\mathbb{R}} \cap M_{\mathbb{R}}} (M_{\mathbb{R}} \times \mathfrak{c}_{\mathbb{R}}) \times_{(M_{\mathbb{R}})_{\mathfrak{s}}} ((M_{\mathbb{R}})_{\mathfrak{s}} \times (N_{\mathbb{R}})_e) \\ &= K_{\mathbb{R}} \times_{K_{\mathbb{R}} \cap M_{\mathbb{R}}} (M_{\mathbb{R}} \times \mathfrak{c}_{\mathbb{R}}) \times_{(M_{\mathbb{R}})_{\mathfrak{s}}} (G_{\mathbb{R}})_e, \end{aligned} \tag{2.33}$$

as real analytic manifold with left $K_{\mathbb{R}}$- and right $(G_{\mathbb{R}})_e$-action; here $(M_{\mathbb{R}})_{\mathfrak{s}}$ acts on $M_{\mathbb{R}}$ by right translation and on $\mathfrak{c}_{\mathbb{R}}$ by conjugation. According to [11, Theorem 5], there exists an isomorphism

$$M_{\mathbb{R}} \simeq (K_{\mathbb{R}} \cap M_{\mathbb{R}}) \times_{(K_{\mathbb{R}})_{\mathfrak{s}}} (\mathfrak{p}_{\mathbb{R}} \cap \mathfrak{m}_{\mathbb{R}} \cap (\mathfrak{m}_{\mathbb{R}})_{\mathfrak{s}}^{\perp}) \times (M_{\mathbb{R}})_{\mathfrak{s}} \tag{2.34}$$

of real analytic manifolds with left $(K_{\mathbb{R}} \cap M_{\mathbb{R}})$- and right $(M_{\mathbb{R}})_{\mathfrak{s}}$-action. Mostow states his decomposition theorem for connected, semisimple groups; the extension to our situation is straightforward. In the decomposition (2.34), $(M_{\mathbb{R}})_{\mathfrak{s}}$ and $(K_{\mathbb{R}})_{\mathfrak{s}}$ act on $\mathfrak{p}_{\mathbb{R}} \cap \mathfrak{m}_{\mathbb{R}} \cap (\mathfrak{m}_{\mathbb{R}})_{\mathfrak{s}}^{\perp}$ by conjugation. We conclude:

$$\mathcal{O} \simeq G_{\mathbb{R}}/(G_{\mathbb{R}})_e \simeq K_{\mathbb{R}} \times_{(K_{\mathbb{R}})_{\mathfrak{s}}} \left((\mathfrak{p}_{\mathbb{R}} \cap \mathfrak{m}_{\mathbb{R}} \cap (\mathfrak{m}_{\mathbb{R}})_{\mathfrak{s}}^{\perp}) \cap \mathfrak{c}_{\mathbb{R}} \right). \tag{2.35}$$

This is equivalent to the statement of the proposition. q.e.d.

3. The instanton flow

In the previous section, we described a flow on a nilpotent orbit $\mathcal{O}$ which retracts the orbit to its core. Kronheimer has constructed a different flow, which also retracts the nilpotent orbit to its core [9]. Let us describe his construction in slightly different language.

We continue with the notation and hypotheses of §2. While we are interested in a nilpotent orbit $\mathcal{O}$ of the real group $G_{\mathbb{R}}$ in the real Lie algebra $\mathfrak{g}_{\mathbb{R}}$, we will work also with the complexified group G, the complexified Lie algebra $\mathfrak{g}$, and the complexification $\mathfrak{s} = \mathfrak{sl}(2,\mathbb{C})$ of $\mathfrak{s}_{\mathbb{R}} = \mathfrak{sl}(2,\mathbb{R})$. In analogy to (2.18), we define

$$(3.1)\qquad \begin{aligned} \operatorname{Hom}(\mathfrak{s},\mathfrak{g}) &= \text{vector space of } \mathbb{C}\text{-linear maps } \Phi : \mathfrak{s} \to \mathfrak{g}, \\ \operatorname{Hom}^{\mathbb{R}}(\mathfrak{s},\mathfrak{g}) &= \{\, \Phi \in \operatorname{Hom}(\mathfrak{s},\mathfrak{g}) \mid \Phi \text{ is defined over } \mathbb{R} \,\}, \\ \operatorname{Hom}^{\theta}(\mathfrak{s},\mathfrak{g}) &= \{\, \Phi \in \operatorname{Hom}(\mathfrak{s},\mathfrak{g}) \mid \theta \circ \Phi = \Phi \circ \theta_{\mathfrak{s}} \,\}, \\ \operatorname{Hom}^{\mathbb{R},\theta}(\mathfrak{s},\mathfrak{g}) &= \operatorname{Hom}^{\mathbb{R}}(\mathfrak{s},\mathfrak{g}) \cap \operatorname{Hom}^{\theta}(\mathfrak{s},\mathfrak{g}). \end{aligned}$$

The Lie bracket can be viewed as a G-equivariant linear map $\wedge^2\mathfrak{g} \to \mathfrak{g}$. In the case of $\mathfrak{s}$, this is an isomorphism for dimension reasons, hence can be inverted to an $SL(2,\mathbb{C})$-equivariant linear map $\mathfrak{s} \to \wedge^2\mathfrak{s}$. Combining the two maps, we get a symmetric bilinear pairing

$$(3.2a)\qquad Q \;:\; \operatorname{Hom}(\mathfrak{s},\mathfrak{g}) \otimes \operatorname{Hom}(\mathfrak{s},\mathfrak{g}) \;\longrightarrow\; \operatorname{Hom}(\mathfrak{s},\mathfrak{g}),$$

which is uniquely characterized by the equation

$$(3.2b)\qquad Q(\Phi_1,\Phi_2)[u,v] \;=\; \frac{1}{2}\left([\Phi_1(u),\Phi_2(v)] - [\Phi_1(v),\Phi_2(u)] \right) \qquad (\, u,v \in \mathfrak{s} \,).$$

Note that

$$(3.3)\qquad Q(\Phi,\Phi) \;=\; \Phi \iff \Phi \in \operatorname{Mor}(\mathfrak{s},\mathfrak{g});$$

here, as in the previous section, $\operatorname{Mor}(\mathfrak{s},\mathfrak{g})$ denotes the set of Lie algebra homomorphisms. The pairing is defined over $\mathbb{R}$, i.e.,

$$(3.4)\qquad Q \;:\; \operatorname{Hom}^{\mathbb{R}}(\mathfrak{s},\mathfrak{g}) \otimes \operatorname{Hom}^{\mathbb{R}}(\mathfrak{s},\mathfrak{g}) \;\longrightarrow\; \operatorname{Hom}^{\mathbb{R}}(\mathfrak{s},\mathfrak{g}),$$

and it is compatible with the Cartan involutions, in the sense that

$$(3.5)\qquad Q \;:\; \operatorname{Hom}^{\theta}(\mathfrak{s},\mathfrak{g}) \otimes \operatorname{Hom}^{\theta}(\mathfrak{s},\mathfrak{g}) \;\longrightarrow\; \operatorname{Hom}^{\theta}(\mathfrak{s},\mathfrak{g}).$$

These properties are immediate consequences of (3.2b).

3.6. Notation. $\mathcal{M}$ is the set of C^∞-maps $\Phi : (0,\infty) \to \operatorname{Hom}^{\mathbb{R},\theta}(\mathfrak{s},\mathfrak{g})$ satisfying the three conditions

a) $\frac{d}{dt}\Phi(t) \;=\; -\,Q(\Phi(t),\Phi(t)),$

b) Φ extends continuously to $[0, \infty)$,

c) $\lim_{t\to\infty}(t\Phi(t))$ exists and lies in $\mathrm{Mor}^{\mathbb{R},\theta}(\mathfrak{s}, \mathfrak{g})$.

For $\Phi_0 \in \mathrm{Mor}^{\mathbb{R},\theta}(\mathfrak{s}, \mathfrak{g})$, we set

$$\mathcal{M}(\Phi_0) = \{\Phi \in \mathcal{M} \mid \lim_{t\to\infty}(t\Phi(t)) = \Phi_0\}.$$

If $C(\mathcal{O})$ is the core of a nilpotent $G_{\mathbb{R}}$-orbit $\mathcal{O} \subset \mathfrak{g}_{\mathbb{R}}$, $\mathcal{M}(C(\mathcal{O}))$ will denote the union of the $\mathcal{M}(\Phi_0)$ corresponding to morphisms Φ_0 whose image $\Phi_0(e)$ under the isomorphism (2.19) lies in $C(\mathcal{O})$.

The conditions b),c) in this definition can be restated in equivalent, but seemingly weaker form – see below.

3.7. Theorem (Kronheimer, [9]). *The space $\mathcal{M}$ has a natural structure of C^∞ manifold. Via the map $\Phi(\,\cdot\,) \mapsto \Phi(0)(e)$, this manifold is $K_{\mathbb{R}}$-equivariantly diffeomorphic to the nilpotent orbit $\mathcal{O}$.*

Strictly speaking, Kronheimer states this result for complex groups. Vergne [17] observed that the statement about real groups formally follows from the result about complex groups by restriction. Kronheimer deduces the manifold structure from general properties of moduli spaces for instantons. The manifold structure also becomes apparent from our results in §5.

To make the transition to Kronheimer's formulation, we attach to each $\Phi \in \mathcal{M}$ a triple of $\mathfrak{g}_{\mathbb{R}}$-valued functions by evaluating $\Phi(t)$ on the triple (2.17b),

$$E(t) = \Phi(t)(e), \quad F(t) = \Phi(t)(f), \quad H(t) = \Phi(t)(h). \tag{3.8a}$$

This triple completely determines the function Φ. The requirement that the values $\Phi(t)$ be compatible with the Cartan involution translates into the condition

$$F(t) = -\theta E(t), \quad H(t) = -\theta H(t). \tag{3.8b}$$

Let us transcribe the conditions a),b),c) in the Definition 3.6. The differential equation (3.6a) becomes

$$\begin{aligned} 2E'(t) &= -[H(t), E(t)], \quad 2F'(t) = [H(t), F(t)], \\ H'(t) &= -[E(t), F(t)]; \end{aligned} \tag{3.9a}$$

the first of these follows from

$$Q(\Phi, \Phi)(e) = \frac{1}{2}Q(\Phi, \Phi)[h, e] = \frac{1}{2}[\Phi(h), \Phi(e)],$$

and similarly for the others. Next,

$$E(t), F(t), H(t) \text{ extend continuously to } [0,\infty), \tag{3.9b}$$

and finally,

$$\begin{aligned} &\text{the limits } E_0 = \lim_{t\to\infty}(tE(t)), \quad F_0 = \lim_{t\to\infty}(tF(t)), H_0 = \lim_{t\to\infty}(tH(t)) \\ &\text{existand satisfy } 2E_0 = [H_0, E_0], \quad 2F_0 = -[H_0, F_0], \quad H_0 = [E_0, F_0]. \end{aligned} \tag{3.9c}$$

In terms of the triple, the map $\Phi \mapsto \Phi(0)(e)$ reduces to evaluating $E(t)$ at zero.

Kronheimer, who works in the context of complex nilpotent orbits, uses a triple of $\mathfrak{g}$-valued functions corresponding to a different basis of $\mathfrak{s}$. Also, he uses the coordinate $x = -\log t$ on $\mathbb{R}$, which gives a slightly different appearance to the differential equation (3.9a) and the "evaluations" $E(t) \rightsquigarrow E_0$ and $E(t) \rightsquigarrow E(0)$.

The $\mathfrak{g}_{\mathbb{R}}$-valued function $\frac{1}{2}H(t)$ is the logarithmic derivative of a C^∞ function $g(t)$ with values in $G_{\mathbb{R}}$ – in other words, $2g(t)^{-1}g'(t) = H(t)$. Since

$$\begin{aligned} \frac{d}{dt}(\operatorname{Ad} g(t)(E(t))) &= \operatorname{Ad} g(t)\left([g(t)^{-1}g'(t), E(t)] + E'(t)\right) \\ &= \operatorname{Ad} g(t)\left([\tfrac{1}{2}H(t), E(t)] + E'(t)\right) \\ &= 0, \end{aligned} \tag{3.10}$$

the curve $E(t)$, for $0 < t < \infty$, stays inside a nilpotent $G_{\mathbb{R}}$-orbit $\mathcal{O}$. The fact that $E(0)$ and $E_0 = \lim_{t\to\infty}(tE(t))$ lie in the same orbit $\mathcal{O}$ is a consequence of Kronheimer's theorem. Because of (3.8b) and (3.9c) – equivalently, because Φ_0 belongs to $\operatorname{Mor}^{\mathbb{R},\theta}(\mathfrak{s},\mathfrak{g})$ – E_0 lies in the core $C(\mathcal{O})$. In particular, then, $E(0) \mapsto E_0$ exhibits $C(\mathcal{O})$ as the strong deformation retract of $\mathcal{O}$. Via the isomorphism (3.7),

$$C(\mathcal{O}) \cong \{\, \Phi_0 \in \operatorname{Mor}^{\mathbb{R},\theta}(\mathfrak{s},\mathfrak{g}) \mid \Phi_0(e) \in \mathcal{O} \,\} \tag{3.11a}$$

corresponds to the $\operatorname{Hom}^{\mathbb{R},\theta}(\mathfrak{s},\mathfrak{g})$-valued functions

$$t \mapsto \Phi(t) =_{\text{def}} \Phi_0\,(1+t)^{-1}, \tag{3.11b}$$

which satisfies the differential equation (3.6a) and takes the value Φ_0 at $t = 0$. There are two simple operations on $\mathcal{M}(\Phi_0)$ as defined in (3.6): for $a \in \mathbb{R}^+$

$$\{\, t \mapsto \Phi(t) \,\} \longrightarrow \{\, t \mapsto a\Phi(at) \,\}, \tag{3.12}$$

which corresponds to scaling on $\mathcal{O}$ under the isomorphism (3.7), and

$$\{\, t \mapsto \Phi(t)\,\} \longrightarrow \{\, t \mapsto a\Phi(a(t+1)-1)\,\}, \tag{3.13}$$

$1 \leq a < \infty$, which induces the homotopy between the identity map $1_{\mathcal{O}}$ and the retraction $\mathcal{O} \to C(\mathcal{O})$; note that (3.13) does act trivially on the functions (3.11b).

The instanton flow is a flow in $\mathrm{Hom}^{\mathbb{R},\theta}(\mathfrak{s},\mathfrak{g})$, the gradient flow of the function $\Phi \mapsto \|\Phi\|^2$ on $\mathrm{Hom}^{\mathbb{R},\theta}(\mathfrak{s},\mathfrak{g})$ [9]. Via the isomorphism $\mathcal{M}(\mathcal{O}) \cong \mathcal{O}$, it corresponds to the retraction (3.13), which is not a gradient flow of a function on $\mathcal{O}$ or $\mathbb{S}(\mathcal{O})$, nor even the flow of a (time independent!) vector field. Curiously, the retraction is induced by a vector field on certain submanifolds of nilpotent orbits, namely those which arise from variations of Hodge structure [13].

The functions $\Phi(\,\cdot\,) \in \mathcal{M}$ are real analytic: for any $t_0 \in (0,\infty)$, the coefficients of the Taylor series of $\Phi(t)$ at $t = t_0$ are polynomials in $\Phi(t_0)$ by repeated differentiation of the equation (3.6a), and the radius of convergence of this Taylor series can be bounded from below by a uniform multiple of $\|\Phi(t_0)\|^{-1}$. In particular, the condition (3.6b) can be replaced by the formally weaker condition

$$\|\Phi(t)\| \;\text{ is bounded on }\; (0,\infty), \tag{3.14}$$

as long as the remaining conditions are maintained. It implies the stronger condition

$$\|\Phi(t)\| \;\text{ extends real analytically to }\; [0,\infty). \tag{3.15}$$

In §5 we shall show that the $\Phi(t)$ are real analytic even at infinity, as functions of the variable $t^{-\frac{1}{2}}$.

4. The normal bundle of the core

The core $C(\mathcal{O})$ of a nilpotent $G_{\mathbb{R}}$-orbit $\mathcal{O} \subset \mathfrak{g}_{\mathbb{R}}$ is a $K_{\mathbb{R}}$-orbit. This fact gives the normal bundle $T_{C(\mathcal{O})}\mathcal{O}$ the structure of $K_{\mathbb{R}}$-homogenous vector bundle. As such, it is associated to the representation of

$$(K_{\mathbb{R}})_\zeta \;=\; \text{isotropy subgroup at } \zeta, \tag{4.1}$$

for any particular $\zeta \in C(\mathcal{O})$, on the quotient

$$[\zeta,\mathfrak{g}_{\mathbb{R}}]/[\zeta,\mathfrak{k}_{\mathbb{R}}] \;\cong\; (T_{C(\mathcal{O})}\mathcal{O})_\zeta . \tag{4.2}$$

In this section, we shall construct a $(K_{\mathbb{R}})_\zeta$-invariant linear complement to $[\zeta, \mathfrak{k}_{\mathbb{R}}]$ in $[\zeta, \mathfrak{g}_{\mathbb{R}}]$. We shall need this construction in subsequent sections.

We identify the base point ζ with the morphism $\Phi_0 \in \mathrm{Mor}^{\mathbb{R},\theta}(\mathfrak{s}, \mathfrak{g})$ which corresponds to ζ via the isomorphism (2.19). To simplify the discussion, we use Φ_0 to identify $\mathfrak{s}_{\mathbb{R}}$ with a subalgebra of $\mathfrak{g}_{\mathbb{R}}$. This physically puts the generators (2.17) into $\mathfrak{g}_{\mathbb{R}}$, with $\zeta = e$. For emphasis,

$$(4.3) \qquad e, f, h \in \mathfrak{g}_{\mathbb{R}}, \quad e = \zeta, \quad \theta e = -f, \quad \theta h = -h.$$

Since $e = \zeta$ and $\theta e = -f$ generate $\mathfrak{s}$,

$$(4.4) \qquad (K_{\mathbb{R}})_\zeta \ \text{ centralizes } \ \mathfrak{s}.$$

The commutation relations of the triple e, f, h imply that h acts semisimply with integral eigenvalues in any finite dimensional representation of $\mathfrak{s}$. Irreducible finite dimensional representations of $\mathfrak{s}$ are uniquely characterized by their highest h-weight, which can be any non-negative integer; the irreducible representation of highest weight r has dimension $r+1$. We set

$$(4.5\text{a}) \qquad \begin{aligned} \mathfrak{g}(r) \ &= \ \mathfrak{s}\text{-isotypic subspace of } \mathfrak{g} \text{ of highest weight } r \ = \\ &\text{linear span of all } \mathfrak{s}\text{-irreducible subspaces of heighest} \\ &\text{weight } r; \mathfrak{g}(r,\ell) \ = \ \ell\text{-weight space of } h \text{ in } \mathfrak{g}(r). \end{aligned}$$

The irreducible $\mathfrak{s}$-module of highest weight r has h-weights $r, r-2, \ldots, -r$, hence

$$(4.5b) \qquad \mathfrak{g} \ = \ \bigoplus\nolimits_{r \geq 0} \mathfrak{g}(r) \ = \ \bigoplus\nolimits_{r\geq 0} \bigoplus\nolimits_{\substack{-r \leq \ell \leq r \\ \ell \equiv r \mod 2}} \mathfrak{g}(r, \ell).$$

The first of these decompositions is $\mathfrak{s}$-invariant, θ-stable, and defined over $\mathbb{R}$.

Recall the notation (3.1). Because of (4.4), $\mathrm{Hom}(\mathfrak{s}, \mathfrak{g})$ contains the $\mathrm{Hom}(\mathfrak{s}, \mathfrak{g}(r))$ as $(K_{\mathbb{R}})_\zeta$-invariant subspaces – invariant with respect to the trivial action on $\mathfrak{s}$ and the natural one on $\mathfrak{g}$. Note that $\mathfrak{s}$ has three natural actions on $\mathrm{Hom}(\mathfrak{s}, \mathfrak{g})$: via the action on $\mathfrak{s}$, via its embedding in $\mathfrak{g}$, and diagonally. The decomposition

$$(4.6) \qquad \mathrm{Hom}(\mathfrak{s}, \mathfrak{g}) \ = \ \bigoplus\nolimits_{r \geq 0} \mathrm{Hom}(\mathfrak{s}, \mathfrak{g}(r))$$

is $\mathfrak{s}$-invariant with respect to all three actions, $(K_{\mathbb{R}})_\zeta$-invariant, θ-stable, and defined over $\mathbb{R}$. The summand corresponding to r is $\mathfrak{s}$-isotypic of highest weight 2 with respect to the first $\mathfrak{s}$-action, and of highest weight r with respect to the second action. Thus

$$(4.7) \qquad \begin{aligned} \mathrm{Hom}(\mathfrak{s}, \mathfrak{g}(r)) \ &= \ \mathrm{Hom}(\mathfrak{s}, \mathfrak{g}(r))(r-2) \oplus \mathrm{Hom}(\mathfrak{s}, \mathfrak{g}(r))(r) \\ &\qquad \oplus \mathrm{Hom}(\mathfrak{s}, \mathfrak{g}(r))(r+2), \end{aligned}$$

with the outer index referring to the $\mathfrak{s}$-type with respect to the diagonal $\mathfrak{s}$-action. This decomposition is also $(K_{\mathbb{R}})_\zeta$-invariant, θ-stable, and defined over $\mathbb{R}$. Note that

$$\text{(4.8)} \qquad \operatorname{Hom}(\mathfrak{s}, \mathfrak{g}(r))(r-2) \;=\; 0 \quad \text{unless} \quad r \geq 2.$$

We write $\operatorname{Hom}^\theta(\mathfrak{s}, \mathfrak{g}(r))$ for the intersection of $\operatorname{Hom}(\mathfrak{s}, \mathfrak{g}(r))$ with $\operatorname{Hom}^\theta(\mathfrak{s}, \mathfrak{g})$, and analogously in the case of the summands in (4.7). Our next statement describes the fiber of the normal bundle $T_{C(\mathcal{O})}\mathcal{O}$ at ζ as $(K_{\mathbb{R}})_\zeta$-module.

4.9. Proposition. *The map* $\operatorname{Hom}(\mathfrak{s}, \mathfrak{g}) \ni \Phi \mapsto \Phi(\zeta) = \Phi(e)$ *is injective on*

$$\mathfrak{d}(\Phi_0) \;=_{def}\; \textstyle\bigoplus_{r\geq 2} \operatorname{Hom}^\theta(\mathfrak{s}, \mathfrak{g}(r))(r-2).$$

The image $\mathfrak{d}(\Phi_0)(\zeta)$ *of* $\mathfrak{d}(\Phi_0)$ *under this map is a* $(K_{\mathbb{R}})_\zeta$*-invariant linear complement to* $[\zeta, \mathfrak{k}]$ *in* $[\zeta, \mathfrak{g}]$*, and is defined over* $\mathbb{R}$.

Let $\mathfrak{d}_{\mathbb{R}}(\Phi_0) = \mathfrak{d}(\Phi_0) \cap \operatorname{Hom}^{\mathbb{R}}(\mathfrak{s}, \mathfrak{g})$ denote the space of real points in $\mathfrak{d}(\Phi_0)$. Then $[\zeta, \mathfrak{g}_{\mathbb{R}}] = \mathfrak{d}_{\mathbb{R}}(\Phi_0)(\zeta) \oplus [\zeta, \mathfrak{k}_{\mathbb{R}}]$, and this identifies $(T_{C(\mathcal{O})}\mathcal{O})_\zeta \cong [\zeta, \mathfrak{g}_{\mathbb{R}}]/[\zeta, \mathfrak{k}_{\mathbb{R}}]$ with $\mathfrak{d}_{\mathbb{R}}(\Phi_0)(\zeta)$ as $(K_{\mathbb{R}})_\zeta$-module.

Proof of 4.9. The evaluation map $\Phi \mapsto \Phi(\zeta) = \Phi(e)$ sends $\operatorname{Hom}^\theta(\mathfrak{s}, \mathfrak{g}(r))(r-2)$ to $\mathfrak{g}(r)$. We can therefore argue one summand at a time. The decompositions (4.6) and (4.7) are defined over $\mathbb{R}$, and $\zeta = e$ is real. This reduces the problem to showing

$$\text{(4.10}a\text{)} \qquad \{\, \Phi(e) \mid \Phi \in \operatorname{Hom}^\theta(\mathfrak{s}, \mathfrak{g}(r))(r-2) \,\} \;\subset\; \operatorname{ad}(e)\mathfrak{g}(r),$$

i.e., the image of the evaluation map lies in the image of $\operatorname{ad}(\zeta)$, and

$$\text{(4.10b)} \qquad \begin{array}{c} \text{for each } \xi \in \mathfrak{g}(r), \text{ there exist } \Phi \in \operatorname{Hom}^\theta(\mathfrak{s}, \mathfrak{g}(r))(r-2) \\ \text{and } \eta \in \mathfrak{k} \text{ so that } \Phi(e) \;=\; [e, \xi + \eta]; \text{in this} \\ \text{situation, } [e, \xi] \text{ uniquely determines } \Phi. \end{array}$$

For the first assertion, note that Φ, which is $(r-2)$-isotypic relative to the diagonal action, has components only in the h-weight spaces corresponding to weights between $2-r$ and $r-2$. The evaluation map is $\mathfrak{s}$-equivariant and e has weight two, so $\Phi(e)$ cannot have a non-zero component in the $(-r)$-weight space. In particular, this forces $\Phi(e)$ to lie in the image of $\operatorname{ad}(e)$.

We write $\xi = \xi_{\mathfrak{k}} + \xi_{\mathfrak{p}}$ with $\xi_{\mathfrak{k}} \in \mathfrak{k}$, $\xi_{\mathfrak{p}} \in \mathfrak{p}$, and combine $\xi_{\mathfrak{k}}$ with η. This transforms (4.10b) into the equivalent assertion

$$\text{(4.11)} \quad \begin{array}{c}\text{for each } \xi \in \mathfrak{p} \cap \mathfrak{g}(r), \text{ there exist } \Phi \in \operatorname{Hom}^\theta(\mathfrak{s}, \mathfrak{g}(r))(r-2) \\ \text{and } \eta \in \mathfrak{k} \text{ so that } \Phi(e) = [e, \xi+\eta]; \text{ in this situation, } [e,\xi] \\ \text{uniquely determines } \Phi.\end{array}$$

The Casimir operator of $\mathfrak{s}$,

$$\text{(4.12)} \qquad \Omega = 2ef + 2fe + h^2$$

acts by the scalar $k^2 + 2k$ on any k-isotypic $\mathfrak{s}$-module. For A in the universal enveloping algebra of $\mathfrak{s}$, we let $A\Phi$ denote the effect of A on Φ, acting via the diagonal $\mathfrak{s}$-action on $\operatorname{Hom}^\theta(\mathfrak{s}, \mathfrak{g}(r))$; $A \circ \Phi$ and $\Phi \circ A$ shall denote the composition of Φ with the action of A on, respectively, the values and arguments. Then $\xi\Phi = \operatorname{ad}(\xi) \circ \Phi - \Phi \circ \operatorname{ad}(\xi)$ if $A = \xi \in \mathfrak{s}$, hence

$$\text{(4.13)} \qquad \begin{aligned} \Omega\Phi &= \Omega \circ \Phi + \Phi \circ \Omega - 4\operatorname{ad}(e) \circ \Phi \circ \operatorname{ad}(f) \\ &\quad - 4\operatorname{ad}(f) \circ \Phi \circ \operatorname{ad}(e) - 2\operatorname{ad}(h) \circ \Phi \circ \operatorname{ad}(h). \end{aligned}$$

Since

$$(r-2)^2 + 2(r-2) - [r^2 + 2r] - [2^2 + 2 \times 2] = -4(r+2),$$

for $\Phi \in \operatorname{Hom}^\theta(\mathfrak{s}, \mathfrak{g}(r))$ the following two conditions are equivalent:

$$\text{(4.14)} \qquad \begin{aligned} &\text{a) } \Phi \in \operatorname{Hom}^\theta(\mathfrak{s}, \mathfrak{g}(r))(r-2), \\ &\text{b) } (r+2)\Phi = \operatorname{ad}(e) \circ \Phi \circ \operatorname{ad}(f) + \operatorname{ad}(f) \circ \Phi \circ \operatorname{ad}(e) \\ &\qquad\qquad + \frac{1}{2}\operatorname{ad}(h) \circ \Phi \circ \operatorname{ad}(h). \end{aligned}$$

To construct a particular $\Phi \in \operatorname{Hom}^\theta(\mathfrak{s}, \mathfrak{g}(r))(r-2)$ amounts to specifying $\Phi(e)$ and $\Phi(h)$ in $\mathfrak{g}(r)$, subject to the following conditions. First, $\Phi(h)$ must lie in $\mathfrak{p}$ since $h \in \mathfrak{p}$, and secondly, the identity (4.14b) must hold when evaluated on either e or h. The θ-equivariance of Φ then forces $\Phi(f) = -\theta\Phi(e)$. The validity of (4.14b) applied to f is automatic since Ω commutes with θ.

Let us suppose that $\Phi \in \operatorname{Hom}^\theta(\mathfrak{s}, \mathfrak{g}(r))(r-2)$, $\xi \in \mathfrak{p} \cap \mathfrak{g}(r)$, and $\eta \in \mathfrak{k}$ are given subject to the condition in (4.11), i.e.,

$$\text{(4.15a)} \qquad \Phi(e) = \operatorname{ad}(e)(\xi + \eta).$$

This implies, and is implied by,

$$\Phi(f) \;=\; \mathrm{ad}(f)(-\xi+\eta), \tag{4.15b}$$

and furthermore, implies

$$\begin{aligned} \mathrm{ad}(h)\eta &= \mathrm{ad}(e)(\Phi(f)) - \mathrm{ad}(f)(\Phi(e)) \\ &\quad - (\mathrm{ad}(e)\mathrm{ad}(f) + \mathrm{ad}(f)\mathrm{ad}(e))\xi. \end{aligned} \tag{4.15c}$$

These identities allow us to express $\Omega\eta$ in terms of $\Phi(e), \Phi(f)$, ξ and the action of $\mathfrak{s}$. Since $\Phi(e), \Phi(f)$, ξ lie in $\mathfrak{g}(r)$ by assumption, $\Omega\eta$ also lies in $\mathfrak{g}(r)$. Thus $\eta = \eta_0 + \eta_1$ with $\eta_0 \in \mathfrak{g}(0) = \ker(\Omega)$ and $\eta_1 \in \mathfrak{g}(r)$. Both $\mathrm{ad}(e)$ and $\mathrm{ad}(f)$ annihilate η_0, so we may as well suppose that $\eta = \eta_1 \in \mathfrak{g}(r)$. For $r = 0$, the right hand sides of (4.15a,b) vanish, and $\mathrm{Hom}^\theta(\mathfrak{s}, \mathfrak{g}(r))(r-2) = 0$, which means that there is nothing to prove. Thus we may assume

$$\eta \in \mathfrak{k} \cap \mathfrak{g}(r), \quad \text{and} \quad r > 0. \tag{4.16}$$

From these hypotheses, we shall conclude

$$\begin{aligned} &\text{a)}\;\; \Phi(h) \;=\; -r\xi + \tfrac{1}{r+2}\mathrm{ad}(h)^2\xi + \tfrac{2}{r+2}\mathrm{ad}(h)\eta, \\ &\text{b)}\;\; \eta \;=\; \frac{1}{r}\,\mathrm{ad}(h)\xi. \end{aligned} \tag{4.17}$$

That, in turn, will imply (4.11).

To establish (4.17), we evaluate (4.14b) on h and use the commutation relations of e, f, h, as well as the identities (4.15a,b):

$$\begin{aligned} (r+2)\Phi(h) &= 2\,\mathrm{ad}(e)(\Phi(f)) - 2\,\mathrm{ad}(f)(\Phi(e)) \\ &= 2\,\mathrm{ad}(e)\mathrm{ad}(f)(-\xi+\eta) \;-\; 2\,\mathrm{ad}(f)\mathrm{ad}(e)(\xi+\eta) \\ &= -\;2\,(\mathrm{ad}(e)\mathrm{ad}(f) + \mathrm{ad}(e)\mathrm{ad}(f))(\xi) \\ &\quad +\; 2(\mathrm{ad}(e)\mathrm{ad}(f) - \mathrm{ad}(e)\mathrm{ad}(f))(\eta) \\ &= -\;(\Omega - \mathrm{ad}(h)^2)(\xi) + \;2\,\mathrm{ad}(h)(\eta) \\ &= -\;(r^2+2r)\xi \;+\; \mathrm{ad}(h)^2\xi + \;2\,\mathrm{ad}(h)(\eta), \end{aligned} \tag{4.18}$$

which is the identity (4.17a). Next, we evaluate (4.14b) on e. We use (4.15a), the commutation relations of e, f, h, and (4.18):

$$\begin{aligned} (r+2)\,\mathrm{ad}(e)(\xi+\eta) &= (r+2)\,\Phi(e) \\ &= -\;\mathrm{ad}(e)(\Phi(h)) + \mathrm{ad}(h)(\Phi(e)) \\ &= -\;\mathrm{ad}(e)(-r\xi + \tfrac{1}{r+2}\,\mathrm{ad}(h)^2\xi + \tfrac{2}{r+2}\,\mathrm{ad}(h)\eta) \\ &\quad + \mathrm{ad}(h)\mathrm{ad}(e)(\xi+\eta). \end{aligned} \tag{4.19a}$$

Note that $\mathrm{ad}(h)\mathrm{ad}(e) = \mathrm{ad}(e)\mathrm{ad}(h) + 2\,\mathrm{ad}(e)$, hence

$$\begin{aligned}(r+2)\,&\mathrm{ad}(e)(\xi+\eta)\\ &= \mathrm{ad}(e)\Big(r\xi - \tfrac{1}{r+2}\,\mathrm{ad}(h)^2\xi - \tfrac{2}{r+2}\,\mathrm{ad}(h)\eta \\ &\qquad + 2(\xi+\eta) + \mathrm{ad}(h)(\xi+\eta)\Big).\end{aligned} \tag{4.19b}$$

We bring all terms to the left and multiply through with $r+2$, to conclude

$$\begin{aligned}&\left(\mathrm{ad}(h)^2 - (r+2)\mathrm{ad}(h)\right)\xi \\ &\qquad + \ \left((r^2+2r) - r\,\mathrm{ad}(h)\right)\eta \ \in\ \ker(\mathrm{ad}(e)).\end{aligned} \tag{4.20}$$

Recall the decomposition (4.5) and write ξ_ℓ, η_ℓ for the components of ξ, η in $\mathfrak{g}(r,\ell)$. Since $h \in \mathfrak{p}$, the Cartan involution interchanges $\mathfrak{g}(r,\ell)$ and $\mathfrak{g}(r,-\ell)$, and

$$\eta_{-\ell} \ = \ -\theta(\eta_\ell), \quad \xi_{-\ell} \ = \ \theta(\xi_\ell), \tag{4.21}$$

since $\mathrm{ad}(h)\eta \in \mathfrak{p}$ and $\mathrm{ad}(h)\xi \in \mathfrak{k}$. The kernel of $\mathrm{ad}(e)$ on $\mathfrak{g}(r)$ is $\mathfrak{g}(r,r)$. This makes (4.20) equivalent to

$$\left((r^2+2r) - r\,\ell\right)\eta_\ell \ = \ -\ \left(\ell^2 - (r+2)\ell\right)\xi_\ell \quad \text{if } \ell \neq r. \tag{4.22}$$

The same identity for $\ell = r$ follows from the case $\ell = -r$ and (4.21). Also, ℓ lies between r and $-r$, so (4.22) is equivalent to $r\,\eta_\ell = \ell\,\xi_\ell$ for all ℓ. That is the assertion (4.17b).

On $\mathfrak{g}(r) \cap \mathfrak{p}$, $r > 0$, $\mathrm{ad}(e)$ is injective, so $[e, \xi]$ determines ξ. From (4.17a,b) and the original hypothesis (4.15a), we now conclude that $[e, \xi]$ completely determines $\Phi(h)$, $\Phi(e)$, and $\Phi(f) = -\theta\Phi(e)$. Thus Φ is indeed uniquely determined. As was pointed out earlier, the existence of Φ comes down to knowing that $\Phi(h)$ and $\Phi(e)$ lie in $\mathfrak{g}(r)$, that $\Phi(h) \in \mathfrak{p}$, and that (4.14b) is satisfied when both sides are evaluated on h and e, respectively. The expression (4.17b) specifies η as element of $\mathfrak{g}(r) \cap \mathfrak{k}$. Since $h \in \mathfrak{p}$, $\mathrm{ad}(h)$ interchanges $\mathfrak{k}$ and $\mathfrak{p}$. Thus (4.17a) exhibits $\Phi(h)$ as lying in $\mathfrak{g}(r) \cap \mathfrak{p}$, as required. The containment $\Phi(e) \in \mathfrak{g}(r)$ follows from (4.15a) and (4.16). The validity of (4.14b) when evaluated on h and e amounts to the two identities (4.18) and (4.19); both hold by construction. This gives us the existence and uniqueness of Φ – in other words, the validity of (4.11). q.e.d.

5. The instanton flow at infinity

In this section we use the proof of the Sl_2-orbit theorem of Hodge theory [13], [1] to show that the flow lines of the instanton flow are real analytic at infinity. In effect, the proof of the Sl_2-orbit theorem produces a real analytic isomorphism between a neighborhood of the core $C(\mathcal{O})$ in a nilpotent orbit $\mathcal{O}$ and a neighborhood of the zero section in the normal bundle $T_{C(\mathcal{O})}\mathcal{O}$. This isomorphism is closely related to Kronheimer's flow. We shall freely use the notation of the earlier sections, in particular that of section 4.

The decompositions (4.5-7) depend on the embedding $\Phi_0 : \mathfrak{s}_{\mathbb{R}} \hookrightarrow \mathfrak{g}_{\mathbb{R}}$ given by any particular choice of $\Phi_0 \in \mathrm{Mor}^{\mathbb{R},\theta}(\mathfrak{s},\mathfrak{g})$; the morphism Φ_0, in turn, was assumed to correspond to some $\zeta \in C(\mathcal{O})$ via the isomorphism (2.19). We shall now let ζ vary over the core, and correspondingly Φ_0 over the inverse image of $C(\mathcal{O})$. In this way, we tacitly regard the decompositions (4.5-7) as depending on ζ, without putting this dependence into the notation. Recall the Definition (3.2) of the pairing Q. We shall need the notion of a Q-polynomial: a function

$$(5.1)\qquad \begin{aligned} &(\Phi_1,\Phi_2,\ldots,\Phi_k) \mapsto P(\Phi_1,\Phi_2,\ldots,\Phi_k) \in \mathrm{Hom}^\theta(\mathfrak{s},\mathfrak{g}),\\ &\qquad \text{with arguments } \ \Phi_1,\ldots,\ \Phi_k \in \mathrm{Hom}^\theta(\mathfrak{s},\mathfrak{g}), \end{aligned}$$

expressible as a finite $\mathbb{C}$-linear combination of monomials in the Φ_ℓ, with Q serving as "multiplication". Note that a real Q-polynomial – i.e., a Q-polynomial with real coefficients – takes values in $\mathrm{Hom}^{\mathbb{R},\theta}(\mathfrak{s},\mathfrak{g})$ whenever the arguments lie in this real subspace.

5.2. Theorem. *Every function $\Phi(t)$ in $\mathcal{M}$ has a convergent series expansion around ∞, in powers of $t^{-\frac{1}{2}}$. Specifically,*

$$a)\qquad \Phi(t) = \Phi_0\, t^{-1} + \textstyle\sum_{k\geq 2} \Phi_k\, t^{-1-\frac{k}{2}} \qquad (t \gg 0),$$

with $\Phi_0 \in \mathrm{Mor}^{\mathbb{R},\theta}(\mathfrak{s},\mathfrak{g})$ and $\Phi_k \in \mathrm{Hom}^{\mathbb{R},\theta}(\mathfrak{s},\mathfrak{g})$ for $k \geq 2$; there exist universal[1] *Q-polynomials with rational coefficients $P_k(\ldots)$, $k \geq 2$, such that*

$$\Phi_k = \Phi_k^k + P_k(\Phi_0,\Phi_2^2,\Phi_3^3,\ldots,\Phi_{k-2}^{k-2}),$$

$$b)\qquad \Phi_\ell^\ell \in \mathrm{Hom}^{\mathbb{R},\theta}(\mathfrak{s},\mathfrak{g}(\ell))(\ell-2) \qquad (\ell \geq 2),$$

$$P_k(\Phi_0,\Phi_2^2,\Phi_3^3,\ldots,\Phi_{k-2}^{k-2}) \in \textstyle\bigoplus_{\ell\leq k-2,\,\ell\equiv k\,(2)} \mathrm{Hom}^{\mathbb{R},\theta}(\mathfrak{s},\mathfrak{g}(\ell)).$$

The polynomial P_k is weighted homogeneous of weight k when the variable Φ_ℓ^ℓ is given weight ℓ, and Φ_0 weight 0. Conversely, any series of this form has a positive radius of convergence, and the resulting $\mathrm{Hom}^{\mathbb{R},\theta}(\mathfrak{s},\mathfrak{g})$-valued function $\Phi(t)$ satisfies the differential equation (3.6a).

[1] i.e., not depending on Φ_0 nor even on $\mathfrak{g}_{\mathbb{R}}$, provided the dimension of $\mathfrak{g}_{\mathbb{R}}$ is bounded.

Loosely paraphrased, the space $\bigoplus_{\ell\geq 2}\mathrm{Hom}^{\mathbb{R},\theta}(\mathfrak{s},\mathfrak{g}(\ell))(\ell-2)$ parameterizes all functions $\Phi(t)$ defined for large positive values of t which satisfy the differential equation (3.6a) and the limiting condition (3.6c). In the preceeding section, we had identified this direct sum with the fiber of the normal bundle $T_{C(\mathcal{O})}\mathcal{O}$ at ζ when the leading coefficient $\Phi_0 \in \mathrm{Mor}^{\mathbb{R},\theta}(\mathfrak{s},\mathfrak{g})$ corresponds to $\zeta \in C(\mathcal{O})$. Note that the power $t^{-\frac{3}{2}}$ gets skipped in the expansion (5.2) – this reflects (4.8).

We shall verify the theorem together with the following companion statement. For any collection of data $\Phi_0 \in \mathrm{Mor}^{\mathbb{R},\theta}(\mathfrak{s},\mathfrak{g})$, $\Phi_\ell^\ell \in \mathrm{Hom}^{\mathbb{R},\theta}(\mathfrak{s},\mathfrak{g}(\ell))(\ell-2)$ for $\ell\geq 2$, and $t>0$, let

$$(5.3) \qquad \Phi(\Phi_0,\Phi_2^2,\dots,\Phi_\ell^\ell,\dots;t)$$

denote the value of the function $\Phi(t)$ in (5.2a,b), provided the analytic continuation of the series is defined at t.

5.4. Theorem. *The assignment*

$$(\Phi_0,\Phi_2^2,\dots,\Phi_\ell^\ell,\dots)\mapsto \Phi(\Phi_0,\Phi_2^2,\dots,\Phi_\ell^\ell,\dots;1)(e)$$

induces a well defined, real analytic, $K_{\mathbb{R}}$-equivariant isomorphism

$$F\,:\,\mathcal{U}\longrightarrow F(\mathcal{U}),$$

between a connected open neighborhood $\mathcal{U}$ of the zero section in the normal bundle $T_{C(\mathcal{O})}\mathcal{O}$ and $F(\mathcal{U})$, an open neighborhood of the core $C(\mathcal{O})$ in the nilpotent orbit $\mathcal{O}$.

Proof of 5.2 and 5.4. We appeal to the results of [1, §6], specifically (6.8-24); these results already appear in [13], in somewhat different language. To begin with, a formal $\mathrm{Hom}(\mathfrak{s},\mathfrak{g})$-valued series $\Phi(t)=\Phi_0\,t^{-1}+\sum_{k>0}\Phi_k\,t^{-1-\frac{k}{2}}$ is a formal solution of the differential equation $\frac{d\Phi}{dt}=-Q(\Phi,\Phi)$if and only if the coefficients Φ_k can be expressed as in (5.2b), (5.5)

with certain specific Q-polynomials $P_k(\dots)$. The terminology "Q-polynomial" is not used in [1]. Rather, the arguments there show that the differential equation translates into the conditions

$$(5.6) \qquad \Phi_k \;=\; \Phi_k^k + A\left(\textstyle\sum_{0<\ell<k} Q(\Phi_\ell,\Phi_{k-\ell})\right) \qquad (\,k>0\,),$$

with $\Phi_k^k \in \mathrm{Hom}(\mathfrak{s},\mathfrak{g}(k))(k-2)$, and with A denoting a particular rational linear combination of projections to the various eigenspaces of the linear map

$$(5.7) \qquad \mathrm{Hom}(\mathfrak{s},\mathfrak{g})\ni T\longmapsto Q(\Phi_0,T).$$

The eigenvalues of this linear map are rational [1, (6.14)], so each projection can be expressed as a rational linear combination of its powers. The coefficient Φ_1 vanishes because of (4.8) and (5.6). Thus, in (5.6), the range of summation is really $2 \leq \ell \leq k-2$. Now, arguing inductively, one finds rational Q-polynomials $P_k(\dots)$, weighted homogeneous of degree k when weights are assigned as in Theorem 5.2, such that

$$\Phi_k = \Phi_k^k + P_k(\Phi_0, \Phi_2^2, \Phi_3^3, \dots, \Phi_{k-2}^{k-2}) \qquad (k > 0). \tag{5.8}$$

The linear map (5.7) preserves the subspaces $\mathrm{Hom}(\mathfrak{s}, \mathfrak{g}(\ell))$, and

$$\begin{aligned} &Q\left(\mathrm{Hom}(\mathfrak{s}, \mathfrak{g}(k_1)), \mathrm{Hom}(\mathfrak{s}, \mathfrak{g}(k_2))\right) \\ &\qquad\subset \textstyle\bigoplus_{0 \leq \ell \leq k_1+k_2, \ell \equiv k_1+k_2 (2)} \mathrm{Hom}(\mathfrak{s}, \mathfrak{g}(\ell)) \end{aligned} \tag{5.9}$$

[1, (6.21)]. Hence

$$P_k(\Phi_0, \Phi_2^2, \Phi_3^3, \dots, \Phi_{k-2}^{k-2}) \in \textstyle\bigoplus_{\ell \leq k-2,\, \ell \equiv k\, (2)} \mathrm{Hom}(\mathfrak{s}, \mathfrak{g}(\ell)), \tag{5.10}$$

again by induction on k. This completes the verification of (5.5). For future reference, we note that

$$\begin{aligned} &\text{for fixed } \Phi_0, \text{ as function of } \Phi_2^2, \Phi_3^3, \dots, \Phi_{k-2}^{k-2} \\ &\text{alone, } P_k(\Phi_0, \Phi_2^2, \Phi_3^3, \dots, \Phi_{k-2}^{k-2}) \text{ has no linear} \\ &\text{and no constant term,} \end{aligned} \tag{5.11}$$

as follows from the homogeneity property of $P_k(\dots)$.

The construction of the P_k readily implies a bound on their size: with Φ_0 kept fixed, there exists a positive constant C such that

$$\|P_k(\Phi_0, \Phi_2^2, \Phi_3^3, \dots, \Phi_{k-2}^{k-2})\| \leq C^k \left(\max_{k \geq 2} \|\Phi_k^k\|\right)^k \tag{5.12}$$

[1, (6.24)]. That, in turn, implies

$$\begin{aligned} &\text{a) the series } \Phi(t) = \Phi_0\, t^{-1} + \textstyle\sum_{k>0} \Phi_k\, t^{-1-\frac{k}{2}} \\ &\quad\ \text{converges if } t^{\frac{1}{2}} > C^{-1} \left(\max_{k \geq 2} \|\Phi_k^k\|\right)^{-1}; \\ &\text{b) } (\Phi_2^2, \dots, \Phi_\ell^\ell, \dots) \mapsto \Phi(\Phi_0, \Phi_2^2, \dots, \Phi_\ell^\ell, \dots; 1) \\ &\quad\ \text{is a well defined, analytic map on some} \\ &\quad\ \text{neighborhood of } 0 \text{ in } \textstyle\bigoplus_{k \geq 2} \mathrm{Hom}(\mathfrak{s}, \mathfrak{g}(k))(k-2). \end{aligned} \tag{5.13}$$

In particular, the map in the statement of Theorem 5.4 is well defined, real analytic when restricted to a small neighborhood of 0 in any fiber $(T_{C(\mathcal{O})}\mathcal{O})_\zeta$ of the normal bundle; Kronheimer's Theorem 3.7 implies that the map takes values in $\mathcal{O}$. Because of (4.9), (5.8) and (5.11), this map sends any sufficiently small neighborhood of 0 in $(T_{C(\mathcal{O})}\mathcal{O})_\zeta$ isomorphically to a real analytic submanifold of $\mathcal{O}$, normal to $C(\mathcal{O})$ at ζ. The Definition (3.2) exhibits Q as $K_\mathbb{R}$-equivariant pairing. We conclude that Q-polynomials are $K_\mathbb{R}$-equivariant as functions of their arguments. The map F is therefore both $(K_\mathbb{R})_\zeta$-equivariant on the fiber at ζ and globally $K_\mathbb{R}$-equivariant. Since $K_\mathbb{R}$ is compact and acts transitively on $C(\mathcal{O})$, F has the properties asserted by Theorem (5.4): $K_\mathbb{R}$-equivariant, real analytic, and real analytically invertible from a small neighborhood of the zero section in $T_{C(\mathcal{O})}\mathcal{O}$ to some neighborhood of $C(\mathcal{O})$ in $\mathcal{O}$.

We now consider a particular curve $\Psi(t)$ in $\mathcal{M}(\Phi_0)$. Then, if $a \geq 1$, the curve $C_a\Psi$, defined by

$$(5.14a) \qquad (C_a\Psi)(t) \;=\; a\Psi(a(t+1)-1),$$

satisfies the three conditions (3.6a-c), with the same limiting morphism Φ_0. In other words, $C_a\Psi \in \mathcal{M}(\Phi_0)$, hence

$$(5.14b) \qquad C_a \;:\; \mathcal{M}(\Phi_0) \longrightarrow \mathcal{M}(\Phi_0) \qquad (\,a \geq 1\,).$$

The condition (3.6c) implies

$$(5.14c) \qquad \lim_{a\to\infty} C_a\Psi(t) \;=\; \Phi_0,$$

for any fixed $t \geq 0$ – recall: $C_a\Psi$, like every curve satisfying (3.6a-c), extends real analytically to a neighborhood of 0. In particular, for a sufficiently large,

$$(5.15) \qquad (C_a\Psi)(0)(e) \in \mathcal{U},$$

with $\mathcal{U}$ having the same meaning as in the statement of Theorem 5.4. Since this theorem has already been proved, there exist

$$\Phi_k^k \in \operatorname{Hom}^{\mathbb{R},\theta}(\mathfrak{s},\mathfrak{g}(k))(k-2),$$

$k \geq 2$, so that

$$(5.16a) \quad \Phi(t) \;=_{\text{def}}\; \Phi_0\, t^{-1} + \textstyle\sum_{k\geq 2}\left(\Phi_k^k + P_k(\Phi_0,\Phi_2^2,\dots,\Phi_{k-2}^{k-2})\right) t^{-1-\frac{k}{2}}$$

converges for $t \gg 0$, extends real analytically to $[1,\infty)$, and satisfies

$$(5.16b) \qquad \Phi(1)(e) \;=\; (C_a\Psi)(0)(e).$$

Because of (5.5), $t \mapsto \Phi(t+1)$ belongs to $\mathcal{M}(\Phi_0)$. By construction, this curve has the same image under the Kronheiner isomorphism as

$C_a\Psi$, and thus must coincide with $C_a\Psi$. We conclude that Ψ can be obtained from Φ by a linear coordinate change, and that $\Psi(t)$ has a convergent series expansion around infinity, in powers of $t^{-\frac{1}{2}}$. Appealing once more to (5.5) and subsequent statements, we conclude that $\Psi(t)$ has the properties asserted in Theorem 5.2. q.e.d.

Theorem 5.4 can be strengthened, as follows. Recall (5.14). A short calculation shows that $C_a \circ C_b = C_{ab}$, hence

(5.17)
$$a \mapsto C_a \text{ induces an action of the multiplicative semigroup } \mathbb{R}_{\geq 1},$$

both on $\mathcal{M}(\Phi_0)$ and on $\mathcal{O} \cong \mathcal{M}(C(\mathcal{O}))$. Using the identification (4.9), we define

$$\begin{aligned} D_a \; : \; T_{C(\mathcal{O})}\mathcal{O} &\longrightarrow T_{C(\mathcal{O})}\mathcal{O}, \\ D_a(\Phi_0) = \Phi_0, \qquad D_a(\Phi_k^k) &= a^{-\frac{k}{2}}\Phi_k^k, \end{aligned} \tag{5.18a}$$

where $\Phi_k^k \in \mathrm{Hom}^{\mathbb{R},\theta}(\mathfrak{s}, \mathfrak{g}(k))(k-2)$. This makes sense for all $a \neq 0$; moreover,

(5.18*b*) $a \mapsto D_a$ defines an action of the multiplicative group $\mathbb{R}^*$,

as can be checked by direct calculation. The map F defined in Theorem 5.4 is $\mathbb{R}_{\geq 1}$-equivariant with respect to the two actions (5.17-18):

5.19. Lemma. *For all $a \geq 1$, $F \circ D_a = C_a \circ F$.*

Proof. If $(\Phi_0, \Phi_2^2, \ldots, \Phi_\ell^\ell, \ldots)$ corresponds to a point in the domain of F, the series (5.16a) converges for large t, extends real analytically to $[1, \infty)$, and

$$\begin{aligned} F(\Phi_0, \Phi_2^2, \ldots, \Phi_\ell^\ell, \ldots) &= \Phi(1)(e) = \tilde{\Phi}(0)(e), \\ &\text{where } \tilde{\Phi}(t) =_{\mathrm{def}} \Phi(t+1). \end{aligned} \tag{5.20}$$

The curve $\tilde{\Phi}(t)$ then belongs to $\mathcal{M}(\Phi_0)$ and corresponds to $F(\Phi_0, \Phi_2^2, \ldots, \Phi_\ell^\ell, \ldots)$ via the Kronheimer isomorphism (3.7). Hence

$$\begin{aligned} C_a(F(\Phi_0, \Phi_2^2, \ldots, \Phi_\ell^\ell, \ldots)) &= (C_a\tilde{\Phi})(0)(e) \\ &= a\,\tilde{\Phi}(a-1)(e) = a\,\Phi(a)(e). \end{aligned} \tag{5.21}$$

On the other hand,

$$(F \circ D_a)(\Phi_0, \Phi_2^2, \ldots, \Phi_\ell^\ell, \ldots) = F(\Phi_0, a^{-1}\Phi_2^2, \ldots, a^{-\frac{\ell}{2}}\Phi_\ell^\ell, \ldots). \tag{5.22}$$

When the Φ_ℓ^ℓ in (5.16a) are replaced by $a^{-\frac{\ell}{2}}\,\Phi_\ell^\ell$, the series $\Phi(t)$ gets transformed into $a\,\Phi(at)$ – this follows from the homogeneity property of the $P_k(\dots)$. Thus

$$(5.23)\quad F(\Phi_0, a^{-1}\Phi_2^2, \dots, a^{-\frac{\ell}{2}}\Phi_\ell^\ell, \dots) \;=\; (a\,\Phi(at)|_{t=1})\,(e) \;=\; a\,\Phi(a)(e),$$

completing the proof of the lemma. q.e.d.

5.24. Corollary. *The map F^{-1} extends to a real analytic isomorphism between the entire nilpotent orbit $\mathcal{O}$ and an open neighborhood of the zero section in $T_{C(\mathcal{O})}\mathcal{O}$.*

Proof. Given $\xi \in \mathcal{O}$, we choose $a > 1$ so large that $C_a\xi$ lies in the domain of F^{-1}, and set

$$(5.25)\qquad F^{-1}(\xi) \;=\; D_{a^{-1}}(F^{-1}(C_a\xi)).$$

This extension of F^{-1} is well defined and one-to-one by (5.17-19), and real analytic by construction. It is also locally invertible, again because of (5.17-19), hence an open map with real analytic inverse. q.e.d.

6. Complex groups and symmetric pairs

In preparation for the next section, in which we discuss the Sekiguchi correspondence, we shall restate the earlier results for symmetric pairs and complex Lie algebras.

There is not so much to say about the complex case – complex Lie algebras can be regarded as real Lie algebras, after all. Let $\mathfrak{g}$ be a complex semisimple Lie algebra. As a matter of general notational convention, we set

$$(6.1)\qquad \mathfrak{g}^{\mathbb{R}} \;=\; \mathfrak{g}, \quad \text{taken as Lie algebra over } \mathbb{R}.$$

In the situation of interest to us, $\mathfrak{g}$ will arise as the complexification of a real semisimple Lie algebra $\mathfrak{g}_{\mathbb{R}}$, in which a Cartan decomposition $\mathfrak{g}_{\mathbb{R}} = \mathfrak{k}_{\mathbb{R}} \oplus \mathfrak{p}_{\mathbb{R}}$ has been specified. The subalgebra

$$(6.2)\qquad \mathfrak{u}_{\mathbb{R}} \;=\; \mathfrak{k}_{\mathbb{R}} \oplus i\,\mathfrak{p}_{\mathbb{R}}$$

is then a compact real form in $\mathfrak{g}^{\mathbb{R}}$, and

$$(6.3)\qquad \mathfrak{g}^{\mathbb{R}} \;=\; \mathfrak{u}_{\mathbb{R}} \oplus i\,\mathfrak{u}_{\mathbb{R}}$$

the Cartan decomposition determined by $\mathfrak{u}_{\mathbb{R}}$. Further notation:

$$(6.4)\qquad \tau : \mathfrak{g} \to \mathfrak{g} \quad \text{is complex conjugation with respect to } \mathfrak{u}_{\mathbb{R}}.$$

In view of (6.3),

$$\tau \text{ is the Cartan involution on } \mathfrak{g}^{\mathbb{R}}. \tag{6.5}$$

We normalize the Killing forms on $\mathfrak{g}$, $\mathfrak{g}^{\mathbb{R}}$, and $\mathfrak{g}_{\mathbb{R}}$ so that they coincide on $\mathfrak{k}_{\mathbb{R}}$. This will allow us to refer to all three by the same symbol B, without ambiguity.

Extension of scalars identifies the space of $\mathbb{R}$-linear maps from $\mathfrak{s}_{\mathbb{R}}$ to $\mathfrak{g}^{\mathbb{R}}$ with the space of $\mathbb{C}$-linear maps from $\mathfrak{s}$ to $\mathfrak{g}$. Also, the Cartan involution on $\mathfrak{s}_{\mathbb{R}}$ equals the restriction to $\mathfrak{s}_{\mathbb{R}}$ of complex conjugation with respect to the compact real form $\mathfrak{su}(2)$ in $\mathfrak{s} = \mathfrak{sl}(2,\mathbb{C})$. This results in the following "dictionary" between the spaces of homomorphisms defined in the preceeding sections and their analogues in the present setting:

$$\begin{aligned} &\mathrm{Hom}^{\mathbb{R}}(\mathfrak{s},\mathfrak{g}) \rightsquigarrow \mathrm{Hom}^{\mathbb{R}}(\mathfrak{s},\mathbb{C}\otimes_{\mathbb{R}}\mathfrak{g}^{\mathbb{R}}) \cong \mathrm{Hom}(\mathfrak{s},\mathfrak{g}) \\ &\mathrm{Hom}^{\mathbb{R},\theta}(\mathfrak{s},\mathfrak{g}) \rightsquigarrow \mathrm{Hom}(\mathfrak{su}(2),\mathfrak{u}_{\mathbb{R}}), \end{aligned} \tag{6.6}$$

and similarly in the case of $\mathrm{Mor}(\mathfrak{s},\mathfrak{g})$. The Lie algebra $\mathfrak{su}(2)$ acts on these spaces, both via the action on the values and diagonally, so the decompositions (4.6-7) have obvious counterparts. Note that the evaluation map

$$\begin{aligned} &\mathrm{Hom}^{\mathbb{R},\theta}(\mathfrak{s},\mathfrak{g}) \ni \Phi_0 \mapsto \Phi_0(e) \qquad \text{corresponds to} \\ &\mathrm{Hom}(\mathfrak{su}(2),\mathfrak{u}_{\mathbb{R}}) \ni \Phi_0 \mapsto \Phi_0(\tfrac{e}{2}-\tfrac{f}{2}) - i\,\Phi_0(\tfrac{ie}{2}+\tfrac{if}{2}) \end{aligned} \tag{6.7}$$

via the translation (6.6). We now let G denote the identity component of $\mathrm{Aut}(\mathfrak{g})$, and $U_{\mathbb{R}} \subset G$ the compact real form determined by $\mathfrak{u}_{\mathbb{R}}$. Then

$$G_{\mathbb{R}} \rightsquigarrow G, \qquad K_{\mathbb{R}} \rightsquigarrow U_{\mathbb{R}} \tag{6.8}$$

completes our dictionary: when the substitutions (6.6-8) are made, the results of the earlier sections – in particular Lemmas 2.11 and 2.19, Theorems 3.7, 5.2, and 5.4, Propositions 2.21, 2.25, and 4.9, Corollaries 2.22, 2.23, and 5.24 – hold in the setting of a complex Lie algebra.

We return to the case of a real semisimple Lie algebra $\mathfrak{g}_{\mathbb{R}} = \mathfrak{k}_{\mathbb{R}} \oplus \mathfrak{p}_{\mathbb{R}}$. As additional datum, we suppose that an involutive automorphism

$$\sigma : \mathfrak{g}_{\mathbb{R}} \to \mathfrak{g}_{\mathbb{R}} \qquad (\sigma^2 = 1_{\mathfrak{g}_{\mathbb{R}}}) \tag{6.9a}$$

is fixed. It induces a pseudo-Riemannian "Cartan decomposition" of $\mathfrak{g}_{\mathbb{R}}$,

$$\mathfrak{g}_{\mathbb{R}} = \mathfrak{h}_{\mathbb{R}} \oplus \mathfrak{q}_{\mathbb{R}}, \qquad [\mathfrak{h}_{\mathbb{R}},\mathfrak{h}_{\mathbb{R}}] \subset \mathfrak{h}_{\mathbb{R}}, \quad [\mathfrak{h}_{\mathbb{R}},\mathfrak{q}_{\mathbb{R}}] \subset \mathfrak{q}_{\mathbb{R}}, \quad [\mathfrak{q}_{\mathbb{R}},\mathfrak{q}_{\mathbb{R}}] \subset \mathfrak{h}_{\mathbb{R}}, \tag{6.9b}$$

with $\mathfrak{h}_\mathbb{R}$ and $\mathfrak{q}_\mathbb{R}$ denoting, respectively, the $(+1)$ and (-1)-eigenspaces of σ. We shall assume that the usual Cartan involution θ preserves this decomposition – equivalently,

$$\theta \circ \sigma \ = \ \sigma \circ \theta. \tag{6.10}$$

When this fails to be the case, it can be brought about by replacing the Cartan decomposition with an appropriate $G_\mathbb{R}$-conjugate. The involution σ lifts to

$$G_\mathbb{R} \ = \ \mathrm{Aut}(\mathfrak{g}_\mathbb{R})^0. \tag{6.11}$$

Let $H_\mathbb{R} \subset G_\mathbb{R}$ be a subgroup lying somewhere between the fixed point group $G_\mathbb{R}^\sigma$ and its identity component,

$$(G_\mathbb{R}^\sigma)^0 \ \subset \ H_\mathbb{R} \ \subset G_\mathbb{R}^\sigma. \tag{6.12}$$

Then $H_\mathbb{R}$ preserves the decomposition (6.9b), and thus acts on the set of nilpotents in $\mathfrak{q}_\mathbb{R}$.

The Lie algebra $\mathfrak{s}_\mathbb{R} = \mathfrak{sl}(2,\mathbb{R})$ and its diagonal subalgebra $\mathfrak{a}_\mathbb{R}$ furnish the simplest non-trivial example of a symmetric pair:

$$\sigma_\mathfrak{s} : \mathfrak{s}_\mathbb{R} \longrightarrow \mathfrak{s}_\mathbb{R}, \qquad \sigma_\mathfrak{s}(\eta) \ = \ \begin{cases} \eta & \text{for } \ \eta \in \mathfrak{a}_\mathbb{R} \\ -\eta & \text{for } \ \eta \in \mathbb{R}e \oplus \mathbb{R}f \end{cases} \tag{6.13a}$$

is the involution, and

$$\mathfrak{s}_\mathbb{R} \ = \ \mathfrak{a}_\mathbb{R} \oplus (\mathbb{R}e \oplus \mathbb{R}f) \tag{6.13b}$$

the non-Riemannian Cartan decomposition; note that $\sigma_\mathfrak{s}$ does commute with the Cartan involution $\theta_\mathfrak{s}$ – cf. (2.17a). The group $S_\mathbb{R} = PSl(2,\mathbb{R})$ and its diagonal subgroup $A_\mathbb{R}$ play the roles of $G_\mathbb{R}$ and $H_\mathbb{R}$. The space $\mathbb{R}e \oplus \mathbb{R}f$ contains five nilpotent $A_\mathbb{R}$-orbits, namely $\mathbb{R}_{>0}\, e$, $\mathbb{R}_{<0}\, e$, $\mathbb{R}_{>0} f$, $\mathbb{R}_{<0} f$, and $\{0\}$.

In the present setting, the roles of the set $\mathrm{Mor}^{\mathbb{R},\theta}(\mathfrak{s},\mathfrak{g})$ and of the vector space $\mathrm{Hom}^{\mathbb{R},\theta}(\mathfrak{s},\mathfrak{g})$ are played by

$$\begin{aligned} \mathrm{Mor}^{\mathbb{R},\theta,\sigma}(\mathfrak{s},\mathfrak{g}) \ &= \ \{\, \Phi_0 \in \mathrm{Mor}^{\mathbb{R},\theta}(\mathfrak{s},\mathfrak{g}) \mid \sigma\circ\Phi_0 = \Phi_0 \circ \sigma_\mathfrak{s} \,\}, \\ \mathrm{Hom}^{\mathbb{R},\theta,\sigma}(\mathfrak{s},\mathfrak{g}) \ &= \ \{\, \Phi \in \mathrm{Hom}^{\mathbb{R},\theta}(\mathfrak{s},\mathfrak{g}) \mid \sigma\circ\Phi = \Phi \circ \sigma_\mathfrak{s} \,\}. \end{aligned} \tag{6.14}$$

Note that the decompositions (4.6-7) induce decompositions of $\mathrm{Hom}^{\mathbb{R},\theta,\sigma}(\mathfrak{s},\mathfrak{g})$, because $\sigma_\mathfrak{s}$ acts trivially on the Casimir operator of $\mathfrak{s}$, and because σ preserves the Φ_0-image of $\mathfrak{s}$ in $\mathfrak{g}$. When Φ_0 lies in

$\mathrm{Mor}^{\mathbb{R},\theta,\sigma}(\mathfrak{s},\mathfrak{g})$, we write $\mathcal{M}^\sigma(\Phi_0)$ for the set of all those functions $\Phi(t)$ in $\mathcal{M}(\Phi_0)$ which take values in $\mathrm{Hom}^{\mathbb{R},\theta,\sigma}(\mathfrak{s},\mathfrak{g})$.

In the following, $\mathcal{O}_{\mathfrak{q}_\mathbb{R}}$ will denote a nilpotent $H_\mathbb{R}$-orbit in $\mathfrak{q}_\mathbb{R}$; there are only finitely many such orbits, and they are invariant under scaling by positive scale factors [16]. The $G_\mathbb{R}$-translates of $\mathcal{O}_{\mathfrak{q}_\mathbb{R}}$ sweep out a nilpotent $G_\mathbb{R}$-orbit $\mathcal{O} \subset \mathfrak{g}_\mathbb{R}$. We use $\mathcal{O}$ to normalize the Killing form, as in (2.9-10). The moment map (2.3) restricts to an $(H_\mathbb{R} \cap K_\mathbb{R})$-equivariant map

$$m : \mathcal{O}_{\mathfrak{q}_\mathbb{R}} \longrightarrow \mathfrak{h}_\mathbb{R} \cap \mathfrak{p}_\mathbb{R}, \tag{6.15}$$

the moment map associated to the $H_\mathbb{R}$-action on $\mathcal{O}_{\mathfrak{q}_\mathbb{R}}$. As in the absolute case, the multiplicative group $\mathbb{R}^+$ acts, by scaling, on $\mathcal{O}_{\mathfrak{q}_\mathbb{R}}$ and on the set of critical points of the function $\zeta \mapsto \|m(\zeta)\|^2$; these actions commute with those of $H_\mathbb{R} \cap K_\mathbb{R}$. By definition,

$$\begin{aligned} C(\mathcal{O}_{\mathfrak{q}_\mathbb{R}}) = \{\, \zeta \in \mathcal{O}_{\mathfrak{q}_\mathbb{R}} \mid \\ \zeta \text{ is a critical point for } \|m\|^2, \text{and } \|\zeta\| = 1 \,\} \end{aligned} \tag{6.16}$$

is the core of the nilpotent orbit $\mathcal{O}_{\mathfrak{q}_\mathbb{R}}$.

6.17. Proposition. *The function $\zeta \mapsto \|m(\zeta)\|^2$ on $\mathbb{S}(\mathcal{O}_{\mathfrak{q}_\mathbb{R}})$ is Bott-Morse. Its set of critical points coincides with the set minima, and consists of exactly one $(H_\mathbb{R} \cap K_\mathbb{R})$-orbit. In particular, the core $C(\mathcal{O}_{\mathfrak{q}_\mathbb{R}})$ is non-empty, and $H_\mathbb{R} \cap K_\mathbb{R}$ acts transitively on it. The map $\Phi \mapsto \Phi(e)$ establishes a $H_\mathbb{R} \cap K_\mathbb{R}$-equivariant bijection*

$$\{\, \Phi_0 \in \mathrm{Mor}^{\mathbb{R},\theta,\sigma}(\mathfrak{s},\mathfrak{g}) \mid \Phi_0(e) \in \mathcal{O}_{\mathfrak{q}_\mathbb{R}} \,\} \;\cong\; C(\mathcal{O}_{\mathfrak{q}_\mathbb{R}}),$$

and, at the point $\zeta = \Phi_0(e) \in C(\mathcal{O}_{\mathfrak{q}_\mathbb{R}})$, identifies

$$\textstyle\bigoplus_{\ell \geq 2} \mathrm{Hom}^{\mathbb{R},\theta,\sigma}(\mathfrak{s},\mathfrak{g}(\ell))(\ell - 2)$$

$(H_\mathbb{R} \cap K_\mathbb{R})_\zeta$-equivariantly with $(T_{C(\mathcal{O}_{\mathfrak{q}_\mathbb{R}})}\mathcal{O}_{\mathfrak{q}_\mathbb{R}})_\zeta$, the fiber at ζ of the normal bundle of the orbit $\mathcal{O}_{\mathfrak{q}_\mathbb{R}}$ along its core. There exists a real analytic, $(H_\mathbb{R} \cap K_\mathbb{R})$-equivariant isomorphism $\mathcal{O}_{\mathfrak{q}_\mathbb{R}} \simeq T_{C(\mathcal{O}_{\mathfrak{q}_\mathbb{R}})}\mathcal{O}_{\mathfrak{q}_\mathbb{R}}$. Lastly, the Kronheimer diffeomorphism (3.7) induces

$$\mathcal{M}^\sigma(C(\mathcal{O}_{\mathfrak{q}_\mathbb{R}})) \;\cong\; \mathcal{O}_{\mathfrak{q}_\mathbb{R}};$$

here $\mathcal{M}^\sigma(C(\mathcal{O}_{\mathfrak{q}_\mathbb{R}}))$ refers to the union of the $\mathcal{M}^\sigma(\Phi_0)$ parameterized by those morphisms Φ_0 which correspond to points in $C(\mathcal{O}_{\mathfrak{q}_\mathbb{R}})$.

These statements are analogous to (2.11), (2.19), (2.21), (4.9), (2.25), and (3.7), but not all them can be deduced directly from those results in the absolute case. We will indicate briefly how to modify the earlier arguments so that they apply in the present situation. As noted in the proof of Proposition 2.21, the fact that $\|m\|^2$ is Bott-Morse on $\mathbb{S}(\mathcal{O}_{\mathfrak{q}_\mathbb{R}})$ is a general property of moment maps for linear actions. By (2.20) and the fact that the moment map m in (6.15) is the restriction of the moment map (2.3) from $\mathcal{O}$ to $\mathcal{O}_{\mathfrak{q}_\mathbb{R}}$, we conclude that the critical set of $\|m\|^2 : \mathcal{O}_{\mathfrak{q}_\mathbb{R}} \to \mathbb{R}_{>0}$ is the intersection of $\mathcal{O}_{\mathfrak{q}_\mathbb{R}}$ with the critical set of $\|m\|^2$ in $\mathcal{O}$. It follows that the core $C(\mathcal{O}_{\mathfrak{q}_\mathbb{R}})$ is the intersection of the core $C(\mathcal{O})$ with $\mathcal{O}_{\mathfrak{q}_\mathbb{R}}$, and $C(\mathcal{O}_{\mathfrak{q}_\mathbb{R}})$ consists precisely of the minima of $\|m\|^2$ on $\mathbb{S}(\mathcal{O}_{\mathfrak{q}_\mathbb{R}})$. Any $\zeta \in C(\mathcal{O}_{\mathfrak{q}_\mathbb{R}})$ can be embedded in a strictly normal S-triple, and by [10] or [16, Lemmas 1.4,1.5] such strictly normal S-triples constitute an $H_\mathbb{R} \cap K_\mathbb{R}$-orbit. This proves the analogues of (2.11) and (2.21). The proof of Lemma 2.19 can now be adapted to establish the bijection $\{\, \Phi_0 \in \mathrm{Mor}^{\mathbb{R},\theta,\sigma}(\mathfrak{s},\mathfrak{g}) \mid \Phi_0(e) \in \mathcal{O}_{\mathfrak{q}_\mathbb{R}} \,\} \;\cong\; C(\mathcal{O}_{\mathfrak{q}_\mathbb{R}})$.

Let us explain next how to modify the proof of Proposition 4.9 in the present setting. We denote the complexifications of $\mathfrak{h}_\mathbb{R}$ and $\mathfrak{q}_\mathbb{R}$ by $\mathfrak{h}$ and $\mathfrak{q}$, respectively. The evaluation map $\Phi \to \Phi(e) = \Phi(\zeta)$ sends

$$\bigoplus_{\ell\geq 2} \mathrm{Hom}^{\theta,\sigma}(\mathfrak{s},\mathfrak{g}(\ell))(\ell-2) \longrightarrow \mathfrak{q}. \tag{6.18}$$

By (4.10a) the image of (6.18) lies in $[\zeta,\mathfrak{g}(\ell)]$ and, because $[\mathfrak{q},\zeta] \subset \mathfrak{h}$, in $[\zeta,\mathfrak{q}]$. This proves the analogue

$$\{\, \Phi(e) \mid \Phi \in \mathrm{Hom}^{\theta,\sigma}(\mathfrak{s},\mathfrak{g}(\ell))(\ell-2) \,\} \;\subset\; [\zeta, \mathfrak{q}\cap\mathfrak{g}(\ell)] \tag{6.19}$$

of (4.10a). It remains to prove the analogue of (4.10b):

$$\begin{gathered}\text{for each } \xi \in \mathfrak{h}\cap\mathfrak{g}(\ell), \text{ there exist } \Phi \in \mathrm{Hom}^{\theta,\sigma}(\mathfrak{s},\mathfrak{g}(\ell))(\ell-2)\\ \text{and } \eta \in \mathfrak{h}\cap\mathfrak{k} \text{ so that } \Phi(e) = [e,\xi+\eta];\\ \text{in this situation, } [e,\xi] \text{ uniquely determines } \Phi.\end{gathered} \tag{6.20}$$

Statement (4.10b) implies the existence of such a $\Phi \in \mathrm{Hom}^{\theta}(\mathfrak{s},\mathfrak{g}(\ell))(\ell - 2)$, an $\eta \in \mathfrak{k}$, and the fact that $[e,\xi]$ uniquely determines Φ. From (4.17b) we conclude that $\eta \in \mathfrak{h}\cap\mathfrak{k}$. Using the defining property $\Phi(e) = [e,\xi+\eta]$ and formula (4.17b) for $\Phi(h)$ one checks readily that $\Phi \in \mathrm{Hom}^{\theta,\sigma}(\mathfrak{s},\mathfrak{g}(\ell))(\ell-2)$.

With the appropriate changes in notation, the proof of Proposition 2.25 carries over almost word-for-word, giving the isomorphism $\mathcal{O}_{\mathfrak{q}_\mathbb{R}} \simeq T_{C(\mathcal{O}_{\mathfrak{q}_\mathbb{R}})}\mathcal{O}_{\mathfrak{q}_\mathbb{R}}$.

Vergne [17] observed that Kronheimer's isomorphism (3.7) restricts to an isomorphism $\mathcal{M}^\sigma(C(\mathcal{O}_{\mathfrak{q}_\mathbb{R}})) \simeq \mathcal{O}_{\mathfrak{q}_\mathbb{R}}$. Indeed, by definition, any $\Phi(\,\cdot\,) \in \mathcal{M}^\sigma(C(\mathcal{O}_{\mathfrak{q}_\mathbb{R}}))$ intertwines σ and $\sigma_\mathfrak{s}$, hence $\Phi(0)(e) \in \mathcal{O}\cap\mathfrak{q}_\mathbb{R}$ – cf.

(6.13b); here $\mathcal{O}$ again denotes the $G_{\mathbb{R}}$-orbit containing $\mathcal{O}_{\mathfrak{q}_{\mathbb{R}}}$. Since $\mathcal{O}_{\mathfrak{q}_{\mathbb{R}}}$ is a union of connected components of $\mathcal{O} \cap \mathfrak{q}_{\mathbb{R}}$, a continuity argument shows that Kronheimer's isomorphism (3.7) restricts to a one-to-one map

$$\mathcal{M}^{\sigma}(C(\mathcal{O}_{\mathfrak{q}_{\mathbb{R}}})) \longrightarrow \mathcal{O}_{\mathfrak{q}_{\mathbb{R}}}. \tag{6.21}$$

To see that it is onto, we consider a particular $\zeta \in \mathcal{O}_{\mathfrak{q}_{\mathbb{R}}}$ and the corresponding $\Phi(\,\cdot\,) \in \mathcal{M}(C(\mathcal{O}))$. The function $t \mapsto \sigma \circ \Phi(t) \circ \sigma_{\mathfrak{s}}$ also satisfies the defining conditions 3.6, and $\sigma \circ \Phi(0) \circ \sigma_{\mathfrak{s}}(e) = -\sigma \circ \Phi(t)(e) = \zeta$. By uniqueness, $\Phi(\,\cdot\,) = \sigma \circ \Phi(\,\cdot\,) \circ \sigma_{\mathfrak{s}}$, hence $\Phi(\,\cdot\,) \in \mathcal{M}^{\sigma}(C(\mathcal{O}_{\mathfrak{q}_{\mathfrak{s}}}))$. Thus (6.21) is surjective, as was to be shown.

7. The Sekiguchi correspondence

The Sekiguchi correspondence in its most general form establishes a bijection between nilpotent orbits attached to certain pairs of commuting involutions [16]. The complete statement and its specialization to the case of interest to us involve substantial notational overhead. For this reason, we discuss only the most important particular case; however, all statements and arguments extend readily[2] to the general case.

We use the notation and conventions of the previous section. In particular, $\mathfrak{g}$ arises as the complexification of $\mathfrak{g}_{\mathbb{R}} = \mathfrak{k}_{\mathbb{R}} \oplus \mathfrak{p}_{\mathbb{R}}$. We let θ denote both the Cartan involution of $\mathfrak{g}_{\mathbb{R}}$ and its extension to $\mathfrak{g}$, and τ complex conjugation with respect to the compact real form $\mathfrak{u}_{\mathbb{R}} \subset \mathfrak{g}$. Then, by construction of $\mathfrak{u}_{\mathbb{R}}$,

$$\tau\zeta = \theta\bar{\zeta} = \overline{\theta\zeta} \qquad (\zeta \in \mathfrak{g}). \tag{7.1}$$

Here $\bar{\zeta}$ refers to the complex conjugate, relative to $\mathfrak{g}_{\mathbb{R}}$. The complexification $\mathfrak{k}$ of $\mathfrak{k}_{\mathbb{R}}$ corresponds to a subgroup

$$K \subset G = \text{identity component of } \operatorname{Aut}(\mathfrak{g}). \tag{7.2}$$

The complex group G also contains

$$G_{\mathbb{R}} = (\operatorname{Aut}(\mathfrak{g}_{\mathbb{R}}))^0 \quad \text{and} \quad U_{\mathbb{R}} = (\operatorname{Aut}(\mathfrak{u}_{\mathbb{R}}))^0, \tag{7.3}$$

as noncompact and compact real form, respectively.

We shall consider nilpotent K-orbits in $\mathfrak{p} = \mathbb{C} \otimes_{\mathbb{R}} \mathfrak{p}_{\mathbb{R}}$ on the one hand, and nilpotent $G_{\mathbb{R}}$-orbits in $i\,\mathfrak{g}_{\mathbb{R}}$ on the other; $\mathcal{O}_{\mathfrak{p}}$ will be the generic

[2]Except for the orientation statements in Theorem 7.20, which needs to be modified when there are no complex and symplectic structures to orient the orbits in question.

symbol for the former, and $\mathcal{O}_{\mathbb{R}}$ for the latter. To avoid trivial exceptions, we always exclude the orbit $\{0\}$. Recall the Definition (2.17b) of the basis $\{e, f, h\}$ of $\mathfrak{s}$.

In the discussion in §2, we can make the trivial substitution of $i\,\mathfrak{g}_{\mathbb{R}}$ for $\mathfrak{g}_{\mathbb{R}}$. Then, as is shown there, $\Phi_0 \mapsto \Phi_0(i\,e)$ induces a $K_{\mathbb{R}}$-equivariant bijection

$$\text{(7.4}a\text{)} \qquad \{\, \Phi_0 \in \operatorname{Mor}^{\mathbb{R},\theta}(\mathfrak{s},\mathfrak{g}) \mid \Phi_0(i\,e) \in \mathcal{O}_{\mathbb{R}} \,\} \;\cong\; C(\mathcal{O}_{\mathbb{R}}),$$

for every nilpotent $G_{\mathbb{R}}$-orbit $\mathcal{O}_{\mathbb{R}} \neq \{0\}$ in $i\,\mathfrak{g}_{\mathbb{R}}$. When we look at all nilpotent $G_{\mathbb{R}}$-orbits in $i\mathfrak{g}_{\mathbb{R}}$ simultaneously, (7.4a) sets up a natural bijection

$$\text{(7.4}b\text{)} \quad \{\,\text{nilpotent } G_{\mathbb{R}}\text{-orbits in } i\mathfrak{g}_{\mathbb{R}}\,\} \;\xrightarrow{\sim}\; \{\, K_{\mathbb{R}}\text{-orbits in } \operatorname{Mor}^{\mathbb{R},\theta}(\mathfrak{s},\mathfrak{g})\,\}$$

We shall argue shortly that the results in §6 imply an analogous statement for nilpotent K-orbits in $\mathfrak{p}$: when $\mathcal{O}_{\mathfrak{p}} \neq \{0\}$ is a nilpotent K-orbit in $\mathfrak{p}$, the assignment $\Phi_0 \mapsto \Phi_0(\frac{h}{2} + \frac{ie}{2} + \frac{if}{2})$ induces a $K_{\mathbb{R}}$-equivariant bijection

$$\text{(7.5}a\text{)} \qquad \{\, \Phi_0 \in \operatorname{Mor}^{\mathbb{R},\theta}(\mathfrak{s},\mathfrak{g}) \mid \Phi_0(h + ie + if) \in \mathcal{O}_{\mathfrak{p}} \,\} \;\cong\; C(\mathcal{O}_{\mathfrak{p}}),$$

which, in turn, determines a bijection

$$\text{(7.5}b\text{)} \quad \{\,\text{nilpotent } K\text{-orbits in } \mathfrak{p}\,\} \;\xrightarrow{\sim}\; \{\, K_{\mathbb{R}}\text{-orbits in } \operatorname{Mor}^{\mathbb{R},\theta}(\mathfrak{s},\mathfrak{g})\,\},$$

in complete analogy to (7.4b). Combining (7.5b) with the inverse of (7.4b), we obtain the *Sekiguchi correspondence*

$$\text{(7.6)} \quad \{\,\text{nilpotent } K\text{-orbits in } \mathfrak{p}\,\} \;\xrightarrow{\sim}\; \{\,\text{nilpotent } G_{\mathbb{R}}\text{-orbits in } i\mathfrak{g}_{\mathbb{R}}\,\}$$

[16], which relates the K-orbit $\mathcal{O}_{\mathfrak{p}}$ to the $G_{\mathbb{R}}$-orbit $\mathcal{O}_{\mathbb{R}}$ precisely when the inverse images of $C(\mathcal{O}_{\mathfrak{p}})$ and $C(\mathcal{O}_{\mathbb{R}})$ in $\operatorname{Mor}^{\mathbb{R},\theta}(\mathfrak{s},\mathfrak{g})$ coincide.

We still need to establish (7.5a). For this purpose, we regard $(\mathfrak{g},\mathfrak{k})$ as symmetric pair over $\mathbb{R}$, with involution θ. We appeal to Proposition 6.17, which needs to be translated into the present setting by means of the "dictionary" (6.6-8). To begin with,

$$\text{(7.7)} \qquad \begin{aligned} \operatorname{Mor}^{\mathbb{R},\theta}(\mathfrak{s},\mathfrak{g}) \;&\rightsquigarrow\; \operatorname{Mor}(\mathfrak{su}(2),\mathfrak{u}_{\mathbb{R}}) \\ &\cong\; \operatorname{Mor}^{\tau}(\mathfrak{s},\mathfrak{g}) \\ &=_{\text{def}}\; \{\, \Phi_0 \in \operatorname{Mor}(\mathfrak{s},\mathfrak{g}) \mid \tau \circ \Phi_0 = \Phi_0 \circ \tau_{\mathfrak{s}} \,\}, \end{aligned}$$

as in (6.6); here $\tau_{\mathfrak{s}} : \mathfrak{s} \to \mathfrak{s}$ stands for complex conjugation with respect to $\mathfrak{su}(2)$. By the same dictionary,

$$\text{(7.8)} \qquad \operatorname{Mor}^{\mathbb{R},\theta,\sigma}(\mathfrak{s},\mathfrak{g}) \;\rightsquigarrow\; \{\, \Phi_0 \in \operatorname{Mor}^{\tau}(\mathfrak{s},\mathfrak{g}) \mid \theta \circ \Phi_0 = \Phi_0 \circ \sigma_{\mathfrak{s}} \,\},$$

since $\theta : \mathfrak{g} \to \mathfrak{g}$ now plays the role of the involution σ. A short calculation in $Sl(2,\mathbb{C})$ gives

$$(7.9) \qquad \sigma_{\mathfrak{s}} = \operatorname{Ad} c \circ \theta_{\mathfrak{s}} \circ \operatorname{Ad} c^{-1}, \qquad \text{with} \quad c = \frac{1}{\sqrt{2}} \begin{pmatrix} 1 & i \\ i & 1 \end{pmatrix},$$

and $\operatorname{Ad} c$ commutes with $\tau_{\mathfrak{s}}$, so (7.7) is equivalent to the assignment

$$(7.10) \qquad \begin{aligned} &\operatorname{Mor}^{\mathbb{R},\theta,\sigma}(\mathfrak{s},\mathfrak{g}) \\ &\rightsquigarrow \{ \Phi_0 \in \operatorname{Mor}^{\tau}(\mathfrak{s},\mathfrak{g}) \mid \theta \circ \Phi_0 \circ \operatorname{Ad} c = \Phi_0 \circ \operatorname{Ad} c \circ \theta_{\mathfrak{s}} \} \\ &= \{ \Phi_0 \circ \operatorname{Ad} c^{-1} \mid \Phi_0 \in \operatorname{Mor}^{\theta,\tau}(\mathfrak{s},\mathfrak{g}) \}. \end{aligned}$$

These morphisms get evaluated on e, as in Proposition 6.17. But $\operatorname{Mor}^{\theta,\tau}(\mathfrak{s},\mathfrak{g}) = \operatorname{Mor}^{\mathbb{R},\theta}(\mathfrak{s},\mathfrak{g})$ by (7.1), and $\operatorname{Ad} c^{-1}(e) = i(\frac{h}{2} - \frac{ie}{2} - \frac{if}{2})$. This gives the correspondence (7.5), with $i(\frac{h}{2} - \frac{ie}{2} - \frac{if}{2})$ in place of $\frac{h}{2} + \frac{ie}{2} + \frac{if}{2}$. Note that nilpotent K-orbits in $\mathfrak{p}$ are invariant under scaling by nonzero complex numbers – this is clear in the case of $\mathfrak{g}_{\mathbb{R}} = \mathfrak{sl}(2,\mathbb{R})$, and follows in the general case by what has been said so far. Since i has absolute value 1, it maps the core of an orbit to itself, so we can drop the factor i. Finally, complex conjugation permutes the nilpotent K-orbits [3] in $\mathfrak{p}$, and this allows us to replace $\frac{h}{2} - \frac{ie}{2} - \frac{if}{2}$ by its complex conjugate.

When the Sekiguchi correspondence relates $\mathcal{O}_{\mathfrak{p}}$ to $\mathcal{O}_{\mathbb{R}}$, various objects attached to the two orbits are naturally isomorphic – the cores because of (7.4a) and (7.5a):

$$(7.11) \qquad \begin{aligned} C(\mathcal{O}_{\mathfrak{p}}) \cong C(\mathcal{O}_{\mathbb{R}}), \quad \Phi_0(\tfrac{h}{2} + \tfrac{ie}{2} + \tfrac{if}{2}) \leftrightarrow \Phi_0(ie) \\ (\Phi_0 \in \operatorname{Mor}^{\mathbb{R},\theta}(\mathfrak{s},\mathfrak{g})). \end{aligned}$$

This isomorphism is $K_{\mathbb{R}}$-equivariant by construction, so the isotropy subgroups of $K_{\mathbb{R}}$ at $\Phi_0(\frac{h}{2} + \frac{ie}{2} + \frac{if}{2})$ and $\Phi_0(ie)$ coincide[4]. Proposition 4.9 identifies the normal space to $C(\mathcal{O}_{\mathbb{R}})$ at the point $\Phi_0(ie) \in C(\mathcal{O}_{\mathbb{R}})$,

$$(7.12) \qquad \begin{aligned} \textstyle\bigoplus_{r \geq 2} \operatorname{Mor}^{\mathbb{R},\theta}(\mathfrak{s},\mathfrak{g}(r))(r-2) \simeq (T_{C(\mathcal{O}_{\mathbb{R}})}\mathcal{O}_{\mathbb{R}})_{\Phi_0(ie)}, \\ \Phi \mapsto \Phi(ie), \end{aligned}$$

[3]There are two equally natural definitions of the Sekiguchi correspondence. They are related by complex conjugation on the side of K-orbits, or alternatively, by multiplication by -1 on the side of $G_{\mathbb{R}}$-orbits. Our choice of the correspondence is dictated by the application in [15].

[4]This can be seen directly: if $k \in K_{\mathbb{R}}$ fixes $\Phi_0(\frac{h}{2} + \frac{ie}{2} + \frac{if}{2})$, it fixes the real and imaginary parts separately, which together generate $\Phi_0(\mathfrak{s})$; similarly, if k fixes $\Phi_0(ie)$, it fixes the Φ_0-image of $\theta(ie) = -if$ and hence also the image of $[e,f] = h$.

equivariantly with respect to the isotropy subgroup of $K_{\mathbb{R}}$ at $\Phi_0(ie)$. The analogue of (4.9) in the symmetric pair case – which is part of Proposition 6.17 – identifies the normal space to $C(\mathcal{O}_{\mathfrak{p}})$ at $\Phi_0(\frac{h}{2}+\frac{ie}{2}+\frac{if}{2}) \in C(\mathcal{O}_{\mathfrak{p}})$,

$$(7.13)\qquad \begin{aligned} \textstyle\bigoplus_{r\geq 2} \mathrm{Mor}^{\mathbb{R},\theta}(\mathfrak{s},\mathfrak{g}(r))(r-2) \ &\simeq\ (T_{C(\mathcal{O}_{\mathfrak{p}})}\mathcal{O}_{\mathfrak{p}})_{\Phi_0(\frac{h}{2}+\frac{ie}{2}+\frac{if}{2})},\\ \Phi &\mapsto \Phi(\tfrac{h}{2}+\tfrac{ie}{2}+\tfrac{if}{2}), \end{aligned}$$

again equivariantly with respect to the isotropy subgroup of $K_{\mathbb{R}}$. The preceding statement involves the same "translation" that we just used to establish (7.5a). Because of (7.12-13), the fiber of the normal bundle $T_{C(\mathcal{O}_{\mathfrak{p}})}\mathcal{O}_{\mathfrak{p}}$ at $\Phi(\frac{h}{2}+\frac{ie}{2}+\frac{if}{2})$ is isomorphic to the fiber of $T_{C(\mathcal{O}_{\mathbb{R}})}\mathcal{O}_{\mathbb{R}}$ at $\Phi(ie)$ – isomorphic as representation space for the common isotropy group. Thus (7.11) extends to a real analytic isomorphism of $K_{\mathbb{R}}$-equivariant vector bundles

$$(7.14)\qquad T_{C(\mathcal{O}_{\mathfrak{p}})}\mathcal{O}_{\mathfrak{p}} \ \simeq\ T_{C(\mathcal{O}_{\mathbb{R}})}\mathcal{O}_{\mathbb{R}}.$$

Appealing to Proposition 2.25 and its analogue for symmetric pairs, as stated in (6.17), we obtain

$$(7.15)\qquad \mathcal{O}_{\mathfrak{p}} \ \simeq\ \mathcal{O}_{\mathfrak{g}_{\mathbb{R}}},$$

a real analytic, $K_{\mathbb{R}}$-equivariant isomorphism between the two orbits.

Vergne [17] deduces the existence of a $K_{\mathbb{R}}$-equivariant diffeomorphism $\mathcal{O}_{\mathfrak{p}} \simeq \mathcal{O}_{\mathfrak{g}_{\mathbb{R}}}$ from Kronheimer's description of nilpotent orbits, as follows. According to (3.7),

$$(7.16)\qquad \mathcal{M}(C(\mathcal{O}_{\mathbb{R}})) \xrightarrow{\sim} \mathcal{O}_{\mathbb{R}}, \qquad \Phi(\,\cdot\,) \mapsto \Phi(0)(ie),$$

is a $K_{\mathbb{R}}$-equivariant diffeomorphism. The analogous statement for symmetric pairs in (6.17), translated as in (7.7-10), gives the $K_{\mathbb{R}}$-equivariant diffeomorphism

$$(7.17)\qquad \mathcal{M}(C(\mathcal{O}_{\mathfrak{p}})) \xrightarrow{\sim} \mathcal{O}_{\mathfrak{p}}, \qquad \Phi(\,\cdot\,) \mapsto \Phi(0)(\tfrac{h}{2}+\tfrac{ie}{2}+\tfrac{if}{2}).$$

Since $\mathcal{O}_{\mathbb{R}}$ and $\mathcal{O}_{\mathfrak{p}}$ are related by the Sekiguchi correspondence, (7.11) implies

$$(7.18)\qquad \mathcal{M}(C(\mathcal{O}_{\mathfrak{p}})) \ =\ \mathcal{M}(C(\mathcal{O}_{\mathbb{R}})).$$

The composition of (7.16-18) gives Vergne's interpretation of the Sekiguchi correspondence.

In the proof of the Barbasch-Vogan conjecture in [15], we were lead to a quite different geometric description of the correspondence. We fix a nilpotent G-orbit $\mathcal{O}$ in $\mathfrak{g}$. Via the isomorphism $\mathfrak{g} \simeq \mathfrak{g}^*$ induced by the Killing form, $\mathcal{O}$ can be viewed as a complex coadjoint orbit. As such, it carries a holomorphic symplectic form $\sigma_{\mathcal{O}}$; in particular, $\mathcal{O}$ has even complex dimension $2k$. The intersection $\mathcal{O} \cap \mathfrak{p}$ decomposes into a union of finitely many K-orbits, all Lagrangian with respect to $\sigma_{\mathcal{O}}$, hence of complex dimension k [8]. Analogously, $\mathcal{O} \cap i\mathfrak{g}_{\mathbb{R}}$ is a union of finitely many $G_{\mathbb{R}}$-orbits, Lagrangian with respect the real symplectic form $\operatorname{Re}\sigma_{\mathcal{O}}$, hence of real dimension $2k$. We enumerate the two types of orbits as

$$(7.19)\quad \mathcal{O} \cap \mathfrak{p} \;=\; \mathcal{O}_{\mathfrak{p},1} \cup \cdots \cup \mathcal{O}_{\mathfrak{p},d} \quad \text{and} \quad \mathcal{O} \cap i\mathfrak{g}_{\mathbb{R}} \;=\; \mathcal{O}_{\mathbb{R},1} \cup \cdots \cup \mathcal{O}_{\mathbb{R},d}.$$

The complex structure orients the orbits $\mathcal{O}_{\mathfrak{p},j}$, which gives meaning to the $[\mathcal{O}_{\mathfrak{p},j}]$ as top dimensional cycles, with infinite support, in $\mathcal{O} \cap \mathfrak{p}$. We had remarked already that $\sigma_{\mathcal{O}}$ restricts to a purely imaginary form on $\mathcal{O} \cap i\mathfrak{g}_{\mathbb{R}}$. Thus $\frac{1}{i}\sigma_{\mathcal{O}}$ defines a symplectic form on the $G_{\mathbb{R}}$-orbits $\mathcal{O}_{\mathbb{R},j}$ – one can check that this gives the same symplectic structure as the identification of $\mathcal{O}_{\mathbb{R},j}$ with a real coadjoint orbit via division by i and the isomorphism $\mathfrak{g}_{\mathbb{R}} \simeq \mathfrak{g}_{\mathbb{R}}{}^*$ induced by the Killing form. We use the symplectic structure to orient the $\mathcal{O}_{\mathfrak{g}_{\mathbb{R}},j}$, and to regard them as cycles $[\mathcal{O}_{\mathfrak{g}_{\mathbb{R}},j}]$ in $\mathcal{O} \cap i\mathfrak{g}_{\mathbb{R}}$. We let $\mathrm{H}_*^{inf}(\ldots,\mathbb{Z})$ denote homology with possibly "infinite supports" (Borel-Moore homology). Then, in view of (7.19),

$$(7.20)\quad \begin{aligned} \mathrm{H}_{2k}^{inf}(\mathcal{O} \cap \mathfrak{p}, \mathbb{Z}) &= \{\textstyle\sum n_j\,[\mathcal{O}_{\mathfrak{p},j}] \mid n_j \in \mathbb{Z}\}, \\ \mathrm{H}_{2k}^{inf}(\mathcal{O} \cap i\mathfrak{g}_{\mathbb{R}}, \mathbb{Z}) &= \{\textstyle\sum n_j\,[\mathcal{O}_{\mathfrak{g}_{\mathbb{R}},j}] \mid n_j \in \mathbb{Z}\}. \end{aligned}$$

There are no relations among the $[\mathcal{O}_{\mathfrak{p},j}]$, respectively $[\mathcal{O}_{\mathbb{R},j}]$, since we are dealing with top dimensional homology. This allows us to view the Sekiguchi correspondence as a specific isomorphism between the two homology groups.

Our description of the Sekiguchi correspondence amounts to a geometric passage between the two homology groups in (7.20). We define a real analytic family of diffeomorphisms

$$(7.21)\quad f_t : \mathcal{O} \longrightarrow \mathcal{O}, \qquad f_t(\zeta) \;=\; \mathrm{Ad}(\exp(t\operatorname{Re}\zeta))(\zeta) \qquad (t \in \mathbb{R});$$

this agrees with the definition in [15, §6], except for the change of variables $s = t^{-1}$. The images $(f_t)_*[\mathcal{O}_{\mathfrak{p},j}]$, $0 \le t < \infty$, of any $[\mathcal{O}_{\mathfrak{p},j}]$ constitute a real analytic family of cycles in the complex orbit $\mathcal{O}$. We argue in [15] that this family of cycles has a limit for $t \to +\infty$ for a priori reasons, and that the limit is a cycle in $\mathcal{O} \cap i\mathfrak{g}_{\mathbb{R}}$. The existence of the limit may seem surprising, since f_t has exponential behavior for large t. At the end of this section, we shall say a few words about the notion of limit of a family of cycles, and about the argument for the existence of the limit in our situation.

7.22. Theorem. *The assignment* $c \mapsto \lim_{t\to+\infty} (f_t)_* c$ *induces the Sekiguchi correspondence, as map from* $\mathrm{H}_{2k}^{inf}(\mathcal{O} \cap \mathfrak{p}, \mathbb{Z})$ *to* $\mathrm{H}_{2k}^{inf}(\mathcal{O} \cap i\mathfrak{g}_{\mathbb{R}}, \mathbb{Z})$. *In other words,*

$$\lim_{t\to+\infty} (f_t)_*[\mathcal{O}_{\mathfrak{p}}] \quad = \quad [\mathcal{O}_{\mathfrak{g}_{\mathbb{R}}}]$$

whenever the K*-orbit* $\mathcal{O}_{\mathfrak{p}}$ *in* $\mathcal{O} \cap \mathfrak{p}$ *and the* $G_{\mathbb{R}}$*-orbit* $\mathcal{O}_{\mathbb{R}}$ *in* $\mathcal{O} \cap i\mathfrak{g}_{\mathbb{R}}$ *are related by the Sekiguchi correspondence.*

This theorem plays a crucial role in our proof of the Barbasch-Vogan conjecture in [15], where it is stated as Theorem 6.3. Its proof splits naturally in two parts. One establishes the existence of the limit and reduces its computation to two geometric lemmas about nilpotent orbits [15, §6]. The second part consists of the proofs of the two lemmas; these proofs occupy the last section of this paper and use the tools developed in the preceding sections.

We had mentioned earlier that our description of the Sekiguchi correspondence carries over to its most general version, which relates nilpotent orbits attached to symmetric pairs defined by commuting involutions [16]. The statement and the various steps of the proof apply in the general case after minimal changes, with one exception: in the absence of complex and symplectic structures, the orbits no longer carry natural orientations and – as far as we know – need not even be orientable. One can deal with this problem by considering the orbits as cycles with values in the orientation sheaf; the isomorphism (7.15) identifies the orientation sheaves of any two orbits related by the Sekiguchi correspondence. When that is done, Theorem 7.22 remains correct as stated.

Let us comment briefly on the meaning of the limit in Theorem 7.22 – for a more detailed discussion of limits of cycles in general, see [14]. When we restrict the family of cycles $\{\, (f_t)_*[\mathcal{O}_{\mathfrak{p}}] \,\}$ to some finite interval $0 \le t \le a$, we obtain a submanifold with boundary in $[0, a] \times \mathcal{O}$, and the boundary consists of the fibers over 0 and a; in this situation, it is natural to think of the fiber over a as the limit of the family as t tends to a from below. What matters here is not the smoothness of the family; it suffices that the total space and the map to the parameter interval $[0, a]$ be Whitney stratifiable. In the case of real algebraic, or more generally, subanalytic families of cycles, Whitney stratifiability is automatic. The family $\{\, (f_t)_*[\mathcal{O}_{\mathfrak{p}}] \,\}$ fails to be subanalytic at $t = +\infty$. It does, however, belong to one of the *analytic-geometric categories* constructed by van den Dries-Miller [2], using recent work in model theory [18], [3]. These analytic-geometric categories generalize the notion of subanalyticity, and share most of the important properties of the subanalytic category, such as Whitney stratifiability. This implies the existence of the limit in Theorem 7.22; in effect, one can argue as if the family were subanalytic even at infinity. By definition, the limit cycle is supported on F_∞, the fiber over $\{+\infty\}$ of the closure of $\{\,(t, f_t(\zeta)) \mid 0 < t < \infty, \zeta \in \mathcal{O}_{\mathfrak{p}}\,\}$ in $[0, +\infty] \times \mathcal{O}$. A fairly simple argument identifies F_∞ as $\mathcal{O} \cap i\mathfrak{g}_{\mathbb{R}}$ [15, §6]. Thus, according to (7.19), the limit cycle can only be an integral linear combination of the $\mathcal{O}_{\mathbb{R},j}$. A normal slice to $\mathcal{O}_{\mathbb{R},j}$ in $\mathcal{O}$, at a generic point of $\mathcal{O}_{\mathbb{R},j}$, intersects $(f_t)_*[(\mathcal{O}_{\mathfrak{p}})]$, for t close to $+\infty$, with an intersection multiplicity m_j not depending on t; here "generic" is to be taken in the sense of the analytic-geometric category to which the family of cycles belongs. Essentially by definition, the intersection multiplicity m_j is the coefficient of $\mathcal{O}_{\mathbb{R},j}$ in the limit cycle. We argue in [15, §6] that the multiplicity m_j can be calculated even at "non-generic" points under certain circumstances. This argument reduces the statement of Theorem 7.22 to the second of the two lemmas in the next section; the first lemma is a crucial ingredient of the proof of the second.

8. Two lemmas

We work in the setting of the complexified Lie algebra $\mathfrak{g} = \mathbb{C} \otimes_{\mathbb{R}} \mathfrak{g}_{\mathbb{R}}$, as in §§6-7. We keep fixed, once and for all, a nilpotent G-orbit $\mathcal{O}$ in $\mathfrak{g} - \{0\}$. Recall the family of real analytic maps

$$(8.1) \qquad f_t : \mathcal{O} \longrightarrow \mathcal{O}, \qquad f_t(\zeta) = \mathrm{Ad}(\exp(t \,\mathrm{Re}\,\zeta))(\zeta) \qquad (t \in \mathbb{R}),$$

defined in (7.21); as was remarked earlier, this agrees with the definition in [15, §5], except for the change of variables $t = s^{-1}$. Note that

$$(8.2) \qquad \mathrm{Re}\,(f_t(\zeta)) \;=\; \mathrm{Re}\,\zeta$$

for all $\zeta \in \mathcal{O}$. It follows that $\{f_t\}$ is a one parameter group of diffeomorphisms.

Because of (6.5) and (7.1), the Definition (2.7) of the moment map translates into

$$(8.3) \qquad m : \mathfrak{g} - \{0\} \longrightarrow i\,\mathfrak{u}_{\mathbb{R}}, \qquad m(\zeta) \;=\; -\,\frac{[\zeta, \theta\bar{\zeta}]}{\|\zeta\|^2}$$

in the present situation. This map is invariant under the action of the maximal compact subgroup $U_{\mathbb{R}} \subset G$. We are interested in the qualitative behavior of $\|m\|^2$ along trajectories $\{f_t(\zeta) \mid t \geq 0\}$ through points $\zeta \in \mathcal{O} \cap \mathfrak{p}$. Thus we consider a particular K-orbit $\mathcal{O}_\mathfrak{p}$ in $\mathcal{O} \cap \mathfrak{p}$ and a point $\zeta \in \mathcal{O}_\mathfrak{p}$. With this choice of ζ kept fixed, we write

$$\begin{aligned} m(f_t(\zeta)) \ = \ m(t) \ = \ m_1(t) + m_2(t) + m_3(t), \quad \text{with} \\ m_1(t) \in \mathbb{R} \cdot \operatorname{Re} \zeta, \quad m_2(t) \in \mathfrak{p}_\mathbb{R} \cap (\operatorname{Re} \zeta)^\perp, \quad m_3(t) \in i\mathfrak{k}_\mathbb{R}. \end{aligned} \tag{8.4}$$

This can be done because $i\,\mathfrak{u}_\mathbb{R} = i\,\mathfrak{k}_\mathbb{R} \oplus \mathfrak{p}_\mathbb{R}$. Our first statement is [15, Lemma 6.28], phrased in terms of the new variable $t = s^{-1}$.

8.5. Lemma. *For $\zeta \in \mathcal{O}_\mathfrak{p}$ as above and $t \in \mathbb{R}$,*

$$\|m_1(t)\|^2 \ + \ \|m_3(t)\|^2 \ \geq \ \|m(0)\|^2.$$

Before embarking on the proof of the lemma, we state the second one. Besides the K-orbit $\mathcal{O}_\mathfrak{p}$ in $\mathcal{O} \cap \mathfrak{p}$, it involves a $G_\mathbb{R}$-orbit $\mathcal{O}_\mathbb{R}$ in $\mathcal{O} \cap i\,\mathfrak{g}_\mathbb{R}$, which may or may not be related to $\mathcal{O}_\mathfrak{p}$ by the Sekiguchi correspondence. We fix a point $\nu \in C(\mathcal{O}_\mathbb{R})$, which can be represented as

$$\nu \ = \ i\,\Phi_0(e), \qquad \text{with} \ \ \Phi_0 \in \operatorname{Mor}^{\mathbb{R},\theta}(\mathfrak{s}, \mathfrak{g}), \tag{8.6}$$

as in (7.4). The choice of Φ_0 gives meaning to the decomposition (4.5) of $\mathfrak{g}$. The space

$$\mathfrak{q}(\nu) \ = \ \textstyle\bigoplus_{r \geq 1} \bigoplus_{\ell < r} \mathfrak{g}(r, \ell) \tag{8.7}$$

is a linear complement to $\operatorname{Ker} \operatorname{ad}(\nu) = \operatorname{Ker} \operatorname{ad}(e)$ in $\mathfrak{g}$, and is defined over $\mathbb{R}$. Thus, for $a > 0$ sufficiently small, the map

$$\begin{aligned} \{ (\xi, \eta) \in \mathfrak{g}_\mathbb{R} \times \mathfrak{g}_\mathbb{R} \ \mid \ \xi, \eta \in \mathfrak{q}(\nu) \cap \mathfrak{g}_\mathbb{R}, \ \|\xi\|, \|\eta\| < a \} \longrightarrow \mathcal{O} \\ (\xi, \eta) \longmapsto \operatorname{Ad}(\exp i\xi \exp \eta)(\nu) \end{aligned} \tag{8.8}$$

sends its domain isomorphically to an open neighborhood of ν in $\mathcal{O}$. In particular, shrinking a further if necessary, we can make

$$N(\nu, a) \ = \ \{ \operatorname{Ad}(\exp i\xi)(\nu) \ \mid \ \xi \in \mathfrak{q}(\nu) \cap \mathfrak{g}_\mathbb{R}, \ \|\xi\| < a \} \tag{8.9}$$

a normal slice to $\mathcal{O}_\mathbb{R}$ in $\mathcal{O}$ at the point ν – in other words, a submanifold of $\mathcal{O}$ that intersects $\mathcal{O}_\mathbb{R}$ at the single point ν, where the intersection is transverse. We remarked in §7 that the orbits $\mathcal{O}_\mathfrak{p}$, $\mathcal{O}_\mathbb{R}$ carry natural orientations: the former as complex manifold, the latter as coadjoint orbit via $\mathcal{O}_\mathbb{R} \ni i\zeta \mapsto \zeta \in \mathfrak{g}_\mathbb{R} \cong \mathfrak{g}_\mathbb{R}{}^*$, hence as canonically symplectic manifold [5]. The orientation of $\mathcal{O}_\mathfrak{p}$ in turn orients the diffeomorphic image $f_t(\mathcal{O}_\mathfrak{p})$. Our next statement is a more specific version of Lemma 6.29 in [15], again phrased in terms of the variable $t = s^{-1}$.

8.10. Lemma. *For a sufficiently small and t sufficiently large in terms of a, the submanifolds $N(\nu, a)$ and $f_t(\mathcal{O}_\mathbb{R})$ of $\mathcal{O}$ meet either exactly once or not at all, depending on whether or not $\mathcal{O}_\mathbb{R}$ is the Sekiguchi image of $\mathcal{O}_\mathfrak{p}$. In the former situation, the intersection is transverse and has intersection multiplicity $+1$ when the orientation of $N(\nu, a)$ and the sign convention for intersection multiplicities are chosen so as to make $N(\nu, a)$ meet $\mathcal{O}_\mathbb{R}$ with multiplicity $+1$ at ν.*

Proof of 8.5. Let us record some observations about nilpotents in $\mathcal{O} \cap \mathfrak{p}$; they will be used not only here but also in the proof of the second lemma. We consider an arbitrary $\zeta \in \mathcal{O} \cap \mathfrak{p}$, which we express as $\zeta = \xi + i\eta$ with $\xi, \eta \in \mathfrak{p}_\mathbb{R}$. In particular, $\operatorname{ad}\xi : \mathfrak{g}_\mathbb{R} \to \mathfrak{g}_\mathbb{R}$ is self-adjoint with respect to the complex extension of the inner product (2.2). Thus

$$\zeta = \sum \zeta_\lambda = \xi + i \sum \eta_\lambda,$$

with λ ranging over $\mathbb{R}$, and

$$\eta_\lambda \in \mathfrak{g}_\mathbb{R}^\lambda = \lambda\text{-eigenspace of } \operatorname{ad}\xi, \tag{8.11a}$$

$$\zeta_\lambda = \begin{cases} \xi + i\eta_0 & \text{if } \lambda = 0, \\ i\eta_\lambda & \text{if } \lambda \neq 0. \end{cases}$$

The nilpotence of ζ implies

$$0 = B(\zeta, \zeta) = B(\xi + i\eta, \xi + i\eta) = \|\xi\|^2 - \|\eta\|^2 + 2i(\xi, \eta).$$

Also, $\mathfrak{g}_\mathbb{R}^\lambda \perp \mathfrak{g}_\mathbb{R}^\mu$ unless $\mu = \lambda$, hence

$$\begin{aligned} \|\xi\|^2 &= \|\eta\|^2 = \sum \|\eta_\lambda\|^2 = \tfrac{1}{2}\|\zeta\|^2, \\ (\xi, \eta) &= (\xi, \eta_0) = 0. \end{aligned} \tag{8.11b}$$

Both ξ and η lie in $\mathfrak{p}_\mathbb{R}$, i.e., the (-1)-eigenspace of θ, hence

$$\theta\eta_\lambda = -\eta_{-\lambda}, \qquad \|\eta_\lambda\| = \|\eta_{-\lambda}\|. \tag{8.11c}$$

All this applies to the point ζ referred to in the statement of the lemma.

[5]The orientation conventions are spelled out in detail in [14, §8]

We calculate $m(t) = m(f_t(\zeta))$, beginning with the Definition (8.3) of the moment map:

$$\begin{aligned} m(t) &= -\frac{[\mathrm{Ad}(\exp(t\xi))\zeta, \theta(\overline{\mathrm{Ad}(\exp(t\xi))\zeta})]}{\|\,\mathrm{Ad}(\exp(t\xi))\zeta\|^2} \\ &= \frac{[\xi + i\sum_\lambda e^{\lambda t}\eta_\lambda, \xi - i\sum_\lambda e^{\lambda t}\eta_{-\lambda}]}{\sum_\lambda e^{2\lambda t}\|\zeta_\lambda\|^2} \\ &= \frac{\sum_{\lambda,\mu} e^{(\lambda+\mu)t}[\eta_\lambda, \eta_{-\mu}] \;-\; i\sum_\lambda \lambda(e^{\lambda t} + e^{-\lambda t})\eta_\lambda}{\sum_\lambda e^{2\lambda t}\|\zeta_\lambda\|^2}. \end{aligned}$$

The imaginary part of this expression equals $m_3(t)$, and $i\eta_\lambda = \zeta_\lambda$ if $\lambda \neq 0$. We conclude:

$$\text{(8.12)}\qquad \begin{aligned} m_3(t) &= -\frac{\sum_\lambda \lambda(e^{\lambda t} + e^{-\lambda t})\zeta_\lambda}{\sum_\lambda e^{2\lambda t}\|\zeta_\lambda\|^2}, \\ \|m_3(t)\|^2 &= \frac{\sum_\lambda \lambda^2(e^{\lambda t} + e^{-\lambda t})^2\|\zeta_\lambda\|^2}{(\sum_\lambda e^{2\lambda t}\|\zeta_\lambda\|^2)^2}. \end{aligned}$$

On the other hand,

$$\begin{aligned} m_1(t) &= \frac{(\mathrm{Re}\; m(t), \xi)}{\|\xi\|^2}\,\xi \;=\; 2\,\frac{B(\mathrm{Re}\; m(t), \xi)}{\|\zeta\|^2}\,\xi \\ &= 2\,\frac{\sum_{\lambda,\mu} e^{(\lambda+\mu)t} B([\eta_\lambda, \eta_{-\mu}], \xi)}{\|\zeta\|^2\,\sum_\lambda e^{2\lambda t}\|\zeta_\lambda\|^2}\,\xi \\ &= -2\,\frac{\sum_{\lambda,\mu} e^{(\lambda+\mu)t} B(\eta_\lambda, [\xi, \eta_{-\mu}])}{\|\zeta\|^2\,\sum_\lambda e^{2\lambda t}\|\zeta_\lambda\|^2}\,\xi \\ &= 2\,\frac{\sum_\lambda \lambda e^{2\lambda t} B(\eta_\lambda, \eta_{-\lambda})}{\|\zeta\|^2\,\sum_\lambda e^{2\lambda t}\|\zeta_\lambda\|^2}\,\xi \\ &= 2\,\frac{\sum_\lambda \lambda e^{2\lambda t}\|\eta_\lambda\|^2}{\|\zeta\|^2\,\sum_\lambda e^{2\lambda t}\|\zeta_\lambda\|^2}\,\xi. \end{aligned}$$

In the numerator, only the summands corresponding to non-zero λ matter, so we can replace $i\eta_\lambda$ by ζ_λ, giving us

$$\text{(8.13)}\qquad \begin{aligned} m_1(t) &= 2\,\frac{\sum_\lambda \lambda e^{2\lambda t}\|\zeta_\lambda\|^2}{\|\zeta\|^2\,\sum_\lambda e^{2\lambda t}\|\zeta_\lambda\|^2}\,\xi, \\ \|m_1(t)\|^2 &= 2\,\frac{(\sum_\lambda \lambda e^{2\lambda t}\|\zeta_\lambda\|^2)^2}{\sum_\lambda \|\zeta_\lambda\|^2\,(\sum_\lambda e^{2\lambda t}\|\zeta_\lambda\|^2)^2}; \end{aligned}$$

in the second line, we have used the equality $\sum_\lambda \|\zeta_\lambda\|^2 = \|\zeta\|^2 = 2\|\xi\|^2$.

At $t = 0$, $m(0)$ is a positive multiple of $-[\zeta, \theta\bar{\zeta}]$, which lies in $i\mathfrak{k}_{\mathbb{R}}$. Thus

$$
\begin{aligned}
m(0) &= -2\,\frac{\sum_\lambda \lambda\zeta_\lambda}{\sum_\lambda \|\zeta_\lambda\|^2},\\
\|m(0)\|^2 &= 4\,\frac{\sum_\lambda \lambda^2\|\zeta_\lambda\|^2}{(\sum_\lambda \|\zeta_\lambda\|^2)^2},
\end{aligned}
\tag{8.14}
$$

as follows from (8.12) with $t = 0$.

To prove the inequality asserted by the lemma, we rewrite it in terms of the expressions (8.12-14) and clear the (positive) denominators. This transforms the inequality into the following equivalent form:

$$
\begin{aligned}
&2(\textstyle\sum_\lambda \|\zeta_\lambda\|^2)(\sum_\lambda \lambda e^{2\lambda t}\|\zeta_\lambda\|^2)^2\\
&\qquad + (\sum_\lambda \|\zeta_\lambda\|^2)^2(\sum_\lambda \lambda^2(e^{\lambda t}+e^{-\lambda t})^2\|\zeta_\lambda\|^2)\\
&\qquad\qquad \geq 4(\textstyle\sum_\lambda \lambda^2\|\zeta_\lambda\|^2)(\sum_\lambda e^{2\lambda t}\|\zeta_\lambda\|^2)^2.
\end{aligned}
\tag{8.15}
$$

The original inequality is homogenous in ζ. So is the inequality (8.15) when one allows only scaling by real numbers and gives λ – which is a typical eigenvalue of $\operatorname{ad}\xi = \operatorname{ad}(\operatorname{Re}\zeta)$ – the same weight as $\|\zeta_\lambda\|$. Thus we are free to renormalize ζ, subject to the condition $2 = \|\zeta\|^2 = \sum_\lambda \|\zeta_\lambda\|^2$. We set $a_\lambda = \|\zeta_\lambda\|^2$. Then $\sum_\lambda a_\lambda = 2$, and $a_0 = \|\xi\|^2 + \|\eta_0\|^2 \geq 1$ by (8.11); also, $a_{-\lambda} = a_\lambda$, again by (8.11). We note that

$$
(e^{\lambda t} + e^{-\lambda t})^2 = e^{2\lambda t} + 2 + e^{-2\lambda t},
$$

and replace $2t$ by t throughout. At this point, the inequality to be proved becomes

$$
\begin{aligned}
(\textstyle\sum_\lambda \lambda e^{\lambda t}a_\lambda)^2 &+ \textstyle\sum_\lambda \lambda^2(e^{\lambda t}+2+e^{-\lambda t})a_\lambda\\
&\geq (\textstyle\sum_\lambda \lambda^2 a_\lambda)(\sum_\lambda e^{\lambda t}a_\lambda)^2,\\
&\text{subject to the conditions } a_\lambda = a_{-\lambda} > 0,\\
&a_0 \geq 1,\ \ \textstyle\sum_\lambda a_\lambda = 2.
\end{aligned}
\tag{8.16}
$$

There must be at least one pair of non-zero indices $\pm\lambda$; otherwise ξ and η would commute, making ζ semisimple – impossible, since ζ is a non-zero nilpotent. One further reformulation of the inequality to be proved: we define

$$
h(t) = \textstyle\sum_\lambda a_\lambda e^{\lambda t}, \qquad t \in \mathbb{R}. \tag{8.17a}
$$

This transforms the inequality into the form

$$
h'(t)^2 + 2h''(t) + 2h''(0) \geq h''(0)h(t)^2, \tag{8.17b}
$$

with the a_λ still subject to the conditions listed in (8.16).

The function $h(t)$ has a globally convergent Taylor series. We can therefore verify (8.17b) by establishing the corresponding inequalities for all derivatives, at $t = 0$, of the expressions on both sides, including the 0-th derivative, of course. The conditions on the a_λ imply, in particular,

$$\text{(8.18)} \qquad \begin{aligned} &\text{a)}\ \ h(0) \ = \ \textstyle\sum_\lambda a_\lambda \ = \ 2; \\ &\text{b)}\ \ 0 < \textstyle\sum_{\lambda \neq 0} a_\lambda \ \leq 1; \\ &\text{c)}\ \ h^{(2k)}(0) \ = \ \textstyle\sum_{\lambda \neq 0} \lambda^{2k}\, a_\lambda, \qquad && \text{for } k > 0; \\ &\text{d)}\ \ h^{(2k+1)}(0) \ = \ 0, && \text{for } k \geq 0. \end{aligned}$$

This gives the inequality at $t = 0$, as an equality, in fact. We still must show that

$$\frac{d^{(k)}}{dt^k}\left(h'(t)^2 \ + \ 2h''(t) \ + \ 2h''(0)\right)|_{t=0} \ \geq \ \frac{d^{(k)}}{dt^k}\left(h''(0)h(t)^2\right)|_{t=0}$$

for all $k > 0$, or equivalently,

$$\text{(8.19)} \qquad \begin{aligned} &\sum_{\ell=0}^{k} \binom{k}{\ell} h^{(\ell+1)}(0)h^{(k-\ell+1)}(0) \ + \ 2h^{(k+2)}(0) \\ &\qquad\qquad \geq \ \sum_{\ell=0}^{k} \binom{k}{\ell} h^{(2)}(0)h^{(\ell)}(0)h^{(k-\ell)}(0). \end{aligned}$$

When k is odd, both sides reduce to zero because of (8.18d). To deal with the even case, we replace k by $2k$, omit the odd derivatives in the two sums, and separate out the summands involving $h(0) = 2$. This reduces the problem to showing that

$$\text{(8.20)} \qquad \begin{aligned} &\sum_{\ell=1}^{k} \binom{2k}{2\ell-1} h^{(2\ell)}(0)h^{(2k-2\ell+2)}(0) \ + \ 2h^{(2k+2)}(0) \\ &\qquad\qquad \geq \ \sum_{\ell=1}^{k-1} \binom{2k}{2\ell} h^{(2)}(0)h^{(2\ell)}(0)h^{(2k-2\ell)}(0) \\ &\qquad\qquad\quad + \ 4h^{(2)}(0)h^{(2k)}(0), \end{aligned}$$

still for $k > 0$.

We shall reorganize the terms on both sides of (8.20) and then compare corresponding terms, using the Chebychev inequality. First the left hand side of (8.20):

$$\begin{aligned}
&\sum_{\ell=1}^{k}\binom{2k}{2\ell-1}h^{(2\ell)}(0)h^{(2k-2\ell+2)}(0) \quad + \quad 2h^{(2k+2)}(0)\\
&\quad = \sum_{\ell=1}^{k}\left\{\binom{2k-1}{2\ell-1}+\binom{2k-1}{2\ell-2}\right\}h^{(2\ell)}(0)h^{(2k-2\ell+2)}(0)\\
&\qquad + \quad 2h^{(2k+2)}(0)\\
&\quad = \sum_{\ell=1}^{k}\binom{2k-1}{2\ell-1}h^{(2\ell)}(0)h^{(2k-2\ell+2)}(0)\\
&\qquad + \sum_{\ell=0}^{k-1}\binom{2k-1}{2\ell}h^{(2\ell+2)}(0)h^{(2k-2\ell)}(0)\\
&\qquad + \quad 2h^{(2k+2)}(0)\\
&\quad = \sum_{\ell=1}^{k-1}\binom{2k-1}{2\ell-1}h^{(2\ell)}(0)h^{(2k-2\ell+2)}(0)\\
&\qquad + \sum_{\ell=1}^{k-1}\binom{2k-1}{2\ell}h^{(2\ell+2)}(0)h^{(2k-2\ell)}(0)\\
&\qquad + \quad 2h^{(2)}(0)h^{(2k)}(0) \quad + \quad 2h^{(2k+2)}(0).
\end{aligned} \tag{8.21}$$

Now the right hand side:

$$\begin{aligned}
&\sum_{\ell=1}^{k-1}\binom{2k}{2\ell}h^{(2)}(0)h^{(2\ell)}(0)h^{(2k-2\ell)}(0) \quad + \quad 4h^{(2)}(0)h^{(2k)}(0)\\
&\quad = \sum_{\ell=1}^{k-1}\left\{\binom{2k-1}{2\ell-1}+\binom{2k-1}{2\ell}\right\}h^{(2)}(0)h^{(2\ell)}(0)h^{(2k-2\ell)}(0)\\
&\qquad + \quad 4h^{(2)}(0)h^{(2k)}(0)\\
&\quad = \sum_{\ell=1}^{k-1}\binom{2k-1}{2\ell-1}h^{(2)}(0)h^{(2\ell)}(0)h^{(2k-2\ell)}(0)\\
&\qquad + \sum_{\ell=1}^{k-1}\binom{2k-1}{2\ell}h^{(2)}(0)h^{(2\ell)}(0)h^{(2k-2\ell)}(0)\\
&\qquad + \quad 4h^{(2)}(0)h^{(2k)}(0).
\end{aligned} \tag{8.22}$$

Matching up corresponding terms on the right in (8.21-22), we see that the inequality (8.20) can be reduced to

$$h^{(2\ell+2)}(0) \;\geq\; h^{(2)}(0)h^{(2\ell)}(0) \tag{8.23}$$

for all ℓ. This is equivalent to

$$\sum_{\lambda\neq 0} \lambda^{2\ell+2} a_\lambda \;\geq\; \Big(\sum_{\lambda\neq 0} \lambda^2 a_\lambda\Big)\Big(\sum_{\lambda\neq 0} \lambda^{2\ell} a_\lambda\Big), \tag{8.24}$$

because of (8.18c). We now appeal to Chebychev's inequality as stated in [5, (2.17.1)], for example:

$$\frac{\sum_{\lambda\neq 0} \lambda^{2\ell+2} a_\lambda}{\sum_{\lambda\neq 0} a_\lambda} \;\geq\; \left(\frac{\sum_{\lambda\neq 0} \lambda^2 a_\lambda}{\sum_{\lambda\neq 0} a_\lambda}\right)\left(\frac{\sum_{\lambda\neq 0} \lambda^{2\ell} a_\lambda}{\sum_{\lambda\neq 0} a_\lambda}\right). \tag{8.25}$$

But $0 < \sum_{\lambda\neq 0} a_\lambda \leq 1$ by (8.18b), so (8.25) implies (8.24), and hence Lemma 8.5. q.e.d.

Proof of 8.10. We express the point ν as in (8.6) and use the morphism Φ_0 to identify $\mathfrak{s} = \mathfrak{sl}(2,\mathbb{C})$ with a θ-stable, conjugation invariant subalgebra of $\mathfrak{g}$. In particular,

$$\nu \;=\; i\,e. \tag{8.26}$$

We must show: for $a > 0$ sufficiently small and $t > 0$ sufficiently large, the equation

$$\begin{gathered} f_t(\zeta) \;=\; \mathrm{Ad}(\exp i\kappa)(i\,e), \\ \text{with } \;\zeta \in \mathcal{O}_{\mathfrak{p}},\;\; \kappa \in \mathfrak{q}(ie)\cap \mathfrak{g}_{\mathbb{R}},\;\; \|\kappa\| < a, \end{gathered} \tag{8.27}$$

has exactly one solution when $\mathcal{O}_{\mathfrak{p}}$ and $\mathcal{O}_{\mathbb{R}}$ are Sekiguchi related, and no solution otherwise.

It is easy to produce a solution when it is supposed to exist. Thus, for the moment, we assume that the two orbits *are* related. Note that the identity $se^{2st} = 1$, with $t \geq 0$, $0 < s \leq 1$, implicitly describes $s = s(t)$ as a decreasing function of t, and $\lim_{t\to\infty} s(t) = 0$. A simple calculation in $SL(2,\mathbb{C})$ shows:

$$\begin{aligned} & f_t(s\,h + i\,s\,e + i\,s\,f) \\ &\qquad = \mathrm{Ad}(\exp(s\,t\,h))(s\,h + i\,s\,e + i\,s\,f) \\ &\qquad = s\,h + i\,e + i\,s^2 f \;=\; \mathrm{Ad}(\exp(i\,s\,f))(i\,e), \end{aligned} \tag{8.28a}$$

and $s(h + i\,e + i\,f)$ lies in the K-orbit related to the $G_{\mathbb{R}}$-orbit of $i\,e$;

in other words, the relation (8.27) with $\zeta = s(h + ie + if)$ and $\kappa = sf$ – which does lie in $\mathfrak{q}(ie) \cap \mathfrak{g}_{\mathbb{R}}$, as required. With little more effort, one checks that

$$\text{(8.28b)}\quad \begin{array}{l} \text{in the case of } (\mathfrak{g}_{\mathbb{R}}, \mathfrak{k}_{\mathbb{R}}) = (\mathfrak{sl}(2,\mathbb{R}), \mathfrak{so}(2)), \text{ with } t > 0, \\ \text{the above solution of the equation (8.27) is the only} \\ \text{solution with the property that } \xi \in \mathbb{R}h \text{ and } \kappa \in \mathbb{R}f. \end{array}$$

In fact, for $(\mathfrak{g}_{\mathbb{R}}, \mathfrak{k}_{\mathbb{R}}) = (\mathfrak{sl}(2,\mathbb{R}), \mathfrak{so}(2))$ and $t > 0$, it is the only solution, even without the additional hypotheses on ξ and κ, as will follow from the arguments below. We shall need to know certain properties of the solution (8.28a):

$$\text{(8.29)}\quad \begin{aligned} m(s\,h + i\,e + i\,s^2 f) & \\ &= (1+s^2)^{-1}\Big((1-s^2)\,h - 2\,i\,s\,e + 2\,i\,s\,f \Big), \\ \|s\,h + i\,e + i\,s^2 f\| &= 1+s^2, \\ \|m(s\,h + i\,e + i\,s^2 f)\|^2 &= 2; \end{aligned}$$

this follows from the description (8.3) of the moment map and another simple calculation.

In the general situation, let us suppose that (8.27) does have a solution, with $a > 0$ sufficiently small and $t > 0$ – the meaning of "sufficiently small" will be specified later. We write $\zeta = \xi + i\eta$, as in (8.11), and we define

$$\text{(8.30)}\qquad s = \frac{\|\xi\|}{\sqrt{2}}.$$

The present meaning of s appears to be different from that in (8.28a); after the fact, we shall see that they agree. Inductively, we shall produce bounds

$$\text{(8.31)}\qquad \|\xi - sh\| < C^k s^k, \quad \|\kappa - sf\| < C^k s^k,$$

for all $k \geq 1$, with some positive constant C which is independent of both k and t. For a small and $\|\kappa\| < a$,

$$\text{(8.32)}\quad \|\xi\| = \|\operatorname{Re} f_t(\zeta)\| \leq \|f_t(\zeta) - i\,e\| = \|\operatorname{Ad}(\exp i\kappa)(i\,e) - i\,e\|$$

is small as well. Thus we can force $Cs < 1$, in which case (8.31) implies $\xi = sh \in \mathbb{R}h$ and $\kappa = sf \in \mathbb{R}f$, hence $\xi, \kappa \in \mathfrak{s}$. But then $\mathfrak{s}$ also contains $\zeta = f_{-t}(\operatorname{Ad}\exp(i\kappa)(ie))$; recall: $\{f_t\}$ is a one-parameter group of diffeomorphisms. Because of (8.28b), our hypothetical solution must coincide with the solution (8.28a) – in particular, no solution exists unless the two orbits are Sekiguchi related.

At this point, we still need to establish the bounds (8.31) and to pin down the nature of the intersection of $f_t(\mathcal{O}_{\mathbb{R}})$ with the normal slice – transverse, with sign +1. The latter is a separate matter, and we shall deal with it last.

To prepare for the verification of (8.31), we re-write the right hand side of (8.27). Since $\operatorname{ad} f(e) = -h$ and $(\operatorname{ad} f)^2(e) = -2f$,

$$\begin{aligned}
\operatorname{Ad}(\exp i\kappa)(ie) &= \sum\nolimits_{\ell\geq 0} \frac{i^{\ell+1}}{\ell!}\left(s \operatorname{ad} f + \operatorname{ad}(\kappa - sf)\right)^{\ell}(e) \\
&= i e + s h + i s^2 f + [e, \kappa - sf] \\
&\quad - \frac{i}{2}\Big((\operatorname{ad}(\kappa - sf))^2 + \operatorname{ad}(\kappa - sf)\operatorname{ad}(sf) \\
&\qquad\qquad + \operatorname{ad}(sf)\operatorname{ad}(\kappa - sf)\Big)(e) \\
&\quad + \sum\nolimits_{\ell>2} \frac{i^{\ell+1}}{\ell!}\left(s \operatorname{ad} f + \operatorname{ad}(\kappa - sf)\right)^{\ell}(e).
\end{aligned} \tag{8.33a}$$

We make a small enough to force $s < 1$ and $\|\kappa - sf\| < 1$. For $k > 2$, $(\operatorname{ad} f)^k(e) = 0$. Thus, when we expand $(s \operatorname{ad} f + \operatorname{ad}(\kappa - sf))^k(e)$ as a sum of monomials, every non-zero term involves at least one power of $\operatorname{ad}(\kappa - sf)$. We can therefore choose $D > 0$ so that

$$\begin{aligned}
&\Big\| \sum\nolimits_{\ell>2} \frac{i^{\ell+1}}{\ell!}\left(s \operatorname{ad} f + \operatorname{ad}(\kappa - sf)\right)^{\ell}(e) \Big\| \\
&\qquad < D\,\|\kappa - sf\| \max(s^2, \|\kappa - sf\|^2).
\end{aligned} \tag{8.33b}$$

Taking the real and imaginary parts of

$$f_t(\zeta) = \xi + i \operatorname{Ad}(t\xi)(\eta) = \operatorname{Ad}(\exp i\kappa)(ie),$$

we find

$$\begin{aligned}
&\text{a)}\quad \|\xi - sh - [e, \kappa - sf]\| \\
&\qquad\qquad < D\,\|\kappa - sf\| \max(s^2, \|\kappa - sf\|^2), \\
&\text{b)}\quad \|\operatorname{Ad}(t\xi)(\eta) - e - s^2 f\| \\
&\qquad\qquad < D\,\|\kappa - sf\| \max(s, \|\kappa - sf\|),
\end{aligned} \tag{8.34}$$

now with a possibly larger value of D.

We remarked already that $\|\xi\|$ and s are necessarily small when a is small. Also, the operator $\operatorname{ad} e$ is injective on the space $\mathfrak{q}(ie)$, which contains both f and κ. Hence $\|[e, \kappa - sf]\|$ can be bounded from below by a positive multiple of $\|\kappa - sf\|$. Using (8.34a), we now conclude:

$$\text{(8.35)} \qquad \|\xi - sh\| \quad \text{and} \quad \|\kappa - sf\| \quad \text{are mutually bounded}$$

when a is sufficiently small. In particular, this makes the two inequalities in (8.31) equivalent to one another. The first holds vacuously when $k = 1$, hence so does the other.

For the inductive step, we assume that (8.31) is satisfied for some $k \geq 1$. Enlarging the constant D in (8.34) if necessary – independently of k – we can arrange

$$\text{(8.36)} \quad \|\xi + i\,\mathrm{Ad}(t\xi)(\eta) - s\,h - i\,e - i\,s^2 f\| \;<\; D\|\kappa - s\,f\| \;<\; C^k D s^k.$$

But $\xi + i\,\mathrm{Ad}(t\xi)(\eta) = f_t(\zeta)$, and $(1+s^2)^{-1}(sh + ie + is^2 f)$ lies in the core $C(\mathcal{O})$; indeed, according to (8.29), $(1+s^2)^{-1}(sh + ie + is^2 f)$ has unit length, and there the function $\|m\|^2$, which is invariant under scaling of the argument, assumes the minimum value 2. Thus (8.36) implies

$$\text{(8.37)} \quad \begin{aligned} \mathrm{dist}(\,(1-s^2)^{-1} f_t(\zeta),\, C(\mathcal{O})\,) \;&\leq\; \Big\| \frac{f_t(\zeta)}{1+s^2} \;-\; \frac{sh + ie + is^2}{1+s^2} \Big\| \\ &< C^k D s^k. \end{aligned}$$

The function $\|m\|^2 : \mathbb{S}(\mathcal{O}) \to \mathbb{R}_{>0}$ is Bott-Morse, with minimal value 2, assumed precisely on the core. Using (8.37) and the invariance of m under scaling of the argument, we find

$$\text{(8.38)} \qquad \|m(f_t(\zeta))\|^2 \;-\; 2 \;<\; C^{2k} D^2 s^{2k},$$

possibly after increasing D, again independently of k. On the other hand, according to Lemma (8.5),

$$\text{(8.39)} \quad \begin{aligned} \|m(f_t(\zeta))\|^2 \;&=\; \|m_1(t)\|^2 \;+\; \|m_2(t)\|^2 \;+\; \|m_3(t)\|^2 \\ &\geq\; \|m_1(t)\|^2 \;+\; \|m_3(t)\|^2 \\ &\geq\; \|m(\zeta)\|^2 \geq\; 2. \end{aligned}$$

Combining (8.38-39), we find

$$\text{(8.40)} \qquad \|m_2(t)\| \;<\; C^k D s^k.$$

The moment map is differentiable, so (8.37) implies a bound on the distance between $m(f_t(\zeta))$ and $m(sh + ie + is^2 f)$,

$$\text{(8.41)} \qquad \|m(f_t(\zeta)) \;-\; m(sh + ie + is^2 f)\| \;<\; C^k D s^k,$$

with a larger D, if necessary. By definition of the $m_j(t)$,

$$(8.42)\qquad m_1(t) \;=\; \operatorname{Re}(m_1(t) + m_3(t)) \;=\; \operatorname{Re}(m(f_t(\zeta)) - m_2(t)).$$

At this point, we can conclude that

$$(8.43)\qquad \| m_1(t) - \frac{1-s^2}{1+s^2}\, h \| \;<\; 2C^k D s^k ,$$

by combining the formula (8.29) for $m(sh + ie + is^2 f)$ with (8.40-42).

Recall that $m_1(t)$ is a real multiple of $\operatorname{Re}\zeta = \xi$ – a positive multiple, as follows from the explicit formula (8.13) in conjunction with (8.11):

$$(8.44)\qquad \xi \;=\; \|\xi\| \,\frac{m_1(t)}{\|m_1(t)\|}.$$

In this formula, we can approximate $m_1(t)$ by $(1+s^2)^{-1}(1-s^2)h$, at the expense of introducing an error term slightly larger than that in (8.43), multiplied by $\|\xi\|$. Since the inner product was normalized by the formula $\|h\|^2 = B(h,h) = 2$,

$$(8.45)\qquad \| \xi - \frac{\|\xi\|}{\sqrt{2}}\, h \| \;<\; 3\|\xi\| C^k D s^k ,$$

provided s is sufficiently small – which, we had seen, can be arranged by making a small. We substitute $\|\xi\| = \sqrt{2}s$ – cf. (8.30) – and choose C at least as large as $3D\sqrt{2}$, giving us

$$(8.46)\qquad \| \xi - s\,h \| \;<\; C^{k+1} s^{k+1}.$$

In view of (8.35) this completes the inductive verification of (8.31). We had remarked already that (8.31) implies the first part of the lemma.

Now let $\mathcal{O}_{\mathbb{R}}$, $\mathcal{O}_{\mathfrak{p}}$ be orbits related by the Sekiguchi correspondence, and ν a point in the core $C(\mathcal{O}_{\mathbb{R}})$. We use the notation (8.26-28); in particular, we again identify $\mathfrak{s}$ with a subalgebra of $\mathfrak{g}$ and the point ν with ie. To shorten formulas, we set

$$(8.47)\qquad \nu_t \;=\; s(h + i\,e + i\,f) \qquad\qquad (0 \le t < \infty),$$

with $s = s(t)$ determined implicitly by $se^{2st} = 1$ as before. Then $\lim_{t\to\infty} s(t) = 0$, $s(0) = 1$, and

$$(8.48)\qquad f_t(\nu_t) \;=\; i\,e + s\,h + i\,s^2 f,$$

as in (8.28). We regard tangent spaces to (real) submanifolds of $\mathfrak{g}$ as vector subspaces of $\mathfrak{g}^{\mathbb{R}}$, i.e., of $\mathfrak{g}$ considered as vector space over $\mathbb{R}$. However, we shall not dwell on the distinction between $\mathfrak{g}$ and $\mathfrak{g}^{\mathbb{R}}$ from now on. We shall show:

(8.49) the limit of vector spaces $\lim_{t\to\infty} (f_t)_* (T_{\nu_t}\mathcal{O}_{\mathfrak{p}})$ exists and equals $T_{ie}\mathcal{O}_{\mathbb{R}}$.

Since $\mathcal{O}_{\mathbb{R}}$ and the normal slice $N(ie,a)$ meet transversely at ie by construction, $f_t(\mathcal{O}_{\mathfrak{p}})$ must then meet $\mathcal{O}_{\mathbb{R}}$ transversely at ν_t for t large, as asserted by the lemma.

The point $(2s)^{-1}\nu_t$ lies in the core $C(\mathcal{O}_{\mathfrak{p}})$. Since scaling by a positive number preserves $\mathcal{O}_{\mathfrak{p}}$, the tangent spaces to $\mathcal{O}_{\mathfrak{p}}$ at ν_t and $(2s)^{-1}\nu_t$ are naturally isomorphic; indeed, they are equal as subspaces of $\mathfrak{g}^{\mathbb{R}}$. Appealing to (4.9) and (6.17), we find

$$
(8.50)\qquad \begin{aligned} T_{\nu_t}\mathcal{O}_{\mathfrak{p}} &= T_{\nu_t}(K_{\mathbb{R}}\cdot\nu_t) \bigoplus \mathfrak{d}_{\mathbb{R}}(\Phi_0)(h+ie+if), \\ \text{with}\quad \mathfrak{d}_{\mathbb{R}}(\Phi_0) &= \textstyle\bigoplus_{r\geq 2} \mathrm{Hom}^{\mathbb{R},\theta}(\mathfrak{s},\mathfrak{g}(r))(r-2). \end{aligned}
$$

We shall apply the differential of f_t separately to the various summands in this decomposition of $T_{\nu_t}\mathcal{O}_{\mathfrak{p}}$ and then take the limit as $t\to\infty$.

The map f_t is $G_{\mathbb{R}}$-invariant by definition. It follows that $(f_t)_*$ maps the tangent space $T_{\nu_t}(K_{\mathbb{R}}\cdot\nu_t)$ isomorphically onto $T_{f_t(\nu_t)}(K_{\mathbb{R}}\cdot f_t(\nu_t))$. Since $f_t(\nu_t)\to ie$, we can let t tend to infinity and conclude

$$
(8.51)\qquad \lim_{t\to\infty} (f_t)_* (T_{\nu_t}(K_{\mathbb{R}}\cdot\nu_t)) = T_{ie}(K_{\mathbb{R}}\cdot ie),
$$

provided the family of $K_{\mathbb{R}}$-orbits $K_{\mathbb{R}}\cdot f_t(\nu_t) = K_{\mathbb{R}}\cdot(ie+sh+is^2f)$ depends smoothly on $s=s(t)$ even at $s=0$. To see this, note that any $k\in K_{\mathbb{R}}$ that fixes $f_t(\nu_t)$ must fix the real and imaginary parts separately, but those generate $\mathfrak{s}$ as Lie algebra. Similarly, if $k\in K_{\mathbb{R}}$ fixes ie, it must fix also $if=-i\theta e$, which together with ie generates $\mathfrak{s}$. The constancy of the isotropy subgroups of $(K_{\mathbb{R}})_{f_t(\nu_t)} = (K_{\mathbb{R}})_{\mathfrak{s}}$ even at $s=0$ implies the smooth dependence of the $K_{\mathbb{R}}$-orbits, hence (8.51).

Recall the decomposition $\mathfrak{g}=\oplus_{r,\ell}\,\mathfrak{g}(r,\ell)$ defined in §4. For $\eta\in\mathfrak{g}(r)$, we let η_ℓ denote the component of η in $\mathfrak{g}(r,\ell)$. We shall need to know:

$$
\begin{aligned}
&\text{a) the map } \Phi \mapsto \Phi\, h \text{ establishes an isomorphism} \\
&\quad \operatorname{Hom}^{\mathbb{R},\theta}(\mathfrak{s},\mathfrak{g}(r))(r-2) \\
&\qquad \simeq \textstyle\bigoplus_{0\le\ell<r} (\mathfrak{g}(r,\ell) + \mathfrak{g}(r,-\ell)) \cap \mathfrak{p}_{\mathbb{R}}; \\
&\text{b) } \Phi \in \operatorname{Hom}(\mathfrak{s},\mathfrak{g}(r))(r-2) \implies (\Phi\, h)_\ell = 0 \quad \text{if } \ell = \pm r, \\
&\quad (\Phi\, e)_{\ell+2} = \frac{-1}{r-\ell}[\,e,\,(\Phi\, h)_\ell\,], \\
&\quad (\Phi\, f)_{\ell-2} = \frac{1}{r+\ell}[\,f,\,(\Phi\, h)_\ell\,].
\end{aligned}
\tag{8.52}
$$

The assertion b) is established in [13, §9], in the arguments[6] leading up to (9.53) in that paper; alternatively, one can deduce b) directly from the identity (4.14b) in the proof of Proposition 4.9. Because of b), the map $\Phi \mapsto \Phi\, h$ is certainly injective on $\operatorname{Hom}^{\mathbb{R},\theta}(\mathfrak{s},\mathfrak{g}(r))(r-2)$, and $\Phi\, h$ has no components in $\mathfrak{g}(r,\pm r)$. But any such Φ respects the Cartan decomposition and real structure, so $\Phi\, h$ lies in $\mathfrak{g}(r)\cap\mathfrak{p}_{\mathbb{R}}$. The space $\mathfrak{g}(r)\cap\mathfrak{p}_{\mathbb{R}}$ is invariant under $(\operatorname{ad} h)^2$, hence splits into the direct sum of the subspaces $(\mathfrak{g}(r,\ell) + \mathfrak{g}(r,-\ell))\cap\mathfrak{p}_{\mathbb{R}}$. Since $(\Phi\, h)_{\pm r} = 0$, the map $\Phi \mapsto \Phi\, h$ in a) does take values in $\bigoplus_{0\le\ell r}(\mathfrak{g}(r,\ell) + \mathfrak{g}(r,-\ell))$. To see the surjectivity of the map, let us fix $\xi \in (\mathfrak{g}(r,\ell) + \mathfrak{g}(r,-\ell))$ for some integer ℓ, $0\le\ell<r$. The formulas in b) are compatible with the Cartan involution and real structure, and consequently determine a unique $\Phi \in \operatorname{Hom}^{\mathbb{R},\theta}(\mathfrak{s},\mathfrak{g}(r))$ such that the formulas in b) hold and $\Phi\, h = \xi$. The criterion (4.14b) and a short computation show that Φ, thus defined, lies in $\operatorname{Hom}^{\mathbb{R},\theta}(\mathfrak{s},\mathfrak{g}(r))(r-2)$. This completes the verification of (8.52).

The definition of the map f_t, coupled with the formula for the differential of the exponential map – see [4, Theorem II.1.7], for example – leads to the formula

$$
\begin{aligned}
&(f_t)_*\zeta \\
&= \operatorname{Ad}\exp(sth)\left(\zeta + \left[\frac{1-e^{-st\operatorname{ad}h}}{s\,t\operatorname{ad}h}(t\operatorname{Re}\zeta),\, s(h+ie+if)\right]\right).
\end{aligned}
\tag{8.53}
$$

We apply this to $\zeta = \Phi(h+ie+if)$, with $\Phi \in \operatorname{Hom}^{\mathbb{R},\theta}(\mathfrak{s},\mathfrak{g}(r))(r-2)$ viewed as tangent vector to $\mathcal{O}_{\mathfrak{p}}$ at ν_t, as in (8.50). To simplify the statement we are about to make, we assume

$$\Phi\, h \in (\mathfrak{g}(r,\ell) + \mathfrak{g}(r,-\ell))\cap\mathfrak{p}_{\mathbb{R}} \qquad (0\le\ell<r). \tag{8.54}$$

[6]The hypotheses "if $r=n$ or $r=n-2$, and if $s=\pm n$, $\pm(n-2)$" in [13, (9.53)] are irrelevant in the present setting; in other words, one should argue as in [13], but with $X_n = Y_n = Z_n = 0$.

In any case, $\mathrm{Hom}^{\mathbb{R},\theta}(\mathfrak{s},\mathfrak{g}(r))(r-2)$ has a basis consisting of linear maps Φ of this type. According to (8.52b), Φe has a nonzero component in $\mathfrak{g}(r,\ell+2)$ – unless $\Phi = 0$, of course – but no components in $\mathfrak{g}(r,j)$ with $j > \ell+2$; similarly, Φf has no components in $\mathfrak{g}(r,j)$ with $j \geq \ell$. The operator $\mathrm{Ad}\exp(sth)$ acts on $\mathfrak{g}(r,\ell)$ as multiplication by $e^{st\ell} = s^{-\ell/2}$, whereas the operator $(\mathrm{ad}h)^{-1}(1 - e^{-st\mathrm{ad}h})$ acts by $\ell^{-1}(1-e^{-st\ell}) = \ell^{-1}(1-s^{\ell/2})$ or $st = -\frac{1}{2}\log s$, depending on whether $\ell > 0$ or $\ell = 0$. Looking at the leading terms, or equivalently the terms involving the lowest power of s, we find

$$(8.55)\qquad \begin{aligned} &(f_t)_*|_{\nu_t}\,\Phi(h+ie+if)\\ &\qquad = \begin{cases} i\,s^{-1-\frac{\ell}{2}}\,\frac{1+\ell(r-\ell)}{\ell(r-\ell)}\,(\Phi e)_{\ell+2} + \dots & \text{if } \ell>0\\ i\,|\log s|\,s^{-1}\,\frac{1}{2r}\,\Phi e \;+ \dots & \text{if } \ell = 0\,; \end{cases} \end{aligned}$$

here $\dots$ refers to lower order terms, and we are using (8.52b) to express $[(\Phi h)_\ell, e]$ as a multiple of $(\Phi e)_{\ell+2}$.

Let us re-state the top line of the identity (8.55) in terms Φe, rather than $(\Phi e)_{\ell+2}$. By (8.54) and (8.52b), if $\ell > 0$,

$$\begin{aligned} \frac{2r}{r+\ell}\,(\Phi e)_{\ell+2} &= \frac{-2r}{(r-\ell)(r+\ell)}\,[e,\,(\Phi h)_\ell]\\ &= \frac{-1}{r-\ell}\,[e,\,(\Phi h)_\ell] - \frac{1}{r+\ell}\,[e,\,(\Phi h)_{-\ell}]\\ &\quad - \frac{1}{r+\ell}\,[e,\,(\Phi h)_\ell - (\Phi h)_{-\ell}]\\ &= (\Phi e)_{\ell+2} + (\Phi e)_{-\ell+2}\\ &\quad - \frac{1}{r+\ell}\,[e,\,(\Phi h)_\ell + \theta((\Phi h)_\ell)]\\ &= \Phi e - \frac{1}{r+\ell}\,[e,\,(\Phi h)_\ell + \theta((\Phi h)_{-\ell})]. \end{aligned}$$

In the next to last line, we have used the fact that θ acts as -1 on Φh and maps $\mathfrak{g}(r,\ell)$ to $\mathfrak{g}(r,-\ell)$. Thus

$$(8.56)\qquad \begin{aligned} (\Phi e)_{\ell+2} &= \frac{r+\ell}{2r}\,\Phi e - [e,\,\eta_\Phi],\\ &\text{with } \eta_\Phi =_{\mathrm{def}} \frac{1}{2r}\,\big((\Phi h)_\ell + \theta((\Phi h)_\ell)\big) \in \mathfrak{k}_{\mathbb{R}}. \end{aligned}$$

Combining (8.55-56), we get

$$
\begin{aligned}
&\lim_{t\to\infty}\left(s^{1+\frac{\ell}{2}}\,(f_t)_*|_{\nu_t}\,\Phi(h+ie+if)\right)\\
&\qquad = \frac{r+\ell+\ell(r^2-\ell^2)}{2r\ell(r-\ell)}\,\Phi(ie)\;+\;\frac{1+\ell(r-\ell)}{\ell(r-\ell)}\,[\,\eta_\Phi,\,ie\,]\\
&\qquad\qquad \text{if } \ell>0,\\
&\lim_{t\to\infty}\left(\frac{s}{|\log s|}\,(f_t)_*|_{\nu_t}\,\Phi(h+ie+if)\right) \;=\; \Phi(ie)\\
&\qquad\qquad \text{if } \ell=0.
\end{aligned}
\tag{8.57}
$$

In analogy to (8.50), we can describe the tangent space to $\mathcal{O}_{\mathbb{R}}$ at the point ie as

$$
\begin{aligned}
T_{ie}\mathcal{O}_{\mathbb{R}} &= T_{ie}(K_{\mathbb{R}}\cdot(ie))\,\bigoplus\,\mathfrak{d}_{\mathbb{R}}(\Phi_0)(ie), \qquad \text{with}\\
\mathfrak{d}_{\mathbb{R}}(\Phi_0) &= \textstyle\bigoplus_{r\geq 2}\operatorname{Hom}^{\mathbb{R},\theta}(\mathfrak{s},\mathfrak{g}(r))(r-2),\\
&\qquad\qquad T_{ie}\mathcal{O}_{\mathbb{R}} \;=\; [\,\mathfrak{k}_{\mathbb{R}},\,ie\,].
\end{aligned}
\tag{8.58}
$$

We have established (8.49); equivalently, there exists a basis $\{\eta_j(t)\}$ of $(f_t)_*(T_{\nu_t}\mathcal{O}_{\mathfrak{p}})$, depending continuously on the parameter t, such that the limits $\tilde\eta_j = \lim_{t\to\infty}\eta_j(t)$ exist and constitute a basis of $T_{ie}\mathcal{O}_{\mathbb{R}}$. This follows from the analogous statement about the tangent spaces of the $K_{\mathbb{R}}$-orbits – which is equivalent to (8.51) – in conjunction with (8.50), (8.52a), (8.57-58), and the non-vanishing of the coefficients of $\Phi(ie)$ in (8.57). We have pointed out already that (8.49) implies the transversality assertion of the lemma.

To pin down the sign of the intersection, it suffices to compare two orientations on $T_{ie}\mathcal{O}_{\mathbb{R}} = \lim_{t\to\infty}((f_t)_*T_{\nu_t}\mathcal{O}_{\mathfrak{p}})$ – on the one hand, the orientation introduced by the symplectic form $\frac{1}{2\pi i}\sigma_{\mathcal{O}}$, on the other, the orientation coming from the complex structure on $\mathcal{O}_{\mathfrak{p}} \simeq f_t(\mathcal{O}_{\mathfrak{p}})$ and the limiting process; the sign of the intersection is the sign which relates the two orientations. We had remarked already that the tangent spaces $T_{\nu_t}\mathcal{O}_{\mathfrak{p}}$ all coincide when we regard them as subspaces of $\mathfrak{g}^{\mathbb{R}}$. In particular, they all coincide with the tangent space at $\nu_0 = h+ie+if$:

$$
T_{\nu_t}\mathcal{O}_{\mathfrak{p}} \;=\; T_{\nu_0}\mathcal{O}_{\mathfrak{p}} \;=\; [\mathfrak{k},\nu_0]. \tag{8.59}
$$

For reasons of continuity, the real 2-form $\operatorname{Im}\sigma_{\mathcal{O}}$ is non-degenerate on $(f_t)_*(T_{\nu_t}\mathcal{O}_{\mathfrak{p}})$ for all large enough values of t. We must show that $(f_t)_*$ is orientation preserving with respect to this symplectic structure on $(f_t)_*(T_{\nu_t}\mathcal{O}_{\mathfrak{p}})$ and the orientation of $T_{\nu_0}\mathcal{O}_{\mathfrak{p}}$ as complex vector space –

equivalently, that $\mathrm{Im}(f_t^*\sigma_{\mathcal{O}})$, for $t \gg 0$, orients the tangent space $T_{\nu_0}\mathcal{O}_{\mathfrak{p}} = T_{\nu_t}\mathcal{O}$ consistently with the complex orientation. In fact, we shall show

$$\text{(8.60)}\quad \begin{array}{ll} \text{a)} & \mathrm{Im}\,\sigma_{\mathcal{O}} \text{ is non-degenerate on } (f_t)_*(T_{\nu_t}\mathcal{O}_{\mathfrak{p}}) \text{ for all } t > 0; \\ \text{b)} & S =_{def} \lim_{t\to 0^+} \left(t^{-1} f_t^*(\mathrm{Im}\,\sigma_{\mathcal{O}})|_{T_{\nu_t}\mathcal{O}}\right) \\ & \text{exists, is non-degenerate,and orients } T_{\nu_0}\mathcal{O} = T_{\nu_t}\mathcal{O} \\ & \text{consistently with the complex structure.} \end{array}$$

That suffices: the 2-forms $f_t^*(\mathrm{Im}\,\sigma_{\mathcal{O}})$, for $t > 0$, are then all nondegenerate on $T_{\nu_0}\mathcal{O}$ and therefore induce the same orientation. Because of b), this orientation agrees with the orientation determined by the complex structure.

We break down the verification of the statement (8.60a) into the following two separate assertions:

$$\text{(8.61)}\quad \begin{array}{ll} \text{a)} & \text{the submanifolds } f_t(\mathcal{O}_{\mathfrak{p}}) \text{ of the complex orbit } \mathcal{O} \text{ are} \\ & \text{Lagrangian with respect to the symplectic form } \mathrm{Re}\,\sigma_{\mathcal{O}}; \\ \text{b)} & (f_t)_*(T_{\nu_0}\mathcal{O}) \cap i\,(f_t)_*(T_{\nu_0}\mathcal{O}) = 0 \text{ for all } t > 0. \end{array}$$

Let us assume this for the moment. If $\zeta \in (f_t)_*(T_{\nu_t}\mathcal{O}_{\mathfrak{p}})$ lies in the radical of the restriction of $\mathrm{Im}\,\sigma_{\mathcal{O}}$ to $(f_t)_*(T_{\nu_t}\mathcal{O}_{\mathfrak{p}})$, (8.61a) allows us to argue

$$\begin{aligned} &\mathrm{Im}\,\sigma_{\mathcal{O}}(\,\zeta,\,(f_t)_*(T_{\nu_t}\mathcal{O}_{\mathfrak{p}})\,) = 0 \\ &\quad\Longrightarrow\quad \sigma_{\mathcal{O}}(\,\zeta,\,(f_t)_*(T_{\nu_t}\mathcal{O}_{\mathfrak{p}})\,) = 0 \\ &\quad\Longrightarrow \sigma_{\mathcal{O}}(\,\zeta,\,i\,(f_t)_*(T_{\nu_t}\mathcal{O}_{\mathfrak{p}})\,) = 0 \\ &\quad\Longrightarrow\quad \sigma_{\mathcal{O}}(\,\zeta,\,(f_t)_*(T_{\nu_t}\mathcal{O}_{\mathfrak{p}}) \oplus i\,(f_t)_*(T_{\nu_t}\mathcal{O}_{\mathfrak{p}})\,) = 0; \end{aligned}$$

at the second step we are using the complex linearity of $\sigma_{\mathcal{O}}$. But (8.61b) and a dimension count imply that $(f_t)_*(T_{\nu_t}\mathcal{O}_{\mathfrak{p}})$ and $i\,(f_t)_*(T_{\nu_t}\mathcal{O}_{\mathfrak{p}})$ span the tangent space of $\mathcal{O}$ at $f_t(\nu_t)$, so ζ lies in the radical of the holomorphic symplectic form $\sigma_{\mathcal{O}}$, forcing $\zeta = 0$. Thus (8.61a,b) do imply (8.60a). At this point, only (8.60b) and (8.61a,b) remain to be proved.

Recall the notation (8.47) and the formula (8.53) for the differential of f_t. We apply this formula to a tangent vector $\zeta \in T_{\nu_t}\mathcal{O}_{\mathfrak{p}}$. Because of

(8.59), we can write $\zeta = [\kappa, \nu_0]$ for some $\kappa \in \mathfrak{k}$, so that

$$
\begin{aligned}
(f_t)_*\zeta &= (f_t)_*[\kappa, \nu_0] \\
&= s^{-1}\left[\operatorname{Ad}\exp(sth)\left(\kappa + \frac{1-e^{-stadh}}{\operatorname{ad}h}\operatorname{Re}[\kappa,\nu_0]\right),\right. \\
&\qquad\qquad \left.\operatorname{Ad}\exp(sth)\,\nu_t\right] \\
&= -\,s^{-1}\left[f_t(\nu_t),\ \operatorname{Ad}\exp(sth)\left(\kappa + \frac{1-e^{-stadh}}{\operatorname{ad}h}\operatorname{Re}\zeta\right)\right].
\end{aligned}
\tag{8.62}
$$

The holomorphic symplectic form $\sigma_{\mathcal{O}}$ is the canonical symplectic form of the complex coadjoint orbit that corresponds to $\mathcal{O}$ when we identify $\mathfrak{g} \simeq \mathfrak{g}^*$ via the Killing form. Thus, for $\zeta_j = [\kappa_j, \nu_0] \in T_{\nu_t}\mathcal{O}_{\mathfrak{p}}$, $j = 1, 2$,

$$
\begin{aligned}
&(f_t^*\sigma_{\mathcal{O}})(\zeta_1,\zeta_2) \\
&= (\sigma_{\mathcal{O}})|_{f_t(\nu_t)}(\,(f_t)_*[\kappa_1,\nu_0],\ (f_t)_*[\kappa_2,\nu_0]\,) \\
&= B\left(\,f_t(\nu_t),\ \left[\,(\operatorname{ad}f_t(\nu_t))^{-1}((f_t)_*[\kappa_1,\nu_0]),\ (\operatorname{ad}f_t(\nu_t))^{-1}((f_t)_*[\kappa_2,\nu_0])\,\right]\,\right) \\
&= s^{-2}\,B\left(\nu_t,\ \left[\kappa_1 + \frac{1-e^{-stadh}}{\operatorname{ad}h}\operatorname{Re}\zeta_1,\ \kappa_2 + \frac{1-e^{-stadh}}{\operatorname{ad}h}\operatorname{Re}\zeta_2\right]\right) \\
&= s^{-2}\,B\left(\nu_t,\ \left[\frac{1-e^{-stadh}}{\operatorname{ad}h}\operatorname{Re}\zeta_1,\ \frac{1-e^{-stadh}}{\operatorname{ad}h}\operatorname{Re}\zeta_2\right]\right) \\
&\quad + s^{-2}\,B\left(\nu_t,\ \left[\kappa_1,\ \frac{1-e^{-stadh}}{\operatorname{ad}h}\operatorname{Re}\zeta_2\right] + \left[\frac{1-e^{-stadh}}{\operatorname{ad}h}\operatorname{Re}\zeta_1,\ \kappa_2\right]\right).
\end{aligned}
$$

Here, in the second line, $(\operatorname{ad}f_t(\nu_t))^{-1}((f_t)_*[\kappa_j,\nu_0])$ is symbolic notation for any element of $\mathfrak{g}$ whose image under $\operatorname{ad}f_t(\nu_t)$ is $(f_t)_*[\kappa_j,\nu_0]$; in passing from the second line to the third, we are using (8.62), the identity $f_t(\nu_t) = \operatorname{Ad}\exp(stadh)(\nu_t)$, and the Ad-invariance of the Killing form; the last step is justified by the perpendicularity of $\nu_t \in \mathfrak{p}$ and $[\kappa_1,\kappa_2] \in \mathfrak{k}$. Next, we use the infinitesimal invariance of B, the relation $\nu_t = s\,\nu_0$, and the relation between ζ_j and κ_j, to conclude

$$
\begin{aligned}
(f_t^*\sigma_{\mathcal{O}})(\zeta_1,\zeta_2) &= (\sigma_{\mathcal{O}})|_{f_t(\nu_t)}(\,(f_t)_*[\kappa_1,\nu_0],\ (f_t)_*[\kappa_2,\nu_0]\,) \\
&= s^{-1}\,B\left(\nu_0,\ \left[\frac{1-e^{-stadh}}{\operatorname{ad}h}\operatorname{Re}\zeta_1,\ \frac{1-e^{-stadh}}{\operatorname{ad}h}\operatorname{Re}\zeta_2\right]\right) \\
&\quad - s^{-1}\,B\left(\zeta_1,\ \frac{1-e^{-stadh}}{\operatorname{ad}h}\operatorname{Re}\zeta_2\right) \\
&\quad + s^{-1}\,B\left(\frac{1-e^{-stadh}}{\operatorname{ad}h}\operatorname{Re}\zeta_1,\ \zeta_2\right).
\end{aligned}
\tag{8.63}
$$

We shall use this formula to verify (8.60b) and (8.61a).

For t near 0, $s(t) = 1-2t+\dots$ and $(\mathrm{ad}h)^{-1}(1-e^{-st\mathrm{ad}h}) = st{\cdot}1+\dots$, hence $s^{-1}(\mathrm{ad}h)^{-1}(1-e^{-st\mathrm{ad}h}) = t\cdot 1+\dots$, and

$$(8.64)\qquad \begin{aligned} &f_t^*(\operatorname{Im}\sigma_{\mathcal{O}})(\zeta_1,\zeta_2)\\ &\quad = \ -t\,B(\operatorname{Im}\zeta_1,\operatorname{Re}\zeta_2) + t\,B(\operatorname{Re}\zeta_1,\operatorname{Im}\zeta_2) + \dots . \end{aligned}$$

We conclude that $S = \lim_{t\to 0^+} t^{-1} f_t^*(\operatorname{Im}\sigma_{\mathcal{O}})$ exists as $\mathbb{R}$-bilinear, alternating form on $T_{\nu_0}\mathcal{O}_{\mathfrak{p}} = [\mathfrak{k},\nu_0]$ and is given by the formula

$$(8.65)\qquad S(\zeta_1,\zeta_2) \ = \ -B(\operatorname{Im}\zeta_1,\operatorname{Re}\zeta_2) + B(\operatorname{Re}\zeta_1,\operatorname{Im}\zeta_2).$$

Let $\{\zeta_j\}$ be a $\mathbb{C}$-basis of $[\mathfrak{k},\nu_0]$, orthonormal with respect to the inner product (2.2). Since θ acts as multiplication by -1 on $[\mathfrak{k},\nu_0]\subset\mathfrak{p}$,

$$(8.66)\qquad \begin{aligned} S(\zeta_j, i\,\zeta_k) &= B(\operatorname{Im}\zeta_j,\operatorname{Im}\zeta_k) + B(\operatorname{Re}\zeta_j,\operatorname{Re}\zeta_k)\\ &= \operatorname{Re}(\zeta_j,\zeta_k) \ = \ \delta_{j,k},\\ S(\zeta_j,\zeta_k) &= -B(\operatorname{Im}\zeta_j,\operatorname{Re}\zeta_k) + B(\operatorname{Re}\zeta_j,\operatorname{Im}\zeta_k)\\ &= -\operatorname{Im}(\zeta_j,\zeta_k) \ = \ 0. \end{aligned}$$

In particular, the nondegenerate alternating bilinear form S orients $[\mathfrak{k},\nu_0]$, viewed as real vector space, in the same way as the complex structure. This establishes (8.60b).

The formula (8.63) and its derivation remain valid if we replace $\nu_t = s\,\nu_0$ by an arbitrary point $\nu\in\mathcal{O}_{\mathfrak{p}}$ and $s\,h = \operatorname{Re}\nu_t$ by $\operatorname{Re}\nu$. We take real parts on both sides, to conclude

$$(8.67)\qquad \begin{aligned} &(\operatorname{Re}\sigma_{\mathcal{O}})|_{f_t(\nu)}(\,(f_t)_*\zeta_1,\,(f_t)_*\zeta_2\,)\\ &\quad = \ B\left(\operatorname{Re}\nu,\left[\frac{1-e^{-t\mathrm{ad}\operatorname{Re}\nu}}{\mathrm{ad}\operatorname{Re}\nu}\operatorname{Re}\zeta_1,\ \frac{1-e^{-t\mathrm{ad}\operatorname{Re}\nu}}{\mathrm{ad}\operatorname{Re}\nu}\operatorname{Re}\zeta_2\right]\right)\\ &\qquad - \ B\left(\operatorname{Re}\zeta_1,\ \frac{1-e^{-t\mathrm{ad}\operatorname{Re}\nu}}{\mathrm{ad}\operatorname{Re}\nu}\operatorname{Re}\zeta_2\right)\\ &\qquad + \ B\left(\frac{1-e^{-t\mathrm{ad}\operatorname{Re}\nu}}{\mathrm{ad}\operatorname{Re}\nu}\operatorname{Re}\zeta_1,\ \operatorname{Re}\zeta_2\right), \end{aligned}$$

for all $\zeta_1, \zeta_2 \in T_\nu \mathcal{O}_\mathfrak{p}$. On the other hand, because of the invariance of B,

$$(8.68)\quad \begin{aligned} &B\left(\operatorname{Re}\nu, \left[\frac{1-e^{-t\operatorname{ad}\operatorname{Re}\nu}}{\operatorname{ad}\operatorname{Re}\nu}\operatorname{Re}\zeta_1, \frac{1-e^{-t\operatorname{ad}\operatorname{Re}\nu}}{\operatorname{ad}\operatorname{Re}\nu}\operatorname{Re}\zeta_2\right]\right) \\ &\quad = B\left(\operatorname{ad}\operatorname{Re}\nu \circ \frac{1-e^{-t\operatorname{ad}\operatorname{Re}\nu}}{\operatorname{ad}\operatorname{Re}\nu}\operatorname{Re}\zeta_1, \frac{1-e^{-t\operatorname{ad}\operatorname{Re}\nu}}{\operatorname{ad}\operatorname{Re}\nu}\operatorname{Re}\zeta_2\right) \\ &\quad = B\left((1-e^{-t\operatorname{ad}\operatorname{Re}\nu})\operatorname{Re}\zeta_1, \frac{1-e^{-t\operatorname{ad}\operatorname{Re}\nu}}{\operatorname{ad}\operatorname{Re}\nu}\operatorname{Re}\zeta_2\right) \\ &\quad = B\left(\operatorname{Re}\zeta_1, \frac{1-e^{-t\operatorname{ad}\operatorname{Re}\nu}}{\operatorname{ad}\operatorname{Re}\nu}\operatorname{Re}\zeta_2\right) \\ &\qquad - B\left(\operatorname{Re}\zeta_1, \frac{e^{t\operatorname{ad}\operatorname{Re}\nu}-1}{\operatorname{ad}\operatorname{Re}\nu}\operatorname{Re}\zeta_2\right) \end{aligned}$$

The operator $\operatorname{ad}\operatorname{Re}\nu$ is skew with respect to B, so $(\operatorname{ad}\operatorname{Re}\nu)^{-1}(1 - e^{-t\operatorname{ad}\operatorname{Re}\nu})$ is the adjoint of $(\operatorname{ad}\operatorname{Re}\nu)^{-1}(e^{t\operatorname{ad}\operatorname{Re}\nu} - 1)$, and

$$(8.69)\quad \begin{aligned} &B\left(\operatorname{Re}\zeta_1, \frac{e^{t\operatorname{ad}\operatorname{Re}\nu}-1}{\operatorname{ad}\operatorname{Re}\nu}\operatorname{Re}\zeta_2\right) \\ &\qquad = B\left(\frac{1-e^{-t\operatorname{ad}\operatorname{Re}\nu}}{\operatorname{ad}\operatorname{Re}\nu}\operatorname{Re}\zeta_1, \operatorname{Re}\zeta_2\right). \end{aligned}$$

Combining (8.67-69), we find that $\operatorname{Re}\sigma_\mathcal{O}$ vanishes identically on $f_t(\mathcal{O}_\mathfrak{p})$. Since $\mathcal{O}_\mathfrak{p}$ has half the dimension of $\mathcal{O}$, this implies (8.61a).

Only (8.61b) remains to be established. Let us assume, then, that $t > 0$. We consider two tangent vectors

$$(8.70)\quad \zeta_j \in T_{\nu_t}\mathcal{O}_\mathfrak{p} = [\,\mathfrak{k}, h+ei+if] \subset \mathfrak{p}, \text{ such that } (f_t)_*\zeta_1 = i\,(f_t)_*\zeta_2.$$

We express the ζ_j in terms of their real and imaginary parts,

$$(8.71)\quad \zeta_j = \xi_j + i\,\eta_j, \qquad \text{with } \xi_j, \eta_j \in \mathfrak{p}_\mathbb{R}.$$

Because $(\operatorname{ad}h)^{-1}(1-e^{-st\operatorname{ad}h})[\xi_j, h] = (e^{-st\operatorname{ad}h} - 1)\xi_j$, the formula (8.53) can be re-written as follows:

$$(8.72)\quad (f_t)_*\zeta_j = \xi_j + i\operatorname{Ad}\exp(sth)\left(\eta_j + \left[\frac{1-e^{-st\operatorname{ad}h}}{\operatorname{ad}h}\xi_j, e+f\right]\right).$$

Our assumption (8.70) on the ζ_j is therefore equivalent to

$$(8.73)\quad \begin{aligned} e^{-st\operatorname{ad}h}\,\xi_1 &= -\eta_2 + \left[e+f, \frac{1-e^{-st\operatorname{ad}h}}{\operatorname{ad}h}\xi_2\right], \\ e^{-st\operatorname{ad}h}\,\xi_2 &= \eta_1 - \left[e+f, \frac{1-e^{-st\operatorname{ad}h}}{\operatorname{ad}h}\xi_1\right]. \end{aligned}$$

We need to separate the components in $\mathfrak{k}_{\mathbb{R}}$ and $\mathfrak{p}_{\mathbb{R}}$. For this purpose, we define

$$\begin{aligned} S &= \frac{\sinh(s\,t\,\mathrm{ad}\,h)}{\mathrm{ad}\,h} = s\,t\cdot 1 + \frac{1}{6}(s\,t\,\mathrm{ad}\,h)^2 + \dots, \\ T &= \frac{1 - \cosh(s\,t\,\mathrm{ad}\,h)}{\mathrm{ad}\,h} \\ &= -\frac{1}{2}s\,t\ \mathrm{ad}\,h - \frac{1}{24}(s\,t\,\mathrm{ad}\,h)^3 - \dots . \end{aligned} \tag{8.74}$$

Even powers of $\mathrm{ad}\,h$ or $\mathrm{ad}(e+f)$ commute with the Cartan involution, whereas odd powers anti-commute; also,

$$e^{-s\,t\,\mathrm{ad}\,h} = 1 - (S + T)\circ \mathrm{ad}\,h = 1 - \mathrm{ad}\,h\circ(S + T). \tag{8.75}$$

Equating $\mathfrak{p}_{\mathbb{R}}$-components in (8.73), we now find

$$\begin{aligned} \eta_1 &= (1 - \mathrm{ad}\,h\circ T)\xi_2 + \mathrm{ad}(e+f)\circ T\ \xi_1, \\ \eta_2 &= -(1 - \mathrm{ad}\,h\circ T)\xi_1 + \mathrm{ad}(e+f)\circ T\ \xi_2, \end{aligned} \tag{8.76a}$$

and the equality of the $\mathfrak{k}_{\mathbb{R}}$-components translates into

$$\begin{aligned} \mathrm{ad}\,h\circ S\ \xi_1 &= -\mathrm{ad}(e+f)\circ S\ \xi_2, \\ \mathrm{ad}\,h\circ S\ \xi_2 &= \mathrm{ad}(e+f)\circ S\ \xi_1. \end{aligned} \tag{8.76b}$$

The latter two equations can be combined into the single complex equation

$$[\,h + i\,e + i\,f,\ S(\xi_1 - i\,\xi_2)\,] = 0. \tag{8.77}$$

We shall use these equations to show that the ζ_j must vanish..

Both ζ_j lie in $[\mathfrak{k}, h+ie+if]$, hence in the image of $\mathrm{ad}(h+ie+if):\mathfrak{g}\to\mathfrak{g}$, and $S(\xi_1-\xi_2)$ lies in the kernel of $\mathrm{ad}(h+ie+if)$ by (8.77). The image and the kernel are each other's annihilator, relative to the Killing form. Thus

$$B(\,S(\xi_1 - i\xi_2),\, \zeta_1 + i\zeta_2\,) = 0.$$

Taking real parts, we find

$$\begin{aligned} 0 &= B(S\xi_1, \xi_1 - \eta_2) + B(S\xi_2, \xi_2 + \eta_1) \\ &= B(S\xi_1, \xi_1) + B(S\xi_1, (1 - \mathrm{ad}h\circ T)\xi_1) \\ &\quad - B(S\xi_1, \mathrm{ad}(e+f)\circ T\xi_2) + B(S\xi_2, \xi_2) \\ &\quad + B(S\xi_2, (1 - \mathrm{ad}h\circ T)\xi_2) \\ &\quad + B(S\xi_2, \mathrm{ad}(e+f)\circ T\xi_1); \end{aligned} \tag{8.78}$$

at the second step, we have used (8.76a) to express the η_j in terms of the ξ_j. The infinitesimal invariance of the Killing form and (8.76b) give

$$\begin{aligned} B(S\xi_1, \mathrm{ad}(e+f)\circ T\xi_2) &= -B(\mathrm{ad}(e+f)\circ S\xi_1, T\xi_2) \\ &= -B(\mathrm{ad}h\circ S\xi_2, T\xi_2) \\ &= B(S\xi_2, \mathrm{ad}h\circ T\xi_2), \end{aligned} \tag{8.79a}$$

and similarly

$$\begin{aligned} B(S\xi_2, \mathrm{ad}(e+f)\circ T\xi_1) &= -B(\mathrm{ad}(e+f)\circ S\xi_2, T\xi_1) \\ &= B(\mathrm{ad}h\circ S\xi_1, T\xi_1) \\ &= -B(S\xi_1, \mathrm{ad}h\circ T\xi_1). \end{aligned} \tag{8.79b}$$

The operators

$$1 - \mathrm{ad}\,h\circ T = \cosh(s\,t\,\mathrm{ad}\,h), \quad \mathrm{ad}\,h\circ T = 1 - \cosh(s\,t\,\mathrm{ad}\,h) \tag{8.80}$$

are series in $(\mathrm{ad}h)^2$, hence symmetric with respect to the Killing form. Thus, combining (8.78-90), we find

$$\begin{aligned} 0 &= B(S\xi_1, \xi_1) + B((1-\mathrm{ad}h\circ T)\circ S\xi_1, \xi_1) \\ &\quad - B(\mathrm{ad}h\circ T\circ S\xi_2, \xi_2) + B(S\xi_2, \xi_2) \\ &\quad + B((1-\mathrm{ad}h\circ T)\circ S\xi_2, \xi_2) - B(\mathrm{ad}h\circ T\circ S\xi_1, \xi_1). \end{aligned} \tag{8.81}$$

The inner product (2.2) agrees with the Killing form on $\mathfrak{p}_{\mathbb{R}}$. Relative to this inner product, $\mathrm{ad}h$ is a symmetric operator, whose eigenspace decomposition diagonalizes S and $\mathrm{ad}h\circ T$. For $t > 0$ – which also makes s strictly positive – the eigenvalues of S and $1 - \mathrm{ad}h\circ T$ are strictly positive, and those of $\mathrm{ad}h\circ T$ non-positive. Thus all terms in (8.81) vanish individually, and $\xi_1 = \xi_2 = 0$. The η_j, which can be expressed in terms of the ξ_j, must vanish also. We have shown that (8.70) forces $\zeta_1 = \zeta_2 = 0$. This completes the verification of (8.61b), and with it, the proof of Lemma 8.10. q.e.d.

References

[1] E. Cattani, A. Kaplan & W. Schmid, *Degeneration of Hodge structures*, Ann. of Math. **123** (1986) 457–535.

[2] L. van den Dries & C. Miller, *Geometric categories and o-minimal structures*, Duke Math. J. **84** (1996) 497–540.

[3] L. van den Dries, A. Macintyre & D. Marker, *The elementary theory of restricted analytic fields with exponentiation*, Ann. of Math. **140** (1994) 183–205.

[4] S. Helgason, *Differential geometry, Lie groups, and symmetric spaces*, Academic Press, 1978.

[5] G. H. Hardy, J. E. Littlewood & G. Pólya, *Inequalities*, Cambridge University Press, Reprint of the 1952 edition, Cambridge Mathematical Library, 1988.

[6] F. Kirwan, *Cohomology of quotients in symplectic and algebraic geometry*, Princeton University Press, Math. Notes, Vol. 31, 1984.

[7] B. Kostant, *The principal three-dimensional subgroup and the Betti numbers of a complex simple Lie group*, Amer. J. Math. **81** (1959) 973–1032.

[8] B. Kostant & S. Rallis, *Orbits and representations associated with symmetric spaces*, Amer. J. Math. **93** (1971) 753–809.

[9] P. Kronheimer, *Instantons and the geometry of the nilpotent variety*, J. Differential Geom. **32** (1990) 473–490.

[10] A. Marian, *On the moment map of a linear group action*, Informal notes.

[11] G. D. Mostow, *Some new decomposition theorems for semi-simple groups*, Amer. Math. Soc. Memoirs **14** (1955) 31–54.

[12] L. Ness, *A stratification of the null cone via the moment map*, Amer. J. Math. **106** (1984) 1281–1325.

[13] W. Schmid, *Variation of Hodge structure: the singularities of the period mapping*, Invent. Math. **22** (1973) 211–319.

[14] W. Schmid & K. Vilonen, *Characteristic cycles of constructible sheaves*, Invent. Math. **124** (1996) 451–502.

[15] ———, *Characteristic cycles and wave front cycles of representations of reductive Lie groups*, Ann. of Math. **151** (2000) 1071–1118.

[16] J. Sekiguchi, *Remarks on nilpotent orbits of a symmetric pair*, J. Math. Soc. Japan **39** (1987) 127–138.

[17] M. Vergne, *Instantons et correspondance de Kostant-Sekiguchi*, C. R. Acad. Sci. Paris **320** (1995) 901–906.

[18] A.J. Wilkie, *Model completeness results for expansions of the ordered field of real numbers by restricted Pfaffian functions and the exponential function*, J. Amer. Math. Soc. **9** (1996) 1051–1094.

Department of Mathematics, Harvard University
Department of mathematics, Northwestern University

Surveys in Differential Geometry, 2000
Vol. VII ©2000, International Press
pp. 625-672

SEIBERG-WITTEN INVARIANTS, SELF-DUAL HARMONIC 2-FORMS AND THE HOFER-WYSOCKI-ZEHNDER FORMALISM

CLIFFORD HENRY TAUBES

1.Introduction

Suppose that $\underline{X}$ is a compact, oriented 4-manifold with $b^{2+} \geq 1$. A symplectic form on $\underline{X}$ is a closed, non-degenerate 2-form whose square provides the given orientation. Little is known by way of sufficient conditions which guarantee the existence of such a form. However, there are smooth, closed forms on $\underline{X}$ which are symplectic off of a disjoint union of embedded circles, with the latter being the vanishing locus of the form. Indeed, if a sufficiently generic Riemannian metric is chosen for $\underline{X}$, then some of the self-dual, harmonic 2-forms on $\underline{X}$ have the aforementioned property. Moreover, the given metric, with such a form, defines a compatible almost complex structure on the complement of the form's zero set. Thus, the complement, $X \subset \underline{X}$, of the zero set of the given closed, self-dual 2-form has a natural pseudoholomorphic geometry, the 'Riemannian pseudoholmorphic geometry'. This geometry seems worthy of study if, for no other reason, then the following:A sufficient condition for the zero set of the form to homologically bound a pseudoholomorphic subvariety in its compliment is for $\underline{X}$ to have non-trivial Seiberg-Witten invariants [16].

Prior to the discovery of the Seiberg-Witten invariants, Hofer introduced [5] and then Hofer, Wysocki and Zehnder [9], [10], [11] (see [6]) systematically developed an elegant formalism for studying a particular

First printed in Asian Journal of Mathematics, 1999. Used by permission.
Author was supported in part by the National Science Foundation.

version of pseudoholomorphic geometry on symplectic manifolds with tubular ends. In particular, the complement, $X \subset \underline{X}$, of the zero set of a form as just described provides a nice example for the Hofer, Wysocki and Zehnder formalism. These pseudoholomorphic geometries on X will be called 'HWZ pseudoholomorphic geometries'. In this regard, it is important to note the the HWZ pseudoholomorphic geometry near the zero set of the given form is not the same as the Riemannian one. In particular, the relationship between the HWZ pseudoholomorphic geometry and the Seiberg-Witten invariants must still be sorted out, and this article provides the first step in doing so with a theorem (Theorem 5.4, below) which implies the following:

- Suppose that $\underline{X}$ has a non-vanishing Seiberg-Witten invariant. Then, there is a finite set of irreducible, HWZ pseudoholomorphic subvarieties in X whose union, with positive integer weights, homologically bounds the zero locus of the given self-dual, harmonic 2-form. This is to say that the weighted union has algebraic intersection number 1 with each linking 2-sphere of the form's zero set.

- Moreover, X has its own Seiberg-Witten invariants from which the Seiberg-Witten invariants of $\underline{X}$ can be computed, and if just the former are non-trivial, then X still has an HWZ pseudoholomorphic subvariety as described in the preceding point.

(1.1)

Note that a Seiberg-Witten based existence proof for pseudoholmorphic subvarieties of compact symplectic manifolds has already been established [17] (but see the revised version in [18] which corrects some arguments in Section 6e of [17]). Moreover, in the case where X is a compact symplectic manifold, the complete Seiberg-Witten invariant of X can be computed completely in terms of the associated pseudoholomorphic geometry (see [19], [20]). This is to say that there is a symplectic invariant, Gr, which is obtained as a count of pseudoholomorphic subvarieties [21] in X and which turns out to be the same as the Seiberg-Witten invariant of X.

In the present context, there is a candidate for a version of Gr which is defined for the cylindrical end manifold $X \subset \underline{X}$, is computable completely in terms of the HWZ pseudoholomorphic geometry of X, and may well be equal to the Seiberg-Witten invariants of $\underline{X}$. This candidate Gr and its relation to the Seiberg-Witten invariants of $\underline{X}$ is the subject of a planned sequel to this article.

By the way, there is some circumstantial evidence to the effect that the pseudoholomorphic geometry of $X \subset \underline{X}$, either Riemannian or HWZ, provides 4-manifold information which goes beyond the Seiberg-Witten invariants (see, e.g. [22]). In this regard, the HWZ pseudoholomorphic geometry may prove the more tractible tool for the study of 4-manifold differential topology. For example, it turns out that the singularities of pseudoholomorphic subvarieties in the HWZ geometry are not hard to classify. In contrast, the singularities of the pseudoholomorphic subvarieties in the Riemannian pseudoholomorphic geometry has only been partly sorted out [23], and may turn out to be very complicated.

The remainder of this article provides the details to (1.1), and is organized along the following lines: Section 2 summarizes the basic features of HWZ pseudoholomorphic geometry, with a special focus on those manifolds which arise as the complement of the zero set of a generic, closed, self-dual 2-form on a compact 4-manifold.This is to say that each end of such a manifold is symplectically concave and is diffeomorphic to $[0, \infty) \times (S^1 \times S^2)$. Section 3 summarizes the Seiberg-Witten story on compact 4-manifolds, and Section 4 summarizes the analgous story for the class of non-compact manifolds 4-manifolds with $[0, \infty) \times (S^1 \times S^2)$ ends. Then, Section 5 points out some of the basic relationships between the Seiberg-Witten story and the HWZ geometry on the class of manifolds under consideration. The results in Section 5 are summarized by Theorems 5.4 and 5.5. The final two sections are devoted to the proofs of these last two theorems.

2. The HWZ pseudoholomorphic geometry

The HWZ geometry is designed for studying symplectic manifolds with contact boundary. The general context for this is described in Hofer [5] and Hofer, Wysocki and Zehnder [9], so attention here will be restricted to the case where the manifold in question is 4-dimensional. With this understood, the purpose of this section is to review the relevant portions of the HWZ geometry.After a general review in the first three subsections, the remaining subsections of Section 2 describe this HWZ geometry in the restricted context that is used in the remainder of this article.

a) Contact boundaries

Let X_0 denote the 4-manifold in question, ω the given symplectic form, and Y a component of the 3-manifold boundary of X_0. The convention here is to orient Y using the restriction to Y of the 3-form

$(\omega \wedge \omega)(v, \cdot, \cdot, \cdot)$ in the case where v is a tangent vector to X_0 along Y which is outward pointing.

The manifold Y is a contact type boundary when there exists a smooth 1-form α on Y such that

- $d\alpha = \iota^* \omega$.
- $\alpha \wedge d\alpha$ *is nowhere zero.*

(2.1)

In this regard, note that $\alpha \wedge d\alpha$ can either agree or disagree with the orientation of Y. In the former case, the boundary is called 'convex' and in the latter case, it is called 'concave'.

In any event, if Y has contact type, then there exists an orientation preserving embedding $\varphi : (0, 1] \times Y \to X_0$ with the following properties:

- $\varphi : \{1\} \times Y \to Y$ is the identity.
- $\varphi^* \omega = du \wedge \alpha \pm u \, d\alpha$ on some neighborhood of $\{1\} \times Y$.

(2.2)

Here, u is the Euclidean coordinate on $(0, 1]$. Also, the $+$ sign is used when Y is convex, and the minus sign when Y is concave. (The concave case will be the case of interest in later sections.)

Write $u = e^{\varepsilon s}$ with $\pm\varepsilon > 0$ depending on whether the contact structure is convex $(+)$ or concave $(-)$. Then,

$$\varphi^* \omega = e^{\varepsilon s}(\varepsilon \, ds \wedge \alpha + d\alpha). \tag{2.3}$$

This form is defined for s non-positive and near zero, but it evidently extends to all positive values of s. This is to say that the form ω extends from X_0 to the noncompact manifold

$$X \equiv X_0 \cup_Y ([0, \infty) \times Y). \tag{2.4}$$

Note that when measured with the product metric on $([0, \infty) \times Y)$, the form in (2.3) either grows or shrinks in size exponentially fast as $s \to \infty$ depending on whether Y is convex or concave.

b) Pseudoholomorphic geometry

In all that follows, assume that all components of ∂X_0 are of contact type. An almost complex structure on X is an endomorphism, J, of TX whose square is -1. Such a J will be called 'ω-compatible' in the case where the bilinear form $\omega(\cdot, J(\cdot))$ defines a Riemannian metric. It proves useful to impose some further requirements on J's restriction to each end $[0, \infty) \times Y$ of X. In particular, the HWZ geometry considers ω-compatible almost complex structures which restrict to $[0, \infty) \times Y$ so that:

- J is invariant under the 1-parameter semi-group of translations $(s, x) \to (s + a, x)$ for $a \geq 0$.
- $J \cdot \partial_s$ is annihilated by $d\alpha$.
- J preserves the kernel of α.

(2.5)

Because the space of ω-compatible almost complex structures is contractible, there is no problem with finding such almost complex structures which also obey the requirements in (2.5). With this last point understood, the almost complex structures henceforth under consideration will be implicitly assumed to satisfy (2.5) as well as being ω-compatible.

c) Pseudoholomorphic subvarieties

A subvariety $C \subset X$ will be called 'pseudoholomorphic' when the following conditions are met:

- C is closed and locally compact.
- There is a non-accumulating set $\Lambda \subset C$ of at most a countable number of points such that $C - \Lambda$ is an embedded submanifold of X whose tangent space is J-invariant.

(2.6)

A pseudoholomorphic subvariety $C \subset X$ will be called an 'HWZ subvariety' when, in addition to (2.6),

$$\int_{C\cap((0,\infty)\times M)} d\alpha < \infty. \tag{2.7}$$

By the way, when integrating either $d\alpha$ or $\varepsilon\ ds \wedge \alpha$ over a domain in a pseudoholomorphic subvariety C, keep in mind that both restrict to C as non-negative 2-forms. This is a consequence of C being pseudoholomorphic for the almost complex structure in (2.5). Here is a simple consequence of this last fact:

Lemma 2.1. *If all boundary components of X_0 are concave, then every pseudoholomorphic subvariety in X satisfies (2.7). That is, all pseudoholomorphic subvarieties are HWZ subvarieties.*

This subsection ends with the.

Proof of Lemma 2.1. It proves useful to make a short, preliminary digression to choose, for each $R \geq 4$, a function σ_R on R with the following properties:

$$
\begin{array}{lll}
 & \bullet\ \sigma_R = 0 & \text{where } s \leq 0, \\
 & \bullet\ \sigma_R = \sigma_1 \ \text{and}\ 0 \leq \sigma_1' \leq 1 & \text{where } 0 \leq s \leq 2, \\
 & \bullet\ \sigma_R = 1 & \text{where } 2 \leq s \leq R, \\
 & \bullet\ -1 \leq \sigma_R' \leq 0 & \text{where } R \leq s \leq R+2, \\
(2.8) & \bullet\ \sigma_R = 0 & \text{where } s \geq R+2,
\end{array}
$$

Thus, σ_R vanishes until $s = 0$, then increases to 1 by $s = 2$, stays equal to 1 until $s = R$ and finally decreases to zero by $s = R + 2$. Moreover, its derivative is nowhere greater than 1 or less than -1.

With the digression now over, remember that $d\alpha$ on C is non-negative as is σ_R; and as $\sigma_R = 1$ where $s \in [2, R]$, the demonstration of an R-independent upper bound for the integral over C of $\sigma_R\, d\alpha$ proves that C is an HWZ subvariety. With this last point understood, remark that $d(\sigma_R\alpha)$ is also integrable over C. Stokes' theorem finds this integral equal to zero, and so

$$
\int_C \sigma_R d\alpha = \int_C -d\sigma_R \wedge \alpha. \tag{2.9}
$$

Thus, it is enough to find an R independent upper bound to the integral over C of $-d\sigma_R \wedge \alpha$.

To achieve the latter task, remark first that $-d\sigma_R \wedge \alpha$ has support in two disjoint sets, the first where $0 \leq s \leq 2$ and the second where $R \leq s \leq R + 2$. Moreover, if all components of ∂X_0 are concave,then $-d\sigma_R \wedge \alpha$. is non-positive on C where $s \geq R$ because σ_R' is non-positive where $s \geq R$ while $-ds \wedge \alpha$ is non-negative on C. Thus,

$$
\int_C \sigma_R d\alpha \leq \int_{C\cap\{0\leq s\leq 2\}} \sigma_1'(-ds \wedge \alpha). \tag{2.10}
$$

As the right-hand side of (2.8) is finite and independent of R, the desired bound follows.

d) The $S^1 \times S^2$ example

The relevant example for this article takes $Y = S^1 \times S^2$. To write the relevant contact form α, take standard spherical coordinates $(\theta, \varphi) \in$

$[0, \pi] \times [0, 2\pi]$ for S^2 and a coordinate $t \in [0, 2\pi]$ for S^1. In terms of these coordinates,

$$\alpha = -(1 - 3\cos^2\theta)dt - \sqrt{6}\cos\theta\sin^2\theta d\varphi. \tag{2.11}$$

A computation finds

$$d\alpha = 6\cos\theta\sin\theta dt d\theta + \sqrt{6}(1 - 3\cos^2\theta)\sin\theta d\theta d\varphi; \tag{2.12}$$

thus $\alpha \wedge d\alpha$ is seen not to vanish:

$$\alpha \wedge d\alpha = -\sqrt{6}(1 + 3\cos^4\theta)dt\sin\theta d\theta d\varphi. \tag{2.13}$$

By the way, take the metric on $S^1 \times S^2$ to be the product of the standard round metrics and introduce the resulting Hodge star operator, $*$, on differential forms. Then α in (2.11) obeys

- $*d\alpha = -\sqrt{6}\alpha.$
- $d^*\alpha = 0.$

(2.14)

This last point is mentioned in as much as it implies that the 2-form

$$\omega \equiv e^{-\sqrt{6}s}(-\sqrt{6}ds \wedge \alpha + d\alpha) \tag{2.15}$$

on $\mathbb{R} \times S^1 \times S^2$ is symplectic, and self-dual with respect to the Hodge star operator from the product metric.

The compact, integral curves in the foliation that is defined by the kernel of $d\alpha$ are of prime importance in the story. In this regard, the kernel of $d\alpha$ is the linear span of the vector field

$$v \equiv -(1 - 3\cos^2\theta)\partial_t - \sqrt{6}\cos\theta\partial_\varphi. \tag{2.16}$$

Thus, all integral curves of v can be parameterized by

$$\tau \to (t = t_0 - \tau(1 - 3\cos^2\theta_0), \theta = \theta_0, \varphi = \varphi_0 - \sqrt{6}\tau\cos\theta_0) \tag{2.17}$$

where t_0, θ_0 and φ_0 are constants. Note that the integral curve in (2.17) is compact if and only if

$$\frac{\sqrt{6}\cos\theta_0}{(1 - 3\cos^2\theta_0)} \in Q \cup \{\infty\}. \tag{2.18}$$

By the way, the kernel of the contact form α in (2.11) is spanned by the vectors

$$\{\partial_\theta, \sqrt{6}\cos\theta\sin\theta\partial_t - (1 - 3\cos^2\theta)(\sin\theta)^{-1}\partial_\varphi\}. \tag{2.19}$$

When considering almost complex structures on $\mathbb{R}\times(S^1 \times S^2)$ which obey (2.5), there is one which is especially useful. To describe this J, it proves convenient to first digress for the purpose of introducing auxilliary functions f and h on $\mathbb{R}\times(S^1 \times S^2)$ as follows:

- $f \equiv e^{-\sqrt{6}s}(1 - 3\cos^2\theta)$.
- $h \equiv e^{-\sqrt{6}s}\sqrt{6}\cos\theta\sin^2\theta$.

(2.20)

In terms of the 'coordinates' (t, f, h, φ), the form ω and the product metric on $\mathbb{R}\times(S^1 \times S^2)$ are as follows:

- $\omega = dt \wedge df + d\varphi \wedge dh$.
- $ds^2 + dt^2 + d\theta^2 + \sin^2\theta d\varphi^2 = dt^2 + g^{-2}(df^2 + \sin^{-2}\theta dh^2) + \sin^2\theta d\varphi^2$.

(2.21)

Here, $g \equiv \sqrt{6}e^{-\sqrt{6}s}(1 + 3\cos^4\theta)^{1/2}$.

With the preceding understood, define J by

- $J \cdot \partial_t = g\partial_f$,
- $J \cdot \partial_\varphi = \sin^2\theta\ g\partial_h$.

(2.22)

It is an exercise to verify that the almost complex structure so defined obeys the constraints in (2.5). (One reason for appreciating this particular J is that it acts as an orthogonal transformation with respect to the standard product metric on $\mathbb{R}\times(S^1 \times S^2)$.)

The final comment on this example concerns orientations for the homology of $S^1 \times S^2$. In this regard, orient $H_3(S^1 \times S^2; \mathbb{Z})$ by requiring $\alpha \wedge d\alpha$ to have negative integral. (This is equivalent to the requirement that $(\omega\wedge\omega)(\partial_s, \cdot, \cdot, \cdot)$ have positive integral.) To orient the 2-dimensional homology, first remark that $d\alpha$ is non-degenerate on the kernel of α, and thus orients this $\mathbb{R}^2$ bundle by requiring $d\alpha$ to be positive on a positively oriented frame. The Euler class of this oriented bundle is twice a generator of $H^2(S^1 \times S^2; \mathbb{Z}) \approx \mathbb{Z}$ and so orients the latter by making this Euler class positive. Use this last orientation to orient the

2-dimensional homology by requiring the oriented generator of the latter to pair with that of H^2 to give 1. Finally, an orientation on H_1 is induced from that on H^2 via Poincaré duality. In less prosaic terms, $S^1 \times S^2$ is oriented by the form $dt \sin\theta d\theta d\varphi$, while $\sin\theta d\theta d\varphi$ orients the S^2 factor and dt orients the S^1 factor.

e) The context

This example arises in the following context: Let $\underline{X}$ be a compact, oriented 4-manifold with $b^{2+} \geq 1$. Put a Riemannian metric, g, on $T\underline{X}$ and Hodge-DeRham theory provides a b^{2+} dimensional vector space of closed, self dual 2-forms. In this regard, note that a 2-form ω is self-dual when

$$\omega \wedge \omega = |\omega|^2 d\,\mathrm{vol}\,. \tag{2.23}$$

Thus, closed, self-dual 2-forms are symplectic except where they vanish.

If the metric is chosen in a suitably generic fashion [7] (from a Baire set of smooth forms), then there exist closed, self-dual 2-forms which vanish transversely as sections of the $\mathbb{R}^3$ bundle of self dual 2-forms. Thus, if ω is such a form, then

$$Z \equiv \omega^{-1}(0) \tag{2.24}$$

is a disjoint union of embedded circles and ω is symplectic on $\underline{X} - Z$.

To consider the behavior of ω near a component circle $Z_0 \subset Z$, note that a neighborhood of Z_0 is diffeomorphic to the product of S^1 with a centered 3-ball $B^3 \subset \mathbb{R}^3$. In particular, a coordinate $t \in [0, 2\pi]$ for S^1 and $x = (x_1, x_2, x_3)$ for B^3 can be chosen so that with respect to these coordinates, the form ω is given by

$$\omega = dt \wedge A_{ij} x_i dx_j + 2^{-1} A_{ij} x_i \varepsilon_{ijk} dx_j \wedge dx_k + o(|x|^2), \tag{2.25}$$

where $A_{ij}(t)$ is a matrix valued function on S^1. Moreover,

- $A_{ij} = A_{ji}$,
- $\sum_j A_{jj} = 0$,
- $\det(A) < 0$.

(2.26)

Here, the first two properties are consequences of the fact that $d\omega = 0$, and the third is a consequence of the fact that ω vanishes transversally along Z_0. (To be precise, this just insures that $\det(A) \neq 0$; the sign is arranged through a choice of orientation for Z_0.)

Now, the fact that A_{ij} is traceless and symmetric and has negative determinant implies that it has, at each $t \in Z_0$, two positive eigenvalues and one negative eigenvalue. Let $\underline{\varepsilon} \to Z_0$ denote the real line bundle whose fiber at each $t \in Z_0$ is the negative eigenspace of the matrix $A_{ij}(t)$. This line bundle can either be orientable or not. In this regard, the following result of Gompf [4] is fundamental:

Lemma 2.4. *The parity (even or odd) of the number of components of Z where $\underline{\varepsilon}$ is orientable equals the parity of $b^{2+} - b^1 + 1$.*

In the case where $\underline{\varepsilon}$ is orientable, the form ω can be modified near Z_0 so that the resulting new form has the following properties:

- The new form is symplectic on $\underline{X} - Z_0$ and agrees with the old form outside of some previously specified neighborhood of Z_0.

- There are coordinates (t, x) for an $S^1 \times B^3$ tubular neighborhood of Z_0 in which the new form is equal to

$$\begin{aligned} dt \wedge (x_1 dx_1 + x_2 dx_2 - 2x_3 dx_3) + x_1 dx_2 \wedge dx_3 \\ -x_2 dx_1 \wedge dx_3 - 2x_3 dx_1 \wedge dx_2 \end{aligned} \tag{2.27}$$

In the case where $\underline{\varepsilon}$ is not orientable, there is a modification of ω near Z_0 so that the first point above holds, and so that the second point with (2.27) holds on a non-trivial, $S^1 \times B^3$ double cover of a tubular neighborhood of Z_0. An equivalent assertion in the $\underline{\varepsilon}$ non-orientable case is the following: The form ω has a modification on the original $S^1 \times B^3$ neighborhood of Z_0 so that the resulting new form obeys the first point in (2.27) and so that there are coordinates (t', x') on some $S^1 \times B^3$ tubular neighborhood of Z_0 in which the new form is equal to the form in (2.27) after the substitutions

- $t = t'/2$,

- $x_1 = x'_1$,

- $x_2 = \cos(t'/2)x'_2 - \sin(t'/2)x'_3$,

- $x_3 = \sin(t'/2)x'_2 + \cos(t'/2)x'_3$.

(2.28)

Now note that $B^3 - \{0\}$ is diffeomorphic to $S^2 \times [0, \infty)$ via the map which sends the centered sphere in B^3 with radius $a > 0$ to $S^2 \times \{s = -2^{-1} \ln a\}$ in $S^2 \times [0, \infty)$. The form in (2.27) pulls back under this diffeomorphism to

$$-d(e^{-s} 2^{-1}(1 - 3\cos^2\theta)dt + e^{-3s/2} \cos\theta \sin^2\theta d\varphi). \tag{2.29}$$

This form is not described by (2.3) for any contact form α on $S^1 \times S^2$. However, it can be modified so that a constant multiple (2.3) is accurate at large s with α given by (2.11). In particular, consider:

Lemma 2.3. *Fix $R > 1$ and there are constants $c_1, c_2 > 0$ with the following significance: Let R' denote either ∞ or a number greater than $2R$. There is a symplectic form ω on $[0, \infty) \times S^1 \times S^2$ which is described by (2.22) on $[0, R/2] \times S^1 \times S^2$, and by c_1 times the form in (2.22) on $[R', \infty)$. Meanwhile, on $[R, R']$, the form is described by c_2 times the form in (2.15).*

(The proof of this lemma is left as an exercise save for the following hint: The numbers $c_{1,2}$ are on the order of e^{-R}. See also [8].)

Lemma 2.3 implies that a disjoint, finite set of embedded circles can be removed from any b^{2+} positive 4-manifold so that the resulting non-compact manifold is described by the *HWZ* formalism. Here, each boundary component is a copy of $S^1 \times S^2$; and after possibly passing to the non-trivial double cover, there are coordinates where the relevant contact form is given in (2.11).

f) A more general context

The subsequent discussions of the *HWZ* geometry takes place on a connected, non-compact manifold X which splits as $X = X_0 \cup ([0, \infty) \times \partial X_0)$, where X_0 is a compact, 4-manifold with boundary where the latter is a disjoint union of some number $N \geq 0$ copies of $S^1 \times S^2$. Furthermore, it will be assumed that X_0 has a symplectic form, ω, for which each boundary component is contact type and concave. Finally, it will be assumed that each boundary component of X_0 is described by at least one of the following points:

- There are coordinates in which the contact form is given by α in (2.11).
- There are coordinates on the non-trivial 2-fold cover in which the pull-back of the contact form is given by α in (2.11).

(2.30)

A component of ∂X_0 will be said to have *orientable z-axis line bundle* when the first point in (2.30) holds. Otherwise, it will be said to have *non-orientable z-axis line bundle.*

Note that (2.15) with (2.30) provides an extension of the symplectic form ω on X_0 to the whole of X. This extension of ω will be implicit in what follows.

The HWZ geometry of X will be defined by a choice of ω-compatible almost complex structure, $J : TX \to TX$ whose restriction to $[0, \infty) \times \partial X_0$ is as follows: When a given component of ∂X_0 has orientable z-axis line bundle, then take J as in (2.30) on the corresponding component of $[0, \infty) \times \partial X_0$. Otherwise, take J so that its lift to the non-trivial double cover of the corresponding component of $[0, \infty) \times \partial X_0$ is given by (2.30).

As the boundaries of X_0 are all concave with respect to the induced contact form, Lemma 2.1 finds all of the pseudoholomorphic subvarieties to be HWZ subvarieties. Moreover, these subvarieties are all reasonably well behaved, as indicated by the following lemma:

Lemma 2.4. *Let X, its symplectic form and its almost complex structure be as described at the beginning of this subsection. Now, let $C \subset X$ be an HWZ subvariety. Then the set $\Lambda \subset C$ of non-manifold points is a finite set at worst; infact, C intersects the complement of a compact subset of X as a properly embedded, disjoint union of cylinders. In particular, this implies that for sufficiently large s, the intersection of C with $\{s\} \times Y$ is transversal and a disjoint union of circles; and, these s-dependent circles in Y converge in the C^∞ topology as $s \to \infty$ to a disjoint union of smooth circles whose tangent lines lie in the kernel of $d\alpha$.*

As this lemma plays only a peripheral role in this article, its proof will be given elsewhere. (Given that the ends of C are embedded cylinders, the implication concerning the intersection of C with $\{s\} \times Y$ for large s follow from Theorem 1.2 in [9].)

The preceding lemma and theorems from HWZ (see [11]) provide a natural topology on the set of HWZ subvarieties in X which makes this set into a reasonable topological space.In particular, a neighborhood of a given HWZ subvariety C in this topological space is homeomorphic to the inverse image of zero for some smooth map between a ball in one finite dimensional Euclidean space to another such Euclidean space. In addition, the components of the space of HWZ subvarieties have natural compactifications as stratified spaces where the extra strata are also spaces of HWZ subvarieties. In short, these spaces of HWZ subvarieties are much like the moduli spaces of pseudoholomorphic subvarieties on compact symplectic manifolds with compatible almost complex structures.

Before preceding to the next subsection, a two part digression is in order to discuss issues which relate to the existence and uniqueness of the coordinates in (2.30).

Part 1. (Existence). Let ν be a concave contact form on $S^1 \times$

S^2. Then ν can be either tight or overtwisted. (A contact form ν is overtwisted or tight if there is or is not an embedded, closed disk which is transversal to the kernel of ν along its boundary, but whose boundary is tangent to the kernel ν.) The contact form α in (2.11) is overtwisted, witness the t = constant disk in S^2 where $\cos^2\theta \geq 1/3$. Thus, both cases of (2.30) require overtwisted contact structure.

Meanwhile, a fundamental theorem of Eliashberg [1] asserts that the overtwisted (concave) contact structures (up to homotopy through contact structures) on a compact, oriented 3-manifold are in 1-1 correspondence with the homotopy classes of oriented, 2-dimensional subbundles of the manifold's tangent bundle. In the case of $S^1 \times S^2$, the latter are classified in part by the degree of the Euler class of the subbundle.In both cases of (2.30), the Euler class in question has minimal (in absolute value) non-zero degree. (This can be verified by examining the zeros of the product of one of the vectors in (2.19) with $\sin\theta$.) Moreover, there are precisely two homotopy classes of 2-plane fields on $S^1 \times S^2$ whose Euler class is $+2$, and these two are not permuted by the diffeomorphisms group of $S^1 \times S^2$. On the otherhand, those with Euler class -2 can be mapped to the corresponding $+2$ classes by an orientation preserving diffeomorphism of $S^1 \times S^2$. Thus, up to homotopy through overtwisted contact structures, any overtwisted contact structure on $S^1 \times S^2$ whose kernel has Euler class with absolute value 2 obeys (2.30).

By way of contrast, another theorem of Eliashberg (see Theorem 4.1.4 in [2]) implies that there are no tight contact structures on $S^1 \times S^2$ whose contact 2-plane field has non-zero degree. Note that there is a unique (up to diffeomorphism) tight contact structure on $S^1 \times S^2$ [2].

The preceding observations directly imply the following:

Lemma 2.5. *Suppose that X_0 is a compact manifold with boundary where each boundary component is diffeomorphic to $S^1 \times S^2$. Let ω be a symplectic form on X_0 for which ∂X_0 is contact type, concave, and such that the kernel of the correponding contact structure has Euler class with absolute value 2 on each component of ∂X_0. Then ω can be homotoped through symplectic forms for which ∂X_0 is contact type so that the resulting form induces a contact structure on ∂X_0 which obeys one of the two points in (2.30) on each component of ∂X_0.*

Part 2. (Uniqueness). Given that a contact form α satisfies (2.30), it is by no means the case that the coordinates which realize (2.30) are unique. Even so, there are certain features of such a coordinate system which are invariant under diffeomorphisms which preserve the form in (2.11). One of these features relates to the closed integral curves in (2.17) of the vector field v in (2.16).In particular, the curves in (2.17) are, but with a single one parameter family of glaring exception, all non-trivial in $H_1(S^1 \times S^2)$. The homologically trivial, closed integral curves are characterized by the condition that $\cos^2\theta_0 = 1/3$. The union of the 1-parameter family of such curves is a pair of embedded tori in $S^1 \times S^2$ whose compliment is the disjoint union of three pieces,

- $a_0 \equiv S^1 \times \{(\theta, \varphi) : -1/3 < \cos^2\theta < 1/3\}$,
- $a_\pm \equiv S^1 \times \{(\theta, \varphi) : \pm\cos^2\theta > 1/3\}$.

(2.31)

With regard to (2.31), note that the component a_0 is the product of the circle with an annulus, while the other two components are the products of the circle with a disk.

The following lemma (the proof is self-evident) concludes the digression:

Lemma 2.6. *Any diffeomorphism of $S^1 \times S^2$ which preserves the form α in (2.11) must map a_0 to itself and either map $a_\pm$ to themselves or to each other.*

g) SW-admissable HWZ subvarieties

Let X with its symplectic form ω be as described in the previous subsection. As every component of ∂X_0 is concave, Lemma 2.1 finds all pseudoholomorphic subvarieties in X to be HWZ subvarieties. Even so, this term will be employed as a reminder that (2.7) is obeyed along with (2.6). Of particular concern are those HWZ subvarieties which are 'SW-admissable', a term which is specified in Defintion 2.9, below. This definition requires a three part, preliminary digression.

Part 1. This first part of the digression presents:

Lemma 2.7. *The contact form on ∂X_0 canonically orients the homology of ∂X_0.*

Proof of Lemma 2.7. This follows from the discussion in the final paragraph of Section 2d.

Remember this lemma when considering the definition of SW-admissable, below.

Part 2. Let $C \subset X$ denote an HWZ subvariety. Then, an irreducible component of C is, by definition, the closure of a component of the complement in C of the set Λ from the second point of (2.6). If C is an HWZ pseudoholomorphic subvariety, then so are its irreducible components. Note that an HWZ subvariety has only finitely many irreducible components. (If not, then the first point of (6) would force most to lie entirely in $(0, \infty) \times \partial X_0$. But, this possibility is ruled out by the fact that $d\alpha$ restricts as a non-negative form with finite total integral.)

Part 3. The symplectic form ω orients psuedoholomorphic subvarieties in X, and so any such variety, C, determines, by restriction, a class $[C] \in H_2(X_0, \partial X_0; \mathbb{Z})$.

Definition 2.8. A *generalized* HWZ subvariety is a finite collection $c \equiv \{(C_\alpha, m_\alpha)\}$ with the C_α's pairwise distinct, irreducible, HWZ subvarieties and the corresponding m_α's non-negative integers. (The integer m_α is called the mulitiplicity of the corresponding C_α). A generalized HWZ subvariety $\{(C_\alpha, m_\alpha)\}$ is called *SW-admissable* when the connecting homomorphism from $H_2(X_0, \partial X_0; \mathbb{Z})$ to $H_1(\partial X_0; \mathbb{Z})$ of the long exact homology sequence for the pair $(X_0, \partial X_0)$ sends $\sum_\alpha m_\alpha [C_\alpha]$ to the sum of the oriented generators of $H_1(\partial X_0; \mathbb{Z})$.

The following lemma offers some perspective on this definition:

Lemma 2.9. *Let C be an HWZ pseudoholomorphic subvariety with two properties: First, the pair $(C, 1)$ is SW-admissable. Second, there exists $s_0 > 0$ such that the intersection of C with each component of $\{s_0\} \times \partial X_0$ is path connected. Then, the following conclusions can be drawn:*

- *C intersects any large, constant s slice of any given component of $[0, \infty) \times \partial X_0$ as a circle which is an oriented generator of the first homology of the corresponding component of ∂X_0.*
- *If a particular component of ∂X_0 has orientable z-axis line bundle,then C intersects the large s portion of the corresponding component of $[0, \infty) \times \partial X_0$ in $[0, \infty) \times a_0$.*
- *If a particular component of ∂X_0 has non-orientable z-axis line bundle, then C intersects the corresponding component of $[0, \infty) \times \partial X_0$ in the image of $[0, \infty) \times a_0$ from the non-trivial 2-fold cover.*
- *Moreover, in this last case: As $s \to \infty$ on C's inverse image in the non-trivial double cover of the component in question the function θ converges to $\pi/2$ while φ converges either to $\pi/2$ or to $-\pi/2$.*

Proof of Lemma 2.9. Given the orientations of the homology of ∂X_0, the lemma is a direct consequence of Theorem 1.2 in [9].

3. The Seiberg-Witten invariants on compact 4-manifolds

This section consists of a summary of some of the relevant properties of the Seiberg-Witten invariants of a compact 4-manifold. These invariants, first introduced by Witten [24], are now discussed in a number of books (see, e.g. [12]) to which the reader is referred for more details.

In this section, X is a compact, connected, oriented 4-manifold with $b^{2+} \geq 1$. Let S denote the set of equivalence classes of Spin^C structures on X. This set is a principal bundle over a point for the additive group $H^2(X;\mathbb{Z})$. After a choice of orientation for the line $\det^+ \equiv \Lambda^{\mathrm{top}}(H^1(X;\mathbb{R})) \otimes \Lambda^{\mathrm{top}}(H^{2+}(X;\mathbb{R}))$, and also $H^{2+}(X;\mathbb{R})$ in the case where $b^{2+} = 1$, the Seiberg-Witten invariants constitute a map

$$SW_X : S \to \Lambda^*(H^1(X;\mathbb{Z})) = \mathbb{Z} \oplus H^1(X;\mathbb{Z}) \oplus \Lambda^2(H^1(X;\mathbb{Z})) \oplus \cdots \tag{3.1}$$

The map SW is defined as an algebraic count of solutions to a certain differential equation defined on X.

a) The Seiberg-Witten equations

The definition of the Seiberg-Witten equations has four parts.

Part 1. Fix a Riemannian metric on X. The latter specifies the principal $SO(4)$ bundle $\mathrm{Fr} \to X$ of oriented, orthonormal frames in TX. By definition, a Spin^C structure is a lift of Fr to a principal

$$\mathrm{Spin}^C(4) = (SU(2) \times SU(2) \times U(1))/\{\pm 1\} \tag{3.2}$$

bundle. In this regard, note that $SO(4) = (SU(2) \times SU(2))/\{\pm 1\}$; and with this understood, the homomorphisms from $\mathrm{Spin}^C(4)$ to $SO(4)$ simply forgets the $U(1)$ factor in (3.2).

In any event, let $F \to X$ denote a lift of Fr to a principal $\mathrm{Spin}^C(4)$ bundle.

Part 2. Associated to F are two canonical $\mathbb{C}^2$ bundles, $S_\pm$. Here, the association is via the representations of $\mathrm{Spin}^C(4)$ to $U(2) = (SU(2) \times U(1))/\{\pm 1\}$ which forget either the first factor of $SU(2)$ or the second. By convention, the projective plane bundles PS_- and PS_+ are the unit sphere bundles in the respective $\mathbb{R}^3$ bundles $\Lambda_\pm$ of anti-self dual and self dual 2-forms. The latter are associated to Fr via the two homomorphisms from $SO(4)$ to $SO(3) = SU(2)/\{\pm 1\}$ which forget one or the other factor $SU(2)$. Note that both $S_\pm$ inherit canonical Hermitian metrics.

There is also an associated $U(1)$ principal bundle, $L \to X$ which is defined via the homomorphism from $\mathrm{Spin}^C(4)$ to $U(1)$ which forgets both factors of $SU(2)$. In this regard, remember that $U(1)/\{\pm 1\}$ is isomorphic to $U(1)$.

As remarked above, the set of Spin^C structures on X is a principal bundle over a point for the group $H^2(X;\mathbb{Z})$. The action of this group on S can be simply described in terms of its effect on the bundles $S_\pm$ and L. Here, remember that $H^2(X;\mathbb{Z})$ is in 1 to 1 correspondence with the set of equivalence classes of complex hermitian line bundles where the correspondence associates a line bundle E to its first Chern class, $c_1(E) \in H^2(X;\mathbb{Z})$. With this understood, remark that when $\mathfrak{s} \in S$ and $e \in H^2(X;\mathbb{Z})$, then the Spin^C structure $e \cdot s$ is characterized by the condition that $S_\pm(e \cdot \mathfrak{s}) = E \otimes S_\pm(\mathfrak{s})$ where $E \to X$ is a complex line bundle with $c_1(E) = e$. Meanwhile, $L(e \cdot \mathfrak{s})$ is characterized by the property that its associated first Chern class equals $c_1(L(\mathfrak{s})) + 2e$.

Part 3. The Seiberg-Witten equations are defined with the help of the Clifford multiplication map

$$cl : TX \to \mathrm{Hom}(S_+, S_-) \tag{3.3}$$

Indeed, cl is a canonical bundle isomorphism between TX_C and $\mathrm{Hom}(S_+, S_-)$ which is defined by viewing the latter bundle as an associated bundle to Fr. The map in (3.3) has the following key property: When $v \in TX$, then $cl(v)^\dagger cl(v)$ and $cl(v)^\dagger cl(v)$ are equal $-|v|^2$ times the identity endomorphism of S_+ and S_-, respectively. Note that cl can also be viewed as a homomorphism

$$\underline{cl} : S_+ \otimes T^*X \to S_-. \tag{3.4}$$

Also required is the extension of cl to

$$cl_+ : \Lambda_+ \to \mathrm{End}(S_+) \tag{3.5}$$

The map cl_+ sends Λ^+ to the traceless, anti-hermitian endomorphisms of S_+. It is defined by the requirement that it send the self-dual projection of $w \wedge w'$ to

$$2^{-1}(cl(w')^\dagger cl(w) - cl(w)^\dagger cl(w')). \tag{3.6}$$

The adjoint of cl_+ maps $S_+^\dagger \otimes S_+$ to the imaginary valued sections of Λ^+. This adjoint will be denoted by cl^+.

Part 4. The data for the Seiberg-Witten equations consists of a pair (A, Ψ), where A is a connection on L and where Ψ is a section of S_+. The Seiberg-Witten equations involve the curvature 2-form F_A of the connection A and its projection, F_A^+, in Λ_+. These equations also involve the covariant derivative ∇_A on sections of S_+ which A induces with the help of the Levi-Civita conection on TX. Indeed, the Levi-Civita connection provides a connection on the principle $SO(3) \times SO(3)$ bundle $\mathrm{Fr}/\{\pm 1\}$. Thus, A and the Levi-Civita connection together provide a connection on the principle $(SO(3) \times SO(3) \times U(1))/\{\pm 1\}$ bundle $\underline{F}$ associated to F. As F is, fiberwise, a 4-fold cover of $\underline{F}$, the connection on $\underline{F}$ induces a unique connection on F. The covariant derivative of the latter connection is ∇_A.

With the preceding understood, the Seiberg-Witten equations read

- $D_A\Psi \equiv \underline{cl}(\nabla_A\Psi) = 0,$
- $F_A^+ = cl^+(\Psi \otimes \Psi^\dagger) + i\mu.$

(3.7)

Here, μ denotes a fixed, favored self-dual 2-form.

b) Properties of the space of solutions to the Seiberg-Witten equations

Fix a Spin^C structure s and so define the principal $U(1)$ bundle $L \to X$ and the $\mathbb{C}^2$ bundle S_+. The set of connections on L is naturally an affine space which is modeled on the space of smooth, imaginary valued 1-forms, $i \cdot C^\infty(T^*X) \subset C^\infty(T^*X) \otimes \mathbb{C}$. This affine structure endows the space of connections, $\mathrm{Conn}(L)$, with the structure of a smooth Frechet space manifold. Meanwhile, the space of sections of S_+ has its linear, C^∞ Frechet space structure.

Now, let $m \subset \mathrm{Conn}(L) \times C^\infty(S_+)$ denote the space of solutions to (3.7) for a given choice of μ. (Thus, m depends on the triple (s, g, μ) of Spin^C structure, Riemannian metric and self-dual 2-form.) Topologize m with the subspace topology.

The space m is always infinite dimensional because the equations in (3.7) are invariant under a certain smooth action on $\mathrm{Conn}(L) \times C^\infty(S_+)$ of the group $C^\infty(X; S^1)$ of smooth maps from X to the circle. Indeed, a map $\eta \in C^\infty(X; S^1)$ acts by sending the pair $c \equiv (A, \Psi)$ of connection on L and section of S_+ to $\eta \cdot c \equiv (A - 2\eta^{-1}d\eta, \eta \cdot \Psi)$. For future reference, note that this action is free except at pairs of the form $(A, 0)$ where the stabilizer is the circle of constant maps to S^1. By the way, such pairs $(A, 0)$ are termed *reducible*. In any event, let M denote the quotient $m/C^\infty(X; S^1)$ which will be viewed as a topological space using the quotient topology.

It also proves useful to introduce the space, $\underline{M}$, which is the quotient of $X \times m$ by the relation $(x, c) \sim (x', c')$ if and only if $x = x'$ and $c = \varphi \cdot c'$ where $\varphi \in C^\infty(X; S^1)$ obeys $\varphi(x) = 1$. Away from reducible points, the obvious projection from $\underline{M}$ to $X \times M$ has fiber S^1.

The following proposition lists some of the salient features of M and $\underline{M}$: (This proposition was known to Witten [24]; and proofs of its assertions can be found in [12].)

Proposition 3.1. *Fix a* Spin^C *structure* $\mathfrak{s}$, *a Riemannian metric* g *and a self-dual 2-form* μ. *Use this data to define the space* M *and* $\underline{M}$. *Then the following are true:*

- *M and $\underline{M}$ are compact.*
- *Each irreducible $c \in M$ has a neighborhood which is homeomorphic to the zero set of a real analytic map between balls about the origin in finite dimensional Euclidean spaces. In particular, the domain ball lies naturally in the kernel of a first order, elliptic operator δ_c and the range of this map is the cokernel of this same operator. Here, the index of δ_c is equal to*

$$(b^1 - 1 - b^{2+}) + 4^{-1}(-\tau_X + c_1(L) \bullet c_1(L)); \tag{3.8}$$

  where τ_X is the signature of X and the symbol '$\bullet$' between a pair of 2-dimensional cohomology classes signifies the value of their cup product on the fundamental class of X.
- *In general, the subspace $M_{\mathrm{reg}} \subset M$ of irreducible orbits where* $\mathrm{cokernel}(\delta_c) = 0$ *is open in M and has the structure of a smooth, orientable manifold whose dimension is given by (3.8). Moreover, an orientation of the line* $\det^+ \equiv \Lambda^{\mathrm{top}}(H^1(X;\mathbb{R})) \otimes \Lambda^{\mathrm{top}}(H^{2+}(X;\mathbb{R}))$ *provides M_{reg} with a canonical orientation. Meanwhile, the the inverse image in $\underline{M}$ of $X \times M_{\mathrm{reg}}$ has the structure of a smooth, principal S^1 bundle.*
- *Suppose that $b^{2+} > 0$. Fix the metric, and there is a Baire set $\mathcal{U} \subset C^\infty(X; \Lambda_+)$ of self dual 2-forms μ for which $M = M_{\mathrm{reg}}$, and so M has the structure of a smooth manifold of dimension given by the number in (3.8). In particular, for $\mu \in \mathcal{U}$, the operator δ_c has trivial cokernel for all $c \in M$.*

By the way, this operator δ_c is defined for any $c \in \mathrm{Conn}(E) \times C^\infty(X; S_+)$ and maps $i \cdot C^\infty(T^*X) \oplus C^\infty(S_+)$ to

$$i \cdot (C^\infty(X) \oplus C^\infty(\Lambda_+)) \oplus C^\infty(S_-).$$

In this regard, δ_c sends a pair $(b, \eta) \in i \cdot C^\infty(TX^*) \oplus C^\infty(S_+)$ to the triple in $i \cdot (C^\infty(X) \oplus C^\infty(\Lambda_+)) \oplus C^\infty(S_-)$ with the components

- $d^*b + 4^{-1}(\eta^\dagger\Psi - \Psi^\dagger\eta)$
- $d^+b - cl^+(\eta \otimes \Psi^\dagger + \Psi \otimes \eta^\dagger)$
- $D_A\eta + 2^{-1}cl(b)\Psi$.

(3.9)

Here are some observations about the preceding assertions:

- The number in (3.8) is either even or odd; its parity is the same as that of $1 - b^1 + b^{2+}$.
- When $b^{2+} > 0, \mu \in U$, and the integer in (3.8) is negative, the proposition asserts that $M = \emptyset$ since there are no negative dimensional manifolds.
- When $b^{2+} > 0$, $\mu \in U$ and (3.8) is zero, then M consists of a finite set of points. In this case, an orientation of M is an association of a sign to each point in M. (Note that a point has a canonical orientation since $H_0(point; \mathbb{Z})$ has a canonical generator.)
- When $b^{2+} > 0$, $\mu \in U$ and (3.8) is positive, let $\underline{c} \in X \times M$ denote the first Chern class of the principal S^1 bundle $\underline{M} \to X \times M$. Then slant product with $\underline{c}$ defines a canonical map, ϕ, from $H_*(X;\mathbb{Z})$ to $H^{2-*}(M;\mathbb{Z})$.

With regard to this map ϕ, note that the image under ϕ of a class $\gamma \in H^1(X;\mathbb{Z})$ has an alternate definition which goes as follows: Choose a map, $\underline{\gamma} : S^1 \to X$, which pushes forward the fundamental class of S^1 to give γ. The association to $c = (A, \Psi) \in \mathrm{Conn}(L) \times C^\infty(X; S_+)$ of the holonomy of $\underline{\gamma}^*A$ around S^1 defines a smooth map, $h_{\underline{\gamma}} : \mathrm{Conn}(L) \times C^\infty(X; S^1) \to S^1$. Then $\phi(\gamma)$ is the same class as the pull-back via h_γ of the generator of $H^1(S^1)$.

c) The Seiberg-Witten invariants

Here is the definition of the invariant SW:

Definition 3.2. Let X be a compact, conected, oriented 4-manifold with $b^{2+} > 0$. Fix an orientation for the line $\det^+$, and, in the case $b^{2+} = 1$, also fix an orientation of the line $H^{2+}(X;\mathbb{R})$. Fix a Riemannian metric on X and a Spin^C structure. Also, fix $\mu \in U$ in (3.7) to define M, but in the case when $b^{2+} = 1$, make the following additional requirement: Let ω denote a non-trivial, closed, self-dual 2-form whose direction provides the orientation for $H^{2+}(X;\mathbb{R})$. Now require that $r \equiv i\int_X \mu \wedge \omega$ be positive and very large. The value of

$$SW \in \mathbb{Z} \oplus H^1(X;\mathbb{Z}) \oplus \Lambda^2 H^1(X;\mathbb{Z}) \oplus \cdots$$

on the given Spin^C structure is computed using M as follows: Let d denote the integer in (3.8).

- If $d < 0$, then $SW = 0$.
- If $d = 0$, then M is a finite set of points and the chosen orientation for det^+ defines a map, ε, from M to $\{\pm 1\}$. With ε understood, then

$$SW = \sum_{c \in M} \varepsilon(c) \in Z. \tag{3.10}$$

- When $d > 0$, then SW has non-zero projection into $\Lambda^p H^1(X;\mathbb{Z})$ only if p has the same parity as $1 - b^1 + b^{2+}$. In this case, SW is defined by its values on the set of decomposable elements in $\Lambda^p(H_1(X;\mathbb{Z})/\mathrm{Torsion})$; and here, SW sends $\gamma_1 \wedge \cdots \wedge \gamma_p$ to

$$\int_M \phi(\gamma_1) \wedge \cdots \phi(\gamma_p) \wedge \phi(*)^{(d-p)/2}, \tag{3.11}$$

where $* \in H_0(X;\mathbb{Z})$ is the class of a point.

The apparent dependence of SW on the Riemannian metric and on μ is spurious as the next proposition asserts:

Proposition 3.3. *Let X be a compact, connected, oriented 4-manifold with $b^{2+} \geq 1$. Then the values of SW on the elements of s are independent of the choice of Riemannian metric and form μ. In fact, SW depends only on the diffeomorphism type of X in the sense that it pulls back naturally under orientation preserving diffeomorphisms. This is to say that if $\varphi : X \to X$ is a diffeomorphism, then φ pulls back the chosen orientation of det^+ (and of H^{2+} when $b^{2+} = 1$), it pulls back $\Lambda^* H^1(X;\mathbb{Z})$ and it pulls back* Spin^C *structures (because metrics pull back). With this understood, then $SW(\varphi * (\cdot)) = \varphi^*(SW(\cdot))$.*

See, e.g. [12] for a proof of this Proposition.

4. The Seiberg-Witten invariants on manifolds with $S^1 \times S^2$ boundaries

There is now a well developed theory of the Seiberg-Witten invariants for manifolds with boundary. Here are a few relevant references:[13], [14],

[15]. In principle, the general story is well understood, though the details may be quite complicated if the boundary is a complicated 3-manifold. Fortunately, the case where the boundary components are all $S^1 \times S^2$'s is fairly simple to describe, and this section provides a description of the salient features.

To begin, let X_0 now denote a connected, compact, oriented 4-manifold with boundary such that $\partial X_0 = \cup_{1 \leq j \leq N} S^1 \times S^2$. And, let S denote the set of equivalence classes of pairs $(\mathfrak{s}, \sigma)$ where $\mathfrak{s}$ is a Spin^C structure on X_0 whose associated line bundle L is trivial on ∂X_0, while σ is a class in $H^2(X_0, \partial X_0; \mathbb{Z})$ which maps to $c_1(L) \in H^2(X_0; \mathbb{Z})$.

After a choice of orientation for certain canonical lines, there is a well defined Seiberg-Witten invariant for X_0 which consitutes a map $SW : S \to \mathbb{Z} \oplus H^1(X_0; \mathbb{Z}) \oplus \Lambda^2 H^1(X_0; \mathbb{Z}) \oplus \cdots$. The definition of this map and a description of its properties occupies the various subsections of this Section 4.

By the way, the map SW which is defined below is very much like the Seiberg-Witten invariant defined in [13] for manifolds with boundary $S^1 \times \Sigma$, where Σ is a surface of genus greater than 1. Indeed, the arguments that justify assertions in this section are almost entirely slightly modified or simplified versions of arguments from [13]. Thus, the proofs of the various propositions and lemma to come will simply refer to the sections in [13] where analogous statements are proved, leaving it to the reader to make the necessary modifications.

a) Geometric preliminaries

To begin, consider X_0 as the complement of an open set in the non-compact manifold (without boundary) X, which is defined by identifying $\partial X_0 \subset X_0$ with $\partial X_0 \times \{0\} \subset \partial X_0 \times [0, \infty)$. Thus, $X \equiv X_0 \cup (\partial X_0 \times [0, \infty))$ is a manifold with 'tubular ends'.

Fix a metric with positive scalar curvature on ∂X_0 and then fix a metric on X which restricts to a neighborhood of $\partial X_0 \times [0, \infty)$ as the product of this standard metric on ∂X_0 and the Euclidean metric on the half line. With such a metric chosen, fix a pair $(\mathfrak{s}, \sigma) \in \mathfrak{s}$ and thus a lift, $F \to X$ of the frame bundle Fr. Use F to define (as in the compact case) the bundles $\Lambda_\pm$ of self and anti-self dual 2-forms, the complex $\mathbb{C}^2$ bundles $S_\pm$, and the principle S^1 bundle L.

Let $\mathrm{Conn}_e(L)$ now denote the space of connections, A, on the bundle L with the property that the norm of the associated curvature 2-form F_A has exponential decay on all ends of X and so that $(2\pi i)^{-1} F_A$ represents the class τ. Meanwhile, let $C_e^\infty(S_+)$ denote the space of sections of S_+ whose norms have exponential decay on all ends of X. To be explicit about this exponential decay condition, fix a function $s : X \to (-1, \infty)$

whose restriction to $\partial X_0 \times [0, \infty)$ is the projection onto the half line factor. Then a pair consisting of a connection A on L and a section Ψ of S_+ are in $\mathrm{Conn}_e(L) \times C_e^\infty(L)$ when there exists $\delta > 0$ such that

$$e^{\delta s}|F_A| + e^{\delta s}|\Psi| \tag{4.1}$$

is bounded on X, and when $(2\pi i)^{-1}F^A$ represents σ.

Note that $C_e^\infty(S_+)$ is a linear Frechet manifold where a neighborhood of zero is labeled by data $(\delta, \varepsilon, n, K)$ which consist of positive numbers δ and ε, a non-negative integer n, and a compact set $K \subset X$. The neighborhood labeled by this data consists of those $\Psi \in C_e^\infty(S_+)$ with the property the following two properties: First, all covariant derivatives of Ψ to order n are bounded by ε on K. Here, the covariant derivatives are defined by some hermitian connection on S_+ which is fixed in advance. Second, $e^{\delta s}|\Psi| < \epsilon$ on X.

Meanwhile $\mathrm{Conn}_e(L)$ is a fiber bundle over $\times_N S^1$ whose fiber is an affine Frechet manifold. With regard to this last point, remember that N denotes the number of components of ∂X_0, and with this understood, the k'th coordinate of this fibering map sends $A \in \mathrm{Conn}_e(L)$ to the limit as s tends to ∞ on the k'th end of X of the holonomy of A around any circle in $S^1 \times S^2$ which generates the latter's first homology. The boundedness of (4.1) for some $\delta > 0$ insures the existence of this limit.

By the way, as the curvature 2-form of each $A \in \mathrm{Conn}_e(L)$ has exponential decay along the ends of X, the 4-form $-(4\pi 2)^{-1}F_A \wedge F_A$ is integrable on X. Moreover, the value of the ensuing integral can be argued to be independent of the particular choice of A from $\mathrm{Conn}_e(L)$ and thus depends only on s. Indeed,

$$-(4\pi 2)^{-1} \int_X F_A \wedge F_A = c_1(L) \bullet c_1(L). \tag{4.2}$$

where the right hand side denotes the evaluation on the fundamental class in $H_4(X_0, \partial X_0; \mathbb{Z})$ of the cup product with itself of any lift of $c_1(L)$ to $H^2(X_0, \partial X_0; \mathbb{Z})$. (In particular, σ is such a lift.)

The point here is that the bilinear pairing, $\bullet$, on $\otimes_2 H^2(X_0, \partial X_0; \mathbb{Z})$ is symmetric, but not perfect and the kernel of this pairing is precisely the image of $H^1(\partial X_0)$ under the connecting homomorphism for the long exact cohomology sequence of the pair $(X_0, \partial X_0)$. Thus, $\bullet$ can be viewed, equivalently, as a non-degenerate pairing on the kernel of the restriction induced homomorphism from $H^2(X_0; \mathbb{Z})$ to $H^2(\partial X_0; \mathbb{Z})$. (The fact is that Poincaré duality provides only a perfect bilinear pairing from $H^2(X_0) \otimes H^2(X_0, \partial X_0)$ to $\mathbb{Z}$.) It is convenient to view the pairing $\bullet$ at times from one or the other of these view points.

In any event, for the time being, view $\bullet$ as a symmetric pairing on $H^2(X_0, \partial X_0; \mathbb{Z})$ and let $b^{2+}(X)$ denote the maximum of the dimensions of those subspaces $V \subset H^2(X_0, \partial X_0)$ on which it is positive definite.

b) Properties of the solutions

The Seiberg-Witten invariant for a fixed pair $(\mathfrak{s}, \sigma) \in S$ is computed via an appropriate count of the solutions $(A, \Psi) \in \text{Conn}_e(L) \times C_e^\infty(S_+)$ of the equation in (3.7) where μ is a fixed self-dual form on X whose norm has exponential decay on the ends of X. (That is, $e^{\delta s}|\mu|$ is bounded on X for some $\delta > 0$.) Thus, of prime interest is the set $m \subset \text{Conn}_e(L) \times C_e^\infty(S_+)$ of pairs (A, Ψ) which obey (3.7) on X. The lemma below lists some basic properties of elements $c \in m$.

Lemma 4.1. *Let m be as just described. Then, there exist constants $\kappa > 0$ and, for each $n \geq 0$, there exists $\zeta_n \geq 1$ with the following significance: Let $(A, \Psi) \in m$. Then*

- $e^{\kappa s}(|F_A| + |\Psi|) \leq \zeta_0$
- *For each $n \geq 1$, $e^{\kappa s}(|\nabla^n F_A| + |(\nabla_A)^n \Psi|) \leq \zeta_n$.*
- *On each end of X, the connection A has exponential decay to some flat connection in the following sense: There is a flat connection A_0 on L's restriction to the given end such that $a \equiv A - A_0$ obeys $e^{\kappa s}|\nabla^n a| \leq \zeta_n$ for all n.*

Proof of Lemma 4.1. The proof is obtained by modifying the arguments in Section 6.4 of [13]. Here, the fact that the scalar curvature of the metric on $S^1 \times S^2$ is positive plays the role played in[13] by the assumption that the solutions to the Seiberg-Witten equations on $S^1 \times \Sigma$ are non-degenerate (See Section 5 of [13]). (The positivity of the scalar curvature also implies that the only solutions to the unperturbed 3-manifold Seiberg-Witten equations for the same metric on $S^1 \times S^2$ have the form $(A, 0)$, where A is flat.) Note that the arguments in Section 6.4 of [13] give directly an exponential decay assertion in the L_1^2 Sobolev norm. However, standard elliptic estimates can be used to prove that there is exponential decay in the C_k norms for all k.

c) The moduli space

The group $C^\infty(X; S^1)$ acts on $\text{Conn}_e(L) \times C_e^\infty(L)$ and this action is smooth when $C^\infty(X; S^1)$ is viewed as a Frechet lie group using the C^∞ Frechet topology. Here, the open neighborhoods of the constant map 1 are indexed by triples (ε, n, K) where $\varepsilon > 0, n \in \{0, 1, \ldots\}$, and $K \subset X$ is compact; and the corresponding open set consists of those maps φ which obey $|\nabla^k \varphi| < \varepsilon$ on K for all $k \in \{0, 1, \ldots, n\}$. Give $m \subset \mathrm{Conn}_e(L) \times C_e^\infty(S_+)$ the subspace topology and then give $M \equiv m/C^\infty(X; S^1)$ the quotient topology.

As in the compact X case, it proves useful to introduce the somewhat larger space $\underline{M}$ which is the quotient of $X \times m$ by the equivalence relation $(x, c) \sim (x', c')$ when $x = x'$ and when $c = \varphi \cdot c'$ where $\varphi \in C^\infty(X; S^1)$ obeys $\varphi(x) = 1$. This space $\underline{M}$ admits an continuous action of S^1 and the quotient space is $X \times M$.

Key properties of M and $\underline{M}$ are described in Proposition 4.2, below. The statement of this proposition uses τ_X to denote the signature of the pairing $\bullet$ on $H^2(X_0, \partial X_0)/\,\mathrm{Image}(H^1(\partial X_0)$ and it uses $H^{2+}(X_0)$ to denote any choice of b^{2+} dimensional subspace of

$$H^2(X_0, \partial X_0)/\,\mathrm{Image}(H^1(\partial X_0)$$

to which $\bullet$ restricts as a positive definite pairing.

Proposition 4.2. *Fix a pair $(\mathfrak{s}, \sigma) \in S$, a Riemannian metric g and a self-dual 2-form μ which exponentially decays on the ends of X. Use this data to define the space M and $\underline{M}$. Then the following are true:*

- *M and $\underline{M}$ are compact. In fact, given the* Spin^C *structure $\mathfrak{s}$, there are only finitely many pairs $(\mathfrak{s}, \sigma') \in S$ with M non empty.*

- *Each irreducible $c \in M$ has a neighborhood which is homeomorphic to the zero set of a real analytic map between balls about the origin in finite dimensional Euclidean spaces. In particular, the domain ball and the range ball lie naturally in the respective kernel and cokernel of a Fredholm operator, δ_c, between separable Hilbert spaces. Here, the index of δ_c is equal to*

$$b^1 - 1 - b^{2+} + 4^{-1}(-\tau_X + c_1(L) \bullet c_1(L)).$$

- *In general, the subspace $M_{\mathrm{reg}} \subset M$ of irreducible orbits where* $\mathrm{cokernel}(\delta_c) = 0$ *is open in M and has the structure of a smooth, orientable manifold whose dimension is equal to the index of δ_c. Moreover, the inverse image in $\underline{M}$ of $X \times M_{\mathrm{reg}}$ has the structure of a smooth, principal S^1 bundle.*

- *Fix attention on an end of X. Then, the assignment to $c = (A, \Psi) \in M$ of the $s \to \infty$ limit of the holonomy of A around the loop $S^1 \times \{\text{point}\} \times \{s\}$ in the given end of X defines a continuous map from M to S^1 which is smooth on M_{reg}.*

- *Suppose that $b^{2+} > 0$. Then, with the metric on X fixed, there is a Baire subset U of choices for μ in (3.7) which have exponential decay on the ends of X and are such that the following hold:*
  - *a) M contains no reducible pairs.*
  - *b) $M = M_{\text{reg}}$.*
  - *c) Fix an end of X and the corresponding map from M to S^1 is generic in the sense that it has at most a finite number of critical points, and each is non-degenerate. Moreover, the critical values of this map can be assumed to miss any previously specified countable set in S^1.*
  - *d) The critical points for the maps to S^1 defined by distinct ends of X are distinct.*
- *An orientation of the line $\Lambda^{\text{top}}H^1(X;\mathbb{R}) \otimes \Lambda^{\text{top}}H^{2+}(X;\mathbb{R})$ canonically orients M_{reg}.*

Proof of Proposition 4.2. Except for the final point about orientations, all of the assertions can be proved by slightly modified versions of arguments from Section 8 of [13]. The final point can be proved by modifying arguments from Section 9.1 of [13].

Two key examples take $X_0 = S^1 \times \underline{B^3}$ and $X_0 = \underline{B^2} \times S^2$, where $\underline{B^p} \subset \mathbb{R}^p$ denotes the closed, unit radius ball centered at the origin. In the $S^1 \times \underline{B^3}$ case, take the metric to be one with non-negative scalar curvature which restricts to a product neighborhood of the boundary $S^1 \times S^2$ as the product of the Euclidean metric on the line with a metric having positive scalar curvature. With such a metric and with $\mu = 0$ in (3.7),

$$M = S^1. \tag{4.3}$$

In fact, M consists solely of reducible pairs $(A, 0)$, where A is pulled back from S^1. In particular, the map from M to S^1 given by the holonomy of A about $S^1 \times \{\text{point}\}$ provides the identification in (4.3).

In the $\underline{B^2} \times S^2$ case, take the product of the round metric on S^2 with a metric on $\underline{B^2}$ which has non-negative scalar curvature and restricts to a product neighborhood of the boundary circle as a flat, product metric. With such a metric and with $\mu = 0$ in (3.7),

$$M = \{\text{point}\}, \tag{4.4}$$

where the point in M is the reducible pair $(A, 0)$ with A a trivial connection.

d) The Seiberg-Witten invariants

Assume now that $b^{2+}(X_0) > 0$. Orient the line

$$\Lambda^{\text{top}} H^1(X_0;\mathbb{R}) \otimes \Lambda^{\text{top}}(H^{2+}(X_0;\mathbb{R});$$

and if $b^{2+} = 1$, also orient $H^{2+}(X_0;\mathbb{R})$. Having done so, the moduli space M can be used to define a map $SW : S \to \mathbb{Z} \oplus H^1(X_0;\mathbb{Z}) \oplus \cdots$ which is suitably invariant under diffeomorphisms of X_0. Indeed, Definition 3.2 translates verbatim to define SW in the present case. This is to say that the value of SW on a element $(\mathfrak{s}, \sigma) \in S$ is computed as follows: First, fix a suitable Riemannian metric on X and then fix μ from Proposition 4.2's set U to define M, but in the $b^{2+} = 1$ case, make the following added requirement on μ: Let ω denote a non-trivial, closed, compactly supported 2-form whose class in $H^2(X_0, \partial X_0;\mathbb{R})$ class has positive square and defines the given orientation for $H^{2+}(X_0)$. Then, require that $r \equiv i\int_X \mu \wedge \omega$ be positive and very large. Note that resulting M is a compact, oriented, smooth manifold of dimension

$$d \equiv \dim M = b^1 - 1 - b^{2+} + 4^{-1}(-\tau_X + c_1(L) \bullet c_1(L)).$$

With this last point understood, set $SW(\mathfrak{s}, \sigma) = 0$ when $d < 0$ as $M = \emptyset$ in this case. In the case when $d = 0$, then M is a finite set of signed points and $SW(\mathfrak{s}, \sigma)$ is the integer which is obtained by summing the signs which are associated to the points of M. Finally, when $d > 0$, then the component of $SW(\mathfrak{s}, \sigma)$ in $\Lambda^p H^1(X; Z)$ is defined by the condition that it send the decomposable element

$$\gamma_1 \wedge \cdots \wedge \gamma_p \in \Lambda^p(H_1(X;\mathbb{Z})/\operatorname{Torsion})$$

to the value of the expression in (3.1).

As just defined, SW has the following properties:

Proposition 4.3. *The values of SW as just defined are independent of the choice of Riemannian metric and form μ subject to the aforementioned constraints on $[0,\infty) \times \partial X_0$. Infact, SW depends only on the diffeomorphism type of X_0 as a manifold with boundary in the sense that it pulls back naturally under orientation preserving diffeomorphisms of the pair $(X_0, \partial X_0)$.*

Proof of Proposition 4.3. The arguments in Section 9.2 of [13] carry over directly. By the way, these arguments do not require a lemma to the effect that the space of oriented diffeomorphisms of $S^1 \times S^2$ is path connected, nor does it require a lemma to the effect that the space of metrics on $S^1 \times S^2$ with positive scalar curvature is path connected. An argument for Proposition 4.3 can be made given only that the space of all metrics on $S^1 \times S^2$ is path connected.

The latter argument is made along the following lines: First, establish a restricted version of Proposition 4.3 which limits the diffeomorphisms under consideration to those whose boundary restriction is isotopic to the identity, and which limits metric variation to that which changes the boundary metric along a path in the Frechet space of positive scalar curvature metrics. Then, establish an appropriate analog of the product formula in Theorem 9.5 of [13] for the resulting restricted Seiberg-Witten invariants. In fact, the analog of Theorem 9.5 from [13] for the restricted Seiberg-Witten invariants can be phrased to read like a modified version of Proposition 4.5, below; with the major modification occuring in the assumptions about $\underline{X}$. In particular, the modified Proposition 4.5 takes $\underline{X}$ diffeomorphic to X_{0+} and X_{0-} diffeomorphic to $[-1,1] \times \partial X_{0+}$. In any event, the analog of Theorem 9.5 from [13] will imply the full invariance of SW as stated in Proposition 4.3.

e) The invariant for X_0 and for compact 4-manifolds

Let $\underline{X}$ now denote a compact, connected oriented, smooth 4-manifold with $b^{2+} \geq 1$, and suppose that φ is an embedding of the disjoint union, Y, of some $N \geq 1$ copies of $(S^1 \times S^2)$ into $\underline{X}$, each of which separates $\underline{X}$. Then, $\underline{X}$ can be written as $\underline{X} \equiv X_{0-} \cup_Y X_{0+}$, where $X_{0\pm}$ are compact, oriented manifolds with boundary Y. In this case, the invariant SW for X can be computed in terms of that for X_{0-} and X_{0+}. The story in the general case is somewhat outside the scope of this paper. However, there are three special cases where the story is quite simple:

- $b^{2+} > 0$ for both $X_{0\pm}$.
- $b^{2+} > 0$ for X_{0+} while $X_{0-} \subset \underline{X}$ is the closure in $\underline{X}$ of a tubular neighborhood of the disjoint union of embedded circles and 2-spheres with self-intersection number zero. Furthermore, at least one of these 2-spheres gives a non-zero class in $H_2(\underline{X};\mathbb{R})$.
- $b^{2+} > 0$ for X_{0+} while $X_{0-} \subset \underline{X}$ is the closure in $\underline{X}$ of a tubular neighborhood of the disjoint union of embedded circles and 2-spheres, where the latter are all inessential in the real, second homology of $\underline{X}$.

(4.5)

The story for the first two cases in (4.5) is simply stated as follows:

Proposition 4.4. *Let $\underline{X}$ be as described above and suppose either the first or the second case in (4.5) holds. Then $SW_{\underline{X}} \equiv 0$.*

To describe $SW_{\underline{X}}$ in the third case of (4.5), digress first to introduce the set $\mathcal{Y}$ of $B^2 \times S^2$ components of X_{0-}. Then, note that the boundary of each $Y \in \mathcal{Y}$ is also a component of the boundary of X_{0+} and thus picks out a distinguished homology class, $\gamma_Y \in H_1(X_{0+};\mathbb{Z})$. (This class is non-zero in $H_1(X_0;\mathbb{Z})/\,\text{Torsion}$ because the core S^2 in Y is non-zero on $H_2(\underline{X};\mathbb{Z})/\,\text{torsion}$.) Digress again to note that the inclusion induced map $\iota : H_1(X_{0+};\mathbb{Z}) \to H_1(\underline{X};\mathbb{Z})$ is surjective.

With these last points understood, consider:

Proposition 4.5. *Suppose that $\underline{X}$ is a compact, oriented 4-manifold with $b^{2+} > 1$ which decompose as $X_{0-} \cup_Y X_0$ where $b^{2+}(X_0) > 0$ and where X_{0-} is a tubular neighborhood of a disjoint union of embedded circles and 2-spheres with zero self-intersection number. Let $\mathcal{Y}$ denote the set of components of X_{0-} which contain these 2-spheres. If $\mathcal{Y} = \emptyset$, set $\kappa \equiv 1$, and otherwise, order the components of $\mathcal{Y}$ and set $\kappa \wedge_{Y\in\mathcal{Y}} \gamma_Y$ where the order of the terms in this exterior product conforms to the chosen ordering of $\mathcal{Y}$. Then,*

$$SW_{\underline{X}}(\iota(\gamma_1) \wedge \cdots \wedge \iota(\gamma_p)) = \pm SW_{X_{0+}}(\kappa \wedge \cdots \wedge \gamma_p),$$

where the $\pm$ here is independent of $\{\gamma_j\}_{1\le j\le p}$.

Proof of Propositions 4.4 and 4.5. The proof of Theorem 9.1 of [13] can be modified in the present context to provide a natural description of the extended moduli space $\underline{M}_{\underline{X}}$ for a suitable metric on $\underline{X}$ in terms of the corresponding moduli spaces $\underline{M}_{\pm}$ for $X_{0\pm}$. In order to write this formula, let $\underline{M}_{\underline{X}}$ denote the extended moduli space $\underline{M}$ for $\underline{X}$. Recalling that the latter comes with a canonical projection to $\underline{X}$, let $\underline{M}_{\underline{X}}|_Y \subset \underline{M}_{\underline{X}}$ denote the subset which lies over Y. Define $\underline{M}_{\pm}|_Y$ analogously. Moreover, use $(\underline{M}_-|_Y \times \underline{M}_+|_Y)|_\Delta \subset \underline{M}_-|_Y \times \underline{M}_+|_Y$ to denote the subspace which lies over the diagonal, $\Delta \subset Y \times Y$. With these definitions understood, the analog here of Theorem 9.1 from [13] asserts that there are certain metrics on $\underline{X}$ for which $\underline{M}_{\underline{X}}|_Y$ has a natural, S^1-equivariant identification with $(\underline{M}_-|_Y \times \underline{M}_+|_Y)|_\Delta$. Here, S^1 acts on the product via the diagonal action. (The metrics in question admit, for some large constant T, an isometric embedding of $[-T,T]\times Y$ into $\underline{X}$ which map $\{0\}\times Y$ to $Y \subset \underline{X}$ via the identity map. Here, $[-T,T]\times Y$ has a metric which is the product of a Euclidean metric on $[-T,T]$ and a positive scalar curvature metric on Y.) The assertions of both propositions follow with a little algebraic topology from this picture of $\underline{M}_{\underline{X}}|_Y$.

f) An important corollary

Proposition 4.6, below, summarizes the features of the map SW which are relevant for the subsequent discussion of the assertions in (1.1).

Note that this proposition is an immediate corollary to Propositions 4.3 and 4.5.

Proposition 4.6. *Let X_0 be as described in Section 4a, and suppose that $(\mathfrak{s}, \sigma) \in S$ has a non-zero value of SW. Let μ be any self-dual 2-form on X with exponential decay on $[0, \infty) \times \partial X_0$. Then, there exists at least one solution to the $((\mathfrak{s}, \sigma), \mu)$ version of (3.7). In particular, if $\underline{X}$ and X_0 are as described in Proposition 4.5 and $\underline{X}$ has non-trivial Seiberg-Witten invariant, then so does X_0 and so there exists $(\mathfrak{s}, \sigma) \in S$ such that the $((\mathfrak{s}, \sigma), \mu)$ version of (3.7) has at least one solution for every choice of exponentially decaying, self-dual 2-form μ.*

5. SW-admissable HWZ subvarieties and the Sieberg-Witten invariants

This section returns to the milieu of Sections 2f and 2g. In particular, suppose that the manifold with boundary X_0 and the corresponding manifold X with its symplectic form ω are as described in the aformentioned parts of Section 2. Likewise, endow X with an ω-compatible almost complex structure, J, which restirct to the components of $[0, \infty) \times \partial X_0$ as follows: If a component of ∂X_0 has orientable z-axis line bundle, take coordinates on this copy of $S^1 \times S^2$ so that the contact form is given by (2.11). Then, use (2.22) to define J. If the component in question does not have an orientable z-axis line bundle, take coordinates on the non-trivial 2-fold cover so that the pull-back of the contact form is given by (2.11). Then, take J so that its pull-back to this same double cover is given by (2.22) in these same coordinates.

Note that J defines the Riemannian metric $g = \sqrt{2}\omega(\cdot, J(\cdot))/|\omega|$, and with respect to this metric, ω is self-dual and J is an orthogonal transformation. Moreover, the metric on $[0, \infty) \times \partial X_0$ is given by the second line in (2.21) in the coordinates on a component or its double cover where J is given by (2.22) and α by (2.11).

The metric here will be used to define the Seiberg-Witten equations on X; thus it plays a role in the definition of SW for X_0. Meanwhile, the almost complex structure will be used, as in Section 2g, to define the SW-admissable subvarieties. The task for this section is to point out a relationship between the map SW on the one hand, and SW-admissable subvarieties on the other.

a) SW-admissable subvarieties and the set S

The relationship between the Seiberg-Witten invariants and SW-admissable subvarieties begins with the fact that a Spin^C structure on

X_0 of the sort which arises in the definition of SW can be canonically assigned to each SW-admissable, generalized HWZ subvariety:

Proposition 5.1. *An SW-admissable, generalized HWZ subvariety $c = \{(C_\alpha, m_\alpha)\}$ canonically defines a* Spin^C *structure $\mathfrak{s}_c$ with associated line bundle L that is trivial over ∂X_0. Moreover, an SW-admissable, generalized HWZ subvariety of the form $(C, 1)$ which satisfies the four points in Lemma 2.9 canonically defines a pair, $(\mathfrak{s}_C, \sigma_C) \in S$.*

The remainder of this section is occupied with the

Proof of Proposition 5.1. To start, fix a Spin^C structure on X_0 whose corresponding line bundle L is trivial on ∂X_0. Introduce the endomorphism cl_+ in (3.5); then $cl_+(\omega)$ is a skew hermitian endomorphism whose eigenvalues are $\pm i \cdot \sqrt{2}|\omega|$. These eigenspaces of $cl_+(\omega)$ decompose S_+ as $L_+ \oplus L_-$, a direct sum of complex line bundles where, $cl_+(\omega)$ acts as $i \cdot \sqrt{2}|\omega|$ on L_+. Note that the line bundle L_+ is non-trivial over any component of ∂X_0; its first Chern class is the oriented generator of the second cohomology of each component of ∂X_0. (The orientation of this cohomology is describe in Lemma 2.7 and the discussion in the final paragraph of Section 2d.)

Meanwhile, duality identifies $H_2(X_0, \partial X_0; \mathbb{Z})$ with $H^2(X_0; \mathbb{Z})$ and so an SW-admissable, generalized HWZ subvariety c defines a canonical element, $e_c \in H^2(X_0; \mathbb{Z})$ which is the Poincaré dual to $\sum_{(C,m)\in c} m[C]$. And, according to Definition 2.8, this class must restrict to each component of ∂X_0 as the oriented generator of the second cohomology of the component in question. Thus, e_c and $c_1(L_+)$ differ by an element in $H^2(X_0, \partial X_0; \mathbb{Z})$ and so one obtains

Lemma 5.2. *Let c denote an SW-admissable, HWZ subvariety. Then, there exists a unique* Spin^C *structure, $\mathfrak{s}_c$, over X_0 with the following properties:*

- *The associated line bundle L is trivial over ∂X_0.*
- $c_1(L_+) = e_c$.

The next task is to find the class $\sigma_C \in H^2(X_0, \partial X_0; \mathbb{Z})$ so that $(\mathfrak{s}_C, \sigma_C)$ are in S under the assumption that C satisfies the four points in Lemma 2.9. For this purpose, introduce the line bundle

$$K \equiv \mathrm{Hom}(L_-, L_+).$$

Now there are four claims to be made about K:

Lemma 5.3. *Define the line bundle K as above. Then K has the following properties:*

- *K admits a canonical section, κ, over ∂X_0.*

- *The zero set of κ is canonically homologous to twice $C \cap \partial X_0$.*

- *The homology from the preceding point with any choice of an extension of κ to a section of K over X_0 can be used with C to canonically define a closed, oriented subvariety in X_0.*

- *The class in $H_2(X_0, \mathbb{Z})$ of the subvariety just mentioned is independent of the extension of κ off of ∂X_0, and its dual maps to $c_1(L)$ in $H^2(X_0; \mathbb{Z})$.*

Accept this lemma on faith for the moment to finish the definition of Proposition 5.1's class σ_C: Take σ_C to be the dual class to the fundamental class in $H_2(X_0; \mathbb{Z})$ of the subvariety from Lemma 5.3's fourth point. Then, the pair $(\mathfrak{s}_C, \sigma_C) \in S$ because σ_c is constructed to be the first Chern class of the complex line bundle $L_+^2 K^{-1}$ and this line bundle is $L = \det(S_+)$.

With the preceding understood, then the proof of Proposition 5.1 is completed with the

Proof of Lemma 5.3. To start, it is important to realize that the $\mathbb{R}^2$ bundle underlying K is isomorphic to the orthogonal complement in Λ_+ to ω. This is because the restriction to this complement of cl_+ produces purely off diagonal endomorphisms of S_+ with respect to the decomposition of S_+ as $L_+ \oplus L_-$. Thus, studying K means studying the orthgonal complement of ω in Λ_+; and this view of K will be used to prove the statements in Lemma 5.3.

To prove the first two statements of the lemma, it proves useful to distinguish in the discussion those components of ∂X_0 with oriented z-axis line bundle from those without. Consider first the case where a component $Y \subset \partial X_0$ has orientable z-axis line bundle. In this case, there are coordinates on Y where α is given by (2.11) and then ω on $[0, \infty) \times Y$ is given by the first line in (2.21). In particular, where $0 < \theta < \pi$, the line bundle K is spanned by the forms

- $dt \wedge \sin^2\theta d\varphi - g^{-2} df \wedge dh$.

- $dt \wedge dh - \sin^2\theta d\varphi \wedge df$.

(5.1)

Take the first form above for the section κ of K. With κ understood, note that $\kappa^{-1}(0)$ on ∂X_0 is the set $S^1 \times \{0, \pi\}$; it is left to the reader to check that the orientations are such that $\kappa^{-1}(0)$ is twice the generator of $H^1(S^1 \times S^2; \mathbb{Z})$. These last observations justify the first two points of Lemma 5.3 for Y.

To obtain the homology in question, use the points in Lemma 2.9 to conclude that when s is large, then C's intersection with $\{s\} \times Y \subset [0, \infty) \times Y$ is a very small distance push-off of the parameterized loop given by (2.17) where the the left hand side of (2.18) is an integer and has positive denominator. (Note that the natural orientations are opposite, however.) Thus, for large s, there is a canonical homotopy between C's intersection with $\{s\} \times Y$ and the orientation reversed version of the loop in (2.17). Meanwhile, the orientation reversed version of this same loop from (2.17) is canonically homotopic to $S^1 \times \{\theta = 0\}$: Indeed, simply decrease the θ coordinate in (2.17) from its given value of θ_0 to 0. Likewise, this loop is canonically homotopic to $S^1 \times \{\theta = \pi\}$ via the homotopy which increases the θ coordinate from θ_0 to π.

Now consider the case where the given component $Y \subset \partial X_0$ has unoriented z-axis line bundle. In this case, there are coordinates on the non-trivial double cover of Y where α pulls back to give the form in (2.11) and ω pulls back to give the form in the top line of (2.21). In particular, where $0 < \theta < \pi$ the pull-back of K is spanned by the forms in (5.1). Save these last observation.

To exploit the preceding, introduce

$$(x_1 = \sin\theta\cos\varphi, x_2 = \sin\theta\sin\varphi, x_3 = \cos\theta)$$

on the non-trivial double cover of Y and use them to define a set of functions (t', x_1', x_2', x_3') on Y itself via (2.28). It then follows from (2.28) and the remarks in the preceding paragraph that the Poincaré dual to the first Chern class of K's restriction to Y is the parameterized curve

$$\tau \to (t' = 2\tau, x_1' = 0, x_2' = \sin(\tau), x_3' = \cos(\tau)). \tag{5.2}$$

Meanwhile, it follows from the final point of Lemma 2.9 that when s is large, the intersection of C with $\{s\} \times Y$ is a very small distance push-off of the circle where $x_1' = 1$. Thus, this intersection is canonically homotopic to the $x_1' = 1$ circle. Meanwhile, the circle in (5.2) is canonically homotopic to twice the $x_1' = 1$ circle. Indeed, consider the homotopy which replaces the right side of (5.2) which sends $(\tau, r) \in S^1 \times [0, 1]$ to $(t' = 2\tau, x_1' = (1 - r^2)^{1/2}, x2' = r\sin(\tau), x_3' = r\cos(\tau))$.

With the first two points of Lemma 5.3 understood, the third point of the lemma follows now from the preceding discussion in an absolutely straightforward manner.

The fourth point of Lemma 5.3 immediately from the first three points.

b) SW-admissable HWZ subvarieties and the Seiberg-Witten invariants

Take X_0 as in the previous subsection and reintroduce the Seiberg-Witten invariant for X_0, $SW : S \to Z \oplus H^1(X_0;\mathbb{Z}) \oplus \Lambda^2 H^1(X_0;\mathbb{Z}) \oplus \cdots$. The following theorem is the key result in this article:

Theorem 5.4. *Let $(\mathfrak{s}, \sigma) \in S$ be a class on which SW is non-zero. Then there exists an SW-admissable, generalized HWZ subvariety c with $\mathfrak{s}_c = \mathfrak{s}$.*

Note that there is no assertion here that about the existence of an SW-admissable, HWZ subvariety C which obeys the four points in Lemma 2.9, and has both $\mathfrak{s}_C = \mathfrak{s}$ and $\sigma_C = \sigma$. Indeed, it is not clear that this must be the case even with reasonable, additional assumptions.

Theorem 5.4 is a generalization of the main theorem in [17], [18] which gives the $\partial X_0 = \emptyset$ version. Theorem 5.4 should also be compared with Theorem 1.2 in [16] which proves an analgous theorem in the more restrictive context of Riemannian pseudoholomorphic geometry. Moreover, the proof of Theorem 5.4 here borrows heavily from that of Theorem 1.2 in [16], while the latter follows many of the lines of the main theorem in [17], [18]. On the other hand, Theorem 1.2 in [16] can be viewed as a corollary to the Theorem 5.4 and Proposition 4.5. Conversely, a special case of Theorem 5.4 can be deduced from Theorem 1.2 in [16].

c) A generalization of Theorem 5.4

With a metric on X as described at the beginning of this section, fix $r \geq 1$ and consider the version of (3.7) where $\mu = 2^{-1} r\omega$. If Ψ in this version (3.7) is replaced by $\sqrt{r}\Psi$, then (3.7) reads

- $D_A\Psi \equiv \underline{cl}(\nabla_A\Psi) = 0,$
- $F_A^+ = r(cl^+(\Psi \otimes \Psi^\dagger) - i2^{-1}\omega).$

(5.3)

Proposition 4.6 implies that the $((\mathfrak{s}, \sigma), r)$ version of (5.3) has a solution for every $r \geq 1$ when $SW(\mathfrak{s}, \sigma) \neq 0$. Thus, Theorem 5.4 is a corollary of:

Theorem 5.5. *Fix $(\mathfrak{s}, \sigma) \in S$ and suppose that there exists an unbounded, increasing sequence $\{r_n\} \in (0, \infty)$ with the property that for each index n, the $r = r_n$ version of (5.3) has a solution, $(A_n, \Psi_n) \in \mathrm{Conn}_e(L) \times C_e^\infty(S_+)$. Then there exists an SW-admissable, generalized HWZ subvariety c with $\mathfrak{s}_C = \mathfrak{s}$. In addition, there is a subsequence* *of* $\{(A_n, \Psi_n)\}$ *(hence relabled by consecutive integers) with the following property: For each* n, *let* α_n *denote the orthogonal projection of* Ψ_n *into eigenspace of* $cl_+(\omega)$ *in* S_+ *with eigenvalue* $i \cdot \sqrt{2}|\omega|$. *Introduce the HWZ subvariety* $C' \equiv \cup_{(C,m)\in c} C$ *Let* $Q \subset X$ *be any compact set and then*

$$\lim_{n\to\infty} [\sup_{x\in C'\cap Q} \operatorname{dist}(x, \alpha_n^{-1}(0)) + \sup_{x\in\alpha_n^{-1}(0)\cap Q} \operatorname{dist}(C', x)] \tag{5.4}$$

exists and equals zero. Finally, there is a constant $\wp$ *which depends solely on the symplectic form and the Riemannian metric of* X_0, *and is such that*

$$\sum_{(C,m)\in c} m \int_C \omega = 2^{-1}[\omega] \bullet c_1(L) + \wp \tag{5.5}$$

6. Estimates for the proof of Theorem 5.5

The argument for Theorem 5.5 is begun in this section and completed in the next. However, before starting, take note of the fact that the argument presented here is a modified version of the proof in [16] of Theorem 2.2 in [16]. (The latter asserts an analog of Theorem 5.5's existence result in the context of Riemannian pseudoholomorphic geometry which plays the role of Theorem 5.5 here.) The proof of Theorem 2.2 in [16] and that given below of Theorem 5.5 can be viewed as having three distinct parts. The first part derives global bounds for various measures of Ψ and F_A. The second part uses the global bounds to obtain stronger estimates on compact domains. Note that these first two parts are more conceptually distinct than chronologically distinct. In any event, the first two parts of the proof occupy Section 6. The final part of the proof occupies Section 7. In the third part of the proof, the bounds on compact subsets of X from this section are used in conjunction with various arguments from [17], [18] to complete the proof of Theorem 5.5. In this regard, note that [17], [18] proves Theorem 5.5 in the case where $\partial X_0 = \emptyset$.

As indicated, the discussion here follows closely the proof of Theorem 2.2 in [16], and so referrals to [16] are frequent.

In the remainder of Section 6 and in Section 7, the implicit assumption is that a pair $(\mathfrak{s}, \sigma) \in S$ has been fixed, that r is large and that (A, Ψ) is a solution to the $((s, \sigma), r)$ version of (5.3).

a) Integral bounds for $|\Psi|^2$.

As might be expected from the title, the purpose of this subsection is to obtain integral bounds for $|\Psi|^2$. The statement of these bounds requires a brief digression to introduce some notation. To start the digression, introduce the characteristic number

$$\varepsilon_\omega(s) \equiv [\omega] \bullet \sigma. \tag{6.1}$$

Here, $[\omega]$ is the class of the symplectic form in $H^2(X_0, \partial X_0)$. In this regard, remember that ω is exponentially decaying along $[0, \infty) \times \partial X_0$ and so canonically defines a class in $H^2(X_0, \partial X_0)$. Also, note that the right hand side of (6.3), though written in terms of $\sigma \in H^1(X_0, \partial X_0)$, depends only on σ's image, $c_1(L)$, in $H^2(X_0)$.

To continue the digression, let g denote the chosen Riemannian metric and let R_g denote g's scalar curvature function and W_g^+ the self-dual part of the Weyl curvature. Then, let R_{g-} denote the minimum of zero and R_g. Thus, R_{g-} has compact support in X_0. Finally, let dvolg denote the volume form of the metric g.

With the digression now over, consider:

Lemma 6.1. *There is a universal constant c with the following significance: Let $(\mathfrak{s}, \sigma) \in S$ be given. Now, suppose that (A, Ψ) solves (5.3) for the given $(\mathfrak{s}, \sigma)$ and for some choice of $r \geq 1$. Then*

- $\int_X (2^{-1/2}|\omega| - |\Psi|^2)^2 d\,\mathrm{vol}_g \leq c r^{-1}(\varepsilon_\omega(s)$ $+ \int_X (|R_g| + r^{-1}|R_g|^2)|\omega| d\,\mathrm{vol}_g$.
- $\int_X |2^{-1/2}|\omega| - |\Psi|^2| d\,\mathrm{vol}_g \leq c r^{-1}(\varepsilon_\omega(s)$ $+ \int_X (|R_g| + |W_g^+|^2)|\omega| + r^{-1}|R_{g-}|) d\,\mathrm{vol}_g$.

Proof of Lemma 6.1. The argument here is almost an exact copy of that which proves Lemma 3.1 in [16]. The only difference occurs in the modification of a particular term which appears in a differential equation for $|\Psi|^2$, the latter being implied by the Weitzenboch formula which writes $D_A^\dagger D_A$ in terms of the Laplacian $\nabla_A^\dagger \nabla_A$. To be precise, note that the Weitzenboch formula used in the proof of Lemma 3.1 of [16] implies that

$$2^{-1} d^* d|\Psi|^2 + |\nabla_A \Psi|^2 + 4^{-1} r|\Psi|^2(|\Psi|^2 - 2^{-1/2}|\omega| + r^{-1} R_g) \leq 0, \tag{6.2}$$

which is the same equation as (3.4) from [16]. Now, the point is that this last inequality holds with R_g replaced by R_{g-} :

$$2^{-1} d^* d|\Psi|^2 + |\nabla_A \Psi|^2 + 4^{-1} r|\Psi|^2(|\Psi|^2 - 2^{-1/2}|\omega| + r^{-1} R_{g-}) \leq 0, \tag{6.3}$$

With (6.3) understood, the subsequent arguments for Lemma 3.1 in [16] can be imported verbatim to prove Lemma 6.1 here. (Note that

R_g can be replaced by R_{g-} in the corresponding Equation (3.4) from [16], but in the latter equation, the distinction is superfluous because [16] considers these equations on compact manifolds.)

b) Pointwise bounds for $|\Psi|^2$

The purpose of this subsection is to obtain pointwise bounds for $|\Psi|^2$. These bounds come via (6.3) with the help of the maximum principle. In particular, since both $|\Psi|$ and $|\omega|$ decay to zero (exponentially fast) as the parameter $s \in [0,\infty)$ gets large on $[0,\infty) \times \partial X_0$, and since $|\omega| \leq \sqrt{2}$, the maximum principle with (6.3) immediately gives the bound

$$|\Psi|^2 \leq 1 + r^{-1}|R_{g-}|. \tag{6.4}$$

The following lemma gives some fine structure to the pointwise behavior of $|\Psi|^2$:

Lemma 6.2. *There is a consant ξ which depends only on the Riemannian metric and which has the following significance: Given $(\mathfrak{s},\sigma) \in S$ and $r \geq 1$, let (A,Ψ) be a solution to the corresponding version of (6.1). Let $\underline{s} \equiv 6^{-1/2}\ln r$. Then,*

- $|\Psi|^2 \leq 2^{-1/2}|\omega| + \xi r^{-1}$ *where $s \leq \underline{s}$.*
- $|\Psi|^2 \leq \zeta r^{-1} e^{-(s-\underline{s})/\sqrt{2}}$ *where $s \geq \underline{s}$.*

(6.5)

The remainder of this subsection contains the

Proof of Lemma 6.2. First, remember that ω is both self-dual and closed, and so $(d^*d\omega)^+ = 0$. The Bochner-Weitzenboch formula for this last equation (see, e.g. Appendix C in [3]) implies that

$$d^*d|\omega| + |\omega|^{-1}|\nabla\omega|^2 \geq -k^+|\omega|, \tag{6.6}$$

where k^+ has compact support on X_0 and is bounded by a universal multiple of $|R_g|$ and $|W_g^+|$. With this last equation understood, introduce $u \equiv |\Psi|^2 - 2^{-1/2}|\omega|$ and then (6.3) and (6.6) together imply that

$$2^{-1}d^*du + (4\sqrt{2})^{-1}r|\omega|u \leq \zeta_1 e^{-\sqrt{6}s}, \tag{6.7}$$

where ζ_1 is a constant which depends only on the Riemannian metric. Here, s has been extended as a smooth function to the whole of X from is original domain of definition, $[0,\infty) \times X_0$. (This extension of the domain of s will be implicit in the subsequent appearances of the function s.)

By the way, the derivation of (6.7) uses (6.4) and the fact that $|\omega|$ and $|\nabla\omega|$ obey bounds on X of the form

- $|\omega| \geq m^{-1}e^{-\sqrt{6}s}$,
- $|\omega| + |\nabla\omega| \leq me^{-\sqrt{6}s}$,

(6.8)

where $m \geq 1$ is a constant.

The next step in the proof of Lemma 6.2, starts with the following observation: Equation (6.7) implies that there exists a constant ξ which depends solely on the Riemannian metric on X and is such that $\underline{u} \equiv u - \xi r^{-1}$ obeys the differential inequality $2^{-1}d^*d\underline{u} + (4\sqrt{2})^{-1}r|\omega|\underline{u} \leq 0$. Since u is negative where s is very large on X, the maximum principle can be invoked with this last equation to prove that $\underline{u} \leq 0$ everywhere on X. That is,

$$|\Psi|^2 \leq 2^{-1/2}|\omega| + \zeta r^{-1}. \tag{6.9}$$

This last bound gives the first point in (6.5).

To obtain the second point, first note that (6.9) and (6.2) together imply that $|\Psi|^2$ obeys

$$d^*d|\Psi|^2 + 2^{-1}|\Psi|^2 \leq \zeta(e^{-\sqrt{6}s} + r^{-1})e^{-\sqrt{6}s}. \tag{6.10}$$

where $s \geq 0$. Here, ζ is a constant which depends only on the Riemannian metric. At the same time, $|\Psi|^2 \leq \zeta r^{-1}$ where $s \geq \underline{s}$, and so the comparison principle can be invoked for (6.10) to establish that

$$|\Psi|^2 \leq \zeta r^{-1}e^{-(s-\underline{s})/\sqrt{2}} \tag{6.11}$$

where $s \geq \underline{s}$. This is the second point in (6.7).

c) Writing $\Psi = (\alpha, \beta)$ and estimates for $|\beta|^2$.

As in the proof of Theorem 2.2 of [16], the next step Theorem 5.5's proof requires the introduction of the components (α, β) of Ψ as follows:

- $\alpha \equiv 2^{-1}(1 + i(\sqrt{2}|\omega|)^{-1}cl_+(\omega))\Psi$,
- $\beta \equiv 2^{-1}(1 - i(\sqrt{2}|\omega|)^{-1}cl_+(\omega))\Psi$.

(6.12)

The claim now is that $|\beta|$ is uniformly small over X. Here is the precise statement:

Proposition 6.3. *There are constant $\xi_1, \xi_2 \geq 1$ which depends only on the Riemannian metric chosen for X and which have the following significance: Let (A, Ψ) be a solution to (5.3) as defined by the chosen* *pair* $(\mathfrak{s},\sigma) \in S$ *and* $r > (\xi_2)^4$. *Let* $R \in (r/\zeta_1, \zeta_1)$. *Then, the* β *component of* Ψ *obeys*

$$|\beta|^2 \leq \xi_2 R^{-1}(2^{-1/2}|\omega| - |\alpha|^2) + \xi_1 R^{-2} \tag{6.13}$$

where $s \leq \underline{s} - 6^{-1/2}\ln R$. *(This is where* $re^{-\sqrt{6}s} \geq R$.*)*

The remainder of this subsection is occupied with the

Proof of Proposition 6.3. Modulo some notational changes, the proof of Proposition 3.3 in [16] proves this proposition. Indeed, the arguments in the latter proof can be followed with minor notational changes to establish the existence of constants ζ_3, ζ_4 which are independent of the data $r, R, (\mathfrak{s},\sigma)$, and (A, Ψ), and which have the following significance: Set $w \equiv (2^{-1/2}|\omega| - |\alpha|^2)$ and then let $u \equiv |\beta|^2 - \zeta_1 R^{-1} w - \zeta_2 R^{-2}$. Also, let u_+ denote the maximum of u and 0. Note that u_+ may only be Lipschitz where it is zero. In any event, where $re^{-\sqrt{6}(s-1)} \geq R$, this u_+, viewed as a distribution, obeys the differential inequality

$$d^* du_+ + \xi^{-1} R u_+ \leq 0. \tag{6.14}$$

where $\xi \geq 1$ is independent of $R, r, (\mathfrak{s},\sigma)$ and (A, Ψ). (Note that (6.8) was used to derive (6.14).)

Meanwhile, where $re^{-\sqrt{6}(s-1)} = R$, the first point in (6.5) and (6.8) imply that $u_+ \leq \zeta_3 r^{-1} R$ with $\zeta_3 \geq 1$ a constant which is independent of $r, R, (\mathfrak{s},\sigma)$ and (A, Ψ). Given this last observation and (6.14), the comparison principle implies that

$$u_+ \leq r^{-1} R \zeta_4 \exp(-\sqrt{R}(\sqrt{6}\ln(r/R) + 1 - s)/\sqrt{\xi}), \tag{6.15}$$

where $re^{-\sqrt{6}s} \geq R$ and where $\zeta_4 \geq 1$ is independent of $r, R, (\mathfrak{s},\sigma)$ and (A, Ψ). This last inequality implies the lemma since $r^{-1}Re^{-\sqrt{R}/\sqrt{\xi}} \leq \zeta R^{-2}$ where ζ can be taken to be independent of r and R.

d) Bounds for the curvature

The purpose of this subsection is to exhibit bounds on the curvature 2-form of the connection A. In this regard, the arguments for these bounds are essentially the same as those which appear in Section 3d of [16] so the discussion will be fairly brief.

The discussion here begins with the self-dual projection, F_A^+, of the curvature. In particular, the second line of (5.3) implies that

$$\begin{aligned} |F_A^+| = r(2\sqrt{2})^{-1}((2^{-1/2}|\omega| - |\alpha|^2)^2 \\ + 2|\beta|^2(2^{-1/2}|\omega| + |\alpha|^2) + |\beta|^4)^{1/2}. \end{aligned} \tag{6.16}$$

This last equality and (6.13) provide the following useful lemma:

Lemma 6.4. *Fix $k \geq 1$ and there is a constant, $\zeta_k \geq 1$, which is independent of the data, $r(\mathfrak{s}, \sigma)$ and (A, Ψ) and which has the following significance: If $r \geq \zeta_k$, then*

$$|F_A^+| \leq r(2\sqrt{2})^{-1}(2^{-1}|\omega| - |\alpha|^2) + \zeta_k \tag{6.17}$$

at all points where $s \leq k$.

Bounds for the anti-self dual part, F_A^-, of the curvature of A are obtained, as in [16], by exploiting a differential equation for the latter which is implied by the fact that the total curvature is a closed 2-form. The following proposition summarizes these bounds:

Proposition 6.5. *Fix the* Spin^C *structure and fix $m \geq 1$. Then, there are constants, $\zeta_m, \zeta_m' \geq 1$, which are independent of the data r, σ and (A, Ψ) and which have the following significance: Take $r \geq \zeta_m'$, and then*

$$|F_A^-| \leq r(2\sqrt{2})^{-1}(1 + \zeta_m r^{-1/2})(2^{-1}|\omega| - |\alpha|^2) + \zeta_m' \tag{6.18}$$

at all points where $s \leq m$.

Proof of Proposition 6.5. Except for some minor notational changes, the proof is essentially identical to the proof of Proposition 3.4 in [16] to which the reader is referred. Note that this argument for Proposition 6.5 provides along the way the amusing integral inequalities given in (6.19), below. Both involve a constant $\zeta \geq 1$ which depends on the given Spin^C structure $\mathfrak{s}$, but which is independent of r, σ, and (A, Ψ). Moreover both inequalities hold only when $r \geq \zeta$. Here are the inequalities:

- $\int_X |F_A|^2 \leq \zeta r$.
- $\int_X (1 + \mathrm{dist}(x, \cdot)^{-2})(|\nabla_A \Psi|^2 + r^{-1}|F_A^+|^2) \leq \zeta$ for any point $x \in X$.

(6.19)

(The proof of the preceding two inequalities is the same as the proof of (3.29) in [16].)

e) Bounds for $\nabla_A \alpha$ and $\nabla_A \beta$

The required bounds for these derivatives are summarized by

Proposition 6.6. *Fix the* Spin^C *structure and fix $m \geq 1$. Then, there are constants, $\zeta_m, \zeta_m' \geq 1$, which are independent of the data r, σ*

and (A, Ψ) *and which have the following significamce: Take* $r \geq \zeta'_m$, *and then*

$$|\nabla_A \alpha|^2 + r|\nabla_A \beta|^2 \leq \zeta_m r(2^{-1/2}|\omega| - |\alpha|^2) + \zeta'_m. \tag{6.20}$$

at all points where $s \leq m$.

Proof of Proposition 6.6. Except for some minor notational changes, the argument is the same as that for Proposition 3.7 in [16].

f) A summary of conclusions from [16] which now apply

The next series of arguments for Theorem 5.5 are borrowed virtually verbatim from the proof of Theorem 2.2 of [16]. The results of these arguments are summarized below, while the reader is referred to the appropriate place in [16] for the proof.

To begin, suppose that $B \subset X$ is a compact set, and consider the energy of B:

$$e_B \equiv (4\sqrt{2})^{-1} r \int_B |\omega||2^{-1/2}|\omega| - |\Psi|^2| d\,\mathrm{vol}_g\,. \tag{6.21}$$

The key feature of e_B is summarized by

Proposition 6.7. *There is a constant* $\zeta \geq 1$, *and given* $m \geq 1$, *there is a constant* $\zeta_m \geq 1$; *and these constants have the following significance:*

Suppose that $r \geq \zeta_m$ *and let* (A, Ψ) *be a solution to the* $((\mathfrak{s}, \sigma), r)$ *version of (5.3). Let* $B \subset X$ *be a geodesic ball with center* x *on which* $s \leq m$. *Let* ρ *denote the radius of* B *and require that* $1/\zeta_m \geq \rho \geq 2^{-1} r^{-1/2}$. *Then*

- $e_B \leq \zeta \rho^2$.
- *If* $|\alpha(x)| \leq (2\sqrt{2})^{-1}|\omega|$, *then* $e_B \geq \zeta_m^{-1}\rho^2$.

This last proposition has various collaries, the most immediate being:

Lemma 6.8. *Given* $m \geq 1$, *there is a constant* $\zeta_m \geq 4$ *with the following significance: Fix* $r \geq \zeta_m$ *and let* (A, Ψ) *be a solution to the* $((\mathfrak{s}, \sigma), r)$ *version of (5.3). Let* $\rho \in (\zeta_m r^{-1/2}, \zeta_m)$. *Then,*

- *Let* Λ *be any set of disjoint balls of radius* ρ *whose centers lie on* $\alpha^{-1}(0)$ *and lie where* $s \leq m$. *Then* Λ *has less then* $\zeta_m^{-1}\rho^{-2}$ *elements.*

- *The set of points in* $\alpha^{-1}(0)$ *which lie where* $s \leq m$ *has a cover by a set* Λ *of no more that* $\zeta_m \rho^{-2}$ *balls of radius* ρ. *Moreover, each ball in this set has its center on* $\alpha^{-1}(0)$. *Finally, the set of concentric balls of radius* $\rho/2$ *is disjoint.*

The preceding lemma can then be used to prove the following refinement of Proposition 6.5:

Proposition 6.9. *Given $m \geq 1$, there are constants $\zeta_m, \zeta'_m \geq 1$ with the following significance: Fix $r \geq \zeta_m$ and let (A, Ψ) be a solution to the $((\mathfrak{s}, \sigma), r)$ version of (5.3). Then, at points of X where $s \leq m$,*

$$|F_A^-| \leq r(2\sqrt{2})^{-1}(2^{-1/2}|\omega| - |\alpha|^2) + \zeta'_m. \tag{6.22}$$

Proof of Propositions 6.7 and 6.9, and Lemma 6.8. These are the respective analogs of Propositions 4.1 and 4.3, and Lemma 4.2 in [16], and the proofs of the latter in Section 4 of [16] carry over with only small notational changes.

The next step in the proof of Theorem 5.5 is also borrowed from [16], this being a description of (A, Ψ) at distances from $\alpha^{-1}(0)$ which are $o(r^{-1/2})$. In particular,the assertion of Proposition 5.2 of [16] holds here with the obvious changes: First, (A, Ψ) is a solution on X to (5.3). Second, instead of choosing $\delta > 0$, choose $m \geq 1$ and restrict the point x to lie where $s \leq m$. Finally, the constant ζ_δ in the statement of Proposition 5.2 of [16] is replaced by a constant $\zeta_m \geq 1$.

With the structure of (A, Ψ) near $\alpha^{-1}(0)$ understood, consider now the behavior from Section 6 of [16] at larger distance from $\alpha_{-1}(0)$. Here, the assertions of Proposition 6.1 and Lemma 6.2 from [16] can be borrowed with only notational changes. The notationally modified assertions are summarized in

Proposition 6.10. *Given $m \geq 1$, there is a constant $\zeta_m \geq 4$ with the following significance: Fix $r \geq \zeta_m$ and let (A, Ψ) be a solution to the $((\mathfrak{s}, \sigma), r)$ version of (5.3). If $x \in X$ is such that $s \leq m$, then*

$$\begin{aligned} &r|2^{-1}|\omega| - |\alpha|^2| + r^2|\beta|^2 + |\nabla_A \alpha|^2 + r|\nabla_A \beta|^2 \\ &\qquad \leq \zeta_m(1 + r\exp[-\sqrt{r}\,\mathrm{dist}(x, \alpha^{-1}(0))/\zeta_m]). \end{aligned} \tag{6.23}$$

Proof of Proposition 6.10. Mimic the proof of Lemma 6.2 in [16].

This last result facilitates the identification of the connection A at distances which are uniformly far from $\alpha^{-1}(0)$. Indeed, at distances from $\alpha^{-1}(0)$ which are $o(1)$, the bounds in (6.17), (6.22) and (6.23) imply that the curvature F_A has an r independent upper bound. This suggests that when r is large, the connection A is close to some fiducial connection, A^0, at such distances from $\alpha^{-1}(0)$. This is indeed the case. To describe this canonical connection, introduce $K \subset \Lambda^+$ to denote the orthogonal complement to the span of ω. As Λ^+ is oriented, so K is oriented by writing $\Lambda^+ = \mathbb{R}\omega \oplus K$. Moreover, Λ^+ has a natural inner product, so does K and thus K can be viewed in a canonical way as a complex line bundle over X. Furthermore, the Levi-Civita connection on TX induces a connection on Λ^+ and thus, by orthogonal projection, a connection on K. The latter is hermitian with respect to the aforementioned complex line bundle structure. Use $\underline{A}^0$ to denote the dual connection on K^{-1}.

To proceed with the defintion of A^0, reintroduce the line bundle $L_+ \to X$ from the proof of Proposition 5.1. The line bundle L_+ enters because the determinant line bundle L for the Spin^C structure is naturally isomorphic to $L = K^{-1}L_+^2$. With this point understood, note that α is a section of L_+ and so α^2 can be viewed as a section of $\mathrm{Hom}(K^{-1}, L)$. In particular, where α is not zero, $\alpha^2/|\alpha|^2$ defines a hermitian identification between K^{-1} and L.

With the previous two paragraphs understood, it can now be stated that the canonical connection A^0 on L is the image of the Levi-Civita induced connection $\underline{A}^0$ on K^{-1} under the identification via $\alpha^2/|\alpha|^2$ of these two bundles.

Having now defined A^0, consider:

Proposition 6.11. *Given $m \geq 1$, there is a constant $\zeta_m \geq 4$ with the following significance: Fix $r \geq \zeta_m$ and let (A, Ψ) be a solution to the $((\mathfrak{s}, \sigma), r)$ version of (5.3). If $x \in X$ is such that $s \leq m$, and* $\mathrm{dist}(x, \alpha^{-1}(0)) \geq r^{-1/2}$, *then*

$$(6.24)\quad |A - A_0| + |F_A - F_{\underline{A}^0}| \leq \zeta_m r^{-1} + \zeta_m r \exp[-\sqrt{r}\,\mathrm{dist}(x, \alpha^{-1}(0))/\zeta_m].$$

Proof of Proposition 6.12. Copy the proof of Proposition 6.1 in [16].

7. Completion of the proof of Theorem 5.5

The proof of Theorem 5.5 is completed here with an analysis of the $n \to \infty$ limit of the sets $\alpha_n^{-1}(0)$ which appear in the statement of Theorem 5.5. The analysis of this limit follows closely the discussion in Section 7 of [16].

a) The curvature as a current

In this section, let $\{r_n\}_{n=1,2,\ldots}$ be an unbounded, increasing sequence of positive numbers such that for each n, the $((\mathfrak{s}, \sigma), r = r_n)$ version of (5.3) has a solution $\{(A_n, \Psi_n)\}$. The difference between the curvature

2-form of the connection A_n and that of the canonical connection $\underline{A}^0$ on K^{-1} defines a current on X, which is to say, a linear functional on the Frechet space of compactly supported, smooth 2-forms. To be precise, the current in question assigns to a smooth, 2-form ν with compact support the number

$$f_n(\nu) \equiv 2^{-1} \int_X i/(2\pi)(F_{A_n} - F_{\underline{A}^0}) \wedge \nu. \tag{7.1}$$

Note that if $m > 0$ is given, then (6.17), (6.22) and (6.23) together provide $\zeta_m > 1$ such that

$$|f_n(\nu)| \leq \zeta_m \sup_X |\nu| \tag{7.2}$$

when $s \leq m$ on the support of ν. This implies, in particular, that the sequence $\{f_n(\cdot)\}$ of linear functionals on the space of compactly supported 2-forms has weak limits and any such limit defines a bounded, linear functional on the space of 2-forms with support on some fixed compact subdomain in X. Choose one such weak limit, denote it by f, and renumber the subsequence of $\{f_n\}$ which converges to f consecutively from $n = 1$.

The current f is integral in the following sense: If ν is a closed 2-form with compact support and with integral periods on $H_2(X; Z)$, then

$$f(\nu) = c_1(L_+) \bullet [\nu] \in Z, \tag{7.3}$$

where $[\nu]$ is the class of ν in $H^2(X, \partial X; Z)$.

b) The support of f

The support of f is described by Lemma 7.1, below. Note that except for notational changes, the proofs of Lemmas 7.1 here and Lemma 7.1 in [16] are the same.

Lemma 7.1. *There is a closed subspace $C' \subset X$ with the following properties:*

- *$f(\nu) = 0$ if ν has compact support on $X - C'$.*
- *Let $B \subset X$ be an open set which intersects C'. Then, there is a 2-form ν with compact support on B and with $f(\nu) \neq 0$.*
- *Fix $m \geq 0$ and the set of points in C' where $s \leq m$ has finite 2-dimensional Hausdorf measure.*

- *Fix $m \geq 0$ and there is a constant ζ_m with the following significance: Let $B \subset X$ be a ball of radius $\rho \leq \zeta_m^{-1}$ and center on C'. Then, the 2-dimensional Hausdorf measure of $C \cap B$ is greater than $\zeta_m^{-1}\rho^2$.*

- *There is a subsequence of $\{(A_n, \Psi_n)\}$ such that the corresponding sequence $\{\alpha_n^{-1}(0)\}$ converges to C' in the following sense: If $Q \subset X$ is any compact set, the following limit exists and is zero:*

$$\lim_{n\to\infty}\Big[\sup_{x\in C'\cap Q} \operatorname{dist}(x, \alpha_n^{-1}(0)) + \sup_{x\in\alpha_n^{-1}(0)\cap Q} \operatorname{dist}(x, C')\Big]. \tag{7.4}$$

With Lemma 7.1 understood, the arguments in Section 7c,d of [16] can be transferred here essentially verbatim to prove

Proposition 7.2. *The set C' from Lemma 7.1 is the image of a smooth, complex curve, C_0, via a proper, pseudoholomorphic map $f : C_0 \to X$. Thus, C' is a pseudoholomorphic subvariety and so an HWZ subvariety. Moreover, there is a positive integer assigned to each irreducible component of C' such that the following is true: Let c denote the corresponding generalized HWZ subvariety. Then c and the current f are related in the following sense: For any compactly supported 2-form ν,*

$$f(\nu) = \sum_{(C,m)\in c} m \int_C \nu. \tag{7.5}$$

(Note that the conclusion here that C' is an HWZ subvariety follows from Lemma 2.1.)

c) An SW-admissable, generalized subvariety

The purpose of this subsection is to prove that the generalized, HWZ pseudoholomorpic variety, c, in Proposition 7.2 is SW-admissable. In this regard, note that (7.3) and (7.5) imply that $e_c = c_1(L)$.

A proof that $\sum_{(C,m)\in c} m[C]$ in $H_2(X_0, \partial X_0; \mathbb{Z})$ maps to the sum of the oriented generators of $H_1(\partial X_0; \mathbb{Z})$ proves that c is admissable. For this purpose, introduce a function χ on $\mathbb{R}$ with total integral equal to 1 and with compact support in $[0, 1]$. Then, for $R \geq 0$, introduce the function χ_R of the parameter s via the formula $\chi_R(s) \equiv \chi(s - R)$. Thus χ_R is a function on X with support where $s \in [R, R+1]$. With χ_R now defined, set $\nu \equiv (2\pi)^{-1}\chi_R ds \wedge dt$.

Now, consider first the case where $Y \subset \partial X_0$ is a component with oriented z-axis line bundle. Identify Y as $S^1 \times S^2$ via coordinates where the contact form is given by (2.11) and ω by (2.21). Then, the image of $\sum_{(C,m)\in c} m[C]$ in $H_1(Y;\mathbb{Z})$ is the number $q \equiv \sum_{(C,m) inc} m \int_C \nu$ times the oriented generator.

Now, the point is that (7.5) identifies this number q as equal to $f(\nu)$, and thus it follows from the definition of f that the number q is also given by (7.1) in the case where n is large. In particular, since the curvature of A_n is exponentially decreasing to zero as $s \to \infty$ on $[0,\infty) \times M$, it follows that the the image of $\sum_{(C,m)\in c} m[C]$ in $H_1(Y;\mathbb{Z})$ is q times the oriented generator, where q is equal to

$$-i/(4\pi) \int_X F_{\underline{A}^0} \wedge \nu. \tag{7.6}$$

On the other hand, this last intergral computes the evaluation of $-2^{-1} c_1(K^{-1})$ on the oriented 2-sphere $\{\text{point}\} \times S^2$ in Y. The latter is half of the evaluation of $c_1(K)$ on this same 2-sphere, and it follows by considering the zeros of the sections of K in (5.1) that this number equals 1, which is the required answer.

A similar argument proves the case for those components of ∂X_0 with unoriented z-axis line bundle.

d) The symplectic area of C

The assertion in (5.5) follows directly from (7.3) and (7.5). In particular, the constant $\wp$ is given by

$$\wp = -2^{-1} \int_X i/(2\pi) F_{\underline{A}^0} \wedge \omega. \tag{7.7}$$

(This integral converges since $F_{\underline{A0}}$ is bounded on $[0,\infty) \times \partial X_0$ while $|\omega|$ decays exponentially fast.)

References

[1] Y. Eliashberg, *Classification of overtwisted contact structures on 3-manifolds*, Invent. Math. **98** (1989) 623–637.

[2] ———, *Filling by holomorphic discs and its applications*, London Math. Soc. Lect. Notes Ser. **151** (1991) 45–67.

[3] D. S. Freed & K. K Uhlenbeck, *Instantons and four-manifolds,* Springer, 1984.

[4] R. Gompf, private communication.

[5] H. Hofer, *Pseudoholomorphic curves in symplectization with applications to the Weinstein conjecture in dimension-3*, Invent. Math. **114** (1993) 515–563.

[6] ———, *Dynamics, topology and holomorphic curves*, Proc. Internat. Congr. Math., Berlin 1998, Vol I, Documenta Math. Extra Volume ICM, 1998, 255–280.

[7] K. Honda, *Harmonic forms for generic metrics*, Preprint.

[8] ———, *Local properties of self-dual harmonic 2-forms on a 4-manifold*, Preprint.

[9] H. Hofer, K. Wysocki & E. Zehnder, *Properties of pseudoholomorphic curves in symplectizations. I*, Ann. Inst. Henri Poincaré **13** (1996) 337–379.

[10] ———, *Properties of pseudoholomorphic curves in symplectizations. II*, Geom. Funct. Anal. **5** (1995) 270–328.

[11] ———, *Properties of pseudoholomorphic curves in symplectizations. III*, Preprint.

[12] J. W. Morgan, *The Seiberg-Witten equations and applications to the topology of smooth 4-manifolds*, Math. Notes, Vol. 44, Princeton Univ. Press, Princeton, 1996.

[13] J. W. Morgan, Z. Szabo & C. H. Taubes, *A product formula for the Seiberg-Witten invariants and the generalized Thom conjecture*, J. Differential Geom. **44** (1996) 706–788.

[14] J. W. Morgan, T. S. Mrowka & Z. Szabo, *Product formulas along T^3 for Seiberg-Witten invariants*, Math. Res. Lett. **4** (1997) 915–929.

[15] P. Osvath & Z. Szabo, *Higher type adjunction inequalities in Seiberg-WItten theory*, Preprint.

[16] C. H. Taubes, *Seiberg-Witten invariants and pseudoholomorphic subvarieties for self- dual, harmonic 2-forms*, J. Top. Geom.

[17] ———, *SW => Gr: From the Seiberg-Witten equations to pseudoholomorphic curves*, J. Amer. Math. Soc. **9** (1996) 845–918.

[18] ———, *SW => Gr: From the Seiberg-Witten equations to pseuooholomorphic curves*, in Proc. First IP Lecture Ser., Vol II, (R. Wentworth ed.), Internat. Press, 2000.

[19] ———, *The Seiberg-Witten and the Gromov invariants*, Math. Res. Lett. **2** (1995) 221–238.

[20] ———, *SW = GR: Counting curves and connections*, J. Differential Geom, to appear, and reprinted in Proc. First IP Lecture Ser., Vol II, (R. Wentworth ed.), Internat. Press, 2000.

[21] ———, *Counting pseudoholomorphic submanifolds in dimension-4*, J. Differential Geom. **44** (1996) 818-893, and reprinted in Proc. First IP Lecture Ser., Vol II, (R. Wentworth ed.), Internat. Press, 2000.

[22] ———, *The geometry of the Seiberg-Witten invariants*, Proc. Internat. Congr. Math., Berlin 1998, Vol II, Documenta Math. Extra Volume ICM 1998, 493–504.

[23] ———, *The structure of pseudo-holomorphic subvarieties for a degenerate almost complex structure and symplectic form on* $S^1 \times B^3$, Geom. Top. **2** (1998) 221–332.

[24] E. Witten, *Monopoles and 4-manifolds*, Math. Res. Lett. **1** (1994) 769–796.

HARVARD UNIVERSITY

SURVEYS IN DIFFERENTIAL GEOMETRY, 2000
Vol. VII ©2000, INTERNATIONAL PRESS
pp. 673-684

UNIFYING THEMES IN TOPOLOGICAL FIELD THEORIES

CUMRUN VAFA

We discuss unifying features of topological field theories in 2, 3 and 4 dimensions. This includes relations among enumerative geometry (2d topological field theory) link invariants (3d Chern-Simons theory) and Donaldson invariants (4d topological theory). (Talk presented in conference on Geometry and Topology in honor of M. Atiyah, R. Bott, F. Hirzebruch and I. Singer, Harvard University, May 1999).

1. Introduction

There has been many exciting interactions between physics and mathematics in the past few decades. Many of these developments on the physics side are captured by certain field theories, known as topological field theories. The correlation function of these theories compute certain mathematical invariants. Even though the original motivation for introducing topological field theories was to gain insight into these mathematical invariants, topological field theories have been found to be important for answers to many questions of interest in physics as well.

The aim of my talk here is to explain certain connections that have been discovered more recently among various topological field theories. I will first briefly review what each one is, and then go on to explain some of the connections which has been discovered between them.

The main examples of topological field theories that have been proposed appear in dimension two [26] known as topological sigma models, in dimension three [27] known as Chern-Simons theory and in dimension four [28] known as topological Yang-Mills theory. The 2d and the 4d topological theory are related to an underlying supersymmetric quantum field theory, and there is no difference between the topological and standard version on the flat space. The difference between conventional supersymmetric theories and topological ones in these cases only arise when one considers curved spaces. In such cases the topological version, is a modified version of the supersymmetric theory on flat space where some of the fields have different Lorentz transformations properties (compared to the conventional choice). This modification of Lorentz transformation properties is also known as twisting, and is put in primarily to preserve supersymmetry on curved space. In particular this leads to having at least one nilpotent supercharge Q as a scalar quantity, as opposed to a spinor, as would be in the conventional spin assignments. The physical observables of the topological theory are elements of the Q cohomology. The path integral is localized to field configurations which are annihilated by Q and this typically leads to some moduli problem which lead to mathematical invariants.

In these theories the energy momentum tensor is Q trivial, i.e.,

$$T_{\mu\nu} = \{Q, \Lambda_{\mu\nu}\}$$

which (modulo potential anomalies) leads to the statement that the correlation functions are all independent of the metric on the curved space, thus leading to the notion of topological field theories (i.e. metric independence).

The case of the 3d topological theory, is somewhat different. In this case, namely the case of Chern-Simons theory, one starts from an action which is manifestly independent of the metric on the 3 manifold, and thus topological nature of the field theory is manifest.

The organization of this paper is as follows: In Section 2 I briefly review each of the three classes of topological theories and discuss how in each case one goes about computing the correlation functions. In section 3 I discuss relations between 2d and 4d topological theories. In section 4 I discuss relations between 2d and 3d topological theories.

2. A brief review of topological field theories

In this section I give a rather brief review of topological field theories in dimensions 2, 3 and 4.

2.1 TFT in $d = 2$: topological sigma models

Topological sigma models are based on $(2,2)$ supersymmetric theories in 2 dimensions. These typically arise by considering supersymmetric sigma models on Kahler manifolds. In other words, we consider maps from 2 dimensional Riemann surfaces Σ to target spaces M which are Kahler manifolds (together with fermionic degrees of freedom on the Riemann surface which map to tangent vectors on the Kahler manifold). The topological theory in this case localizes on holomorpic maps from Riemann surfaces to the target:

$$X: \qquad \Sigma \to M$$

$$\overline{\partial} X = 0$$

If we get a moduli space of such maps we have to evaluate an appropriate class over it. This class is determined by the topological theory one considers (for precise mathematical definitions see [5]. Also there are two versions of this topological theory: coupled or uncoupled to gravity. Coupling to gravity in this case means allowing the complex structure of Σ to be arbitrary and looking for holomorphic curves over the entire moduli space of curves. The case coupled to gravity is also sometimes referred to as 'topological strings'.

A particularly interesting class of sigma modelds both for the physics as well as for mathematics, corresponds to choosing M to be a Calabi-Yau threefold, and considering topological strings on M. In this case the virtual dimension of the moduli space of holomorphic maps is zero. If this space is given by a number of points, the topological string amplitude just counts how many such points there are, weighted by $e^{-k(.)}$ where $k(.)$ is the area of the holomorphic map (pullback of the Kahler form integrated over the surface) times λ^{2g-2}, where g denotes the genus of the Riemann surface and λ denotes the string coupling constant. More generally the space of holomorphic maps will involve a moduli space. This space comes equipped with a bundle with the same dimension as the tangent bundle (the existence of this bundle and the fact that its dimensions is the same as the tangent bundle follows from the fact that the relevent index is zero). Topological string computes the top Chern class of such bundles again weighted by $e^{-k(.)}\lambda^{2g-2}$. These have to be defined carefully, due to singularities and issues of compactifications, and lead in general to rational numbers. The sum of these numbers for a given class $v \in H_2(M, \mathbf{Z})$ and fixed genus g, which we will denote by $r_{g,v}$, is known as Gromov-Witten invariant. We thus have the full partition function of topological string given by

$$F(\lambda, k) = \sum_{v \in H_2(M,\mathbf{Z})} r_{g,v} e^{-k(v)} \lambda^{2g-2}$$

here k denotes the Kahler class of M. Even though the numbers $r_{g,v}$ are not integers, it has been shown, by physical arguments that F can also be expressed in terms of other *integral* invariants [8]. These integral invariants are related to certain aspects of cohomology classes of moduli of holomorphic curves *together with flat bundles.*

These invariants associate for each $v \in H_2(M, \mathbf{Z})$ and eash positive (including 0) integer s a number $N_{v,s}$ which denotes the 'net' number of BPS membranes with charge in class v and 'spin' s (for precise definitions see [8]). Then we have

$$F(\lambda, k) = \sum_{n>0, v\in H_2(M,\mathbf{Z})} \frac{1}{n} N_{v,s} e^{-nk(v)} [2Sin(n\lambda/2)]^{2s-2} \tag{1}$$

For all cases checked thus far the Gromov-Witten invariants $r_{g,v}$ has been shown to be captured by these simpler integral invariants $N_{v,s}$ through the above map. In particular the checks made for constant maps [6] and for contribution of isolated genus g curves to all loops [21] as well as some low genus computations for non-trivial CY 3-folds [14] all support the above identification.

Let us illustrate the above results in the case of a simple non-compact Calabi-Yau threefold, which we will later use in this paper. Consider the total space of the rank 2 vector bundle $O(-1) + O(-1) \to \mathbf{P}^1$. This space has vanishing c_1, and is a non-compact CY 3-fold. In this case the only BPS state is a membrane wrapping $\mathbf{P}^1$ once. This state has spin $s = 0$. If we denote the area of $\mathbf{P}^1$ by t, then we have from (1)

$$F = \sum_{n>0} \frac{1}{n[2Sin(n\lambda/2)]^2} e^{-nt} \tag{2}$$

For this particular case this has also been derived using the direct definition of topological strings in [6], [21].

2.2 Topological field theory in 3d: Chern-Simons theory

The 3d topological theory we consider is Chern-Simons theory, which is given by the Chern-Simons action for a gauge field A:

$$S_{CS} = \frac{k}{4\pi} \int_M Tr[AdA + \frac{2}{3} A^3]$$

where M is a 3-manifold and k is an integer which is quantized in order for $exp(iS)$ to be well defined. As is clear from the definition of the above action, S does not depend on any metric on M and in this sense the theory is manifestly topological (i.e., metric independent).[1]

[1] At the quantum level there is a metric dependence which can be captured by a gravitational Chern-Simons term [27] [2].

Thus the partition function of Chern-Simons theory gives rise to topological invariants for 3-manifolds for each group G. In other words

$$Z_M(G) = exp(-F_M(G)) = \int \mathcal{D}A exp[iS_{CS}]$$

where A is a connection on M for the gauge group G and the above integral is over all inequivalent G-connections on M. The simplest way to compute such invariants is to use the relation between Hilbert space of Chern-Simons theory on a Riemann surface Σ and the chiral blocks of WZW model on Σ with group G and level k. For example the partition function on S^3 can be computed by viewing S^3 as a sum of two solid 2-tori, which are glued along T^2 by an order 2 element of $SL(2,\mathbf{Z})$ on T^2. In this way the partition function gets identified with

$$Z_{S^3}(G,k) = S_{00}(G,k)$$

where $S_{00} = \langle 0|S|0\rangle$ is a particular element of the order 2 operation of $SL(2,\mathbf{Z})$ on chiral characters, and is well studied in the context of WZW models. In particular for $G = SU(N)$ it is given by:

$$\begin{aligned} Z_{S^3}(SU(N),k) &= exp(-F) \\ &= e^{i\pi N(N-1)/8}\frac{1}{(N+k)^{N/2}}\sqrt{\frac{N+k}{N}} \\ &\quad \cdot \prod_{j=1}^{N-1}(2\sin\frac{j\pi}{N+k})^{N-j}. \end{aligned} \tag{3}$$

One can also consider knot invariants: Consider a knot γ in M and choose a representation R of the group G and consider the character of the holonomoy of A around the knot γ, i.e.,

$$P[\gamma,R] = Tr_R P\exp(i\int_\gamma A)$$

By the equation of motion for Chern-Simons theory, which leads to flatness of A, we learn that the above operator only depends on the choice of the knot type and not the actual knot[2]. One then obtains a knot invariant by computing the correlation function

$$< \prod_i P[\gamma_i,R_i] >= \int \mathcal{D}A \prod_i P[\gamma_i,R_i]\exp(iS_{CS})$$

Again these quantities can be computed by the braiding properties of chiral blocks in 2 dimensional WZW models and leads in particular to HOMFLY polynomial invariants for the knots.

[2]In the quantum theory one also needs to choose a framing for the knot.

2.3 Topological field theories in 4-dimensions

If one consider $N = 2$ supersymmetric Yang-Mills, with an unconventional spin assignments, one finds a topological field theory. The partition function is localized on the moduli space of instantons and the observables of this theory are given by intersection theory on the moduli space of instantons. More precisely each d-cycle on the four manifold M will lead to a $4-d$ cohomology element on the moduli space of instantons (obtained by integrating out $\int \mathrm{Tr} F \wedge F$ over the corresponding cycle on the universal moduli space of instantons), and the wedging of the cohomology classes gives rise to the observables in Donaldson theory. This does not depend on the metric in M (except when $b_2^+(M) = 1$) but will depend on the choice of smooth structure on M.

The computations in this case can be done for many choices of M by finding an equivalence of this theory and a simpler abelian theory. In this case studying the moduli space of non-abelian instantons gets replaced with the study of an abelian system known as the Seiberg-Witten equation. The relevant geometry for the case of $SU(N)$ Yang-Mills is captured by a certain geometric data related to a Jacobian variety over an $N-1$ dimensional family of genus $N-1$ Riemann surface, known as Seiberg-Witten geometry [22]. For topological field theory aspects and how the Seiberg-Witten geometry leads to computation of the topological correlation functions see [25], [19].

There is another topological theory in 4 dimensions which has been studied [24] and is related to twisting the maximal supersymmetric gauge theory in 4 dimensions. This theory computes the Euler characteristic of moduli space of instantons. In particular for each group G and each complex parameter q one considers

$$Z_M(G) = q^{-c(M,G)} \sum_k q^k \chi(\mathcal{M}_k)$$

for some universal constant c (depending on M and G), where k denotes the instanton number and $\chi(\mathcal{M}_k)$ denotes the euler characteristics (of a suitable resolution and compactification) of $\mathcal{M}_k$, the moduli space of anti-self dual G-connections with instanton number k on M. Moreover, according to Montonen-Olive duality conjecture one learns that the above partition function is expected to be modular with respect to some subgroup of $SL(2, \mathbf{Z})$ acting in the standard way on τ where $q = exp(2\pi i\tau)$. For certain M (such as $K3$) the above partition function has been computed and is shown to be modular in a striking way. For recent mathematical discussion on this see [11] and references therein.

3. Connections between 2d ↔ 4d TFT's

There are three different links between 4 dimensional and 2 dimensional TFT's that I would like to discuss. In all three links the common theme is that the moduli space of instantons are mapped to moduli space of holomoprhic curves on appropriate spaces.

3.1 Topological reduction of $4d$ to $2d$

The simplest link between the two theories involves studying the 4d TFT on a geometry involving the product of two Riemann surfaces $\Sigma_1 \times \Sigma_2$, which was studied in [4]. In the limit where Σ_1 is small compared to Σ_2 one obtains an effective theory on Σ_2 which is the topological sigma model with target space given by moduli space of flat connections on Σ_1, in case one considers $N = 2$ topological field theories in 4 dimensions or the Hitchin space associated with Σ_1 if one considers $N = 4$ topological field theories. This is natural to expect because studying light supersymmetric modes in either case gives rise to the corresponding space of solutions, which thus behaves from the viewpoint of the space Σ_2 as a target space. In particular the moduli space of 4d instantons get mapped to moduli space of holomorphic maps for these target spaces. Thus quantum cohomology rings of moduli of flat connections on a Riemann surface, which are encoded in 2d topological correlation functions capture the corresponding topological correlation functions of the 4 dimensional $N = 2$ theory. Similarly in the $N = 4$ case the reduction to 2 dimensions yields a sigma model on the Hitchin space (which can also be viewed as a Jacobian variety). In this context the Montonen-Olive duality of $N = 4$ theory gets mapped to mirror symmetry of this 2d sigma model (by a fiberwise application of T-duality to Jacobian fibers).

3.2 A more subtle 2d ↔ 4d link

For the $N = 2$ topologically twisted theory, an important role is played by the Seiberg-Witten geometry, which is an abelian simplification of the non-abelian gauge theory. This geometry is a quantum deformation of the classical one, due to pointlike four dimensional instantons. This geometry was first conjectured based on consistency with various properties of $N = 2$ quantum field theories and its deformation to $N = 1$ quantum field theories with mass gap, where plausible properties of $N = 1$ theories were assumed.

With the recent advances in our understanding of string theory, the same 4-dimensional gauge theories have been obtained by considering particular geometries where strings propagate in. This procedure is known as geometric engineering of QFT's (see [14], [12] and references therein). These geometries involve a non-compact Calabi-Yau threefold geometry which is a blow up of a geometry with some loci of A-D-E singularities (locally modelled by $\mathbf{C}^2/G$ where G is a discrete subgroup of $SU(2)$), giving rise to the corresponding gauge theory in 4 dimensions. Depending on the detailed structure of singularities one can obtain various interesting gauge groups and various matter representations.

It turns out that in this description of gauge theory, the guage theory instantons are mapped to stringy instantons, which are just worldsheet instantons. Thus being able to compute worldsheet instantons, i.e., counting of holomorphic curves in these target geometries, captures the geometry of 4-dimensional gauge theory instantons. Counting of holomrophic curves is precisely what the (A-model) topological string computes and thus in this way the geometry of vacua of 4-dimensional gauge theory gets mapped to solving topological amplitudes in 2d. This in turn can be done by using (local) mirror symmetry. For a physical derivation of mirror symmetry and some references on this subject see the recent work [10]. In this way, not only the Seiberg-Witten geometry has been rederived, but also other geometries which describe other $N = 2$ systems with various kinds of gauge groups and intricate matter representations have been obtained [15].

3.3 $N = 4$ Yang-Mills on elliptic surfaces and $2d$ topological theories

If we consider an $N = 4$ supersymmetric $SU(N)$ topological theory on an elliptic surface, with base B, the stable bundles get mapped to spectral covers of B on a dual elliptic surface M (where the Kahler class of the elliptic fiber is inverted). This uses the fact that in the limit of small tori, the stable bundles become flat fiberwise and flat bundles on tori are related to points on the dual tori. See [7], [3] for a discussion of how this arises. In particular a rank N stable bundle with instanton number k gets mapped to a spectral curve which is a holomorphic curve wrapping the base N times and the elliptic fiber k times. Thus the topological $N = 4$ amplitude on M, denoted by $Z_M(SU(N))$ which computes the Euler characteristic of moduli space of $SU(N)$ instantons on M gets mapped to computing Euler characteristic of moduli space of holomorphic curves (together with a flat bundle) which in turn is captured by genus zero topological string amplitudes, and can be computed using mirror symmetry. This idea has been implemented in great detail for the case of rational elliptic surface (also known as "half $K3$") [15], [18], [17]. The results for the case of rank 2 and its implications for the Euler characteristic of moduli space of instantons on rational elliptic surface has been confirmed using rigorous mathematical methods in [30].

4. Connections between 2d $\leftrightarrow$ 3d TFTs

Over two decades ago 't Hooft conjectured that $SU(N)$ gauge theories with large N look alot like string theories. In particular the partition function for these theories can be organized in terms of Riemann surfaces where each Riemann surface is weighted with N^{χ} where χ denotes the Euler characteristic of the Riemann surface. In particular the low genera dominate in the large N limit. The weight factor N^{χ} follows simply from the combinatorics of Feynman diagrams, and the Riemann surface can be identified with the Feynamn diagrams where the would be holes have been filled.

The main difficulty in the conjecture of 't Hooft is to identify precisely which string theory one obtains. In the past few years for serveral interesting gauge theories and in particular some in 4 dimensions the corresponding string theory has been identified [1]. Even though it has not been possible to actually compute the string theory amplitudes in these cases, due to the complicated background strings propagate in, there has been mounting evidence for the validity of the identification. One would like to have a similar conjecture in a setup which is more computable. An ideal setup for this is topological guage theories, and in particular the topological Chern-Simons theory.

If we consider $SU(N)$ Chern-Simons theory on S^3 in the limit of large N, one could hope to get a string theory. It has been conjectured in [9] that this is indeed the case. In particular it has been conjectured that $SU(N)$ Chern-Simons theory at level k on S^3 is equivalent to topological string with target being a non-compact Calabi-Yau threefold which is the total space of $O(-1)+O(-1) \to \mathbf{P}^1$, where the (complexified) size of $\mathbf{P}^1$ is given by $t = 2\pi i N/(k+N)$ and the string coupling constant $\lambda = \frac{2\pi i}{N+k}$. This is a natural conjecture in the following sense: The Chern-Simons theory on S^3 can itself be viewd as an open string theory with target T^*S^3 [29] By open string we mean considering Riemann surfaces with boundaries, where the boundaries are mapped to S^3. The geometry $O(-1) + O(-1) \to \mathbf{P}^1$ can be obtained from the T^*S^3 geometry by shrinking S^3 to zero size and blowing $\mathbf{P}^1$ instead. This kind of transition is also very similar to what is observed to happen in the other cases where large N string theory description was discovered [1]. In fact one can determine [9] the map of the parameters t and λ given above using this picture (and recalling the metric dependence anomaly in Chern-Simons theory).

This conjecture has been checked at the level of the partition function (which we have briefly reviewed for both the Chern-Simons theory on S^3 and for $O(-1)+O(-1)\to \mathbf{P}^1$ in section 2). The implications of this conjecture for knot invariants has been explored in [20] and provides a reformulation of knot invariants in terms of integral invariants which again capture the degeneracy of spectrum of (BPS) particles in the corresponding string theory. This involves considering a Largrangian submanifold which intersects T^*S^3 along the knot and following it through the transition to $O(-1)+O(-1)\to \mathbf{P}^1$ where it corresponds to a Lagrangian submanifold. The corresponding computation on the topological string side will now involove Riemann surfaces with boundaries, where the boundary can lie on this Lagrangian submanifold in $O(-1)+O(-1)\to \mathbf{P}^1$. The results for the unknot [20] as well as the integrality properties of the torus knots [16] are in perfect agreement with the conjecture.

5. Conclusions

We have seen some intricate relations among topological theories in 2, 3 and 4 dimensions and in some ways these connections parallel the discovery of duality symmetries in superstring theories (see [23] for a review of some mathematical aspects of string dualities). These topological examples provide a simpler version of superstring dualities, which one could hope to understand more deeply and which might provide a hint as to how to think about dualities in general.

This research was supported in part by NSF grants PHY-9218167 and DMS-9709694.

References

[1] O. Aharony, S. S. Gubser, J. Maldacena, H. Ooguri & Y. Oz, *Large N field theories, string theory and gravity,* Phys. Rept. **323** (2000) 183.

[2] S. Axelrod & I. Singer, *Chern-Simons perturbation theory. II,* J. Differential Geom. **39** (1994) 173.

[3] M. Bershadsky, A. Johansen, T. Pantev & V. Sadov, *On four dimensional compactifications of F-theory,* Nucl. Phys. **B505** (1997) 165.

[4] M. Bershadsky, A. Johansen V. Sadov & C. Vafa, *Topological reduction of 4D SYM to 2D sigma models,* Nucl. Phys. **B448** (1995) 166.

[5] D.A. Cox & S. Katz, *Mirror symmetry and algebraic geometry*, Math. Surveys and Monographs **68** (ASMS, 1999).

[6] C. Faber & R. Pandharipande, *Hodge integrals and Gromov-Witten theory,* math. AG/9810173.

[7] R. Friedman, J. Morgan & E. Witten, *Vector bundles over elliptic fibrations*, alg-geom/9707004.

[8] R. Gopakumar & C. Vafa, *M-Theory and topological strings. I,II,* hep-th/9809187, hep-th/9812127.

[9] ———, *On the gauge theory/geometry correspondence*, hep-th/9811131.

[10] K. Hori & C. Vafa, *Mirror symmetry*, hep-th/0002222.

[11] M. Kapranov, *The elliptic curve in the S-duality theory and Eisenstein series for Kac-Moody groups*, math.AG/0001005.

[12] S. Katz, P. Mayr & C. Vafa, *Mirror symmetry and exact solution of 4D* $N = 2$ *gauge theories,* Adv. Theor. Math. Phys. **1** (1998) 53.

[13] S. Katz, A. Klemm & C. Vafa, *Geometric engineering of quantum field theories,* Nucl. Phys. **B497** (1997) 173.

[14] bysame, *M-Theory, topological strings and spinning black holes,* hep-th/9910181.

[15] A. Klemm, P. Mayr & C. Vafa, *BPS states of exceptional non-critical strings,* hep-th/9607139.

[16] J. M. F. Labastida & M. Marino, *Polynomial invariants for torus knots,* hep-th/0004196.

[17] J. A. Minahan, D. Nemeschansky, C. Vafa & N. P. Warner, *E-Strings and* $N = 4$ *topological Yang-Mills theories,* Nucl. Phys. **B527** (1998) 581.

[18] J. A. Minahan, D. Nemeschansky & N. P. Warner, *Partition functions for BPS states of the non-critical* E_8 *string,* Adv. Theor. Math. Phys. **1** (1998) 167.

[19] G. Moore & E. Witten, *Integration over the u-plane in Donaldson theory,* hep-th/9709193.

[20] H. Ooguri & C. Vafa, *Knot invariants and topological strings,* hep-th/9912123.

[21] R. Pandharipande, *Hodge integrals and degenerate contributions,* math. AG/9811140.

[22] N. Seiberg & E. Witten, *Electric-magnetic duality, monopole, condensation, and confinement in N=2 supersymmetric Yang-Mills theory,* Nucl. Phys. **B426** (1994) 19.

[23] C. Vafa, *Geometric Physics,* Proc. ICM-98, hep-th/9810149.

[24] C. Vafa & E. Witten, *A Strong coupling test of S-duality,* Nucl. Phys. **B431** (1994) 3.

[25] E. Witten, *On S-duality in Abelian gauge theory,* hep-th/9505186.

[26] ———, *Topological sigma model,* Comm. Math. Phys. **118** (1988) 411.

[27] ———, *Quantum field theory and the Jones polynomial,* Comm. Math. Phys. **121** (1989) 351.

[28] ———, *Topological quantum field theory,* Comm. Math. Phys. **117** (1988) 353.

[29] ———, *Chern-Simons gauge theory as a string theory,* hep-th/9207094.

[30] K. Yoshioka, *Euler characteristics of $SU(2)$ instanton moduli spaces on rational elliptic surfaces,* Comm. Math. Phys. **205** (1999) 501.

JEFFERSON PHYSICAL LABORATORY, HARVARD UNIVERSITY

SURVEYS IN DIFFERENTIAL GEOMETRY, 2000
Vol. VII ©2000, INTERNATIONAL PRESS
pp. 685-696

NONCOMMUTATIVE YANG-MILLS THEORY AND STRING THEORY

EDWARD WITTEN

I review recent work on the relation between string theory and Yang-Mills theory on noncommutative spaces. In the long wavelength limit, string theory has a conventional string perturbation expansion in terms of ordinary Yang-Mills theory with small, α'-dependent corrections. On the other hand, in a certain limit, where the B-field is effectively large, the stringy excitations drop out and the string theory admits a systematic description in terms of non-commutative Yang-Mills theory. Compatibility of the two decriptions rests on a surprising mathematical fact: though the gauge group of noncommutative Yang-Mills theory is different from the conventional Yang-Mills gauge group, the equivalence relations generated by the two groups are the same, modulo a change of variables. Open string field theory might offer a systematic framework for describing open strings in terms of noncommutative associative algebras, with all of the excited string states included, but this description has not yet been useful.

Lecture at the Differential Geometry Conference, Harvard (May, 1999).

In this lecture, I will describe recent results with N. Seiberg [8] aiming to systematically describe the role in string theory of "noncommutative Yang-Mills theory" in the sense of A. Connes. Noncommutative Yang-Mills theory was first shown to give the solution of a string theory problem by Connes, Douglas, and Schwarz in the context of matrix model compactification on a torus [2]. There have been many subsequent contributions. Some of the new contributions that are most directly relevant to today's lecture are the work of Nekrasov and Schwarz [6] on instanton solutions of noncommutative Yang-Mills theory, the work of Schomerus

[7] on open strings in a background B-field, and Kontsevich's work on deformation quantization [4] together with its quantum field theory interpretation by Cattaneo and Felder [1]. See also [3] for a recent review. In the course of describing some of these results, I will also try to explain how they relate to a wider picture.

First of all, I think that the main reason that string theory will interest mathematicians in the long run is that at its core it is based on a new kind of geometry, a successor to Riemannian geometry. But this is very hard to convey, for two reasons. First of all, physicists do not really understand the new geometry in any systematic way. Second, the pieces of the story that we do understand are based on quantum field theory constructions that are very difficult and typically inaccessible mathematically.

Roughly speaking, in the new stringy geometry, the role of the Einstein equations is played by the requirement that a certain two-dimensional quantum field theory should be conformally invariant. This gives equations that are, in a certain limit, asymptotically close to the Einstein equations, but do not coincide with them. Both parts of this assertion are important.

If two-dimensional conformal invariance did not give the Einstein equations to very high accuracy under ordinary conditions, string theory would be in trouble, as a theory of nature. For Einstein's theory is certainly very successful experimentally.

On the other hand, as Einstein's theory apparently cannot be quantized, there is a need for a new theory that reduces to it in a suitable limit. If two-dimensional conformal invariance gave the Einstein equations on the nose, string theory would fall short of providing this new theory.

String theory in fact gives equations that differ from those of Einstein's theory in a very characteristic way. Einstein's classical equations are invariant under rescalings of length. If g denotes the spacetime metric and t is a positive real number, then the Einstein equations in vacuum are invariant under $g \to tg$. So, for example, classical black holes can come in any size.

In string theory, this scale invariance is lost. There is a characteristic length scale $\sqrt{\alpha'}$; in the most straightforward way of trying to relate string theory to the real world, this length is about 10^{-32} cm. (The value is found by using the string theory formulas for the fine structure constant and Newton's constant.) For objects much bigger than this, the Einstein equations are a good approximation. For small objects, they are not.

By analyzing the conditions for two-dimensional conformal invariance, one can make a systematic expansion of the equations in powers of the curvature. The expansion reads schematically

$$(1)\quad 0 = R_{IJ} + \alpha' R_{IKLM} R_J{}^{KLM} + (\alpha')^2 D_T R_{IKLM} D^T R_J{}^{KLM} + \dots$$

where only a few illustrative terms have been written, and the ellipses denote terms of higher order in α'. The generic term on the right-hand side of (1) is of the form $(\alpha')^s$ (for some integer $s \geq 0$) times a polynomial in the Riemann tensor and its covariant derivatives that is homogeneous of degree $2s+2$ (here the Riemann tensor is considered to be of degree two, and a covariant derivative to be of degree one). In the small curvature limit, the equation is dominated by the leading term $0 = R_{IJ}$. For simplicity, I have here considered only the vacuum Einstein equations and their stringy extension. One can also incorporate matter; in fact, on the string theory side, one is forced to do so, and the matter takes a very definite form.

It is important for our story that the corrections to the Einstein equations that appear in equation (1) are ordinary, local, covariant terms. Einstein omitted them from his theory primarily on grounds of simplicity, but otherwise they obey most of his criteria. (The one general criterion formulated by Einstein that the corrections violate is, I believe, that they contain higher derivatives while Einstein looked for second order PDE's. As I have tried to explain elsewhere [10] in a lecture that was in a similar spirit to the one I am giving today, the higher derivative terms indicate that some additional "fast" variables have been averaged out of the equations. This is an important part of the story, but one that I will not describe today.)

Though the string theory corrections to the Einstein equations are usually negligible for large objects, for small objects these corrections are typically big. A relatively simple example is a Calabi-Yau threefold X. To use such a threefold for physical applications, one takes spacetime to be $\mathbf{R}^4 \times X$ (where $\mathbf{R}^4$ is intepreted as four-dimensional Minkowski space). When X is large compared to the stringy scale $\sqrt{\alpha'}$, it can be treated by classical Ricci-flat Kähler geometry, but when X becomes small, the classical description breaks down and wild things happen, such as mirror symmetry. It is very difficult to give a full account of all of the strange things that happen in string theory for Calabi-Yau threefolds.

I am going to talk today about a case where we can come closer to understanding what is going on in the stringy regime where the familiar classical equations fail. This will be the problem of Yang-Mills instantons on $\mathbf{R}^4$. How does this problem arise?

Gauge fields, and therefore instantons, can be incorporated in string theory in several different ways (which are nowadays often related to each other by nonperturbative dualities). In a previous lecture [10], I considered instantons mainly from the point of view of the heterotic string, but today we will use the older approach where gauge fields are associated with open strings while gravity is associated with closed strings. The most significant known physical application of the discussion is to D-branes in Type II superstrings. For today, all that one needs to know about D-branes is that a D-brane corresponds to a submanifold Y of spacetime, and that we will be doing gauge theory on Y. Moreover, for our purposes we can take Y to be a copy of $\mathbf{R}^4$ with its flat metric. $\mathbf{R}^4$ is linearly embedded in the spacetime, which for today's lecture we can take to be a flat $\mathbf{R}^{10}$.

Like the Einstein equations, the Yang-Mills equations receive corrections in string theory which are unimportant for large objects but very important for small ones. If F is the Yang-Mills curvature, $*$ the Hodge star operator, and D the gauge-covariant extension of the exterior derivative, the classical Yang-Mills equations read $0 = D * F$, or equivalently $0 = D^I F_{IJ}$. The stringy extension of these equations reads schematically

$$0 = D^I F_{IJ} + \alpha'[F_{KL}, D_J F^{KL}] + \dots \tag{2}$$

with higher order terms that are local, gauge invariant polynomials in F and its covariant derivatives, multiplying suitable powers of α'. As always, for large objects, the stringy corrections are small, and for small objects, they are large.

The classical instanton equations, in particular, are scale-invariant, so a classical Yang-Mills instanton can have any size. For a large instanton, the classical Yang-Mills equations are a good approximation; for a small instanton, they are not. So far, this is the usual story. The specific problem of instantons on $\mathbf{R}^4$, however, has some additional features.

The flat metric on $\mathbf{R}^4$ is, of course, essentially unique. However, the problem of string instantons on $\mathbf{R}^4$ depends not only on this flat metric but on an additional microscopic parameter $\theta \in \wedge^2\mathbf{R}^4$; I will say a word about its origin later. The self-dual projection of θ will be called θ^+.

If $\theta \neq 0$, then the rotation symmetry of $\mathbf{R}^4$ is broken to a subgroup. Thus, the case $\theta = 0$ is most similar to the classical instanton problem. Indeed, one can show using the hyper-Kähler structure of $\mathbf{R}^4$ that if $\theta^+ = 0$, then the instanton moduli space is the same in string theory as in classical Yang-Mills theory. The string theory instantons of size $\leq \sqrt{\alpha'}$ are not well approximated by classical instantons, but they have the same moduli space, if $\theta^+ = 0$. I will let $\mathcal{M}^\theta_{k,n}$ denote the string theory moduli space of based instantons on $\mathbf{R}^4$ of rank N and instanton number k for given θ.

In particular, the stringy instantons of $\theta^+ = 0$ have the familiar "bubbling" singularities that bedevil Donaldson theory. In "bubbling," an instanton becomes small and collapses to a delta function. Oddly, the term "bubbling," which was certainly coined long before instantons were studied in string theory, seems particularly appropriate in this stringy situation. Our instantons are supported on $\mathbf{R}^4 \subset \mathbf{R}^{10}$, but an instanton that shrinks to a point in $\mathbf{R}^4$ can literally "bubble away" into the higher dimensional world. The bubbled instanton is a point-like object (called technically a "-1-brane") in $\mathbf{R}^{10}$. The bubbling phenomenon in string theory is described by the ADHM construction of instantons.

If $\theta^+ \neq 0$, the instanton moduli space is modified from what it is in classical gauge theory. For $\theta^+ \neq 0$, there is a "no bubbling theorem," which is proved by using the fact that there is an energetic barrier to separating the -1-brane from $\mathbf{R}^4$. The barrier exists because a state with such a separated -1-brane would not be supersymmetric. Hence, for $\theta^+ \neq 0$, the moduli space $\mathcal{M}^\theta_{k,N}$ lacks the bubbling singularities. As a result, in fact, $\mathcal{M}^\theta_{k,N}$ is smooth if k and N are relatively prime. $\mathcal{M}^\theta_{k,N}$ still inherits a hyper-Kähler structure from the hyper-Kähler structure of $\mathbf{R}^4$, and it is independent of θ in the limit that the instantons are extremely large.

What hyper-Kähler manifold has those properties? According to the ADHM construction of instantons, the classical instanton moduli space is a "hyper-Kähler quotient" $\mu^{-1}(0)/G$, where μ is the hyper-Kähler moment map for a linear action of $G = U(k)$ on a flat hyper-Kähler manifold $\mathbf{R}^{4k^2+4kN}$. The relevant action of G preserves a hyper-Kähler structure on $\mathbf{R}^{4k^2+4kN}$, and μ is the associated hyper-Kähler moment map. μ takes values in $S = \wedge^{2,+}\mathbf{R}^4 \otimes \mathbf{g}$, with $\mathbf{g}$ the Lie algebra of G.

Because the center of G is $U(1)$, there is a natural embedding of $\wedge^{2,+}\mathbf{R}^4$ in S. A hyper-Kähler manifold that lacks bubbling singularities and is smooth if $(k, N) = 1$ is $\mu^{-1}(\theta^+)/G$, for nonzero $\theta^+ \in \wedge^{2,+}(\mathbf{R}^4)$. Taking $\theta^+ \neq 0$ does not change the behavior of the big instantons, but it eliminates bubbling for small instantons. This is what we want. The hyper-Kähler manifold $\mu^{-1}(\theta^+)/G$ has been studied mathematically as a partial desingularization of the usual instanton moduli space on $\mathbf{R}^4$ [5]. But what sort of objects does it parametrize?

This old question, which has been with us since the discovery of the ADHM construction of instantons and the hyper-Kähler quotient construction of hyper-Kähler manifolds, was neatly answered by Nekrasov and Schwarz [6]. They identified $\mu^{-1}(\theta^+)/G$ as the "moduli space of instantons on noncommutative $\mathbf{R}^4$" with the given θ^+.

To describe the appropriate notion of gauge theory on noncommutative $\mathbf{R}^4$, we begin with a bivector $\theta \in \wedge^2\mathbf{R}^4$. (The definition of the theory will depend on an arbitrary bivector. It can be shown, for instance via the ADHM construction, that the instanton moduli space depends only on the self-dual projection θ^+ of θ.) θ determines a Poisson bracket of functions on $\mathbf{R}^4$:

$$\{f,g\} = \sum_{i,j=1}^{4} \theta^{ij}\frac{\partial f}{\partial x^i}\frac{\partial g}{\partial x^j}. \tag{3}$$

One can deform the algebra of functions on $\mathbf{R}^4$ to an associative algebra $\mathcal{A}$, with multiplication $*$, such that

$$f * g - g * f = \{f,g\} + \dots \tag{4}$$

where the ellipses denote terms that in a suitable sense are small. This notion is often captured by introducing a formal deformation parameter $\hbar$, and writing $f*g-g*f = \hbar\{f,g\}+O(\hbar^2)$. For our present purposes, though, it is more pertinent to consider the behavior under scaling of $\mathbf{R}^4$. If we set $f_t(x) = f(x/t)$, keeping f fixed as $t \to \infty$, then $\{f_t, g_t\} \sim 1/t^2$. The property of the $*$ product that we want, apart from associativity, is that

$$f_t * g_t - g_t * f_t = \{f_t, g_t\} + O(1/t^4). \tag{5}$$

The $*$ product with these properties is essentially unique (up to automorphism of the algebra $\mathcal{A}$) and can be described by a very explicit formula:

$$f * g(x) = \left.\left(\exp\left(\frac{1}{2}\sum_{i,j}\theta^{ij}\frac{\partial}{\partial y^i}\frac{\partial}{\partial z^j}\right) f(y)g(z)\right)\right|_{y=z=x}. \tag{6}$$

Now let us move on to gauge theory, which we will formulate in the most elementary possible way. A gauge field, in the rank one case, is given by a "one-form"

$$A = \sum_{i=1}^{4} A_i dx^i, \tag{7}$$

where the A_i are elements of the algebra $\mathcal{A}$. The gauge-covariant exterior derivative is $D = d + iA$. The gauge transformation law is the statement that under an infinitesimal gauge transformation, $\delta D = i[D, \epsilon]$, with $\epsilon \in \mathcal{A}$. We get for noncommutative gauge fields of rank one

$$\delta A = d\epsilon + iA * \epsilon - i\epsilon * A. \tag{8}$$

For rank N gauge fields, one would use the same formulas, with A_i and ϵ regarded as elements of $\mathcal{A} \otimes \mathrm{Mat}(N)$, where $\mathrm{Mat}(N)$ is the algebra of $N \times N$ complex matrices. The gauge-covariant curvature is

$$\widehat{F} = \frac{1}{2} \sum_{i,j} F_{ij} dx^i \wedge dx^j, \tag{9}$$

where

$$F_{ij} = \partial_i A_j - \partial_j A_i + iA_i * A_j - iA_j * A_i. \tag{10}$$

The instanton equation is

$$\widehat{F}^+ = 0, \tag{11}$$

where $\widehat{F}^+$ is the self-dual projection of $\widehat{F}$. Nekrasov and Schwarz showed that solutions of this equation can be obtained by an ADHM construction, and that the moduli space of solutions so obtained is $\mu^{-1}(\theta^+)/G$.

This gives an interpretation of the deformed hyper-Kähler quotient, but is it what we want for string theory? So far, I have described two theories that both have classical Yang-Mills theory as a limiting approximation. In fact, in each case, the deformation has small effects for large objects, and large effects for small objects. In string theory the characteristic length, above which the theory reduces to classical Yang-Mills theory, is $\sqrt{\alpha'}$. In the case of the noncommutative Yang-Mills theory, a similar role is played by $\sqrt{|\theta|} = (\theta, \theta)^{1/4}$, where $(\ ,\ )$ is the natural inner product on bivectors in $\mathbf{R}^4$, and we take a fourth root because (θ, θ) has dimensions of $(\text{length})^4$. If the functions f and g have characteristic scale of variation much greater than $\sqrt{|\theta|}$, then the Poisson bracket $\{f, g\}$ is small and the noncommutative Yang-Mills theory reduces to ordinary Yang-Mills theory.

So far, so good. There is a rough parallel between these two theories with α' corresponding to $|\theta|$. But if we probe a bit more closely, we find what at first sight appears to be an insuperable obstacle to matching up these two theories. Both string theory and noncommutative Yang-Mills theory can be systematically expanded in powers of $(\text{length})^{-1}$. In the string theory case, the general form of the expansion is schematically indicated in (2). In noncommutative Yang-Mills theory, one obtains an analogous expansion by expanding in powers of the Poisson bracket. In each case, one is expanding in powers of a quantity (α' or $|\theta|$) with dimensions of length squared.

But there appears to be a crucial difference between the two. In the string case, the expansion involves more and more complicated terms that are written in the standard framework of classical gauge invariance. The higher order corrections involves increasingly complicated terms that are all written in the standard framework. For noncommutative Yang-Mills theory, by contrast, the expansion seems to involve a change in the rules: it involves an expansion of the multiplication law (in the definition of the curvature $\widehat{F}$) in powers of the Poisson bracket. These two expansions sound very different. How can they agree?

Here we meet a surprise, described more fully in [8]. These is a sense in which these two types of expansion do agree. To draw out the essential issue in the sharpest way, consider the case of gauge fields of rank one. Let us contrast two types of gauge fields and gauge invariances. In the first case, we have a gauge field A that takes values in the space $\mathcal{B}$ of classical rank one connections. In the second case, the gauge field A' takes values in the space $\mathcal{B}'$ of noncommutative rank one connections. The respective gauge invariances are:

(A) Classical abelian gauge invariance: $\delta A_i = \partial_i \epsilon$.

(B) "Non-commutative" gauge invariance: $\delta A'_i = \partial_i \epsilon' + i A'_i * \epsilon' - i\epsilon' * A'_i$.

These infinitesimal gauge transformation laws generate group actions. The two groups involved are in fact different. The first is abelian and the second is non-abelian. No change of variables will establish an isomorphism between an abelian group and a nonabelian one.

It seems, therefore, that it is impossible for these two types of gauge theory to be equivalent. But that is not the right conclusion. To do physics with gauge theory, we do not need to know what the gauge group is; we only need to be able to identify its orbits. In other words, we need to know when two gauge fields should be considered equivalent. We need the equivalence relation that is generated by the infinitesimal gauge invariances, but we do not need to make a particular choice of generators of this equivalence relation.

It turns out that, though no change of variables could convert the commutative group (A) into the noncommutative group (B), there is a change of variables that maps one equivalence relation into the other. To identify only the equivalence relation, and not the group, one has more flexibility in the change of variables. A change of variables that would map one group into the other would take the general form

$$\epsilon \to \epsilon'(\epsilon, d\epsilon, \dots) \tag{12}$$

$$A \to A'(A, dA, \dots). \tag{13}$$

Here, in other words, one transforms the group generator ϵ to a new group generator ϵ' which (in a formal series expansion in powers of θ) can be a general local functional of ϵ and its derivatives. But ϵ' is independent of A: to show that two groups are isomorphic, one should establish an isomorphism that is independent of any details about the space that the groups act on. Likewise, in claiming an equivalence between commutative and noncommutative gauge theory, one would want a mapping between the two spaces $\mathcal{B}$ and $\mathcal{B}'$ of connections, so $\widetilde{A}$ should be a function of A and its derivatives, independent of ϵ. Of course, a mapping of the type (12) does not exist; an abelian group cannot be equivalent to a nonabelian one.

To show not that the two gauge groups are the same, but only that the two equivalence relations are the same, modulo a change of variables, one has more freedom. For this, one looks at a change of variables of the form

$$\epsilon \to \epsilon'(\epsilon, d\epsilon, \dots; A, dA, \dots) \tag{14}$$

$$A \to A'(A, dA, \dots). \tag{15}$$

There is no change in the second equation: we want to define a definite map from $\mathcal{B}$ to $\mathcal{B}'$, so A' depends on A only and not on ϵ. The change is in the first equation: ϵ' may depend on A as well as ϵ, as we are not aiming to identify the two gauge groups, but only the orbits they generate in $\mathcal{B}$ and $\mathcal{B}'$. Existence of a change of variables of the form (14) from the classical to the noncommutative theory has the following implication: if A is a classical gauge field, g a classical gauge transformation, and A^g the transform of A by g, then the corresponding noncommutative gauge fields A' and $(A^g)'$ are gauge equivalent in the noncommutative sense, but the gauge transformation g' that establishes this equivalence will generally depend on A as well as g. Such a transformation from classical to noncommutative gauge invariance does exist, and can be found in a completely elementary way once one is persuaded to look for it [8].

Thus, the general framework of classical gauge invariance is equivalent to the general framework of noncommutative gauge invariance. The question is thus not which of these is correct in describing a given problem, but which is more useful. In particular, in string theory, one wants to know which framework for describing the corrections to Yang-Mills theory is more convenient in a given situation.

The answer to this question turns out to be as follows. String theory has both α' and θ. It can be usefully described as noncommutative Yang-Mills theory in a certain limit in which effectively $|\theta| >> \alpha'$. For $|\theta| \leq \alpha'$, the noncommutative Yang-Mills framework is still perfectly correct, but does not appear to be particularly useful.

For a hint of how this comes about (for more detail see [8]), we will finally have to look at the two dimensional quantum field theories that stringy geometry actually comes from. The action for a string with worldsheet Σ is

$$S = \frac{1}{4\pi\alpha'}\int_\Sigma g_{IJ} dX^I \wedge *dX^J - \frac{i}{2}\int_\Sigma \sum_{I,J} B_{IJ} dX^I \wedge dX^J. \tag{16}$$

Here X^I, $I = 1, \ldots, 4$ are coordinates on $\mathbf{R}^4$ that we use to describe a map $X : \Sigma \to \mathbf{R}^4$; g_{IJ} is the flat metric on $\mathbf{R}^4$; and $*$ is the Hodge star, using a conformal structure on Σ. The B-field for our purposes is a two-form with constant coefficients B_{IJ}.

This theory leads in general to the full complexity of string theory. There is, however, a limit of the theory in which the excited states of the strings drop out and the string theory can be described systematically in terms of noncommutative Yang-Mills theory. This is the limit in which, by taking g/α' to zero with fixed B (or by scaling things in various other ways to get a similar result) the second term in the action dominates, so that the action reduces to

$$S' = -\frac{i}{2}\int_\Sigma \sum_{I,J} B_{IJ} dX^I \wedge dX^J = -i\int_\Sigma X^*(B). \tag{17}$$

Actually, to be more precise, this limit does not exist for closed strings, for indeed if Σ is a closed surface, then S' always vanishes, since the two-form B is exact. However, if Σ has a boundary, S' is nontrivial. For the important case that Σ is a disc, S' is a functional only of the boundary values of X. If B is nondegenerate, then S' is the usual action functional for maps of a circle (namely $\partial\Sigma$) to the symplectic manifold $\mathbf{R}^4$ with symplectic form B, and is hence intimately connected with quantization of particle motion on $\mathbf{R}^4$. Note that I said "particle motion" rather than "string motion": in the limit that the full action functional S reduces to S', the strings effectively reduce to particles on $\mathbf{R}^4$, and that is why things become simple. This is also tied up with the fact that S' has more symmetry than S: it does not depend on the conformal structure of Σ, and so is invariant under arbitrary diffeomorphisms of Σ.

At any rate, in the limit that S' dominates, the string theory can be analyzed systematically [8] in terms of noncommutative Yang-Mills theory, with the noncommutativity parameter being the bivector $\theta = B^{-1}$. Actually, this limit is closely related to the content of many important recent papers. An example in which S' dominates is the limit of toroidal compactification (small area with fixed period of B) studied in the original application [2] of noncommutative Yang-Mills theory to string theory. Also, as Cataneo and Felder explain [1], the action they use in reinterpreting Kontsevich's results on deformation quantization reduces in the symplectic case to S'.

Thus, in this limit, which one can think of roughly as $|\theta| >> \alpha'$, the string theory remains nonclassical but can be described in great detail in terms of noncommutative Yang-Mills theory. The simplicity of this limit is tied with the fact that the characteristic excited states of the string drop out, and the conformal action S is replaced by the topological action S'. Is there a systematic framework, which somehow reduces to this description in the relevant limit, for using noncommutative, associative algebras to study the full-fledged string theory, with all the excited string states?

String field theory provides such a framework, at least for the open strings [9], but is regrettably messy. Here one looks not at functions on spacetime, but at functions on the path space of spacetime (suitably enriched with ghosts), and one defines a multiplication law for such functions using a gluing law for the paths. This description includes all of the stringy degrees of freedom, and is based on an elegant concept with an abstract Chern-Simons action $\int \left(A * QA + \frac{2}{3} A * A * A\right)$. But it is messy in detail and not much useful in practice. Indeed, the limit I have sketched, in which the stringy excitations drop out and the string theory can be described via noncommutative Yang-Mills theory, is the only known limit in which the open string field theory reduces to something nonclassical yet tractable.

But the purpose of the open string field theory, or whatever replaces it, should be precisely to incorporate the excited string states in the noncommutative framework. Many mathematicians and physicists have felt that the messiness of open string field theory comes from trying to shoehorn the more elegant two-dimensional worldsheet quantum field theory into an associative algebra framework that does not naturally fit. It has, in particular, been suggested that one should use an A_∞ algebra rather than an ordinary associative algebra, but this suggestion has not yet been accompanied by a suggestion of how to use an A_∞ algebra to write a Lagrangian.

This work was supported in part by NSF Grant PHY-9513835 and the Caltech Discovery Fund.

References

[1] A. S. Cattaneo & G. Felder, *A path integral approach to the Kontsevich quantization formula,* math.QA/9902090.

[2] A. Connes, M. R. Douglas & A. Schwarz, *Noncommutative geometry and matrix theory: compactification on tori,* JHEP **9802:003** (1998) hep-th/9711162.

[3] M. Douglas, *Two lectures on D-geometry and noncommutative geometry,* hep-th/9901146.

[4] M. Kontsevich, *Deformation quantization of Poisson manifolds,* q-alg/9709040.

[5] H. Nakajima, *Resolutions of moduli spaces of ideal instantons on* $\mathbf{R}^4$, Topology, Geometry, and Field Theory, (eds. K. Fukaya, M. Furuta, T. Kohno, and D. Kotschick), World Scientific, 1994).

[6] N. Nekrasov & A. Schwarz, *Instantons on noncommutative* $\mathbf{R}^4$ *and (2,0) superconformal six-dimensional theory,* Commun. Math. Phys. **198** (1998) 689, hep-th/9802068.

[7] V. Schomerus, *D-Branes and deformation quantization,* JHEP **9906:030** (1999), hep-th/9903205.

[8] N. Seiberg & E. Witten, *String theory and noncommutative geometry,* hep-th/9908142.

[9] E. Witten, *Noncommutative geometry and string field theory,* Nucl. Phys. **B268** (1986) 253.

[10] ———, *Small instantons in string theory,* Prosp. Math., (ed. H. Rossi), Amer. Math. Soc. **111** (1998).

California Institute of Technology, Pasadena
Institute for Advanced Study, Princeton